Greenhouse Design and Control

Greenhouse Design and Control

Pedro Ponce, Arturo Molina,
Paul Cepeda & Esther Lugo
Tecnologico de Monterrey, Campus Ciudad de Mexico, Mexico City, Mexico

Brian MacCleery
National Instruments, Austin, TX, USA

CRC Press
Taylor & Francis Group
Boca Raton London New York Leiden

CRC Press is an imprint of the
Taylor & Francis Group, an **informa** business

A BALKEMA BOOK

First published in paperback 2024

First published 2015
by CRC Press/Balkema
4 Park Square, Milton Park, Abingdon, Oxon, OX14 4RN

and by CRC Press/Balkema
2385 NW Executive Center Drive, Suite 320, Boca Raton FL 33431

CRC Press/Balkema is an imprint of the Taylor & Francis Group, an informa business

© 2015, 2024 Taylor & Francis Group, LLC

Publisher's Note
The publisher has gone to great lengths to ensure the quality of this reprint but points out that some imperfections in the original copies may be apparent.

Library of Congress Cataloging-in-Publication Data

Ponce Cruz, Pedro.
 Greenhouse design and control / authors, Pedro Ponce, Arturo Molina, Paul
Cepeda & Esther Lugo, InstitutoTecnologico de Monterrey, Ejidos de Huipulco, Mexico,
Brian MacCleery, National Instruments, Austin, TX, USA. — First edition.
  pages cm
 Includes bibliographical references and index.
 ISBN 978-1-138-02629-2 (hardback) — ISBN 978-1-315-77155-7 (ebook)
1. Greenhouses–Design and construction. I. Title.
 SB416.P66 2014
 635.9'823—dc23
                                                        2014016067

ISBN: 978-1-138-02629-2 (hbk)
ISBN: 978-1-03-292086-3 (pbk)
ISBN: 978-0-429-22732-5 (ebk)

DOI: 10.1201/b17391

Typeset by MPS Limited, Chennai, India

Pedro

To my charming and lovely Norma who is my inspiration every single day of my life.

Arturo

To my wife Silvia and lovely kids Julio and Monse. In memory of my mom Rosita.

Paul

I dedicate this work to my father in heaven and my beautiful family: Francisco, Ana María, María José, Rosa, Ruby, David, Carlos, Nathan, whose support, patience and love are the search engine for a better future. Selvia, you're part of this, thank you for living through and sharing hours of work, effort and dedication, encouraging me in difficult times and celebrating the best of times. To my colleagues and co-authors, thank you for the opportunities provided. Finally, this is also for all the people not mentioned here and who have been part of my growth and achievements.

Esther

To the ITESM CCM for all the support for this research that has become part of this book.
To my tutor Dr. Pedro Ponce for the opportunity to participate in this research work.
To my parents, brothers, sister and cousin, for supporting me.
To my boyfriend for the support, time and ideas for improving this research.
To my friends for the time and help they have given me to perform my job.

Brian

To Eva Jane MacCleery, greenhouse for my spirit.

Table of contents

List of figures

List of tables

Preface

Since weather conditions are changing dramatically around the planet, new forms of cultivation are needed more than ever. A greenhouse is one of the best options for improving the autonomous cultivation process; thus, researchers are looking for ways of improving the complete greenhouse performance. Because protected agriculture can be defined as an integrated engineering topic, a cluster of engineering areas are needed for greenhouse design. To establish the most favorable environmental conditions, it is necessary to control conditions inside the greenhouse, so the controller and structure have to be robust. The structures of greenhouses that are chosen are selected depending on the environmental factors that must be controlled, to what degree they must be controlled, and the cost of controlling them in relation to the value of the crop being produced.

Greenhouse technology has opened novel opportunities in different research fields, such as control systems, mechanical structures, digital sensors, wireless communications and many more. Although there are greenhouse structures and controllers for solving internal and external greenhouse disturbances, the optimal crop conditions could be improved by designing better greenhouse structures and controllers. Each environmental parameter to be controlled increases crop production cost, hence the paramount importance of excellent environmental control design. The goal is a controlled environment structure that allows for the control of those parameters at the level of precision required depending on the crop.

On the other hand, monitoring and controlling greenhouse climate control can be designed with intelligent systems in combination with non-linear control, where control actions are mainly calculated against internal and external disturbances. This kind of control methodologies generates better response. Greenhouses are the key factors for keeping optimal climate conditions. If these variables are optimally regulated, the crop growth performance will increase dramatically. Non-autonomous greenhouses require great effort in order to produce good quality crops, hence autonomous greenhouses have been installed in many parts of the word in order to attain optimal growth of crops.

This book shows the complete state of art regarding greenhouse controllers and structures and it also presents a basic greenhouse prototype based on LabVIEW for tomatoes in order to give a clear description about greenhouse design and control.

We would like to thank Tecnológico de Monterrey campus ciudad de Mexico and National Instruments Austin Texas for supporting this research.

Pedro
Arturo
Paul
Esther
Brian

How to read this book

The flowchart below is a guide to reading the book in a fast track method; just select the topic you want to learn more about and follow the diagram.

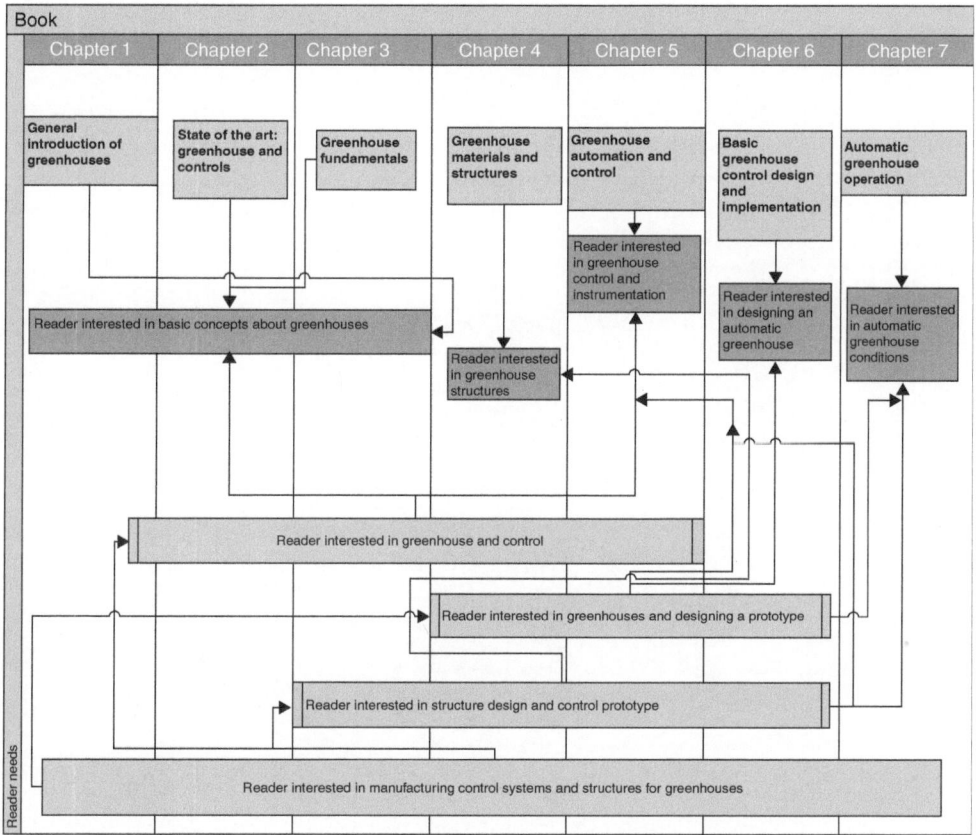

All the LabVIEW programs presented in the chapters can be downloaded at the book's website http://www.crcpress.com/product/isbn/9781138026292
The password of the zip programs is BOOK-GREENHOUSE

Chapter 1

Introduction

The agriculture sector faces the daunting challenge of providing adequate food and other necessities for a growing world population, which is projected to be nine billion by 2050. There is limited scope for the expansion of arable land, and the emerging threat to agriculture from climate change in the form of unpredictable weather, floods, and other disastrous events makes the task of providing enough food for the global population even more challenging (Clements et al., 2011). Since 1950 agricultural researchers have applied the scientific principles of genetics and breeding in an aggressive effort to improve crops grown primarily in less developed countries. The effort typically was accompanied by collateral investments to develop or strengthen the delivery of extension services, production inputs and markets, and to develop physical infrastructural features such as roads and irrigation (IAASTD, 2009).

Agriculture is understood to be the cultivation of plants that serve as food, provender, oils and industrial products. This activity involves the use of different groups of plants: grasses, vegetables, flowers comestible and ornamental, legumes, oil-seeds, industrial crops, medicinal plants and others. All these plant groups are grown in soil and open fields, some of which can be planted with protective covers, hence the term "Protected Agriculture".

One of the most significant difficulties faced by designers of greenhouses is the construction of a structure that allows the correct regulation of the environmental conditions, which is the main idea of developing the present book. The structure, automation systems and control should be closely linked; if the structure or automation systems are not properly installed, the control strategy could be the best but it will perform as the worst and vice versa (Cepeda et al., 2013).

In theory any plant can be grown in a protected environment, however, economic profitability, management and production cycles, have determined that mainly vegetables and cut flowers are grown this way. That is why the concept of "Protected Horticulture" is more frequently mentioned than "Protected Agriculture" (Cedillo & Calzada, 2012). As well as production schemes in open fields, agriculture is practiced in a wide variety of modified systems (Juárez et al., 2011; AMCI, 2008), among which include:

Micro-tunnels – These are small structures built with arches on which are placed plastic covers. Due to their small size it is not possible for people to work inside them so the work is done from the outside. Micro-tunnels are approximately one meter in

Figure 1.1 Micro-tunnels (Courtesy of Grupo Xurde, www.grupoxurde.es).

Figure 1.2 Macro-tunnel (Courtesy of BIOqualitum S.A. de C.V.)

height and width is consistent with the groove where the plant is being developed. Once it reaches a certain height, the micro-tunnel is removed to allow the plant to reach to its full height. The work is the same as in an open field system.

Macro-tunnels – These structures, mainly formed by arches, do not have the appropriate width and height to be considered greenhouses but they allow room for people to perform tasks inside the structure. They are 4 to 5 m (meters) in width and 2 to 3 m in height. They have the advantage of being easy to construct but the disadvantage of retaining less heat than a greenhouse, due to their low volume.

Shade houses – Shade houses are used in large areas and mostly with commercial crops. The sowing either occurs on the ground or in a hydroponic system. The shading mesh is used where solar radiation is very high, it is recommended for dry climates as it does not offer protection from the rain. Dimensions may vary according to

Figure 1.3 Shade House (Courtesy of ININSA Greenhouses, www.fabricanteinvernaderos.com/english/).

available space in terms of width and length, height depends on the crop, but this usually is from 3 to 5 m.

Greenhouses – A greenhouse is a metal frame agricultural building used for cultivation and/or protection of plants, covered with translucent plastic film which does not allow the passage of rain inside and has the aim of reproducing the most suitable conditions for the growth and development of crops established inside with some independence of the external environment, and whose dimensions allow the work of the people inside. Greenhouses can have a total plastic enclosure and side vents.

The changes in lifestyle and the accompanying increased needs brought on by new consumer habits has led to increased consumption of fruits and vegetables. Currently these products enjoy wide acceptance by consumers, largely due to their beneficial health aspects. The new needs presented by consumers demand products of good quality and taste. The increase in demand for these products has caused companies to incorporate new production techniques such as automated greenhouses in production fields, getting better quality products, and a long shelflife (Lucero & Sánchez, 2012). Since agriculture is still one of the most important economic sectors in many developing countries, providing employment and the poor's main source of income, it is not surprising that most developing countries are interested in technologies for adapting agriculture to climate change (Clements et al., 2011).

The greenhouse effect, first proposed by Jean Baptiste Joseph Fourier in 1824, is related to the old gardening glass buildings; its meaning is linked to the overheating of the planet's surface due to the atmosphere (Balas & Balas, 2008). A building designed for the protection of growing plants (protected agriculture) out of season against excessive cold or heat was called greenhouse or glasshouse, hence the term greenhouse effect.

Figure 1.4 Greenhouses.

The greenhouse structure protects the crops from the rain and wind allowing solar radiation to pass through the walls making heat exchange with the exterior environment possible. By the middle of the 19th century, all the techniques needed for successful greenhouse gardening had been developed; the greenhouse went from being a mere refuge from a hostile climate to a controlled environment, adapted to the needs of particular plants (Greenhouses, 2011).

By the late 1970s, it was possible to build a small greenhouse with automatic controls for less than one third of the cost of a small car. Today a variety of greenhouses are available, many of them built with glass or plastic enclosed in a framed structure, modular in form to permit easy expansion. Automated greenhouses are commonly used for the production of fruits, vegetables, flowers and any other plants that require special environmental conditions (Tiwari, 2003).

The best examples of the evolution and development of greenhouses are to be found in the Netherlands, with steel and glass greenhouses; and Spain, particularly in the region of Almería, with wooden greenhouses and tubular profiles covered with flexible plastic films.

The Netherlands, in half a century, increased its greenhouse area from 30 to 6,946 hectares. Considering only the period 1968–1994, the increase was 142 hectares per year for a period of 26 years. Spain, in the same period, experienced a greenhouse increase at a rate of more than 1,400 hectares per year. Currently it is estimated that there are over 30 thousands hectares of land occupied by greenhouses in Almería. Figure 1.5 shows the Almería region for protected agriculture viewed from a satellite.

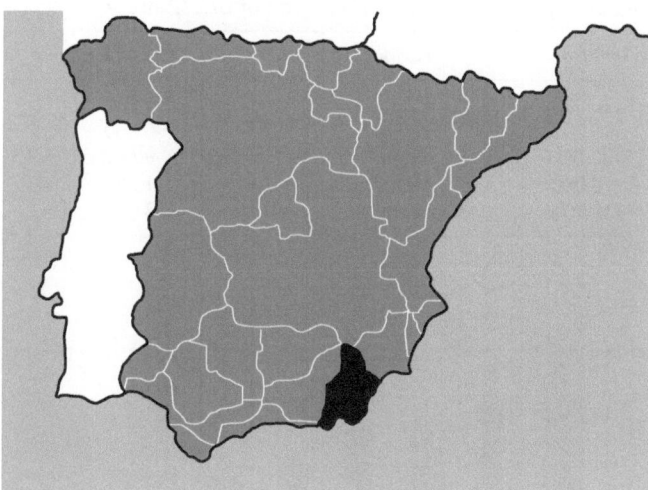

Figure 1.5 Almería, Spain.

In the Far East Japan is one of the most developed countries in terms of greenhouse use. Protected cultivation began in the late 19th century with the construction of glass greenhouses for the production of grapes. By the year 1982, it is reported that greenhouses covered an area of a little more than 27,000 hectares, of which nearly 26,000 were of plastic. Since 1975, plasticulture in Japan led to an important extension of fruit trees under protected structures and for 1993 it was estimated that 10,800 hectares of greenhouses were for fruit production, surpassing the greenhouse area for cultivation of ornamental plants and vegetables (Bastida T., 2011).

Currently China is the country with the largest area of greenhouses, with more than 2 million hectares. The Netherlands is the country with the most developed technology in the sector and Spain hosts the largest concentration of greenhouses in the Almería region. Experts suggest however that new surface growth has stopped in recent years, and that has been replaced by conversion of low technological greenhouses to modern structures.

Although statistics are not very reliable and differ considerably from one source to another, the estimated global area occupied by greenhouses in 1980 was 100,000 hectares, which had increased to 450,000 hectares by 1998; an annual growth close to 20%. Asia accounts for 66% of the area, Europe for 26%, while in America and Africa both account for 4%. For the year 1992, it was reported that worldwide about 280,000 hectares of greenhouse were with plastic covers, of which in Europe some 127,000 were concentrated, in Asia another 140,000 hectares were plastic covererd, while in the Americas another 13,000 hectares were covered with flexible plastic. By the middle of the 1990s, it was estimated that there were about 300,000 hectares of greenhouses and other micro-tunnels, most of them with plastic cover.

A study conducted in Europe between 1996 and 1997 indicated that of the more than 93,000 hectares of greenhouse in the region, 74% had plastic covers and 26% of them had glass covers. Glass covered greenhouses represented 98% of the total in the

Table 1.1   Global Status of Plasticulture by Region – End of Last Century in hectares.

Region	Padded	Floating Covers	Micro-tunnels	Greenhouses
Asia	350 000	12 000	192 960	192 000
Europe	380 000	40 000	90 000	86 080
America	200 000	3 150	9 000	10 000
Africa	15 000	ND	11 050	27 000
Totals	945 000	55 150	303 010	315 080

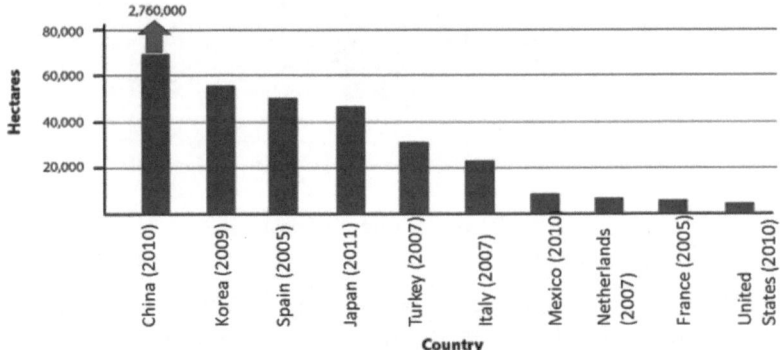

Figure 1.6   Total Area in Major Greenhouse Production Countries.

Netherlands and Denmark, Belgium 95%, Germany (West) with 90%, Switzerland 86%, England 85% and Austria with 80% (Bastida T., 2011).

By the beginning of the 21st century, the global area of protected crops was around a million hectares; with China accounting for an estimated 700,000 hectares, another 80,000 cultivated in South Korea, in addition to Europe and America.

Thus in the last decades, plasticulture has penetrated almost all areas of agricultural activity. In addition to the greenhouses, shade houses, tunnels and quilts also used plastic products for almost all systems irrigation and horticultural product packaging, among other activities where plastics are widely used (Bastida T., 2011).

The total areas in major greenhouse production countries are shown in Figure 1.6 (University of Arizona & CEAC, 2012).

If we consider the entire surface of the globe, greenhouses are concentrated in two geographic areas: in the Far East (China, Japan and Korea) are grouped 80% of the greenhouses in the world, and in the Mediterranean Basin about 15%. Growth is slow in Europe, but in Africa, America and the Middle East growth ranges from 15 to 20% annually. Notably, China has grown from 4,200 hectares in 1981 to over two million today (30% annually). Excluding China, the global area of greenhouses is estimated around 406,000 hectares (Bastida T., 2011).

The current development of greenhouses presents a wide variation of types that allow us to compare the different national situations and regions, as well as the influence of the elements on greenhouse design. Therefore, where we find the more

Figure 1.7 Protected Crops in the World: over 3 million hectares.

severe weather conditions is also where we find the more sophisticated greenhouses and sophisticated protection facilities. Thus, in the countries of central and northern Europe prevail glass greenhouses with high technology managed by computers and powerful productive, as in the Netherlands; while the Mediterranean, North Africa and the Americas are dominated by plastic greenhouses and low technology.

The control of environment and irrigation systems in greenhouses has received considerable attention in these last years. The main purpose of a greenhouse is to improve the environmental conditions in which plants are grown. Greenhouses provided with the appropriate equipment can further improve the environmental conditions by means of climate and irrigation control (Javadikia et al., 2009). These improvements have become, in recent years, the means for achieving controlled production yielding higher quality produce and improved economic benefits (Trabelsi et al., 2007).

There are several tools for improving and controlling the climatic and irrigation conditions in a greenhouse, such as:

- Natural ventilation: open-and-close side curtains and roof, as well as high greenhouses structures.
- Forced ventilation: exhaust fans and air circulator.
- Heating systems: heat pump, convection tubes.
- Shade: bleaching the ceiling, shade and mesh screens.
- Water evaporation, adiabatic cooling: wet wall and mist.
- Irrigation: drip, hydroponics.

The use of climate controllers such as alarms and standard programs in greenhouses allows manipulation of the different climatic factors (ventilation, heating, thermal screens or shade, humidification, recirculating). In advanced greenhouse automation systems, the regulation of indoor environment is focussed on creating an appropriate micro-climate for the intensification of plant growth and reduction of the final cost (Kolokotsa et al., 2010). Crop development, in its different stages of growth, is conditioned by environmental factors; the most relevant factors or variables considered and controlled in a hydroponic automated greenhouse are temperature, relative humidity, light intensity, carbon dioxide (CO_2), and electrical conductivity/pH of the nutrient solution in the irrigation system.

For a plant to perform its task it is vital that the combination of these factors are within minimum and maximum limits, outside of which a plant ceases metabolism and dies. Temperature is the most important parameter to consider in the greenhouse environment, since it is the most influential variable on the growth and development of plants. Normally the optimum temperature for plants is between 10°C and 20°C (InfoAgro, 2006).

The use of automatic controllers allows early planting and harvesting, so that produce can reach the market when prices are favorable, and the product's quality is also improved to the extent that it is the best option amongst competitors on the market. Today automatic controllers are not only used as a tool to streamline the climatological crop management, but also to obtain metrics that help the new generation of producers to make accurate decisions for better management of the plants. This has created the different production systems that are increasingly independent of external climatic factors and human error.

Greenhouse production, often called Controlled Environment Agriculture (CEA), usually entails hydroponics, as a cultivation technique for controlling the crop's nutrients. Today, hydroponics is possibly the most implemented method for crop production in the agricultural industry.

Globally the scope of hydroponics is constantly increasing. According to ISOSC (International Society for Soilless Culture) more than 25,000 hectares of greenhouses in the world are hydroponic. The primary countries in hydroponic greenhouse cultivation are the Netherlands, Spain, France and Japan. Figure 1.8 shows the world map with the

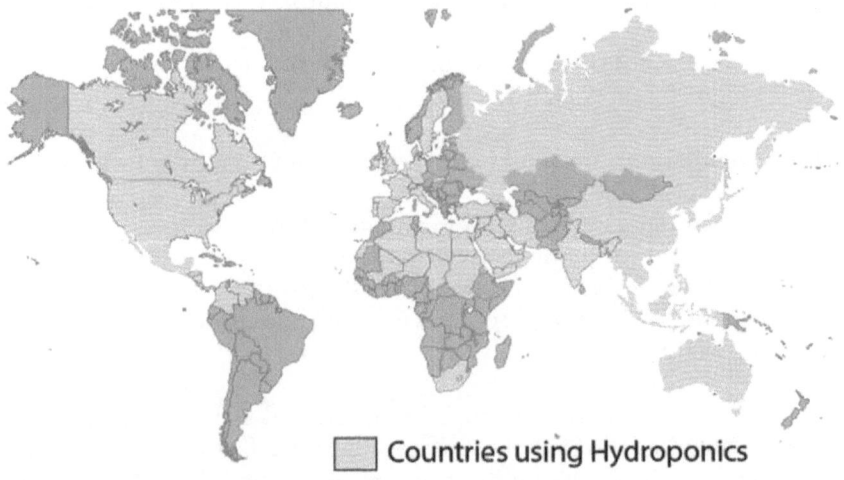

☐ Countries using Hydroponics

Figure 1.8 Countries using Hydroponics.

main countries that use hydroponics highlighted. Among these countries, besides the ones already mentioned, are: the United States, Mexico, Canada, Italy, Great Britain, Ireland, Germany, Sweden, Russia, Israel, the Middle East, China, India, Southeast Asia, the West Indies, the countries in or bordering the Sahara Desert, Venezuela, Colombia, parts of Central America, South Africa and Australia.

A hydroponic food presents unsurpassed features of purity and quality for human consumption. The mineral salts added to the water or irrigation are the same salts crops extract from the soil. Nutritional formulas are created using natural 100% soluble mineral salts. The formulas also use organic chelating trace elements. The nutritional value of hydroponic produce for humans is significantly superior, in most cases, to that of products obtained by conventional methods (soil cultivation). The explanation for this is that a product of soil-less cultivation is supplied during its entire life, if there is a well-crafted and balanced nutritional formula, with an adequate and optimum nutrient ratio such that the plant or the fruit complete their requirements effectively. This way the fruit or plant has everything it must have for a correct and healthy diet.

Furthermore, the progressive deterioration of greenhouse soils and horticultural production areas in general, due to increasingly widespread exhaustion, fungal contamination and salinization, is forcing farmers to opt for hydroponics as a solution. On the other hand, it is now essential to implement techniques that will lead to an economization of increasingly scarce water; hydroponics, given its high mechanization, leads to the consumption of only the amount of water that is needed, minimizing any losses. This combined with increased productivity and quality achieved by using hydroponics allows a greater product yield with minimum consumption of water and fertilizers.

Hydroponic culture is possibly the most intensive method of crop production in today's agricultural industry. In combination with the greenhouses, hydroponics is high-tech and capital intensive. Yet, hydroponics requires only basic agricultural knowledge (Moreno et al., 2011). Some advantages of this technique are the following (H. Ponce, 2009):

- No contamination from machinery.
- Intensive production to whit more growing per year.
- No erosion of the soil.
- Helpful for growing in cities.
- Resources conservation.
- Crop rotation is not necessary.
- No fertilizers for growing.
- Avoid great quantities of waste.

The most important crops grown under protected agriculture and hydroponics are shown in Table 1.2; this table also presents a comparison between growing using both soil and hydroponics where yield figures are strikingly in favor of hydroponics.

Protected agriculture is booming, using different levels of technology and focusing primarily on the production of vegetables. The technological level of the production systems working on protected environments are divided into three groups (Cedillo & Calzada, 2012):

Low technology – They are totally environment dependent, using simple technologies similar to those used in open field cultivation. The produce has passive ventilation (overhead and side ventilation), no heating, and is grown on a substrate. The average cost of this type of structure is $25–$30 dollars per square meter. In this case, it is very important to know the variations between day- and night-time temperatures because of the lack of heating and cooling.

Medium technology – Corresponds to modular structures with semi-climate control, programmed irrigation and soil cultivation or hydroponic. Usually productivity and quality is higher than the low technology level above. The producer uses a combination of both technology levels, high and low, and the temperature control is usually very simple. Medium technology encompassed both passive cooling systems and active, so there are models of structures that have heating and others that do not. The average cost of a system can be between $30 and $100 per square meter.

High technology – This level includes facilities with automated climate control (greater independence from the outdoor weather), computerized irrigation, CO_2 injection and soil cultivation or hydroponics. For this purpose they have sensors and devices operating the irrigation and ventilation systems, shading meshes for the lighting control and cultivation substrates. Generally based on the response of the plant to the environment. With these systems, the producer can optimize plant growth and maximize production and fruit quality. Because of these highly developed features, the average cost is the highest of all, from $100 to $150 dollars per square meter.

Greenhouse technologies with controllable environments will trigger the following achievements:

- To ensure production quality – commercial objectives by setting quality standards.
- To achieve highest productivity.
- To control calendars of production – manage the beginning and completion of production.
- To save energy – low cost of operation.

The main problem that should be taken into account for the greenhouse control is the complex interactions between the inside and outside parameters. Conventional

Table 1.2 Comparison between Two Growing Methods.

Crop (No. of harvests a year using hydroponics)	Yield using Soil (tons per hectare at harvest time)	Yield using Hydroponics (tons per hectare at harvest time)
Lettuce (10)	52	300–330
Tomato (2)	80–100	350–400
Cucumber (3)	10–30	700–800
Carrot	15–20	55–75
Potato	20–40	120
Peppers (3)	20–30	85–105
Cabbage (3)	20–40	180–190

control systems are not suitable for these type of applications because a greenhouse model shows non-linear behavior on many points (Fourati & Chtourou, 2007). Hence, it is not possible to achieve an accurate mathematical model, which is the reason why this work will focus on combining non-linear control methods, such as Sliding Modes and Feedback/Feed-forward Linearization, with artificial intelligence techniques such as Fuzzy Logic and Neural Networks. The latter do not need a mathematical model (Oduk & Allahverdi, 2011).

Nowadays, protected agriculture has continually lagged because of scarce and slow technological advances, and unskilled labor. The implementation of intelligent greenhouses to improve the quality and quantity of crops, allows a precise regulation of the environmental conditions inside the greenhouse in relation to the life-cycle of the crop within the control rules (Cepeda et al., 2010).

1.1 GREENHOUSE MOTIVATION

A look through history reveals that growth in agriculture has tended to be the antecedent of wider economic development. From the Industrial Revolution that began in England in the 18th century and spread to other countries, through to more recent examples of China or Vietnam, agriculture advances have always been the precursor to the rise of industry and services. In many poor developing countries, agriculture remains one of the most important activities, and in turn for the global economy, because it depends on the primary power of millions of people. Inadequate infrastructure, incomplete markets and a large presence of subsistence producers are frequent characteristics of these economies. Strategies to promote economic growth must be firmly anchored in agriculture. Increasing productivity in the sector is a necessary condition for resources to migrate towards non-agricultural activities, thus gradually diversifying the economy (FAO, 2012b).

In 2002, the World Bank and the FAO (Food and Agriculture Organization of the United Nations) initiated the IAASTD (International Assessment of Agricultural Knowledge, Science and Technology for Development). The objective of the IAASTD was to assess the impacts of past, present and future agricultural knowledge, science and technology on (IAASTD, 2009):

- Reduction of hunger and poverty,
- Improvement of rural livelihoods and human health (nutrition), and
- Equitable, socially, environmentally and economically sustainable development.

Recent food crises and the growing concern about climate change global agriculture have placed a priority on the international agenda. Governments, international organizations and civil society groups gathered at the summits of the Group of Eight (G8), the Group of Twenty Finance Ministers and Central Bank Governors (G20) and Rio +20 held in 2012, reiterating its common commitment to ensure the promotion of a sustainable future economically, socially and environmentally for our planet and for present and future generations. Investment in agriculture is one of the most effective ways to promote agricultural productivity, reduce poverty and improve environmental sustainability. The transition to sustainable agriculture will not be possible without significant new investments to protect and enhance the efficient use of natural

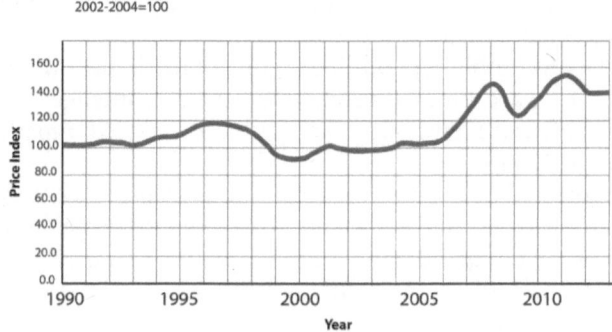

Figure 1.9 FAO Food Price Index.

resources and reduce losses at all stages of production, processing and consumption (FAO, 2012a).

After one decade of stability in global food markets, the period after 2007 was marked by an ongoing increase in prices for main agricultural products. The international Food Price Index[1], which is tracked by the FAO, increased by 55.3% between 2007 and 2011 as seen in Figure 1.9 (Ruíz-Funes & Smith, 2012).

Several studies conclude that the main cause of price increases in agricultural products are:

- The world's economic growth, especially in emerging countries such as China and India.
- An increase in the per capita consumption of meat and dairy products, the production of which requires an intensive use of feed grains.
- The reduction of the agricultural product inventory.
- The dollar devaluation.
- The expansion of bio-fuel production in Europe and the United States.
- "Panic" buying by some importing countries.
- The reallocation of investment portfolios to raw materials future markets, in many cases with speculative purposes.
- A slowdown in the growth of global agricultural production.
- The conversion of productive land for use in non-agricultural activities.
- The increase in water opportunity cost.
- Adverse weather phenomena in major production regions caused by climate change.
- Export restrictions imposed by major producer countries in certain periods.
- The rise in price of oil and other fuels, which increases agricultural production costs.

---

[1]**Food Price Index:** Consists of the average of 5 commodity group price indices weighted with the average export shares of each groups for 2002–2004: in total 55 commodity quotations considered by FAO as representing the international prices of the food commodities noted are included in the overall index.

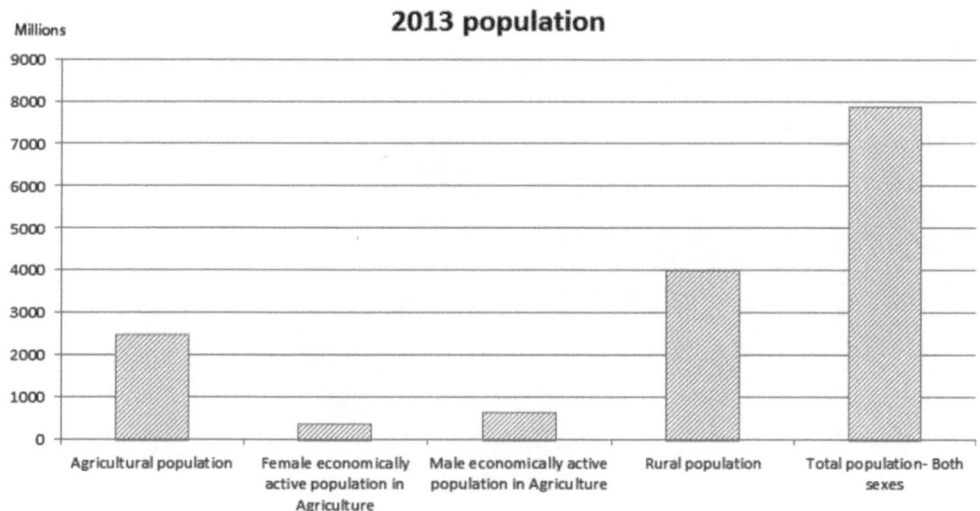

Figure 1.10 Actual worldwide population involved in agriculture.

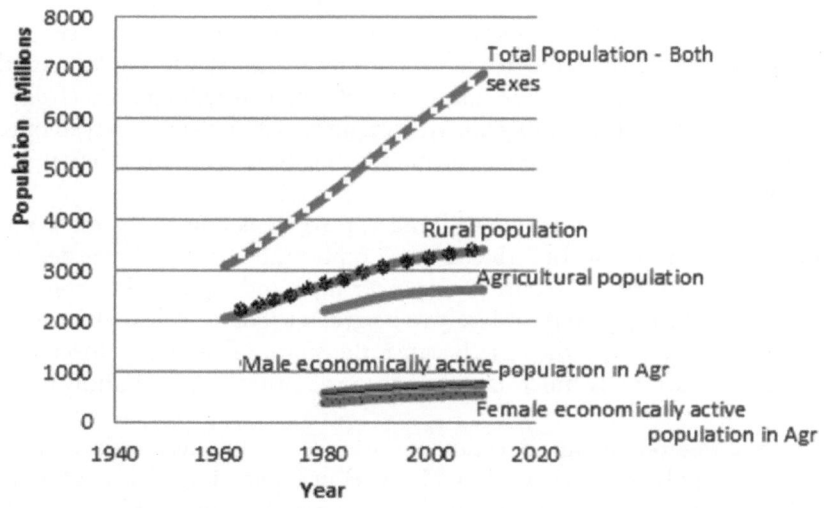

Figure 1.11 World's Population Dynamics.

The world's population has nevertheless doubled since year 2013 to around 6.9 billion, and is projected to increase considerably over the next decades as depicted in Figure 1.11. Figures 1.10 shows the population changes. Well over half of the developing world's population of 3.4 billion people live in rural areas. Of them, roughly 2.6 billion derive their livelihoods from agriculture (FAO, 2012b).

Approximately two-thirds of the world's agricultural added value is generated in developing countries, and in many of them the agricultural sector contributes as

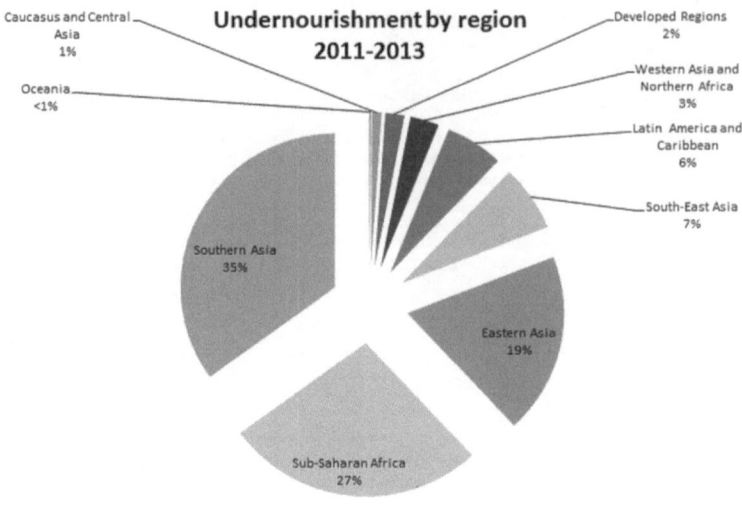

Figure 1.12 Undernourishment in 2011–2013, by Region.

much as 30% to the Gross Domestic Product[2] (GDP) and is a source of employment for two-thirds of the labor force. According to the World Bank, growth in the agricultural sector can be up to 3.2 times more effective at reducing US$1/day poverty than growth in other sectors. Importantly, agriculture can provide a haven of resilience against global economic and financial turmoil, often more effectively than other sectors (FAO, 2012b).

Among the many determinants of hunger, poverty is one of the most important. Not simply a lack of income or reduced consumption, poverty includes deprivation in health, education, nutrition, security, empowerment and dignity. The role of agricultural growth in reducing poverty is likely to be greater than its role in driving economic growth. This is likely to be the case because the share of the labor force that works in the agriculture sector is much larger than agriculture's contribution to economic output.

About 842 million people are estimated to have been undernourished (in terms of dietary energy supply) in the period 2011–2013 in Figure 1.12. This value represents 12% of the global population, or one in eight people. A major reason that people may not have access to food even when enough is produced is that there is no guarantee that a market economy will generate a distribution of income that provides enough for all to purchase the food needed (FAO, 2012b). Table 1.3 represents the population suffering from malnutrition since the last 20 years.

---

[2]**Gross Domestic Product:** The monetary value of all the finished goods and services produced within a country's borders in a specific time period, usually calculated on an annual basis. It includes all of private and public consumption, government outlays, investments and exports less imports that occur within a defined territory.

Table 1.3 Undernourished Population.

Number and Percentage of Undernourished Persons		
2011–2013	842 million	12%
2008–2010	878 million	13%
2005–2007	907 million	14%
2000–2002	957 million	15%
1990–1992	1015 million	19%

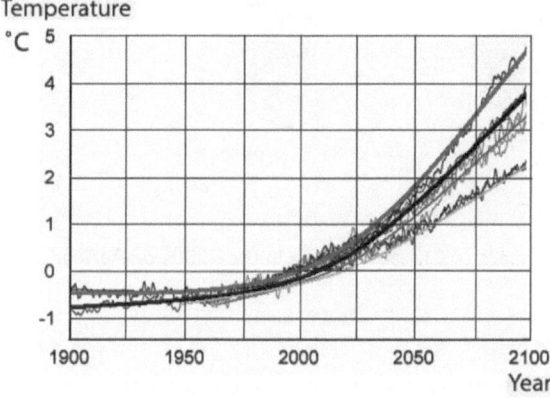

Figure 1.13 Temperature Rising Predictions.

The global climate change that is affecting the planet is due to the release of greenhouse gases, which have increased significantly through massive use of fossil fuels. The root causes of this problem are the generation and consumption of energy in the form of coal or oil, automotive transport, and energy-intensive industrial processes (IAASTD, 2009). As a result of global climate change is expected the occurrence of extreme temperatures, water shortages and floods. The Earth's temperature has increased between 1850 and 2010 at a rate of 0.5°C per century, but that mark increased to 0.7°C since 1900, to 1.3°C from 1950 and 1.8°C during the past 35 years. The last two decades are among the warmest since records of temperatures have been kept (ASERCA, 2012). The poorest and most food-insecure regions around the globe are the most vulnerable to effects of climate change. Figure 1.13 shows projection curves calculated by several prediction centres for the temperature rise over the next 90 years.

Climate change will significantly impact agriculture, farmers' incomes, and food security by increasing water demand, limiting crop productivity and reducing water availability in areas where irrigation is most needed. Global atmospheric temperature is predicted to rise by approximately 4°C by 2080 (Cline, 2007), as depicted in Figure 1.14 consistent with a doubling of atmospheric CO_2 concentration (FAO et al., 2012). Also, the presence of numerous insects and mites harmful to crops may be increasing because of the global warming (ASERCA, 2012).

Under present trends, by 2030, maize production in Southern Africa could decrease by 30% while rice, millet and maize in South Asia could decrease by 10%. By 2080,

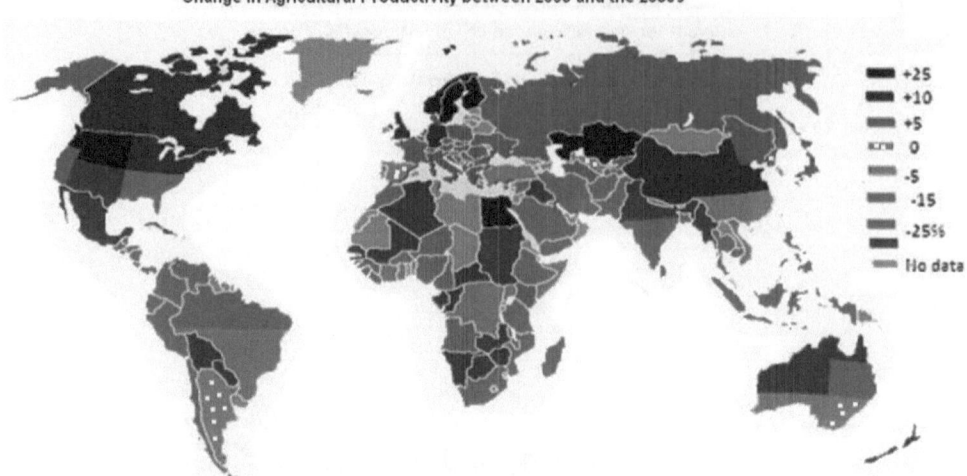

Change in Agricultural Productivity between 2003 and the 2080s

Figure 1.14 Impact of Climate Change in the 2080s on Agricultural Productivity.

yields in developing countries could decrease by 10% to 25% on average while India could see a drop of 30% to 40%. By 2100, when the population is expected to double, rice and maize yields in the tropics are expected to decrease by 20%–40% because of higher temperatures without accounting for the decrease in yields as a result of soil moisture and water supplies stressed by rising temperatures.

Future warming of around 3°C (by 2100, relative to 1990–2000) could result in increased crop yields in middle and high latitude areas, but in low latitude areas, yields could decline, increasing the risk of malnutrition. A similar regional pattern of net benefits and costs could occur for economic (market-sector) effects. Warming above 3°C could result in crop yields falling in temperate regions, leading to a reduction in global food production.

Climate change is expected to impose additional constraints on agriculture, but also brings some opportunities, as it will affect land and water availability (OECD, 2012). High temperatures accelerate the decomposition of organic matter and increase the pace of development of other processes occurring in the soil and can affect fertility. The rhythms of root growth and decomposition of organic matter decreases significantly in dry soils, reducing ground cover and makes the latter more vulnerable to wind erosion. For its part, the increased rainfall can also cause significant erosion of soil on the mountain slopes (ASERCA, 2012).

Furthermore, climate change in the case of many countries has caused open field cultivation to suffer through the impacts of economic relations, particularly in trade agreements. Although in this case, these agreements benefit the country by providing world markets and stimulate demand for products and services, they also cause damages to smallholders, as they are required to continue the investment needed for seeds, fertilizers, machinery and labor (Explorando México, 2012).

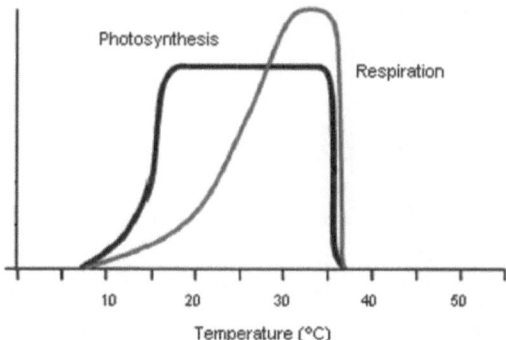

Figure 1.15 The Temperature Effect on Photosynthesis and Respiration.

Table 1.4 Problem in Crops by Climate Fluctuations.

Temperature	
Over 35°C	– Immature pollen, no fertilization, abortion, flower drop, affects fruit growth
Below 12°C	– Affects plant growth
Over 35°C/Below 12°C	– No fertilization, fruits take on a yellow hue
Relative Humidity	
Over 80%	– Pollen cakes, disease incidence increases, fruit cracks
Below 60%	– Pollen dehydrated preventing pollination
Light	
Low light	– Low growth, flowering, pollination and maturation

For example, for the cultivation of such vegetables as tomatoes, chillies and similar, the optimum temperature would be 24° to 27°C during the day and 14–17°C overnight. The daily temperature being over 30°C or nights over 20°C, has the following negative effects: low quantity and pollen fertility, fewer blooms per plant, fewer flowers bloom, the pistil elongates and leaves the flower, asymmetric blooms, longer time to first flowering, long inter-nodes, thin stems and poorly formed red pigment (lycopene). Optimal temperatures keep the sugar production in photosynthesis above consumption in breathing. The remaining sugar is used for growth and production. It may be noted in Figure 1.15, above 27°C the plant consumes more sugar than it produces and at 35°C the stomata close, which is reflected in a drastic drop in both processes (Guy, 2012). Table 1.4 shows the most common problems in crops due to climate variations.

At present, more than 1.5 billion hectares of the globe's land surface (about 12%) is used for crop production (arable land and land under permanent crops) as shown in

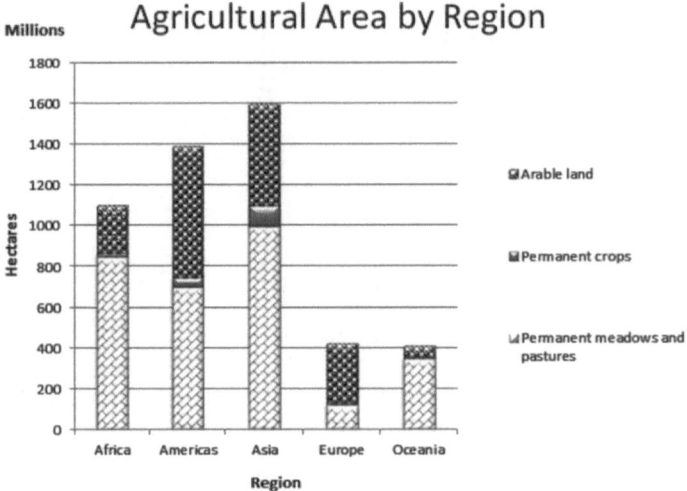

Figure 1.16 Composition of Agricultural Area by Region.

Figure 1.16. According to FAO, there is little scope for further expansion of agricultural land. Despite the presence of considerable amounts of land potentially suitable for agriculture, much of it is covered by forests, protected for conservation reasons, or employed for urban settlements (FAO, 2012b).

Water availability and its distribution may also be profoundly affected. While warming may extend the frontier of agriculture in higher-latitude areas (both northern and southern hemispheres), it is anticipated that key agricultural systems will have to cope with new temperature, humidity and water stress. This makes the need to increase the efficiency of land and water use even more urgent.

So far, land and water management systems have been able to meet the rapidly rising demands placed on them. This was made possible through gains in yields thanks to increased use of inputs, technology and irrigation. World agricultural production has grown between 2.5 and 3 times over the last 50 years while the cultivated area (permanent cropland and arable land) has grown by only 12%. More than 40% of the increase in food production came from irrigated areas, which have doubled in surface. These outcomes underscore the steady trend toward precision agriculture and commercialization of all types of food and industrial crops (FAO, 2012b).

Agriculture, though, has a complex relationship with natural resources and the environment. While agriculture is a major user of land and water it must also maintain the quantity and quality of those resources in order to stay viable. Considering the last issues which point to an increase in food demand in the coming years, the challenge for the farmers lies in increasing the supply rapidly and efficiently.

Thus, protected agriculture comes as a necessity in food production in regions with agro-climatic problems, initially in areas with frost and low temperatures, mainly in the Netherlands; however, this type of agriculture was also developed in areas with water shortages as the case of Israel, and also evolved in hot and wet regions such as Colombia, so this technology was developed to protect all crops from inclement weather, pests and diseases, and climate change on the planet. In Mexico and other countries like United States, France, Canada, Italy, Spain, Germany and China, among others, the development and evolution of this agriculture came later. With protected agriculture the crop is provided by its optimal requirements of temperature, relative humidity, soil moisture, ventilation, light and solar radiation, carbon dioxide, oxygen and fertigation, with the aim of achieving the most optimal performance and quality of crops (ICAMEX, 2011). Thereby it allows to minimize restrictions that poor weather cause to crops (SAGARPA, 2012).

In recent years, horticultural crops have presented a tendency towards obtaining early or out of season production, under conditions other than those practiced in traditional open field cultivation. This trend has created the need to use different elements, tools, materials and structures for crop protection with the aim of obtaining better quality products (Juárez et al., 2011).

Generally, protected horticulture has the following benefits over conventional agriculture (Cedillo & Calzada, 2012, SAGARPA, 2012):

- Generation of jobs: The occupancy rate in protected agriculture is higher than in other production systems, with the advantage that in most cases occupation is permanent and not temporary. For growing vegetables in greenhouses, on average it takes 5 to 10 workers per hectare work throughout the year on a permanent basis. For the production of cut flowers and ornamental plants 10 to 15 people per hectare are needed, also permanently. An agronomist can serve about five hectares.
- It provides the plants with greater protection, especially against the weather, pests, and diseases.
- It allows the use of techniques like hydroponics, irrigation technologies, balanced nutrition and climate control.
- It yields higher quality crops.
- It increases the production per unit area (up to 5 times compared to open field (Tomato: 350 ton/ha with protected agriculture vs. 70 ton/ha in open field).
- It can be harvested at any time of the year and permanently taking the advantage of market windows for competitive prices.
- The obtained products are healthier, decreasing pesticide application.
- Average water savings of 50%. In tomato saving is up to 77% (in open field 89 liters per kilo produced is used while in hydroponics 20).
- It can incorporate poor soils or growing in urban areas through the use of substrates.

Despite all the advantages of protected agriculture, the world is behind in taking up this technology. Managing proportions, traditional agriculture has about 1.5 billion hectares worldwide, while protected agriculture are estimated just over 3 million hectares as equivalent to 0.2% of total agricultural production.

Given the problems of climate change, water and land deteriorated, projects about expanding protected agriculture are in continuous development. It is necessary for governments around the world to encourage this type of investment in the coming years.

Traditional Over Protected Agriculture

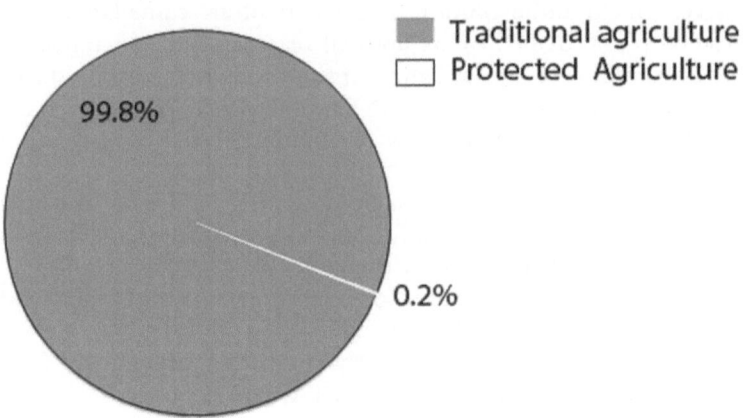

Figure 1.17 Traditional Over Protected Agriculture.

1.1.1   The state of North American protected agriculture

The United States, Canada, and Mexico have developed major greenhouse industries. All the large commercial greenhouses in the United States and Canada use active climate control and hydroponics, and many U.S. and Canadian growers would like to define a greenhouse tomato as one grown in that type of greenhouse. Although some greenhouse growers in Mexico have similar technology levels, other growers use greenhouses with lower technology systems, perhaps without fully active climate control, hydroponics, or both. Lower technology systems are less costly than high-technology greenhouses, but they produce lower yields and a less consistent product.

More recently, as the industry has expanded in Mexico, heterogeneity in production methods has increased. Shade houses are becoming more common in Mexican export-oriented field production regions, and it is becoming increasingly difficult to distinguish greenhouse and shade house produce in the marketplace. Growers in the United States and Canada view the growth of lower technology greenhouses and shade houses in Mexico with some alarm. Higher expected year-round production volumes in Mexico portend greater competition in all seasons, and continued downward pressure on prices. The historical development of the North American vegetable sector is shown in Table 1.5 and Figure 1.18 illustrates the distribution of protected agriculture (Cook & Calvin, 2005).

The major crops grown in North America includes tomato, cucumbers and peppers. Tomato is the main crop grown.

Much of the U.S. greenhouse tomato industry began in the North-East in the early 1990s, with production in the same months as Canadian producers. Eventually, several producers moved west and south, lured by the prospect of producing tomatoes year-round and capturing a slice of the high-priced winter market. The four largest greenhouse tomato firms in the United States are now located in Arizona, Texas,

Table 1.5 Vegetable Development in North America (Hectares).

Year	Canada	U.S.	Mexico	Total
1991	n/a	n/a	50	?
1992	361	n/a	n/a	?
1997	n/a	370	350	?
2002	876	395	1,520	2,791
2004	941	400	2,700	4,041
2011	1,343	529	15,000	16,872

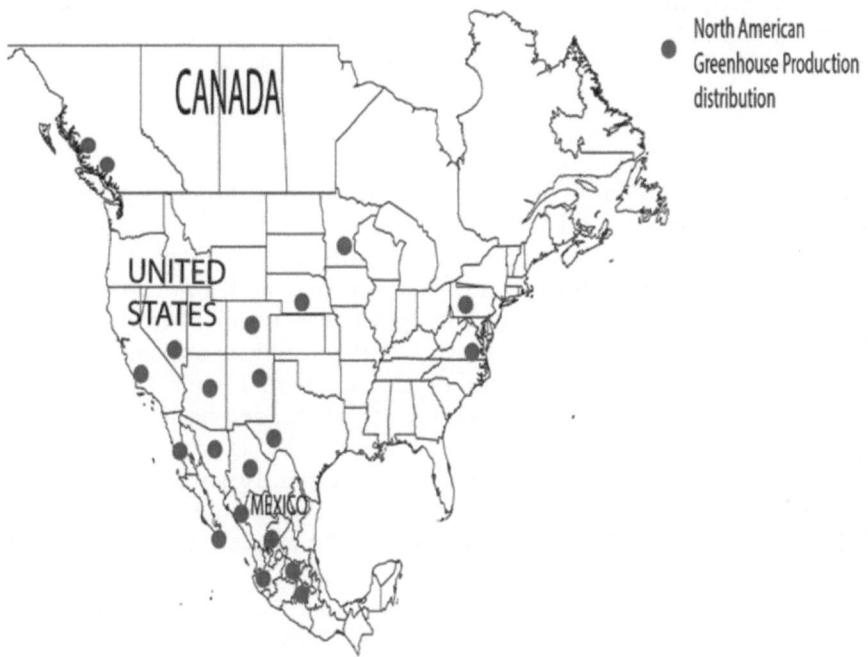

Figure 1.18 North American Greenhouse Production Distribution.

Colorado, and coastal southern California, and account for 67% of domestic production. Smaller greenhouses are located throughout the United States but these are frequently seasonal producers and local marketers. The profitable winter market helps the year-round U.S. producers withstand the very low prices during the summer season when Canadian volume inflates supplies. However, south-western greenhouses face special challenges posed by the summer heat and often need expensive cooling systems to produce high-quality tomatoes. Furthermore, expanding winter production in Mexico will likely reduce greenhouse tomato prices and increase competitive pressure on year-round U.S. growers.

The Mexican greenhouse tomato industry is the fastest growing in North America and the most varied. In Mexico, large field tomato grower-exporters in Sinaloa on the north-west coast and the Baja California peninsula are experimenting with protected culture, either shade houses or greenhouses, near their field operations. In contrast, U.S. field tomato growers usually have no connections to the greenhouse industry. This gives Mexican growers a foot in both camps and potentially reduces market- and other types of risk. Because of its hot, humid summers, Sinaloa, the principal fresh field tomato-exporting region in Mexico and a leading greenhouse exporter, is a winter producer only. Growers there have less incentive to invest in the highest technology greenhouses because the limited shipping season reduces the return on investment. Nevertheless, the technology levels and yields in coastal areas are improving, with more growers moving into mid-level technology systems to improve yields, quality, and marketing (Cook & Calvin, 2005).

The next subsection shows the Mexican panorama in terms of poverty, under-nourishment and natural resources and why Mexico, as part of North America, has becoming the leader in vegetable production and exports.

Mexico

Of the three North American countries that grow greenhouse vegetables, Mexico continues to see the most rapid growth.

Good weather and access to natural resources make Mexico an ideal place for producing vegetables. But Mexican production is lagging behind compared to other countries because they have greater technological advances and better business models applied to their greenhouses (Cepeda et al., 2010). Linking actors such as research centres, universities and non-governmental organizations to support the development of improved products to withstand the climatic onslaught, and pests, as well as improve the yield per hectare, are some alternatives to combat the lag in this sector. Since 2001, SAGARPA has awarded various supports for Protected Agriculture. In 2009, the federal government launched the National Strategy for Protected Agriculture, recognizing the benefits and profitability of this activity in the agricultural sector (SAGARPA, 2012).

Mexican agriculture is one of the largest in the world and has become a very important part of the country's economy as well. Mexico is the world's fourth most mega diverse country. It harbors 10% of the world's known species and ranks first in number of reptiles, second in mammals, fourth in amphibians and fifth in plants. It is also one of the eight centers of origin of edible plants such as avocado, squash, beans, tomato, maize, potato and papaya, among others (Ruíz-Funes & Smith, 2012).

Nowadays, Mexico faces a double policy challenge. First, managing the duality of its rural development with investments in infrastructure and innovations to facilitate the transfer of technological knowledge and technical assistance services to farmers while, at the same time, focusing public policy on poverty alleviation. Second, engaging in consistent agricultural and environmental objectives, implementing policies that enhance sustainability and efficient land and water management. This is the case of the PESA (Proyecto Estratégico de Seguridad Alimentaria México), in collaboration with the FAO, which provides support to small farms and farm households in highly marginal rural areas (OECD, 2012).

Mexico is the twelfth largest economy in the world, and a large country in terms of population, 117 million, and land area, 196 million hectares, as seen in Figures 1.19

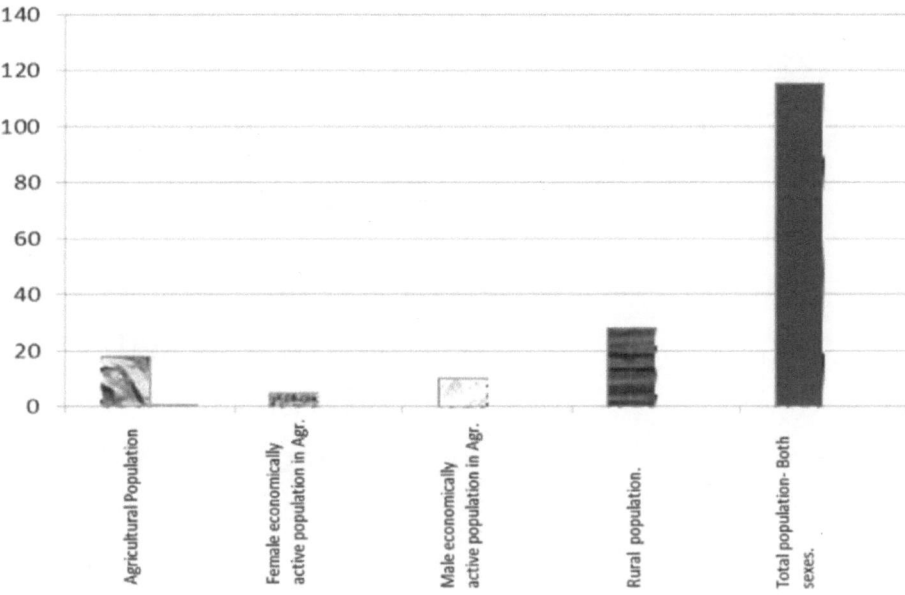

Figure 1.19 Actual Mexican Population.

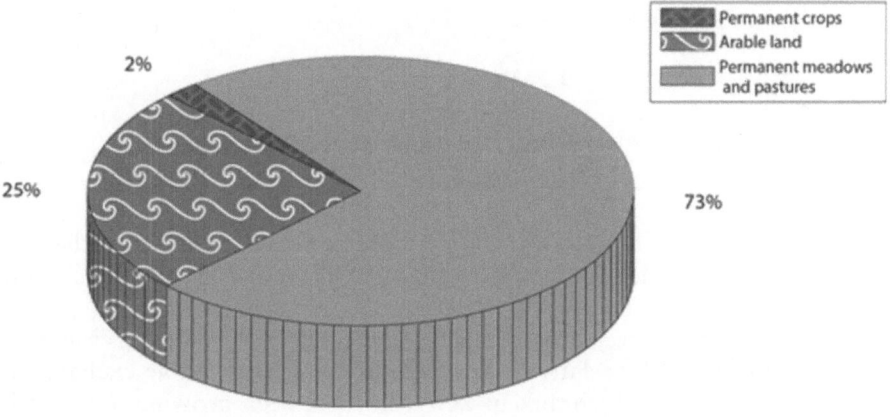

Figure 1.20 Composition of Agricultural Area in Mexico.

and 1.20, respectively (FAOSTAT, 2013). Figure 1.19 shows the current population of Mexico deploying the agricultural and rural population from the total number; then, Table 1.6 presents the statistics of Mexican land distribution while Figure 1.20 illustrates the portions of land intended for agricultural activities.

According to World Bank figures, Mexico had in 2010 a poverty incidence rate of 51.3%. According to the National Council for Evaluation of Social Development Policy 46.2% of the Mexican population (54 million) was in poverty (CONEVAL,

Table 1.6 Land Resources in Mexico.

Land Resources		
Item	Value	Unit
Country area	196438	1000 ha
Agricultural area	103166	1000 ha
Forest area	64647	1000 ha
Other land	28625	1000 ha

Figure 1.21   Mexico: Main Macroeconomic Indicators, 1995–2011.

2011a), of which 11.7 million people were in extreme poverty, not having the resources to acquire the food needed for a healthy life (CONEVAL, 2011b). These figures are alarming.

After some years of monetary instability in the mid 1990s, the Mexican economy had been characterized by relatively low inflation and more stable exchange rate in recent years. The economy shrank in 2009, but has been growing at a yearly rate of 4–5% in 2010 and 2011. The agricultural sector contributes 3.6% to GDP but employs 12.7% of people. Mexico is a net agro-food importer, and its share of agro-food import in total imports is 7%. Arable land represents 24% of total agricultural land, and irrigated land around 6%. There are two forms of land tenure in Mexico: private land and social property (ejidos). This later represents half of the territory of Mexico and, despite recent reforms, its sale requires approval from the Ejido assembly (OECD, 2012).

In Mexico, protected agriculture has existed for over 100 years, but only in the 1990s has its production become important to the consumer. In 1980, it was reported that 300 hectares (ha) had this production system, in 1999 this grew to near 721 ha,

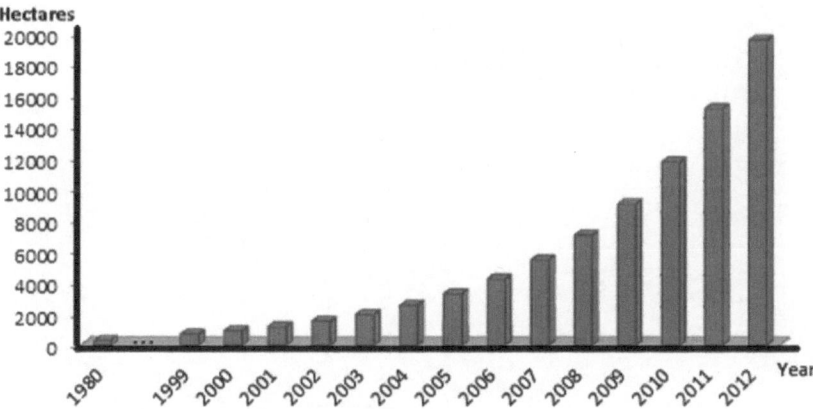

Figure 1.22 Panorama of Protected Agriculture in Mexico.

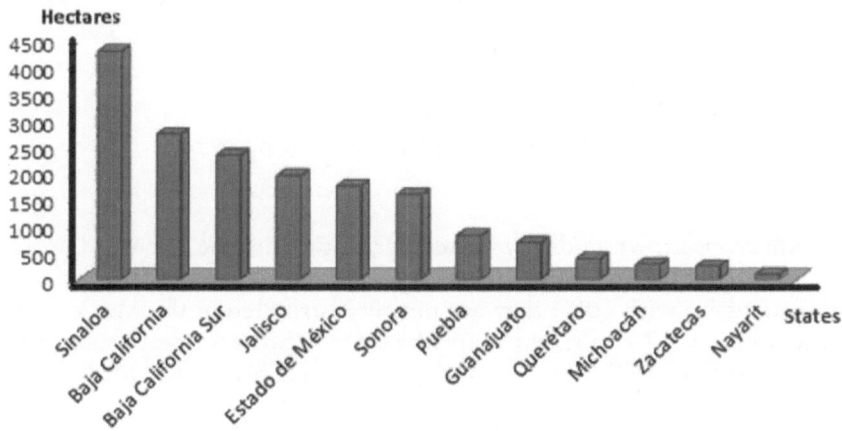

Figure 1.23 Greenhouse Sector Development in Mexico.

and in 2009 10,000 ha in Mexico used protected agriculture. This system of production has shown strong growth in recent years (20% to 25% annually), which has led to inconsistencies in the number of hectares currently set. SAGARPA (Secretaría de Agricultura, Ganadería, Desarrollo Rural, Pesca y Alimentación), in 2010 reported 11,759 ha while the AMHPAC (Asociación Mexicana de Horticultura Protegida A.C.) censused in the same year 15,300 ha (Juárez et al., 2011). The growth trend of protected agriculture through the years in Mexico is presented in Figure 1.22.

Today, Mexico has about 20 thousand hectares under protected agriculture of which approximately 12,000 are greenhouse and the other 8,000 correspond to shading mesh and macro tunnel systems mainly. More than 50% of the surface with protected agriculture is concentrated in four states: Sinaloa (22%), Baja California (14%), Baja California Sur (12%) and Jalisco (10%) (SAGARPA, 2012); the rest of the surface is distributed as seen in 1.23.

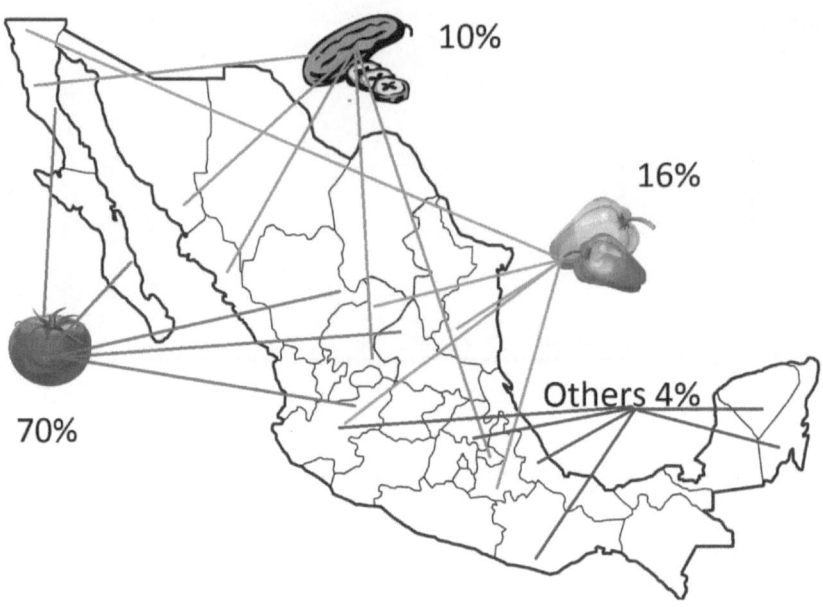

10%

16%

Others 4%

70%

Figure 1.24 Protected Agriculture Crops in Mexico.

The main crops grown under protected agriculture are the tomato (*Lycopersicon esculentum* Mill) with 70%, pepper (*Capsicum annuum* L.) with 16%, and cucumber (*Cucumis sativus* L.) with 10%; they are mostly distributed in the Mexican territory as depicted in Figure 1.24. In recent years, the diversification of crops such as papaya, strawberry, habanero chilli, flowers, among others has been intensified (SAGARPA, 2012).

Mexico participates as worldwide producer and exporter of organic food. It also ranks among the countries positioned in international markets. It is curious to observe that the main crops grown under protected agriculture (tomatoes, peppers and cucumbers) put Mexico in the top 15 ranking of worldwide producers, these commodities are in the top 10 of the most exported products in Mexico and they set Mexico in the first place of the top 5 exporting countries, as shown in Table 1.7, 1.8 and 1.9, respectively.

Specifically, protected agriculture technology in Mexico ranges from low to medium and from medium to high technology. Given the acreage, 79% of protected agricultural systems are high technology, 17% with medium technology and 5% with low technology (SAGARPA, 2008). About 86% of the production units are less than 0.5 ha, 11.5% from 0.51 to 5 and 2.5% have more than 5 hectares. This indicates that most of the farmers have very small production units, limiting their access to technology, training and technical assistance, as well as greater access to more demanding markets (P. Ponce, 2011). For example, the production yields of tomatoes in low technology is about 120 ton/ha, in medium technology ranges from 200 to 250 ton/ha, and in the high technology leads to get up to 600 ton/ha (Cedillo & Calzada, 2012).

Table 1.7 Top 15 Producing Countries by Commodities.

Top 15 Producers

Tomatoes		Chillies and Peppers		Cucumber	
Area	Production	Area	Production	Area	Production
China	46876088	China	15001503	China	45711326
USA	12858700	Mexico	2335560	Iran	1811630
India	12433200	Turkey	1986700	Turkey	1739190
Turkey	10052000	Indonesia	1332360	Russia	1161870
Egypt	8544990	USA	932580	USA	880530
Italy	6024800	Spain	873100	Ukraine	860100
Iran	5256110	Egypt	655841	Spain	666300
Spain	4312700	Nigeria	500000	Egypt	631408
Brazil	4106850	Algeria	383030	Japan	587800
Mexico	2997640	Netherlands	365000	Indonesia	547141
Uzbekistan	2347000	Korea	310462	Mexico	477366
Russia	2049640	Tunisia	304000	Poland	442212
Ukraine	1824700	Ghana	294100	Netherlands	435000
Nigeria	1799960	Italy	293647	Iraq	431868
Greece	1406200	Romania	243493	Uzbekistan	390000

Table 1.8 Top 10 Exported Products in Mexico.

Top 10 Exports

Product	Export Quantity	Export Value	Unit ($/ton)
Beer of Barley	1646096	1790500	1087.73
Wheat	1136317	276341	243.19
Tomatoes	1136299	1210757	1065.53
Sugar Refined	648340	359614	554.67
Chillies and peppers, green	608644	562593	924.34
Vegetables fresh nes	601127	363931	605.41
Watermelons	554410	287681	518.9
Cucumbers and gherkins	491368	261483	532.15
Lemons and limes	464514	189817	408.64
Beverage Non-Alc	417980	239464	572.91

Table 1.9 Top 5 Exporting Countries by Commodities.

Top 5 Exporters

Tomatoes		Chillies and Peppers		Cucumber	
Area	Quantity	Area	Quantity	Area	Quantity
Mexico	1509616	Mexico	644560	Mexico	498822
Netherlands	943119	Spain	446299	Spain	449395
Spain	738773	Netherlands	433868	Netherlands	397126
Turkey	574279	EU(12)ex.int	245645	Iran	242714
EU(12)ex.int	491636	EU(15)ex.int	188077	EU(12)ex.int	202770

Since climatic conditions dictate what kind of technology is needed, most producers use shade houses or basic plastic greenhouses. In the case of Mexico, medium to higher technology can be found in northern states like Sinaloa and Baja California or in the central states like Querétaro. The central states have the advantage of producing year-round, whereas northern states produce mainly during the winter season (Flores & Ford, 2010).

The greenhouse cultivation has always yielded primier productions, quality and higher performance, at any time of the year, while allowing lengthening of the growing season, resulting in turn in better prices. This increase in the value of products allows the farmer to invest in the establishment technology improving the greenhouse structure and systems, which are then translated into improved yields and quality of the final product. The lack of technology, infrastructure, training and organization of human capital, increases the vulnerability of the production structure in the social, economic and climatic phenomena above (Moreno et al., 2011).

Chapter 2

State of the art: Greenhouse and controls

Historians do not report the beginning of growing plants that were under cover, but Kenneth Lemmon presented in The Covered Garden book in 1963 that Plato in his *Phaedo* mentions that plants were grown under cover. One of the first references of glass used in agriculture was in 1385 in Bois le Duc France, where they grew flowers in glass pavilions facing south (Tiwari, 2003).

The greenhouse industry as we know it today probably originated in Holland during its "Golden Age", the 1600s. In the early 1600s, only rudimentary conditions could be imposedusing glass covers. By the middle of the 17th century the Netherlands provided half of the world's shipping, Amsterdam was the world's leading commercial city and the Dutch standard of living was the world's highest. The royal courts of Europe at this time had a taste for elegance and the means to afford it, and spring flowers in winter and fruit out of season were much sought after. The productive capacity of the large middle class, unique to the Netherlands, and the trade channels of the merchant segment soon gave birth to what is today the largest greenhouse industry in the world. Grapes were grown along rock walls in western Holland under cover; these greenhouses conserved the energy of the sun during the winter and yielding an early crop.

The French named their greenhouses orangeries because they used them primarily to protect orange trees from freezing. This also meant huge structures to house the size and number of trees they cultivated. The greenhouse at the Palace of Versailles is

Figure 2.1 Dutch Style Greenhouse (Courtesy of Journal American Rhododendron Society, Editor: Dr. Glen Jamieson).

Figure 2.2 The Orangerie at Versailles (Courtesy of Horacio Andres Galacho).

Figure 2.3 English Greenhouse (Courtesy of Anai Alicia Valencia Lazcano).

a perfect example of an early greenhouse. Built in the 17th century, the sheer size and elaborateness shows the love and care they gave their trees. Later, pineapples became popular and pineries were built as well, using the same principles.

As the invention of Victorian majolica pushed the development of greenhouses, the British abolition of the glass tax in 1847 and window tax in 1851 took place. The vast English upper and middle classes had entered an era of ostentatious wealth where conspicuous consumption was the fashion of the day and nothing said wealth to the Victorians like the luxury of the home conservatory. New exotic species of plants, flowers and fruits were appearing from the far corners of the vast British empire and the only way to enjoy these things were in a greenhouse.

From these origins to the present, there has been a steady evolution of controls (Nelson, 1998). Until the last century farmers and growers relied upon tradition and experience to judge the state of their soils, crops and animals before making the many, varied and often complex operational management decisions that are needed for successful food production. The beginnings of the automatic control era has its founding around the 1930s as is shown in the following time-line:

1930s Instrumentation of field machinery began when tractor manufacturers followed the automobile industry in fitting electrical gauges to their machines. At the same time, the spread of farm electrification led to the widening use of electrical instrumentation and control equipment for heating, ventilation and illumination of animal houses, crop stores and greenhouses (Cox, 1987). Early greenhouse control was as simple as pulling a chain to open or close a vent, turning a valve to control heat or irrigation, or throwing a switch to activate a pump or fan. Over the years this evolved as greenhouse systems themselves became more complex and more reliable (NGMA, 1998a).

1940s–50s During the first half of the 20th century, the manual control was common and widely used; in this type of control there is a person in charge of checking the temperature inside the greenhouses in order to activate the control equipment by hand to maintain the desired temperature. Analog sensors as thermometers were used. This mode of operation entailed large deviations above and below the desired temperature (Nelson, 1998). At this moment, electronic equipment began to appear on the farm making its first major impact by facilitating the measurement of moisture in cereals during and after harvest. Since then, farm electronics has slowly but steadily extended into many spheres of crop and animal production, greatly assisted in recent years by the development of microelectronics (Cox, 1987).

1960s Early automated control consisted of independent thermostats, humidistats, and timers. Even these simple devices allowed major advances in efficiency and product quality and made grower's lives simpler. However, many of these control devices and methods cannot deliver the level of automation and efficiency needed in today's dynamic, competitive environment. The thermostat is inexpensive and simple to install but it has poor accuracy, low energy efficiency and each actuator in the heating and cooling system require its own thermostat.

The common problems experienced with using several independent thermostats and timers to control a greenhouse led to the development of early electronic analogue controls, also known as "step" controls. These devices made a major contribution to improving the growing environment and increasing efficiency by combining the functions of several thermostats into a single unit with a single temperature sensor. The step controller is a bit expensive but it could manage several output connections such as two stages of heating and three stages of cooling, giving more accuracy (NGMA, 1998a).

1970s–80s As operating costs increased, and greenhouse systems became increasingly complex, the demand grew for increased control capability. The computer revolution of the late 70s/early 80s created the opportunity to meet the needs of improved control (NGMA, 1998a). Microprocessors were used for automating greenhouses because they could handle several input and output connections; some microprocessors can control up to 20 devices, more than step controllers. At this time the on/off control strategy was fully used and developed but due to its poor accuracy and the establishment of computers, the PID (Proportional-Integral-Derivative) controllers were implemented. These type of controllers show less fluctuations and a faster response getting the set point but the operation range of these controllers limits their functionality because not every state can be reached, so they have to be implemented for specific purposes with limited ranges (Nelson, 1998).

Also, one of the first attempts to control the temperature of a greenhouse was made in 1970 (Bowman & Weaving, 1970). A light-modulated greenhouse control system is a practical possibility and light-modulated temperature regimes appear to be profitable. From the practical point of view, there is considerable scope for the application of knowledge and skill in the choice of environmental programs. Apparatus has been described for measuring the CO_2 assimilation rate of whole plants in response to various environmental factors, and from such measurements practical environmental programs may be derived. During the course of any one day, more accurate temperature control would be achieved by the use of a more responsive heating system, e.g. pipes heated directly by steam, or by hot air systems. Although the apparatus described has been used only for the modulation of air temperature, it is suitable for the modulation of any environmental factor which can be sensed or controlled by electrical means, e.g. CO_2 concentration or relative humidity.

Over the past 40 years many attempts have been made to develop electronic equipment and control techniques for farm use but most of these developments have failed to gain acceptance in farming and the reasons have been more often economic than technical. The engineer who works with the agricultural engineering industry needs to appreciate the special problems that are posed by farming, to judge how and where electronics is likely to succeed commercially (Cox, 1987). Nowadays, computers with multipurpose microprocessors and embedded systems with specific purpose microprocessor are widely used, adding features like data recording and forecasting (Nelson, 1998).

Consequently there have been dramatic improvements in control technology. All the mechanisms just presented for controlling environmental conditions inside the greenhouses have been recently used. Today, computerized control systems are the standard for modern greenhouses, with continued improvements as the technology advances. High technologies are using more advanced control systems in which artificial intelligence is involved. The evolution along recent years of the different control techniques applied to greenhouses is presented below:

1991 A digital control algorithm was proposed by Davis & Hooper (1991) in order to improve a Venlo-type greenhouse heating control (heating pipe + valve). Experimentally estimated parameters of the pipe and internal air temperature transfer functions were used to design the digital controller. At first, it was designed as an open loop control which was found to be unsuitable when the greenhouse had insulation; then, the final algorithm developed included feedback of pipe and air temperature, and an addition of the integral control. The last proved to work well with both insulated and non-insulated greenhouses giving results consistently superior to the commercial algorithm, even when the new algorithm had a much longer interval between measurements. The algorithm used 10 minute sampling and adjustment interval, compared with the 1 minute (or less) by the commercial controllers. Greenhouse experiments demonstrated the effectiveness of filtering the greenhouse temperature measurements to reduce the valve movement.

1996 The main purpose of the research project of Putter & Gouws (1996) was to develop an optimal controller to create an ideal greenhouse environment controlling the main parameters such as temperature, relative humidity, light intensity and carbon dioxide concentration. This control problem was classified by the authors as non-linear and multi-variable. Discrete PID controller and Fuzzy Logic techniques were developed and a supervisory expert system was implemented in conjunction with the controller. The two control techniques were applied to an experimental greenhouse measuring $5\,m \times 3\,m \times 2\,m$. The main purpose of the expert system was to serve as a user friendly link between the user and the control system, where the user could select from a database previously created the name of the plant species and the stage of development or he can create a new one. Both control techniques proved to be successful in their specifics areas, where the discrete PID was most certainly the cheaper, but limited solution and Fuzzy Logic the more expensive, yet sophisticated option. The application of PID controller for environmental control in a greenhouse would be more acceptable for systems with relatively small variations in external conditions. The Fuzzy Logic controller was developed to handle drastic variations in environmental factors, involving the high interdependence between the control parameters.

The same year Occhipinti & Nunnari (1996) proposed a methodology dealing with the use of artificial intelligence techniques in the modeling and control of some climate variables within a greenhouse. The non-linear physical phenomena governing the dynamics of temperature and humidity on such systems are difficult to be modeled and controlled using traditional techniques. The objective of this work was to obtain some preliminary results in order to demonstrate the validity of the use of Multi-Input-Multi-Output (MIMO) Fuzzy Logic controllers in the field of greenhouses. A comparative analysis was carried out for both traditional *bang-bang* controllers and a MIMO Fuzzy Logic controller where this last gave the best performance in terms of precision, energy and also robustness.

1998 A framework was proposed by Caponetto et al. (1998) for the development of "human-based" Fuzzy Logic controllers in modern greenhouses. The use of Fuzzy Logic for the regulation of climate variables like temperature and humidity in artificially conditioned greenhouses represents a powerful way to minimize the heating energy spending, which is the more important aspect of greenhouse climate control. Modeling a greenhouse from a physical point of view requires a large computer effort due to the intrinsic complexity of the system and of the phenomena involved. A comparative analysis carried out for both traditional on/off or *bang-bang* and Fuzzy Logic controllers showed that the latter gave the best performance in terms of precision, energy and robustness.

2000 The subject of Caponetto et al.'s 1998 work was further developed and continued by Caponetto et al. (2000) to give some results in order to demonstrate the validity of the use of the artificial intelligence techniques in the field of greenhouse climate control. The Fuzzy Logic controller designed in 1998 and a newly developed distributed PID controller were used; both techniques were optimized through Genetic Algorithms to improve the their performance. There were less improvements compared to the "human-based" Fuzzy Logic controllers, justifying the robustness of Fuzzy Logic in directly transferring human expertise into automatic control laws. Finally, the distributed PID controller when compared to a common *bang-bang* and a non-distributed PID controllers, showed better results.

2001 An example of a coupled, non-linear controller for greenhouse air temperature and humidity was developed by Albright et al. (2001). Simplifying methods were used to obtain a linearized, uncoupled control model with significant lag times in the two controlled variables. An example of tuning, utilizing PDF (Pseudo-Derivative Feedback Controllers), was given, and the results of simulations of step changes in set points and disturbances, and the effect of parameter uncertainty, were explored. The simulations suggest that the resultant control system is robust, stable, and responds appropriately to disturbances and parameter uncertainties.

2002 The study of Fuzzy Logic controllers for greenhouse climate control was continued by Lafont & Balmat (2002). Since 1991, a classical controller on/off has been implemented in their experimental greenhouse, which generated a great number of data files. Knowing that the conventional techniques of regulation are difficult to implement in this type of system (multi-variable, non-linear, non-stationary) where the interdependence of temperature and hygrometry with the meteorological disturbances are strong, the authors developed two types of multi-variable Fuzzy controller, a basic Fuzzy controller and an optimized Fuzzy controller with a significant number of inputs and outputs. The basic controller allows to take into account external disturbances but its actual use poses a problem because the derivative variations are not available; indeed, it is not reasonable to add these inputs. On the other hand, the optimized Fuzzy controller has a structure which is easier to implement. It gives good results thanks to the introduction of variation senses of inputs, yet, it does not take into account external disturbances. The results showed a similar behavior between the Fuzzy controllers but the optimized was simpler; these controllers only had good performance half of a day. The authors conclude that techniques like mathematical fusion, decentralized structure and hierarchical organization will permit to develop a complete Fuzzy controller combining the advantages of the ones presented.

2003 A method presented by Pasgianos et al. (2003) for decoupling a highly nonlinear and coupled system proved to be very effective in meeting formal requirements for climate control of greenhouses such as set point tracking and disturbance rejection. The PCG (pre-compensator and command generator) block computes set point trade-offs based on psychrometric properties and actuator limits and costs to provide optimized set points that will allow the Feedback/Feed-forward controller to operate without hunting or chattering. The Feedback/Feed-forward controller achieves global input/output linearization and decoupling. Finally, the outer PI feedback controller compensates for model mismatch and deviations from expected disturbances. Simulations illustrated the efficiency and good performance of the proposed non-interacting control scheme.

2005 The work of Chao-gang et al. (2005) used a Proportional-Integral plus a Feed-Forward (PI + FF) method for controlling the heating pipe temperature by adjusting the electric valve. Many coefficients must be identified in this method. Most greenhouse operators set the coefficients based on their experience but facing the challenge of determining the proper values. Hence, a greenhouse temperature prediction model using Radial Basis Function (RBF) Neural Networks is implemented to propose a way to tune the PI + FF coefficients.

By the same year (Piñón et al., 2005), proposed an approach based on two different control schemes: Feedback Linearization (FL) and standard linear Model Predictive Control (MPC). The work deals with the problem of following trajectory for the optimum temperature behavior inside a greenhouse, which has been modeled as a time-invariant non-linear system, linearized by Feedback Linearization. The reason for introducing the combination of control strategies, MPC + FL, was to increase the computing efficiency by linearizing the plant and by reformulating the MPC problem on the new linearized coordinates. Various simulations tests were applied to gain some insight into the performance and efficiency of the MPC + FL approach as compared to a Non-linear Model Predictive Control (NLMPC) approach. The results showed very similar performance between them but a reduction in computational load by the use of MPC + FL was obtained in more than four orders of magnitude.

2006 A rough sets based on Fuzzy Logic control method was proposed by Fang et al. (2006) for greenhouse temperature control. The theory of rough sets was used to distinguish the essential factors from those negligible in terms of decision making for the greenhouse temperature. The overall control rules, on which based the construction of the Fuzzy Logic controller, were extracted from the pool of data collected in a Venlo-type intelligent glasshouse located in Zhejiang. The simulation results indicated good feasibility and validity of employing rough sets based Fuzzy Logic control for greenhouse temperature control.

Also Pucheta et al. (2006) came up with a Neuro-dynamic programming-based (NDP) optimal controller to guide the growth of tomato seedling crops by manipulating the environmental conditions in a greenhouse. The NDP enables the system to learn how to make good decisions by observing its own behavior, and to improve its actions with a built-in mechanism that uses reinforcement signal. This type of design methodology relies on the Neural Networks capability for learning non-linear functions and for solving certain problems that arise whenever massive parallel computation is required. The Neuro-controller considered the dynamic behavior of the crop-greenhouse system and the climate data of San Juan, Argentina, for July 1999. The control law obtained was simple, and minimizes operative costs involved along the evolution of the control process. The simulation studies showed that the performance of the proposed controller encourages its implementation in greenhouse systems for commercial production.

Another research done by (Pucheta et al. (2006) developed a crop growth guidance methodology in a greenhouse, where the environment variables CO_2 and temperature were manipulated, whereas the remaining variables (e.g., irrigation, humidity, pesticides, PAR radiation) were presented in the greenhouse proposal and given within suitable levels and dosages. The optimal control problem was stated by considering a non-linear system, with a non-quadratic performance index, with restrictions on both state and manipulated variables. A main premise for the optimization process was to know the climate pattern along the system evolution, a limiting condition that was met for the studied case. The approach attempted to minimize the operational costs and to meet the delivery date for the crop. The results show the feasibility of implementing a direct control of the environment variables straightforwardly on the crop, according to the business objectives of the producer.

On the other hand Nachidi et al. (2006) derived a Takagi-Sugeno (T-S) Fuzzy model from a given non-linear dynamic model of an empty greenhouse system, since this non-linear system can be readily obtained. Then, the concept of Parallel Distributed Compensation (PDC) was used to design a Fuzzy controller from the greenhouse T-S Fuzzy model. The results showed that the robust Fuzzy controller effectively achieves the desired climate conditions in a greenhouse, which shows the importance of the use a T-S Fuzzy model in the regulation of a very complex process with high non-linearity such as a greenhouse climate.

2007 The work of Fourati & Chtourou (2007) used an Elman Neural Network to emulate the direct dynamics of a greenhouse. The obtained model was implemented in a closed loop control using a multilayer Feed-Forward Neural Network. This last was trained to emulate the inverse dynamics of the greenhouse and then used as a non-linear controller with feedback state to provide the control actions for the process. The simulation results showed that Neural Networks strategies give good performances when controlling complex process such as greenhouses.

2008 The authors Bennis et al. (2008) address the modeling and control problems of the air temperature and humidity of an experimental greenhouse located at the University of South Toulon-Var (France). The control objective aimed to ensure a favorable inside micro-climate for the crop development minimizing production costs. The proposed regulation was based on H_2 robust control design involving a discrete linear control model of the process, obtained by an offline parametric identification technique. Evaluation of control performance was achieved through a benchmark physical model derived from energy balance for the temperature and water mass balance for the humidity. Simulations results showed a good performance of the proposed controller despite the high interaction between the process variables and the external meteorological conditions. One can expect much better results if the greenhouse was equipped with suitable and sufficient power actuators.

Another development was proposed by Shihua et al. (2008) which introduced a new technique which they called Adaptive Fuzzy Controller. In a greenhouse control, the main control factor is temperature, humidity, CO_2 and light intensity. The control of CO_2 concentration is simple, when concentration is high ventilation is reduced and when concentration is low CO_2 is sprayed; a reason why a traditional PID controller was used. On the other hand, the light intensity control is basically non-existent, so the main focus was over temperature and humidity control. These two variables were controlled by regulating the amount of steam heating, air speed and volume of water vapor and, skylights openings. This is a multi-variable control system and it presents lagged behaviors. The strong coupling between environmental conditions and the desire to find an accurate mathematical model brought some difficulties, so the controller was divided into two parts: an adaptive forecast predictor and the Fuzzy Logic controller.

2009 An irrigation controller based on a Fuzzy Logic methodology based on Mamdani controller is presented by (Javadi-Kia et al., 2009). The developed Fuzzy Logic controller can effectively estimate the amount of water uptake of plants in different depth using the reliable irrigation model, evapo-transpiration functions, environmental conditions of the greenhouse, soil type, type of plant and another factors affecting the irrigation of the greenhouse. The proposed system, when compared to simple on/off and an on/off with hysteresis controllers, found that the Fuzzy Logic controller had more ability in terms of less oscillations, operating points, energy savings, reliability and low cost.

At the same time Javadikia et al. (2009) developed a Fuzzy Logic controller based on Mamdani for greenhouse climate regulation. This work was compared to a conventional on/off controller, like the previously presented system. Both controllers were used with the greenhouse model in simulation to understand their performance and affection on the control parameters such as temperature and air humidity inside the greenhouse. The results showed that the proposed Fuzzy Logic controller is very user friendly, easy to design, highly adaptable and quick to perform.

Other researchers Zhou et al. (2009), introduced a Fuzzy control system based on multi-factor control of greenhouse. Temperature, light intensity and CO_2 concentration were the control parameters. The greenhouse environmental system was a complex large-scale system, so it was difficult to set up a systematic mathematical model, as it is very difficult to achieve control using the classic control or modern control theory. According to its characteristics, the Fuzzy control does not need the mathematical model of the controlled object, and it suits the control for non-linear, time-varying and time-delay systems. The Fuzzy Logic controller was applied to Programmable Logic Controller (PLC) procedures as the core of the environmental control system.

2010 Two Fuzzy Logic controllers were developed by Kolokotsa et al. (2010) embodying the expert knowledge of agriculturists and indoor environment experts. These controllers consisted of Fuzzy P (Proportional) and PD (Proportional-Derivative) control using desired climatic set points. The factors being monitored were the greenhouse's indoor illumination, temperature, relative humidity, CO_2 concentration and the outside temperature. Output actuation included: heating units, motor-controlled windows, motor-controlled shading curtains, artificial lighting, CO_2 enrichment bottles and water fogging valves. The system was tested in a greenhouse located in MAICh (Mediterranean Agronomic Institute of Chania). The test of the controller through modeling showed that the set points can be reached. The foremost characteristic was the universality of this control and its application in any cultivation regime with different environmental set points based on the crop requirements.

On the other hand, a design of an optimal controller for saving energy and precise prediction of greenhouse temperature and humidity is needed to compensate for the negative effects of inherent delay time of the greenhouse climate. Therefore Yousefi et al. (2010) proposed a hybrid Neuro-Fuzzy approach based on fuzzy clustering to model the greenhouse climate built upon the experimental data. In the first stage, the nearest neighborhood method generates the necessary Fuzzy rules automatically. Then, the cluster centers were used as the initial condition for the applied Neural Network trained and optimized using the Self-Organized Feature Mapping (SOFM) algorithm. Intelligent methods seem to be the most proper choices for the modeling of systems like greenhouses because the properties of universal approximation, they can model non-linear systems with trained data by arbitrary fitness. The simulation results showed the efficiency of proposed model.

Consecutively Yingchun & Yue (2010) argued that the greenhouse control technology was still unpopular on Chinese farms due to its cost and poor control performance, a reason why they developed a MIMO Fuzzy Logic controller based on the agronomists experiences for controlling temperature and humidity. The control system was embedded into an ATMAGE16 microcontroller. The system proved to be simple, cheap, high-precision and convenient, and it has practical potential for popularizing greenhouse application enhancing the effectiveness of the greenhouse industry.

Lastly Hu et al. (2010) presented a hybrid control scheme combining RBF Neural Networks with a conventional PID controller for the greenhouse climate control. A model of non-linear thermodynamic laws between the system variables affecting the greenhouse climate was formulated. It was anticipated that the combination of such techniques would take advantage of the simplicity of PID controllers and the powerful capability of learning, adaptability and tackling non-linearity of RBF Neural Networks. The presented Neuro-PID control was validated through simulations of set point tracking and disturbance rejection. The results showed that the proposed controller has good adaptability, strong robustness and satisfactory control performance by tracking square wave trajectory and being compared with conventional PID control methods.

2011 A solution to the problem of controlling the minimum temperature in greenhouses using controllers developed from non-linear models of the system was discussed and applied by Nachidi et al. (2011). The proposal started from a simple non-linear model obtained from energy balances, which enables a T-S model to be derived with only two Fuzzy rules, which represent precisely the non-linear model. As a simplified model is used, the resulting controller is simple to implement in nocturnal temperature control system for greenhouses. Moreover, this controller ensures stability and performance by using a specific design based on Linear Matrix Inequalities (LMIs). Compared with other control techniques that use Fuzzy rules, such as Mamdani-type Fuzzy Logic controllers, the proposed approach makes it possible to derive Fuzzy models directly from a physical description of the plant, which reduces the number of Fuzzy rules needed and gives simple stability results. The controller was applied to a greenhouse situated in El Ejido (Almería, South East of Spain). When tests were carried out, the designed T-S controller clearly showed how particular minimum temperature could be maintained during winter nights when heating was necessary, reducing actuator commutations and the fuel consumption.

The advantages of the Fuzzy Logic control over traditional control systems in greenhouse automation is presented by (Oduk & Allahverdi, 2011). Because of not having good results with traditional control techniques on the systems that have complex and non-linear form-like power systems, the usage of Fuzzy Logic controllers has been highlighted. The designed automation control system is intended to use six input parameters (temperature, air humidity, light intensity, soil humidity, CO_2 concentration and wind speed). A Fuzzy Expert system has been designed by running the heating, cooling, ventilation, lighting, irrigation and shading systems which are the output parameters. The performance indicators obtained from the simulation results were successful; they showed a minimization of the human errors and less energy consumption.

Another development by Chunfeng & Yonghui (2011) considered the characteristics of time-delay, non-linear and difficulties to establish a precise mathematical model in greenhouse climate; they proposed a control system based on predictive-PID cascade control. The real-time feedback correction technology was used to correct the uncertainty timely and effectively, which is caused by factors such as model mismatch, time-varying and environmental interference. It combines a cascade control structure with predictive control algorithm, which suppresses the main interference in the inner loop by a high frequency PID control, while utilizes Dynamic Matrix Control (DMC) to achieve good tracking performance and robustness in the outer loop. The DMC-PID simulations results demonstrated that prediction PID control overcome the shortness of the traditional PID and predictive control, it not only had a small overshoot, rapid response, good stability, but also had a slight steady-state error.

A Fuzzy irrigation control system based on a FPGA (Field Programmable Array) for greenhouses was developed by Gómez-Melendez et al. (2011). This irrigation control system is made up of two modules. One module named climate module is used to determine when and how much nutrient solution should be applied to the crop. The second defined as the nutrition module is in charge of pH control, nutrient solution dosage, and monitoring of the electrical conductivity. The module is based on a closed loop control system that employs Fuzzy Logic in order to control the multi-variable irrigation system with non-linear conditions related to the variables already mentioned. The system showed a potential to save a significant amount of water and nutrients making it a very economical fertigation control option compared to conventional systems. The Fuzzy control is adaptable, simple, and easily implemented and therefore represents an excellent tool to be applied to the optimization of fertigation systems. According to the experimental analysis, the system showed excellent performance and should allow for optimal fertigation control in a variety of greenhouses with crops that have different nutritional needs.

Finally Cepeda et al. (2011) presented the analysis, design and development of three intelligent open controls regulating temperature, relative humidity, and nutrients electrical conductivity. These controllers were implemented in the sustainable greenhouse located at Tecnológico de Monterrey Campus Ciudad de México. The intelligent controllers are based on a Fuzzy C-Means T-S control in order to deal with the nonlinear environmental conditions and variables inside the greenhouse. Fuzzy C-Means method is used to achieve a generator of Fuzzy rules from real data to predict the behavior of the actuators involved in the regulation of the variables to guarantee the optimal conditions inside the greenhouse for the optimal crops growth. The intelligent controllers are based on an open loop control system that could be used under the boundaries of different crops. Finally, experimental results showed a good performance but the system delays and the actuators efficiency triggered some errors in tracking set points.

2012 Hybrid systems are heterogeneous dynamical systems whose behavior can be defined by interacting continuously and by having discrete changes in their dynamics. A controller for this kind of system must ensure that the closed loop system meets some requirement, regardless of what the plant does. A design strategy based on stated stability and quadratic stability theorems for hybrid systems was proposed by Rajaoarisoa et al. (2012). It is well known that the agricultural greenhouse is a very complex system, composed of elements that can interact and exchange energy between them and with their environment. So, the setting and tuning of greenhouse climate controllers is by no means an easy or standard procedure. The closed loop system depends not only on the input control, it depends particularly on the external input disturbance dynamics. If the controller rejects correctly the disturbance effect, then it can ensure that the state feedback closed loop system becomes stable or quadratically stable. The results showed the feasibility of the approach for a class of hybrid system.

A greenhouse climate model with a Feedback/Feed-forward compensator for linearization, decoupling, and disturbance compensation was presented by Gurban & Andreescu (2012). The equivalent model was reduced to integral plus dead time decoupled processes suitable for temperature and humidity control of greenhouse climate. A comparison study of associated PI/PID controllers employing different tuning techniques were applied, such as Ziegler-Nichols rules, internal model control, closed loop transfer function coefficients matching, direct synthesis based design and specification of desired control signal. Simulation tests with system responses to set point step and ramp changes, and disturbance step changes, were analyzed and compared. The results showed that the smallest settling time was obtained by Ziegler-Nichols PID tuning rules, and the smallest overshoot by internal model control. Simulation tests showed that in the case of a set point change, the decoupling can be lost since the control variables are limited by actuators (ventilation rate and fog debit). Simulation results proved very good disturbance rejection by the compensator based on Feedback/Feed-forward linearization and decoupling.

And last Chouchaine et al. (2012) proposed an approach of Feedback Linearization control based on a T-S Fuzzy system integrating it in an adaptive Fuzzy scheme to control the temperature inside a greenhouse. The authors used a Fuzzy C-Means method to find the input membership functions and a recursive weighted least-squares method to estimate the consequences. A thermal model was developed from a real greenhouse located in a semi-urban environment in north Tunisia. Simulation results showed that the controller is able to reject disturbances and maintain the output in the desired trajectory during the day. However, if the outside temperature exceeds the reference output, the control action reacts with the maximum of its value and it can only bring inside temperature in the same level as the outside. Anyway, the proposed adaptive Fuzzy control provides acceptable results and this application can lead to productive consequences in agriculture.

Currently, unlike the stable markets of the past decades, farmers must adopt innovative techniques to deal with the globalization of markets, the changing needs and demands of customers, the rapid evolution of technology and the importance of respecting the environment.

Chapter 3

Greenhouse fundamentals

The world scenario has been changing from plentiful to limited resources owing to exponential population growth. This exerts a continuous pressure on land and agriculture and demands a radical change in agricultural practices in years to come. Sustainable ecological environment principles will be the guiding line in determining the desirability of certain agricultural practices over the other.

In contrast to many other vocations, agriculture is not a profession which is chosen voluntarily by many. Its importance in the future shall lie with the introduction of new agricultural technologies; a sense of pride has to be associated with farming for the recurrence of a green revolution (Tiwari, 2003).

A greenhouse is a structure that provides the most suitable micro-climate for the maximum plant growth during off-season. Air temperature is the most dominant parameter affecting plant growth. Transparent covering of the greenhouse allows the short wave solar radiation to enter but is partially opaque to the long wave radiation resulting in the greenhouse effect. In cold climates, higher inside air temperature is desirable during all hours for maximum plant growth; this can be achieved by keeping the greenhouse closed for maximum greenhouse effect or by using any suitable heating system. Whereas, in a composite climate, greenhouse effect is desirable only for a brief winter period (2–3 months) but for the rest of months, excess heat from the greenhouse must be removed by using any suitable cooling system.

The inside air temperature of a passive greenhouse directly depends on the ambient air temperature, the solar radiation, the overall heat transfer coefficient, the covering material and the wind velocity. Total solar radiation received by a greenhouse at a particular time and location also depends on its shape as well as its orientation which ultimately determines the inside temperature (Sethi, 2009).

A greenhouse can be built to either of two basic designs: (i) attached to a house, generally known as solarium and (ii) freestanding or greenhouse. The free standing greenhouses have general advantages. The free standing can also be used for crop drying. The classification of greenhouse system has been given in Figure 3.1.

Moreover, greenhouses can be also classified based on working principles as seen in Figure 3.2 (Tiwari, 2003):

Passive – These systems are normally those, which do not require mechanical energy for moving fluids for their operation. Fluids and energy move by virtue of temperature gradients established by the absorption of radiation. The greenhouse act as

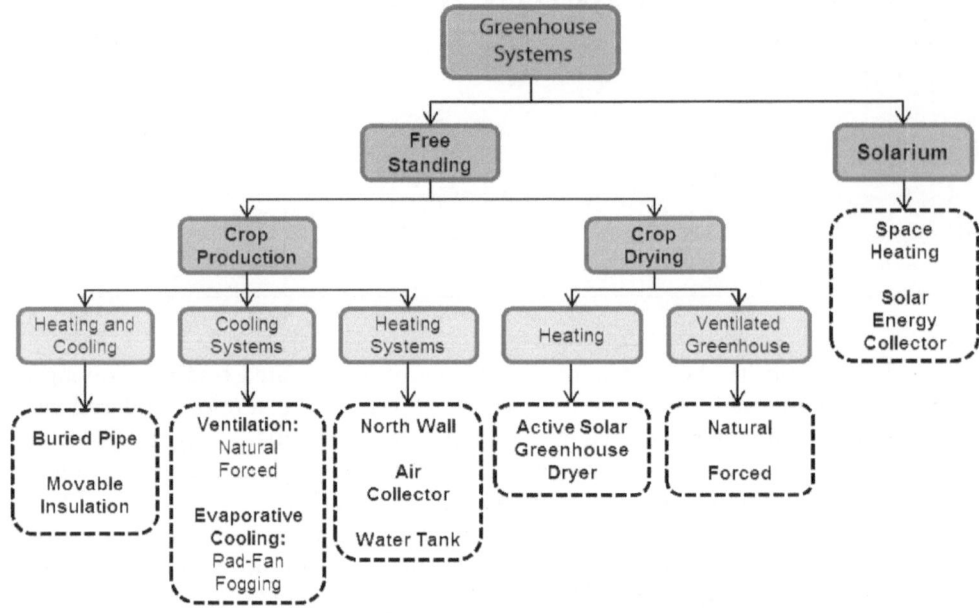

Figure 3.1 Classification of Greenhouse Systems.

a collector, as the glazed area,walls, and roofs are used for collecting, storing, and distribution of solar energy in the greenhouse by natural processes of convection, radiation, and conduction. Passive systems purely depend on architectural design that can be used to maximize solar gain in the winter (and minimize them in the summer) to reduce heating (and cooling) loads.

Active – In these greenhouses there is an external thermal energy available either from conventional fuel or solar energy through a collector panel is fed inside the greenhouse. These greenhouses use fans and pumps with the help of mechanical energy to move the working fluid in the system.

Planning before construction is vital when building any greenhouse structure. A short time spent weighing design factors can make large differences in profits and convenience of operation over the lifetime of a structure. Economic and business considerations such as capital availability, interest rates, and whether the operation is wholesale or retail will influence the size and type of building needed. Government regulations and taxing policies can influence the design of the structure. In urban areas, building codes and zoning laws may prohibit certain types of operations, or specify details that must be included in the building design (Bucklin, 2012).

The choice of crops grown in the greenhouse is based on physical size of the structure and economic feasibility of particular type of crop. As a result of high value horticultural crops have been more popular in greenhouse cultivation. A list of commonly grown crops inside greenhouses is mentioned in Table 3.1 (Tiwari, 2003).

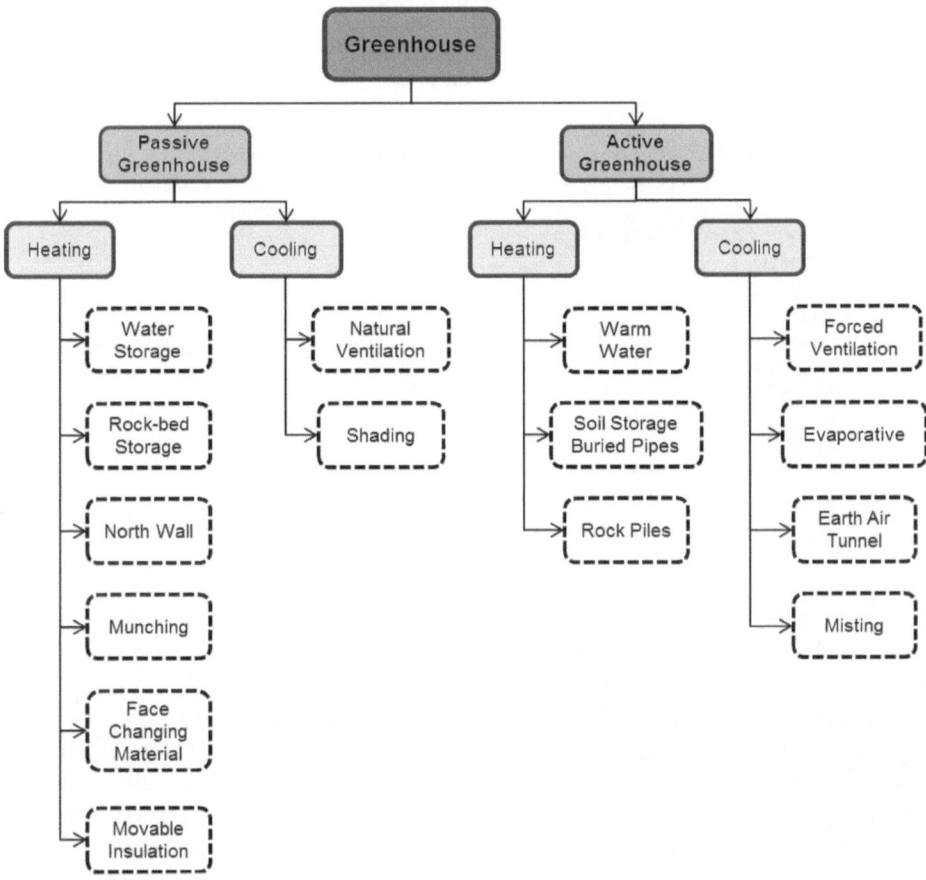

Figure 3.2 Classification of Greenhouses based on Working Principles.

Lastly, the accessibility to needed commercial services should be considered along with the availability of shipping facilities and main roads. The supply of utilities like electricity, water and fuel must be considered, and telephone communications and communications between buildings must be planned (Bucklin, 2012).

This chapter presents the fundamental greenhouse theory with the aim of giving the basics about the entire process of construction of an automated greenhouse. Important details such as design, orientation, supply utilities, designs, materials, cooling and heating systems, and control systems, among others, are covered in the next chapters.

3.1 GREENHOUSE GENERALITIES AND CONSTRUCTION

A greenhouse permits the maintenance of a micro-climate favorable to crops being considered for growing. Its structure should be light though also able to withstand

Table 3.1 Commonly Grown Greenhouse Crops.

Vegetables	Fruits	Ornamental Crops	Other
Tomato	Strawberries	Roses	Tobacco
Cucumber	Grapes	Poinsettias	Nurse
Lettuce	Citrus	Potted Plants	Chrysanthemums Plants
Onion	Melons	–	–
Cabbage	–	–	–
Beans	–	–	–
Peas	–	–	–
Spinach	–	–	–
Egg Plant	–	–	–
Peppers	–	–	–
Squash	–	–	–
Chilies	–	–	–

certain forces for a given design. Greenhouses are considered to be semi-permanent structures and are designed for a service life of 25 years. A greenhouse structure must support loadssuch as its own weight, wind, snow and transmit the maximum of sunlight (Tiwari, 2003). For an intelligent greenhouse containing high value equipment and crops, they should not be designed for a life-cycle less than 10 years.

To avoid structure damages, the stress limits should be considered, which are associated with a structure collapse that could potentially endanger human life. The ultimate limit states, which may require consideration, are: loss of balance of the structure and loss of load bearing capacity due to breakage, shakiness, fatigue, excessive tensions and deformations. Regarding the durability, it has been established that the protection against corrosion must not damage the structure integrity at least during the greenhouse life-cycle (Muñoz et al., 2013).

Greenhouses around the world must conform to same standards of design and construction as those used for homes and other small buildings. During the past four decades, greenhouses have been redesigned by growers, entrepreneurs and plant enthusiasts. Configurations have been numerous with different super-structural materials. Electrical conduits, plastic pipes, rolled metals and fiberglass reinforced plasters have been used for greenhouse frameworks along with standard materials such as steel pipes, wood and aluminum (Tiwari, 2003).

3.1.1   Site selection

A greenhouse complex should be located in such a way that it is well connected with markets, both for its supply and sale of its produce. The proposed greenhouse site should have adequate communication means i.e. telephone, fax, etc. There should be sufficient availability of good quality water and electricity (Tiwari, 2003).

The service building and the greenhouses should be on the same level for easy movement of personnel and materials and to permit maximum automation. Thus, the building site should be as level as possible to reduce the cost of grading. The site should be well drained. Because of the extensive use of water in greenhouse operations, providing a drainage system is always advisable. Where drainage is a problem, it is wise to install drainage tile below the surface prior to constructing the greenhouses. It is also advisable to select a site with a natural windbreak, such as a treeline or hill, on the north and northwest sides. In regions where snow is expected, trees should be 100 ft (feet) (30.5 m) away from the greenhouses in order to keep back snowdrifts. To prevent overshadowing of the crop, trees located on the East, South, or West sides should be set back at a distance of 2.5 times their height (Nelson, 1998). The site should ensure room for future expansion of the facility to make the system economically viable (Tiwari, 2003).

3.1.2 Orientation

In general, two criteria relevant to greenhouse orientation are (Tiwari, 2003):

- The light level in the greenhouse should be adequate and uniform for crop growth.
- The prevailing winds should not adversely affect either the structure or the operation of the facility.

Single greenhouses (free standing and single span) located above 40°N latitude in the northern hemisphere should be built with the ridge running East to West so that low-angle light of the winter sun can enter along a side rather than from an end where it would be blocked by the frame trusses. Below 40°N latitude, the ridge of single greenhouses should be oriented from north to south, since the angle of the sun is much higher. Gutter-connected greenhouses or multi-span (greenhouses connected to another along their length) at all latitudes should be oriented north to south. This north-south arrangement avoids the shadow in a greenhouse that would occur from the greenhouse lying immediately south of it in an East-West arrangement. Although the north-south orientation has a shadow from the frame trusses, it is much smaller than the shadow that would be cast from a whole greenhouse oriented to the south (Nelson, 1998).

Various researchers have used different greenhouses along East-West and North-South orientations for raising off-season vegetable or ornamental plants as seen in Table 3.2 (Sethi, 2009).

Prevailing wind direction of the site in study also influences the orientation. In naturally ventilated greenhouses, the ventilation should open in the wind side. The effect of wind can be checked by constructing fences of varying heights or growing trees and shrubs for wind breaks. A solid wind break, which causes turbulence, is much less effective than one which allows a small amount of wind to pass through it.

On the other hand, a greenhouse facility is severely affected by the shading caused by surrounding terrain, buildings, and plant material. Hence, greenhouses should not be constructed near large trees, buildings or other obstruction. As general rule, no objects taller than 3.3 m should be within 9 m of the greenhouse in either East, West or South, west or south direction (Tiwari, 2003). The magnitude of the shadows depends on the angle of the sun and thus on the season of the year (Nelson, 1998).

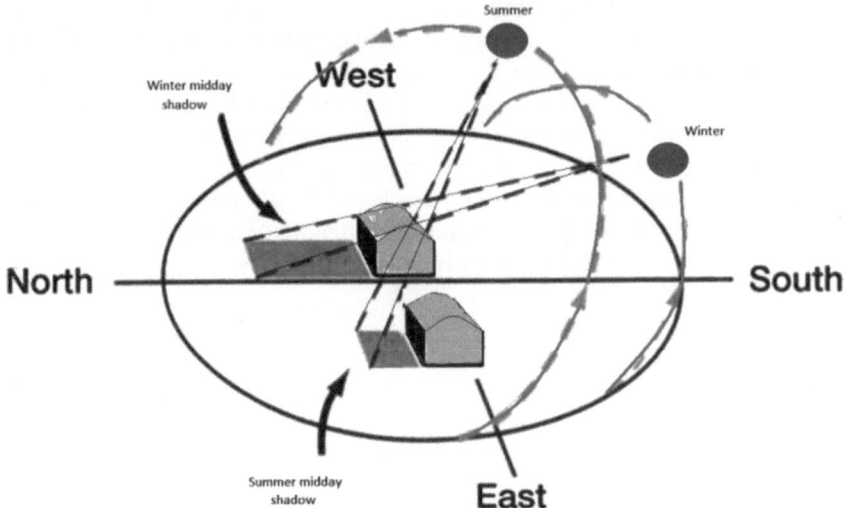

Figure 3.3 Location Priorities for Greenhouse based on Ambient Sunlight.

Table 3.2 Different Greenhouse Orientation by Researchers.

Shape	Orientation	Area (m²)	Location	Researcher
Arch	East-West	179	–	Gauthier et al.
Quonset	East-West	79	Quebec, CAN	Bernier et al.
Even Span	East-West	24	Chandigarh, IND	Sethi & Sharma
Quonset	North-South	100	Ludhiana, IND	Sethi & Gupta
Uneven Span	East-West	20	Delhi, IND	Dutt et al.
Vinery	East-West	–	ISR	Albright et al.
Even Span	East-West	1000	Agrinion, GRE	Santamouris et al.

3.1.3   Plan layout

A greenhouse facility includes space for storage of inputs, sales, works, etc., in addition to the production area. A general layout of the entire facility is given in Figure 3.4 (Tiwari, 2003); it should be prepared keeping the following factors under consideration:

- The work space also called head house should be located to the north of the production area, if possible to counter any kind of shadow from it.
- Storage/receiving area, sale area should have a separateaccess point.
- Customers and visitors should not be allowed to walk through the production area.
- The wind breaks should be at least 30 m away from the north and west of the greenhouse structure.

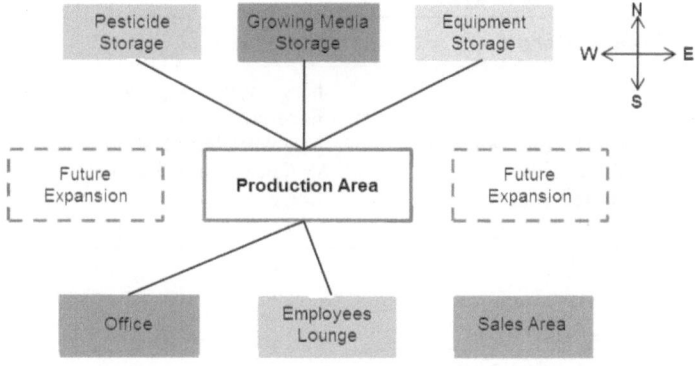

Figure 3.4 Plan Layout of a Greenhouse Along with other Support Facilities.

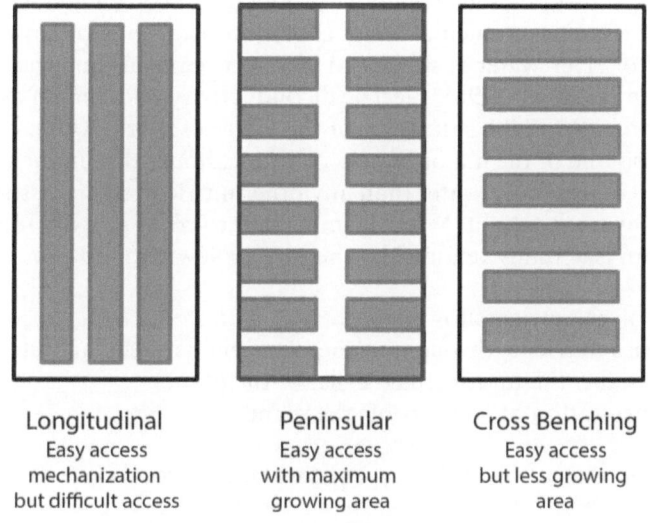

Figure 3.5 Benches and Beds Arrangements.

Production area: benches and bed arrangements

The first choice in growing plants is whether to grow them in raised benches or in ground beds; hydroponics can be performed both ways. If the crop is of moderate height, raised benches can be used; however, these benches should be located close to ground keeping the plants at a practical level for disbudding, spraying, and harvesting. Taller plants are grown in ground beds to minimize height (Nelson, 1998). Common arrangements of benches and beds are as depicted in Figure 3.5.

Table 3.3 Watering Effects.

Under Watering	Over Watering
Wilt	Large, soft new growth
Stress	Excessively tall
Reduced photosynthesis	May wilt under high light
Smaller leaves	Will not ship or last well
Shortened internodes	Oxygen depletion in root-zone
"Hard"	

3.1.4 Availability of water

Water is one of the most frequently overlooked resources in the establishment of a greenhouse business. Before a site is purchased, the available water source should be tested for quality. Water quantity is equally important, since as much as 20 liter can be applied to $1\,m^2$ of growing area in a single application. The amount of water requirement depends upon the area under production, crop and climate. Well water is the desired source, since municipal water is often too costly and may contain harmful fluoride. Pond or river water is subject to disease organisms and may require expensive sterilization (Nelson, 1998). Electrical conductivity and and an assay of the total soluble salts can indicate the suitability of the water (Tiwari, 2003).

Water is also one of the major factors of a greenhouse that influence crop productivity. A plant requires more water than any other nutrient. Leafy tissue contains 95% or more water by fresh weight. Water is important to plant growth and requires much more water than is actually retained by the plant (Newman, 2013).

- Water dissolves nutrients for uptake.
- Water carries nutrients throughout the plant via its vascular system.
- Water maintains the turgor (filled cells) of the plant.
- Evapo-transpiration of water cools the plant.

Some of the effects of under or over watering greenhouse plants are listed in Table 3.3.

The irrigation of greenhouse crops is one of the most critical of all production practices. And yet it is frequently overlooked and taken for granted. Often, watering is given to junior employees, yet irrigation mistakes frequently account for the greatest loss in crop quality. To provide conditions for optimum plant growth it is essential to become familiar with the factors that influence soil moisture (Newman, 2013).

3.1.5 Availability of electricity

All the electrical appliances such as motor fans, lights and other measuring instruments need assured electrical power for uninterrupted operation. Therefore, it becomes vital to ensure availability of electrical power on the site of construction (Tiwari, 2003). Table 3.4 gives a general requirement of power for different sizes of greenhouses.

Table 3.4 Electrical Power Requirement for Greenhouses of Different Size.

Greenhouse Size (m²)	Electrical Power Requirement (kW)
500	15
500–2000	24
2000–3000	36
3000–4000	48
4000–8000	96
8000–12000	145

Several standards exist around the world for electrical power distribution. AC (Alternate Current) electric power can be delivered at differing voltage and frequency. The electrical supply for greenhouses (actuators, central control) depends on the zone around the world.

Canada, Mexico and the United States all use a 100–127 V (Volts), 60 Hz (Hertz) electrical system (as do some other central and South American countries). Most of Africa, Asia and Europe use a 220–240 V, 50 Hz electrical system.

Renewable energies

Agriculture is one of the sectors with the greatest potential for using alternative energy sources and, in this sense, there are many farm experiences that prove the viability and the effectiveness of autonomous electrification and pumping powered by renewable energies.

Greenhouses are productive systems characterized by an intensive and efficient use of primary resources as wind or solar energy. In adittion the processes involved in configuration and function of indoor climate conditioning and the contribution of photosynthetically active radiation (PAR) for plants, can spend important amount of energy which can be minimized by using renewable sources of energy (Pérez & Sánchez, 2012).

In any case, alternative arrangements should be made to supply electricity to greenhouses. Solar energy has been the most used and tested technology as a natural source to obtain energy for greenhouses.

Renewable energy facilities will depend on the region in which the greenhouse is located. For areas where solar radiation is considerably high the optimum solution is to install a photovoltaic system, but when the greenhouse is located in an area with high air flow a wind turbine system has the preference. Greenhouse facilities can have hybrid systems (solar + wind) in order to magnify the electricity generating system.

The main advantages and disadvantages of using renewable energy are presented below.

Advantages:

• Do not consume fuel.
• Long life (from 15–20 years).

Figure 3.6 Greenhouse   Solar   Panel   System   (Courtesy   of   ININSA   Greenhouses, www.fabricanteinvernaderos.com/english/).

Figure 3.7 Wind farm.

- Minimal environmental impact.
- Low operating and maintenance costs.

Disadvantages:

- Relatively high initial investment.
- Limited service access.
- The electric loads needed to power a greenhouse can be calculated as shown in Table 3.5.

From the above, the greenhouse would need an installation of solar panels or wind turbines that generate around 4.96 kW at any time of day.

Table 3.5 Electric Loads for a Common Hi-Tech Greenhouse.

System	Actuator	Unit Power (W)	Total (W)
Passive Ventilation	3 Motors	370	1110
Active Ventilation	2 Motor (Fan)	370	740
Shading System	1 Motor	370	370
Heater	1 Heater	2000	2000
Fog System	1 Pump	370	370
Irrigation System	1 Pump	370	370
		Total	**4960**

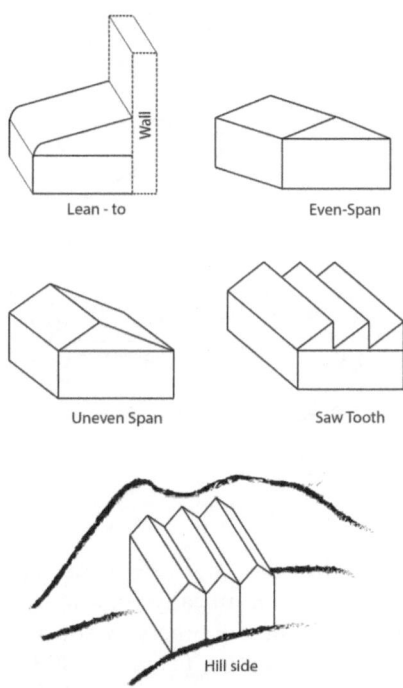

Figure 3.8 Greenhouse forms still in Use.

3.1.6 Structure design

Greenhouses are available in different shapes and sizes suitable for different climatic zones prevailing in the world. Each zone requires different shapes for providing favorable climatic conditions for the growth of plants. The greatest amount of insulation possible, covering of maximum ground area for the least cost and a structurally sound facility are some of the criteria for development of several types of greenhouse. Many greenhouse structures as seen in Figure 3.8 such as solarium ("lean-to", attached to a

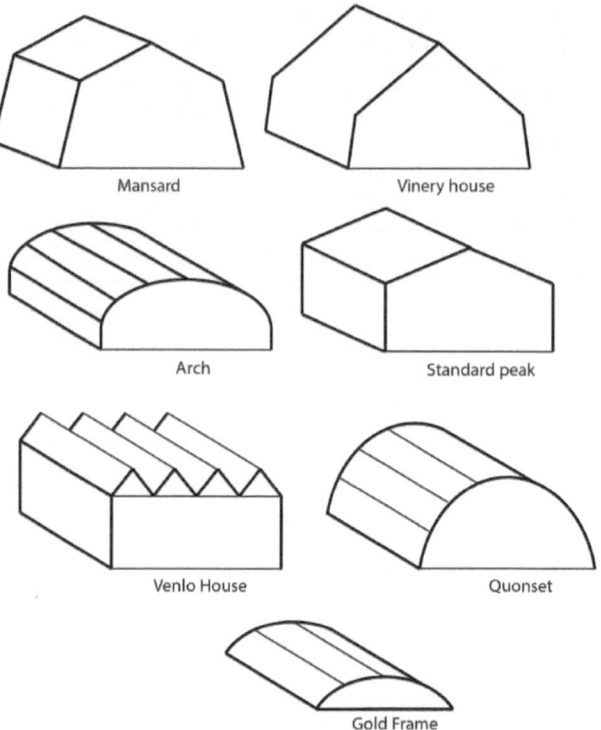

Figure 3.9 Basic Structures of Greenhouse.

house), even and uneven span, "hillside" and saw tooth types are still found throughout the world. Some of these are economically impractical and would not meet the requirements of a controlled environment; the saw tooth, however, has been use in floriculture industry of developed countries (Tiwari, 2003). The designs showed next are accepted by both European and American standards for greenhouse construction.

Modern designs have followed from these first approaches. Free standing greenhouses are commonly of two types which are shown in Figure 3.9: peak roof (A-frame) or arch/curvilinear. These structures are supported by themselves, i.e. no external support is provided. Arch roof structures are developed not because of light transmission considerations, but due to economic factors; these can be constructed for approximately 25% less cost than a peak roof structure. The arched roof is easily adaptable to both rigid and flexible covering material (Tiwari, 2003).

Sometimes the free standing peak or arch structures are combined to form "ridge and furrow" facility or multi span, shown in Figure 3.10. This type of configuration is applicable to most of the commercial greenhouses used for floriculture and vegetable production. They are less expensive to build, conserve ground area, and require less heating cost per ground area compared to stand-alone structures (Kacira, 2013).

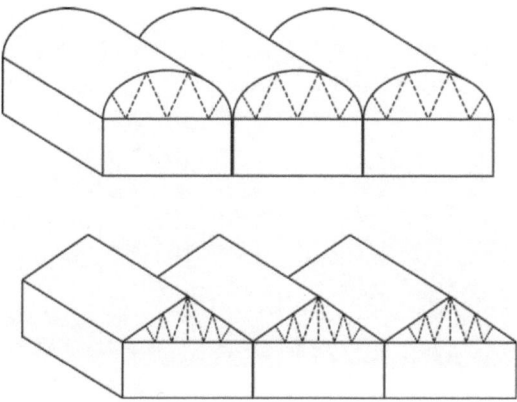

Figure 3.10 Ridge and Furrow Configuration of Greenhouse Structures.

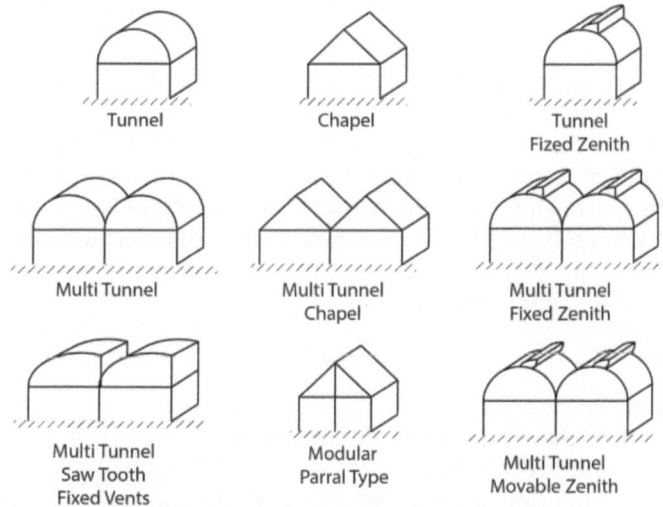

Figure 3.11   Greenhouse Structure Designs based on Mexican Standard.

Greenhouse certification according to Mexican Standard NMX-E-255-CNCP-2008

The greenhouse structure designs according to the Mexican standard are given in Figure 3.11. Note that the labels differ from the above but the geometric shapes are the same; some of these designs add a zenith or roof vent for achieving better ventilation.

Designs description

There are some greenhouse classifications according to different criteria (e.g. material for construction, cover material type, roof features, etc.). However, it is preferred

Figure 3.12 Arch Roof Greenhouse (Multi-Span) (Courtesy of ININSA Greenhouses, www.fabricanteinvernaderos.com/english).

to list the most important ignoring some features for classification. Among the most common types of greenhouses in the world are (Bouzo & Gariglio, 2009):

Arch roof – or tunnel, it is characterized by the shape of its housing and its all-metal structure. The use of this type of greenhouse is spreading due to its greater ability to control the micro-climate, its resistance to high winds and rapid installation with prefabricated structures. The brackets are galvanized iron pipes spaced 5 m × 8 m or 3 m × 5 m apart. The maximum height of such greenhouses is between 3.5 m and 5 m. Side walls adopt heights from 2.5 m to 4 m. The width of these greenhouses is between 6 m and 9 m and they allow multi-span configuration. Ventilation is through lateral and roof windows.

Advantages:

- High transmittance of sunlight.
- Good indoor air volume (high thermal inertia).
- Good resistance to winds.
- Totally free interior space providing easy displacement, mechanized tillage, crops driving, etc.
- Construction of medium to low complexity (due to the availability of prefabricated elements).

Disadvantages:

- High cost.
- They face ventilation difficulties if they are built in multi-span and there is no roof vent system.

Standard peak – or chapel, is one of the oldest structures, used in forced crop cultivation. The slope of the roof is variable according to radiation and rainfall (usually varying between 15° and 35°). Width dimensions vary between 6 and 12 m (even higher) for variable length. The height of the lateral range between 2.0 m to

2.5 m and 3.0 m to 3.5 m the ridge (also built lower than those indicated, but are not recommended). The ventilation of these greenhouses in single span have no difficulties, becoming more difficult when these are gutter-connected.

Advantages:

- Construction of low to medium complexity.
- Use of inexpensive materials depending on the area (eucalyptus poles and timbers, pine, etc.).
- Side ventilation is very easy. It is also easy to install roof windows.
- Suitable for both covering materials flexible and rigid.
- It has great facilities for draining rainwater.

Disadvantages:

- Ventilation problems with gutter-connected greenhouses.
- It has less enclosed volume than curved greenhouses with the same peak height.
- Larger number of elements that reduce light transmittance (greater shading).
- Internal support elements hinder the movement and location of crops.

Saw tooth – A variation of chapel greenhouses, which was first used in areas with very low rainfall and high levels of radiation were greenhouses that had a single roof tilted at angles ranging from 5° to 15°. The lateral coupling of such originated the greenhouses known as "saw tooth". The need to evacuate water from precipitation determined an inclination in collection areas from the middle towards both ends.

Advantages:

- Construction of medium complexity.
- Excellent ventilation which differs from multi-span chapel greenhouses.
- Use of inexpensive materials depending on the area.

Disadvantages:

- Shading much greater than chapel due to greater number of supporting structural elements.
- Low volume of air enclosed (for the same peak height) than chapel.

Modular Parral type – These greenhouses are originated in the province of Almería (Spain), made up of poles and wires called "parral" which are a modified version of the structures used to grow table grape vines. Currently, there is a modern version of the original built with galvanized pipes as indoor supports, the use of posts remain for lateral tension holding wind loads. These greenhouses generally have a ridge height of 3.0 m to 3.5 m, the width is variable, ranging within 20 m or more depending on the length. The slope is almost nonexistent, or in areas with high rainwater risk, usually is between 10° to 15°, which represents a lateral height of about 2.0 m to 2.3 m. It is vented only through lateral openings.

Advantages:

- Low cost construction.
- Large volume of air enclosed (good behavior depending on the thermal inertia).

Figure 3.13 SawTooth Design (Courtesy of Murat Kacira, www.ag.arizona.edu).

- Negligible incidence of roof elements in the interception of light.
- High wind load resistance.

Disadvantages:

- Poor ventilation.
- High risk of breakage by heavy precipitation due to the low drainage capacity.
- Construction of high complexity (requires specialized personnel).
- In areas of low radiation, the low sloped roof represents a low uptake of sunlight.

Venlo house – These are glass greenhouses where the panels rest on the rainwater collection channels; they are generally used in Northern Europe. The width of each module is 3.2 m and the spacing between posts in the longitudinal direction is about 3 m. These greenhouses have no side windows (maybe because in Holland there are not many demands for ventilation). Instead, it has roof windows, opening alternating (one to one side and next to the other) whose dimensions are 1.5 m long and 0.8 m wide.

Advantages:

- The better thermal performance due to the type of material used: glass, and rigid materials currently.
- High degree of control of environmental conditions.

Disadvantages:

- High cost.
- The transmittance is affected, not because of the cover material, but by the large number of supporting elements due to the weight of the cover material.
- Being a rigid material, lasting several years, their light transmission is affected by dust, algae, etc.

Figure 3.14 Venlo House (Courtesy of Duijnisveld & zn b.v., www.duijnisveld.nl).

Greenhouse design by climate

Not all greenhouses are designed equal. A design that works well in a cool climate with long cold winters, snowfall, low light and high winds will not be the best design for a humid, tropical climate with variable light intensity. Different greenhouses are characterized by the level of protection from the outside environment they can offer and the capability they can provide growers to control the inside environment to a specific set of conditions. The level of protection required depends on the type of crop being grown and the local climate. The objective with building any greenhouse is to find a design that will allow the grower to overcome the most limiting climatic problems in their particular area and obtain the maximum growth rates possible from their crops. The needs of greenhouse structures due to different climatic zones that can be found around the world are (Morgan, 2012):

Dry tropical or desert climates – The main environmental threats are high winds carrying dust or sand, which can blast both crops and greenhouses. Simple tents with poles constructed with high-tensile steel wires to form a basic framework over which a single layer of fine insect mesh is stretched and secured around the edges. Humidity can be increased by fogging or misting, which also acts to reduce temperatures. More advanced hi-tech, computer controlled and air conditioned structures are also in use in climates like this.

Subtropical desert and Mediterranean climates – A structure that can be heated but still maintains a cool environment in summer is necessary. In this type of climate a suitable structure is the "pad and fan" cooled plastic greenhouse with top vents and heating. Along with shading over the outside of the greenhouse, this produces an ideal environment during dry summer conditions.

Humid tropical climates – Good tropical greenhouse designs can be as simple as a rain cover or plastic roof with open or roll-up sides covered with insect mesh. In larger greenhouses, the structure is best designed with a "saw tooth" roof layout which allows good venting of the hot air inside the greenhouse on clear days. Heating

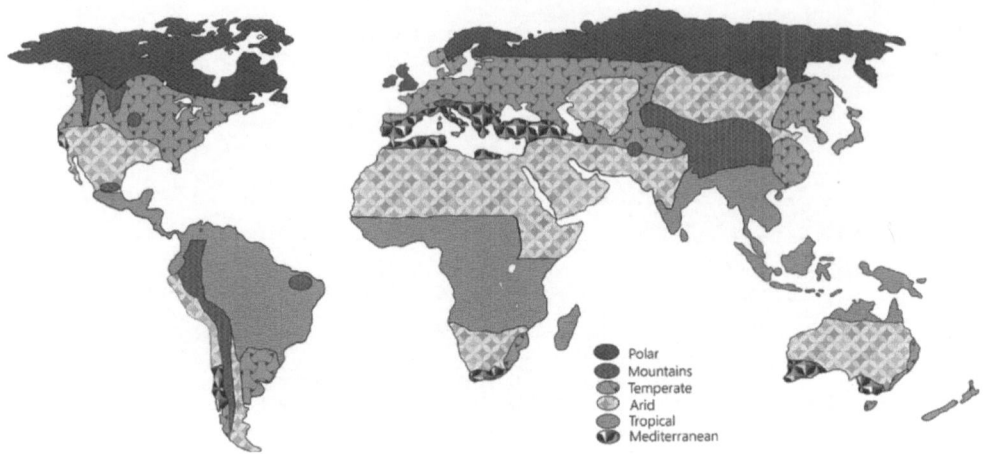

Legend:
- Polar
- Mountains
- Temperate
- Arid
- Tropical
- Mediterranean

Figure 3.15 World Agro-Climatic Zones.

and insulation aren't required and vents can remain open. Misting systems and air-movement fans can be used to cool the environment inside this type of structure and movable thermal screens can be employed to reduce incoming sunlight on bright, cloudless days and pulled back to allow maximum light penetration under overcast conditions. High winds from tropical typhoons or hurricanes can be a major risk in this climate.

Temperate climates – Efficient heating of the air inside the greenhouse and insulating and maintaining this heated air is the main consideration. Growers wanting year-round high growth rates and maximum yields in these environments usually select greenhouses featuring fully clad side walls, roof and side vents, allowing large ventilation areas and computer control of environmental equipment such as heaters, shade or thermal screens, fogging and vents. Temperate zone greenhouse designs often make use of plastic cladding "twin skins" where the space between the two layers of plastic is inflated, offering improved insulation and better environmental control.

Cold temperate climates – Greenhouses for this type of environment need solid walls and strongly constructed, comparatively steep solid roofs to carry snow loads that would collapse plastic film structures. These greenhouses are often double insulated by installing plastic film on the inside walls and positioning retractable thermal screens across the eaves at stud height. To prevent heat loss, vents are often kept closed during the winter months.

The world agro-climatic zones are shown in Figure 3.15. At this point, Table 3.6 shows the most suitable greenhouse structure designs taking into account the above information.

Most modern hydroponic greenhouses for all climates these days feature a stud height of at least 3.05 m and sometimes much more. Regardless of the type or design of the greenhouse or what crop is being grown, a tall greenhouse structure provides

Table 3.6 Most Suitable Greenhouse Designs by Climate.

Climate Type	Suitable Greenhouse Design
Dry tropical/desert climates	Arch roof & standard Peak
Subtropical desert/ Mediterranean climates	Modular Parral Type
Humid tropical climates	Saw tooth, arch roof & uneven span
Temperate climates	Arch roof & standard peak
Cold temperate climates	Venlo house

a better environment for plants and a larger buffer against minor changes in external temperatures. The resulting improved capacity for air movement is a necessary aspect of modern greenhouse cropping that has been shown to benefit numerous crops by improving transpiration and reducing disease. The volume of air that needs to be heated in cooler climates can be reduced by pulling thermal screens across the greenhouse roof at night and heating only under the screen, this creates a large insulation layer above the screen and under the greenhouse roof, thus slowing the rate of heat loss through the cladding (Morgan, 2012).

For improved crop production and quality, a careful selection of greenhouse structure, glazing, and climate control system is required. All greenhouses should be designed properly to withstand all possible load factors for safety and proper functionality purposes. The National Greenhouse Manufacturers Association (NGMA) publishes standards that gives guidance for determining design loads for greenhouses (Kacira, 2013).

3.1.7 Standards for the construction

Little attention has been paid to the role of a greenhouse's structural strength in relation to a more competitive product. The geographical location has a definite bearing on structural needs and greenhouse design also. Areas prone to snow must consider ice and snow load, those adjacent to sea are concerned with wind factors. Whereas high rainfall areas must be interested in the rain intensity in view of greenhouse design.

Greenhouse manufacturers could conceivably manufacture greenhouse designed for 10–15 different total load conditions, but they cannot afford to customize each job and therefore retain only two to three design based on load factors needs (Tiwari, 2003). There are several standards for greenhouse construction around the world but most significant are the European standard (EN-13031-1) and the American standard (ANSI A 58.1-1972) provided by the National Greenhouse Manufacturers Association (NGMA). Attached to these, there are also standards for Latin American countries such as Mexico which has the standard NMX-E-255-CNCP-2008 for the construction of greenhouses. These three mentioned standards will be described below.

Greenhouse: Design and construction EN-13031-1

The European standard EN-13031-1 defines a greenhouse as a structure for growing and/or to provide plant and yield protection, optimizing the solar radiation

Table 3.7 Greenhouse Classification according to EN-13031-1.

2* Classification	Minimum Design Working Life		
	15-year	10-year	5-year
Class A	A15	A10	–
Class B	B15	B10	B5

transmission under controlled conditions, to improve the crop environment and whose dimensions allow people to work inside. EN-13031-1 gives rules for structural design and construction of greenhouse structures for the professional production of plants and crops.

According to the standard greenhouses should be designed by verifying that no relevant limit state is exceeded. The relevant limit states to be considered depend on the class of the greenhouse. On the other hand, a greenhouse should be classified in accordance with a minimum design working life (5, 10 or 15 years) and the tolerance to frame displacements of the cladding systems as seen in Table 3.7.

Greenhouses are divided into two classes, A and B, depending on the tolerance to possible frame displacements (UNE EN-13031-1, 2002):

Class A greenhouses – Greenhouses in which the cladding system is not tolerant to frame displacements, resulting from the design actions. Class A greenhouses should be designed by considering serviceability limit states (SLS) as well as ultimate limit states (ULS).

Class B greenhouses – Greenhouses in which the cladding system is tolerant to frame displacements, resulting from the design actions. Class B greenhouses may be designed by considering the ultimate limit states only.

Recommendations: Glass-covered greenhouses have to be designed according to A15, sophisticated plastic film covered multi-span greenhouses according to B15, and simple plastic film tunnels and shade houses according to B10 and B5.

The *serviceability limit states* correspond to states beyond which no longer fulfil the criteria specified in the proposed service. These include (Muñoz et al., 2013):

- Deformations that can affect the appearance or effective use of the structure (including the malfunction of machines or services) or may cause damage to non-structural elements.
- Vibration that can cause discomfort to persons, damage to the building or its contents or limiting their functional efficacy.

The *ultimate limit states* are those associated with the collapse of a structure which endangers human life. The ultimate limit states which may require consideration, include (Muñoz et al., 2013):

- Loss of equilibrium of the structure or part of it, considered as a rigid body.
- Loss of bearing capacity, due for example to: break, unsteadiness, fatigue or other agreed limit state, such as excessive stress and strain.

The main loads or actions to be considered are (UNE EN-13031-1, 2002) (Muñoz et al., 2013):

Permanent load – or dead load, self-weight of structural and non-structural elements, excluding the installations even if they are permanently present.

Permanently-present installation load – These loads are caused by systems installed on the greenhouse structure. These may include: heating, lighting, shading, irrigation, ventilation, insulation and cooling.

Wind load – Action imposed on the structure by wind.

Snow load – Considered in regions with snowfall.

Crop load – Considered where structures support crops. Where crops are suspended on separate horizontal wires the horizontal tensile forces transmitted to the structure have to be taken into assessment.

Concentrated vertical load – Corresponds to the loads produced by the weight of the people when performing maintenance and repair operations.

Incidentally-present installation load – Correspond to variable loads caused by temporary mobile equipment such as cleaning equipment, rails etc.

The different actions should be considered in combinations which are given in the standard. The European standard EN-13031-1 is valid for the EU (Europe Union). It can be taken as an example also for other regions outside the EU.

National Greenhouse Manufacturers Association (NGMA) Standards

Uniform building codes are used as a guideline for construction requirement. Some state government departments have modified the codes providing leniency; others have adopted them as a minimum standard. Structures will withstand a certain weight applied to them either vertically or horizontally before they collapse. The weight imposed, commonly referred to as load includes the following (ANSI A 58.1-1972); they differ from the European standard in some names and definitions (NGMA, 1996):

Dead load – Weight of permanent components like heaters, water pipeline, and all fixed service equipment.

Live load – Weight superimposed by use: framing inside greenhouse for misting, climbing of green plant, mounting of measuring instruments, handing baskets, shelves, etc.

Wind load – Load caused by wind from any horizontal direction. They are determined by correlation of mean recurrence internal and the fastest speed for 25, 50 and 100 years. All permanent structures require wind loads to be at least calculated from 50-year recurrence means unless the structure has no human occupants. Where there is negligible risk for human life, a 25 near mean recurrence interval may be used. Wind loads are based on height of building above ground (greenhouse will fall in the less than 10 m category) versus the surrounding terrain such as in cities or hilly areas, suburban or open country.

Snow load – Greenhouse structures should also be designed for snow load conditions. Snow load designs are based on climatologic data of particular location. Once again, because a greenhouse has a negligible use over the span of a human life, a 25-year recurrence interval may be used. Snow load on greenhouse depends upon

Table 3.8 Minimum Values of Greenhouse Design Loads.

Load	Minimum (kg/m²)
1. Dead Load	
Pipe frame, double PE	10
Buss frame, lapped glass	25
Hanging baskets, crops	20
such as tomatoes	
and cucumber	
2. Live Load	25
3. Wind Load	Depending upon prevailing wind speed
4. Snow Load	75
(heated greenhouse)	

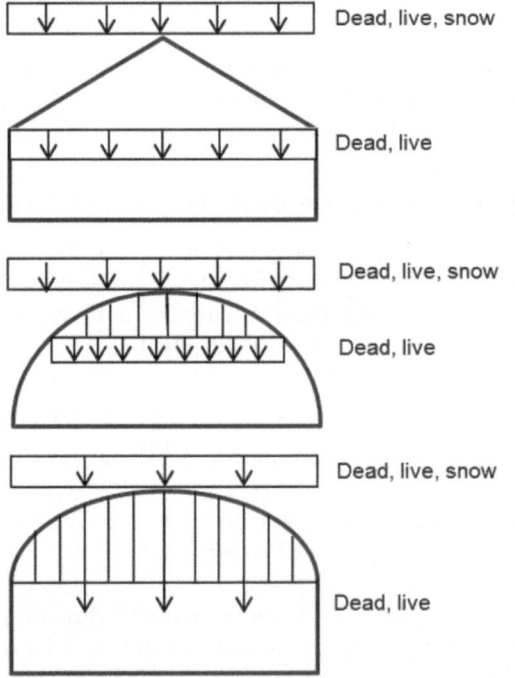

Figure 3.16 Design Loads on Different Greenhouse Structures.

roof slope, snowfall on the ground, single span or multi-span, and weather heated or unheated. A snow depth of 7.5 cm (centimeters) (wet) or 30 cm (dry) is equivalent to 2.5 cm depth of water.

Table 3.8 gives the minimum design values for load on greenhouse and Figure 3.16 shows the load effect on different greenhouse structures.

As the uniform building code does not fit completely to the greenhouse structures, the NGMA was established in 1959 in the US. NGMA adopted a standard in 1975 for greenhouse construction.

Design components are supposed to carry the following loads (NGMA, 1996):

i.   Dead load.
ii. Live load minimum 700 Pa (Pascals) on vertical horizontal projected area.
iii. Wind load minimum 950 Pa on vertical projected. In designing for the above loads, the loads may be considered to act in any of the following combination:

   a.   Dead load plus live load.
   b.   Dead load plus wind load plus 1/2 live load.
   c.   In addition to above, roof bars, purlins and rafters shall be capable of carrying a minimum 4700 N (Newtons) concentrated load at the center of any span.

In the case of Mexico, on July 8th 2008, the Ministry of Economy, through the Directorate General of Standards, published in the Official Journal of the Federation, the Declaration of validity of the Mexican Standard NMX-E-255-CNCP-2008 for Greenhouses Design and Construction Specifications (AMCI, 2008).

Mexican Standard for Design and Construction of Greenhouses NMX-E-255-CNCP-2008.

Highlights of the Standard to consider (AMCI, 2008):

Materials used in general Structures

The materials of the structures should be economical, lightweight, strong and slender; should form little bulky structures to avoid shadows on the same plant, easy construction, maintenance and preservation, modifiable and adaptable to future growth and expansion structures.

- Foundation anchors, columns, bows, arrows, stringers and reinforcements: square or round tubular galvanized steel profile based on a G-90 layer on both sides. Metallic zinc based on the welding seam. Different sections.
- Gutters and fastener profile. Galvanized steel sheet based on a G-90 coating on both sides, various gauges.
- Cable. Galvanized Steel G-90 layer, several measures.
- Wires. Galvanized low carbon steel G-90 various calibers.
- Spring fastener. Galvanized high carbon steel.
- Screws. High strength galvanized G-5 several measures.

The concrete should have a resistance $f'c = 150 \, kg/cm^2$ to manufacture bases for foundations.

Ventilation

For effective ventilation it is recommended that the ventilation area be approximately equal to 15% to 30% of the floor area occupied by the greenhouse. The level of cooling is enhanced when the curtains of the side walls are included in the total ventilation area.

Norms of Materials

Regarding to meet the standards of materials, for steel used in the structure of a greenhouse, builders must meet the following specifications according to the manufacturer:

- Square or round profile steel manufactured according to standard NMX-B-009, steel grade 30 (Fy $= 2320$ kg/cm^2); cold rolled.
- The coating of these profiles should be heat galvanized zinc-aluminum, AZ-90 coating (0.90 Oz/ft$^2 = 274$ gr/m$^2 = 0.0015$ in, according to NOM-B-469, ASTM-792), which the material must provide resistance to corrosive environments.
- The structures must have 5 to 6 kg/m^2 galvanized steel. The columns must be at least 2″ and the arches of 1 3/4″.
- Covers with Polyethylene caliber 720 treated against UV II, different percentages of shade and color.

3.1.8 Greenhouse costs

Greenhouse cultivation is capital-intensive technology both in terms of initial investment as well as operating cost (heating and cooling methods) when compared to the traditional open field cultivation. It is therefore, essential that the agriculture produce from the greenhouse is able to not only offset the higher cost of cultivation but also register adequate profits. This consideration alone should be the criteria for selection of suitable crops for cultivation in greenhouse at any given geographical location.

For an agriculture enterprise based on greenhouse technology, it is necessary to work out its economic viability so that the users may know the relevance and can use the area under their command to their greatest advantage. Greenhouse crops can be the most profitable if the cultivation can be carried out within a period of 3 to 4 months, a short duration compared to open field cultivation.

The cost analysis of greenhouse technology mainly depends on the following factors (Tiwari, 2003):

- Initial investment for construction of greenhouse.
- Initial cost of heating and cooling systems (ground air collector, earth air heat exchanger, evaporative cooling and misting, etc.).
- Operating cost of heating and cooling systems.
- Annual maintenance cost, which includes cost of seeds, preparation root media, irrigation, labors, etc.
- Finally annual cost of crops produced inside the greenhouse.
- Life of greenhouse system and its salvage value.

The environment inside a greenhouse should be maintained to an optimum level for a given crop for higher yield particularly during off-season either by passive or active methods. Greenhouse costs are dependent on the percentage of technology involved in its construction. Table 3.9 presents some cost approximations for different types of greenhouse technologies; these represent the initial investment for a greenhouse that lasts over 10 years.

Broadly speaking, the calculation of return on investment could be calculated in the following way for a hi-tech greenhouse producing tomatoes in one hectare.

Table 3.9 Greenhouse Technologies Costs.

GH Tech.	Structure & Glazing	Canopy	Root	Culture Method	Cost ($/m²)
High	Steel or aluminum Glass, PE or PC	Forced vent. Evap. cooling Hot water pipe CO₂ enrichment Shading Energy blanket	Soil-less subs. Rockwool Coir Drip irrig. with full automated control EC control	High wire culture Fully computerized fertigation Recirculation hydroponics	100–200+
Medium	Steel Double PE or rigid plastic	Passive/active cooling (vents + pad/fan) With/without air heating Basic level of computer control	Soil/soil-less substrate Drip irrig. Some control	High wire culture Longer season Usually computerized fertigation	30–100
Low	Wood or steel Single layer PE film	Passive cooling roof and side wall vents No heating	Soil or tezontle sand Drip irrig. with manual control	Med-high wire	25–30

The total cost of the greenhouse will be 150 \$/m² × 10000 m² (1 Ha) which equals \$1,500,000 of initial investment. Let's say in that area of production the grower can have 40,000 tomato plants and each plant will give around 8 fruits per harvest. In a hi-tech greenhouse tomatoes can be harvested three times a year. Thus in a year the grower might collect around 960,000 tomatoes, each one weighing about 250 gr for a total of 240,000 kg. If the price of a Kilogram is \$2.00 then the grower will earn \$480,000 after a year, which indicates that the producer will have a return in investment in about 3.5 years. This calculation does not contemplate fixed or variable costs, only the initial investment.

Chapter 4

Greenhouse materials and structures

A greenhouse is designed in such a way that it can be verified that it does not exceed the load limits at any time. This is determined according to the type of greenhouse and the materials used. For an intelligent greenhouse containing high value equipment and crops, they should not be designed for a life-cycle less than 10 years.

Regarding the durability, it has been established that the protection against corrosion must not damage the structure integrity at least during the greenhouse life-cycle.

The most common materials used for the structure, cladding and floor are:

Structure materials

There are several frame types being used to construct greenhouses. Depending upon the needs some are better than others.

Galvanized Steel – Mechanical properties of high resistance to stresses. It can withstand the onslaught of both wet and extremely dry climates. Ensures structural integrity against natural disasters as high speed winds. Usually greenhouses whose structure is made of this material use polyethylene or glass claddings.

Wood – Low cost and versatility. The types of wood most used are pine and maple. Steel and wood structures possess concrete foundations to ensure structural integrity against natural disasters.

Low Carbon Steel AISI 1010 – Mechanical properties very similar to those of galvanized steel. It is cheaper and it has less environmental impact than galvanized steel. Galvanized steel has a chemical protection which makes it more durable when exposed.

High-Strength Low-Alloy HSLA 340 – It has mechanical properties that favor the greenhouse structural integrity. It is used in some tubes, bases and joints of the structure as well as screws and nails. Not recommended for use in agricultural production areas. Low durability compared to galvanized steel.

Aluminum – It is very light and useful in countless industrial processes. Not used to build structures due to having poor resistance to stresses. It is advisable to use near places where food is produced.

Cladding materials

Greenhouse designs vary widely with the type of greenhouse being the principal decision facing growers. The cladding will drastically affect the amount of sunlight reaching the crop. The cladding will also determine heat loss of the structure. The most common materials are:

Low Density Polyethylene (PE) – Low cost. Resistant to extreme weather conditions, molding ability and it is light hence it does not cause large loads on the structure. Not readily degradable so its use must be measured. PE film only lasts about 2 years. Clear PE is used for growing most plants, but white PE can be used to reduce light and heat for growing lowlight plants or for propagation.

Glass – The traditional greenhouse covering against which all others are judged. Good-quality glass is an attractive, very transparent, and formal (in appearance) covering material. High cost. It is not as malleable and adjustable as polyethylene. It is heavier so the structure would be subject to greater stresses. For use in non-aggressive environments due to its fragility.

Polyvinyl Chloride PVC – The most economical and used choice. Widely used due to its versatility, malleability and mechanical properties. It is lightweight so it does not generate large stresses in the structure. Not environmentally friendly, slow to degrade.

Vinyl Sheet – Heavier than polyethylene, more durable and considerably more costly. If made with an ultra-violet inhibitor it can last as long as five years. Also, like polyethylene, it has electrostatic properties that attract dust, which clouds the sheeting and therefore cuts down the transmission of light.

Polyester – The best known polyester film is mylar. In the 5-mm thickness used for greenhouses it has the advantages of being lightweight, strong enough to resist damage by hail, it is unaffected by extreme temperatures and has light-transmission characteristics quite similar to glass. Mylar is, however, expensive. It will not be so effective when used on poorly built frames that are rocked by wind.

Fiberglass Reinforced Panels FRPs – Rigid plastic panels made from acrylic or polycarbonate that comes in large corrugated or flat sheets. They are durable, retain heat better than glass does, and are lightweight. The greatest advantage of fiberglass is its exceptionally high resistance to breakage, means it should last between 10 and 15 years. This panels are especially attractive where light intensity is high.

Acrylic Semi-Rigid – Usually flat acrylic panels are ideal for greenhouses because of their strength, light weight, resistance to sunlight and good light-transmission characteristics. They do scratch easily, but apart from this their principal disadvantage is their very high cost. However, acrylic is worth the outlay as it will give good service for many years.

Floor materials

A foundation is one of the most important parts of the greenhouse. No matter where the foundation lies, it must be level and square. The materials of greenhouse floors range from bare ground to concrete and some examples are:

Standard Concrete – Regular concrete will endure about 2500 pounds per square inch. This mix is appropriate for heavy loads such as soil-mixing areas and locations in the greenhouse where heavy equipment is used.

Porous Concrete – Allows drainage preventing puddling and still provides a barrier for weed control. Properly cured porous concrete will have a capacity to endure 600 pounds per square inch of surface. A four-inch floor of this mixture will adequately endure light equipment and personnel.

Gravel or Dirt Floors – They are inexpensive but often not worth the initial savings. Floors will be mud with frequent irrigations and will generally appear unacceptable. In fact, these muddy and unstable floors are a problem in terms of the cost and pollution to the extent that they have prompted the search for substitutes in the shape of eco-materials.

Eco-materials

Some of the materials used for building a greenhouse could not be environment-friendly due to their long period of degradation inducing air pollution. Eco-materials appear to handle this situation, these are the materials which combine less environmental burden in production with high recyclability, and realize more effective utilization of material. Materials should be friendly not only to the nature but also to mankind.

Eco-materials can be referred to the materials used for the life-cycle product design developed in order to protect the environment. There are seven elements of eco-efficiency such as:

- Reducing the material requirements for goods and services.
- Reducing the energy intensity of goods and services.
- Reducing toxic dispersion.
- Enhancing material recyclability.
- Maximizing sustainable use of renewable resources.
- Extending product durability.
- Increasing the service intensity of goods and services.

The eco-materials can be classified into four main categories:

- Non-linear source materials.
- Materials for ecology and environmental protection.
- Materials for society and human health.
- Materials for energy based on two main criteria as their sources and functions.

Some examples for eco-materials are showed in the next list. As it can be seen, some combinations of eco-materials may serve for the greenhouse structure and floor in future developments.

Greenhouses are required to allow high light transmittance, low heat consumption, sufficient ventilation efficiency, adequate structural strength and good overall mechanical behavior, low construction and operating costs.

Greenhouses are designed taking into account safety, serviceability, general structural integrity and suitability. The location, size and frame design of the greenhouse determines the type of material and structure form to be used during the construction of the greenhouse. The components of greenhouse structures; poles, ridges, belts, and downspouts. The greenhouses' mainstay could be made of glass, whose mission is to secure the cover to protect the crop. Besides protection, they support any loads, such as

Table 4.1   List of Eco-materials Applications.

Categories	Examples
Recycled	Eco-cement, coal ash concrete, marine block.
Renewable	Wood based materials, biodegradable plastic made of a vegetable base.
Material for efficiency	Wear-resistant metals and alloys, pre-paint steels, corrosion resistant steel and alloy.
Material for easy disposal or recycle	Biodegradable plastics, functionally graded material, colorbetos which replaces asbestos.
Hazardous free materials	Chromium-free steel, heavy metals free polyesters.
Materials for reducing human health impacts	Vibration damping steel sheet, sound proof panels.
Materials for energy efficiency	Ultra-light steel, high magnetic induction steel sheets, highly endothermic steel.
Materials for green energy	Selective transparent glass, highly durable sealing sheets for solar batteries.

wind, rain, snow, or irrigation facilities staking, to prevent problems like infiltration, minimal shade during the day, and reduce the problem of nightly temperature drops (usually weather). The ventilation, the shape and the material of the greenhouse (which will be described in this chapter) are factors to be taken into account. It is important to mention that the resilience of the greenhouse, is not only based on the materials it is built of, but also on its design. Among consideration are the structures used, the stresses to which the parties submit (there must be tension and no compression), the braces of these structures should be longitudinal and tangential to counteract the horizontal force of the wind, and have good anchor roofing materials, etc. In respect to generating loads due to the irrigation, refrigeration and other types of equipment, these should not be below 15 kg/m². When staking crops overloads of 14–16 kg/m² should be considered.

Based on these necessities, standards have been created that seek to unify the design, the construction and the production of greenhouses; in Mexico this is the NMX-E-255-CNCP-2008 standard, in Europe the UNE-EN13031-1 and in American the standard (ANSI A 58.1-1972) that is provided by the National Greenhouse Manufacturers Association (NGMA).

Based on the need of having greenhouses for optimum crop, in this chapter a description of the most common used materials to build the structure of the greenhouse will be presented, as well as covers and the type of material required to place on the ground to create the foundations (FAO, 2002, Service 2012). There is also information about eco-materials, which are now becoming more economically and environmentally feasible in greenhouses. In addition there is an analysis of the mechanical properties of the materials used in greenhouses, this information is provided with the aim of determining the behavior that the galvanizing steel has under certain wind speeds and thus to predict if the material is a good choice to build a greenhouse.

4.1 STRUCTURAL MATERIALS

Using the proper type of material to construct a structure is important for the resilience and efficiency of the finished product. Harsh weather conditions (snow, rain, hail, wind or hot weather) are the most common cause of failure of greenhouse structures.

The structure should be able to take all the necessary types of loads; dead, live, wind and snow. The foundation, columns and trusses are designed accordingly. The main loads, which have to be taken into account in the greenhouse design, are (Elsner et al., 2000):

1   Dead load or permanent load.
2   Imposed loads (crop loads).
3   Installations.
4   Snow load.
5   Wind load.
6   Seismic load.

This classification is based on the estimation of the lifetime of the greenhouse from the economic and technical point of view.

To avoid structural damages, stress limits should be considered, which are associated with a structure collapse endangering human life (Muñoz et al., 2002). The ultimate limit states, which may require consideration, are the following: loss of balance of the structure and loss of load bearing capacity due to breakage, shakiness, fatigue, excessive tensions and deformations.

Regardless of the type of material used, the stress limits will be analyzed mechanically in a plane structure, which is a set of discrete elements connected to each other in a rigid way, which is designed to withstand the external forces applied and transmit them to the foundation, without which there would be an excessive breakage or deformation of the material.

There are several frame types being used to construct greenhouses. Depending upon the needs some of them are better than others (Juárez-López et al., 2011, Kumar Jaypuria 2008, Yuste Pérez 2008):

Galvanized Steel: Has mechanical properties of high resistance.

- Advantages:

It can resist both the damage of humid and extremely dry climates.
It ensures structural integrity due to natural disasters such as high-speed winds.
The roof is easier to be built in a rounded shape or in a pointed arch (chapel shaped like).
It can be used with polyethylene coatings or glass.

- Disadvantages:

High cost.
*Should be avoided if it is at all possible the contact of the film with the structural elements heated by solar radiation.

Wood: The pine and maple are the most commonly used, combined with steel tubes and foundations of cement or concrete to ensure their structural integrity in case of natural shocks of great intensity.

- Advantages:

A wood frame provides natural insulation for the greenhouse.
The wood absorbs the rays of the sun during the day and holds in the heat at night.

- Disadvantages:

It may be needed to treat the wood to prevent wood rot, and the chemicals out-gassing from the pre-treated wood can be toxic to the plants that are grown inside the greenhouse.

Low carbon steel (AISI 1010): Contains less carbon than other steels and is much easier to cold-form due to their soft and ductile nature.

- Advantages:

Cheaper than the galvanised steel.
Less impact on the environment

- Disadvantages:

It is neither externally brittle nor ductile due to its low carbon content. It has lower tensile strength and is malleable.

High-strength low alloy Steel (HSLA 340): The steels in the HSLA range are suitable for structural parts. Since they generally have lower carbon than the SS steels, their ductility is superior.

- Advantages:

It is recommended to be used in some pipes and in the foundations.
It is recommended to be used in some basic junctions of the structure as well as in the screws and nails.

- Disadvantages:

Do not use a material such as this in areas where there is planted land or agricultural production as it cannot be left outside for a long time.

Aluminum: Round or square tubing is used, depending on the covering material to be used on it.

- Advantage:

It is a strong material and will not rust, and is lightweight. Drilling holes in it is fairly easy.

- Disadvantage:

Fiberglass panels or polycarbonate sheets could be fastened to the structure.

In terms of structural materials, approximately 50% of the greenhouses have metallic structures, 30% is made of wood and the rest is of mixed structures, understood as the combined use of laminated wood, metallic profiles, etc.

All materials are good when matched to the size, shape and location of the finished greenhouse structure and the most adequate structure will be the best to avoid sudden temperature changes and allow great ventilation. That is to say, that it will provide the best conditions of cultivation.

4.1.1 Structural conformation

Another important factor to have a good structure, in addition to materials, is to determine the structural conformation or external profile (Gasso-Busquets & Solomando-Valderrabano 2011):

Modular parral: It is used in areas of low rainfall. The structure of these greenhouses is composed of two parts: a vertical and horizontal structure.

Multitunel greenhouses (Raspa and amagado): Its structure is similar to the Parral type but it varies the shape of the cover. It increases the maximum height of the greenhouse in the ridge, which ranges between 3 and 4.2 m, forming what it is known as scraped. In the lowest part, known as threatened, the mesh cover is joint to the floor by iron winds and forks that allow to place the gutters for rainwater drainage. The height of the threatened range goes from 2 to 2.8 m of the bands between 2 and 2.5 m.

Asymmetric: Differs from the scraped and threatened types in that it increases of the surface in the exposed south face, with the aim of increasing its capacity for taking up solar radiation. For this reason, the greenhouse is oriented East-West, parallel to the sun's apparent path. The tilt of the cover should allow the solar radiation to impact perpendicularly on the cover at the solar noon during the winter solstice, the time where the sun reaches its lowest point. This angle must be close to 60°, but this causes great inconvenience due to the instability of the structure caused by strong winds. Therefore the angle has to be set between the 7 and 9° in the south face and between the 15 and 23° in the north face.

Standard peak or chapel: The greenhouses of simple chapel have the roof forming one or two inclined planes, according to a water or gable. If the inclination of the planes of the roof is greater than 25°, it offers no drawbacks in the evacuation of the rain water. Ventilation is by the front and side windows. When it comes to structures formed by uniting several ships, in the absence of windows, the ceiling skylights provide the ventilation.

Double chapel: Consist of two naves juxtaposed. Ventilation is better than in other types of greenhouse, due to the zenithal ridge of the two steps that form the juxtaposition of the two ships; these ventilation openings usually remain open constantly and generally have mosquito mesh. In addition they also have a vertical vent in the front and side walls. This type of greenhouse is not widely used due to its construction, which is more difficult and expensive than the simple chapel or gabled type of greenhouse.

Arch roof: Has a greater capacity for the control of climatic factors, great resistance to strong winds and speed of installation due to the use of prefabricated structures. These structures have some variants for example butterfly, Chinese hat, with zenithal ventilation type, etc.

Regardless of the selection of the structure type, the dimensions should be considered, for example the average height of a greenhouse characterizes its volume; a large greenhouse volume results in a slow response of the indoor environment to changes of the external weather conditions. Therefore, higher greenhouses exhibit smaller fluctuations in their indoor microclimate (Elsner et al., 2000b). Higher greenhouses have increased energy consumption and are more demanding in terms of structural stability due to larger wind loads. The height of a greenhouse is optimized with respect to these two competing factors: light transmittance of greenhouses and ventilation.

Considering the variable of light transmittance and ventilation as a reference to optimize the greenhouse, one must also select the best cover, based on the climatic needs of the area and type of crop.

4.2   CLADDING MATERIALS

The characteristics of the cladding material determine the quality of the light transmitted into the greenhouse, and radiation transmittance can be improved qualitatively and quantitatively (Castilla and Hernandez 2007). The cladding material must have characteristic such as: strength, consistency, durability, manufacturing quality control, safety, transmission of solar radiation and energy conservation. Greenhouse designs vary widely and choosing the correct type of greenhouse is an important decision faced by growers. The glazing will drastically affect the amount and type of sunlight that reaches the crop.

The glazing will also determine the heat loss of the structure. When selecting a cladding or cover for a greenhouse, it is important to consider the next question (Giacomelli 2002):

How much energy (light) does it let into the greenhouse, and how much energy (heat) will go out?
What are the purchase, installation, and maintenance costs?
How well can the grower manage the environment which is imposed by the glazing to produce a quality, salable product for profit?

These questions can be answered with the experience of the grower, the crop produced, the glazing, the local outside environment, and the greenhouse environmental control systems. There are also other variables that affect the cover used for the crop, such as transmittance, dust and dirt, the gases and also salt (Zabeltitz 1990):

High transmittance of visible light, photosynthetic active radiation PAR, with wavelengths of 400–700 nm.
Low transmittance of long-wave radiation, FIR, in the range of wave lengths from 3,000–20,000 nm.
Low reduction of light transmittance by global radiation (long duration of life).
Low ageing due to UV-radiation.
No drop-wise condensation on the inside of the roof, but condensation as a liquid film (No-Drop properties).
Low accumulation of dirt and dust due to the type reduced light transmittance.
Endurance against wind.

In order to solve the current necessities there are 3 types of coverings typically used for greenhouses: glass, plastic films and rigid plastic panels (Quiminet.com, 2011, AMCI 2008, Yuste Pérez, M. P. 2008):

Glass:
Glass is the traditional covering. It is available in many designs to blend with almost any style or architecture. Glass greenhouses may have slanted sides, straight sides and eaves, or curved eaves.

Glass: The glass that is used as a cover in the greenhouses is always the printed glass or glass "cathedra". Its thickness varies from 2 to 6 mm, and the plates measure approx., 60 cm. Its thickness is measured in gauges and sold in coils in variable widths (from 80 cm up to 12 m and thicknesses from 200 to 1200 gauges, i.e., from 0.05 to 0.3 mm).

- Advantage:

Good-quality glass is an attractive, very transparent, and formal (in appearance) covering material.
It is very strong (tensile strength).

- Disadvantage:

It is subject to shattering and can become brittle with age.
It is also very expensive.
Its weight requires sturdier framing support than is required with other covering materials.

Plastic films:
The use of plastic is increasing in the building of greenhouses due to its cost per square foot. It can be heated as satisfactorily as glass greenhouses, crops that grow under plastic are of equal quality to those grown under glass, it is considered a temporary structure and usually carries a low assessment rate for tax purposes, or may not be taxed at all. Plastic structures can be made of polyethylene (PE), polyvinyl chloride (PVC), copolymers of these materials, and other readily available clear films. Plastic film is the most applied cladding material in countries with tropical, subtropical and arid climates (Zabeltitz 1990).

Polyethylene (PE): Permits passage of much of the reradiated heat energy given off by the soil and plants inside the structure. When it is used an Ultraviolet-inhibited polyethylene, it is lasts longer than regular polyethylene. It has an inhibitor that prevents the rapid breakdown caused by ultraviolet light. UV-inhibited polyethylene is available in 2- and 6-mil thickness up to 40 feet wide and 100 feet long. Therefore, a polyethylene type loses heat more quickly than glass both during sunny periods and after sunset. This is an advantage during the day and a disadvantage at night. These films are reported to reduce 20% of the heat loss from the greenhouse and have become common in today's industry, especially in Europe.

A newly developed polyethylene film in Israel has been designed to allow very low levels of UV light to be transmitted. There is good evidence that UV blocking films have an adverse effect on flying insects such as *Bemisia tabacci*, aphids and thrips (Merle 2001).

- Advantage:

It is low in cost and lightweight.
It also stands well in fall, winter, and spring weather, and lets through plenty of light for good plant growth.

- Disadvantage:

Constantly exposed to the sun deteriorates during the summer and must be replaced each year.
Ultraviolet light energy causes polyethylene to break down.
Low density polyethylene (PE): This is a good choice for home-built greenhouses.
Clear PE is used for growing most plants White PE can be used to reduce light and heat for growing lowlight plants or for propagation. In the majority of the polyethylene films for greenhouse, the point of yielding occurs around a 2.6–3% of stretching. The most secure point of tension in the middle of the point of yielding is between a 1.3–1.5% of the length of the film that it is taut.

- Advantage:

Less structural support is required.
It costs much less than other materials.
Easy to adapt to all types of structures.
Great flexibility.
It resists well the mechanical effects of the wind and hail and is tear resistant.
Does not deteriorate by chemical substances that can be used inside the greenhouse.
Easy to incorporate additives that improve their performance

- Disadvantage:

PE film only lasts about 2 years.
Degrades more quickly than other materials.
It is a poor conductor of heat.
Any deficiency in their placement may cause tears in the plastic.
It is degraded by UV radiation and Oxygen
Polyvinyl Chloride (PVC): Can be used to construct a Quonset type greenhouse frame.

- Advantage:

It is readily available, fairly cheap and easy to work with.
Up to one inch diameter can be bent to form the hoops over which the Polyethylene film can be stretched.
Ends can be constructed from either plywood or fiberglass sheets.
Retains more heat at night, thus avoiding a possible temperature inversion. *Greater resistance to cracking and tearing.
Ages more slowly.
It can be transparent, translucent, photo-selective, bluish, and black.

- Disadvantage:

Fiberglass sheets can also be screwed to the hoops to form a cover over the frame.

Little used (only 2% of all the covering materials are PVC). You have a greater greenhouse effect than that of the PE, due to the fact that it is a poorer heat conductor than this one.

The moisture condenses very little.

Dust quire fixed on the surface.

Vinyl sheet: Used in ends of the greenhouse so customers could have a nice, clear view of the facilities. Not intended for use as covering for the top of the greenhouse. The clear vinyl siding is supposed to last about 4 years.

- Advantage:

It is heavier than polyethylene, more durable and considerably more costly if made with a ultra-violet inhibitor.

It can last as long as five years.

- Disadvantage:

Like polyethylene, it has electrostatic properties that attract dust, which clouds the sheeting and therefore cuts down the transmission of light.

Rigid plastic panels:

Polyester: The best known of the polyester films is Mylar, with a 5-mm thickness it is used for greenhouses construction; it has the advantages of being light.

- Advantage:

It is strong enough to resist damage by hail, it is unaffected by extreme temperatures and has light-transmission.

Similar characteristics to glass.

- Disadvantage:

It is however expensive.

It will not be so effective when used on poorly built frames that are rocked by wind.

Fiberglass Reinforced Panels (FRPs): Are rigid plastic panels made from acrylic or polycarbonate that come in large corrugated or flat sheets, their greatest advantage is their high resistant to breakage. This factor, coupled with its good resistance to ultra-violet, means it should last between 10 and 15 years.

- Advantage:

Are durable, retain heat better than glass does, and are lightweight.

The panels are either flat or corrugated.

The light admitted to the greenhouse is soft and shadow less.

- Disadvantage:

Make sure that it is not exposed to flame or extreme heat, because it burns readily and rapidly.

Acrylic Semi-rigid: Usually flat acrylic panels are ideal for greenhouses because of their strength, light weight, resistance to sunlight and good light-transmission characteristics.

- Advantage:

Strength.

Light weight.

Resistant to sunlight and good light-transmission.
Will give good service for many years.
It is resistant to snow, strong winds and even hail impacts.

- Disadvantage:

Scratches easily.
It is very expensive, but can last 20 years, so one manufacturer claims it is cost effective over the long run, especially in commercial greenhouses where heating costs can be reduced up to 30%.
Polymethyl methacrylate (PMMA OR PMM): A transparent and rigid plastic, PMMA is often used as a substitute for glass in products such as shatterproof windows, skylights, illuminated signs, and aircraft canopies.

- Advantages:

Has great transparency.
Allows almost all UV rays.
It has high opacity to the nocturnal radiation.
High breaking strength, tear and aging.

- Disadvantage:

Its broadcasting power is almost zero.
Scratches easily and loses optical qualities.
Polypropylene: Has demonstrated certain advantages in improved strength, stiffness and higher temperature capability over polyethylene.
It has been very successfully applied to the forming of fibers due to its good specific strength.

- Advantage:

It is manufactured with a special non-stick treatment to avoid the dust that accumulates on the surface.
Easy to handle in positioning and fixing to the structure.

- Disadvantage:

It is transparent to long wavelength infrared radiation.
Due to its surface that has a high degree of porosity, it can cause rain water dripping, if the structure does not have sufficient slope.
Ethylene-vinyl-acetate or EVA: It is a thermic film necessary for places where the night temperature drops below the optimum temperature necessary for the crop.

- Advantage:

It has a higher thermal effect than with PE.
It is more flexible and more impact-resistant than PE, as well as more resistant to tearing.
More broadcasting power than PE.

- Disadvantage:

High dilation that results in rain water bags and break wind.

But independently from the material, when choosing a glazing, it is important to consider the following requirements Smith, S. (2002):

Fire resistance
Hail rock resistance
Guaranteed life span
Energy efficiency
Security
Ease of application
Light transmittance

Fiberglass covers some of these requirements, it is one of the best light diffusers; polyethylene, polycarbonates, acrylics and glass follow roughly, but they are not the best options because it is necessary to consider durability, aesthetics, strength and cost.

Finally it is important to mention that there is no ideal glazing; according to researchers such as Smith (2002) and Zabeltitz (1990), most glazings developed for greenhouses allow satisfactory growth but have advantages and disadvantage:

The light loss depends on the cladding material and on the region (industrial or rural).
The influence of dust on light transmittance can be higher with No-Drop film than with standard PE film.
Cladding materials with special spectral transmittance can have various effects on greenhouse climate, plant behavior and pest control.
Photo selective plastic films and screens will be used as cladding materials for green- and screen houses as well as insect screens in greenhouses.

Furthermore, based on the types of materials and structures, as well as the covering materials a selection material can be performed that will serve to raise the floor and bases that would support the entire structure. This is very important as it helps on the safety and durability of the structure.

4.3   FOUNDATIONS AND FLOOR MATERIALS

A foundation is one of the most important parts of the greenhouse. No matter where the foundation lies, it must be leveled and square.

According to the Construction Specifications and Regulations SNiP II-B.1-62 (Kiselev, 1975), the depth of the foundations on heaving soils depends on the depth of freezing of the soils and is taken to be no less than the calculated value of the latter. However, for lightweight structures with shallow foundations the construction measurements taken with respect to the depth of the foundations do not provide stability for the buildings, since under the effect of the tangential forces of frost heaving considered non uniform, the vertical displacement of the foundation occurs in time as a consequence of the accumulation of residual heaving deformations during annual freezing and thawing of the soils. The concrete foundation of a greenhouse should fulfill the following requirements (Zabeltitz 1990):

1   It should safely sustain and transmit the loads of the greenhouse to the ground.
2   The footing of the foundation should rest on undisturbed soil at a depth of about 500–600 mm below the ground surface.

The materials for greenhouse floors go from bare ground to concrete: some examples are (Schnelle & Dole 1990, Smith, 1992):

Standard Concrete: Regular concrete will endure about 2500 pounds per square inch.

- Advantage:

This mix is appropriate for heavy loads such as soil-mixing areas and locations in the greenhouse where heavy equipment is used.

- Disadvantage:

It will not drain properly.

Porous concrete: Allows for drainage, will help prevent paddling, and still provides a barrier for weed control. Concrete will have a capacity to endure 600 pounds per square inch of surface.

- Advantage:

A four-inch floor of this mixture will adequately endure light vehicle traffic and personnel.

- Disadvantage:

Often not worth the initial savings.

Gravel or dirt floors: Use only with some fine rocks to stabilize the floor. These surfaces drain well and are great on hot days.

- Advantage:

It can be watered down and the heat of the day will cause evaporative cooling.
It is cheap to purchase and easy to install.

- Disadvantage:

Floors will be chronically muddy with frequent irrigations and will generally appear unacceptable, particularly in retail operations.

These muddy, unstable floors will be a liability because of the risk of customers falling and injuring themselves. But all floors have a problem: the cost and pollution they produce had led to look for others solutions such as eco-materials.

4.3.1 Connections and clamps

All steel components of the greenhouse structure should be connected by screws or clamps (Zabeltitz 1990), as it is important for the wind resistance of the greenhouse structure and the stable connection of the steel components. The clamps must not slide on the tubes, but have to be tightened firmly. After the mounting has been finished, it has to be checked whether all clamps, screws and bolts are screwed and fixed tightly.

4.3.2 Leaks

In the greenhouse structure leaks must be avoided, wherever they occur at doors, ventilation openings, plastic-film fastenings, etc., for the following reasons (Zabeltitz 1990):

1   Solar energy will be stored during the daytime and will keep the air temperature in unheated greenhouses some degrees above outside temperature at night.
2   If there are holes in the plastic film caused by the installation of the fastening clips, those holes are the starting point for damage to the plastic film caused by wind forces.
3   Leaks in the structure, vents and insect screens are not permissible when: the integrated production and protection (IPP) system is used, when there are useful insects inside the greenhouse and when pest insects must be kept out.
4   Leaks in gutters and cladding material cause rainwater penetration, crop flooding and disease infestation.

4.3.3 Windbreak

The relationship between windbreak structures and their function has been summarized as follows (Heisler and Dewalle 1988):

The horizontal extent of wind protection is generally proportional to windbreak height.
The wind speed reduction is related to the open area of the windbreak.
Very dense barriers are less effective than medium porous barriers for wind speed reduction of 10–30% at larger distances.
Height growth of a natural windbreak may be more important than density when areas as large as possible have to be protected.
Natural barriers with width less than height and a steep side produce a larger wind reduction over a greater distance than very wide windbreaks or streamlined windbreaks in cross-section.
Tree windbreaks lose less effectiveness in oblique winds than thin artificial wind screens.
Turbulent wind flow decreases with the increasing of the open area of the windbreak.

4.3.4 Insect screens

Used in front of the ventilation openings and doors to keep useful insects inside and to prevent pest insects from penetrating the greenhouse. Criteria for the choice of insect screens are (Zabeltitz 1990):

The species of insects to be screened out.
The influence on the greenhouse climate.
The UV stability and the mechanical durability (thickness of threads).
The cost in comparison to the economic value of the crop.

A disadvantage of insect screens is the reduction of the ventilation efficiency with influence on temperature and humidity, as well as the reduction of light transmittance.

The main factors of characterization of insect screens are: the porosity, the ratio of the open area against the total area of the screen, the mesh or hole size, the thread dimension (woven or knitted), The light transmittance, the color and its influence on pest behavior.

On the other hand it is important to mention that in addition to the different materials used in structures, claddings and foundations materials, there are new developments that are moving materials towards the improvement of their mechanical properties and to radiation selectivity, quantity and quality (Theorem Ambient, 2002), which were designed to be sustainable and avoid environmental pollution without decreasing their quality or life span. For example in greenhouses components, in terms of covers you can find the following selection: photo-selective plastics, antivirus films, botrytis films, photodegradable films, multilayers, Drip, Biodegradable Films, Films for solarisation. With these kind of materials it is intended to develop longer-life span materials (with no reduction of their properties), one of the best options is to reduce at source the use of plastics, and use instead bamboo structures and foundations for hydraulic cement, all these alternatives belonging to the eco-materials.

4.4 ECO-MATERIALS

Eco-materials create less environmental burden as they are produced with high recyclability, and create a more effective utilization of the materials. These are also often produced with sustainable materials as described in research by (Cornell University, 2010, VOX, G. et al., 2010) among others uses of this materials they are already being used in greenhouses.

The concept of eco-materials was proposed in Japan one year before the Rio summit in 1992 (Nowosielsky et al., 2007), through a discussion among materials scientists and materials engineers.

It can be referred to the materials used for the life cycle products designed and developed in order to protect the environment. There are very important and have seven elements of eco-efficiency such as:

- Reduce the material requirements for goods and services.
- Reduce the energy intensity of goods and services.
- Reduce toxic dispersion.
- Enhance material recyclability.
- Maximize the sustainable use of renewable resources.
- Extend the products durability.
- Increase the service intensity of goods and services.

The eco-materials can be classified into four main categories as (Nowosielsky et al., 2007):

Nonlinear source materials.
Materials for ecology and environmental protection.
Materials for society and human health.
Materials for energy based on two main criteria as their sources and functions.

Table 4.2 Example of Eco-Materials.

Categories	Examples
Recycled Materials	Eco-cement, coal ash concrete, marine block.
Renewable Materials	Wood based materials, biodegradable plastic made of a vegetable base.
Material for efficiency	Wear resistant metals and alloys, pre-paint steels, corrosion resistant steel and alloy.
Material for Waste treatment	Membranes for exhausted gas separation, ion-exchange resins, microbial enzymes, absorbent materials for oil and grease removal.
Materials for reduction of environmental load	Catalysis and biological membrane materials for fuel cells, carbon-fiber composites, photo-catalysis coating materials for construction.
Material for Ease Disposal or Recycle	Biodegradable plastics, functionally graded material, colorbetos which replaces asbestos.
Hazardous free materials	Chromium-free steel, heavy metals free polyesters.
Materials for reducing human health impacts	Vibration damping steel sheet, sound proof panels.
Materials for energy efficiency	Ultra-light steel, high magnetic induction steel sheets, highly endothermic steel.
Materials for green energy	Selective transparent glass, highly durable sealing sheets for solar batteries.

Each property of an engineering material has a characteristic range of values. Some examples for eco-materials are showed in Table 4.2 (Nowosielsky et al., 2007):

They can be divided into three categories based on the relation between a material's properties and its role in improving the environment (Nowosielsky et al., 2007):

Functional materials for environmental protection: Materials properties are optimized to improve each environmental problem.

Materials supporting low-emission systems: Materials properties are needed to support environmentally benign systems.

Materials of strategic substitution for an environment friendly social system; materials are used for a given property but society demands that they have lower environmental burden.

As can be seen, this combination of eco-materials may serve for the structure, gladding and the floor. Table 4.3 shows some examples of eco-materials used in greenhouses (Emerson, R. W. 2009, CSIC 2013, ALECO 2009):

These materials are currently being used, but still need to have a greater dissemination of the benefits they offer to the greenhouses.

As it can be seen there is a large amount of materials to build a greenhouse, which implies the know how to choose the most suitable for cultivation. The following explains how to make the selection of the optimal materials to build a greenhouse.

4.5 MATERIAL SELECTION FOR A GREENHOUSE

Material selection involves using a material database to select the best material for a product. By following selection techniques, the most appropriate is selected from all known materials in order to satisfy product requirements and goals.

Table 4.3 Eco-materials application in greenhouses.

Type	Description	Advantages	Disadvantages
Bamboo greenhouses	Bamboo has twice the compression strength of concrete and roughly the same strength-to-weight ratio of steel. Bamboo poles are treated with non-toxic borates to prevent termite and powder post beetle infestations as well as decaying fungi.	Captures carbon dioxide Bamboo poles are able to withstand strong winds and earthquakes.	Bamboo is not easily available.
Natural origin Biodegradable plastic.	Synthesis of the polyester polialeurato. It is not used solvents or catalysts, which allows savings in raw materials, a reduction of waste to remove and the obtaining of a material of higher purity.	Decreases the collection time. Minimum environmental impact. Do not use toxic solvents. It is a perfect substitute for the current plastic.	Cost
Biodegradable Plastic potato and corn starch	They respected the size, width, height, thick and even color of commercial plastics such as polyethylene.	Always will be 100% biodegradable made with biodegradable corn starch, wheat and potatoes.	The cost can be double or triple in the manufacturing process.

The material selection method that is most widely used in practice is the method developed by Ashby (1999). This process begins with a database containing all known materials, after screening and ranking techniques to reduce the number of feasible materials based on product geometry and loading conditions. After the first candidates have been selected, local load conditions combined with design requirements lead to the final material choice.

The basic question is how we go about selecting a material for a given part. This may seem like a very complicated process until we realize than we are often restrained by choices we have already made. Some criteria that can be made to choose a material are:

Availability and price: (cost of the material, labor and equipment, and maintenance thereof).

Specialization of workers and equipment needed.

Safety.

Speed of manufacturing.

Consumption (natural water and energy) resources.

The material properties (mechanical stiffness, strength, ductility; technology: physical, chemical and mechanical), modulation, composition and adaptation to the environment in which it was located.

Conditions about the temperature, water, acoustic and optical abilities.

Reaction to water, weather and fire.

Include all processes, techniques, systems and equipment required for the manufacture, shaping, transport, placement and application of materials.

Determine the feasibility of incorporating the materials and products in a particular work.

When talking about choosing materials for a component, we take into account many different factors. These factors can be broken down into the following areas:

First we need to look at the function of the product – product analysis.
Characteristics of the crop.
Material Properties.
Material Cost and Availability.
Processing.
Environment.

Once these variables are covered, four basic steps should also be considered to perform the material selection:

1) Translation: express design requirements as constraints and objectives.
2) Screening: eliminate materials that cannot do the job.
3) Ranking: find materials that best do the job.
4) Supporting info: handbooks, expert systems, web, etc.

So if it is desireable to use a novel or unusual material, the choice must be made early in the design process. Then it can do the detailed design work using the correct material properties.

The availability of the necessary structural materials and accessories in a standard form can strongly support the development of efficient greenhouse designs. In several cases, however, the incompatibility of the structural components undermines the intended target of a cheaper greenhouse by increasing the building costs (Elsner et al., 2000b).

4.6 MATERIAL PROPERTIES OF A TUNNEL TYPE GREENHOUSE

The material properties are a set of characteristics that make a material behave in a certain way due to external stimuli such as climate or forces. These properties can be divided into physical, chemical and ecological, being the physical one the most important for the job, as the mechanical properties are the ones that describe the behavior of materials when subjected to external forces actions.

To better explain the material properties, these will be analyzed in an actual greenhouse (Tunnel fixed zenith ventilation greenhouses), and it will be studied the principal variables of materials such as: forces, moments, restrictions and others, with the aim to determine the behavior of a given material that will appear under certain conditions of wind.

4.6.1 Tunnel fixed zenith ventilation greenhouses

An orientated greenhouse is placed North–South in a mountain tropical climate with a wind speed of 30 km/hr (This information is taken from the National Meteorological Service, but it bears mentioning that the velocity of the wind changes with each season of the year).

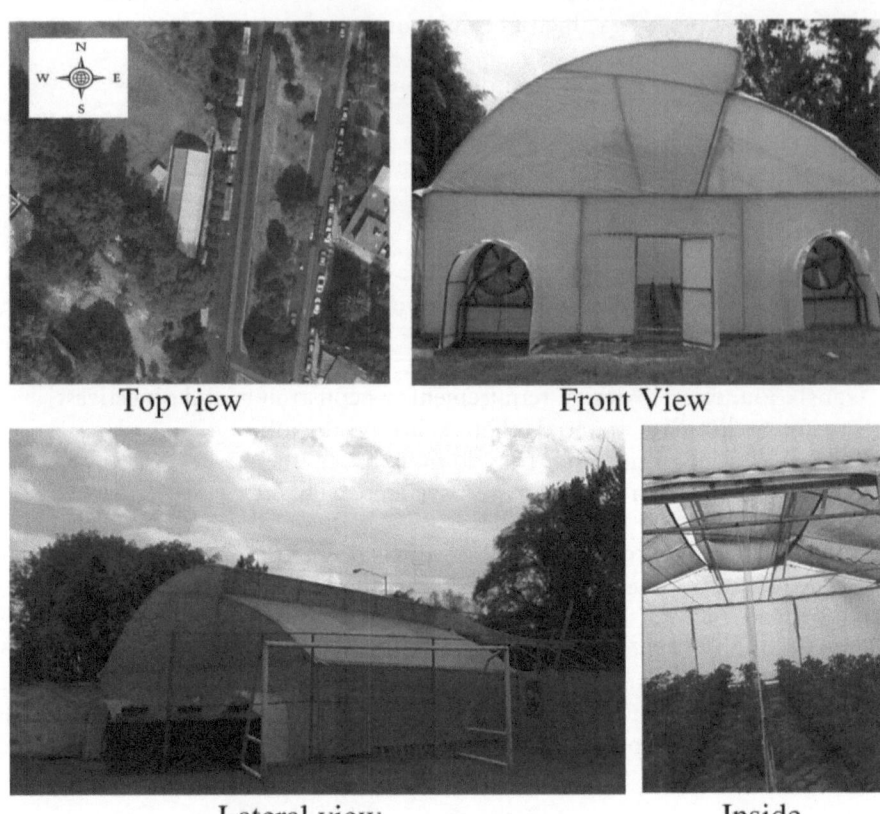

Top view　　　　　　　Front View

Lateral view　　　　　　　Inside

Figure 4.1 Tunnel Fixed Zenith Greenhouses.

The design of the structure is a tunnel with two side vents and a zenith, facilitating the air exchange and a passive temperature regulation. Its dimensions are 10m of length by 10 m of width, which results in a crop area of $100\,m^2$. The dimensions of the construction are responsible for the stability of the greenhouse. The design of the structure is modular (Figure 4.1).

The materials used in the Greenhouse Structure are:

Arches: galvanized pipes 2" – schedule 30 standard ASTM A53.
Posts: galvanized pipes 2" – schedule 30 standard ASTM A53.
Stringers: galvanized pipes 1" – schedule 30 standard ASTM A53.
Treated plastic UV II 720 caliber.
Galvanized surrounds OKI-PET.
Gateway: angles 1/8 × 11/4.
Bushings 11/4, winch, stringers nipples, poles rods.
Cocks N°10.
Screws 3/8 hex head.
High strength screws 2" and 4".

Welding paint.

Galvanized gutters $15 \times 10 \times 15$.

Expansion anchors for attaching to concrete.

Computing techniques used in systems modeling for greenhouse.

For a greenhouse structure, the force analysis is the most used due to the results obtained for the structures, although it is also possible to obtain the following results on a computer analysis (Kacira, 2012):

Provide a detailed analysis of the greenhouse.

Evaluate the interaction between the crop and its surrounding climate.

Save time and cost on analyzing various designs, configurations and climate control strategies.

For example the force method of analysis can be applied to any structure subject to loading or environmental effects (Kendirli, 2006). The degree of statically indeterminacy is determined in this method, in the same way the determination of the displacements in the released structure most be calculated. These displacements are required at the same location and in the same direction.

Cooper et al. (1983) describes in his research a computer method based on the modeling of the transient performance of the greenhouses, that presented details of the mathematical models for each component and listed the assumptions used with each, the simulated performances of a number of different greenhouse types in winter and summer are presented and an analysis of the simulated crop yields, energy flows and temperatures indicates that the model is simulating the expected trends in greenhouses.

Traditional concepts of object oriented techniques (OOT) in the component-based software development (CBSD) branch of software engineering (such as design based on models, specification of components, standards, the use of component libraries, and the reuse of design structures, also known as pattern design), as well as the concepts of assembly or composition of the components, are all common techniques in the design of greenhouses. Modeling greenhouses by using the CBSD paradigm involves the use of modeling standards such as DPS (distributed problem solving) and CAD (computer assisted design) techniques.

The formers are useful in engineering processes that require design strategies using decomposition for the creation of the object product. Other OOTs used for modeling complex systems are techniques based on extensible markup language (XML) from the W3C, and unified modeling language (UML) devised by the OMG which nowadays allows the structure and functioning of a system under study to be understood and described. In CAD techniques, and computer graphics in general, it is used a change between coordinate systems to facilitate and reduce the calculation of a particular graphic element by eliminating one of its dimensions (Z) and work in 2D (Iribarne et al., 2007).

Another technique currently used is the wind analysis using simulators, which can be verified wind loads in a specific place, the study of the speed and the force of the wind in 2D and 3D, defines the maximum loads and stockings anywhere in the model, among others, with the aim of defining for example a greenhouses practically guide, positioning passive ventilation systems and a structure design based in maximum loads.

Figure 4.2 The home screen of Inventor®.

In works such as those presented by Bartzanas et al. (2002), Rouboa & Monteiro (2007), Couto et al. (2008), N. et al. (2012), Flores-Velázquez et al. (2012), Tadj et al. (2013) tools such as CFD from INVENTOR® and Fluent from ANSYS® were used to simulate the winds and their behavior in different circumstances.

Based on the information provided on the greenhouse, an analysis can be performed using the program INVENTOR® from AUTODESK® to obtain the principal mechanical information and basic results about the dynamics of the wind.

4.6.2 Simulation in the Inventor® program

Autodesk Inventor® provides software for 3D mechanical design, simulation, creating tools and design communication that enables profitable workflow design, Digital Prototyping to design and manufacture products in less time generating manufacturing documentation directly from validated 3D digital prototypes reducing errors and allowing changesto be made to orders before manufacturing, and offers rapid and accurate production of drawings directly from the 3D model. Among its features are: parametric solid design of highly complex assemblies handling more than 1000 components, creating presentations and cuts, automatic generation of production drawings, customizable management technology, 3D module welding, bending modulus Reed, mechanical module for animation and photo realism, etc. What follows will be a description of how to use this software with application to the wind and mechanical analysis of a greenhouse, using the tool as a basis inventor and CFD program from the same platform, fluid analysis, in this case the wind.

The first step is to get to know the environment Inventor®, for this, the home screen is as follows (Figure 4.2).

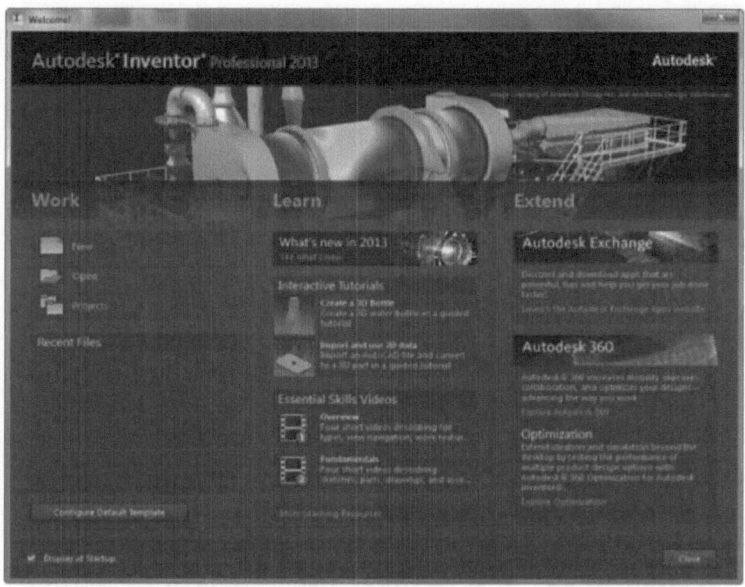

Figure 4.3 Selecting screen of work, learn or extend, Inventor®.

Then in the next screen that appears (Figure 4.3), select an existing file or create a new one.

In the next screen (Figure 4.4) you can set the basic configuration of the template or you can do so when you are already working within the Inventor® environment.

The Inventor® template defines the type of file to be used, such as; component files, assemblies, presentations and manufacturing drawings, it also sets the unit system for modeling, refining, text styles, dimensions, and tables notes, types of welding, modeling preferences, flat feet and frames, among other features that can be pre-configured to support the user in its standardization and accuracy.

Since it was installed, Inventor® creates three shrines: Default, Metric and English, but the user can create any number of inserts required. Each folder created inside the Templates folder, creates an additional tab in the New File window and each is saved within the file created in the folder and will become a user template.

To start a sketch, a new file is opened; a sketch tool is selected and then it begins to draw in the Inventor® window (Figure 4.5). While it is automatically plotting it applies certain rules allowing the required stroke. For that reason the following screen is created and it is necessary begin in a 2D sketch.

When you start working on the piece, the following screen appears which presents all the commands that can be used to create the piece, for example: lines, splines, circles, ellipses, arcs, rectangles, polygons, or points. You can cut corners, extend or apply radios or chamfers, and compensate and project geometry of other features. When working with the figure in the work plane it shows some tools to begin the piece

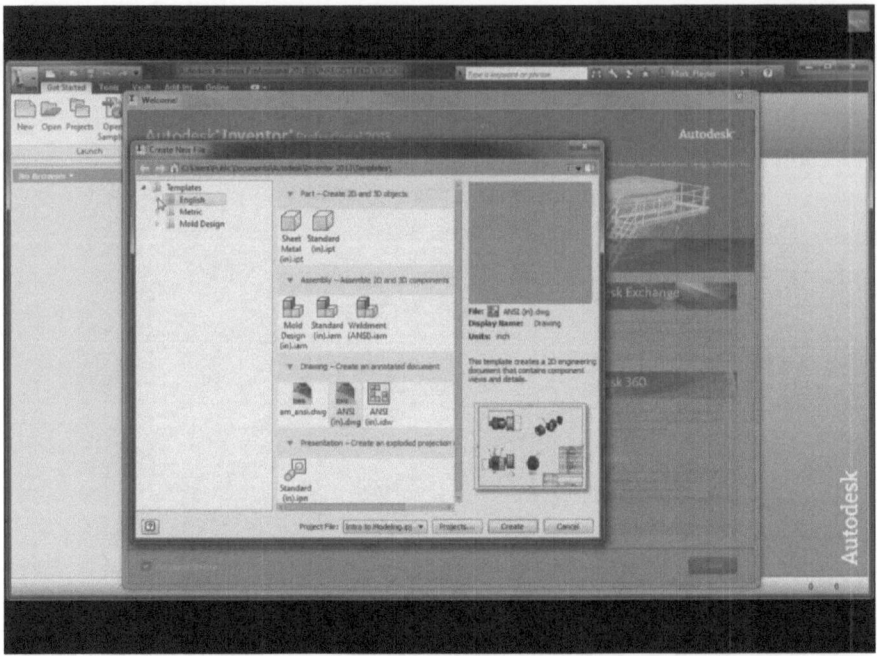

Figure 4.4 Configuration of the template, Inventor®.

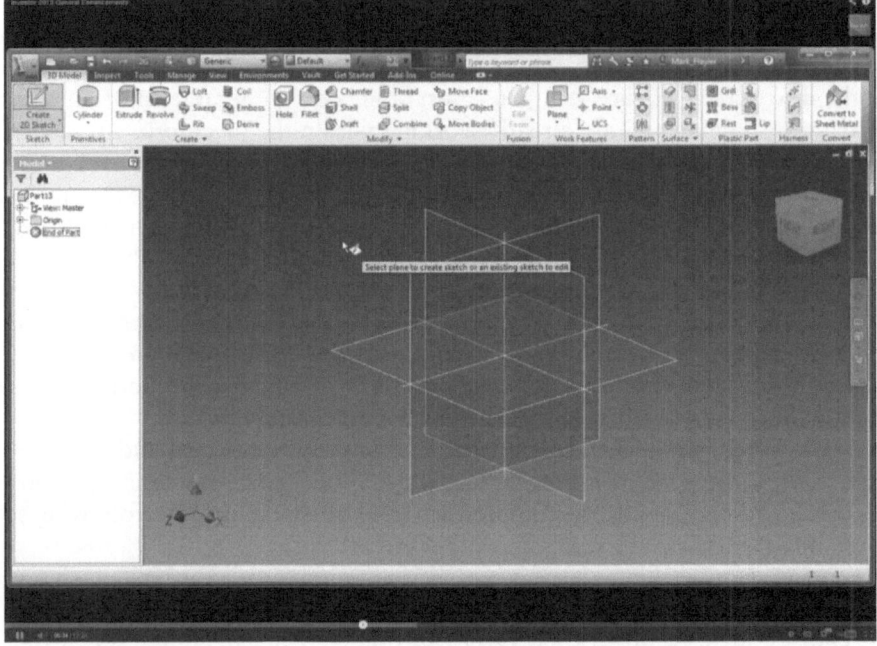

Figure 4.5 First screen of the sketch in Inventor®.

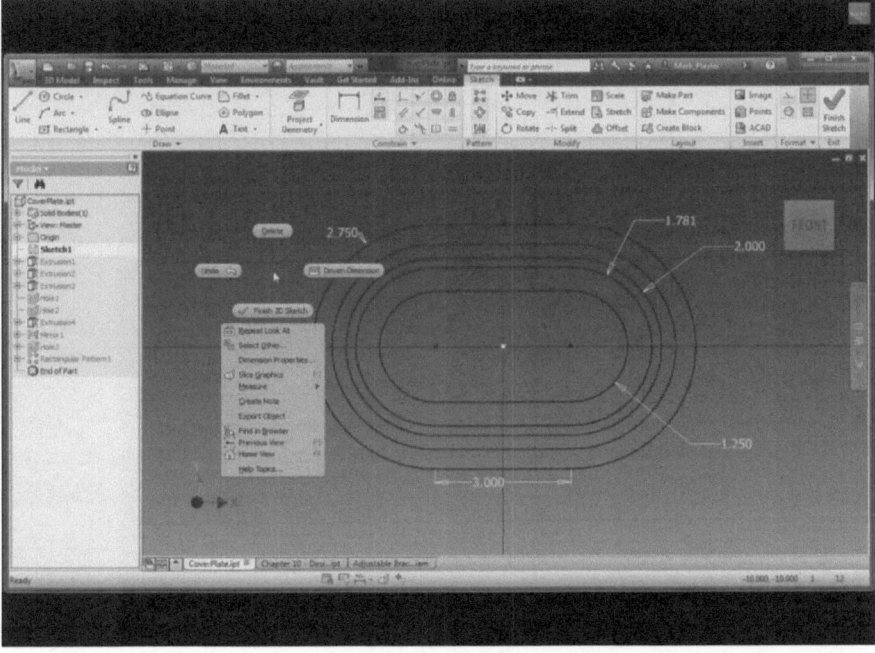

Figure 4.6 Example of the auxiliary tools, Inventor®.

construction; the advantage of using this program is that it is very intuitive and to every action to be performed there are alternative tools to enhance the piece (Figure 4.6).

All projects begin with a sketch that has a starting point whether you are creating a sketch or a profile. A sketch is the profile of a Feature (3D operation) and any geometry (such as a sweep path or axis of rotation) that are required to create a 3D object. All sketch geometry is created and fixed in the sketch environment, using the tools in the tool pane.

On the other hand, it is important to mention that there are 5 basic methods to generate features based (solid, three-dimensional objects or surfaces) in Inventor®: Extrude, Revolve, Sweep, Loft and Coil, Helicoid. In addition to the creation of these objects the Work Features that are used for the abstract construction geometry (auxiliary) can be used when the one that already exists in a model is insufficient for creating and positioning new features that are needed. To fix the position of the work features, you can restrict several complementary tools that help create increasingly complex parts such as: holes, shells, fillets, chamfers, face drafts, and threads. Inventor® allows you to create pieces with maximum detail, for which it relies on tools that will create more complex geometries such as Move Face, Face Draft Split, Thicken/Offset, Decal or Emboss.

Inventor® provides important information to validate model geometry, using various analysis tools. For more basic information you can consult the following link: http://usa.autodesk.com.

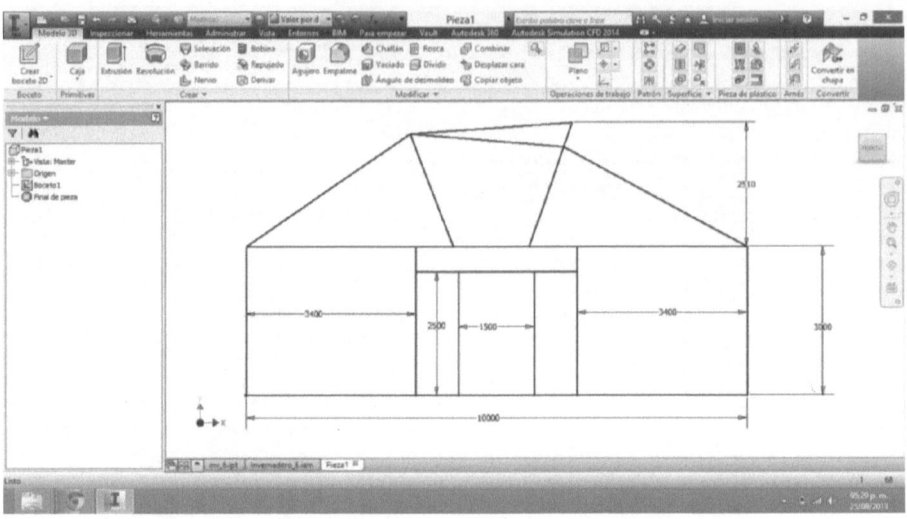

Figure 4.7 Sketch with general characteristics of the structure.

Subsequent to getting to know and how to use the basic commands, you can start with the construction of a structure, which initially takes place in 2D and then moves to 3D projects. To develop a 3D model of a structure, it is necessary to take into account that this is formed by united frames using rails, then the first step is to create the sketch containing the general characteristics of the structure, in this case the dimensions and restrictions are specified in mm (Figure 4.7).

Later levels are used to determine the depth of the structure and create a 3D sketch (Figure 4.8 in mm) which will join the frames using rails.

Once the sketch is generated for the structure the beams shall be inserted, within the program it is possible to determine characteristics such as: type of rule, shape, size and material, in this case it has been selected a line of $1/2 \times 0.145$ in galvanized steel (Figure 4.9).

To perform a study and an analysis of the mechanical properties of the structure it is necessary to know:

General data of the structure – Type of structure that is analyzed.

Nomenclature and coordinates of the nodes that make up the structure that is analyzed.

Nomenclature and incidents of the bars in the structure.

Location and nomenclature of the flat elements.

Geometry of the elements – Number of dimensions that the bars and the flat elements have.

Properties of the materials – which include the type of material that constitutes each elements as well as its modulus of elasticity, coefficient of Poisson, its density, its resistance, among others.

Features of the connections with the foundation – This is, if there are under rides or some degree of freedom in these connections.

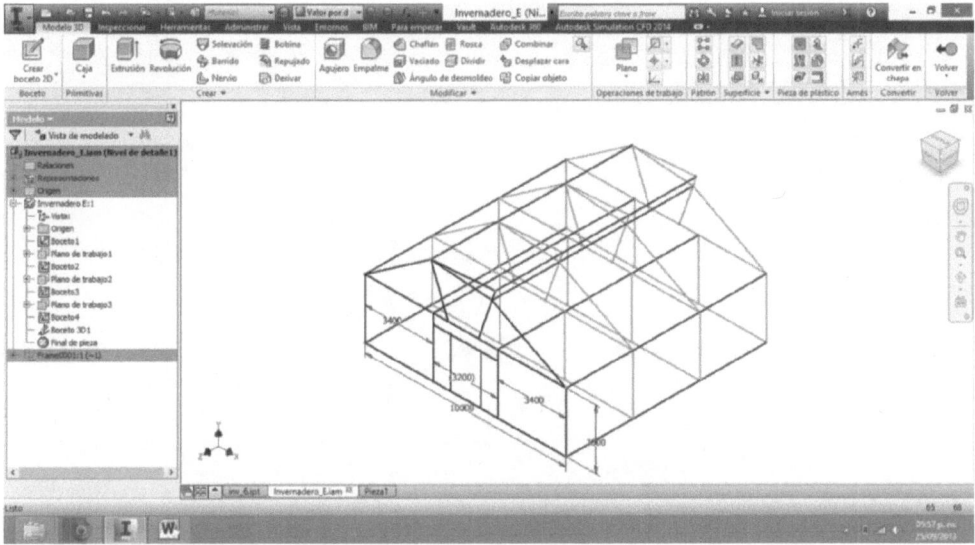

Figure 4.8 3D sketch.

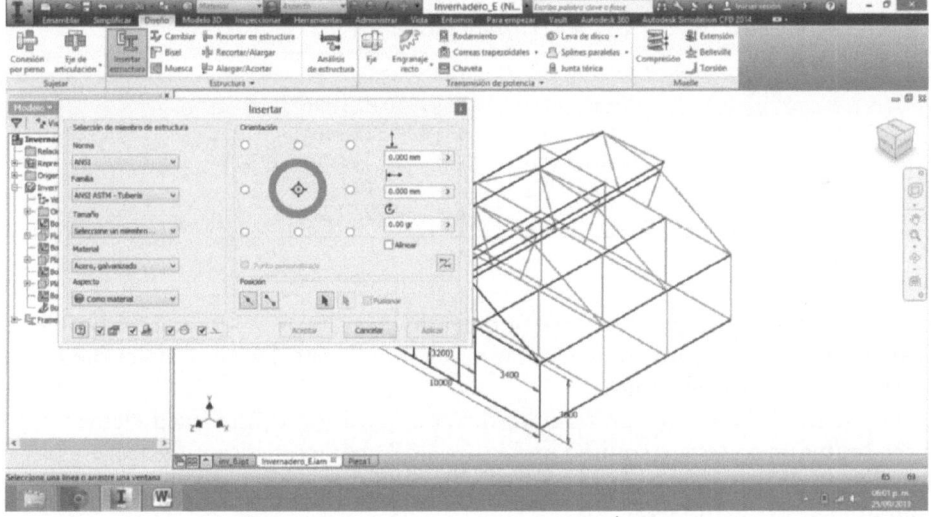

Figure 4.9 Approach characteristics of the structures.

4.6.3 Characteristics of the loads that are applied to the structure

Considering number 8 of the list, and by applying the wind factor to the structure there are two types of load: permanent loads and overload operation. The permanent loads are the main cover material, ranging from 7 to 15 kg/m² when glass is used, and

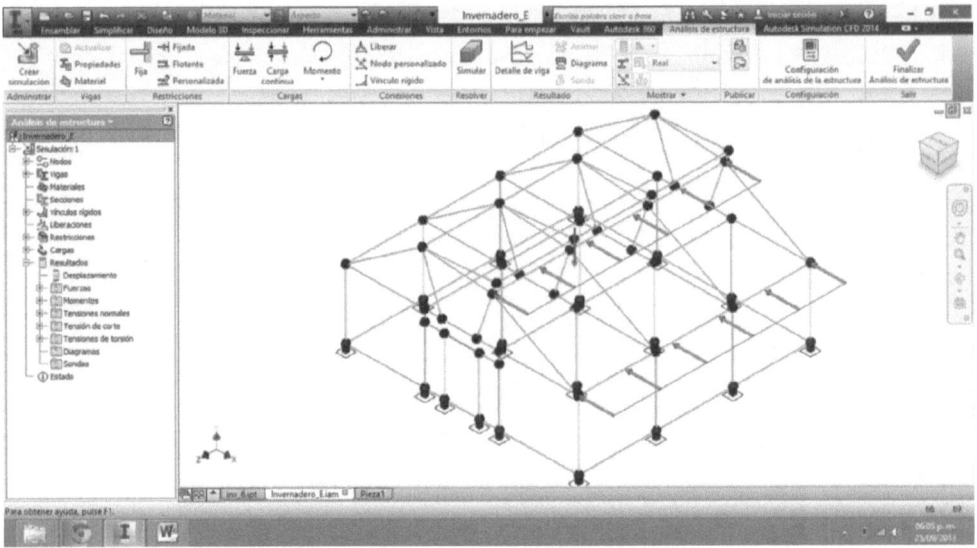

Figure 4.10 Starting the structural analysis.

Table 4.4 Physical properties.

Mass	1673.712 kg
Area	5939950.823 mm^2
Volume	656404.323 mm^3
Gravity center	x = 4934.348 mm
	y = 2760.104 mm
	z = −4899.912 mm

up to a negligible burden in the case of polyethylene sheets, operating on heavier ones considering the wind, snow or general weather conditions that may affect the structure and the roof of the greenhouse.

Figure 4.10, here you can observe the configuration of the fixed elements, loads on the beams or nodes as required and other load characteristics, this process has been completed and it is possible to start with the respective analysis. The wind factor is being used as the load factor.

In the analysis of the results it is possible to find in a general way the behavior that the structure will present before the actual loads are applied, in the same way, it is possible to study the reactions of each element. As expected the upper beams and the area close to the point of load are the points showing the more severe effects and those points should be handled very carefully during their design and construction.

Once these processes have been finished, the results obtained are applied in the following steps.

Decision of the physical properties of the complete structure of the greenhouse (Table 4.4).

Table 4.5 Material.

Name	Galvanized steel	
General	Mass density	7.850 g/cm^3
	Elasticity limit	207.000 MPa
	Maximum tensile strength	345.000 MPa
Tension	Young modulus	200.000 GPa
	Coefficient of Poisson	0.300 su
	Modulus cutting	0.0000120 su/c
Thermic tension	Expansion coefficient	53.000 W/(mK)
	Thermal conductivity	0.450 J/(kg K)
	Lower yield strength of the material (Fy)	2460 N
	Minimum breaking stress in tension (Fu)	4220 N

Table 4.6 Transversal section.

	Section area (a)	515.778 mm^2
Geometrical properties	Width of section	48.260 mm
	High of section	48.260 mm
	Section of Centroid (x)	24.130 mm
	Section of Centroid (y)	24.130 mm
Mechanical properties	Moment of inertia (Ix)	128987.802 mm^4
	Moment of inertia (Iy)	128987.802 mm^4
	Module of rigidity of torsion (J)	257975.605 mm^4
	Module of section (Wx)	5345.537 mm^3
	Module of section (Wy)	5345.537 mm^3
	Module of torsion section (Wz)	10691.074 mm^3
	Area of Shear reduced (Ax)	259.060 mm^2
	Area of Shear reduced (Ay)	259.060 mm^2

Note: These properties are general, and are in the datasheet of every material. This information is supported with ASTM A53 Standard.

Table 4.7 Girders model.

Nodes	112
Girders	56
Round tubes	56

The physical characteristics are given according to the material used in the strucrure (Table 4.5).

The first analysis shows the mechanical properties in the girders (Table 4.6).

All nodes are specified in the complete structure of the greenhouse (Table 4.7).

Here is shown the initial load conditions that the greenhouse will have; in this case a wind speed factor of 30 km/hr is applied, to complement the following information (Table 4.8).

To perform the analysis applying the wind factor to a structure, it is necessary to know the characteristics of the place where the greenhouse is located and because this case study was conducted in the city of Mexico, all calculations described were

Table 4.8 Load conditions.

Load type	Continuous load
Magnitude	10 N/m
Coordinates of girders system	No
Plane angle	180.00°
Angle in plane	90.00°
Qx	−0.010 N/mm
Qy	0.000 N/mm
Qz	0.000 N/mm
Discrepancy	0.000 mm
Length	10000.000 mm
Final magnitude	0.010 N/mm

Table 4.9 Initial conditions to perform the analysis of wind structures.

Name	Classification	Description
Importance	group B	Structures for a moderate degree of safety are recommended in adjudicating these to generate a low loss of human lives and cause damage of intermediate magnitude, if wind fails to endanger other structures.
Response to the wind action	Type 2	Structures whose slender or small size of their cross section makes them particularly sensitive to short bursts, and whose long natural periods favor the occurrence of significant oscillations. Considered in this type, there are buildings with slenderness, defined as the ratio between the height and the minimum plan size greater than 5, or critical period greater than 1 second.
Category field by roughness	R3	Typical urban and suburban area. The site is surrounded by buildings predominantly medium and low altitude or wooded areas.
Topography factor	Normal Ft	Land practically flat.
Factor topography and roughness	Roughness T3	Virtually flat terrain, open fields, absence of major topographical changes, with slopes less than 5% (normal).

conducted as provided in the Complementary Technical Standards for wind Design based on the Official Gazette of the Federal District (2004). To begin the analysis it should be considered Table 4.9.

The first option is used to obtain the design speed since in this case the static wind is applied on a structure or component thereof, the following equation is used to define it:

$$VD = FTR \ F\alpha \ VR \qquad (4.1)$$

where: FTR is the correction factor that takes into account the local conditions related to the topography and roughness of the terrain around the site rudeness; $F\alpha$ is the factor that takes into account the variation of the velocity against the height; VR is the regional speed of the area that corresponds to the site where the structure will be built.

For this greenhouse FTR = C as z < 10 where z is the height above the natural ground to apply to the wind design. C is the roughness coefficient, in this case it is type 3 FTR = C = 0.881; VR II is a local data and has a regional speed of 32 m/s. Land category must be R3 $\alpha = 0.156$ and $\delta = 390$. Therefore:

$$VD = (0.881)(0.15632)(32) = 4.406 \, m/s = 15.86 \, km/hr$$

This is the optimal design value, but for analysis and testing purposes, based on the national meteorological system, the rate of the season that was used is 30 km/hr, which corresponds to moderate wind characteristics that cause to start rocking small tree branches and small peaks of water to form waves on the lakes.

The next variable is the wind pressure that flows over a given construction, pz, in Pa (kg/m²), it is obtained by considering its form, and it is given by the following expression:

$$pz = 0.47 \, Cp \, VD^2 \tag{4.2}$$

where: Local pressure is the coefficient Cp, which depends on the shape of the structure; VD design velocity at height z.

$$pz = (0.47)(0.795)(15.86)^2 = 94.006 \, Pa$$

And as consequence of having applied the force of the wind to the structure, a force and a couple of reactions are obtained in each of the restrictions that are formed by joining the girders to form the structure. In figures 12 to 23, the main behaviors of girders and structures are shown, indicating by colors the magnitude of the applied forces and the reaction of the structure (Beer 2004, Fitzgerald 2007).

Displacement

It is a basic element in the process of structural design; as it is associated with: the structure inelastic deformations and non-structural elements, the structure overall stability and the damage to the non-structural elements (figure 4.11). This is manifested in the degrees of freedom that exist in the structure and are presented through the translation and rotation of the nodes. To use the displacement method components must be calculated numerically; cutting diagrams, moment and axial load. This method is explained in Kaveh (2006). A considerable number of calculations are needed to use this method, but since it is found in most of the applications, it can be used with any software.

As noted the critical part is located on the arches of the greenhouses since the force is being applied in the nodes of the connections of the database structures.

Normal stress

It is the relationship between the applied force and the area of the section on which it acts. In other terms it is the load that acts per unit area of the material (Figure 4.12).

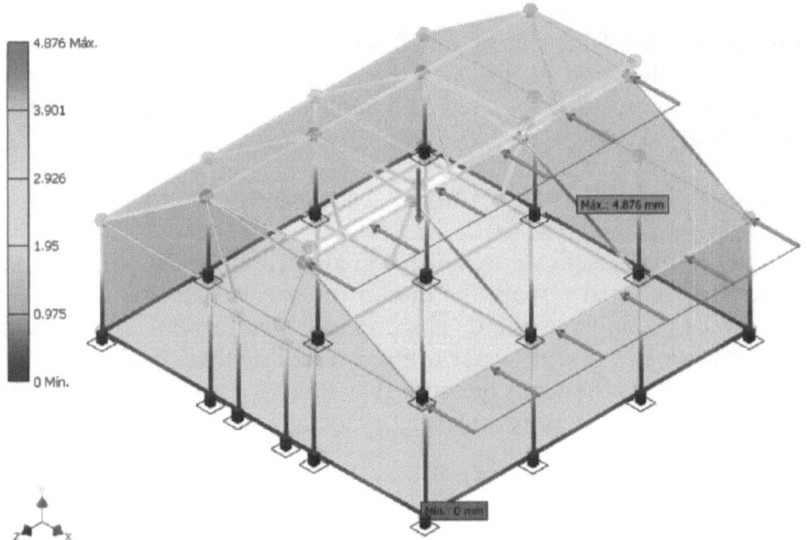

Figure 4.11 Displacement in the nodes of the greenhouses in mm.

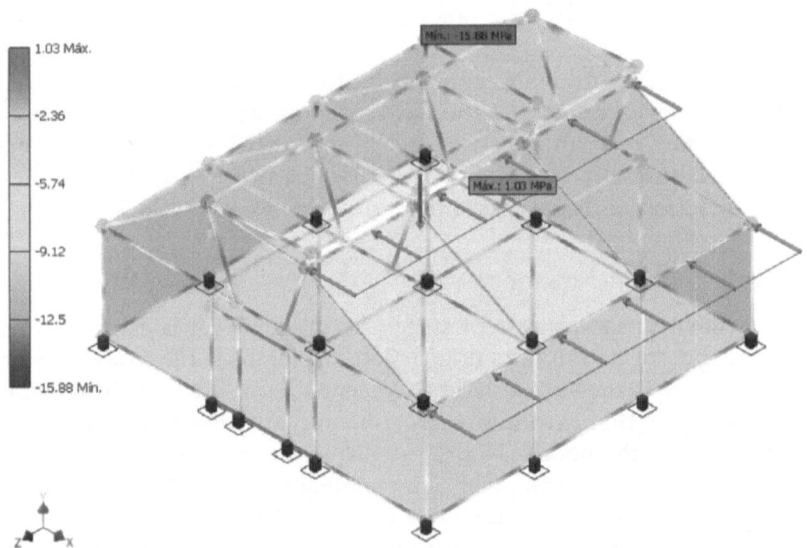

Figure 4.12 Normal stresses in the greenhouse, MPa.

As can be seen, a normal stress impact to the structure, although this has very low values, which are apparent only seen in small sections of the girders, the stress is present in every one of the axes and consequently there will be one on which more force will be applied. In table 4.17 it is shown the summary of the results.

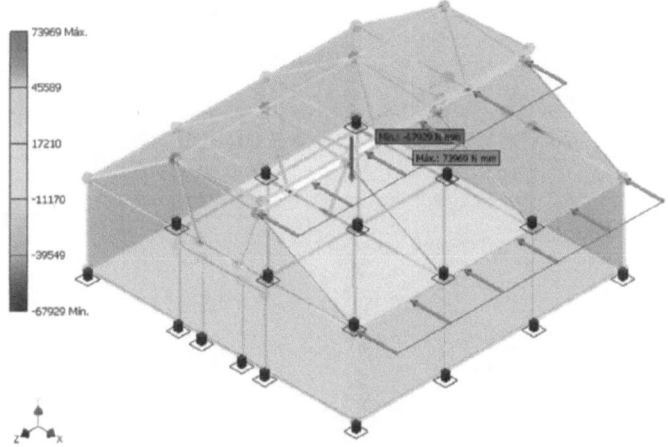

Figure 4.13 Moment of force MX in N mm.

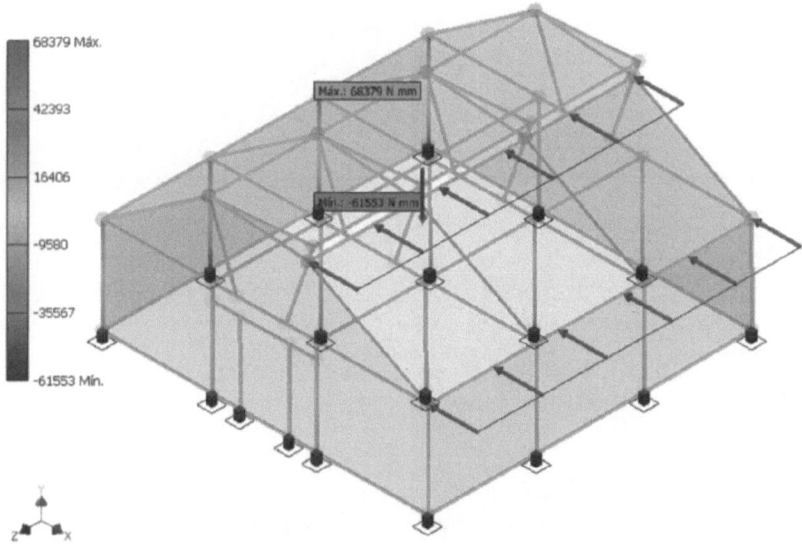

Figure 4.14 Moment of force My in N mm.

Moment

It is the tendency of a force to turn around any axis. The magnitude of the effect of the force rotation around an axis is the intensity of the moment. The units of the moment are the force and the distance. In Figures 4.13, 4.14, and 4.15, it is possible to see that in the Y axis the moment is more visible in the girders on the top of the greenhouses, but the result is minimal.

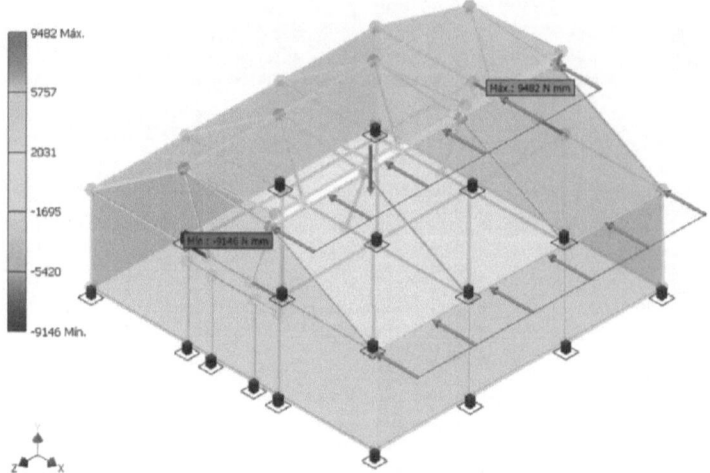

Figure 4.15 Moment of force Mz in N mm.

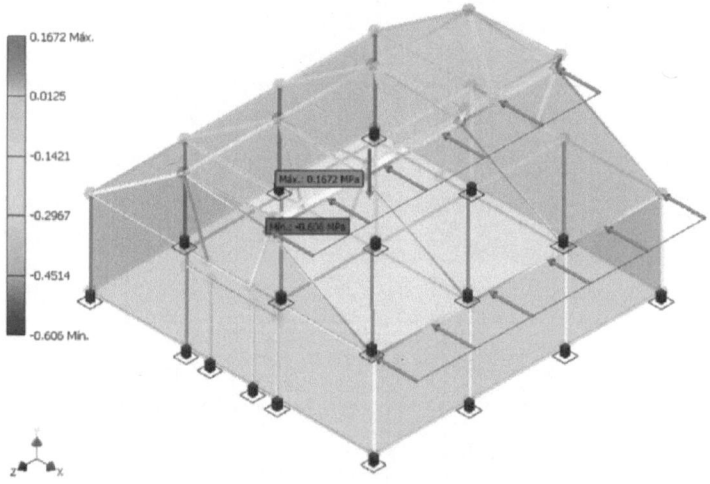

Figure 4.16 Tension cut off Tx in Mpa.

As can be seen the moment is not very representative in the greenhouse, as the most visible expression is found only on the y axis, and the value is not significant and does not represent damage to this.

Tension cut off

There are produced in a body when the applied loads tend to cut or slide a part of the same with respect to another, and are presented in items subjected to direct cutting (Figure 4.16 and 4.17).

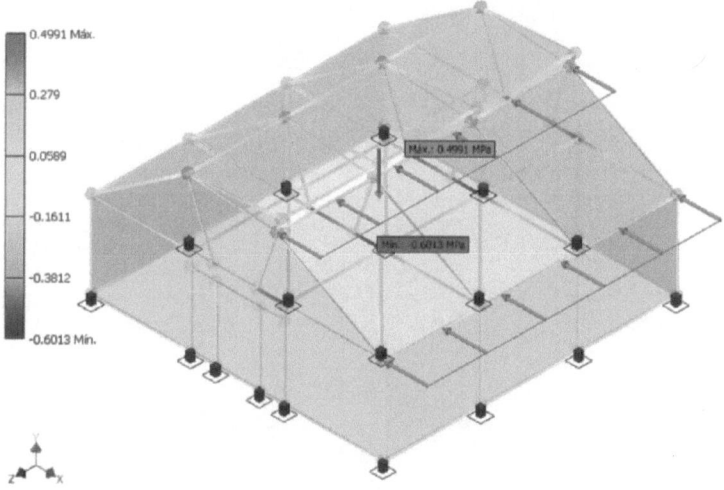

Figure 4.17 Tension cut off Ty in Mpa.

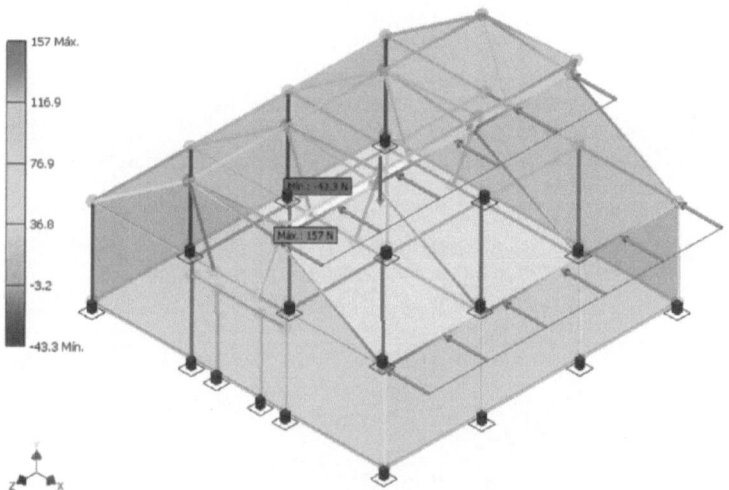

Figure 4.18 Forces in Fx in N.

As can be seen the tension cut off has more consequences in the x-axis, especially in the girders supporting the structure. To solve this problem you will have to place girders with a more reinforced material. This is a direct affectation because it is found on the same direction in which the force acts on these and therefore the resistance that must be countered is greater.

Reaction Forces are the forces applied to a beam or other structure when it rests against something. The structure needs a force that acts at a single point on a structure and it is depicted by a single arrow on the diagrams (Figure 4.18, 4.19 and 4.20).

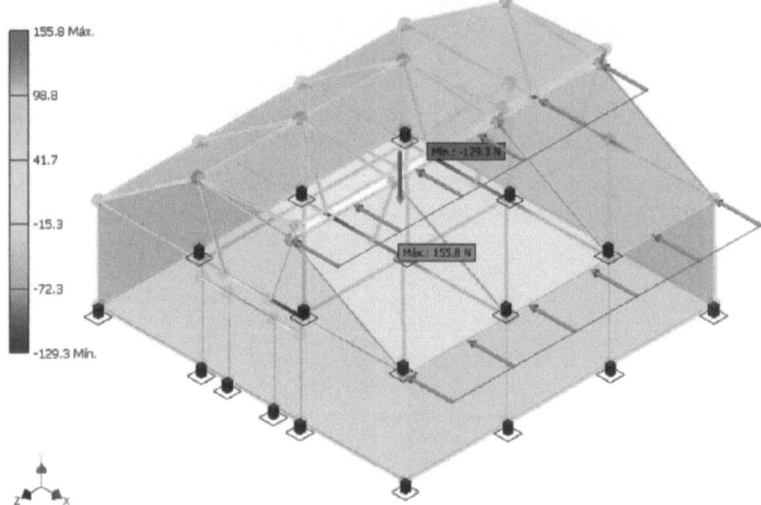

Figure 4.19 Forces in Fy in N.

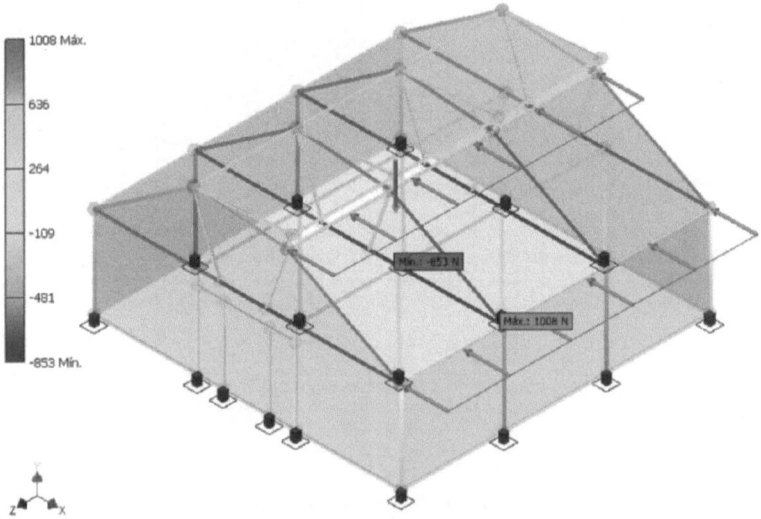

Figure 4.20 Forces in Fz in N.

In this case, the most critical result is found in Z, the client needs to pay attention to this axis in the greenhouses, e.g. the union between the girders needs bolstering.

Torsional stress

The type of load that tends to twist a bar around its longitudinal axis occurs when time is applied on the longitudinal axis of a constructive element or mechanical prism,

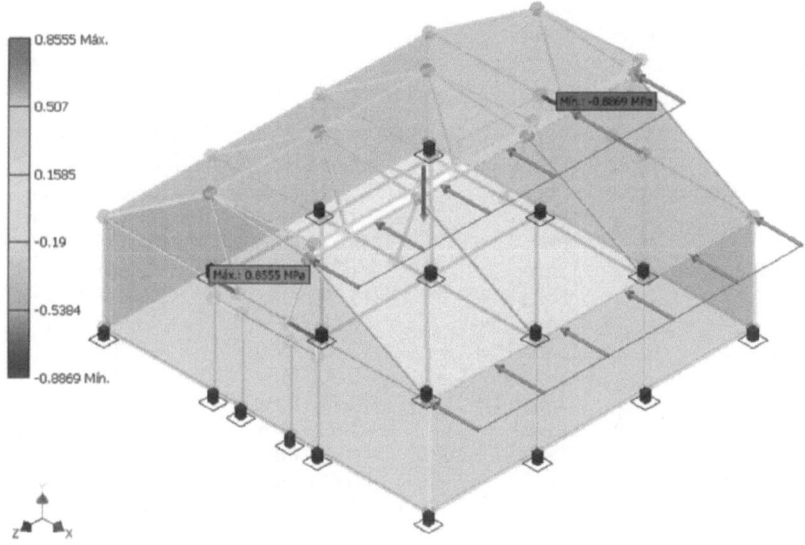

Figure 4.21 Torsional stresses in Mpa.

as shafts or elements where there is a predominant dimension over the other two, although it is possible to find it in various situations (Figure 4.21).

As the force is applied, the torsional stress does not represent any damage to the greenhouse.

These series of figures have a main function to show the possible damage that could occur to the structure according to the force applied, in this case the wind.

One of the advantages that the program Inventor® has, is the power to get an overview of the numerical results that shows the same variables as the figures, linking in this way both to ensure a proper interpretation of the results.

Table 4.10 shows the minimum and the maximum values that are obtained by applying a force to the greenhouse type Tunnel with zenith.

The purpose of this analysis is to determine how effective is the construction by applying the force of a wind of 30 km/hrs, but it should be noted that in order to have optimal results for the construction, it is necessary to make simulations with changes of wind speed, direction of the greenhouse and in any given moment a change in the material, in order to cover the design with the needs of the client.

To perform a critical analysis and to do the second point of the analysis with the CFD, it has also been placed tables 4.10 to 4.17, with the results of the most used greenhouses around the world, such as: Butterfly, Sawtooth, Multi Tunnel, Modular Parral, Tunnel, Chapel and Chinese Hat.

Based on the data obtained from the ASTM A53 standard and comparing the results obtained in the simulation by inventor, corresponding to Fy, the value of the force is very small which leads to the conclusion that the material used is very resistant to the force and wind speed being used for the simulation, as a standard the value of

Table 4.10 Results of force applied in a Tunnel Fixed Zenith greenhouse.

Name: Zenith	Tunnel	Fixed Minimum	Maximum
Displacement		0.000 mm	4.876 mm
Reaction Forces	Fx	−43.311 N	156.994 N
	Fy	−129.285 N	155.762 N
	Fz	−853.434 N	1008.298 N
Momentum	Mx	−67928.711 N mm	73968.671 N mm
	My	−61552.943 N mm	68379.112 N mm
	Mz	−9145.710 N mm	9482.123 N mm
Normal stress	Smax	−1.717 MPa	15.661 MPa
	Smin	−15.885 MPa	1.027 MPa
	Smax(Mx)	0.000 MPa	13.837 MPa
	Smin(Mx)	−13.837 MPa	−0.000 MPa
	Smax(My)	−0.000 MPa	12.792 MPa
	Smin(My)	−12.792 MPa	0.000 MPa
	Saxial	−1.955 MPa	1.655 MPa
Tension cut off	Tx	−0.606 MPa	0.167 MPa
	Ty	−0.601 MPa	0.499 MPa
Torsional stress	T	−0.887 MPa	0.855 MPa

Table 4.11 Results of force applied in a Butterfly greenhouse.

Name: Butterfly		Minimum	Maximum
Displacement		0.000 mm	45.908 mm
Forces	Fx	−230.322 N	229.480 N
	Fy	−414.229 N	263.253 N
	Fz	−955.201 N	1299.625 N
Momentum	Mx	−304255.550 N mm	267876.276 N mm
	My	−423370.223 N mm	456595.175 N mm
	Mz	−22547.028 N mm	22032.190 N mm
Normal stress	Smax	−2.355 MPa	88.916 MPa
	Smin	−89.024 MPa	1.202 MPa
	Smax(Mx)	0.000 MPa	56.918 MPa
	Smin(Mx)	−56.918 MPa	−0.000 MPa
	Smax(My)	−0.000 MPa	85.416 MPa
	Smin(My)	−85.416 MPa	0.000 MPa
	Saxial	−2.520 MPa	1.852 MPa
Tension cut off	Tx	−0.886 MPa	0.889 MPa
	Ty	−1.016 MPa	1.599 MPa
Torsional stress	T	−2.061 MPa	2.109 MPa

$Fy = 2460$ N and in the greenhouse was used the value $Fy = 155,762$ N. The displacement is negligible since a peak that begins to damage the structure is 5 mm, which will appear depending on the force applied, as is shown in the example presented.

These tables summarize the maximum values for each greenhouse, showing which ones offer the best conditions and which the minimum. This is shown in table 4.18.

Table 4.12 Results of force applied in a Sawtooth greenhouse.

Name: Sawtooth		Minimum	Maximum
Displacement		0.000 mm	10.824 mm
Forces	Fx	−331.958 N	407.950 N
	Fy	−401.216 N	394.935 N
	Fz	−384.940 N	757.200 N
Momentum	Mx	−353323.918 N mm	237780.576 N mm
	My	−208798.405 N mm	271493.092 N mm
	Mz	−8957.221 N mm	11409.255 N mm
Normal stress	Smax	−1.209 MPa	66.535 MPa
	Smin	−67.004 MPa	0.377 MPa
	Smax(Mx)	0.000 MPa	66.097 MPa
	Smin(Mx)	−66.097 MPa	−0.000 MPa
	Smax(My)	0.000 MPa	50.789 MPa
	Smin(My)	−50.789 MPa	−0.000 MPa
	Saxial	−1.468 MPa	0.746 MPa
Tension cut off	Tx	−1.575 MPa	1.281 MPa
	Ty	−1.524 MPa	1.549 MPa
Torsional stress	T	−1.067 MPa	0.838 MPa

Table 4.13 Results of force applied in a Multi Tunnel greenhouse.

Name: Multi tunnel		Minimum	Maximum
Displacement		0.000 mm	4.888 mm
Forces	Fx	−383.018 N	436.621 N
	Fy	−283.781 N	327.602 N
	Fz	−193.877 N	825.011 N
Momentum	Mx	−139457.113 N mm	186867.772 N mm
	My	−250522.735 N mm	194233.670 N mm
	Mz	−4144.139 N mm	4025.153 N mm
Normal stress	Smax	−1.420 MPa	46.998 MPa
	Smin	−46.951 MPa	−0.036 MPa
	Smax(Mx)	0.000 MPa	34.958 MPa
	Smin(Mx)	−34.958 MPa	−0.000 MPa
	Smax(My)	−0.000 MPa	46.866 MPa
	Smin(My)	−46.866 MPa	0.000 MPa
	Saxial	−1.600 MPa	0.376 MPa
Tension cut off	Tx	−1.685 MPa	1.478 MPa
	Ty	−1.265 MPa	1.095 MPa
Torsional stress	T	−0.376 MPa	0.388 MPa

As can be seen the selected material, in this case the galvanized steel, is perfectly designed for building structures, since in all cases it is found that the maximum force that supports in an ideal location for the force applied is minimal compared to the rule of this type of steel (2460 N). With respect to the movements, which are reflected in the translation and rotation of the nodes of the structure, it can be seen that the butterfly type is the one that allows an increased movement before being damaged, which is also reflected by the moment of inertia. On the other hand, the Tunnel greenhouses

Table 4.14 Results of force applied in a Modular Parral greenhouse.

Name: Modular	Parral	Minimum	Maximum
Displacement		0.000 mm	7.011 mm
Forces	Fx	−596.521 N	431.376 N
	Fy	−558.448 N	386.000 N
	Fz	−796.545 N	935.831 N
Momentum	Mx	−107184.651 N mm	131318.841 N mm
	My	−155034.786 N mm	141205.529 N mm
	Mz	−23195.536 N mm	22586.602 N mm
Normal stress	Smax	−1.507 MPa	41.382 MPa
	Smin	−39.075 MPa	0.833 MPa
	Smax(Mx)	0.000 MPa	24.566 MPa
	Smin(Mx)	−24.566 MPa	0.000 MPa
	Smax(My)	−0.000 MPa	29.003 MPa
	Smin(My)	−29.003 MPa	0.000 MPa
	Saxial	−1.814 MPa	1.544 MPa
Tension cut off	Tx	−1.665 MPa	2.303 MPa
	Ty	−1.490 MPa	2.156 MPa
Torsional stress	T	−2.113 MPa	2.170 MPa

Table 4.15 Results of force applied in a Tunnel greenhouse.

Name: Tunnel		Minimum	Maximum
Displacement		0.000 mm	6.242 mm
Forces	Fx	−135.687 N	133.386 N
	Fy	−71.287 N	67.523 N
	Fz	−1115.728 N	1447.388 N
Momentum	Mx	−75652.091 N mm	80652.321 N mm
	My	−94181.246 N mm	91531.392 N mm
	Mz	−4688.606 N mm	4396.278 N mm
Normal stress	Smax	−2.606 MPa	17.629 MPa
	Smin	−20.916 MPa	1.736 MPa
	Smax(Mx)	0.000 MPa	15.088 MPa
	Smin(Mx)	−15.088 MPa	−0.000 MPa
	Smax(My)	0.000 MPa	17.619 MPa
	Smin(My)	−17.619 MPa	−0.000 MPa
	Saxial	−2.806 MPa	2.163 MPa
Tension cut off	Tx	−0.515 MPa	0.524 MPa
	Ty	−0.261 MPa	0.275 MPa
Torsional stress	T	−0.411 MPa	0.439 MPa

with ventilation or the Tunnel Fixed Zenith ones are not very recommended in areas with wide open spaces that can have radical climates.

On the other hand it is important to say that the structural static analysis is performed because the loads are permanent, which is the same as to say that they do not change over time.

Finally Inventor's CFD software is used that presents the summary of the static results, for the main variables that determine the behavior of the material.

Table 4.16 Results of force applied in a Chapel greenhouse.

Name: Chapel		Minimum	Maximum
Displacement		0.000 mm	6.935 mm
Forces	Fx	−324.314 N	413.760 N
	Fy	−309.148 N	304.677 N
	Fz	−654.029 N	820.445 N
Momentum	Mx	−111063.185 N mm	146534.172 N mm
	My	−142280.657 N mm	185057.571 N mm
	Mz	−6803.750 N mm	6228.139 N mm
Normal stress	Smax	−1.400 MPa	37.045 MPa
	Smin	−37.855 MPa	0.288 MPa
	Smax(Mx)	0.000 MPa	27.412 MPa
	Smin(Mx)	−27.412 MPa	−0.000 MPa
	Smax(My)	0.000 MPa	34.619 MPa
	Smin(My)	−34.619 MPa	−0.000 MPa
	Saxial	−1.591 MPa	1.268 MPa
Tension cut off	Tx	−1.597 MPa	1.252 MPa
	Ty	−1.176 MPa	1.193 MPa
Torsional stress	T	−0.583 MPa	0.636 MPa

Table 4.17 Results of force applied in a Chinese hat greenhouse.

Name: Chinese hat		Minimum	Maximum
Displacement		0.000 mm	4.535 mm
Forces	Fx	−205.384 N	318.760 N
	Fy	−347.958 N	318.104 N
	Fz	−1575.609 N	2089.251 N
Momentum	Mx	−159362.249 N mm	152082.655 N mm
	My	−128073.781 N mm	126381.397 N mm
	Mz	−6232.038 N mm	6675.515 N mm
Normal stress	Smax	−3.810 MPa	31.249 MPa
	Smin	−36.153 MPa	2.457 MPa
	Smax(Mx)	0.001 MPa	29.812 MPa
	Smin(Mx)	−29.812 MPa	−0.001 MPa
	Smax(My)	0.000 MPa	23.959 MPa
	Smin(My)	−23.959 MPa	−0.000 MPa
	Saxial	−4.051 MPa	3.055 MPa
Tension cut off	Tx	−1.230 MPa	0.793 MPa
	Ty	−1.228 MPa	1.343 MPa
Torsional stress	T	−0.624 MPa	0.583 MPa

4.6.4   CFD simulator by Inventor®

The CFD simulator, that is a software by Inventor, aims to predict the rheological behavior of one or several substances, in this case the wind; the analysis of the ventilation provides a detailed overview of the movement of the air and its impact on the climate inside the greenhouse.

Table 4.18 Abstract of results over greenhouses.

Characteristics		Maximum	Greenhouse Name	Minimum	Greenhouse Name
Displacement		45.908 mm	Butterfly	4.53 mm	Chinese Hat
Force	Fx	436.621 N	Multi Tunnel	133.386 N	Tunnel
	Fy	394.935 N	Saw tooth	67.523 N	Tunnel
	Fz	2089.251N	Chinese Hat	757.200 N	Saw tooth
Momentums	Mx	267876.276 N mm	Butterfly	73968.671 N mm	Tunnel Fixed Zenith
	My	456595.175 N mm	Butterfly	68397.112 N mm	Tunnel Fixed Zenith
	Mz	22586.602 N mm	Modular Parral	4025.153N mm	Multi Tunnel
Normal tension	Smax	88.916 MPa	Butterfly	15.661 MPa	Tunnel Fixed Zenith
	Saxial	3.055 MPa	Chinese Hat	0.376 MPa	Multi Tunnel
Tension cut off	Tx	2.303 MPa	Modular Parral	0.167 MPa	Tunnel Fixed Zenith
	Ty	2.156 MPa	Modular Parral	0.275 MPa	Tunnel
Torsional stress	T	2.170 MPa	Modular Parral	0.388 MPa	Multi Tunnel

To perform an analysis with the CFD of Inventor in the greenhouse, you have to go through several stages: begin building the geometry of a real greenhouse, on that stage it is identified the areas where the wind circulates, then it is inserted in a fluid space, as is the wind tunnel and finally it is obtained the computational model.

The advantages of the CFD are that it predicts the properties of the fluid with great detail in the domain studied, it aids to the design and produce prototypes and to provide quick solutions avoiding costly experiments; you obtain a display and an animation of the process in terms of the variables in the fluid. The disadvantages are that it requires users with extensive experience and specialized training, consumes resources of hardware and software that require significant investments, in some cases, the computational cost is high.

The process of construction and simulation of a model by computational fluid dynamics comprises three stages that can be performed in series: 1) preprocessing, 2) solution and 3) post-processing. Preprocessing is dedicated to building the geometry. It also determines the shape and the dimensions with which they will work during the simulation. Then it is important to establish the structural materials that form the walls of the geometry and the exterior.

The analysis considers the following as a single structure cover:

Structure material: Galvanized steel.
Gladding material: Low density polyethylene.
Wind velocity: 30 km/hrs.
Windward wind.
Temperature: 19.85°C, it is the base of the program CFD of Autodesk®.

This analysis is carried out in three dimensions from a real greenhouse as described above, the geometry shown in Figure 4.22 is constructed with these dimensions 10 m × 10 m × 5.5 m, in which the fluid areas are divided and identified, then it is inserted in a fluid space (7 m × 14 m × 20 m) in which the computational model is obtained.

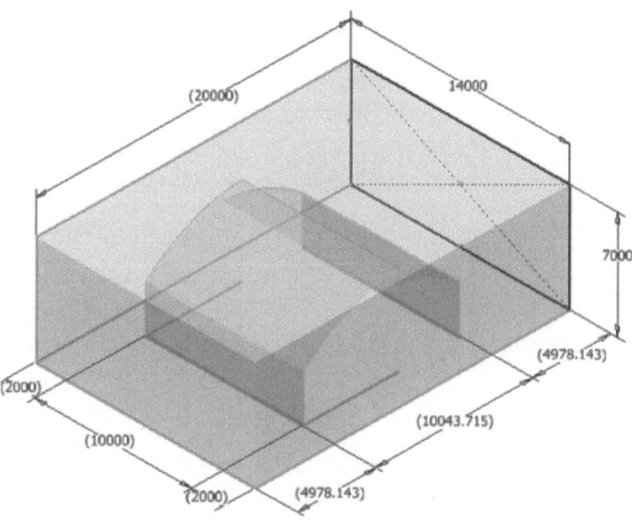

Figure 4.22 Dimensions of the greenhouse inside a Winter Tunnel.

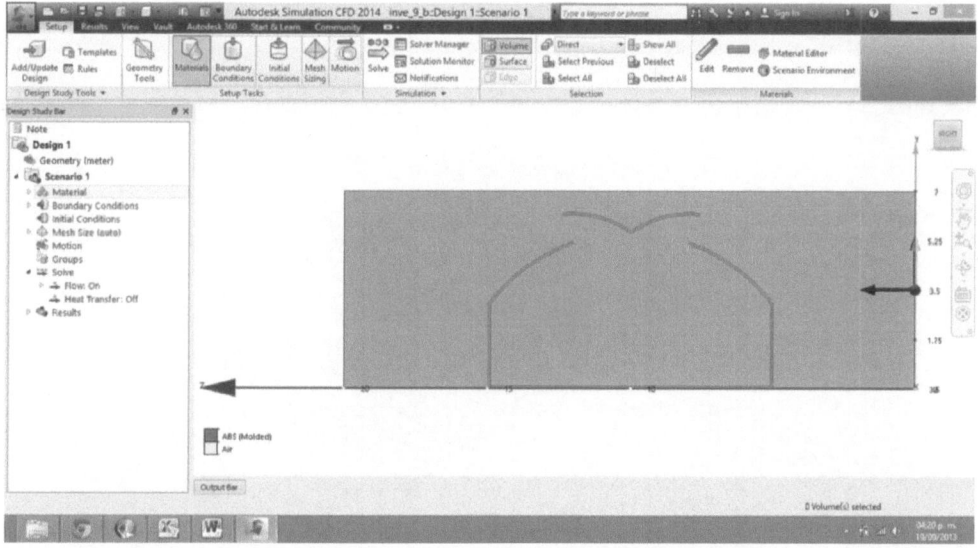

Figure 4.23 Selection of the materials and the wind flow.

To perform the necessary analysis to establish what materials will be used for the structure and the type of fluid to be used, it is necessary to specify the speed with which it will move in the wind tunnel (Figure 4.23).

To establish the boundary conditions (Figure 4.24), it is necesary to determine the wind direction, as well as the velocity over the cube wall where the fluid will begin

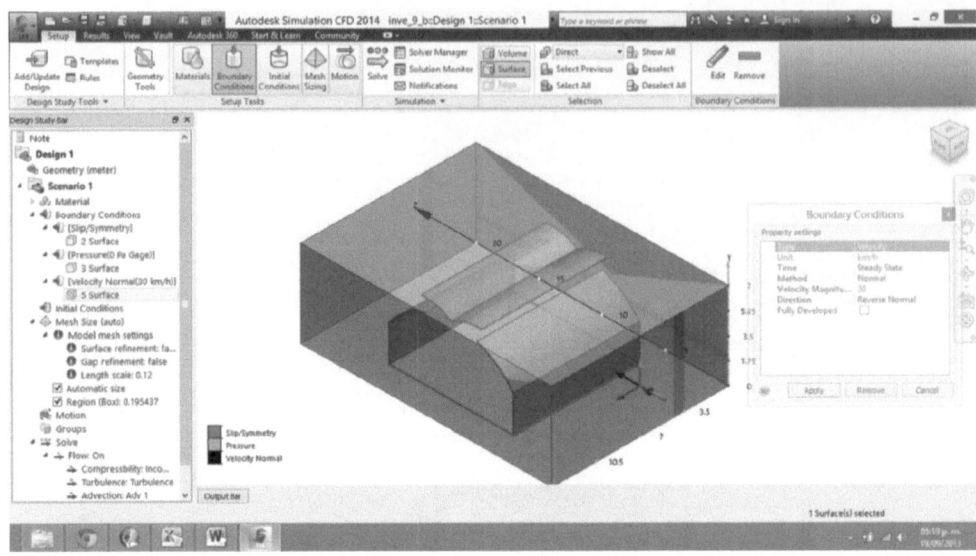

Figure 4.24   Boundary conditions in greenhouses.

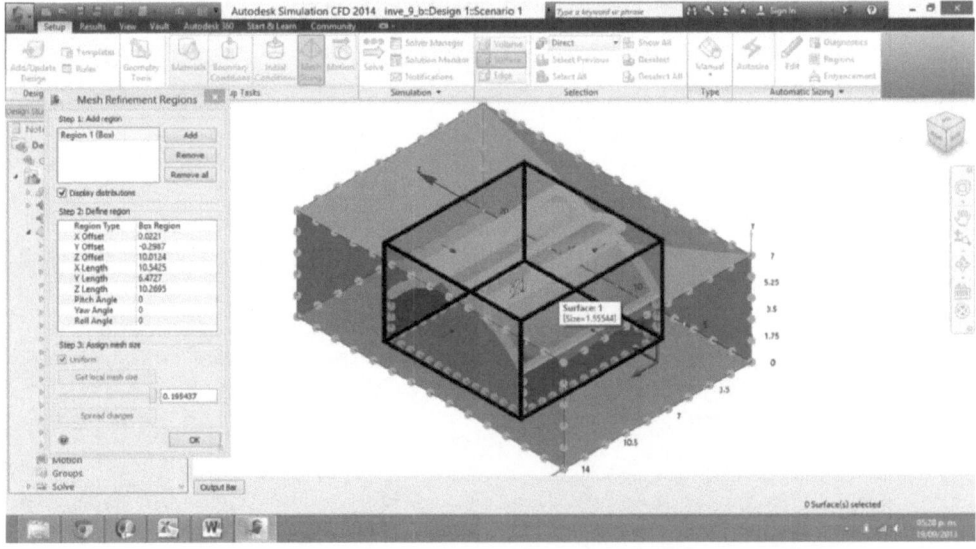

Figure 4.25   Creation of elements in the structure of the greenhouses.

the movement; in the parallel wall the flow of the wind conditions the simetry, the last one in the opposite wall the flow is established without pressure.

Subsequently it begins with the creation of the elements for a solution of the system, in this case it creates a general structure and another in the region of interest in order to obtain greater accuracy in the results (Figure 4.25).

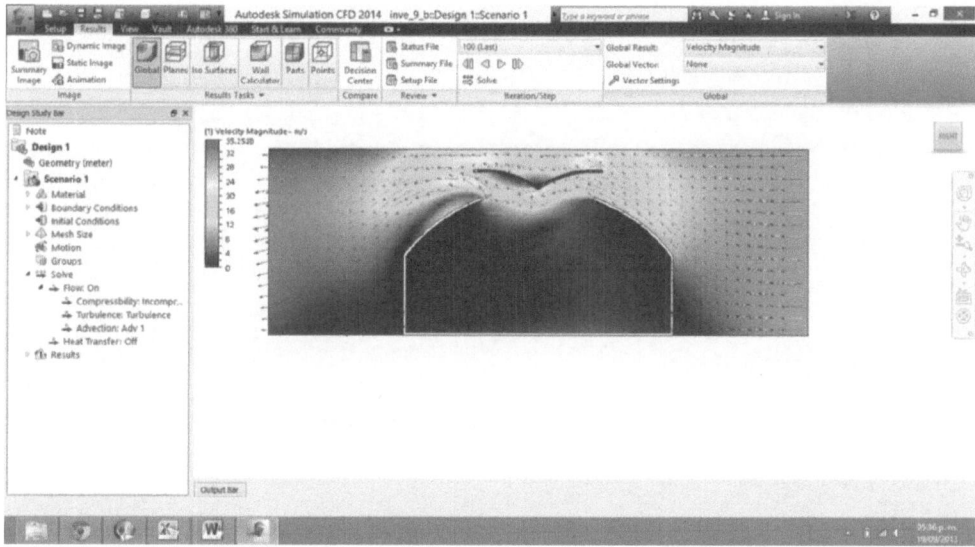

Figure 4.26 Post processing of the information with respect to the force applied in the greenhouses.

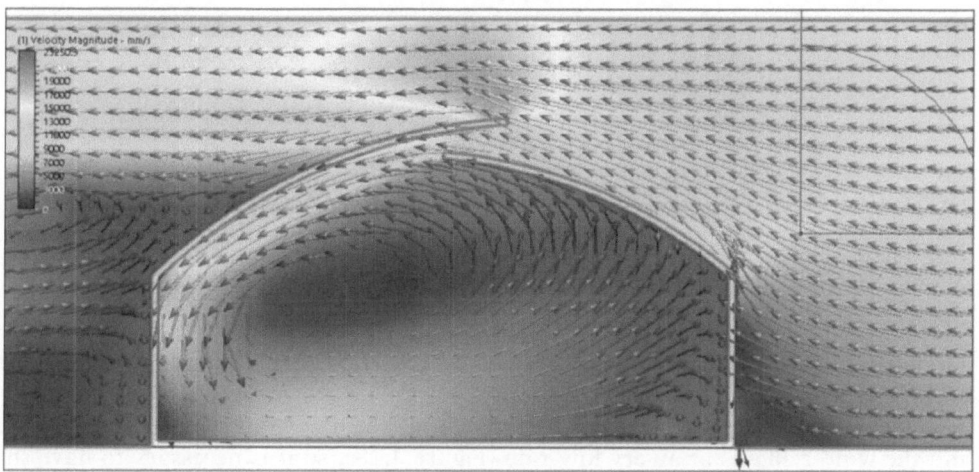

Figure 4.27 Wind analysis for a Tunnel Fixed Zenith greenhouse.

After fulfilling these conditions, you obtain the solution and post processing of the information. In this analysis the velocities of the wind are determined throughout the length of the greenhouse, as well as its behavior within.

Since the flow is determined as the air inside the greenhouse, the analysis is performed on each Inventor structure to analyze which one is the best answer due to their shape and material (Figure 4.26).

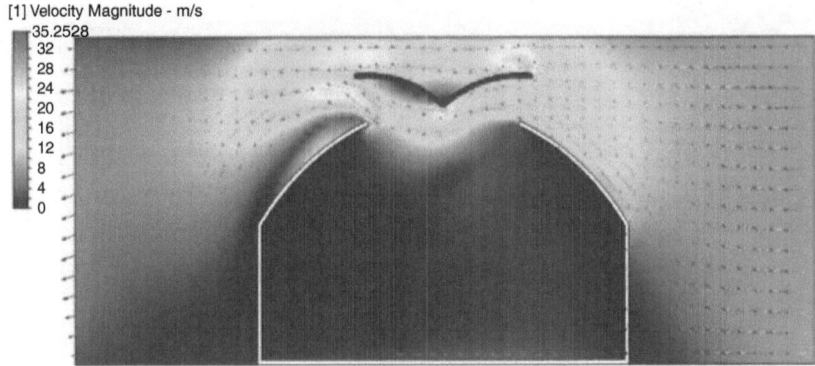

Figure 4.28 Wind analysis for a Butterfly greenhouse.

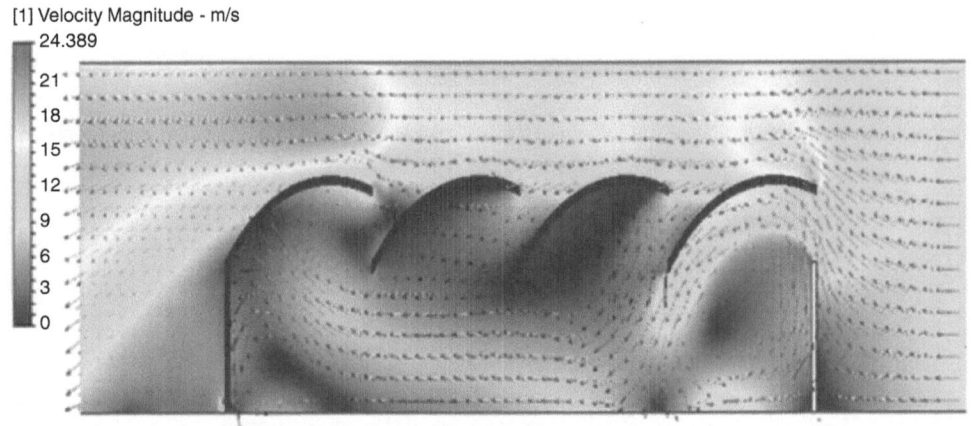

Figure 4.29 Wind analyses for a Sawtooth greenhouse.

Finally the results are shown in Figures 4.27 to 4.34 that show the wind response in some types of greenhouses.

There is no air exchange inside the greenhouse, as the wind only rotates within it and the wind remains at a very low rate (Figure 4.28), so it is necessary to have side vents and an appropriate fan.

The heat exchange occurs, because the cold air that enters pushes the warm air that is inside the greenhouse, benefiting the temperature from this action. To improve the quality of air, side vents and fans are placed inside, facilitating the recirculation according to Figure 4.29.

As is noted the wind runs at a certain speed, which declines from one greenhouse to another, moving the hot air that exists within. The speed of the wind, decreases with the mesh that exists between them (Figure 4.30).

As is noted there is only one air inlet, so it is necessary to add fans to facilitate the exit of the air according to Figure 4.31.

[1] Velocity Magnitude - m/s

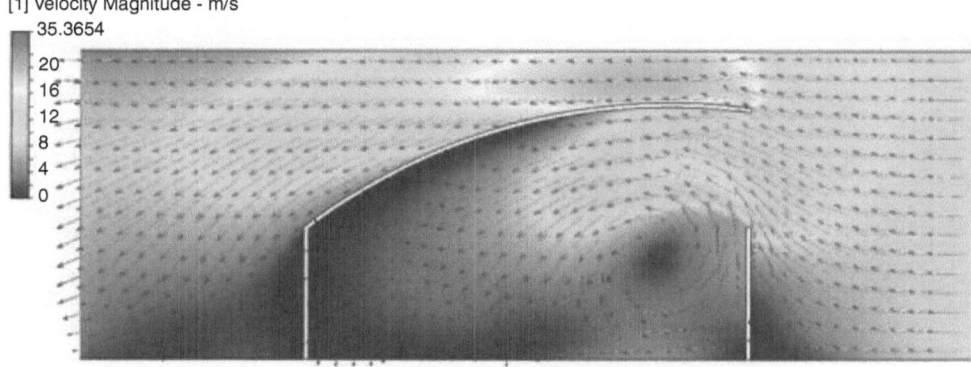

Figure 4.30 Wind analysis for a Multi Tunnel greenhouse.

[1] Velocity Magnitude - m/s

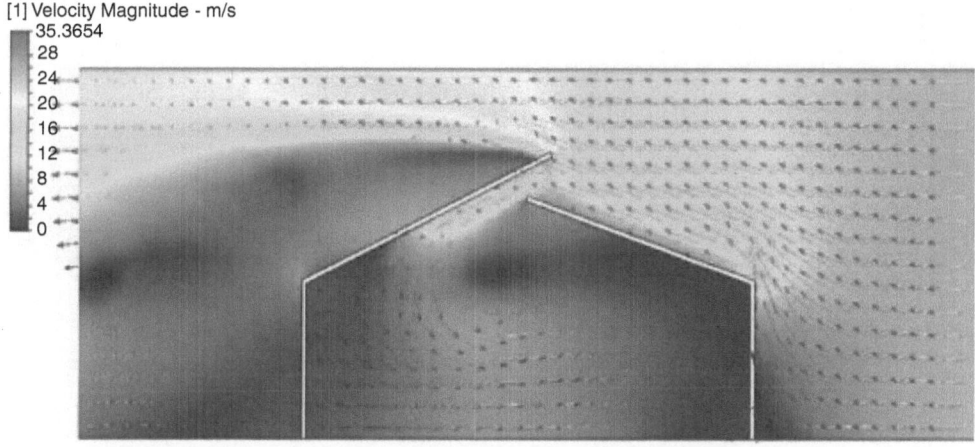

Figure 4.31 Wind analysis for a Modular Parral greenhouse.

As is noted there is only one air inlet, so it is necessary to add fans to facilitate the exit of the air (Figure 4.32).

In this figure there is heat exchange using side vents (Figure 4.33).

As noted there is only one air inlet, so it is necessary to add fans to facilitate the exit of the air (Figure 4.34).

There is no exchange of heat in this greenhouse; there is a need for side vents.

In these figures you have the rate that is in the air movement, showing the advantages and disadvantages of each of the structure type, in Table 4.19 a summary of these results is shown to compare and determine which one is the best air movement to enter the greenhouse.

As can be seen the one that has a faster speed is the butterfly, so this greenhouse should be considered to be used in places where a better air circulation is required to control the internal temperature.

[1]Velocity Magnitude - m/s

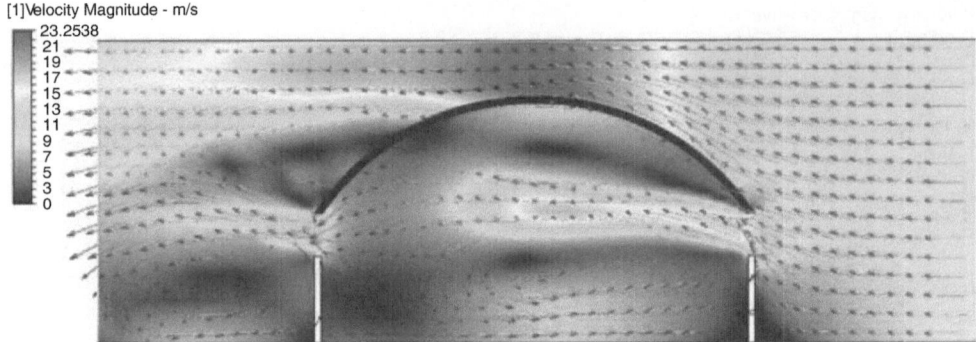

Figure 4.32 Wind analysis for a Tunnel greenhouse.

[1]Velocity Magnitude - m/s

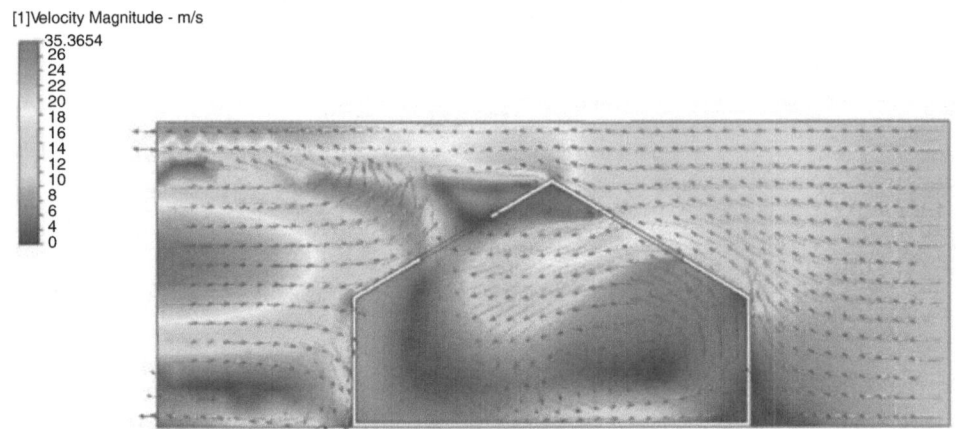

Figure 4.33 Wind analysis for a Chapel greenhouse.

[1]Velocity Magnitude - m/s

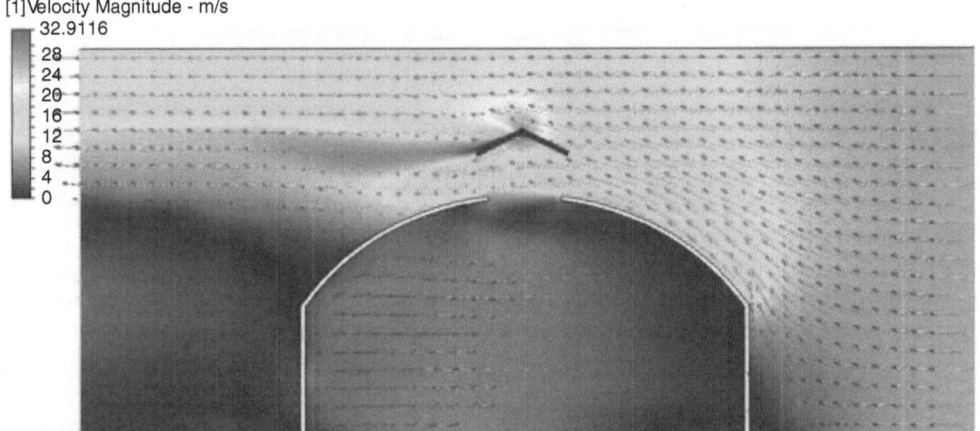

Figure 4.34 Wind analysis for a Chinese hat greenhouse.

Table 4.19 Velocities of the air in greenhouses.

Type of greenhouse	Velocities of the air(m/s)
Tunnel Fixed Zenith	15
Butterfly	21
Sawtooth	15
Multi Tunnel	16
Modular Parral	16
Tunnel	17
Chapel	20
Chinese hat	16

In the examples of air recirculation in the different types ofgreenhouses shown in Figures 4.27 to 4.34, only one is using natural ventilation in which the air that enters through the vents and the zenith to perform the air exchange, on which the ventilation idea is based on performing a mixing of air in order to maintain the volume of air in the greenhouse at an uniform temperature between the temperature of the soil, the one that the crop generates and the environment.

For a real homogenization of the temperature it is necessary to use forced ventilation through the use of ventilators, for which it is necessary to perform an analysis to determine the heat exchange inside the greenhouse and perform the necessary control strategy. Furthermore the results should be analyzed to validate the mathematical analysis that is necessary for the boundary conditions to initialize the model, which can be proposed using the experimental results of air temperature, wind speed and heat flux. However the main factors that determine the air and temperature exchange are the magnitude and direction of the wind, the greenhouse structure, position and type of windows (Figure 4.35).

With the knowledge of these variables it is possible to control the microclimate inside the greenhouse. There are authors like Kacira (2012) who through graphs have shown the behavior of variables such as temperature and wind to determine the type of control to be applied.

4.7 GREENHOUSE CONSTRUCTION STANDARDS

On July 8, 2008, the Ministry of Economy, through the Directorate General of Standards, published in the Official Journal of the Federation, the Declaration of validity of the Mexican Standard NMX-E-255-2008 CNCP for Greenhouses Design and Construction Specifications.

The following subsections are highlights of the published standard.

4.7.1 Materials used in the structures

The materials of the structures should be economical, lightweight, strong and slender; should form little bulky structures, to avoid shadows on the same plant, of easy construction, maintenance and preservation, modifiable and adaptable to future growth and expansion of the structures.

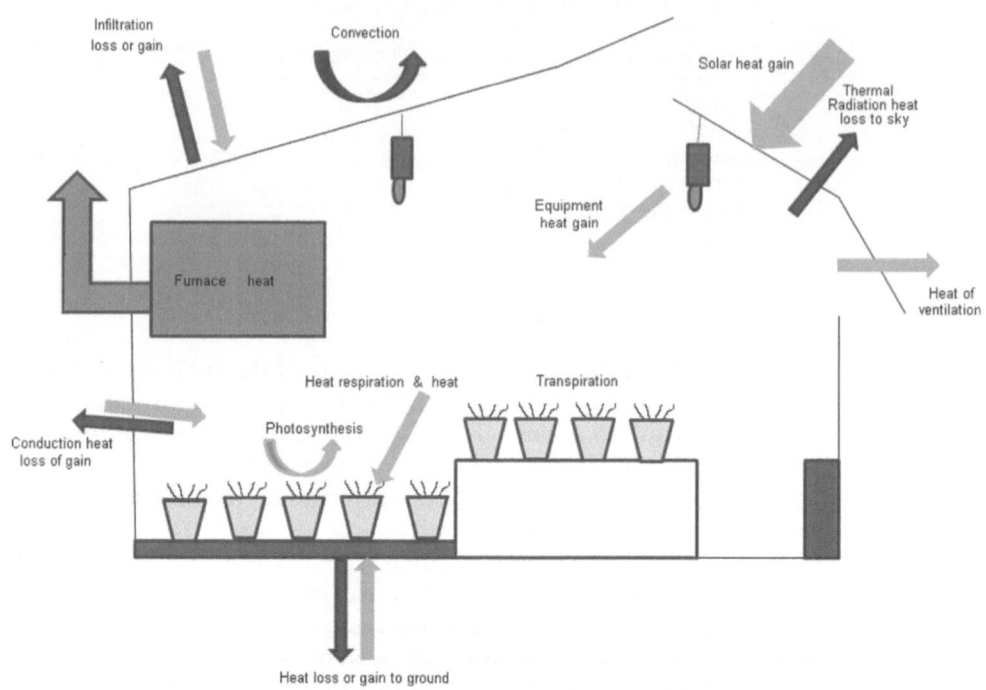

Figure 4.35 Exchanges of fluxes in greenhouse environment.

Foundation anchors, columns, bows, arrows, stringers and reinforcements: Of square or round tubular profile galvanized steel base G-90 layer on both sides. Metallic Zinc based on the welding seam. Different sections.

Gutters and fastener profile. Galvanized steel sheet based on a G-90 coating on both sides, various gauges.
Cable. Galvanized Steel G-90 layer, several measures.
Wires. Low carbon steel galvanized G-90 various calibers.
Spring fastener. Galvanized high carbon steel.
Screws. High strength galvanized G-5 several measures.
The concrete should have a resistance f'c = 150 kg/cm² for manufactured bases.

4.7.2   Norms of materials

In order to meet the standards of materials for the steel used in the structure of a greenhouse, the following specifications must be met according to the manufacturer:

Square or round profile steel manufactured according to standard NMX-B-009, grade 30 steel (Fy = 2,320 kg/cm²); cold rolled.
The coating of these profiles should be zinc-aluminum heat galvanized, AZ-90 coating (0.90 Oz/ft² = 274 gr/m² = 0-0015 in, according to NOM-B-469, ASTM-792), in which the material must provide resistance to corrosive environments.

The structures must have 5 to 6 kg/m^2 steel. The columns must be at least 2″ and the arches of 1 3/4″ of galvanized steel (ASTM A53).

Covers of Polyethylene caliber 720 treated against UV II, different percentages of shade and color.

4.7.3 Activities for greenhouse construction

The most important issues to consider before constructing greenhouses are the location, available budget and potential crops. Other considerations include orientation, airflow, shading, windbreaks, drainage, soil quality, weeds or pests.

The general variables to construct greenhouses are shown in figure 4.36 (Rutledge 1998, Alabama & Auburn Universities 2013):

Selecting and preparing the site: Select a site that is in full sun and slightly elevated so there will be good air movement and ventilation. Locate the site near a clean water supply. The floor can be either leveled or sloped if trickle irrigation is planned. Some systems require leveled floors to install the growing troughs suitable for nutrient delivery across the roots.

Width considerations: Determine the width of the greenhouse by adding the widths of the plant benches and the walks. The width must be so that it allows easy passage for visitors who may not be used to walking between rows of plants. Should a disabled individual confined to a wheelchair desire access, special design specifications should be kept in mind. Access walks to the greenhouse need to be at least four feet wide (Schnelle et al. 1990). To determine the length of the greenhouse, divided the total number of plants to be grown in the greenhouse by the density at which they can be spaced per linear foot of bench. Calculate the number of plants that could be grown per linear foot of bench space. The height of the greenhouse depends on the desired height of the eaves. The eave height, the distance from the sidewall to the center of the greenhouse, and the roof pitch will determine the height of the green-house at the center. Most experts say that at least 6,000 sq. ft. (557.4 m^2), is the preferred growing space. At least 10,000 sq. ft. (929 m^2), of the greenhouse space is needed for a full-time operation.

Type of structure: The greenhouses need to be designed to provide the necessary safety, serviceability, general structural integrity and suitability. The structure should be able to take all the necessary dead, live, wind and snow loads. The foundation, columns and trusses are to be designed accordingly. There are some styles that can be chose by their characteristics such as: short duration of life (structures manufactured in wood, with untreated poles and without foundations, have a life span of only 3–5 years), for a longer life span structures can be manufactured with treated timber profiles or steel tubes. Those structures need more investment for the structure itself, but they are more cost-effective and economical in the long run. On the other hand, the available space and the cost usually have a large impact on the choice of size for a greenhouse, for example if it is more than 30 meters long it should be built with zenithal opening, in these cases, the opening zenithal must be oriented in the opposite direction to the direction of the wind. The roof of the greenhouse will be given a 30% inclination to facilitate the water dropping, a product of the condensation of the transpiration from plants and soil evaporation, it should drop down onto the sides and not on the crops.

Gussets: Greenhouses to be covered in PE usually do not require an extensive foundation, but the support posts must be set in concrete footings. Attached greenhouses and those covered with glass should have a strong concrete or concrete block foundation that extends below the frost line according to local building codes; concrete block, stone, or brick are the most popular materials used, but shingle, clapboard, and asbestos rock have also been used. Choose the type that works best with the overall architectural scheme. In many places where drainage is adequate, a solid floor is not necessary. Walkways can be constructed of concrete for easy movement of equipment and people, especially if a person is disabled or in a wheelchair. Brick filled with sand, flagstone, or stepping stones can be used for decorative walks. Gravel under the benches keeps the walkways free of debris and reduces weeds.

Painting: Corrosion is a problem, especially in areas of industrial air pollution and near the sea. Modern alloys are much more resistant so corrosion is only likely to occur in areas of very high industrial pollution, which are not widespread. Galvanizing can also be broken down by an electrolytic reaction when the alloy and the steel members touch. When repainting it may be necessary to strip back and re-primer if the paint is blistered or cracked as moisture is rapidly absorbed once the skin of the paint is broken. The structures must be built with materials that do not produce plenty of shade within the greenhouses, particularly if they are in the Southern zone where there is less light. Painted in white color reflects the light, while the dark (blackened timber) absorbs it. The painting also allows a better conservation of the wood.

Covering material: There are several covering materials to choose from (as showed in the second point of this chapter), each of which has its own advantages and disadvantages. Common covering materials are glass, polyethylene film, fiberglass reinforced panels, and double layer structured panels.

Installed controls: Automatic controls are important in greenhouses. Lights, fans, pumps, heaters and mist systems must be turned on and off at prescribed times. Time clocks, photocells, thermostats and other automatic controls are available commercially. Individual controls or combinations of controls provide interval control as desired. A thermostat can turn the heater on when the temperature drops to a certain point. Humidistats are available to regulate humidifiers automatically. Automatic ventilators, controlled by a thermostat, open the vents and turn on the fans. The water requirements of plants vary so much of this segment is going to require a very close attention. Automatic controls are costly and you may want to add some of them after you get started.

A summary of the points discussed in this chapter is shown in Tables 4.20 to 4.23 (Mississippi State University 2012, Zabeltitz 1990), information such as: mechanical advantages, weather characteristics, type of crop or expansion of the greenhouses is provided.

Considering these variables, a summary of the variables above mentioned and some more has been developed in Table 4.24 by (Ortiz Vertiz 2013), to determine which is the most profitable greenhouse for the farmer based on the type of crop.

As can be seen, it cannot be generalized or said which is the best greenhouse, but Table 4.24 serves to give an idea of the most relevant features to be considered for the construction and selection of a greenhouse.

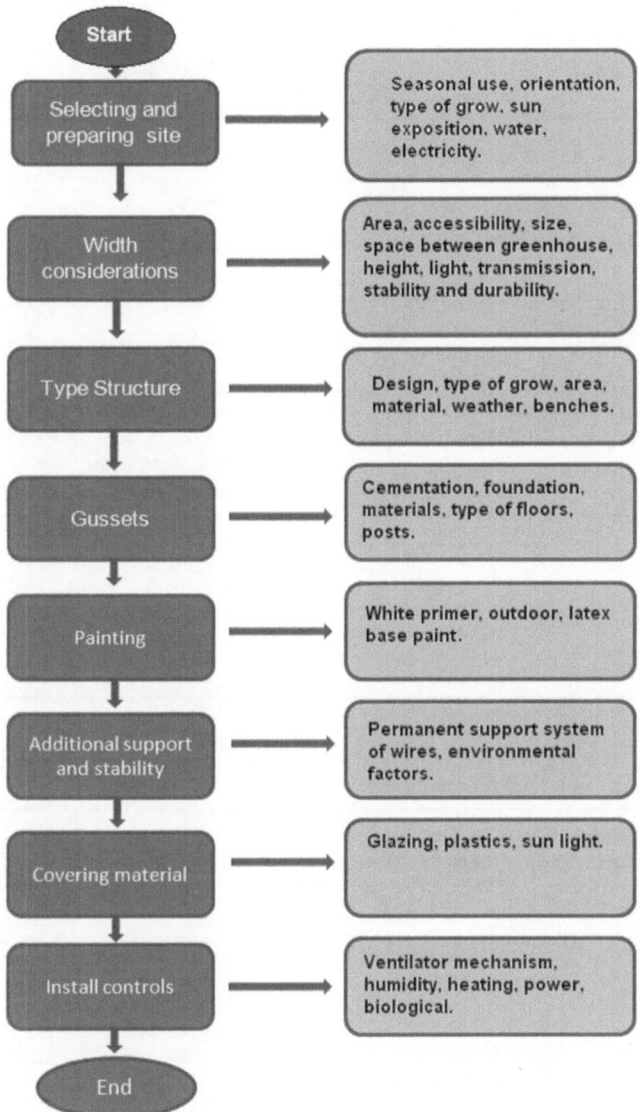

Figure 4.36 Activities for greenhouse construction.

4.8   STRUCTURES AND ENVIRONMENTAL CONTROL

Based on the characteristics of the greenhouses, the gardeners select the specific structure but sometimes other options are needed, such as the environmental control. The mechanical, electrical, electronic and intelligent control of a greenhouse contemplates the physical structure, ventilation, humidification system, irrigation system and nutrient supply to its respective automation, the latter being composed of sensors, actuators

Table 4.20 Mechanical Characteristics of the structures according to the type of greenhouse.

Characteristics (Mechanical advantages)	Tunnel Fixed Zenith	Butterfly	Sawtooth	Multi-tunnel	Modular Parral	Tunnel	Chapel	Chinese hat
Displacement	x					x		
Force		x	x					x
Momentums	x				x			
Normal tension	x							x
Tension cut off					x			
Torsional stress					x			

Table 4.21 Weather characteristics to construct greenhouses.

Characteristics (Weather)	Tunnel Fixed Zenith	Butterfly	Sawtooth	Multi-tunnel	Modular Parral	Tunnel	Chapel	Chinese hat
Dry tropical or desert climates	x	x		x		x	x	x
Subtropical desert and Mediterranean climates	x	x		x		x	x	x
Humid tropical climates			x		x		x	
Cold temperate climates							x	

Table 4.22 Type of crop for each greenhouse.

Characteristics (Types of crops)	Tunnel Fixed Zenith	Butterfly	Sawtooth	Multi-tunnel	Modular Parral	Tunnel	Chapel	Chinese hat
Tomato	x	x	x	x	x	x	x	x
Peppers		x					x	
Cucumbers		x	x			x		
Lettuce	x	x			x	x		
Flowers					x			
Eggplant	x	x				x		
Strawberries					x			
Chilies							x	
Beans	x	x						

Table 4.23 Possible creation of greenhouses in battery.

Characteristics (settlement in battery)	Tunnel Fixed Zenith	Butterfly	Sawtooth	Multi-tunnel	Modular Parral	Tunnel	Chapel	Chinese hat
It has excellent ventilation	x	x	x	x				x
Problems with ventilation					x	x	x	

Table 4.24 Characteristics of some greenhouses.

Characteristics	Parral or Classic	Post and Rafter	Asym-metrical	Tunnel or semi circular	Venlo	Chapel or double chapel
Construction	x	x	x			
Low cost of operation Good use of rainwater in dry periods	x					
Good use of light in the winter			x			
Good deal of the light inside the greenhouse	x			x	x	
Good sealing to the rain and wind		x	x	x	x	
Good ventilation			x	x		x
It has great facilities to evacuate rainwater						x
High thermal inertia due to its unit volume			x			
It is easy to build, easy installation and maintenance				x		x
Placement adaptable for all types of plastic on the cover						x
Structures with few obstacles		x		x		
Allows installation of roof ventilation downwind		x	x	x		x
Allows connection of several ships in battery						x
Their great adaptability to terrain geometry	x					
Increased unit volume and therefore a greater thermal inertia		x				
Surface greater = heat losses through the cover		x	x			
Too much specialization in construction and maintenance	x					
Brightness differences between slopes		x				
Difficulty in tillage structure by Shape	x					x
Abundance of structural elements means less light transmission					x	x
Installation of roof windows is quite difficult	x					
Bad ventilation	x					x
Shed very small due to the complexity of its structure					x	
No rainwater advantage		x	x	x		
Risk of collapse by rainwater bags	x					
Little tight dripping rainwater and air through the holes in the deck joints, which favors the proliferation of fungal diseases.	x					
Little or nothing advisable in rainy places	x					
Low volume of air	x					
Vulnerability to the winds and aging fast installation	x					
It is difficult to change a plastic cover		x	x			
High cost				x	x	x

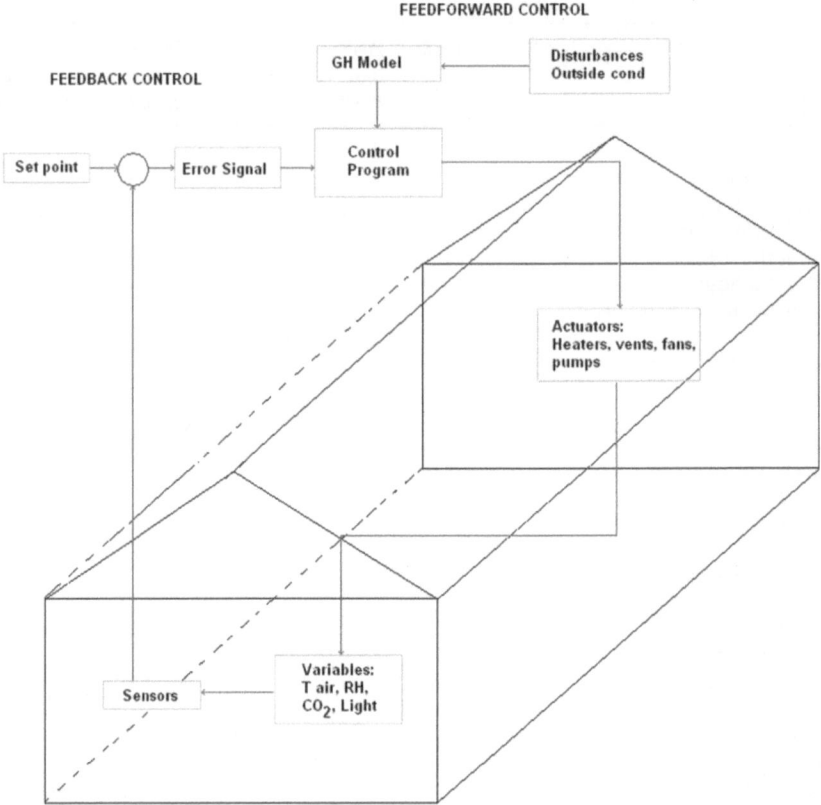

Figure 4.37 Environmental controls inside greenhouses.

and a controller form to control the loop to regulate the microclimate variables that need to be considered inside a greenhouse.

On the other hand, for the control inside the greenhouse (Figure 4.37), the hydroponic is an option, because it includes high-density maximum crop yield, crop production where no suitable soil exists, a virtual indifference to ambient temperature and seasonality, more efficient use of water and fertilizers, minimal use of land area, and suitability for mechanization and disease control.

For this reason and the necessities, the combination of climate conditions and the structure of the greenhouses determine the most appropriate shape of the greenhouse and control. Based on this Morgan (2012) defined some types of controls applied in greenhouses:

Control by defined time without feedback (CDTWF): The main objective of this control system is to keep the interior temperature around a base parameter. The relationship between the opening of windows and the interior temperature is not linear. Based on dates and times a schedule is established and it performs the opening of vents, the lifting of meshes and all the necessary equipment for the control of climate.

Table 4.25 Control types for dry tropical or desert climates.

Dry tropical or desert climates

Characteristics: Greenhouses are very important for tropical regions because they avoid the disadvantages of open-air production and increase the yield and quality remarkably. The main environmental threats are high winds carrying dust or sand, which can blast both crops and greenhouses. Low humidity levels allow the effective use of evaporative cooling, which is the main feature of cropping in this kind of dry, arid climate. Air movement is essential inside the greenhouses to maintain good levels of transpiration within the crop. More advanced high-tech, computer controlled and air conditioned structures are also in use in climates like this.

Control type	Greenhouses	Recommendations
PID, FUZZY	Tunnel Fixed Zenith Butterfly Multi Tunnel Tunnel Chapel Chinese hat	Crop must have protection from rain, wind, and too high global radiation. Need cladding material impermeable to rain. UV-stabilized plastic film with long duration of life. The relation between the greenhouse volume to the ground floor area should be as large as possible. Large ventilation openings with more than 60% opening related to floor area covered by insect screens to block insect entry physically and optically. Spectrally modified cladding material, to reduce heat load by blocking NIR transmittance. Forced extraction ventilation when temperatures exceed the permissible maximum.

Table 4.26 Control types for subtropical desert and Mediterranean climates.

Subtropical desert and Mediterranean climates

Characteristics: Very efficient ventilation for hot periods is a prerequisite for good climate. Usually, they have forced ventilation by fans in combination with evaporative cooling. Along with shading over the outside of the greenhouse, this produces an ideal environment during dry summer conditions. As temperatures drop, which can occur during the night, even in summer, the moist air can be vented through the top vents and the interior of the greenhouse can be heated.

Control type	Greenhouses	Recommendations
CDTWF, ON-OFF, PID, Fuzzy	Tunnel Fixed Zenith Butterfly Multi Tunnel Tunnel Chapel Chinese hat	A structure that can be heated but still maintains a cool environment in summer is necessary. Fans used in conjunction with adequate air-intake systems may enhance temperature control over natural ventilation systems.

It is therefore necessary to rely on computers to measure pH meter and electrical conductivity, and to operate equipment such as electrical pollinators, irrigation equipment etc.

ON-OFF: The users make periodic trips through the greenhouses during the night, checking the temperature in each greenhouse and controlling it by opening or closing

Table 4.27 Control types for humid tropical climates.

Humid tropical climates

Characteristics: Good tropical greenhouse designs can be as simple as a rain cover or plastic roof with open or roll-up sides covered with insect mesh. In larger greenhouses, the structure is best designed with a "saw tooth" roof layout which allows good venting of the hot air inside the greenhouse on clear days. Heating and insulation aren't required and vents can remain open. High winds from tropical typhoons or hurricanes can be a major risk in this climate. Since temperatures are warm during the day and night for much of the year, heating and insulation aren't required and vents can remain open, covered with insect mesh.

Control type	Greenhouses	Recommendations
CDTWF, ON-OFF	Sawtooth Modular Parral Chapel Tunnel Fixed Zenith Butterfly Chinese hat	Misting systems and air-movement fans can be used to cool the environment inside this type of structure and movable thermal screens can be employed to reduce incoming sunlight on bright, cloudless days and pulled back to allow maximum light penetration under overcast conditions. Have even a simple irrigation programming device and the proportion of computerized irrigation systems. Fans used in conjunction with adequate air-intake systems may enhance temperature control over natural ventilation systems.

valves of heating pipes as required. During the day, employees open or close ventilators by hand to maintain the temperature. Hence the temperatures are manually controlled throughout the day during the cropping season.

PID: It is the most complete controller available and the most resorted to since it provides a quick response, a control signal that tends to provide stability to the system and a minimum steady state error. The PID controller is an important control tool for industrial processes and only three gains have to be tuned.

Fuzzy: This controller has the task of managing and monitoring the weather conditions inside the greenhouse, temperature and relative humidity, making decisions autonomously (without an operator) of what to do considering the optimal conditions programmed. These decisions are obtained through monitoring the internal and external sensors.

In all the cases, the variables to be analyzed and controlled are the temperature, relative humidity (RH), CO_2 enrichment, radiation and many others that are necessary to generate the fundamental conditions for successful protected agriculture. Based on this information, tables 4.25 to 4.29 (Zabeltitz, C. 1990, Morgan, L. 2012) show the characteristics of the weather depending on the zone, the control type, greenhouses type and recommendations to help the crop and to obtain the best one.

As you can see the different types of control seek to meet the needs of the climate of each area and ensure food production through water savings, reduced use of agrochemicals and the space required to grow crops while ensuring high quality thereof.

Table 4.28 Control types for temperate climates.

Temperate climates

Characteristics: Efficient heating of the air inside the greenhouse and insulating and maintaining this heated air is the main consideration. Growers wanting year-round high growth rates and maximum yields in these environments usually select greenhouses featuring fully clad side walls, roof and side vents, allowing large ventilation areas and computer control of environmental equipment such as heaters, shade or thermal screens, fogging and vents. Temperate zone greenhouse design often makes use of plastic cladding "twin skins" where the space between the two layers of plastic is inflated, offering improved insulation and better environmental control. Shade screens and whitewash are the major existing methods used to reduce the income of solar radiation; greenhouse ventilation is an effective way to remove the extra heat through air exchange between inside and outside.

Control type	Greenhouses	Recommendations
ON-OFF, PID, FUZZY	Tunnel Fixed Zenith Butterfly Multi Tunnel Tunnel Chinese hat	Requires mechanical ventilation equipment to operate properly in horizontal position. Fans used in conjunction with adequate air-intake systems may enhance temperature control over natural ventilation systems. The main steel materials are all hot dip galvanized to offer excellent anti-corrosion property. The nozzles of the fog system should be located at the highest possible position inside the greenhouse to allow water evaporation before the water drops to crop or the ground.

Table 4.29 Control types for cold temperate climates.

Cold temperate climates:

Characteristics: Greenhouses for this type of environment need solid walls and strongly constructed comparatively steep solid roofs to carry snow loads that would collapse plastic film structures. These greenhouses are often double insulated by installing plastic film on the inside walls and positioning retractable thermal screens across the eaves at stud height. To prevent heat loss, vents are often kept closed during the winter months.

Control type	Greenhouses	Recommendations
On-OFF, PID, FUZZY	Venlo	Need to do perfectly closing ventilation system and very close glazing systems just like the adjustment of double glass and bar double plates in the gables. The glasshouse frames are constructed with hot galvanized steel and aluminum alloy extruded profiles. Shading, heating, cooling, irrigation, climate control systems can be installed if requested cover material: float glass, tempered glass, insulating glass.

Analyzing the information it is possible to show that not all strategies may be applicable in all of the greenhouses, but by analyzing the variables of each site and reviewing the structure corresponding to each weather a good crop can be obtained.

Researchers need to analyze different control theories to determine which one is the most proper for their projects according to their specific requirements of greenhouse climate control systems. In the next chapter, more explanations will be provided.

Chapter 5

Greenhouse automation and control

The design of environmental control systems for plants is complicated owing to the interaction of many environmental variables affecting growth and production. The clear necessity of matching food production to the needs of the world's population requires the accurate knowledge of the factor limiting primary production. Therefore, the environmental factors that limit agricultural production and physiological responses of plants to environmental stresses need to be identified and understood. The fluctuating and interacting natural environment makes it nearly impossible to analyze the effects of various climatic factors on plant behavior. The most appropriate solution for this problem is to move research into a plant growth structure with partial or complete environmental control, then this well maintained chamber is named as controlled environment greenhouse or automated greenhouse (Tiwari, 2003).

A hi-tech greenhouse, which is embodied in Figure 5.1, is suitable for any climate. Crops which are sensitive to environmental conditions should be grown in hi-tech greenhouses. Some of the advantages of hi-tech greenhouses are (Tiwari, 2003):

- The computer always knows about the functioning of individual system, and if programmed properly can coordinate this system to provide the optimum environment.

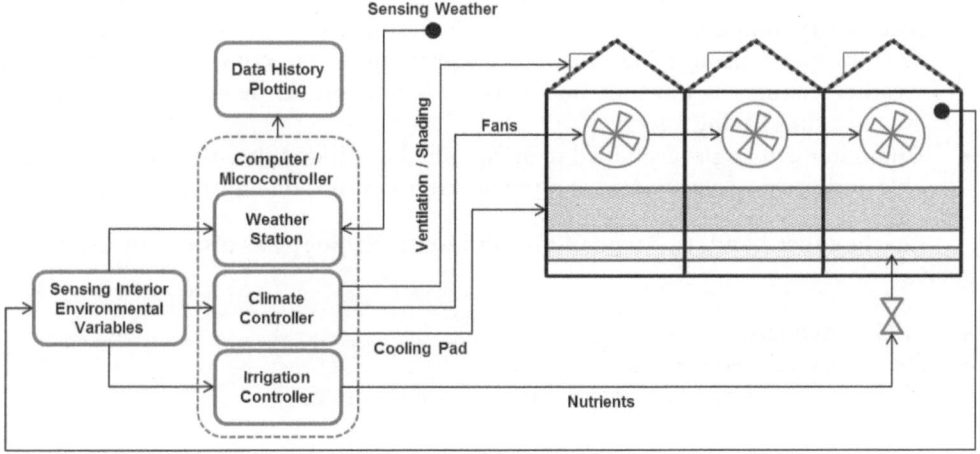

Figure 5.1 Diagram of a Typical Hi-Tech Greenhouse.

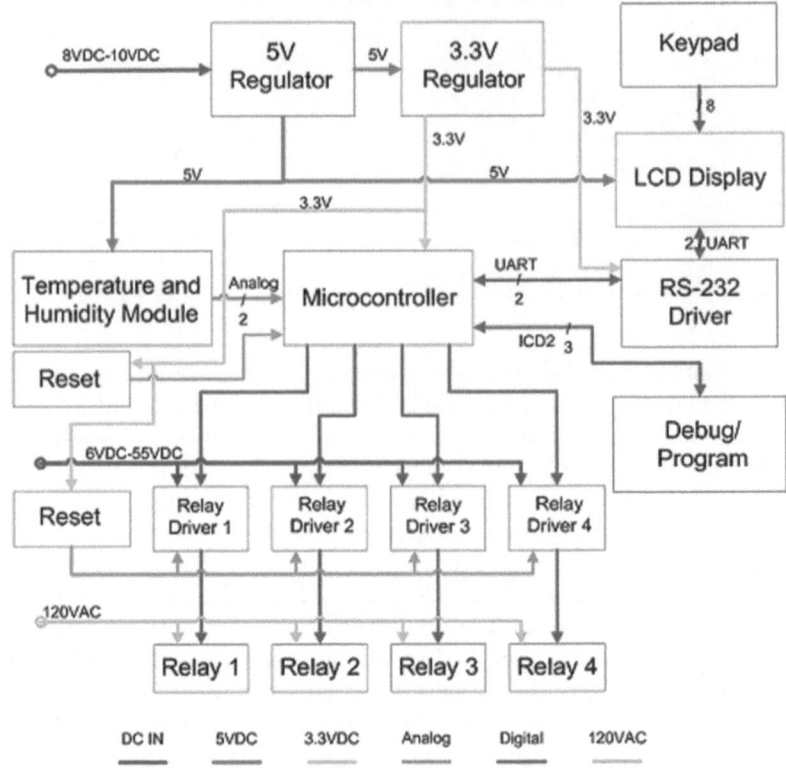

Figure 5.2 Greenhouse Control Design.

- Computer can record environmental data which can be displayed, stored, processed to provide history of cropping period.
- A high speed computer can control several remotely located greenhouses. By placing the computer in a central area, the results can be monitored frequently.
- With proper sensing apparatus and programming, the computer can anticipate weather change in advance and make adjustments in heating and ventilation systems, thus saving energy.
- Computer can be programmed to sound an alarm if conditions become unacceptable and warn of sensor and equipment failure, if not managed properly.

On the other hand, this type of greenhouse technology has also some drawbacks such as:

- High initial cost.
- Requires qualified operators.
- High maintenance, care and precautions are required.

Figure 5.2 presents a typical greenhouse control design in terms of the electric/electronic signals used as well as the control devices.

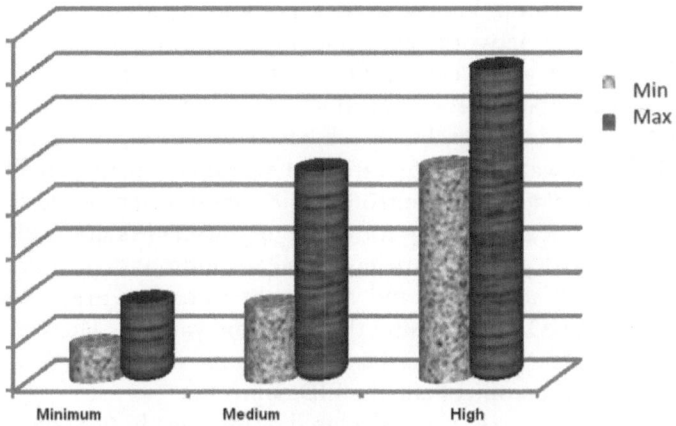

Figure 5.3 Yields of Tomato Production in Greenhouse Technology.

The importance of having a well-controlled micro-climate within a greenhouse is to maintain optimum conditions to plants so they could grow with high quality as color, robustness, healthiness and size, among others. In addition, quantity of crop is also increasing if all conditions are prepared for good growth cultivation. For example, Figure 5.3 shows the quantities of cultivated tomatoes in different greenhouse technology regimes (H. Ponce, 2009).

5.1   REAL TIME CONTROL SYSTEM AND GREENHOUSE

A control system, basically, consists of sensors, actuating devices (i.e. heating/cooling systems) and a controller which can be an operator or a computer. This section will introduce the first two parts of a control system, describing the instrumentation and automation; in other words the measuring variables and instruments, and control devices.

Since novel control techniques were available to climate control for greenhouses in horticulture, non-convention controllers were used. The main limitation was the digital system for running all the information that was needed, so different digital systems were introduced for implementing greenhouse controllers. Nowadays, real time systems are the most powerful tool for implementing complex control laws. However, a lot of controllers do not need a real time system for running because the constant time of the greenhouse is slow.

Advanced digital systems allow implementing complex laws for digital controllers in a short period of time, so it is important to understand some of the basic controllers that are used in controller implemented greenhouses. One of the best options for implementing controllers is to use real time systems based on a real time operative system. A real-time operating system is able to reliably execute programs with specific timing requirements, which is important for the greenhouse system. The key component needed to build a real-time system is a special real-time operating system (RTOS).

For instance, when an artificial neural network controller is used in a greenhouse, it is very useful to run this topology in a real time system with parallel characteristics. There are artificial neural networks controllers that have been applied and tested in different greenhouse meteorological conditions. The most important neural networks controllers that can be applied for controlling a greenhouse are: Albus's cerebellar model articulation which is a neural network based on part of the brain known as the cerebellum (CMAC), Kawato et al.'s hierarchical neural network controller and Psaltis et al.'s multilayered neural network controller. The last one offers a feed-forward neural network which has an ideal topology for deploying in FPGAs and runs on real time system. The greenhouse depends on meteorological conditions, thus the greenhouse's disturbances are difficult to predict and model. In fact, they are non-linear, multi-input multi-output (MIMO) systems and they are time-varying. These are some of the reasons that make it difficult to describe a greenhouse with analytic models and to control them with classical controllers like PID controller.

In the case of control laws that are calculated by the greenhouse non-linear multiple input multiple output coupled model in real time, running a measurement or control program on a standard PC with a general-purpose OS installed (such as Windows) is unacceptable. At any time, the OS might delay execution of a user program for many reasons: running a virus scan, updating graphics, performing system background tasks, and so on. For programs that need to run at a certain rate without interruption (for example, a temperature control system based on non-linear multiple input multiple output coupled model), this delay can cause system failure.

LabVIEW is a graphical programming environment used by engineers and scientists to develop control systems using graphical icons and wires that resemble a flowchart. It offers integration with thousands of hardware devices and provides hundreds of built-in libraries for advanced control, analysis, and data visualization. The LabVIEW platform is scalable across multiple targets and OSs, and, in the case of CompactRIO, LabVIEW can be used to access and integrate all of the components of the reconfigurable I/O (RIO) architecture. The real-time controller contains a processor that reliably and deterministically executes LabVIEW Real-Time applications and offers multirate control, execution tracing, onboard data logging, and communication with peripherals. Additional options include redundant 9 to 30 VDC supply inputs, a real-time clock, hardware watchdog timers, dual Ethernet ports, up to 2 GB of data storage, and built-in USB and RS232 support.

Operative Systems are optimized to run many processes and applications at the same time and provide other features like rich user interface graphics. In contrast, real-time OSs are designed to run a single program with precise timing. Specifically, real-time OSs help you implement the following:

- Perform tasks within a guaranteed worst-case timeframe
- Carefully prioritize different sections of the program
- Run loops with nearly the same timing on each iteration (typically within microseconds)
- Detect if a loop missed its timing goal

In addition to offering precise timing, real-time computing systems can be set up to run reliably for days, months, or years without stopping. This is important not only for

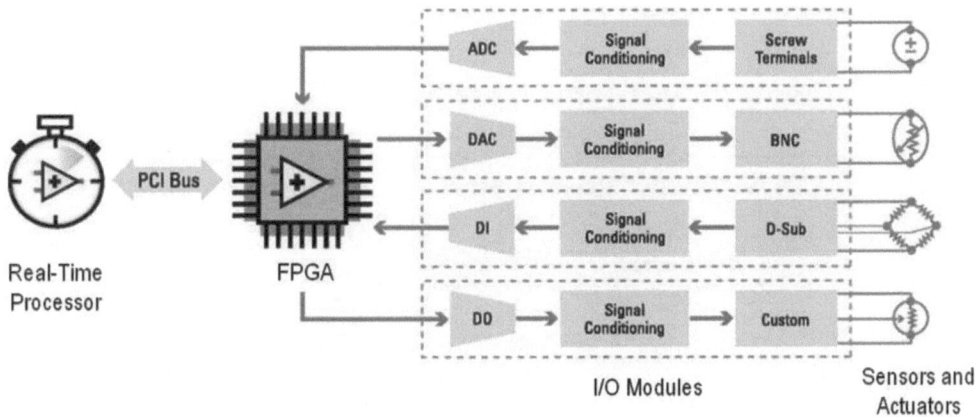

Real-Time Processor   PCI Bus   FPGA

I/O Modules   Sensors and Actuators

Figure 5.4 Embedded System Architecture (*Source:* National Instruments).

the control of greenhouses that operate 24/7 but also for any part of the greenhouse application where downtime is costly. A "watchdog" feature is also typically included in real-time systems to automatically restart an entire computer if the user program stops running. Furthermore, hardware used in a real-time system is often designed to be rugged to sustain harsh conditions for long periods.

CompactRIO is a rugged, reconfigurable embedded system containing three components: a processor running a real-time operating system (RTOS), a reconfigurable field-programmable gate array (Field-Programmable Gate Array (FPGA) Technology), and interchangeable industrial I/O modules. The real-time processor offers reliable, predictable behavior and excels at floating-point math and analysis, while the FPGA excels at smaller tasks that require high-speed logic and precise timing. Often CompactRIO applications incorporate a human machine interface (HMI), which provides the operator with a graphical user interface (GUI) for monitoring the greenhouse's state and setting operating parameters.

FPGAs offer a highly parallel and customizable platform that can perform advanced processing and control tasks at hardware speeds. An FPGA is a programmable chip composed of three basic components: logic blocks, programmable interconnects, and I/O blocks.

The logic blocks are a collection of digital components such as lookup tables, multipliers, and multiplexers where digital values and signals are processed to generate the desired logical output. These logic blocks are connected with programmable interconnects that route signals from one logic block to the next. The programmable interconnect can also route signals to the I/O blocks for two-way communication to surrounding circuitry. For more information on FPGA hardware components see Figure 5.8 and Figure 5.9.

FPGAs are clocked at relatively lower rates than CPUs and GPUs, but they make up for this difference in clock rate by allowing to create specialized circuitry that can perform multiple operations within a clock cycle. Combine this with their tight integration with I/O on NI RIO devices and you get much higher throughput, determinism,

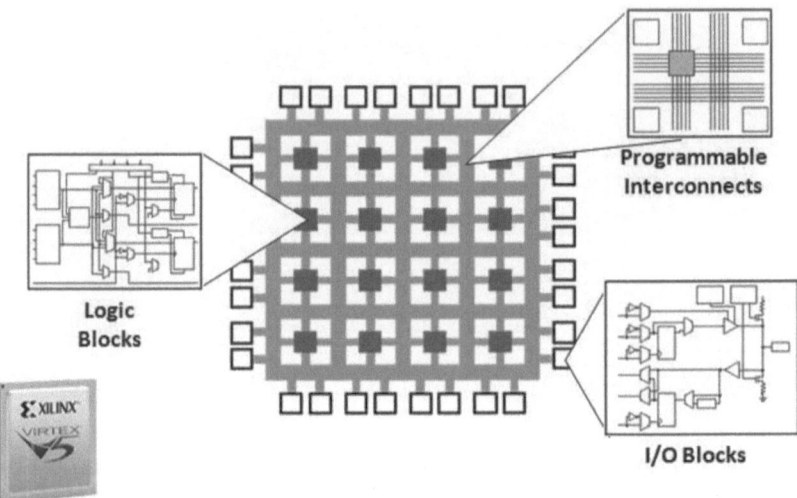

Figure 5.5   FPGA configurable logic and I/O blocks (*Source:* National Instruments).

and faster response times to tackle high-speed streaming, digital signal processing (DSP), control, and digital protocol applications, than what you could accomplish with a processor-only based solution.

The FPGA enables programs on the real-time controller to access I/O with less than 500 ns of jitter between loops. It can also directly program this FPGA to further customize the system. Because of the FPGA speed, this chassis is frequently used to create controller systems that incorporate high-speed buffered I/O, fast control loops, or custom signal filtering. For instance, using the FPGA, a single chassis can execute more than 20 analog proportional integral derivative (PID controller-conventional controllers) control loops simultaneously at a rate of 100 kHz. Additionally, because the FPGA runs all code in hardware, it provides the highest reliability and determinism, which is ideal for hardware-based interlocks, custom timing and triggering, or eliminating the custom circuitry normally required with nonstandard sensors and buses. When a cascade control loop for greenhouses is designed, the time for executing the control law is a critical variable.

Because LabVIEW FPGA VIs are synthesized down to physical hardware, the FPGA compile process is different from the compile process for a traditional LabVIEW for Windows or LabVIEW Real-Time application. When writing code for the FPGA, you write the same LabVIEW code as you do for any other target, but when you select the **Run** button, LabVIEW internally goes through a different process. First, LabVIEW FPGA generates VHDL code and passes it to the Xilinx compiler. Then the Xilinx compiler synthesizes the VHDL and places and routes all synthesized components into a bitfile. Finally, the bitfile is downloaded to the FPGA and the FPGA assumes the personality you have programmed. This process, which is more complicated than other LabVIEW compiles, can take up to several hours depending on how intricate your design is.

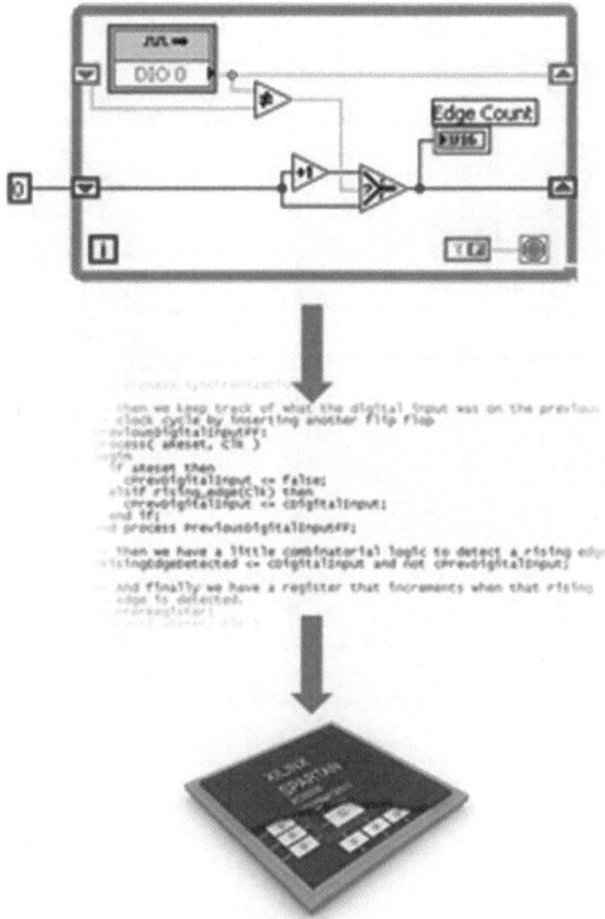

Figure 5.6 LabVIEW FPGA compiler (*Source:* National Instruments).

Depending on the complexity of your LabVIEW FPGA application for the greenhouse, you may want to quickly write a program and compile it down to hardware, or you may want to leverage the built-in simulator to debug, test, and verify your code without having to compile to hardware every time you make a change.

1   Design a software architecture (*covered in Introduction and Basic Architectures*)
2   Implement LabVIEW FPGA code
3   Test and debug LabVIEW FPGA code
4   Optimize LabVIEW FPGA Code
5   Compile LabVIEW FPGA code to hardware
6   Deploy your system

It is recommended that you develop your VIs in simulation mode, by right-clicking **FPGA Target** and selecting **Execute VI on ≫ Development Computer with Simulated**

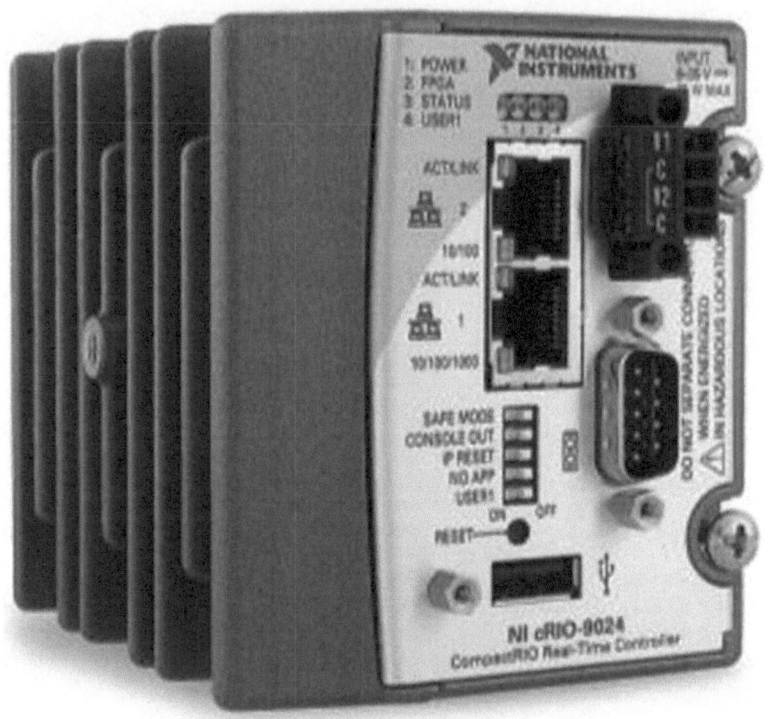

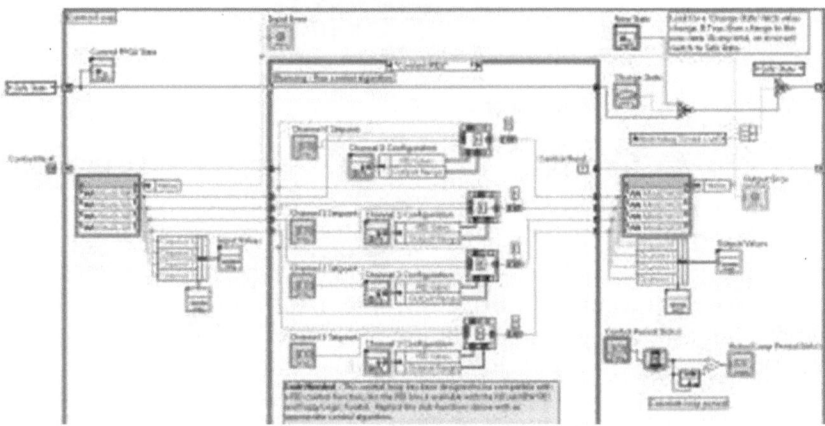

Figure 5.7   NI cRIO-9024 Real-Time Controller and PID code for FPGA (*Source*: National Instruments).

I/O. By taking this approach, you can quickly iterate on your design and have access to all of the standard LabVIEW debugging features. If you need to access real-world I/O, you will need to change the execution mode to **Execute VI on ≫ FPGA Target.**

The reconfigurable FPGA chassis is the center of the embedded system architecture. The RIO FPGA is directly connected to the I/O for high-performance access to

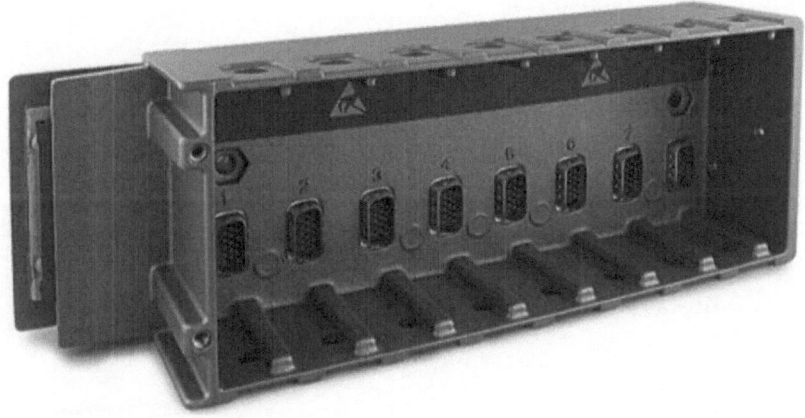

Figure 5.8 Reconfigurable FPGA Chassis (*Source:* National Instruments).

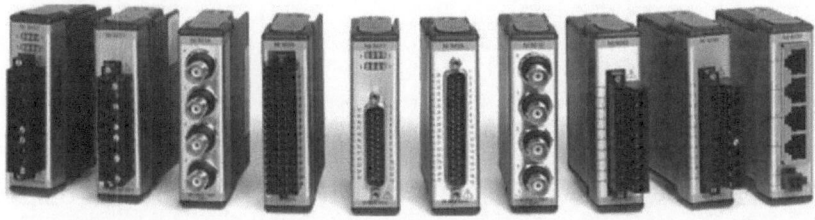

Figure 5.9 There are more than 50 NI C Series I/O modules for CompactRIO to connect to almost any sensor or actuator (*Source:* National Instruments).

the I/O circuitry of each module and timing, triggering, and synchronization. Because each module is connected directly to the FPGA rather than through a bus, you experience almost no control latency for system response compared to other controller architectures. By default, this FPGA automatically communicates with I/O modules and provides deterministic I/O to the real-time processor. Out of the box, the FPGA enables programs on the real-time controller to access I/O with less than 500 ns of jitter between loops. It can also directly program this FPGA to further customize the system. Because of the FPGA speed, this chassis is frequently used to create controller systems that incorporate high-speed buffered I/O, fast control loops, or custom signal filtering. For instance, using the FPGA, a single chassis can execute more than 20 analog proportional integral derivative (PID) control loops simultaneously at a rate of 100 kHz. Additionally, because the FPGA runs all code in hardware, it provides the highest reliability and determinism, which is ideal for hardware-based interlocks, custom timing and triggering, or eliminating the custom circuitry normally required with nonstandard sensors and buses.

I/O modules contain isolation, conversion circuitry, signal conditioning, and built-in connectivity for direct connection to industrial sensors/actuators. By offering a variety of wiring options and integrating the connector junction box into the modules, the CompactRIO system significantly reduces space requirements and field-wiring

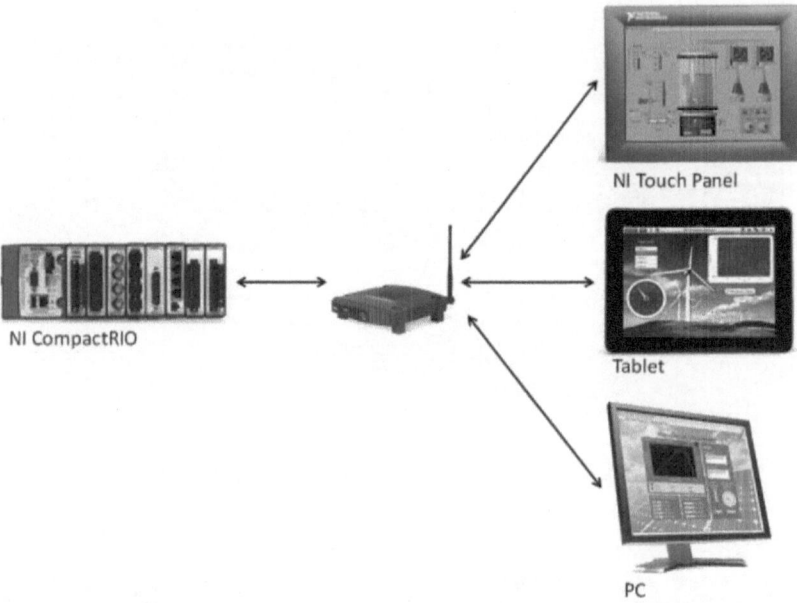

Figure 5.10 (1 Host to 1 Target) Configuration (*Source:* National Instruments).

costs. You can choose from more than 50 NI C Series I/O modules for CompactRIO to connect to almost any sensor or actuator. Module types include thermocouple inputs; ±10 V simultaneous sampling, 24-bit analog I/O; 24 V industrial digital I/O with up to 1 A current drive; differential/TTL digital inputs; 24-bit IEPE accelerometer inputs; strain measurements; RTD measurements; analog outputs; power measurements; controller area network (CAN) connectivity; and secure digital (SD) cards for logging. Additionally, It can build your own modules or purchase modules from third-party vendors. With the NI cRIO-9951 CompactRIO Module Development Kit, you can develop custom modules to meet application-specific needs. The kit provides access to the low-level electrical CompactRIO embedded system architecture for designing specialized I/O, communication, and control modules. It includes LabVIEW FPGA libraries to interface with your custom module circuitry. In the case of greenhouses those modules can be used for controlling the greenhouse conditions.

The simplest embedded system consists of a single controller running in a "headless" configuration. This configuration is used in applications that do not need an HMI except for maintenance or diagnostic purposes. However, a majority of control and monitoring applications requires an HMI to display data to the operator or to allow the operator to send commands to the embedded system. A common configuration is 1:1, or 1 host to 1 target. The host can be an NI Touch Panel, a tablet PC with NI Data Dashboard, or a PC, as shown in Figure 5.10. The HMI communicates to the CompactRIO hardware over Ethernet through a direct connection, hub, or wireless router. This kind of configurations are useful for monitoring and controlling a greenhouse.

The next level of system capability and complexity is either a 1:N (1 host to N targets) or N:1 (N hosts to 1 target configuration). The host is typically either a desktop

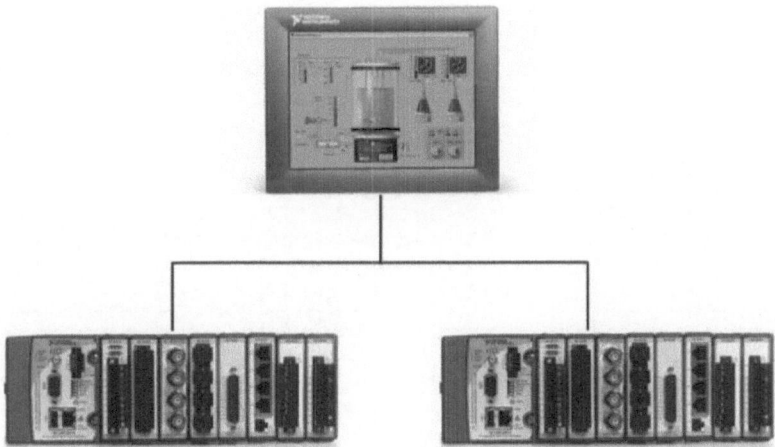

Figure 5.11 A 1:N (1 host to N targets) configuration is common for systems controlled by a local operator (*Source:* National Instruments).

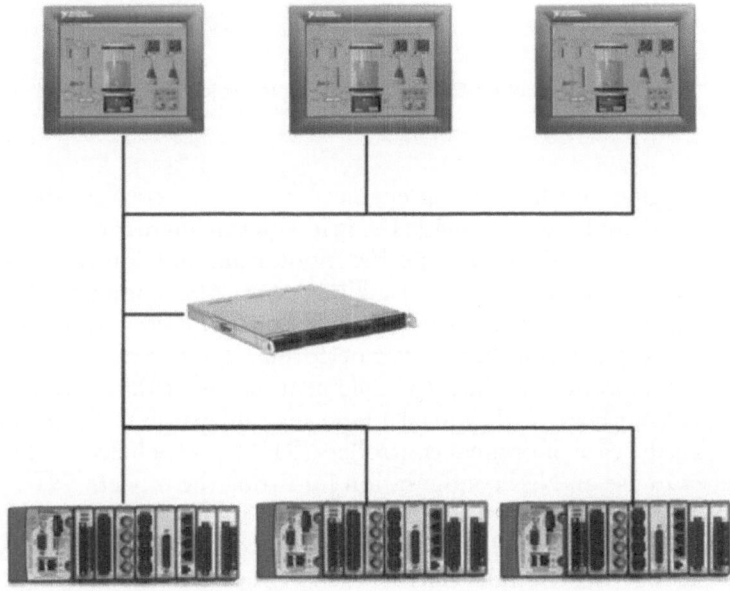

Figure 5.12 Distributed Machine Control System (*Source:* National Instruments).

PC or an industrial touch panel device. The 1:N configuration, shown in Figure 5.11, is typical for systems controlled by a local operator. The N:1 configuration is common when multiple operators use an HMI to check on the system state from different locations.

Complex greenhouse control applications may have an N:N configuration with many controllers and HMIs (Figure 5.12). They often involve a high-end server that

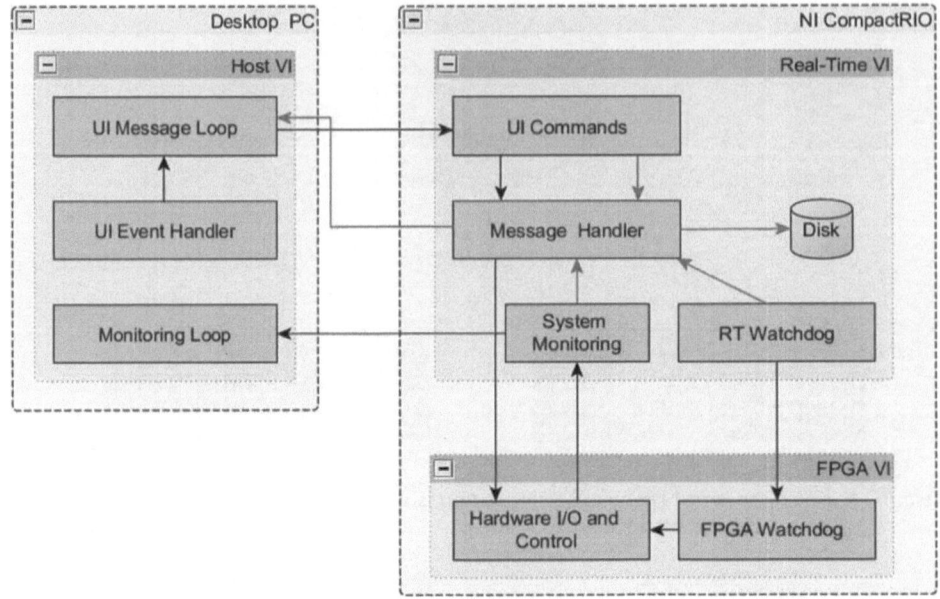

Figure 5.13 LabVIEW FPGA Control on CompactRIO Sample Project Architecture (*Source*: National Instruments).

acts as a data-logging and forwarding engine. This system configuration works for physically large or complex greenhouses. Using it, you can interact with the greenhouse from various locations or distribute specific monitoring and control responsibilities among a group of operators. For example, climate control in commercial greenhouses is based on heuristic rules that can be implemented on N:1 configuration but there are alternative methods which can show better performace, in this case different configuration as N:N configuration can help to implement this algorithms. Some of them are difficult to implement because they need a laborius calibatrion to fit the model in the controller. This is the case of optimal controllers (RHOC) which need to fit the model, hence it is better to use an N:N configuration for fitting the model.

The LabVIEW FPGA Control on CompactRIO sample project implements deterministic, hardware-based control of a plant. The control algorithm, which was written with the LabVIEW FPGA Module, runs on the FPGA inside the CompactRIO device. It sends commands and set-point changes to the FPGA from the user interface, running on a desktop computer, by way of the real-time controller in the device. This controller also monitors the status of the application, such as CPU load and memory usage. This sample project has the following features:

- High-performance control—The control loop can run faster than 10 kHz and features four control algorithms operating in parallel, all with minimal jitter.
- Hardware-based control—Running the control algorithm and safety logic on the FPGA provides maximum reliability.

- User interface with headless option—The user interface VI interacts with the CompactRIO device and displays data. This VI can connect and disconnect from the device at any time without affecting the control loop.
- Error handling—The application reports and logs all errors from the CompactRIO device and then shuts down on critical errors.

A greenhouse is developed with a set of agricultural environment features by including several I/O components (sensors and actuators) and control devices (SCADA).

Greenhouses also require a safety environment which can be strictly controlled by humans or machines for providing the best condition for the growth of plants. Wire and wireless controls and sensor networks in agriculture have been implemented in different greenhouse around the planet. The information of I/O variables is of great importance since it allows the farmer to perform the most suitable operations in order to improve the growth of plants and productivity and low cost.

The LabVIEW Supervisory Control and Data Acquisition System sample project demonstrates how to implement a supervisory control and data acquisition (SCADA) system with scalable architecture for building systems with many I/O points. This sample project has the following features:

- SCADA server—The SCADA server manages I/O points in the system, logs data and alarms in the historical database, and regularly archives data from the logging database to the archive database.
- Administrator console—The administrator console allows the system administrator to configure and manage the SCADA server.
- Client—The client allows the operators to monitor the system status and I/O points. The operator can view the values of the I/O points, alarms, and historical trends in different detailed views.
- Simulated CompactRIO system and simulated PLC-based system—A simulated CompactRIO system and a simulated PLC-based system are included in this sample project. The systems demonstrate how to connect to third-party devices. You can replace the simulated systems with a real CompactRIO system and a real PLC-based system.
- Scalability—You can use this sample project as a starting point to scale to a large system by adding a large number of I/O points or adding many subsystems in this sample project.

The sample project has VIs running on three different targets: the desktop PC, the CompactRIO real-time target, and the CompactRIO FPGA target. The FPGA target features two tasks running in parallel: one executes the control loop and one executes the Watchdog Loop. On the real-time target, tasks are receiving commands from the desktop, monitoring the health of the real-time system, handling messages from all the other tasks, and monitoring the Watchdog Loop. The desktop PC features tasks responsible for sending and receiving commands over the network to the real-time target, handling actions on the user interface, and updating the user interface with data from the real-time target.

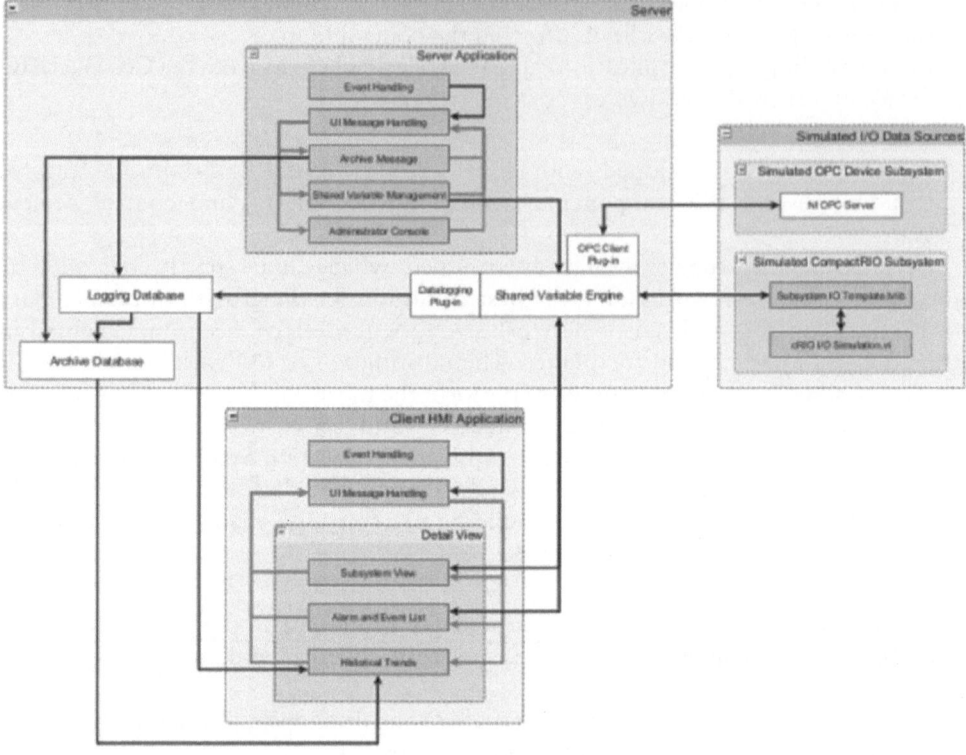

Figure 5.14 LabVIEW Supervisory Control and Data Acquisition *Sample Project Architecture* (*Source*: National Instruments).

To create your own application based on the LabVIEW FPGA Control on CompactRIO sample project, click the Create Project button in the Getting Started Window. This launches the Create Project dialog.

5.2 ADVANCED CONTROL SYSTEMS FOR PORTABLE GREENHOUSES

At this point, a variety of greenhouse controllers have been developed with hydroponics for achieving high quality cultivation. However, people involved in maintaining greenhouses even in good conditions have to turn to experts from time to time. And even experts dependent on their knowledge from practice and experience. Given that experience is finite, the implication is thatover time entropy will exert itself, resulting in a decrease in efficiency, low quality and quantity of cultivation.

Hydroponics is the solution for providing nutrients to the crop, but it does not regulate the environment in which the crop develops. The growth highly depends on the environmental conditions like temperature, light intensity and relative humidity. In order to regulate these variables, it is necessary to isolate the environment in which the crops are going to be harvested.

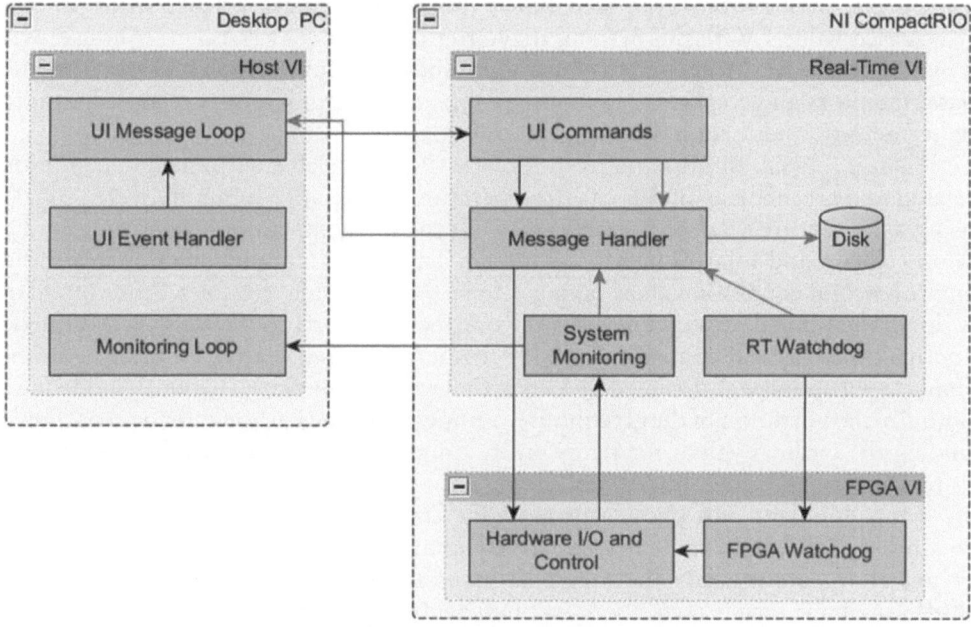

Figure 5.15 Data Communication Diagram (*Source*: National Instruments).

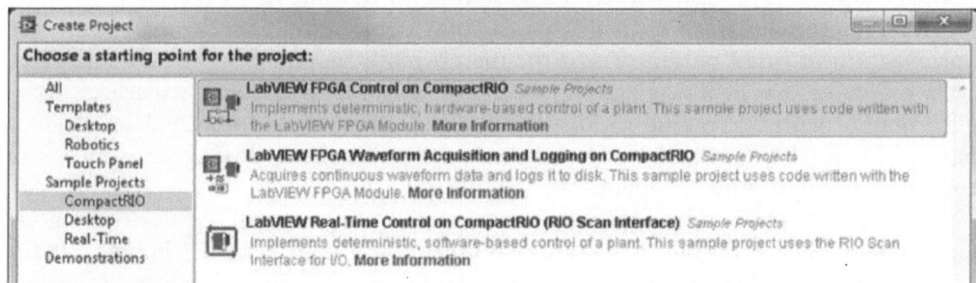

Figure 5.16 Sample Projects by CompactRIO (*Source*: National Instruments).

Therefore, control systems have been used in regulating some environmental variables and controlling nutriments. Typical controllers used into hydroponic greenhouses are classical control systems and, to a smaller degree, intelligent control systems. The last technological systems based on artificial intelligence techniques as fuzzy logic and neural networks have helped to increase efficiency in cultivation.

Related works have been developed. In this way, Shihua, Shiyan, and Limei proposed an adaptive intelligent greenhouse control system dividing the entire system into two tasks: a fuzzy control and forecasting. It uses five triangular membership functions and controls temperature and humidity variables with a proportional fuzzy control system. Carbon oxide control is implemented with a classical PID because time delay is small and accuracy for them is reasonable. In addition, they model a forecasting or observer of the greenhouse (plant) to obtain a prediction of the next real sensed values. With this system, they offer control over several greenhouses with a master computer. Moreover, they use a graphical user interface that draws historical data, monitors and controls greenhouse parameters as temperature, relative humidity, light intensity, and carbon dioxide.

Salazar, López, and Rojano from Mexico City modeled environmental conditions related to a greenhouse. In this case, temperature and relative humidity were modeled using a 3-layer artificial neural network with outside temperature, relative humidity, wind velocity and solar radiation, as inputs. This model permits prediction of inside environmental condition values taking into account outside variables. After modeling is done, the trained artificial neural network model is used for estimating parameters and makes a classical control system for temperature and relative humidity. Another similar application was designed by Linker that extends the neural network modeling to optimize the operation of the greenhouse. Temperature and carbon dioxide is optimized finding the optimal values for temperature range and carbon dioxide concentrations as function of changing in weather.

Fang, Junqiang and Jiaoliao proposed a fuzzy control for temperature based on rough sets. The latter is an emerging theory in artificial intelligence where incomplete or uncertain data is used. The rough set based fuzzy logic control uses the rough set theory to obtain conclusions about incomplete data in temperature databases and then uses this information for generating a set of rules in the fuzzy inference system inside the fuzzy control. This control is used for regulating the greenhouse temperature and its advantage is that high-dense uncertainty in data can be used. Additionally, Fang, Jiaoliao, Libin and Hongwu described a method for optimizing a fuzzy control system for greenhouse temperature. In this work, they proposed a genetic algorithm based optimization of fuzzy parameters in the membership function of the control system. They discussed that genetic algorithms can operate in high dimensional search spaces with nonlinearities, avoiding local minima and finding the optimal values for fuzzy parameters at the current set-point. In addition, this methodology economizes the energy costs of greenhouse operation.

Additionally, new industrial greenhouses are building around the world using high technologies. For example, a sustainable greenhouse complex was inaugurated on May 2009 in Los Angeles, California, United States. A high technology is used for controlling environmental variables inside the greenhouse. Moreover, this greenhouse uses green engineering as photovoltaic cells for generating 1.2 MW of electricity for self-maintaining. This 20 acre greenhouse will produce around 482 tons of tomatoes per acre. In constrast, Xiamen Wondland Environmental Engineering in China offers a greenhouse intelligent control system that controls irrigation taking into account weather conditions measured in a station. An international company on horticulture and building automation Priva offers high technology on intelligent controller systems. They use artificial intelligence for controlling environmental conditions. Additionally, offers sensors and a variety of accessories for automated greenhouses.

However, not only automated greenhouses have been developed. Portable automated greenhouses are also being brought on the market. This type of greenhouse is gaining in popularity because of demands for fresh and organic food. One of the most common portable greenhouses is the AeroGarden. It regulates light and nutriments

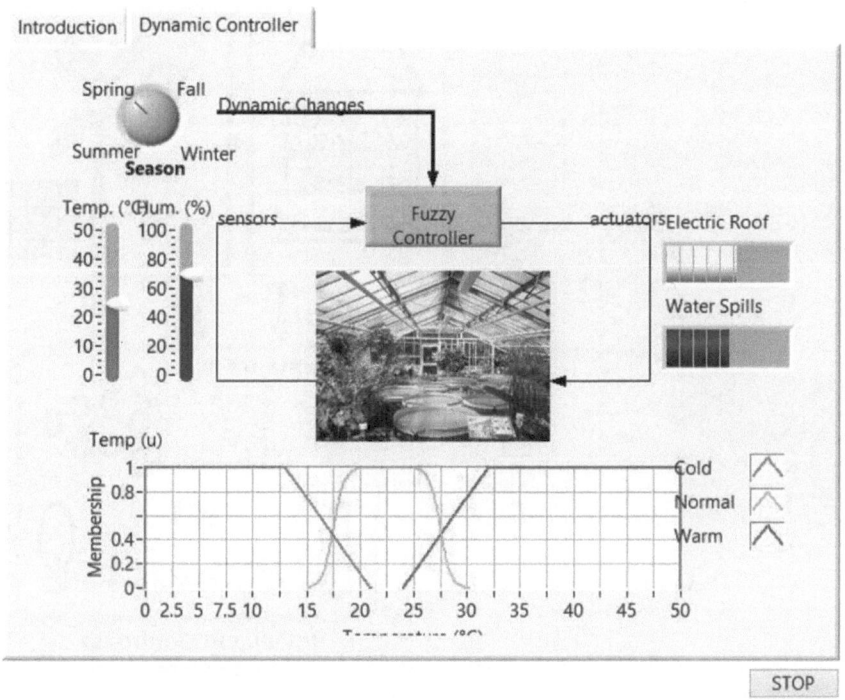

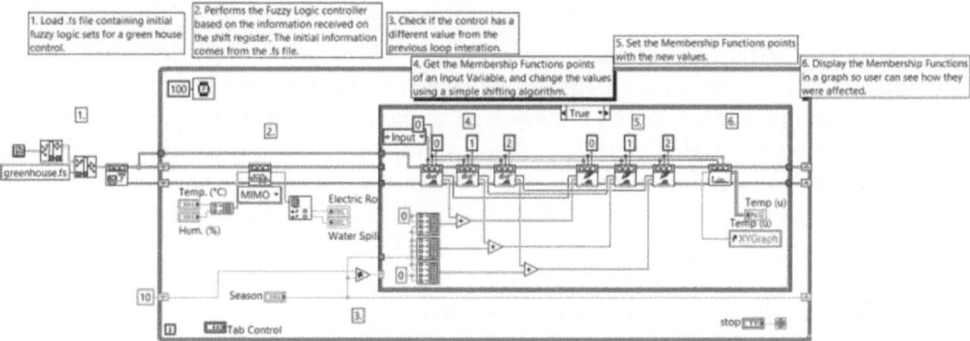

Figure 5.17 Shows the fuzzy logic controller for a greenhouse that is an example of LabVIEW program (*Source:* National Instruments).

supply with a simple on/off control and a temporizer. Additionally, several types of crop can be grown only selecting a predetermined program function. Another portable greenhouse was patented by SIEMENS. This product controls light, nutriments supply, air ventilation, and carbon supply. It is also regulates them with a temporizer.

Portable greenhouses are automated with raw technologies without controlling environmental conditions efficiently. At Tecnologico de Monterrey campus Ciudad de México, a portable intelligent greenhouse based on fuzzy logic and neural networks has been designed in order to get better response related to obtain good quality in

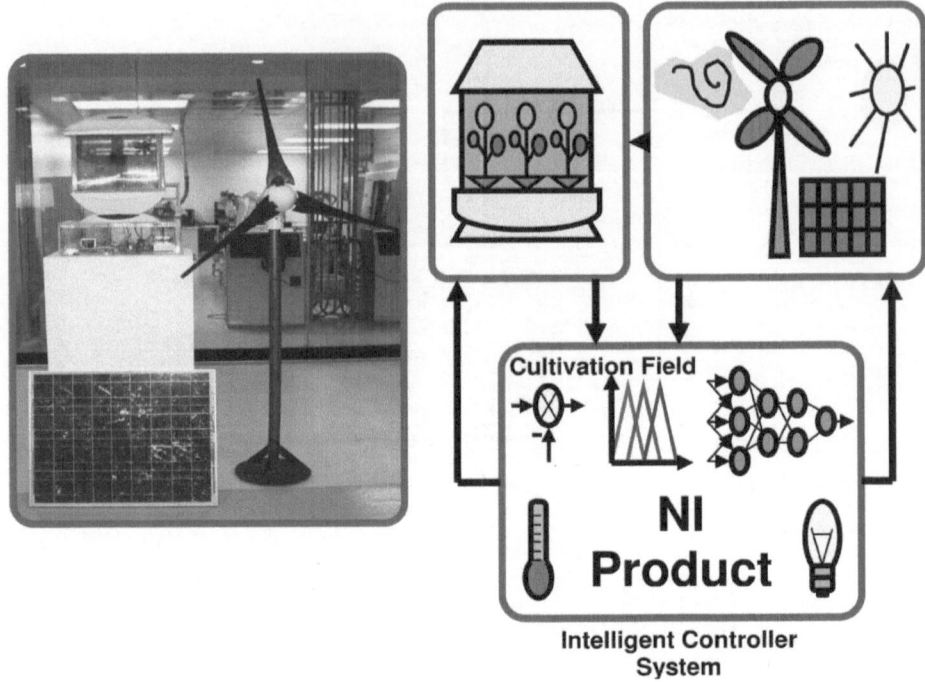

Figure 5.18 Portable Greenhouse concept and experimental one using solar and wind energy.

cultivation and reduce the usage of energy in the whole system. Temperature, relative humidity, light intensity, and nutriments supply are regulated to create a micro-climate for crops. Finally, the proposed system was implemented in a greenhouse prototype specially designed for this purpose. Also, simulation and validation was done on LabVIEW platform. LabVIEW includes an editor for Fuzzy Logic inside of the toolkit PID and Fuzzy Logic. It also has a basic example about greenhouse controller using Fuzzy Logic.

5.2.1 Basic portable greenhouse description

In order to validate an intelligent control system that can be implemented in a conventional greenhouse, a portable hydroponic greenhouse prototype is implemented to produce organic harvests with high quality and low cost. The system consists in a plastic structure thermally isolated with polycarbonate walls; it also has a ventilation system and a heat exchanger to regulate the temperature inside the greenhouse. It has lamps used to regulate the light intensity. A humidifier is used to regulate the relative humidity. All of these elements are controlled using a computer interfaced to the greenhouse with a NI 6211-USB data acquisition (DAQ) target. Additionally, this system counts with sensors that allow monitoring these conditions. Finally, the intelligent system is programmed in the graphical language LabVIEW that manages the DAQ target for measuring and generating signals, and processes the intelligent control. Figure 5.18 shows the schematic diagram of the prototype.

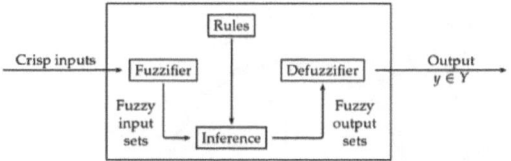

Figure 5.19 Block diagram of a fuzzy control system.

5.2.2 Advanced control systems

Agriculture and horticulture are sciences that have thousands of years of development and evolution behind them. Thus, there is a surfeit of precise information on the environmental and nutritional conditions that plants need to grow. All this information can be moved to a fuzzy controller. This is a branch of artificial intelligence that regulates systems based on linguistic rules in some degree that are modeled by fuzzy logic sets (Mamdani system).

The most important characteristic of a fuzzy control system is the capacity that it has to store the knowledge of experts. Actually, it can handle on different set points instead of using one like in conventional control. This kind of intelligent system is represented in the block diagram of Figure 5.19 It consists of three blocks: fuzzification, inference, and defuzzification.

Fuzzification evaluates crisp input values and assigns a membership value (among 0 and 1) to a linguistic term or fuzzy set. The function that represents the degree of an input belonging to the fuzzy set is called membership function.

Inference process is a set of rules written down as (5.1):

Ri: if x1 is Bi1 and ... and xn is Bin then yi is Di

Ri: if x2 is Bi2 and ... and xn2 is Bin2 then yi2 is Di2   (5.1)

where, Di is a set of output fuzzy sets in the case of a Mamdani fuzzy controller. This procedure returns a consequent fuzzy value of a set of fuzzy values as premises. Finally, defuzzification process is a mapping of fuzzy values to crisp values. Several methods are commonly used as detailed on different works. In this paper, a center of gravity is used for a Mamdani inference system. The crisp output value $y_{i=1,...,q}$ is defined as it is shown in equation (5.2) center of mass

$$y_i = \frac{\sum_h \mu_B(y_h) \cdot y_h}{\sum_h \mu_B(y_h)} \qquad (5.2)$$

Neural networks

Artificial neural networks are mathematical models based on nervous system in order to learn and they can improve their behaviour with training or historical experience.

The basic concept in these networks is the neuron. In Figure 5.20 is presented a neural model. It has an input vector x of n input signals. Each input signal is then connected

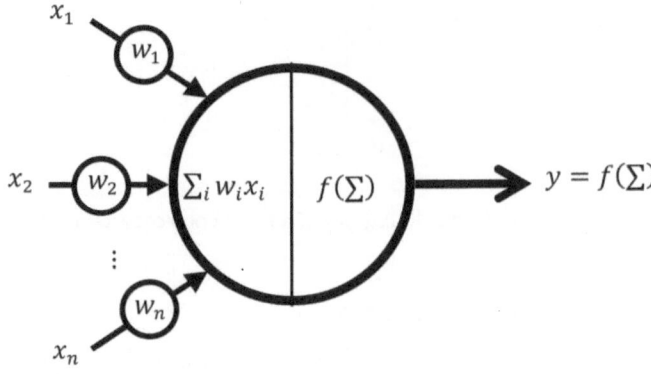

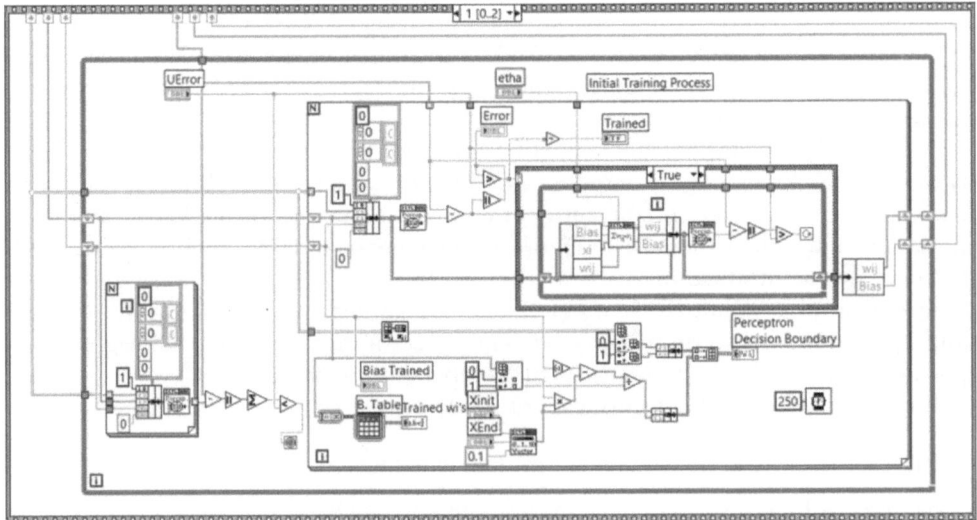

Figure 5.20 Neuron model in artificial neural networks and LabVIEW perceptron program.

to a weight value. A weight vector w is also defined with n elements. Each weighted signal is summing. In fact, as human neurons do, artificial neurons produce an effect related to a set of input stimuli. Thus, artificial neurons have an activation function (e.g. sigmoid, linear, hyperbolic tangent) that describes this relationship. The output signal is then a crisp value *y*. A mathematical model for a neuron is described in (5.3).

$$y = f(w \cdot x) = f\left(\sum_i w_i x_i\right) \tag{5.3}$$

When a set of neurons are interconnected to produce a response of some input signals, this is known as neural network. Depending on their applications, artificial neural networks can be classified as: supervised networks, unsupervised networks, competitive or self-organizing networks, and recurrent networks.

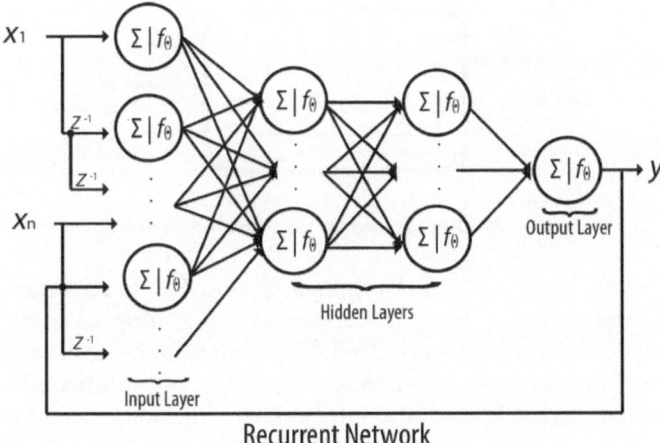

Figure 5.21 Dynamical neural network structure and LabVIEW Artificial Neural Network program (ICTL).

In control, artificial neural networks can be used for modeling the plant or used as a controller system. The approach presented in this work is based on modeling the behavior of the whole system (i.e. cultivation area and environmental conditions on it) using neural networks because it has several nonlinear conditions that cannot be seen as an analytic mathematical equation. In addition, modeling of the micro-climate under actuator's response is also done for knowing the behavior of the greenhouse under controlled circumstances. In both cases, nonlinear information is treated and these data are changing in time. Thus, recurrent or dynamical neural networks are used to face the chaos data problem.

If an array of neurons is located on a layer and several layers are interconnected and input propagation has only one direction, then a feed-forward multilayer neural network is presented. Moreover, a dynamical neural network is the same as the latter, but at the input layer some inputs come from current input signals, and others are delays of these input signals. Additionally, some current output signals, and delays of these signals are also presented in the input layer. Figure 5.21 shows this scheme.

As known, neural networks need to be trained for particular data information or modeling. Thus, the backpropagation algorithm with momentum parameter (5.4) is implemented.

$$w_{ik}^q \leftarrow w_{ik}^q + \Delta w_{ik}^q + \alpha(w_{ik}^q - w_{ik\_last}^q)$$ (5.4)

5.3 CONTROL IMPLEMENTATION

The intelligent control system is implemented in LabVIEW and based on the Intelligent Control Toolkit for LabVIEW (ICTL). In addition, a NI USB-6211 data acquisition (DAQ) target is used for acquiring and generating signals. It is a device optimized for good accuracy at fast sampling rates. It has 16 analog input ports, 2 analog output

Figure 5.22 Representation of a fuzzy-PD control.

ports, 8 digital input lines, 8 digital output lines, and 2 counter/timers. Digital triggering is also offered with this device. No external power supply is required. Actually, these software and hardware are optimized for simulation and validation.

Three control systems were designed for regulating environmental conditions, i.e. temperature and light intensity. Additionally, relative humidity is also controlled with an on/off controller to keep it in a range of 50%RH to 70%RH.

In order to understand the proposal, outputs of the system are values among 0% and 100% that represent the duty cycle of the pulses that will be sent to the actuators: a heating resistance, a pair of 12V DC fans and two 75 W lamps. These output variables will be represented by:

- DR: resistance duty cycle.
- DF: fan duty cycle.
- DL: lamp duty cycle.

Two intelligent control techniques were adopted for controlling environmental conditions (i.e. fuzzy and neural network controllers) in order to see both performances. Responses were compared for decision-making designing purposes.

On the other hand, fuzzy-PD controllers were adopted for temperature and light intensity regulation. Let a temperature controller be used to exemplify the design of the fuzzy controllers.

Consider a constant value of temperature (TR) that will be used as a reference to the system. Based on this value, the error can be obtained by (5.5). From this equation, the expression for the error derivative is (5.6), see Figure 5.22.

These signals (5.5) and (5.6) are used as inputs for the fuzzy controller. As it was mentioned, the output of the controller is the duty cycle. Figure 5.23 shows a typical output signal for the proposed control system. In the same way, light intensity fuzzy-PD controller is implemented.

where,

$$\varepsilon = T_R(k) - T(k) \tag{5.5}$$

$$\Delta\varepsilon(k) = \varepsilon(k) - \varepsilon(k-1) \tag{5.6}$$

$$u(k) = K_P\varepsilon(k) + K_D\Delta\varepsilon(k) \tag{5.7}$$

k is the current sample

Additionally, as hydroponics uses mineral nutrient solutions instead of agricultural land, mineral nutrients are introduced into the water supply. These essential

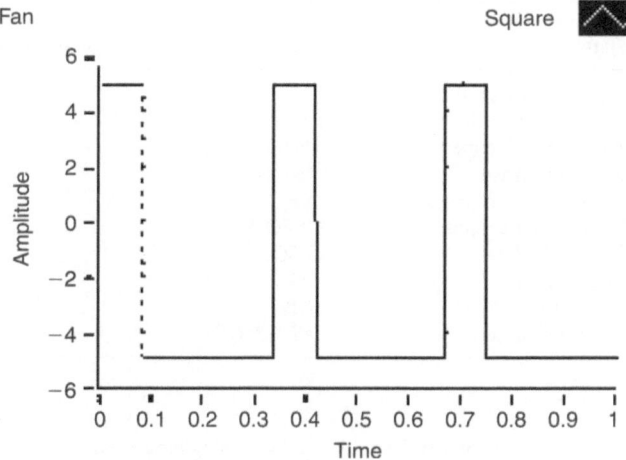

Figure 5.23 Duty cycle signal as output in the proposed intelligent controller.

Table 5.1 Nutriments Formula for Premature germination.

Name	Formula	Quantity (gr)
Calcium Nitrate	$Ca(NO_3)_2$	90
Magnesium Sulfate	$MgSO_4$	30
Monopotassium phosphate	KH_2PO_4	20
Potassium Nitrate	KNO_3	35
Potassium Sulfate	K_2SO_4	15
Iron Sulfate	$FeSO_4 7H_2O$	10
Manganese Sulfate	$MnSO_4 4H_2O$	1
Boricacid	H_2BO_3	0.5
Zinc Sulfate	$ZnSO_4 7H_2O$	0.5
Copper Sulfate	$CuSO_4 5H_2O$	0.5

minerals are provided using static or dynamic formulas. On one hand, static formulas are those that do not change the composition throughout the plant's growing process. On the other hand, dynamic formulas are those that change the proportion of various nutrients during the plant's production process. This helps the plant to have a better development.

In this greenhouse system, nutriments supply controller uses dynamic formulas. Three general steps are presented following:

1   In the first stage of the plant, a general formula is used for premature germination and strengthening of the plant in its first growing period (Table 5.1).
2   In the second step, flowering period uses the formula presented in Table 5.2.
3   The formula in Table 5.3 is developed for increasing flowering achieving a better consistency and preservation of fruits.

Table 5.2 Nutriments Formula for Flowering.

Name	Formula	Quantity (gr)
Potassium Nitrate	KNO_3	
Calcium Chloride	$CaCl_2 6H_2O$	
Manganese Sulfate	$MnSO_4 4H_2O$	
Sodium Nitrate	$Na(NO_3)$	
Ammonium Chloride	$(NH_4)Cl$	
Calcium phosphate	$Ca(PO_3)$	
Copper Sulfate	$CuSO_4 5H_2O$	
Zinc Sulfate	$ZnSO_4 7H_2O$	
Boricacid	H_2BO_3	
Ferrous Sulfate	$FeSO_4 7H_2O$	

Table 5.3 Nutriments Formula for Increasing Flowering.

Name	Formula	Quantity (gr)
Sodium Nitrate	$Na(NO_3)$	100
Calcium Chloride	$CaCl_2 6H_2O$	150
Potassium Nitrate	KNO_3	75
Magnesium Sulfate	$MgSO_4$	50
Ammonium Chloride	$(NH_4)Cl$	20
Copper Sulfate	$CuSO_4 5H_2O$	1
Ferrous Sulfate	$FeSO_4 7H_2O$	15
Calcium phosphate	$Ca(PO_3)$	7
Zinc Sulfate	$ZnSO_4 7H_2O$	1
Manganese Sulfate	$MnSO_4 4H_2O$	5

These formulas are introduced to a fuzzy control system for obtaining a smooth dynamical formula for supplying nutriments to plants in the greenhouse.

Dynamical neural networks were used for modelling the behavior of the plant. This means that temperature, relative humidity, and light intensity variables inside the greenhouse were modelled with dynamical neural networks, taking into account the control action values. In this way, a complete understanding (of how variables response to signal controls) of the greenhouse is found.

Additionally, an inverse plant model was adopted for controlling internal conditions and for obtaining the reference values for temperature, relative humidity, and light intensity.

A first stage for acquiring historical data was done. Then, a dynamical neural network for plant behavior was implemented. Finally, an inverse model of the plant was done with neural networks.

5.3.1   Portable greenhouse results

For validating the intelligent control system, all fuzzy and neural network controllers, the relative humidity time-table controller, and the look-up table nutriments

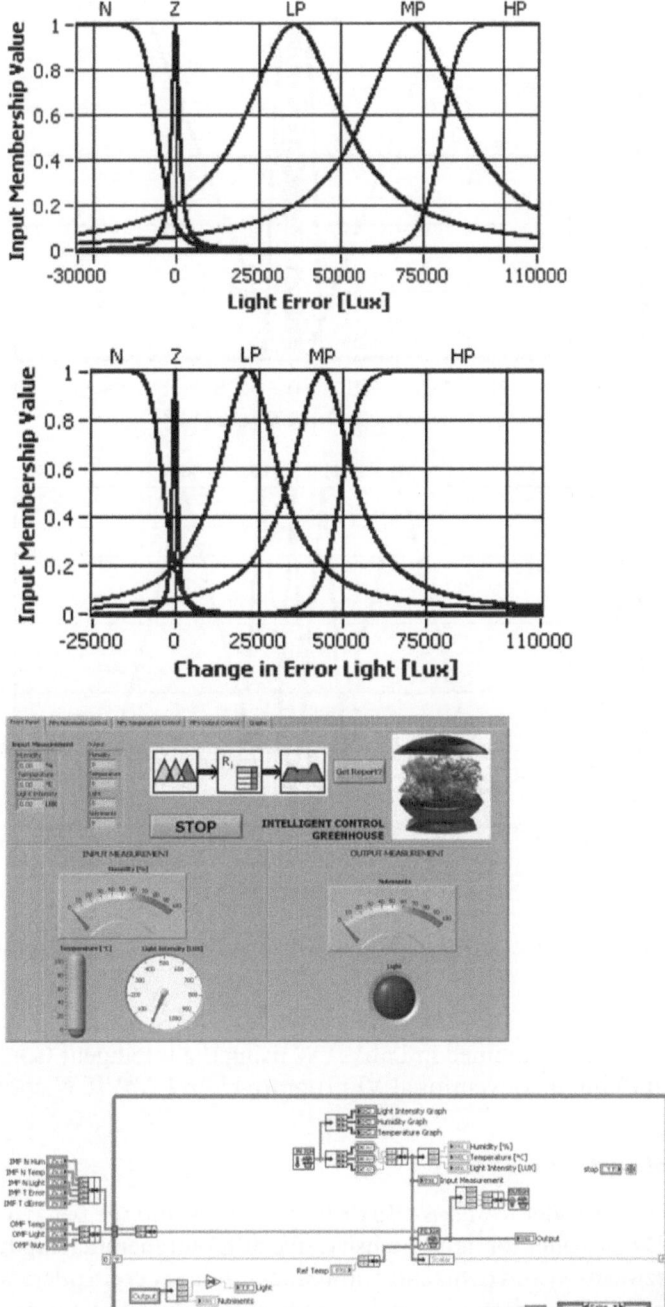

Figure 5.24 Five bell membership functions for input signals and LabVIEW program.

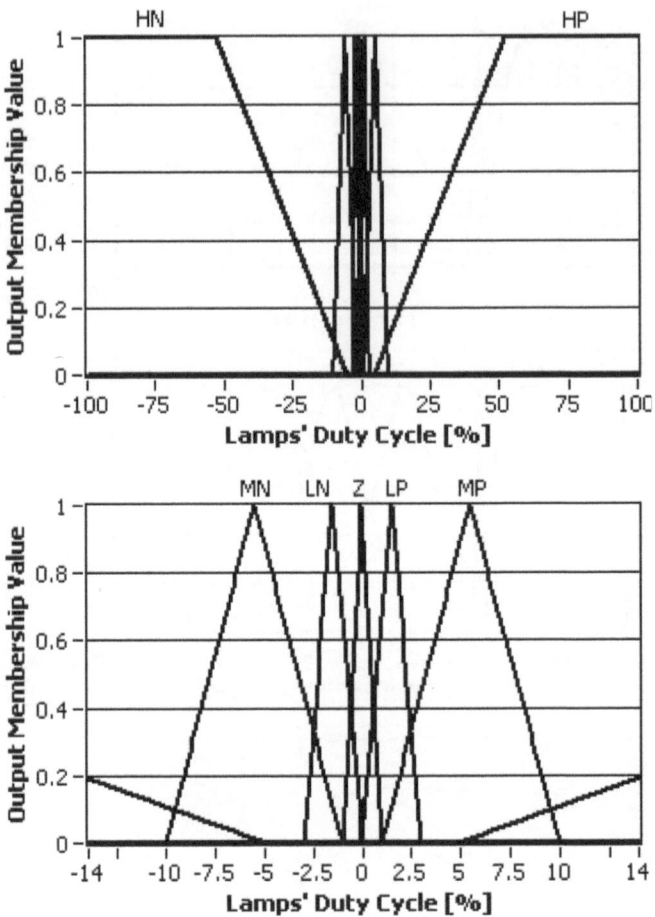

Figure 5.25 Seven triangular membership functions for output signal.

supply system were programmed in LabVIEW using the Intelligent Control Toolkit for LabVIEW (ICTL) but a conventional VI program from LAbVIEW could be used.

Fuzzy control system response

Light Intensity Controller Response. In order to understand the behavior and response of the fuzzy-PD controller for light intensity, five different membership function configurations at fuzzification and defuzzification blocks in fuzzy controllers were developed.

In the first configuration, five bell membership functions for error and change in error inputs were designed. These fuzzy labels are: *negative* (N), *zero* (Z), *low positive* (LP), *medium positive* (MP), and *high positive* (HP). Figure 5.24 depicted these functions. Seven triangular output membership functions (Figure 5.25) were used for obtaining the crisp duty cycle value for lamps: *high negative* (HN), *medium*

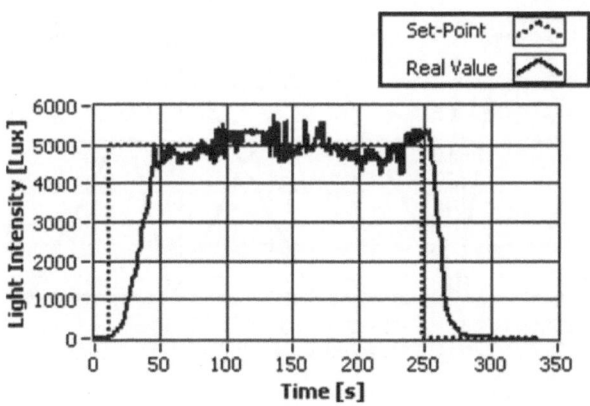

Figure 5.26 Step response of the 5-bell MF's of the fuzzy-PD controller.

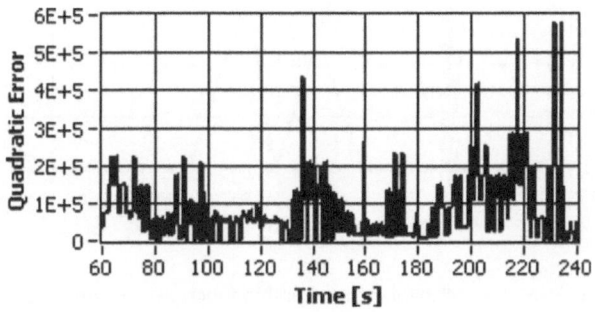

Figure 5.27 Quadratic error plot for the step response of the 5-bell MF's of the fuzzy-PD controller.

negative (MN), *low negative* (LN), *zero* (Z), *low positive* (LP), *medium positive* (MP), and *high positive* (HP). The step response of this fuzzy-PD is shown in Figure 5.26 with a set-point of 5,000 Lux. A quadratic error analysis is done for this response in which three picks were found as seen in Table 5.4. Those picks correspond to instants when the controller failed (Figure 5.27).

In the second configuration, five triangular membership functions for error and change in error inputs were designed. The same fuzzy sets as previously are: *negative* (N), *zero* (Z), *low positive* (LP), *medium positive* (MP), and *high positive* (HP). Figure 5.28 draws these triangular functions. The same output membership functions were used (Figure 5.25). In the same way, this fuzzy controller is excited with a step signal. Current response is shown in Figure 5.29 with a set-point of 5,000 Lux. A quadratic error analysis found two picks as seen in Table 5.4. Figure 5.30 shows the quadratic error analysis.

The third configuration corresponds to a seven bell membership functions for error and change in error inputs. Actually, seven fuzzy sets were designed (Figure 5.31). These are: *low negative* (LN), *medium negative* (MN), *high negative* (HN), *zero* (Z),

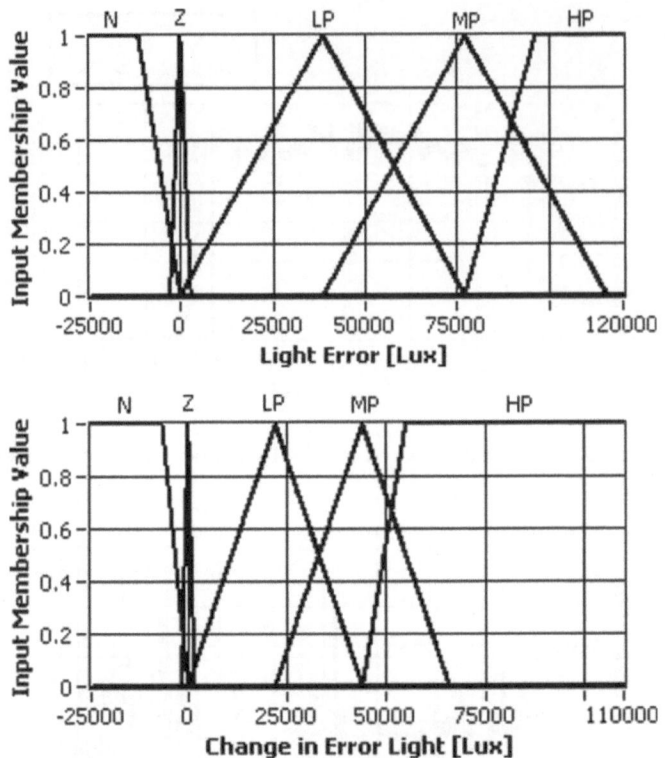

Figure 5.28 Five triangular membership functions for input signals.

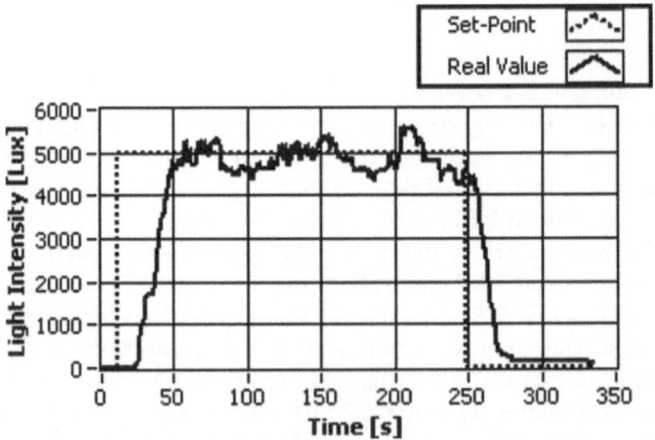

Figure 5.29 Step response of the 5-triangular MF's of the fuzzy-PD controller.

Table 5.4 Quadratic Error Analysis on Interesting Variables.

Configuration	Time [s]	Light [Lux]	Error* [Lux]	Error^2
5-bell MF's	136	5,661	661	436,921
	202	4,354	646	417,316
	217.5	4,268	732	535,824
	231.5	4,240	760	577,600
	234	4,240	760	577,600
5-triangular MF's	80	5,514	514	264,196
	154	5,544	544	292,936
	186.5	4,468	532	283,024
	205	5,749	749	561,001
	232	4,497	503	253,009
7-bell MF's	60	5,457	457	208,849
	67.5	5,471	471	221,841
	114.5	5,441	441	194,481
	176	5,368	368	135,424
	226.5	4,572	428	183,184
7-triangular MF's	101	4,468	532	283,024
	139	4,583	417	173,889
	157	4,612	388	150,544
	211	4,583	417	173,889
	228	4,354	646	417,316
7-singleton MF's	67	4,612	388	150,544
	87	5,309	309	95,481

*Considering a set-point of 5,000 Lux

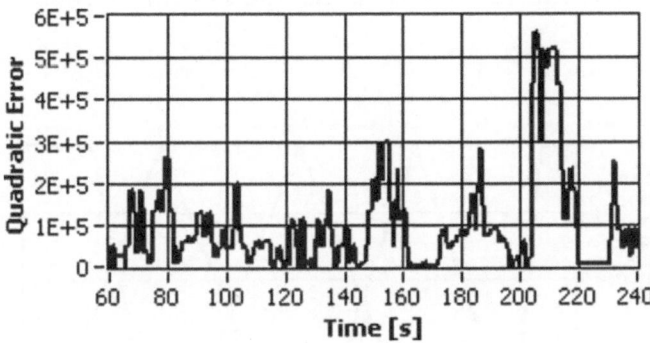

Figure 5.30 Quadratic error plot for the step response of the 5-triangular MF's of the fuzzy-PD controller.

low positive (LP), *medium positive* (MP), and *high positive* (HP). Five triangular output membership functions were used as seen in Figure 5.32. They are: *nothing* (N), *few* (F), *medium* (Me), *high* (H), and *maximum* (Ma) The step response of this fuzzy-PD is shown in Figure 5.33 with a set-point of 5,000 Lux. A quadratic error analysis is done for this response and results are found in Figure 5.34.

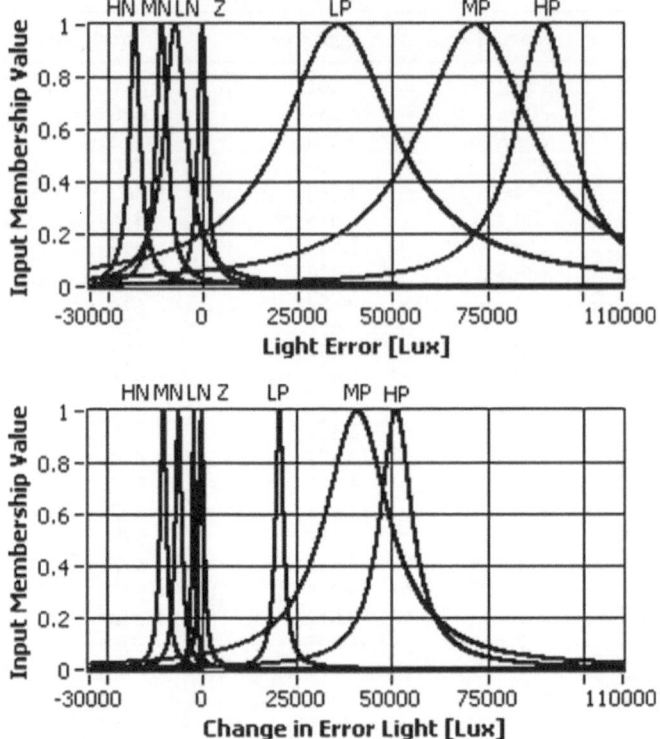

Figure 5.31   Seven bell membership functions for input signals.

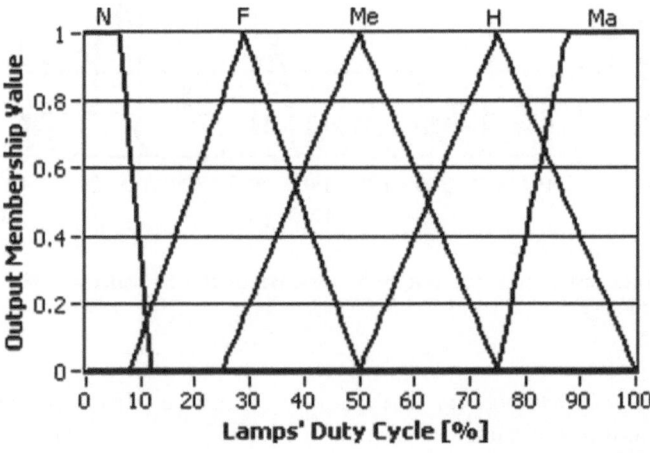

Figure 5.32   Five triangular membership functions for output signal.

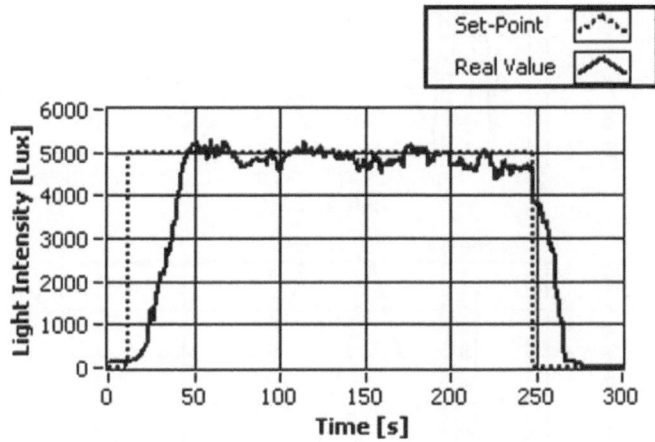

Figure 5.33 Step response of the 7-bell MF's of the fuzzy-PD controller.

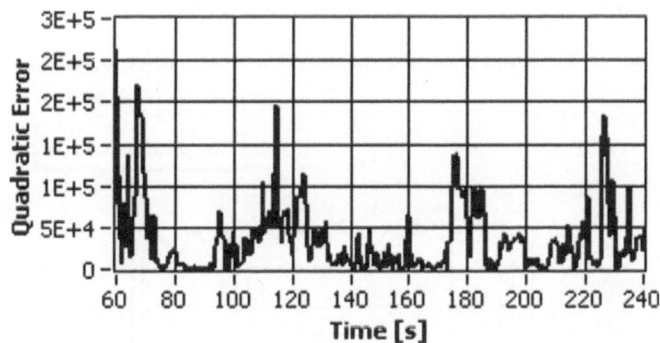

Figure 5.34 Quadratic error plot for the step response of the 7-bell MF's of the fuzzy-PD controller.

In the fourth configuration, seven triangular membership functions for error and change in error inputs were tuned (Figure 5.35). The output membership functions are the same as in the first case (Figure 5.25). The step response of this fuzzy-PD is depicted in Figure 5.36. Additionally, a quadratic error analysis is done (Figure 5.37).

Finally, the output membership functions were exchanged with seven singleton membership functions for the lamp duty cycle output, see Figure 5.38. As expected, the step response in not too accurate (Figure 5.39). Again, the quadratic error analysis is done and is shown in Figure 5.40. Finally, Figure 5.41 shows the comparison between responses of the different fuzzy-PD configurations implemented.

Temperature Controller Response. Also, different tests were done for fuzzy-PD temperature controllers. In the first case, it uses five input bell membership functions as seen in Figure 5.42. Fuzzy sets are: *negative* (N), *zero* (Z), *low positive* (LP), *medium*

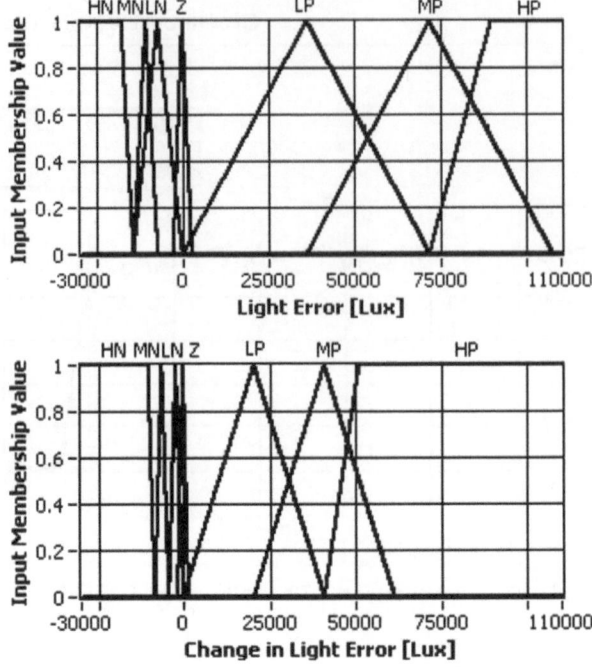

Figure 5.35 Seven triangular membership functions for input signals.

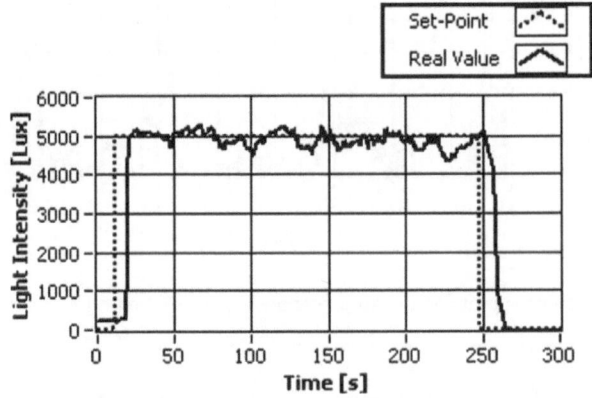

Figure 5.36 Step response of the 7-triangular MF's of the fuzzy-PD controller.

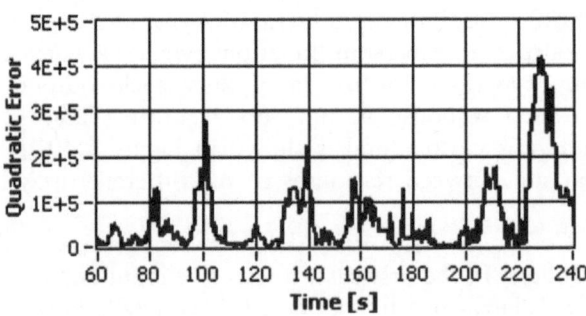

Figure 5.37 Quadratic error plot for the step response of the 7-triangular MF's of the fuzzy-PD controller.

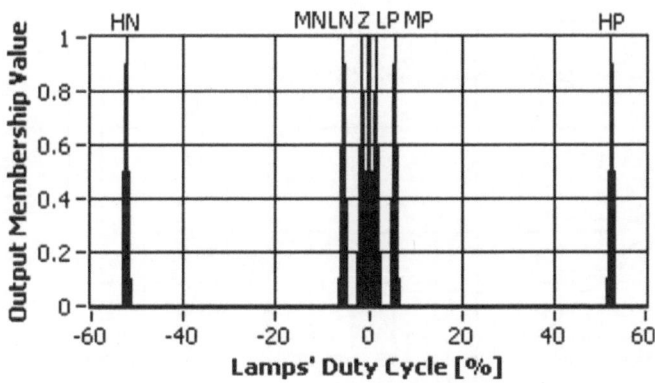

Figure 5.38 Seven singleton membership functions for output signal.

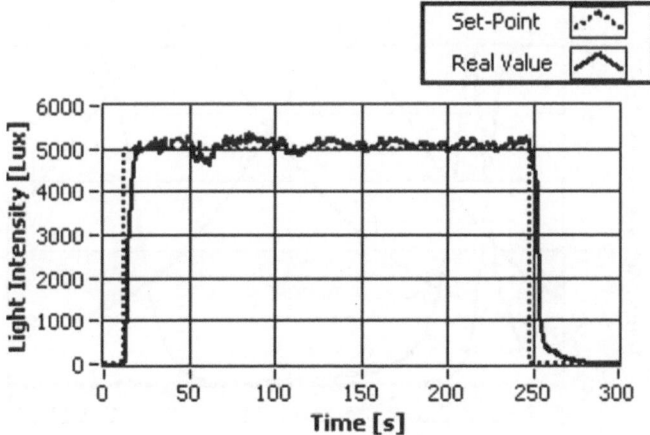

Figure 5.39 Step response of the 7-triangular input MF's and 7-singleton output MF's of the fuzzy-PD controller.

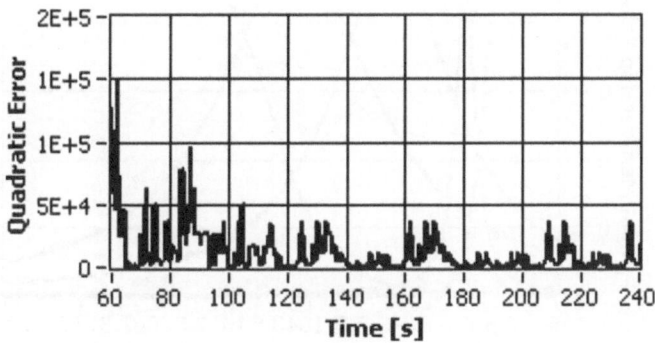

Figure 5.40 Quadratic error plot for the step response of the 7-triangular input MF's and 7-singleton output MF's of the fuzzy-PD controller.

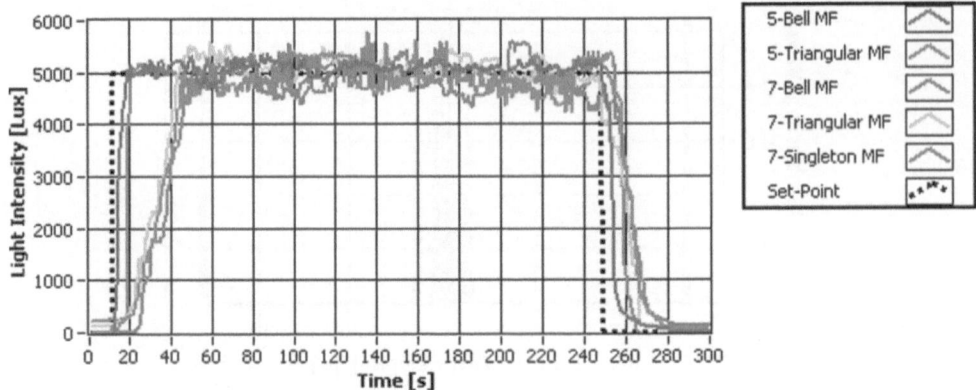

Figure 5.41 Comparison between the five fuzzy-PD configurations for the light intensity controller.

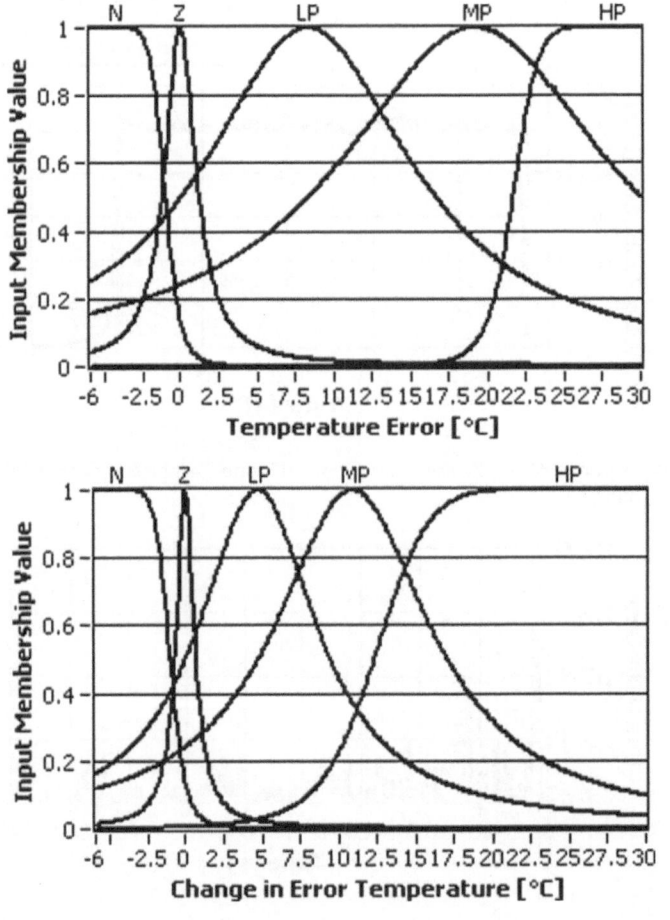

Figure 5.42 Seven triangular membership functions for input signals.

positive (MP), and *high positive* (HP). However, two outputs are related. Five triangular membership functions represent the heating resistance's duty cycle output, and three triangular membership functions represent the fan speed (duty cycle output). Figure 5.43 shows these fuzzy sets in which, for resistance output are *nothing, few, little, medium,* and *high*; and for fan velocities are *low, medium,* and *high*.

In order to emulate the internal temperature behavior, a ramp response for the fuzzy-PD controller is obtained as seen in Figure 5.44. A quadratic error analysis is done. It can be seen from the quadratic error plot (Figure 5.45) that during the rising time the error is very significant from zero to around the 400 seconds when the controller had several perturbances in its action. Table 5.5 shows these statistics.

During the falling time, the error decreased in comparison with the rising time, at the end of the test the error reached almost 0.56°C. In this case overshooting is presented, with the maximum 1.2°C upper, in the steady state set-point. It follows from the fact that the greenhouse is thermally isolated and losses in energy are rather lower.

In the second case, it uses five input triangular membership functions as seen in Figure 5.46. Fuzzy sets are: *negative* (N), *zero* (Z), *low positive* (LP), *medium positive* (MP), and *high positive* (HP). The same output membership functions were

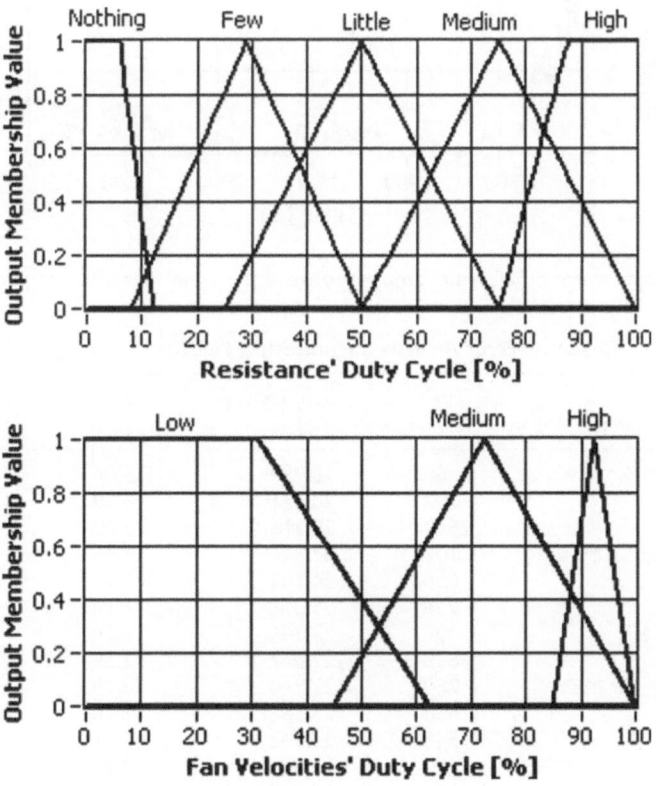

Figure 5.43 Triangular membership functions for output signals.

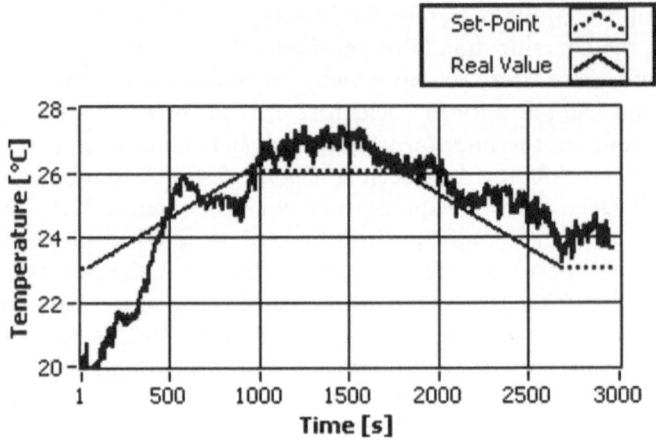

Figure 5.44   Ramp response of the 5-bell MF's fuzzy-PD temperature controller.

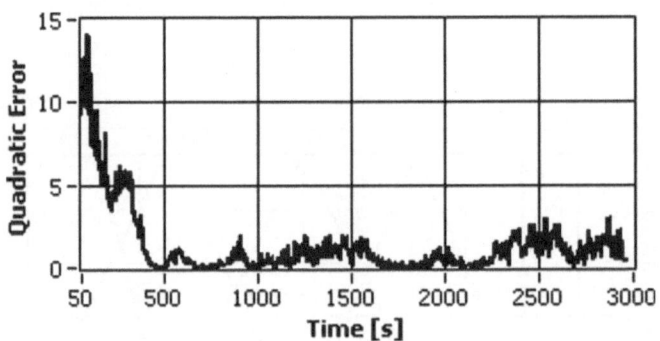

Figure 5.45   Quadratic error plot for the ramp response of the 5-bell fuzzy-PD temperature controller.

Table 5.5   Quadratic Error Analysis on Interesting Picks.

Configuration	Time [s]	Temperature [°C]	Error [°C]	Error²
5-bell MF's	60	18.984	3.04	9.22
	102	20.451	1.89	3.58
	2512	25.216	2.60	6.74
5-triangular MF's	102	22.344	1.40	1.96
	656	24.205	0.97	0.94
	956	25.015	1.14	1.31
7-bell MF's	400	23.346	2.24	5.01
	2688	22.027	2.39	5.72
7-triangular MF's	280	22.940	0.19	0.04
	466	23.566	1.18	1.39
	548	23.842	1.11	1.22
	1950	24.511	0.81	0.65

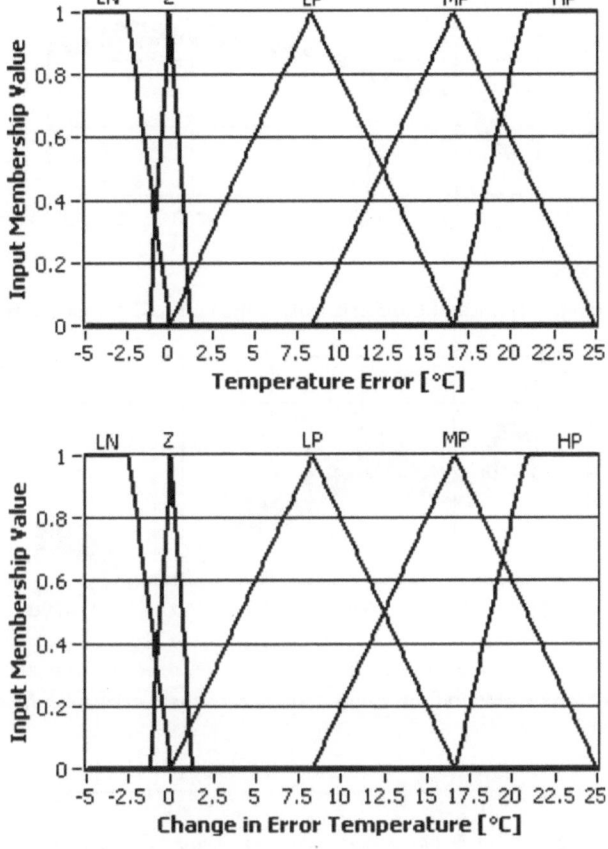

Figure 5.46 Five triangular membership functions for input signals.

used (Figure 5.45). The response of a ramp is shown in Figure 5.47 and the quadratic error analysis is shown in Figure 5.48. Interesting points in the analysis are shown in Table 5.5. In this case, the response is undesirable because the plant cannot track the reference at any point.

The third configuration consisted of using seven input bell membership functions as shown in Figure 5.49 in which fuzzy sets are: *high negative* (HN), *medium negative* (MN), *low negative* (LN), *zero* (Z), *low positive* (LP), *medium positive* (MP), and *high positive* (HP). The same output membership functions were used (Figure 5.45). The response of a ramp is shown in Figure 5.50 and the quadratic error analysis is shown in Figure 5.51. Table 5.5 resumes this analysis. As seen in the response, the plant with the controller cannot track the reference in the rising and falling range time. In addition, in the steady state set-point, it has an overshooting of 0.4°C.

The last configuration is done using seven triangular membership functions for the inputs. In this case, the fuzzy sets are: *high negative* (HN), *medium negative* (MN), *low negative* (LN), *zero* (Z), *low positive* (LP), *medium positive* (MP), and *high positive*

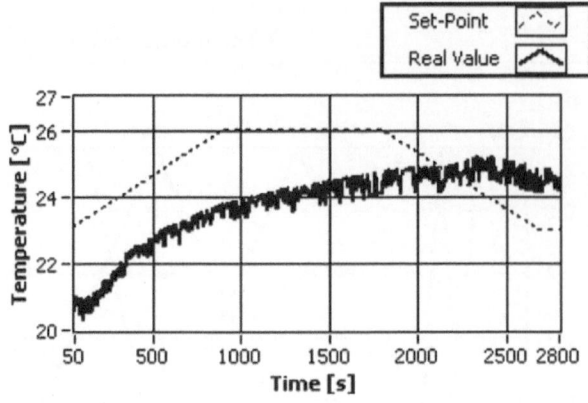

Figure 5.47 Ramp response of the 5-triangular MF's fuzzy-PD temperature controller.

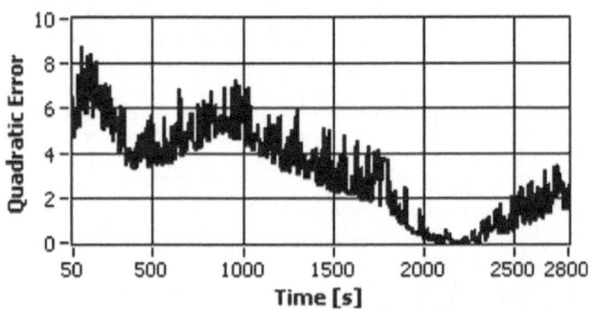

Figure 5.48 Quadratic error plot for the ramp response of the 5-triangular fuzzy-PD temperature controller.

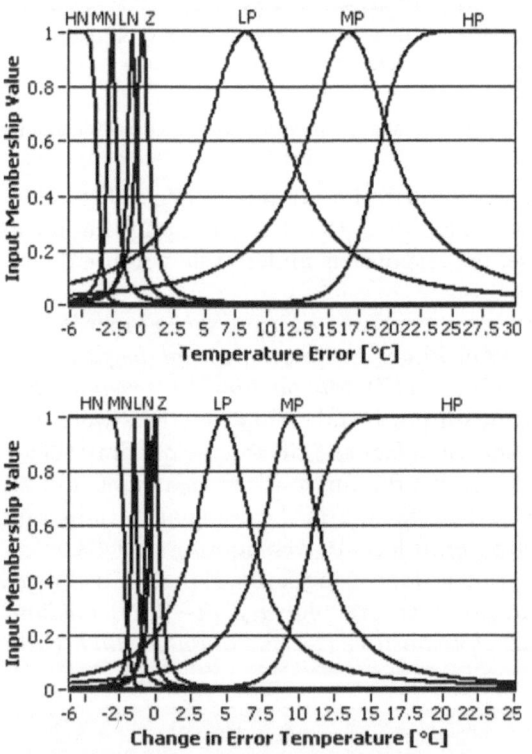

Figure 5.49 Seven bell membership functions for input signals.

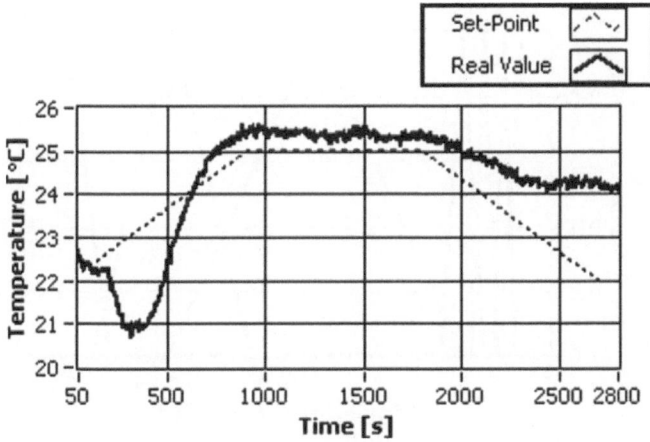

Figure 5.50 Ramp response of the 7-bell MF's fuzzy-PD temperature controller.

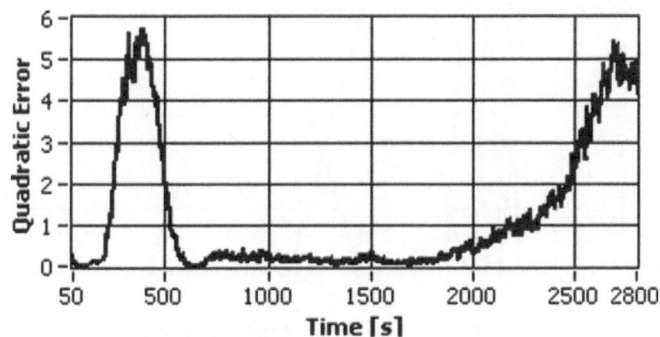

Figure 5.51 Quadratic error plot for the ramp response of the 7-bell fuzzy-PD temperature controller.

(HP). They can be seen in Figure 5.52. In addition, the response of this configuration is shown in Figure 5.53 and the quadratic error analysis is depicted in Figure 5.54. Interesting variables in the quadratic error analysis are summarized in Table 5.5. As seen, this configuration has the best performance and quadratic error is in 1°C range.

As can be seen, four configurations were depicted. Comparison between these responses is shown in Figure 5.55. In fact, the better response is found when seven triangular membership functions for inputs were used.

Dynamical neural network control system response

Two dynamical neural network controllers were developed for regulating temperature, and light intensity conditions inside the greenhouse.

For each controller, two models were depicted. In this way, the first modeling contemplates signal corrections as inputs and real sensed variables as outputs.

17,500 historical data points were obtained from the greenhouse prototype when random control signals were fired. Each sample is picked up every two seconds.

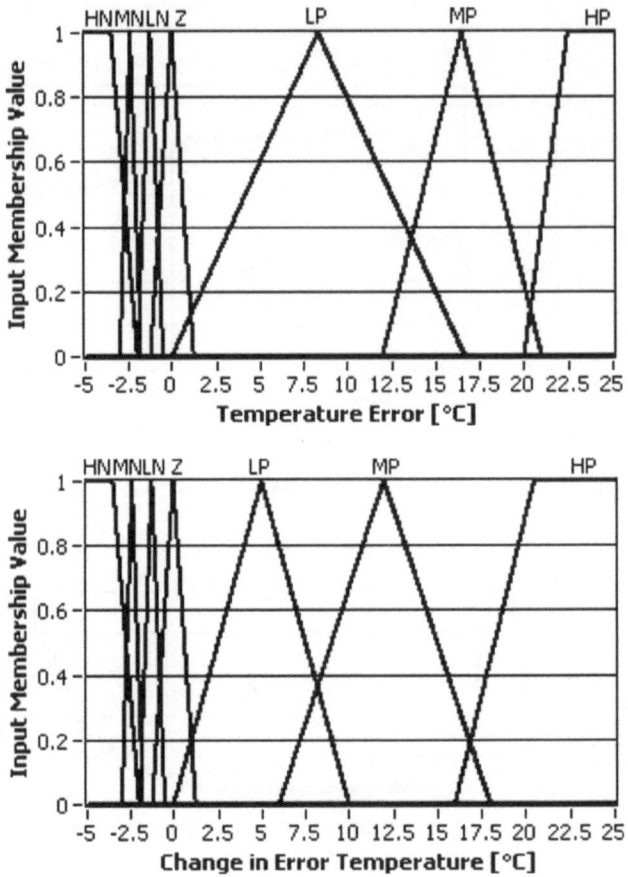

Figure 5.52   Seven triangular membership functions for input signals.

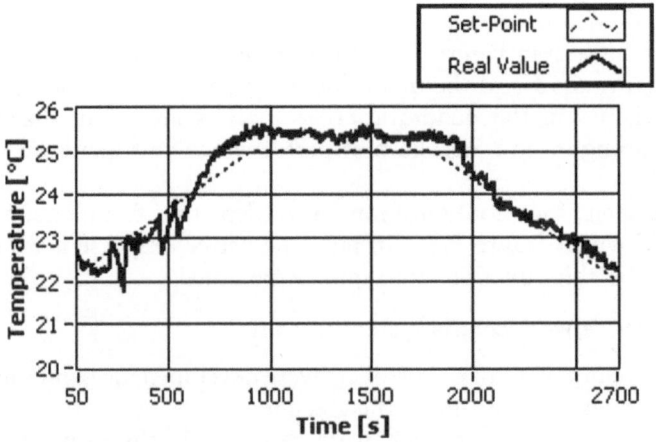

Figure 5.53   Ramp response of the 7-triangular MF's fuzzy-PD temperature controller.

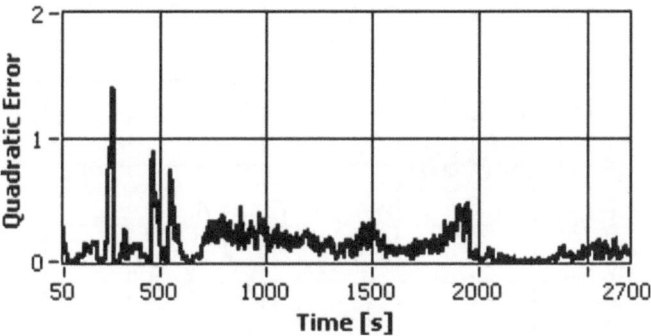

Figure 5.54 Quadratic error plot for the ramp response of the 7-triangular fuzzy-PD temperature controller.

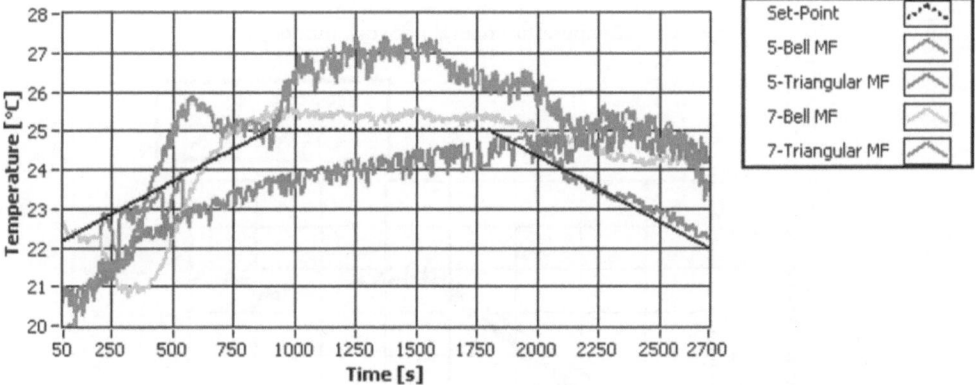

Figure 5.55 Comparison between the four fuzzy-PD configurations for the temperature controller.

Additionally, all sensed signals as temperature, relative humidity, and light intensity were registered, too.

In Figure 5.56 is seen the model of the plant with inputs: *resistance duty cycle, fan duty cycle, water pump activation signal*, and *humidifier activation signal*. Also, as output is *temperature*. This neural network was trained with a backpropagation algorithm with a learning rate of 0.03 and momentum parameter of 0.001, two delays in inputs and two delays in the output were asked, with 50 neurons in the hidden neuron. The total epochs for training were 9,880 taking into account 200 samples per batch. Additionally, the model of the inverse plant (the controller) was obtained in the same manner but using inputs as outputs, and vice versa. The response of the controller is shown in Figure 5.57, where temperature is related. In addition, a quadratic error analysis was performed (Figure 5.58).

In the same way, the model of the plant taking into account the light intensity was done with *lamp duty cycle* as input and *light intensity* as output (Figure 5.59). Two delays at inputs and two at outputs were asked, with 50 neurons in the hidden

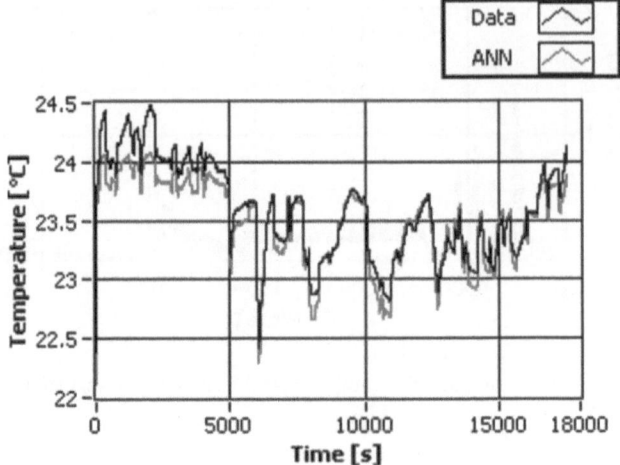

Figure 5.56 Temperature neural network model plant.

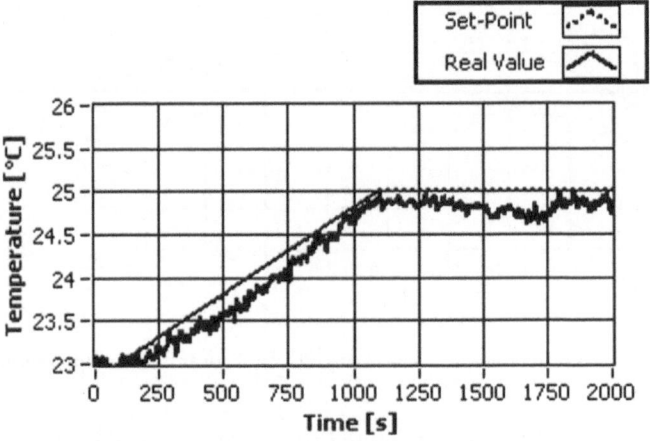

Figure 5.57 Response of the temperature controller.

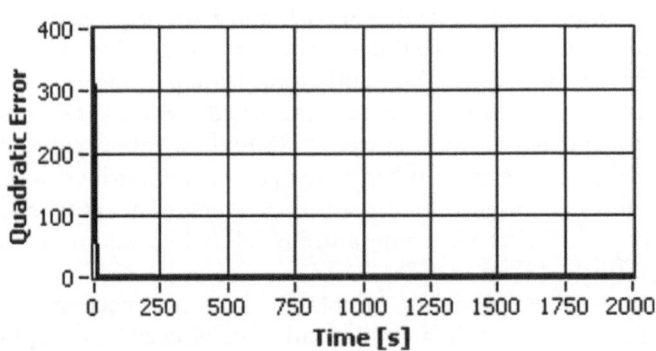

Figure 5.58 Quadratic error plot for the ramp response of the temperature neural network controller.

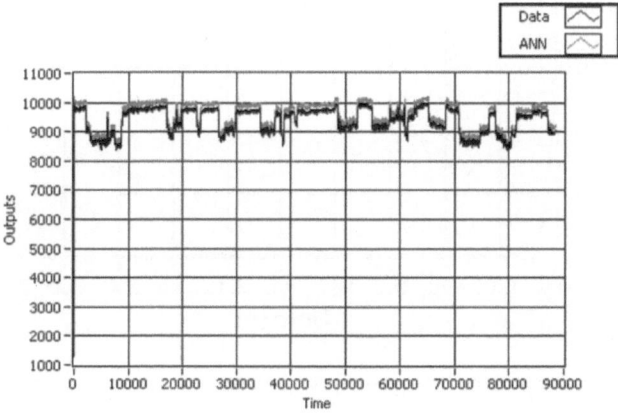

Figure 5.59 Light intensity neural network model plant.

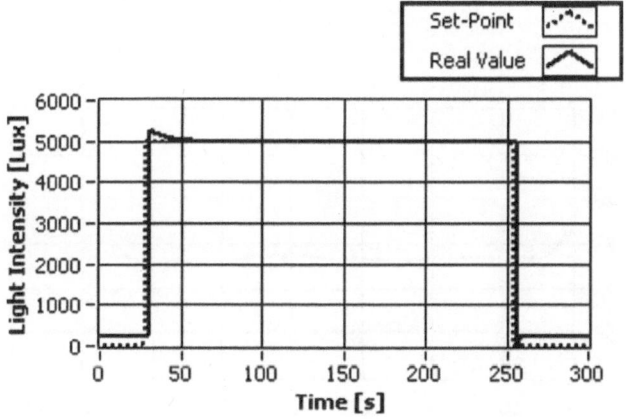

Figure 5.60 Response of the light intensity controller.

neuron. A backpropagation algorithm with learning rate 0.01 and momentum parameter 0.001 were used. Additionally, a sigmoid activation function was used in every model. Furthermore, 200 samples per batch were used for this training. 10,000 epochs were required to find a 0.015 of maximum error between the historical data and the neural network modeling. In the same way, the inverse plant for control the light intensity condition was done with inputs as outputs, and vice versa. In Figure 5.60 is shown the response of the light intensity value. 88,000 historical data points were registered.

As seen, set-point is reached with high accuracy, no error is found in the step response except for the rising time. Figure 5.61 shows the quadratic error analysis.

Relative humidity controller response

This controller was developed for maintaining a relative humidity between 50%RH and 70%RH. This bandwidth was selected because the major varieties of plants that can be grown in greenhouses need a relative humidity variation similar to 50–70%RH.

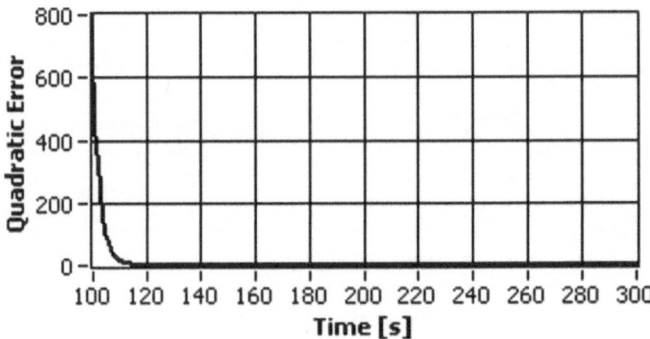

Figure 5.61   Quadratic error plot for the step response of the light intensity neural network controller.

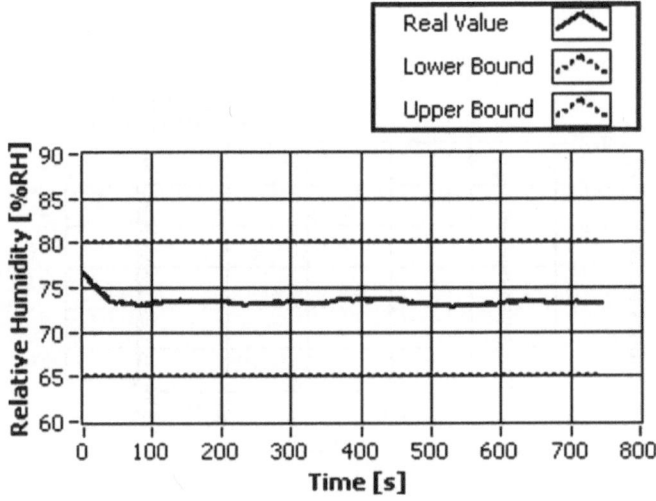

Figure 5.62   Response of the relative humidity on/off controller.

Figure 5.62 shows the response of the on/off controller. Actually, because LabVIEW simulation and validation are optimized, bandwidth can also be changed for others, if needed.

Taking into account the previous analysis done with fuzzy logic and neural network techniques, controllers for light intensity and temperature have some advantages. First, these techniques can handle high nonlinearities in the greenhouse in which several conditions are participating and cannot be decoupled. Second, expert systems, as fuzzy logic, depend on how well the behavior is understood and the experience of the engineer designer. As seen, the number of membership functions, their shapes and their locations in the universe of discussion are related in the response. In contrast, neural network models depend on the experimental data affecting directly through the system (the greenhouse exposed to the environmental variables).

The following can be observed: if experience is needed, fuzzy systems are better for controlling systems. In terms of knowing how the response is acting depending on several variables, fuzzy controllers are well determined. However, if the number of variables increases, rules grow exponentially, making them intractable. On the other hand, neural networks are better for making a generalization of systems (the control system do not depend on specific environment or geographical situation) and then these controllers can be used in any place, with the condition of monitoring before using them. Actually, an algorithm is needed for modeling the behavior of the greenhouse. Then, the neural control can be effectively used.

One advantage of neural networks is that the model can be adapted for the current environmental states and adjust its parameters (weights) for a better response. In comparison with fuzzy controllers where membership functions are tuned for some specific set points, neural networks can be generalized and adapt to other circumstances.

5.4 CONSTITUENTS OF CONTROLLED ENVIRONMENT FOR CONVENTIONAL GREENHOUSE

One of the main problems in designing greenhouses is how to build a structure that allows a controlled environment. It is necessary to understand the ideal micro-climate of plants before attacking the problem. In that sense, there are five components of this climate: temperature, relative humidity, light, CO_2 and root media (Ponce & Ponce 2009); the importance of these variables is summarized in Table 5.6.

Table 5.6 Importance of Controlling Environmental Conditions.

Variable	Impact on Crop
Temperature	It affects photosynthesis, respiration, translocation, transpiration, pigment formation, reproduction and all basic physiological plant processes. It depends on diurnal fluctuation requirements, season and geographical location, age of plants, stage of growth.
Relative Humidity	It impacts on leaf area development and stomatal conductance, interfering with photosynthesis process. High levels of relative humidity produces yield loss (e.g. tomato crop), and leads to fungal diseases. Optimal humidity performances well vapor pressure gradient.
Light Intensity	It affects photosynthesis. If it is higher, then photosynthesis would be better. If light energy is in the infrared range, it produces weak and long plants. Photoperiod is also needed. If a photoperiod is 12–18 hours in range, plant growth increases. Dark period is needed for assimilating photosynthesis process.
CO_2	It is harmful for the photosynthetic process. A plant leaf seeks to combine molecules of CO_2 with water in the presence of sunlight to form carbohydrates and oxygen. CO_2 enrichment if needed, is carried out in lighting duration. It can ultimately be translated in terms of growth, yield and quality.
Root Media	It directly impacts on growth behavior of the plant. It contains chemical substances (nutrients) for feeding the crop. Depending on the stage of growth, optimal nutrients vary. If there is not sufficient nutrients, plants grow weak and unhealthy.

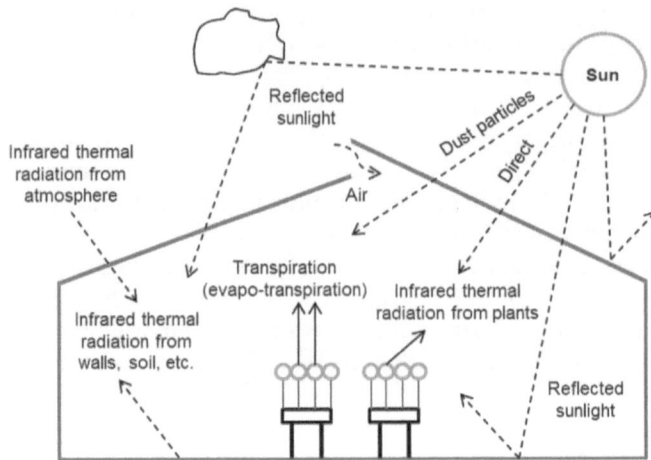

Figure 5.63 Energy Exchange between Greenhouse and Surroundings.

Temperature

Temperature is a measure of the level of heat present. All crops have a temperature range in which they can grow. Below this range, processes necessary for life stop; ice forms within the tissue, tying up water necessary for life processes; and cells are possibly punctured by ice crystals. At the upper extreme, enzymes become inactive and again essential processes stop (Nelson, 1998). Temperature control in greenhouses affects growth and development processes directly, as well as other processes such as nutrient uptake, disease resistance, and pest development and activity.

Air temperature affects development through promoting more rapid leaf expansion (and thereby thinner leaves). If the canopy has not closed, larger individual leaves intercept more light and growth is faster. For example, it is found that leaf expansion is sufficiently increased when air temperature is raised from 24°C to 27°C that light interception, photosynthesis, and thereby growth accelerated by approximately 5% in lettuce (Albright et al., 2001).

To ensure that photosynthesis exceeds respiration, plants are grown in cool temperatures at night to keep the respiration rate down and in warm temperatures by day to enhance photosynthesis. As a general rule, greenhouse crops are grown at a day temperature 3°C to 6°C higher than the night temperature on cloudy days and 8°C higher on clear days. The night temperature of greenhouse crops is generally 7°C to 21°C.

A rule by F.F. Blackman states that the rate of any process that is governed by two or more factors will be limited by the factor in least supply. Photosynthesis is a good case in point. It is dependent on heat (temperature), light, CO_2 and other factors. Blackman's law is well illustrated in Figure 5.64. In the lowest curve, the rate of photosynthesis began to plateau at about 40 kLux, regardless of whether the temperature was at 20° or 30°. The 300 ppm (parts per million) level of CO_2 (0.03% CO_2) became a limiting factor at that point. When the temperature was held at 20° and the CO_2 level was increased

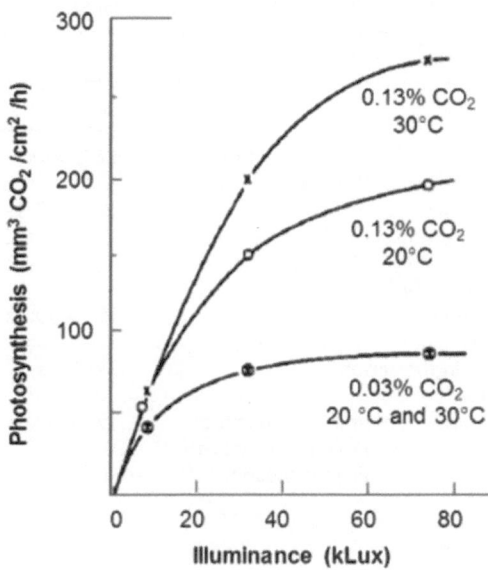

Figure 5.64 Effects of CO_2, Light and Temperature on Photosynthesis in Cucumber.

Table 5.7 Temperature Requirements for Different Crop Species.

Optimum Temperature Range (°C)		
Crop	Day	Night
Tomato	21–27	13–16
Pepper	23–27	16–18
Eggplant	22–27	17–22
Cucumber	20–25	16–18
Melon	25–30	18–21
Watermelon	23–28	17–20

to 1300 ppm (0.13% CO_2), the rate of photosynthesis increased. Then the 20° temperature became the limiting factor, because the increase in temperature to 30° at the same 1300 ppm CO_2 level brought about another increase in photosynthesis (Nelson, 1998).

For temperature management is important to know the needs and constraints of the cultivated species. It also should clarify the following concepts of temperatures, indicating the target values to be considered for the proper functioning of the culture and its limitations (InfoAgro, 2006). Table 5.7 presents some temperature requirements for different crops.

It is important to be clear that the temperatures shown in Table 5.7 are wide ranges while the crops sometimes need shorter temperature ranges or specific values depending on their phenological stage. Table 5.8 gives specific temperature ranges depending on the development stage of a tomato plant including its duration.

Table 5.8 Temperature Requirements for Tomato by Phenological Stage.

Optimum Temperature Range (°C)

Phenological Stage			Time	Day	Night
Initial	Nursery	Rapid absorption	12 Hours	25–29	
		Repose	40 Hours		
	Germination	Rise	20–27 Days	18	
		Roots		22–25	
		First leaves		12	
	TRANSPLANT				
Vegetative	Development		25–30 Days	18–23	16–18
	Flowering		30–60 Days		
Reproductive	Pollen grain formation		8–12 Days	20	
	Pollination			15.5-32	13–24
	Fecundation	Pollen germination	49–63 Days	21	
		Pollen tube growth		20–25	
		Egg fecundation		20–25	
	Fruit set			23–26	14–17
	Fruit ripening			15–22	
	HARVEST				

Relative humidity

The relative humidity inside the greenhouse is related to various processes such as: damping temperature changes, increase or decrease of the transpiration, the growth of tissues, pollen viability for greater percentage of ovarian fertilization of flowers and in the development of diseases and pests.

Relative humidity is a function of both its water content (absolute humidity – ratio of moisture in 1 kg of air) and temperature. This last and relative humidity are commonly measured air properties, highly coupled through nonlinear thermodynamic laws (psychrometric) (Pasgianos et al., 2003). Temperature is inversely related to relative humidity, so that at high temperatures, air increases the capacity to contain water vapor and therefore decreases the relative humidity. At low temperatures, the relative humidity increases (InfoAgro, 2006). The moist air interactions of plant environments are complex, but fortunately there seem to be few detrimental effects of permitting relative humidity (or vapor pressure deficit) within an established plant canopy to vary over a wide range.

Low relative humidity (high vapor pressure deficit, e.g., in excess of 1–2 kPa) leads to reduced plant growth, presumably by causing stomata closure to conserve water. High relative humidity (low vapor pressure deficit, e.g., 0.3 kPa) can reduce transpiration, limit calcium uptake, induce physiologic disorders in some plant species, and promote fungal diseases and insect infestations (Albright et al., 2001). Each species has an ideal humidity for vegetating in perfect condition, Table 5.9 shows some of them.

Light intensity

The growth rate of a closed plant canopy is closely related to photosynthesis and thereby PAR. Light saturation is a condition where plant processes become noticeably less efficient in their use of the incident radiation, which occurs at relatively high light

Table 5.9 Relative Humidity Requirements for Different Crop Species.

Crop	Optimum Relative Humidity Range (%)
Tomato	60–80
Pepper	50–60
Eggplant	50–60
Cucumber	70–90
Zucchini	65–80
Lettuce	60–80
Melon	65–75
Watermelon	65–75
Roses	14–16
Chrysanthemums	20–25

Table 5.10 Light Levels Requirements for Some Crop Species.

Crop	Optimum Light Intensity Range (kLux)
Tomato	10–40
Cucumber	15–40
Lettuce	12–30
Eggplant	65–85
Melon	80–100
Watermelon	80–100
Roses	80–100
Chrysanthemums	75–95

levels. Greenhouses often transmit little more than half of the solar radiation incident upon their exterior surfaces. Thus, light saturation is less likely to occur in greenhouses than outdoors. Before light saturation is significant, the photosynthetic response to light is relatively linear. Thus, one could expect growth rate to be approximately linear with respect to the daily light integral intercepted by the crop.

Light is the only source of energy for plant growth, and the major effect of light on plants is through photosynthesis (Albright et al., 2001). The light intensity varies from place to place but it generally varies from zero at the beginning of the day to about 100 to 150 kLux (1000 lumen/m^2) around noontime. Light intensity on cloudy days is quite low which leads to poor photosynthetic process. Optimum light intensity for a plant is 32 kLux, but below 3.2 kLux and above 129 kLux are not ideal for the plant (Tiwari, 2003).

Solar radiation transmittance needs major attention while designing and constructing a greenhouse. The transmittance of various greenhouse designs with East-West and North-South orientation is shown in Figure 5.65 for December and June. It is clear that light transmittance is higher for East-West orientation in winter and lower in summer than north-south orientation; also, greenhouses with arched roofs have better transmittance than greenhouses with a pitched roof of 25° slope (Tiwari, 2003).

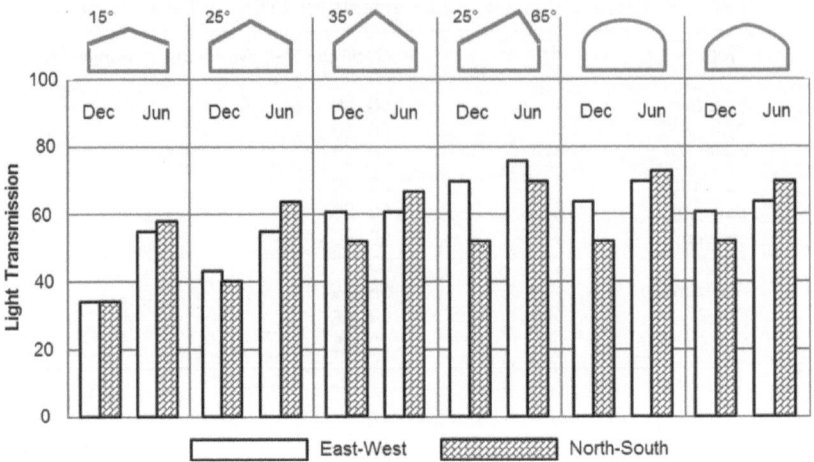

Figure 5.65 Solar Radiation for Different Greenhouse Designs and Orientations.

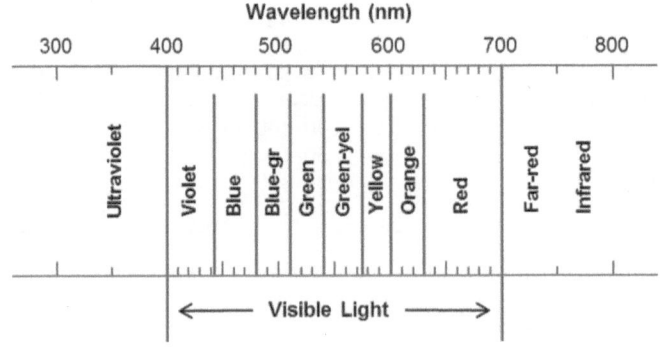

Figure 5.66 Types of Radiant Energy.

During peak summer, some protection from the high intensity of light is needed because it will raise the temperature of the leaf and causes sunburning. Hence, some type of shading screen, either over the greenhouse or inside greenhouse is provided.

Not all light is useful in photosynthesis. Light is classified according to its wavelength in nanometers (nm), which can be seen in Figure 5.66. This classification is referred to as "quality". Plants use visible light which is in the range of 400 to 700 nm; this range is most commonly referred to PAR. Ultraviolet (UV) light is harmful to plants while far-red and infrared have no influence on photosynthesis (Nelson, 1998).

Figure 5.67 shows the rates of photosynthetic activities occurring under different qualities of light. There are peaks in the blue and red bands where photosynthetic activity is higher. When blue light is supplied to plants, growth is shortened, hard and dark in color. When plants are grown in red light, growth is soft and internodes are long, resulting in tall plants. It is clearly shown that all visible light qualities are readily utilized in photosynthesis (Nelson, 1998).

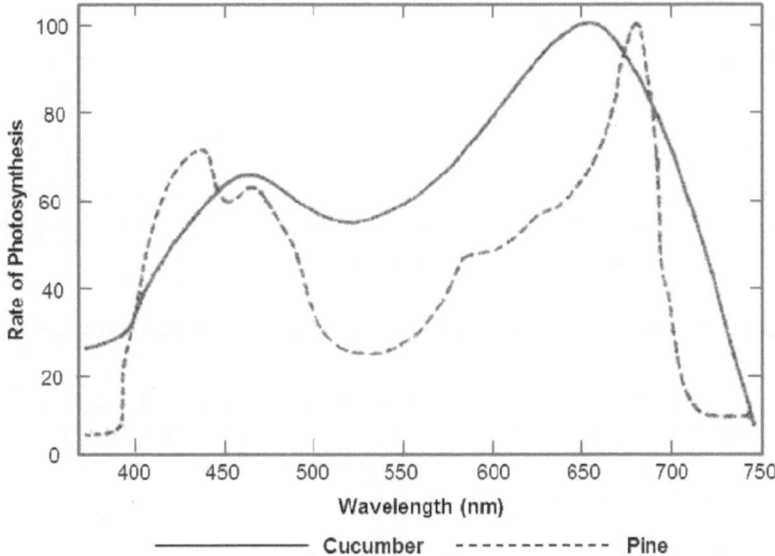

Figure 5.67   Rates of Photosynthetic Activity under Different Qualities of Light.

During the dark seasons of the year, light intensity is below optimum for most crops in most greenhouse production areas of the world. This situation can be rectified by using supplemental lighting in the greenhouse to increase the rate of photosynthesis. In that case, there exist three categories of lamps (H. Ponce, 2009):

Incandescent – These are generally not used because they generate excessive heat, poor radiation, and produce far-red and infrared wavelengths not desirable for plants. In terms of energy they are very inefficient. Incandescent lights are low cost. The luminous efficacy of a typical incandescent bulb is 16 lumen/watt.

Fluorescent – They are the most common lamps used in growing rooms because they offer a variety of wavelengths that can be used for different situations. Contrasting to incandescent, fluorescent lamps are more efficient and they do not produce much heat. Fluorescent lamps offer around 40 to 60 lumen/watt.

High intensity discharge (HID) – These lamps covert electrical energy into radiation very efficiently. They are more used in Europe than America. HID lamps have a higher cost than fluorescent but need less maintenance.

Figure 5.68 summarizes the light spectrum of lamps comparing radiation between the wavelengths produced by them.

Carbon dioxide

Carbon is an essential plant nutrient and is present in the plant in greater quantity than any other nutrient. About 40% of the dry matter of plants is composed of carbon. Plants obtain carbon from CO_2 in the air. For the most part, CO_2 gas diffuses through the stomatal openings in leaves when they are open. Once inside the leaf, carbon

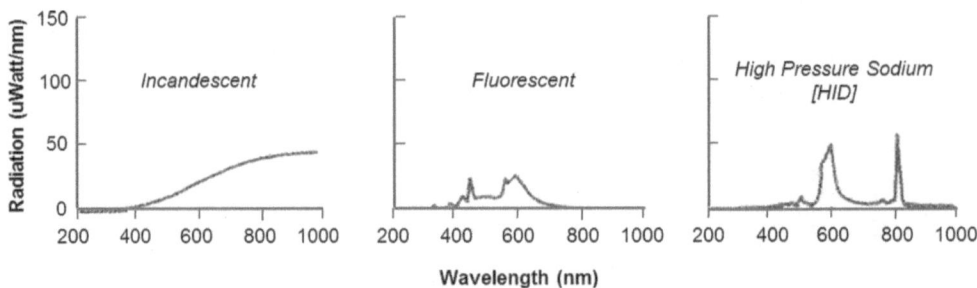

Figure 5.68 The Spectrum of Light Emissions of Lamps for Use in Greenhouses.

from CO_2 moves into the cells, where, in presence of light and water, it is used to make carbohydrates (sugars) and oxygen. The process whereby CO_2 is utilized by the plant is photosynthesis and it occurs in the green chloroplasts within cells. The process is summarized in the following equations:

$$CO_2 + Water + Energy\ from\ Sunlight \rightarrow Carbohydrate + Oxygen \qquad (5.8)$$

$$3CO_2 + 6H_2O[Photosynthesis\ Process]Sunlight\ C_6H_{12}O_6 + 3O_2 \qquad (5.9)$$

Air, on average, contains slightly more than 0.03% CO_2. The average level at the present time is 345 ppm. Due to combustion and deforestation, the level of CO_2 has been increasing 1 or 2 ppm per year since 1880, when the average level was about 294 ppm. The CO_2 level will also be higher in areas such as swamps and riverbeds, where large quantities of plant material are decomposing. Microorganisms feeding upon plant or animal remains respire CO_2 gas, much as humans do when they use plant and animal derived foods. Plants also produce CO_2 gas through the process of respiration which is opposite to photosynthesis.

A CO_2 level of 300 ppm is sufficient to support plant growth. Most plants, however, have the capacity to utilize greater concentrations of CO_2 and, in turn, attain more rapid growth (Nelson, 1998). Concentrations up to three to five times ambient levels show advantage, but diminishing returns are seen at higher values (Albright et al., 2001). This genetic capability apparently stems back to primordial times, when plants adapted to CO_2 levels 10 to 100 times the level that currently exists (Nelson, 1998). Furthermore, the actual benefits of carbon dioxide enrichment may be uneconomical when other factors (such as light level) are limiting. Suggested concentration levels have been established for many crops as seen in Table 5.11, but the temporal dynamics of carbon dioxide assimilation are not yet well quantified (Albright et al., 2001).

As the amount of energy from sunlight increases, the supply of CO_2 must increase if the process is not to be restricted or limited. The general relationship between CO_2 concentration, light intensity and photosynthesis is illustrated in Figure 5.69 (Tiwari, 2003).

In the closed field conditions like greenhouses, the enclosed air may have a CO_2 concentration of 1000 ppm because respired CO_2 remained trapped overnight. As the sunlight becomes available, photosynthesis process begins and CO_2 from the

Table 5.11 CO$_2$ Requirements for Some Crop Species.

Crop	Optimum CO$_2$ Range (ppm)
Tomato	1000–2000
Cucumber	1000–3000
Lettuce	1000–2000
Roses	1000–2000
Chrysanthemums	400–1200

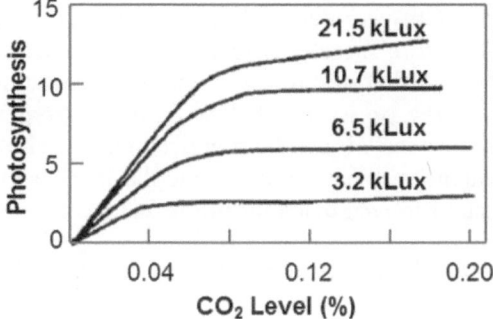

Figure 5.69 Relationship between CO$_2$, Light and Photosynthesis in Wheat.

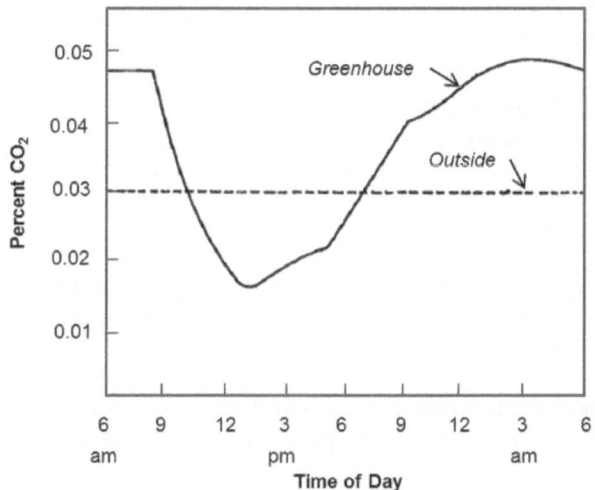

Figure 5.70 Variation of CO$_2$ Concentration in Greenhouse.

greenhouse air gets depleted. Owing to this, the CO$_2$ level in greenhouses goes below even 300 ppm much before noon. Obviously, if the greenhouse air does not receive additional CO$_2$ from some other sources, the plant would become CO$_2$ deficient. A typical graph of daily variation of CO$_2$ in a greenhouse is shown in Figure 5.70 (Tiwari, 2003).

CO_2 enrichment activity, if needed, is carried out in lighting duration. The enrichment is seldom practiced during ventilation. Accordingly, CO_2 enrichment is not done during summer unless the greenhouse is being cooled with a complete closed loop air circulation system. Depending on local climatic conditions, the period of CO_2 enrichment is generally restricted from October to March. Different enrichment methods are (Tiwari, 2003):

Combustion – A hydrocarbon such as natural gas, paraffin oil or kerosene, when burnt in the presence of sufficient oxygen, the CO_2 and H_2O (water) is produced as a result.

Liquid CO_2 – When dry CO_2 is filled in bottles and tanks under high pressure, it liquifies. The CO_2 gas from these pressurized tanks is released with the help of a set of regulating valves so that gas at low pressure is spread inside the greenhouse.

Solid CO_2 – CO_2 under low pressure and low temperature gets solidified and it is popularly known as "dry ice". It can be practiced for enrichment of CO_2 in greenhouse system if the quantity to maintain a particular level, is known. Air circulation system will be required for even distribution of CO_2.

The amount of CO_2 uptake by leaves depends upon several factors:

- Plant species and variety.
- Temperature.
- Radiation intensity.
- Wind velocity.
- Water stress.
- CO_2 concentration in air.
- Resistance to CO_2 diffusion through the stomates.
- Previous history of plant.
- Leaf area.

Root medium

Almost all of the vegetables found on the market are produced either directly or indirectly in open field soil. However, soil itself isn't necessary for plant growth, only some of its constituents. Field soil serves two basic purposes:

- It acts as a reservoir to retain nutrients and water.
- It provides physical support for the plant through its root system.

Artificial means can also provide these important requirements for plant growth with equal (and sometimes better) growth and yield results compared to field soil, although at substantially greater expense. Well-drained, pathogen-free field soil of uniform texture is the least expensive medium for plant growth, but soil does not always occur in this perfect package. Some soils are poorly textured or shallow, and provide an unsatisfactory root environment because of limited aeration and slow drainage. Pathogenic organisms are a common problem in field soils. When adverse conditions are found in soil and reclamation is impractical, some form of soil-less culture may be justified (Johnson, 2013).

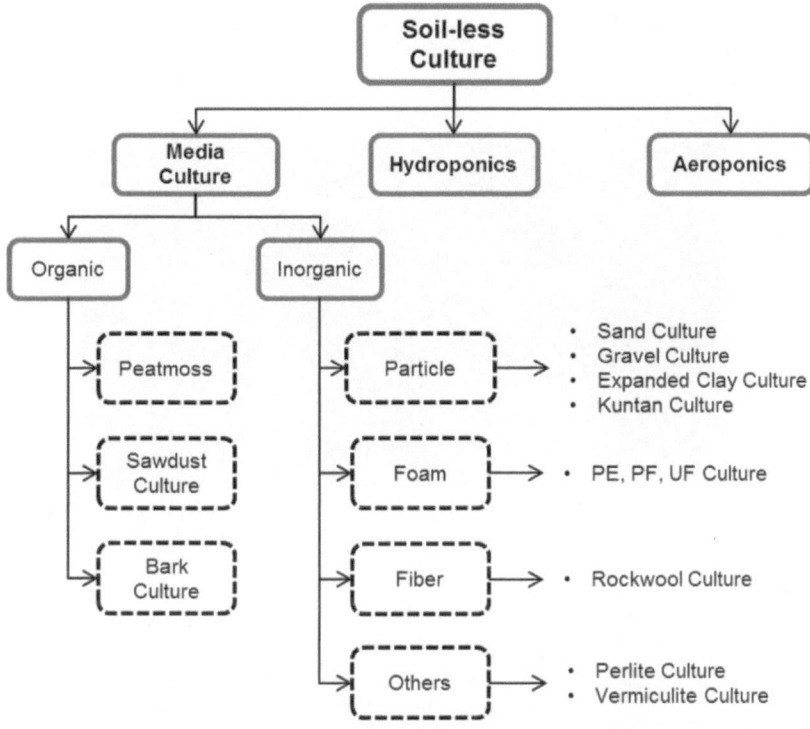

Figure 5.71 Different Soil-less Cultivation Methods.

Table 5.12 Signal Transmission Standards.

Analog	Digital
4–20 mA	Discrete
0–5 V	(switch, relay, alarm)
0–10 V	
3–15 Psig	Pulse train
0.2–1 Bar	(modulated signal)

Soil-less culture is a cultivation technique by which plants are grown away from the soil. Plants are cultivated in containers filled with several possible growing media. If these media are solid, the method is called *media culture*. If no medium is present and the plant roots are bathed in circulated nutrient solution, the method is called *hydroponics*. If no medium is present and plant roots get their nutrients by frequent spraying or misting, the method is called *aeroponics*. Figure 5.71 shows this classification of soil-less cultivation methods (Ronen, 2013).

The limited volume of medium and water availability generally causes rapid changes in the status of water and nutrients. Changes in the medium solution such as pH, electrical conductivity (EC) and nutrients level, should be monitored for the efficient use of water and nutrients.

Figure 5.72 Hydroponic Cultivation Method.

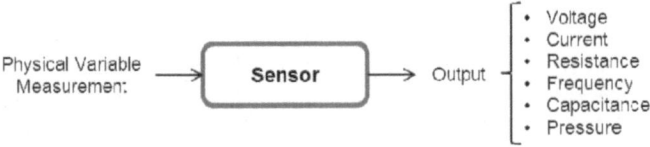

Figure 5.73 Sensor Phenomenon.

5.5 SENSORS FOR GREENHOUSE MONITORING

Sensor/transducers are devices that convert physical variable to an electrical signal generally for processing, control or display. Good crop management depends on having the right information to make necessary decisions. In the past, the grower has been the greenhouse sensor and control system, checking conditions and adjusting equipment settings as needed to optimize crop growth (Perdigones et al., 2004).

The automation and continuous operations in commercial greenhouses have advocated the scope and the use of individual sensor and sensor system for automatic control and measurement of the environmental parameters discussed previously. These environmental factors influence the quality and productivity of plant growth. Hence, continuous monitoring of these parameters gives valuable information to the grower to better understand how each factor affects the quality and the rate of plant growth, and how to maximize crop yield (Tiwari, 2003).

Sensor technologies have made an enormous impact on the modern day industries. There are thousands of sensors available on the market ready to be attached to a sensing platform. Medium and high technology greenhouses make use of a range of sensors which link into automated control systems. The information is used to control heating, venting, fans, screens, nutrient dosing, irrigation, carbon dioxide supplementation and

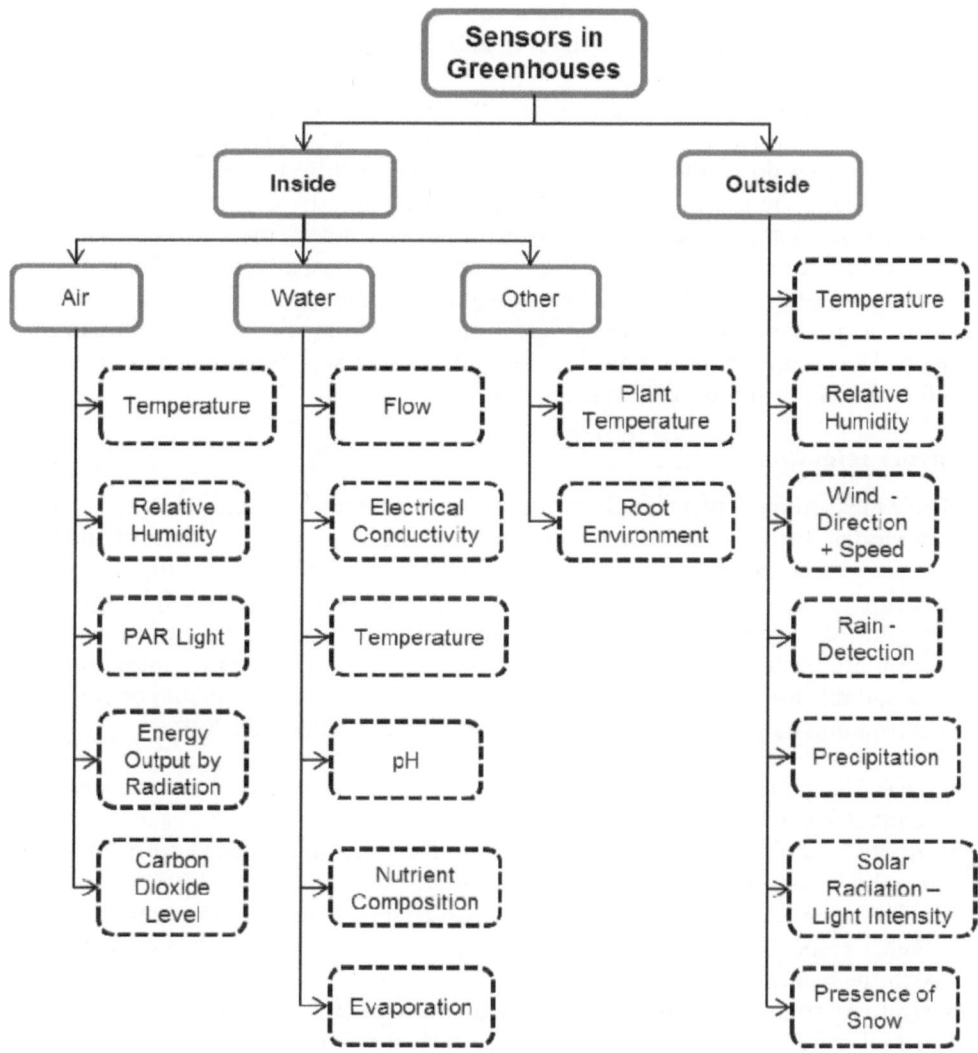

Figure 5.74 Use of Sensors in Greenhouses.

fogging or misting systems (Quan et al., 2011). Figure 5.74 shows a complete scenario about all the possible sensors that can be used in a hi-tech greenhouse, but the main ones are: *inside* – temperature, relative humidity, light, carbon dioxide, EC and pH; *outside* – temperature, relative humidity, light, wind direction, wind speed and rain detection.

Monitoring growing conditions is essential. Even without automated control of the production system, it is not possible to make the right decisions about the crop without having the right information. Temperature and relative humidity need be monitored in every greenhouse. Light levels should be checked at least periodically to make sure covering materials are performing adequately, but ideally light levels need to be checked on a regular basis in order to know the optimal temperature regime for the crop. The EC and pH of both the feed and drain solutions should be monitored in every hydroponic system.

As new technologies are appearing, the new tendency respect to sensors are the use of wireless sensors. In modern greenhouses, several measurements points are required to trace down the local climate parameters in different parts of the big greenhouse to make the greenhouse automation work properly. Cabling would make the measurement system expensive and vulnerable. Moreover, the cabled measurement points are difficult to relocate once they are installed. Thus, a Wireless Sensor Network (WSN) consisting of small size wireless sensor nodes equipped with radio and or several sensors, is an attractive and cost-efficient option to build the required measurement system. The predominant standards commonly used in WSN communications include: ZigBee, 802.15.4 and 6LoWPAN.

Sensors selection

When a measurement of a non-electrical quantity (temperature, light, etc.) is converted to an electric form, a sensor must selected (or a suitable combination of them) the conversion to be carried out. The first step in the selection process is to clearly define the nature of the quantity to be measured. This also includes knowledge of the range of magnitudes and frequencies that are expected. When the problem of measurement has been established, it must examine the fundamental principles of operation of the sensor suitable for the type of measurement. The following points should be considered in determining its appropriateness for a measurement (Dieck, 2000):

Range – The range should be large enough to cover all expected magnitudes of the quantity being measured.

Sensitivity – To obtain a significant data, the sensor should produce a sufficient output signal due to a change in the measured input.

Electrical signal characteristics – The electrical characteristics (output impedance, frequency response and time response) of the sensor output signal must be compatible with the measurement system.

Physical environment – The selected sensor must be able to withstand the environmental conditions to which they may be subject while making measurements. The temperature, humidity, and corrosive chemicals can damage some sensors but not others.

Operating errors – The errors inherent in the operation of the sensor or caused by environmental conditions must be sufficiently small or controllable such that are not significant in measurements.

Once the sensor is selected and incorporated into the design of the measurement system, the following points must be observed to increase the accuracy of the measurements (Dieck, 2000):

- The sensor output is calibrated with respect to any known pattern while using under the conditions in which it will work. This calibration should be performed regularly, at least once for each change of season.

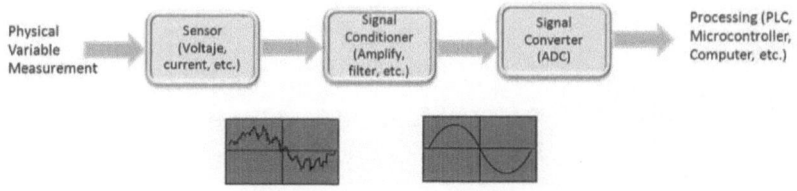

Figure 5.75 Sensor Signal Conditioning.

- Changes in environmental conditions sensor should be recorded continuously. Following this procedure the measurements obtained can be corrected later taking into account changes in environmental conditions.
- Errors can be reduced by controlling the sensor environment artificially locking it in a box or cabinet at a controlled temperature, and isolating it from vibrations and external shocks.

Specifically, greenhouses are challenging measurement environments. Constant high humidity, risk of condensation, potential spray irrigation, dust and dirt, and constant exposure to solar radiation are all factors of a challenging environment. Only instruments designed to work in harsh environments will survive in a greenhouse. Other important factors to consider besides those already mentioned are (Vaisala, 2011):

- Required accuracy and long-term stability.
- Instrument degree of protection IP65/NEMA4 minimum.
- Operating range in high relative humidity.
- Capability to recover from condensation.
- Solar shield for the temperature and humidity sensors.
- Potential wear and tear of moving parts.

Sensors signal conditioning

Hardly an electronic designer connects the sensors directly to the processing unit of a system, since the sent signal is usually very weak or contains unwanted noisy components, therefore stages of signal conditioning are needed. This means that all data acquisition system requires analog electronic circuits based on operational amplifiers (op-amps) to adjust the electrical signals produced by the sensors to be compatible with analog to digital conversion systems (ADC) (Dieck, 2000).

Signal conditioners perform a variety of general-purpose conditioning functions to improve the quality, flexibility and reliability of your measurement system such as:

- Protection to avoid damage to the next element, such as a microprocessor as a result of high voltage or current.
- Converting a signal on a suitable signal type. Would be the case when there is a need to convert a signal to a dc voltage or current.
- Obtaining the appropriate level of the signal.

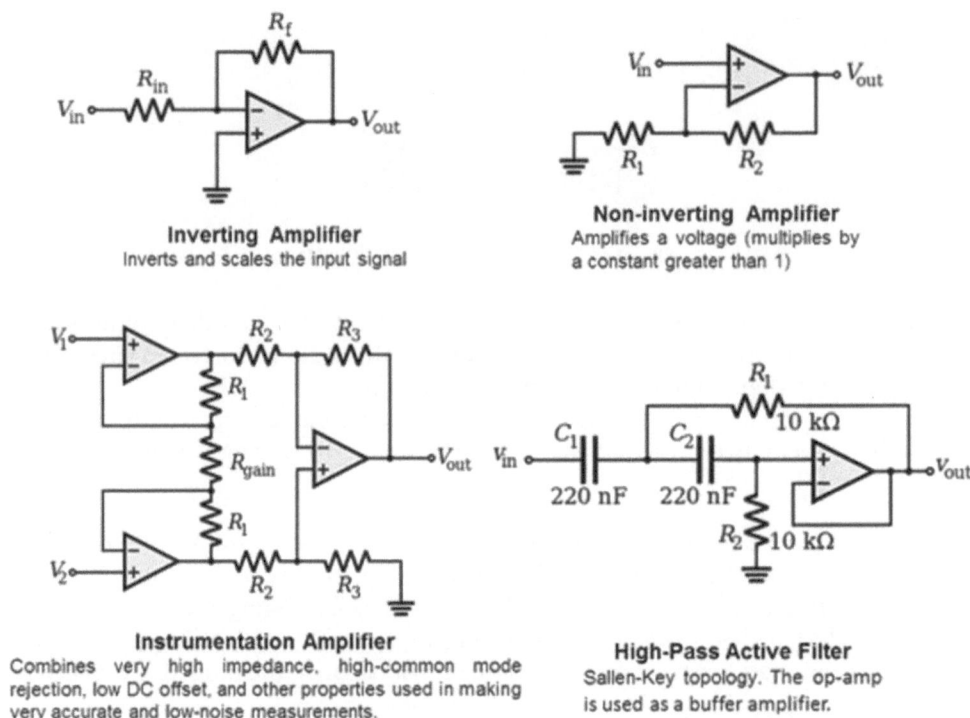

Figure 5.76 Some Op-Amp Configurations in Signal Conditioning.

- Eliminating or reducing noise. For example, to remove noise in a signal using filters.
- Signal handling, for example, convert a variable into a linear function.

An op-amp is a DC-coupled high gain electronic voltage amplifier with a differential input and, usually, a single-ended output. An op-amp produces an output voltage that is typically hundreds of thousands of times larger than the voltage difference between its input terminals.

The op-amp is the solid foundation of any system of electronic conditioning in instrumentation systems and data acquisition. With op-amps encapsulated in integrated circuits can perform many "electronic maneuvers" to condition, convert, edit, adjust and process analog signals from the sensors. One can say that the op-amp is the quintessential processor of analog signals (Dieck, 2000). Some op-amp configurations for different applications of signal conditioning are displayed in Figure 5.76.

Sensors location

Sensor location can be a significant problem in production facilities. For example, it is not uncommon to find a temperature sensor located in a drafty area of a room or where sunlight may strike it for some time during the day. Obtaining representative environmental measurements is not straightforward for many reasons. Examples include the location of radiation sensors (impact of moving shadows in greenhouses, spectral responses of sensors exposed to non-solar radiation), aerial environment (temperature or humidity in stratified zones, poorly mixed zones, inability to measure what is wanted, such as within a crop canopy), and sensor technology and maintenance (calibration, drift, noise immunity) (Albright et al., 2001).

Select a location for the sensor that well represents the greenhouse climate. Temperature, relative humidity and light sensor inside the greenhouse should be placed level with the growing tip of the crop or just above the canopy. Placing these sensors near the door, with shadows, close to the roof, near irrigation water sprays or close to heating pipelines does not represent the climate around the plants and it will not give you the information needed for producing an optimal crop. CO_2 sensor should not be placed near a vent or exhaust duct (Vaisala, 2011). On the other hand, EC and pH sensors must be submerged and in contact with the nutrient solution.

Outside the greenhouse, the sensors or the weather station should be located near the greenhouse where it may represent the outside air climate that could get into the greenhouse. It is necessary to utilize solar shields for the temperature and humidity sensors protecting them against the solar radiation. If the greenhouse is covered by trees, which is uncommon, or barely used, the weather station located outside the greenhouse should be also under the trees. This weather station should be put at a certain height depending on the height of the greenhouse; the wind speed and direction could be different 2 m above the greenhouse than at the level of the vent system.

Sensors description

A description of different kinds of sensors based on their working principle is given below. Advantages and disadvantages of them are also mentioned. The sensors described below can be categorized as commercial sensors, some of which are used widely in industrial processes.

Temperature measurement

Temperature is one of the most widely used sensing technologies in the world of sensors. Four different families of temperature sensors are available on the market. Depending on the application one sensor may be more suitable than the other.

Thermocouples – A thermocouple is a junction between two wires of dissimilar metals. The point of contact between the wires generates a voltage that is proportional to the temperature. Thermocouples are suitable for measuring over a large temperature range, up to 2300°C. They are less suitable for applications where smaller temperature differences need to be measured with high accuracy. For such applications RTDs and thermistors are more appropriate. Applications include kilns, gas turbine exhaust, diesel engines, and other industrial processes (Quan et al., 2011). Some thermocouples are shown in Table 5.13.

Advantages:

- Wide temperature range.
- Relatively cheap.

Table 5.13　Some Thermocouple Types.

Type	Positive Material	Negative Material	Accuracy	Range (°C)
B	Pt, 30%Rh	Pt, 6%Rh	0.5% >800°C	50 to 1820
E	Ni, 10%Cr	Cu, 45%Ni	0.5% or 1.7°C	−270 to 1000
J	Fe	Cu, 45%Ni	0.75% or 2.2°C	−210 to 1200
K*	Ni, 10%Cr	Ni, 2%Al, 2% Mn, 1%Si	0.75% or 2.2°C	−270 to 1372
N*	Ni, 14%Cr, 1.5%Si	Ni, 4.5%Si, 0.1%Mg	0.75% or 2.2°C	−270 to 1300
R	Pt, 13%Rh	Pt	0.25% or 1.5°C	−50 to 1768
S	Pt, 10%Rh	Pt	0.25% or 1.5°C	−50 to 1768
T*	Cu	Cu, 45%Ni	0.75% or 1.0°C	−270 to 400

*Most commonly used thermocouple types.

- Highly accurate.
- Minimal long-term drift.
- Fast response time.

Disadvantages:

- Non-linear.
- Low output signal (mV).
- Vulnerable to corrosion.
- Calibration may be tedious and difficult.

Resistance temperature detectors (RTDs) – RTDs are basically temperature sensitive resistor devices with a sensing range of −200°C to 500°C. The resistance increases with temperature. Most RTD elements consists of a length of fine coiled wire wrapped around a ceramic or glass core. RTDs are widely used in many industrial applications such as air conditioning, food processing, textile production, processing of plastics, micro-electronics and exhaust gas. The materials used to construct RTDs are generally conductors such as copper, nickel, molybdenum or platinum. Of these is platinum which give better performance. A common sensor is the Pt100 (platinum RTD with R = 100 Ω at 0°C).

Advantages:

- Linear over a wide temperature range.
- Relatively accurate.
- Good stability and repeatability at high temperature.

Disadvantages:

- Low sensitivity.
- Higher cost than thermocouples.
- Vulnerable to shock and vibration.

Thermistors – Similar to RTDs, thermistors are also temperature dependent resistor devices. Thermistors are not as accurate or stable as RTDs but they are easier to wire, cost less and almost all automation panels accept them directly. Thermistors

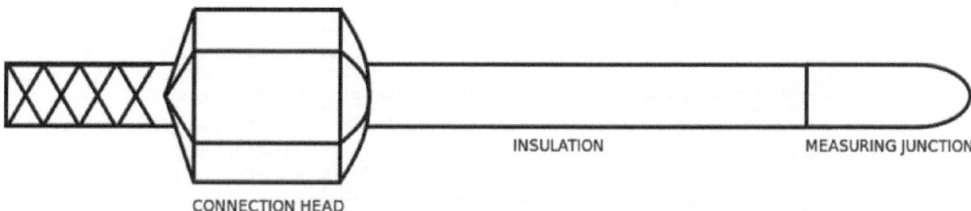

INSULATION   MEASURING JUNCTION

CONNECTION HEAD

Figure 5.77 Thermocouples.

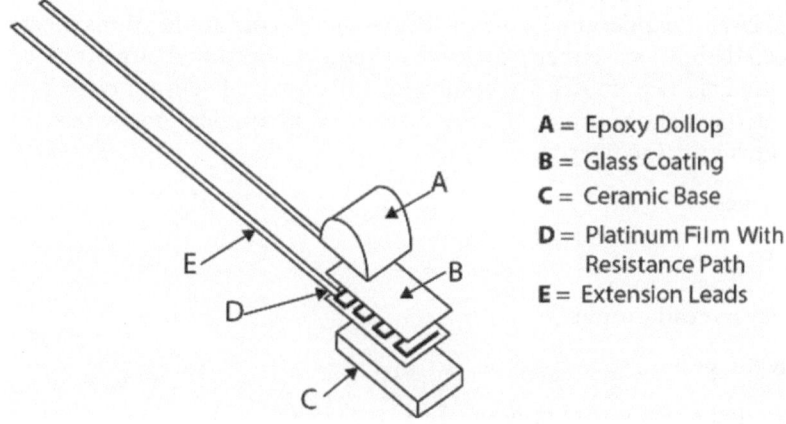

A = Epoxy Dollop
B = Glass Coating
C = Ceramic Base
D = Platinum Film With
Resistance Path
E = Extension Leads

Figure 5.78 Platinum Thin Film RTD.

are made of semiconductor materials with resistivity that is especially sensitive to temperature. Their applications are automotive, modern digital thermostats, 3D printers, food handling and processing industry, and consumer appliance industry (Quan et al., 2011). Thermistors typically achieve a higher precision within a limited temperature range, typically −40°C to 260°C.

Advantages:
- Highly sensitive.
- Low cost.
- Accurate over small temperature range.
- Good stability.

Disadvantages:
- Non-linear.
- Self heating.
- Limited temperature range.

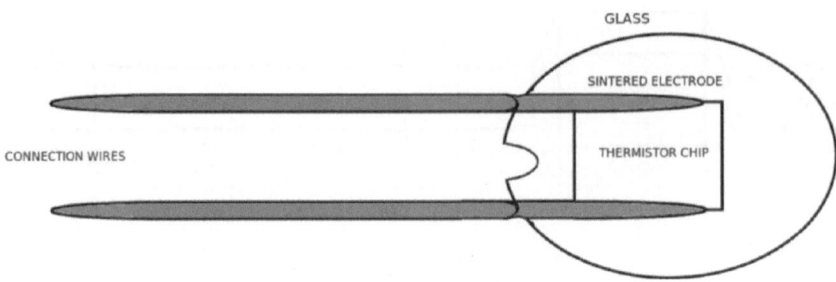

Figure 5.79 Thermistors.

Integrated circuit temperature sensors (ICs) – In low cost applications most of the sensors stated above are either expensive or require additional circuits or components to be used. However ICs are complete, silicon-based sensing circuits with either analogue or digital output. ICs are often used in applications where the accuracy demand is low (Quan et al., 2011).

Advantages:

- Low cost.
- Excellent linearity.
- Easy to read output.

Disadvantages:

- Limited temperature range ($-55°$C to $150°$C).
- Self heating.
- Fragile.
- Slightly less accurate when compared to the other types.

Humidity measurement

When it comes to humidity sensing technology, there are three types of humidity sensors: capacitive, resistive and thermal conductivity humidity sensors.

Capacitive humidity sensors (CHSs) – CHSs consist of a substrate on which a thin film of polymer or metal oxide is deposited between two conductive electrodes. The sensing surface is coated with a porous metal electrode to protect it from contamination and exposure to condensation. The substrate is typically glass, ceramic, or silicon. The changes in the dielectric constant of a CHS are nearly directly proportional to the relative humidity of the surrounding environment. CHSs are widely used in industrial, commercial, and weather telemetry applications (Quan et al., 2011).

Advantages:

- Able to function in high temperature environments.
- Near linear voltage output.

Figure 5.80 Capacitive Humidity Sensors (Courtesy of Ingeniería MCI Ltda).

- Wide relative humidity range.
- High condensation tolerance.
- Reasonable resistance to chemical vapors and contaminants.
- Minimal long-term drift.
- High accuracy.
- Small size.
- Low cost.

Disadvantages:

- Limited sensing distance.
- Sensor interface can be tedious and difficult.

Resistive humidity sensors (RHSs) – RHSs measure the changes occurred in electrical impedance of a hygroscopic medium such as conductive polymer, salt or treated substrate. RHS consists of noble metal electrodes either deposited of a substrate by photo resist techniques or wire-wound electrodes on a plastic or glass cylinder. These sensors are suitable for use in control and display products for industrial, commercial and residential applications (Quan et al., 2011).

Advantages:

- Faster response time than CHSs.
- Near linear voltage output.
- High accuracy.
- Small size.
- Low cost.
- Wide relative humidity range.

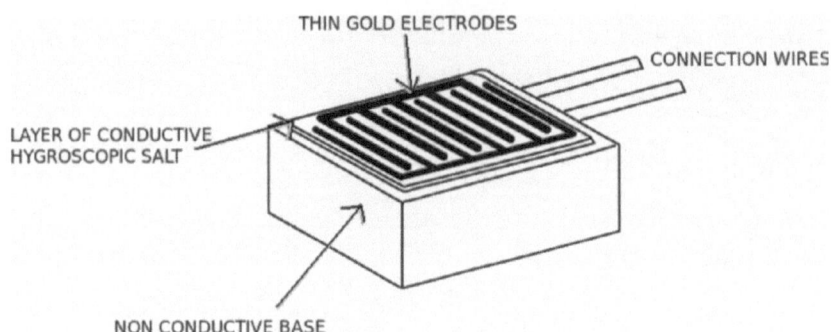

Figure 5.81 Resistive Humidity Sensors.

Disadvantages:

- Lower operating temperature when compared to CHSs.
- Sensitive to chemical vapors.
- Low tolerance against contaminants.
- Low condensation tolerance.

Thermal conductivity humidity sensors (TCHSs) – TCHSs measure the absolute humidity by quantifying the difference between the thermal conductivity of dry air and that of air containing water vapors. TCHS consists of two matched Negative Temperature Coefficient (NTC) thermistor elements in a bridge circuit; one is hermetically encapsulated in dry nitrogen and the other is exposed to the environment. These sensors are suitable for applications such as kilns for drying wood, machinery for drying textiles, paper, chemical solids, pharmaceutical production, cooking and food dehydration.

Advantages:

- Durability.
- Able to operate in high temperature environments.
- Excellent immunity to many chemical and physical contaminants.
- High accuracy.
- High condensation tolerance.

Disadvantages:

- Responds to any gas that has thermal properties different from those of dry nitrogen; this affects the measurements.

Light intensity measurement

Light from the sun is responsible for nearly all life on the earth. Sunlight fuels the process of photosynthesis where plants convert carbon dioxide and water into carbohydrates. Plants use light in the range of 400 to 700 nm. Monitoring PAR is important to ensure that plants are receiving adequate light for photosynthesis. Some of the

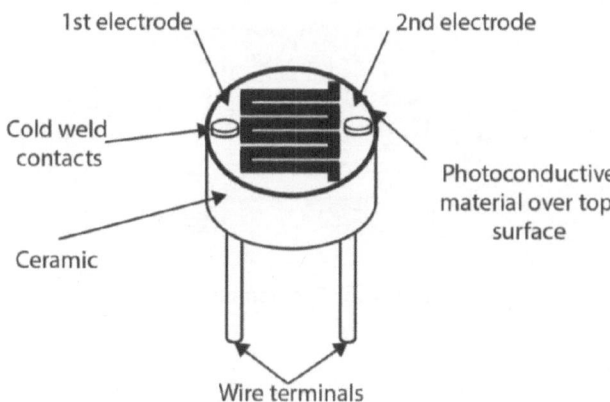

Clear coating over entire top surface

1st electrode

2nd electrode

Cold weld contacts

Photoconductive material over top surface

Ceramic

Wire terminals

Figure 5.82 Light Dependent Resistors.

popular light sensors on the market that can be used for environmental monitoring applications.

Photometric sensors – These sensors measure visible radiation or light as seen by the human eye.

Advantages:

- Highly sensitive.
- Good stability.
- Fast response time.
- Low temperature dependency.
- Excellent linearity.
- Small size.

Disadvantages:

- Expensive.
- These sensors are mostly used to measure indoor lighting conditions.

Light dependent resistors (LDRs) – Similar to photometric sensors, LDRs measure visible light as seen by the human eye. A LDR is basically a resistor; the internal resistance increases or decreases dependent on the level of light intensity impinging on the surface of the sensor (Quan et al., 2011).

Advantages:

- Very cheap.
- Fast response.
- Linear output.
- Small in size.

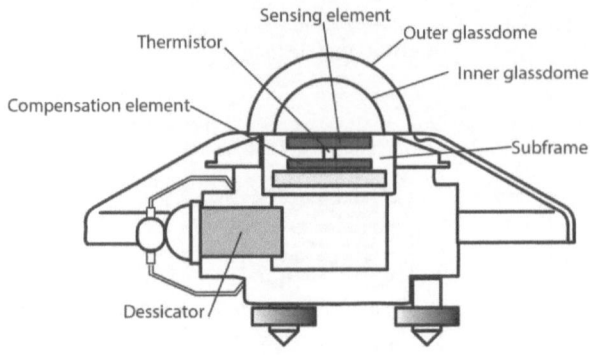

Figure 5.83 Pyranometer.

Disadvantages:

- Like photometric sensors, LDRs are mostly used to measure indoor lighting conditions.

Pyranometers – These measure total solar radiation. The sensor is composed of a silicon photovoltaic detector mounted in a miniature head. The sensor output current is directly proportional to the level of solar radiation. These sensors are commonly used for agriculture, meteorology and solar energy applications (Quan et al., 2011).

Advantages:

- Highly accurate.
- Excellent linearity.
- Good stability.
- Fast response time.

Disadvantages:

- Bulky.
- Expensive.

Quantum sensors – Quantum sensors measure the PPFD (Photosynthetic Photon Flux Density) of the PAR. They are the most popular types of light sensors used in agriculture and environmental industries (Quan et al., 2011).

Advantages:

- Very sensitive.
- Fast response.
- Highly accurate.
- Excellent linearity.
- Good stability.
- Small in size.

Disadvantages:

- Expensive.

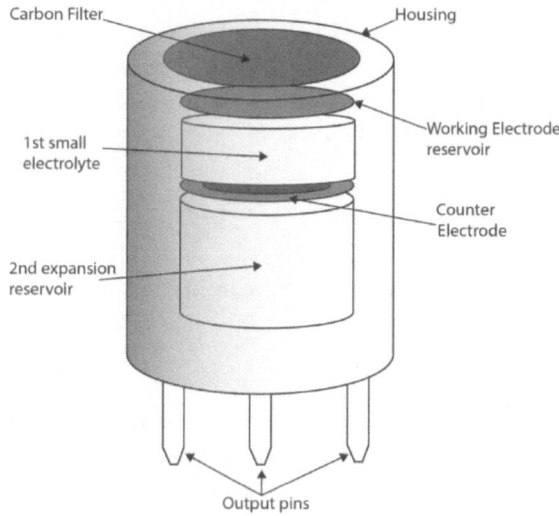

Figure 5.84 Electrochemical CO_2 Sensors.

Carbon dioxide measurement

Measuring CO_2 is important in monitoring indoor air quality and many industrial processes. Two types of CO_2 detectors are available to measure the CO_2 level in the surrounding environment.

Electrochemical CO_2 sensors – The CO_2 sensitive element consists of a solid electrolyte formed between two electrodes, together with a heater substrate. By monitoring the changes in the electromotive force generated between the two electrodes, it is possible to measure CO_2 gas concentration (Quan et al., 2011).

Advantages:

- Cheap.
- Accurate.
- Real-time sensing.
- High tolerance against contaminants.
- Small in size.

Disadvantages:

- Require a significant amount of power because they operate at high temperature.

Non-dispersive infrared (NDIR) CO_2 sensors – NDIR are spectroscopic sensors that detect CO_2 in gaseous environments by its absorption characteristics. The key components are an infrared source, a light tube, an interference (wavelength) filter, and an infrared detector. The gas is pumped or diffused into the light tube and the electronics measures the absorption of the characteristic wavelength of light (Quan et al., 2011).

Advantages:

- High speed.
- Real-time sensing.
- Low power consumption.
- High contamination tolerance.
- Small in size.

Disadvantages:

- Carbon monoxide (CO) often coexists with CO_2 and absorbs a similar wavelength range as CO_2 which results in inaccurate readings.
- Very expensive.

Electrical conductivity measurement

An EC sensor measures the concentration of ions in a solution; the more ions there are, the more conductive the solution is. This is basically because more ions means there are more particles to carry electrons. Electrons are fundamental components of electricity, making the solution more conductive to current when a voltage is applied. It is possible to differentiate the various EC sensors due to the metering used, amperometric or potentiometric.

Amperometric – This method applies a known potential difference (V) to two electrodes and measures the current (I) passing through them. According to Ohm's law: $I = V/R$, where R is resistance. Therefore, the higher the current, the greater the conductivity. The resistance, however, depends on the distance between the two electrodes and surfaces, which can vary due to possible deposits of salts or other materials (electrolysis). Therefore, the amperometric method is recommended for solutions with low concentration of dissolved solids, usually up one gram per liter (approximately $2000\,\mu S/cm$).

Potentiometric – The four ring potentiometric method is based on the induction principle and eliminates common problems associated with amperometric method as the effects of polarization. The two outer rings apply an alternating voltage and induce a voltage loop in the solution. The two inner rings measure the induced voltage drops in the current loop, which depends on the conductivity of the solution. Using the method of four rings is possible to measure conductivity ranges up to $200000\,\mu S/cm$ and $100\,g/l$. This is the most effective method for measuring the electrical conductivity of solutions (InfoAgro, 2010).

pH measurement

The pH sensor includes a measuring electrode, a reference electrode, and sometimes a temperature sensor for compensation. A typical modern pH probe usually combine these components into one device called a combination pH electrode.

A pH sensor is essentially a battery where the positive terminal is the measuring electrode and the negative terminal is the reference electrode. The measuring electrode, which is sensitive to the hydrogen ion, develops a potential (voltage) directly related to the hydrogen ion concentration of the solution. The reference electrode provides a stable potential against which the measuring electrode can be compared. The primary function of a pH sensitive electrode is to produce a voltage of $59.1\,mV = 1\,pH$ at $25°C$.

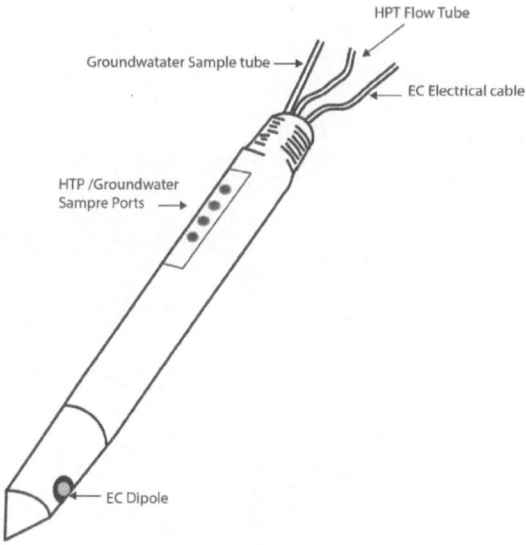

Figure 5.85 Electrical Conductivity Probes.

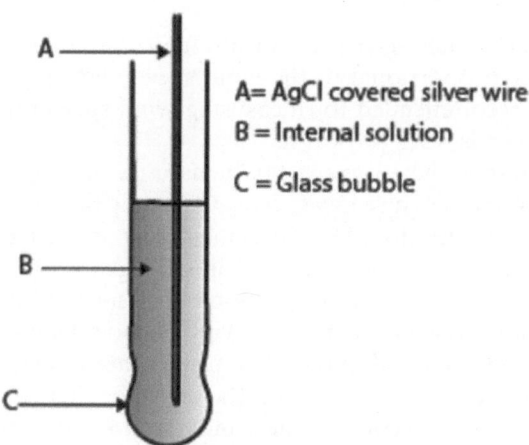

A= AgCl covered silver wire

B = Internal solution

C = Glass bubble

Figure 5.86 pH Probes.

The measuring electrode is usually glass and quite fragile. Glass electrodes have been utilized in a wide range of applications from pure research, control of industrial processes, to analyze foods, cosmetics and comparison of indicators of the environment and environmental regulations. Recent developments have replaced the glass with more durable solid-state sensors (pH Info, 2005).

Wind direction and speed measurement

Cup anemometers – The hemispherical cups are attached to a rotating shaft driven by air currents, so the wind speed becomes the rotation of the shaft. To convert it into

Figure 5.87   Cup Anemometer and Vane.

an electrical signal a tacho-generator or alternator are often used. The generators produce a voltage proportional to the wind speed. This type of anemometers are widely used and recommended for measuring wind speed outside the greenhouse, usually reaches speeds near 100 km/h.

Hot wire anemometers – These sensors can be used to measure the wind speed inside the greenhouse and to measure low speeds. Generally their use is limited to research studies concerning ventilation. Hot wire anemometers are based on evaluating the cooling that occurs in a resistor inserted in the stream of fluid. If an air flow is passed through, this will cool in proportion to the speed of the fluid.

Vanes – These determine the direction of the wind outside. Vanes are formed by an arm which rotates about a vertical axis which incorporates a deflector. When there is a current of air, wind exerts a pressure on the deflector that causes the arm to rotate and the end points the direction of the wind. As in the cup anemometers, a variation in the position of the arm or shaft induces an electrical magnitude. A common method is the potentiometer. The vertical axis about which rotates the vane arm is connected to a precision potentiometer, so that the output signal of the potentiometer is proportional to the angle formed with the vane north (Perdigones et al., 2004).

Rain measurement

Pluviometer – Liquid precipitation gauge that transforms rainfall into electrical pulses, each pulse generally corresponding to 0.1 or 0.2 mm (depending on type of rain gauge) precipitation. They can be found in different volumes.

Rain detector – Not It detects the presence or absence of rain. Raindrops will be responsible for closing the electric circuit that carries the sensor to give the signal. The main application for greenhouses is to connect or disconnect a device (e.g. a

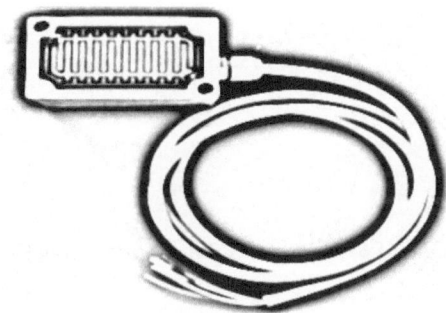

Figure 5.88 Rain Detector.

gear motor to open or close the vents) depending on the presence or absence of rain. Can also be used to detect the presence of condensation (Perdigones et al., 2004).

Image processing in crop inspection

Color and size of post-harvested crops are characteristics that researchers have taken as an area of opportunity to inspect the quality of them using image processing methods, where a camera is the sensing device. The color is visually considered one of the most important parameters in the definition of quality of any food. Its measurement has always been of concern to the food industry and food engineering research. The use of image processing techniques for color and quality assessment of food products require an absolute color calibration technique based on a common interchange format for color data and a knowledge of which features from an image can be best correlated with product quality.

Rapid advances in hardware and software for digital image processing have motivated several studies on the development of computer vision systems (CVS) to evaluate the quality of diverse raw and processed foods. Color imaging analysis not only offers a methodology for specification of uneven coloration but it can also be applied to the specification of other attributes of total appearance. Computer vision systems are also recognized as the integrated use of devices for non-contact optical sensing, and computing and decision processes to receive and interpret automatically an image of a real scene. The technology aims to replace human vision by electronically perceiving and understanding an image.

Image processing and image analysis are the core of computer vision with numerous algorithms and methods capable of objectively measuring and assessing the appearance quality of several agricultural products. In image analysis for food products, color is an influential attribute and powerful descriptor that often simplifies object extraction and identification from an image and that can be used to quantify the color distribution of non-homogeneous samples.

5.6 AUTOMATION SYSTEMS

Greenhouses provide a shelter in which a suitable environment is maintained for plants. Solar energy from the sun provides sunlight and some heat, but you must provide a

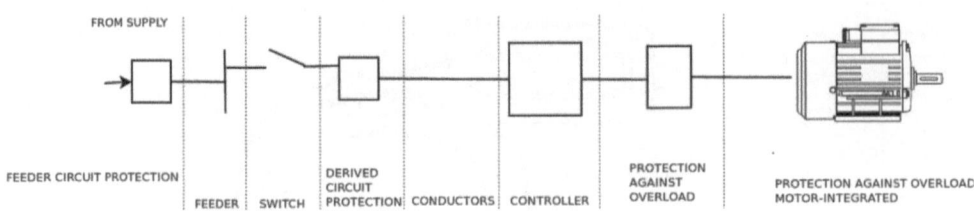

FROM SUPPLY

FEEDER CIRCUIT PROTECTION

FEEDER   SWITCH

DERIVED
CIRCUIT
PROTECTION   CONDUCTORS   CONTROLLER

PROTECTION
AGAINST
OVERLOAD

PROTECTION AGAINST OVERLOAD
MOTOR-INTEGRATED

Figure 5.89 Electric Motor Derived Circuit Elements.

system to regulate the environment in your greenhouse. This is done by actuators; these are the technology devices that generate an action in order to achieve a specific goal. Following this, the main actuator systems are ventilation and cooling technologies, heating arrangements and irrigation systems; these last includes hydroponics and this will be described separately.

Actuator systems are mainly composed of electrical devices, specifically by electrical motors or pumps. Therefore, before diving into the description of each actuator systems, it is first necessary to make a short mention of some control elements of AC electric motors.

Control elements of AC electric motors

In all industrial electrical installations where electric motors appear, the task isn't just bringing energy to them, also requires means connection and disconnection, and also control them depending on the specific application for which they were selected.

These elements in the way they appear in the technical standards for electrical installations are shown in Figure 5.89.

The driver or controller of an electric motor is a device that is normally used to start a motor that is going to play a certain behavior in normal operating conditions, and to stop when required. The controller can be a simple switch to start and stop the motor, it can be also a station button to start the engine locally or remotely. It can also be a device to start the motor by steps or reversing its direction of rotation or using signals from the control elements such as temperature, pressure, liquid level or other physical change that requires starting or stopping a motor and clearly give a greater degree of complexity to the control circuit.

A fixed-speed can be achieved by these types of controllers. If the need leads to vary the motor speed, more complex devices are available, variable speed controllers for AC motors are provided with a range of different power inverter, variable-frequency drive or electronic commutator technologies. Thus, a variable-frequency drive (VFD is a type of adjustable-speed drive used in electro-mechanical drive systems to control AC motor speed and torque by varying motor input frequency and voltage. Examples of applications include fans, centrifugal blowers, centrifugal pumps, propeller pumps, turbine pumps, agitators, and axial compressors.

The variable frequency drive controller is a solid state power electronics conversion system consisting of three distinct sub-systems: a rectifier bridge converter, a direct current (DC) link, and an inverter.

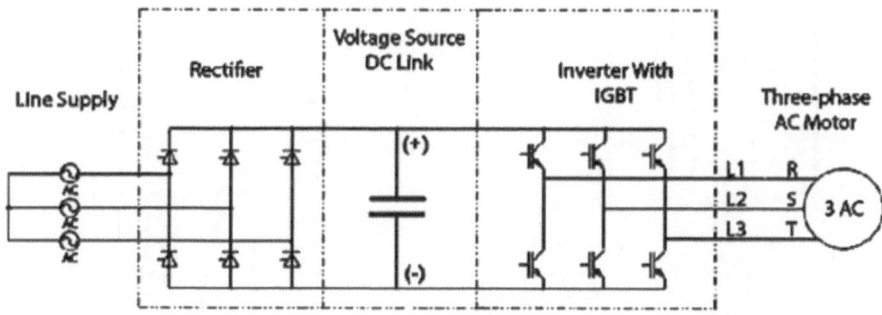

Figure 5.90 Variable Frequency Drive Components.

VFD Principle of Operation – The rectifier in a VFD is used to convert incoming ac power into direct current (DC) power. One rectifier will allow power to pass through only when the voltage is positive. A second rectifier will allow power pass through only when the voltage is negative. Two rectifiers are required for each phase of power. Since most large suppliers are three phase, there will be a minimum of six rectifiers used. After the power flows through the rectifiers it is stored on a DC bus. The DC bus contains capacitors to accept power from the rectifier, store it, and later deliver that power through the inverter section. The final section, the inverter, contains transistors that deliver power to the motor. The Insulated Gate Bipolar Transistors (IGBTs) is a common choice. The IGBT switch on and off several thousand times per second and precisely control the power delivered to the motor. The IGBT uses a Pulse-Width Modulation (PWM) method to generate a current sine wave at the desired frequency to the motor. Varying the frequency of a VFD controls the motor speed.

Each control circuit for simple or complex it may be, it is composed of a certain number of basic components connected together to meet a particular behavior. The principle of operation of these components is the same and its size varies depending on the size of the motor to be controlled. Even when the range of components for the control circuits is large, the main electrical control elements are shown below (Enríquez, 1994; Enríquez, 1999).

- Switches.
- Circuit breakers.
- Drum type switches.
- Button stations.
- Control relays.
- Magnetic contactors.
- Fuses and relays.
- Pilot lamps.
- Level, limit and other switches.
- Resistors, reactors, auto-transformers, transformers and capacitors.

Figure 5.91 shows a basic configuration of an electric motor connection considering control and protection elements. For a deeper knowledge it is necessary to review the references cited in this subsection.

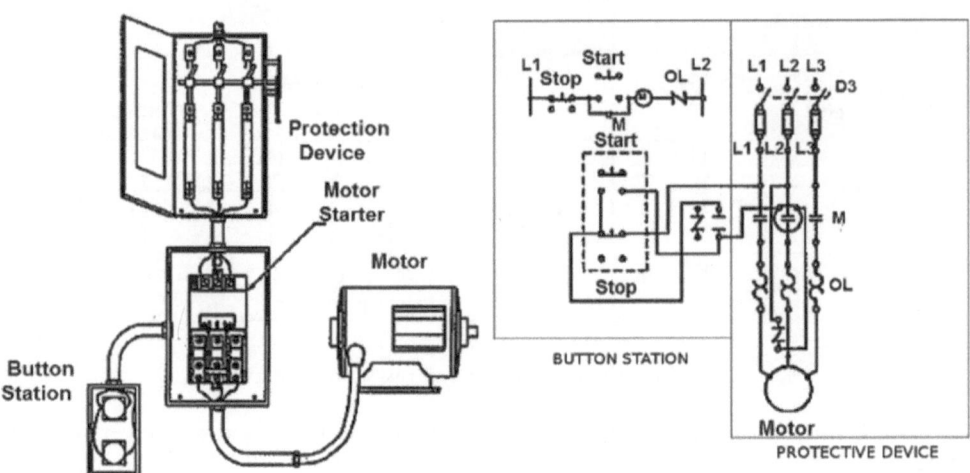

Figure 5.91 Electric and Physical Diagram of Basic Motor Elements.

Electric motors operate on three different physical principles: magnetic, electrostatic and piezoelectric. By far the most common is magnetic. In magnetic motors, magnetic fields are formed in both the rotor and the stator. The product between these two fields gives rise to a force, and thus a torque on the motor shaft. One, or both, of these fields must be made to change with the rotation of the motor. This is done by switching the poles on and off at the right time, or varying the strength of the pole.

The main types are DC motors and AC motors, the former increasingly being displaced by the latter. AC electric motors are either asynchronous and synchronous. Once started, a synchronous motor requires synchronism with the moving magnetic field's synchronous speed for all normal torque conditions. In synchronous machines, the magnetic field must be provided by means other than induction such as from separately excited windings or permanent magnets.

Ventilation and cooling technologies

Greenhouses require two distinctly different forms of cooling, one for summer and the other for winter. Most localities, with the general exception of those in higher elevations, experience periods of summer heat that are adverse to greenhouse crops. Temperatures inside a conventional ventilator-cooled greenhouse can be 11°C higher than those outside, in spite of open vents. This system of cooling is known as *passive cooling* or *natural ventilation* (Nelson, 1998). Ventilation is the process of allowing the fresh air to enter into the enclosed area by driving out the air with undesirable properties. In the greenhouse context, ventilation is essential for reducing temperature, replenishing CO_2 and controlling relative humidity. Ventilation requirements for greenhouses vary greatly, depending on the crop grown and the season of production. Usually greenhouses that are used seasonally employ natural ventilation only (Radha & Igathinathane, 2007).

Due to the inability of passive systems to cool down the greenhouse in hot summers, *active cooling* systems were developed to reduce the excess-heat problem. These systems are in essence evaporative cooling systems; they are based on the process of heat absorption during the evaporation of water. The two evaporative cooling systems in use today are *fan-and-pad* and *fog*. On the other hand, sometimes winters are excessively cold hence is necessary to temper the air temperature before it reaches the plant zone. Two active winter cooling systems that have been developed to solve this problem are *convection-tube cooling* and *horizontal airflow (HAF) cooling*.

Good temperature control is vital for a productive greenhouse, and this in turn means that the cooling system needs to be effective and efficient. It is important to make sure that the system is designed properly and working well.

Passive ventilator cooling

In the tropics, the sides of greenhouse structures are often left open for natural ventilation. With the advent of plastics and its use in greenhouses, provision of passive or natural ventilation is a challenge, especially in the absence of exhaust fans. In natural ventilation, the heated air becomes less dense and rises up. This warm air moves out and allows the dense cool air to flow into the greenhouse. Prevailing winds above certain level also aid in the creation of additional natural ventilation.

Until the 1950s, all greenhouses were cooled by passive air movement through ventilators. Vents were located on both roof slopes adjacent to the ridge and also on, both side walls of the greenhouse. The ventilators on the roof as well as those on the side were of area, each about 10% of the total roof area. During winter cooling phase, the south roof ventilator was opened in stages to meet cooling needs. When greater cooling was required, the north ventilator was opened in addition to the south ventilator. In summer cooling phase, the south ventilator was opened first, followed by the north ventilator. Air entered through the side ventilators. As the incoming air moved across the greenhouse, it was warmed by sunlight and by mixing with the warmer greenhouse air. With the increase in temperature, the incoming air becomes lighter and rises up and flows out through the roof ventilators. This sets up a chimney effect that, in turn, draws in more air from the side ventilators creating a continuous cycle as seen in Figure 5.92. This system did not adequately cool the greenhouse. On hot days, the interior walls and floor were frequently injected with water to help cooling (Nelson, 1998).

Another method of ventilation is to roll up the sides, allowing air to flow across the plants which is shown in Figure 5.93. The amount of ventilation on one side, or both sides, may be easily adjusted in response to temperature, prevailing wind and rain. During periods of excessive heat, it may be necessary to roll the sides up almost to the top. If insects, especially those that are vectors for virus diseases, are prevalent the open vent areas must be covered with a fine mesh. The holes must be large enough to permit free flow of air; meshes with small holes blocks air movement and cause a build up of dust. Such ventilation systems on plastic greenhouses are only effective on free standing greenhouses and not on multi-span greenhouses. The vents can be operated with a modernized vent thermostat for automatic climate control or by a computer system. The purpose of the side curtain and roof ventilator system or the

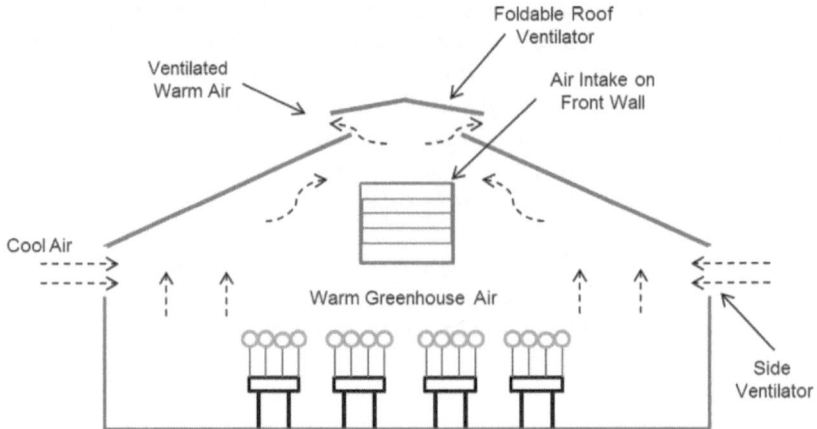

Figure 5.92 Chimney Effect in General Ventilation.

Figure 5.93 Roll Up Side Greenhouse Ventilation.

retractable roof is to replace high energy consuming fan and pad cooling systems. These passive cooling systems work well in hot and cold climates, but the main limitation is the tolerance of the crop to full light intensity when the roof is opened (Radha & Igathinathane, 2007).

Fan-and-pad cooling

Pad-and-fan systems, shown in Figure 5.94, consist of exhaust fans at one end of the greenhouse and a pump circulating water through and over a porous pad installed at the opposite end of the greenhouse. If all vents and doors are closed when the fans operate, air is pulled through the wetted pads and water evaporates. The air will be at its lowest

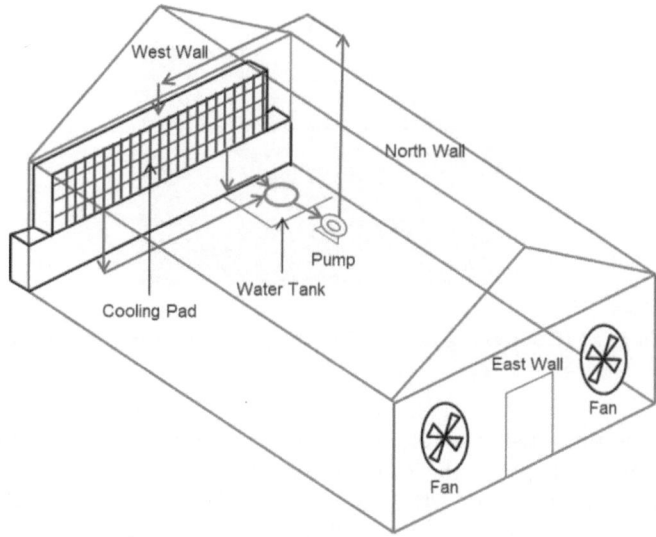

Figure 5.94 Pad-and-Fan Cooling System and Water Trickling Arrangement.

temperature immediately after passing through the pads. As the air moves across the greenhouse to the fans, the air picks up heat from plants and soil and the temperature of the air gradually increases. A temperature gradient across the greenhouse results, with the pad side being coolest and the fan side warmest (Tiwari, 2003).

Pad and fans should be placed either end walls or side walls of the greenhouse and the distance between them is important. A distance of 30 to 61 m is the best. The size of the exhaust fan should be selected to achieve proper temperature difference and good circulation; fans should not be spaced more than 7.6 m apart. If the pad to fan distance is less, then there is less opportunity time for the flowing air to cool the surroundings; whereas with very large distance uniform cooling is not possible as fans may not pull enough air through the pads. To achieve a given degree of cooling, more amount of air is required when pad to fan distance is less and vice versa. So the velocity of incoming air is to be modified accordingly (Radha & Igathinathane, 2007).

Aspen and coated cellulose are common pad materials that usually have a life span of one to three years. The required evaporative pad area depends on the pad thickness. For the typical, vertically mounted four-inch thick pads, the required area in ft^2 can be calculated by dividing the total greenhouse ventilation fan capacity in CFM by the 250 which is the recommended air velocity through the pad. For six-inch thick pads, the fan capacity should be divided by the 350. A general rule says that approximately 1 ft^2 of pad is needed for 20 ft^2 of floor area.

The recommended minimum pump capacity is 0.5 and 0.8 GPM per linear foot of pad for the four and six-inch thick pads, respectively. The recommended minimum sump tank capacity is 0.8 and 1 gallon per ft^2 of pad area for the four and six-inch pads, respectively. For evaporative cooling pads, the estimated maximum water usage is 20–30 GPM per 100 ft^2 of pad area. Approximately 10% (or 0.005 GPM per 1000

Figure 5.95 Cooling Pad (Courtesy of Dr. A.J. Both, Department of Environmental Sciences, Rutgers University, USA).

CFM of air flow when using water with a salt concentration of less than 700 ppm, or three times as much when the salt concentration is as high as 1500 ppm) of the returning water should be bled off to prevent salt buildup on the pads. Salt buildup reduces the efficiency of the pads (Both, 2008). Water for the pads should be clean and low in mineral content to prevent clogging and coating of the pads. Algae growth in the pads can reduce the effectiveness of the system and result in deterioration of the pads. The addition of an algaecide to the water supply will help in control.

Fog cooling systems

A second alternative of evaporative cooling is fog cooling. The cost of installation of a fog cooling system relative to a pad-and-fan system range from less, when pure water is available, to more when extensive filtering and chemical treatment is needed. However, the cost of electrical power is much less for the fog cooling system. Fog cooling involves dispersion of water particles in greenhouse air, where they extract heat from the air as they evaporate. The speed of evaporation of water, consequently, the rate of cooling of air increase proportionately as water droplet size decreases.

Mist droplets are in the range of 1000 microns (0.04 inch) in diameter. If a cup of water were converted to mist, it would have 400 times as much surface area and would evaporate 400 times faster than the same water left in the cup. Mist droplets are large and will settle out of air, wetting surfaces of plants, soil, and people. In contrast, fog droplets that are 40 microns or smaller (0.0016 inch), the surface area and rate of evaporation is 10 thousand times greater than the same volume of water in a cup. These droplets stay suspended in air while they evaporate to cool the air without condensing out on surfaces. Greenhouse fog cooling systems that can convert 99.5% of the water in the system to 40 microns or smaller, with an average droplet size less than 10 microns (0.0004 inch). These droplets evaporate at 40 thousand times the speed with which water evaporates from a cup. With such a rapid evaporative response, air can be cooled at nearly 100% efficiency. Pumps used to provide fog droplets can typically operate at 1000 psi (6.9 MPa) and possibly at pressures up to 1500 psi (10.3 MPa) (Nelson, 1998).

Fog nozzles can be installed throughout the greenhouse, resulting in a more uniform cooling pattern compared to the pad-and-fan system. The recommended spacing is approximately one nozzle for every 50–100 ft^2 of growing area. The water usage per

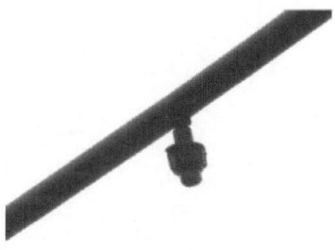

Figure 5.96   Fog Cooling Nozzle.

nozzle is small: approximately 1–1.2 GPH (gallons per hour). In addition, the water needs to be free of impurities to prevent clogging of the small nozzle openings (Both, 2008).

Greenhouses equipped with exhaust-fan cooling lend themselves well to fog cooling. A line of fog nozzles is installed just inside the inlet ventilators. Exhaust fans on the opposite wall draw outside air in through the open ventilators and then through the fog, where it is cooled. Only about half of the exhaust-fan capacity, 4 to 5 CFM/ft^2 of floor area, of fan-and-pad systems is used. If there were no more to the system, air would rise in temperature as it crossed the greenhouse, as happens in a fan-and-pad system. To prevent this, a second row of fog nozzles in installed further inside the greenhouse, parallel to the first row.

Various control systems are used for fog cooling. Timers provide the simplest form of control. The circuit continues through a recycle timer that is typically set to apply fog from 30 seconds to 4 minutes out of each cycle of 1 to 20 minutes.

Advantages cited by greenhouse firms that have installed fog cooling include the following (Nelson, 1998):

- There is less electrical consumption, since the sum of the wattage of the fog pump and exhaust fans is less than that of the exhaust fans and pad water pumps in the fan-and-pad system.
- Heat across the greenhouse is controlled.
- Cooler average temperatures can be achieved across the greenhouse.
- The system is a good substitute for the mist system in cutting-propagation greenhouses, where it uses less water and causes less disease.

Convection-tube cooling

When the temperature set point for winter greenhouse cooling is reached, three events are activated simultaneously as illustrated in Figure 5.97. An exhaust fan, located anywhere in the greenhouse, is turned on to create a vacuum. A louver is opened in a gable, through which cold air enters in response to the vacuum. A pressurizing fan in the end of the clear polyethylene convection tube turns on to pick up cool air entering the louver, since the end of the convection tube is separated 1 or 2 ft (0.3 to 0.6 m). Cold air under pressure in the convection tube shoots out of holes on either side of the convection tube in turbulent jets. The cold air mixes with warm greenhouse air

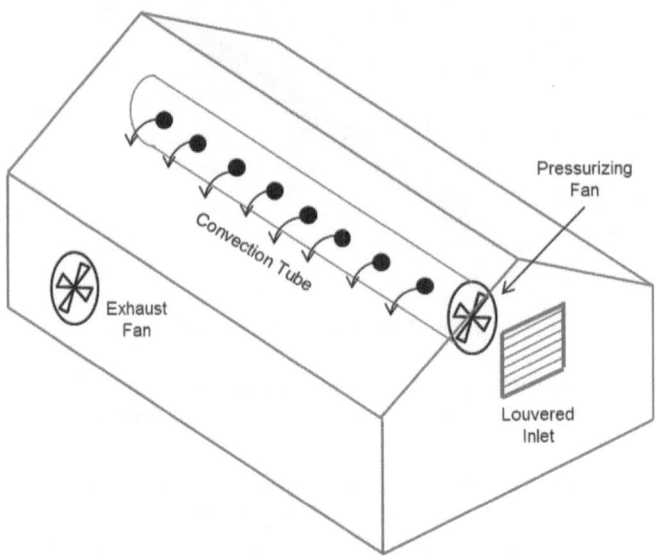

Figure 5.97 Convection-Tube Winter Cooling System.

well above plant height. The cooled mixture air, being heavier, gently falls to the floor, cooling the plant area.

The pressurizing fan must be capable of moving at least the same volume of air as the exhaust fan. If it moves less, excess incoming cold air will drop to the ground at the point of entry and cause a cold spot. When cooling is not required, the inlet louver closes, and the pressurizing fan continue recirculating air within the greenhouse.

Various published specifications call for as little as 1.5 CFM and as much of 4 CFM of air to be exhausted for each square foot of floor area. The high-capacity system costs more to set up but can be operated earlier in the fall and later in the spring to extend the winter cooling season. This can be an advantage, since frosts usually occur during these extension periods. A high-capacity convection-tube system eliminates the necessity of switching back and forth between the summer fan-and-pad and the winter fan-tube systems at these times (Nelson, 1998).

When convection-tube ventilation is used, standard conditions specify a maximum inside temperature of 8°C above the outside temperature. The temperature inside the greenhouse can become adversely high on a winter day when the sun is shining, even though the outside temperature is below the desired level. The convection-tube cooling system is designed to reduce the internal temperature to within 8°C of the outside temperature. If a lower inside temperature is required, cold air must be introduced into the greenhouse at a higher rate.

Convection tubes are conventionally oriented from end to end in the greenhouse. Each convection tube can be used to cool up to 30 ft (9.1 m) of greenhouse width, although it is desirable to use two tubes for greenhouses 30 ft wide. One tube placed down center of the greenhouse will cool greenhouses up to 30 ft in width. Greenhouses 30 to 60 ft (9.1 to 18.3 m) wide are cooled by two tubes placed equidistantly across the greenhouse. Holes along the tube exist in pairs on the pairs on the opposite vertical sides. The holes vary in size according to the volume of greenhouse to be cooled (Nelson, 1998).

Horizontal airflow (HAF) cooling

A more recent system for establishing uniform temperature in greenhouses is the horizontal airflow (HAF) system developed at the University of Connecticut. This system uses small horizontal fans for moving the air mass instead of convection tubes.

   Minimum and maximum airflow velocities for this system are 50 and 100 ft/min (0.25 and 0.5 m/s). Below this level, airflow is erratic, and uniform mixing air cannot be assured. A velocity of 50 ft/min causes slight leaf movement on plants with long leaves, such as tomato. This system should move air at 2 to 3 CFM/ft^2 of floor space. Fans of 1/30 to 1/15 HP (31 to 62 W) and a blade diameter of 16 in (41 cm) are sufficient. Commercial, continuous-duty motors used be used. With approximately one fan per 50 ft (15.2 m) of greenhouse length, fans should be aimed directly down the length of the greenhouse and parallel to the ground. The first fan should be located no closer than 10 to 15 ft (3.1 to 4.6 m) from the end of the greenhouse; the last should be placed, 40 to 50 ft (12.2 to 15.2 m) from the end toward which it is blowing.

   Cold air entering through the louvers high in the gables of the greenhouse is picked up in the air-circulating pattern of the HAF fans and is distributed throughout the greenhouse. Specifications for the HAF system are shown in Figure 5.98 and are described below (Nelson, 1998).

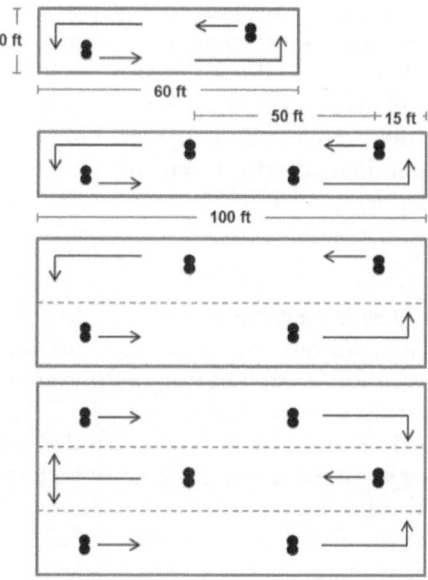

Figure 5.98 Fan Arrangements for HAF System.

- For individual or one span greenhouses, install two rows of fans along the length of the greenhouse, each row one-quarter of the width of the greenhouse in from the side wall. The row of fans one side of the greenhouse should blow air opposite to the direction of the row of fans on the other side of the greenhouse to form a circulating pattern. Fans should be 2 to 3 ft (0.6 to 0.9 m) above the plants.
- For multi-span greenhouses, install a row of fans down the center of each greenhouse. If the block contains an even number of greenhouses, move air down to one greenhouse and back in the adjacent one. In this way, each pair of greenhouses has a circulating air pattern. Connecting gutters must be sufficiently high to permit air movement beneath them. If the block contains an odd number of greenhouses, move air in the same direction in the first and third greenhouses and back the opposite way in the second house.

Heating arrangements

Greenhouse heating is required in cold weather conditions, wherein the entrapped heat is not sufficient for plant growth during the nights. The heating system must provide heat to the greenhouse at the same rate at which it is lost by conduction, infiltration, and radiation. There are three popular types of heating systems for greenhouses.

- Unit heaters.
- Central heaters.
- Radiant heaters.

There is a fourth possible type of system, but it has gained almost no place in the greenhouse industry: solar heating system. Solar heating is still too expensive to be a viable option. Several alternatives to conventional heating methods are heat pumps, biomass systems, and co-generation systems. A comparison of these to conventional systems indicates that bio-mass and co-generation systems offer substantial benefits throughout the state, while the feasibility of heat pumps depends on electricity rates and, hence, on the geographic location of the greenhouse. Only the heat pump alternative will be treated in the present. Heat or warm air is distributed from the unit heaters by one of two common methods: the convection-tube method or HAF, similar to winter cooling systems (Nelson, 1998).

Unit heater system

The most common and least expensive is the unit heater system. In this system, warm air is blown from unit heaters that have self-contained fireboxes. Heaters are located throughout the greenhouse, each heating a floor area 2000 to 6000 ft^2 (186 to 558 m^2). These heaters consist of three functional parts, namely, firebox, metal tube heat exchanger, and heat distribution fan. Fuel is combusted in a firebox to provide heat. The heat is initially contained in the exhaust, which rises through the inside of a set of thin walled metal tubes on its way to the exhaust stack. The warm exhaust transfers heat to the cooler metal walls of the tubes. Much of the heat is removed from the exhaust by the time it reaches the stack through which it leaves the greenhouse. A fan in the back of the unit heater draws in greenhouse air, passing it over the exterior side of the tubes and then out the front of the heater to the greenhouse environment

Figure 5.99 Unit Heater used with Convection-Tube Air Distribution System (Courtesy of American Plants Products & Services, Inc.).

again. The cool air passing over hot metal tubes is warmed and the air is circulated (Radha & Igathinathane, 2007).

Unit heaters are the most commonly used form of heating due to the following reasons (NGMA, 1998):

- They provide the air circulation needed.
- They can be combined with ventilation systems and waste heat applications.
- They can provide uniform bench top/under temperatures.
- They are comparably the least expensive.
- They provide quick response to temperature changes.
- They are easy to install.
- They offer inexpensive expansion for additions.
- They provide snow load protection which facilitates solar gain and plant growth.

Central heat system

A second type of system is central heating system, which consists of a central boiler that produces steam or hot water, plus a radiating mechanism in the greenhouse to dissipate the heat. A central heating system can be more efficient than unit heaters, especially in large greenhouse ranges. In this system, two or more large boilers are in a single location. Heat is transported in the form of hot water or steam through pipe mains to the growing area, and several arrangements of heating pipes in greenhouse are possible as illustrated in Figure 5.100 (Radha & Igathinathane, 2007).

The heat is exchanged from the hot water in a pipe coil on the perimeter walls plus an overhead pipe coil located across the greenhouse or an in-bed pipe coil located in the plant zone. Some greenhouses have a third pipe coil embedded in a concrete floor. A set of unit heaters can be used in the place of the overhead pipe coil, obtaining heat from hot water or steam from the central boiler (Radha & Igathinathane, 2007).

Radiant heat system

In this system, gas is burned within pipes suspended overhead in the greenhouse. The warm pipes radiate heat to the plants. Low intensity infrared radiant heaters can save

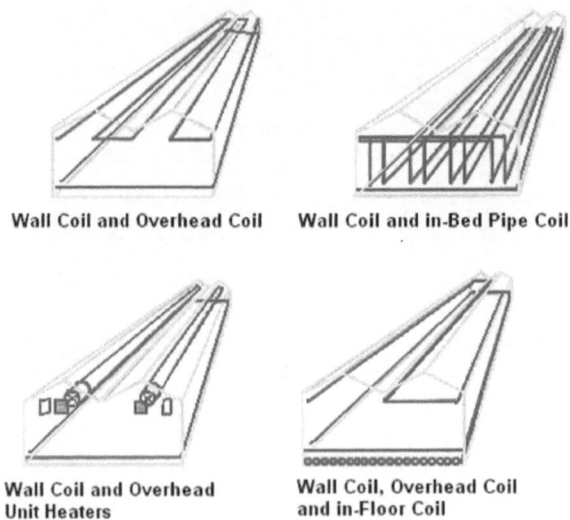

Wall Coil and Overhead Coil   Wall Coil and in-Bed Pipe Coil

Wall Coil and Overhead   Wall Coil, Overhead Coil
Unit Heaters   and in-Floor Coil

Figure 5.100 Arrangement of Heating Pipe Coils.

30% or more of fuel compared to conventional heaters. Several of these heaters are installed in tandem in the greenhouse. Lower air temperatures are possible since only the plants and root substrate are heated directly by this mode of heating (Radha & Igathinathane, 2007).

Conversion from fuel to heat energy occurs in the same way as in a boiler or unit heater. A small burner heats the air in a combustion chamber. The hot air is then distributed in a round steel tube which generally runs down the length of the greenhouse, near the peak. The heat energy is then transferred directly to the plants and the growing surface through electro-magnetic waves traveling at the speed of light. The plants and growing surfaces then absorb this energy and convert it into heat, thus warming the plants and soil. Infrared systems are easy to install in many applications and can provide an environment with warm dry leaves (NGMA, 1998).

Solar heating system

Solar heating systems are found in hobby greenhouses and small commercial firms. Both water and rock energy storage systems are used in combination with solar energy. The high cost of solar heating systems discourages any significant use by the greenhouse industries.

Solar heating is often used as a partial or total alternative to fossil fuel heating systems. Few solar heating systems exist in greenhouses today. The general components of solar heating system, shown in Figure 5.101 are collector, heat storage facility, exchanger to transfer the solar derived heat to the greenhouse air, backup heater to take over when solar heating does not suffice, and set of controls.

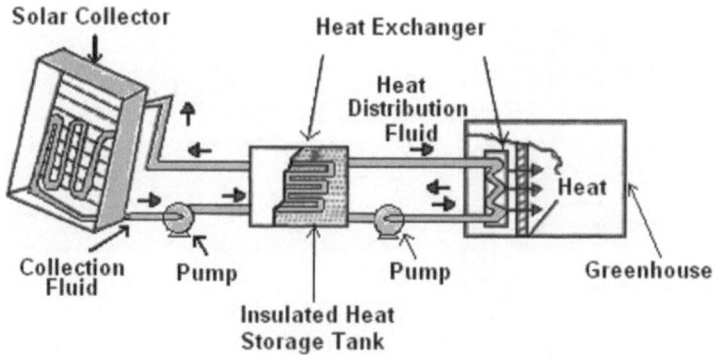

Figure 5.101 A Typical Solar Heating System.

Various solar heat collectors are in existence, but the type that has received greatest attention is the flat-plate collector. This consists of a flat black plate (rigid plastic, film plastic, sheet metal, or board) for absorbing solar energy. The plate is covered on the sun side by two or more transparent glass or plastic layers and on the backside by insulation. The enclosing layers serve to hold the collected heat within the collector. Water or air is passed through or over the black plate to absorb the entrapped heat and carry it to the storage facility. A greenhouse itself can be considered as a solar collector. Some of its collected heat is stored in the soil, plants, greenhouse frame, floor, and so on. The remaining excessive heat not required for plant growth is therefore vented to the outside. The excess vented heat could just as well be directed to a rock bed for storage and subsequent use during a period of heating. Collection of heat by flat-plate collectors is most efficient when the collector is positioned perpendicular to the sun rays at solar noon. Based on the locations, the heat derived can provide 20 to 50% of the heat requirement (Radha & Igathinathane, 2007).

Heat pump

Heat pumps use electricity to transfer heat from the outside environment to the inside of the greenhouse. An organic fluid or refrigerant such as dichlorodifluoromethane (R-12) or 1,1,1,2-tetrafluoroethane (R-134a) is passed through a heat exchanger, where it absorbs energy from the outside environment and vaporizes. This vapor is compressed and then passed through a second heat exchanger (condenser) inside the greenhouse. The vapor condenses and releases heat. The high-pressure liquid returns to a reservoir and then through an expansion valve back to the evaporator. This operating principle is illustrated in Figure 5.103.

Electricity drives the compressor, fans, and pumps in the heat pump system. The thermal energy source can be the outside air, groundwater, surface water, the soil, or direct solar energy. Heat from the pump can be produced in the form of heated air or hot water. Because most heat pumps use electricity, their widespread adoption could induce a load shift for electric utilities, increasing electricity demand at night and in the winter, and reducing natural gas consumption.

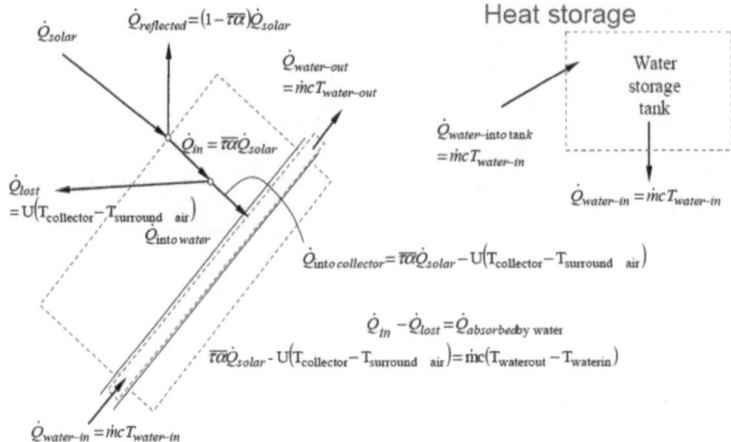

Figure 5.102 Basic Theory: Heat Collection.

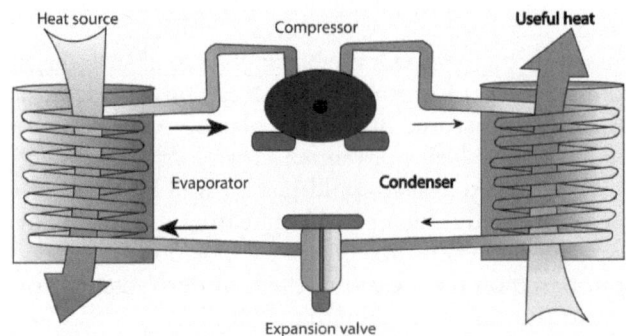

Figure 5.103 Heat Pump Schematic.

Heat pumps save energy when compared with conventional systems only if operated at a high coefficient of performance (Jenkins, 1985).

Irrigation systems

A well-designed irrigation system will supply the precise amount of water needed each day throughout the year. The quantity of water needed would depend on the growing area, the crop, weather conditions, the time of year and whether the heating or ventilation system is operating. Water needs are also dependent on the type of soil or soil mix and the size and type of the container or bed. Watering in the greenhouse most frequently accounts for loss in crop quality. Though the operation appears to be the simple, proper decision should be taken on how, when and what quantity to be given to the plants after continuous inspection and assessment. Since underwatering (less frequent) and overwatering (more frequent) will be injurious to the crops, the rules of watering should be strictly adhered to. The following are the three important rules of application of irrigation water (Radha & Igathinathane, 2007):

Rule 1: Use a well-drained substrate with good structure – If the root substrate is not well drained and aerated, proper watering cannot be achieved. Hence substrates with ample moisture retention along with good aeration are indispensable for proper growth of the plants. The desired combination of coarse texture and highly stable structure can be obtained from the formulated substrates and not from field soil alone.

Rule 2: Water thoroughly each time – Partial watering of the substrates should be avoided; the supplied water should flow from the bottom in case of containers, and the root zone is wetted thoroughly in case of beds. As a rule, 10 to 15% excess of water is supplied. In general, the water requirement for soil based substrates is at a rate of 20 liter/m^2 of bench, 0.3 to 0.35 liter per 0.165 m diameter pot.

Rule 3: Water just before initial moisture stress occurs – Since overwatering reduces the aeration and root development, water should be applied just before the plant enters the early symptoms of water stress. The foliar symptoms, such as texture, color, and turgidity can be used to determine the moisture stresses, but they vary with crops. For crops that do not show any symptoms, color. Feel and weight of the substrates are used for assessment.

A more precise method to determine the time of irrigation is to use a sensing device, such as tensiometer, an evaporation simulator, a soil conductivity meter for indicating the soil moisture levels.

Some systems are designed on fixed interval between irrigation. The predetermined frequency being such that maximum water requirement of the plants are met. The length of interval can be increased during periods of low transpiration rates and decreased during periods of high evaporation and thus high transpiration. Porous soils which have high infiltration rate and low water holding capacity require small and frequent applications if losses from deep percolation are to be avoided. Soils of much better water holding capacities can be given larger amount of water less frequently. In general it can be said that shallow root crops (vegetables) are likely to be sensitive, specially trees will be less affected.

It has been suggested that the minimum time of application based on providing water for the plant when it can use it and allowing for an adequate factor of safety, would be 6–10 hours per day for porous soil and 10–18 hours per day for soils of good water holding capacity. For peak periods, the duration could be increased to 20–22 hours (Tiwari, 2003).

Several irrigation water application systems, such as hand watering, channel irrigation, perimeter watering, sprinkler irrigation, boom watering, and trickle/drip irrigation which are currently in use will be discussed below.

Hand watering

Hand watering is still employed where the number of units is small, the labor cheap or there is a large diversity in unit size and species. Water is spread manually as and when need arises. One end of flexible hose pipe is connected to a water supply and the other is used for irrigating the plants. This system of irrigation has some disadvantages (Tiwari, 2003):

- It is labor intensive, thus not economically viable in developed countries where labor is expensive.
- Great risk of applying too little water or of waiting too long between waterings.
- Hand watering requires considerable time and is very boring.
- It is usually performed by inexperienced employees, who may be tempted to speed up the job or put it off to another time. Good watering requires considerable training and reliable labor.
- Overhead flooding, and free of moving water, breaks down soils structure leading to crust formation and reduce infiltration and oxygen supply.
- The end of hose is one of the best ways to spread disease.

Automatic watering is rapid and easy and is performed by the grower himself. Where hand watering is practiced, a water breaker should be used on the end of the hose. Such a device breaks the force of the water, permitting a higher flow rate without washing the root substrate out of the bench or pot. It also lessens the risk of disrupting the structure of the substrate surface (Radha & Igathinathane, 2007).

Channel irrigation

This is the system which has been used since ancient times and still in operation in many countries. Water is fed from main channel into a series of details which are created when plants are grown in ridges. The flow of water from the main channel into gullies can be controlled by using polyethylene siphon tubes. A simple S-shaped siphon is hooked over the end of the conduit at any place along its length, making it simple to disperse the water where it is needed (Tiwari, 2003).

Perimeter watering

Perimeter watering system can be used for crop production in benches or beds. A typical system consists of a plastic pipe around the perimeter of a bench with nozzles that spray water over the substrate surface below the foliage.

Either polyethylene or PVC pipe can be used. While PVC pipe has the advantage of being very stationary, polyethylene pipe tends to roll if it is not anchored firmly to the side of the bench. This causes nozzles to rise or fall from proper orientation to the substrate surface. Nozzles are made of nylon or a hard plastic and are available to put out a spray arc of 180°, 90° or 45°. Regardless of the types of nozzles used, they are staggered across the benches so that each nozzle projects out between two other nozzles on the opposite side. Perimeter watering systems with 180° nozzles require one water valve for benches up to 30.5 m in length. For benches over 30.5 m and up to 61.0 m, a water main should be installed on either side, one to serve each half of the bench. This system applies 1.25 liter/min/m of pipe. Where 180° and 90° or 45° nozzles are alternated, the length of a bench serviced by one water valve should not exceed 23 m (Radha & Igathinathane, 2007).

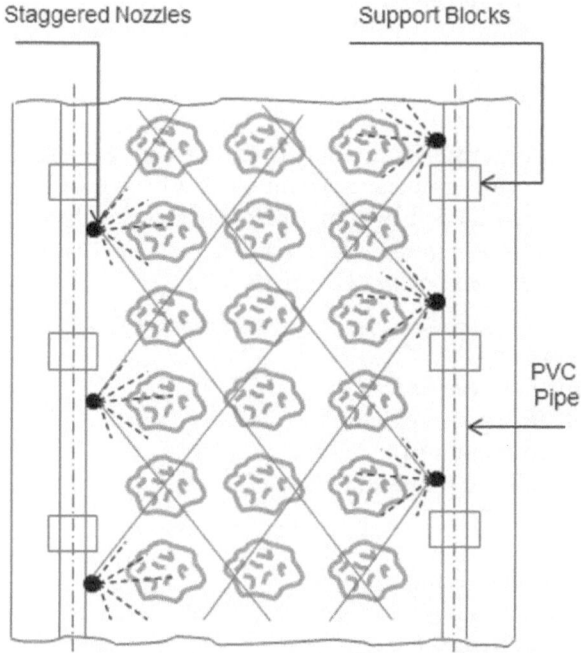

Staggered Nozzles   Support Blocks

PVC
Pipe

Figure 5.104 Schematic Diagram Perimeter Watering System.

Figure 5.105 Sprinkler Irrigation System.

Sprinkler irrigation

Sprinkler irrigation system consists of pumping water through a pipe and rotating heads so as to apply water to the soil in a manner similar to that received by natural rainfall. This is the most popular system in humid region. The power required to lift a given quantity of water depends on the length of time the pump is required. As power is defined as the rate of doing work and can be determined as the rate of doing work and can be determined by the given expression:

$$P_m = \frac{\dot{Q} \varrho g h}{\eta} \tag{5.10}$$

where $\dot{Q}$ is the discharge rate in m^3/s, ϱ is the density of water, h is the head in meters and η coefficient of efficiency. About 0.1 kWh or 0.035 liter of gas or 0.05 liter of petrol is required to deliver $1\,m^3$ of water at a height of 25 m. The service pressure is generally kept low (1.5–5 kg/cm^2). This system also has some advantages and drawbacks as mentioned below (Tiwari, 2003).

Advantages:

- Land leveling is not necessary.
- Drainage problem are decreased.
- Erosion is kept to minimum.
- Fewer special skills are required.

Disadvantages:

- The initial investment is high.
- Power consumption is high.
- More labor is required to move the pipe.
- Wind prevents a uniform distribution, making it often necessary to irrigate in nights.
- Evaporation losses of water are higher than with other methods of irrigation.

Boom watering

Boom watering can function either as open or a closed system, and is used often for the production of seedlings grown in plug trays. Plug trays are plastic trays that have width and length dimensions of approximately 0.30×0.61 m, a depth of 13 to 38 mm, and contain about 100 to 800 cells. Each seedling grows in its own individual cell. A boom watering system generally consists of a water pipe boom that extends from one side of a greenhouse bay to the other. The pipe is fitted with nozzles that can spray either water or fertilizer solution down onto the crop. The boom is attached at its center point to a carriage that rides along rails, often suspended above the center walk of the greenhouse bay. In this way, the boom can pass from one end of the bay to the other. The boom is propelled by an electrical motor. The quantity of water delivered per unit area of plants is adjusted by the speed at which the boom travels (Radha & Igathinathane, 2007).

Trickle or drip irrigation

The system of trickle or drip irrigation was originally developed in the Neger Desert in Israel because irrigation by sprinkling the sand with salt gave unsatisfactory results. It is now used in many countries and a whole range of different commercial systems is available. The technique is gaining popularity in those areas where water supplies are inadequate.

Figure 5.106 Boom Watering (Courtesy of Cherry Creek Systems, www.cherrycreeksystems.com).

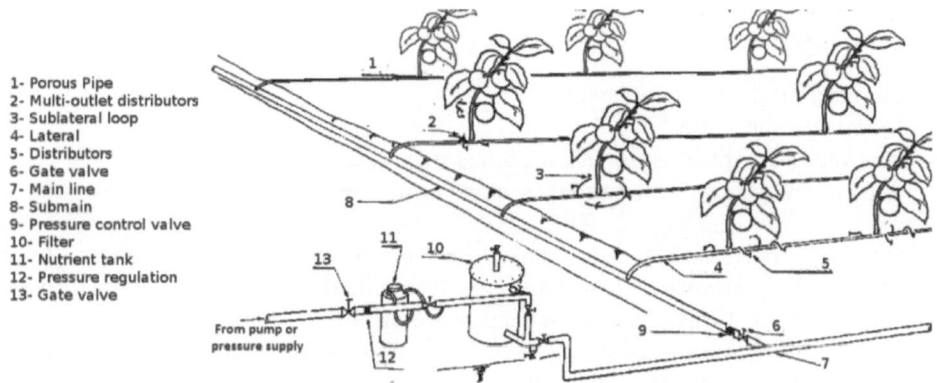

1- Porous Pipe
2- Multi-outlet distributors
3- Sublateral loop
4- Lateral
5- Distributors
6- Gate valve
7- Main line
8- Submain
9- Pressure control valve
10- Filter
11- Nutrient tank
12- Pressure regulation
13- Gate valve

Figure 5.107 Basic Components of a Trickle Irrigation System.

The distribution pressure of the water is reduced from 1.0 to 1.75 kg/cm^2 to a few meter head at the point where it is fed to plant at a very low rate as a trickle or series of drips. There are many systems by which this can be accomplished.

- By use of porous plastic tube.
- By a micro-tube from a main feeder.
- By perforated tube.
- By means of dripper/emitter device fixed to a feed pipe.

The quantity of water can be controlled so that optimum results are obtainable with any particular crop under different climatic conditions. It is essential to use filtered water to avoid blockage of the feeding system. The basic layout and components of a drip irrigation system are shown in Figure 5.107.

For crops grown at wide spacing such as orchards, tube fitted with drippers are used. Lay-flat hose systems are used for crops grown fairly closely together such as vegetables in open and flowers in greenhouse.

As each system has its merits and demerits, so drip irrigation method is also attributed with:

Advantages:

- The roots are never flooded and always have access to air and water.
- Increase in crop yields, sometimes as much as 100%, 20–40% more than with channel irrigation and 10-20% more than sprinkler irrigation method.
- Increase of crop quality and size.
- Water consumption reduced by 25–50%.
- Labor cost reduced by 25%.
- Absence of water on leaves means fewer diseases.
- Possibility of using water with high solid constant since salts are not deposited at root level.

Disadvantages:

- Higher initial cost investment.
- Relatively short life.
- Blockage perforations, holes and pores may occur.
- Attacks by rodents and animals seeking water in dry regions.

Since only a part of soil volume is needed, determination of the amount of application per tickle irrigation cycle is unique. It is determined as for sprinkling or flooding, much water may be lost by excessive deep percolation.

The maximum amount of water to be applied in each irrigation cycle depends on desired depth of wetting, the capacity of soil to hold available moisture, the level of moisture depletion allowed or desired, and the proportion of area or volume of soil wetted. The volume of water applied per irrigation cycle can be determined by multiplying the total surface area to be irrigated by the depth of application per irrigation.

The maximum root depth to be considered and typical plant spacing of some crops which are commonly grown under drip irrigation are given in Table 5.14 (Tiwari, 2003).

5.7   ADVANCED PROTECTED SYSTEMS – HYDROPONICS

Hydroponics is a technology for growing plants in nutrient solutions with or without the use of an artificial medium, such as sand, gravel, vermiculite, rock wool, peat moss and sawdust to provide mechanical support. Liquid hydroponic systems have no other supporting medium for the plant roots, whereas aggregate systems have a solid medium for support. Hydroponic systems are further categorized as open system when the nutrient solution is delivered to the plant roots and is not reused and closed system when the surplus solution is recovered, replenished, and recycled.

Table 5.14 Rooting Depth and Spacing for Use in Design of Drip Irrigation System.

Crop	Root Depth (m)	Plant Spacing (m)	Row Spacing (m)
Tomatoes	1.0–1.2	1	1–2
Vegetables	0.3–0.6	1	1–2
Citrus	1.0–1.2	3–6	5–7
Fruit trees	1.0–2.0	2–8	4–8
Grapes	1.0–3.0	2–3	2–4
Bananas	0.4–0.6	2–3	2–3

Hydroponic culture is possibly the most intensive method of crop production in today's agricultural industry. In combination with greenhouses or protective covers, it is highly technology oriented and capital intensive. It is also highly productive, conservative of water and land, and protective of the environment. Since regulating the aerial and root environment is a major concern in greenhouses, production takes place inside enclosures designed to control air and root temperatures, light, water, plant nutrition, and adverse climate. During the last 12 years, there has been increasing interest in hydroponics or soil-less techniques for producing greenhouse horticultural crops. The future growth of hydroponics depends greatly on the development of production systems that are cost competitive with open field agriculture.

While hydroponics and Controlled Environment Agriculture (CEA) are not synonymous, CEA usually accompanies hydroponics. The principal advantages of hydroponic CEA are:

- High density.
- Maximum crop yield.
- Crop production where no suitable soil exists.
- Virtual independence to ambient temperature and seasonality.
- More efficient use of water and fertilizers.
- Minimal use of land area.
- Suitability for mechanization.
- Effective disease control.

A major advantage of hydroponics, as compared to the open field agriculture is the isolation of the crop from the underlying soil, which often has problems of disease, salinity, poor structure and drainage. The costly and time-consuming tasks of soil sterilization and cultivation are not necessary in hydroponic systems and a rapid production of crops is readily achieved. Because of the precise control over the environment and balanced supply of plant nutrients, the maximum potential yield is assured in hydroponic culture. Studies have shown that the yield of tomatoes in hydroponic CEA, shown in Figure 5.108, is 375 million ton/ha/year when compared to 100 million ton/ha/year in open field.

Figure 5.108 Tomato Production using Hydroponics.

The principal disadvantages of hydroponics relative to conventional open field agriculture are the high costs of capital and energy inputs, and the high degree of management skills required for successful production. Capital costs may be especially excessive if the structures are artificially heated and evaporatively cooled by fan-and-pad systems, and have systems of environmental control that are not always needed in the tropics. Workers must be highly competent in plant science and engineering skills. Studies of prices have shown that only high quality, garden type vegetables, such as tomatoes, cucumbers, and specially lettuce can provide break even or better revenues in hydroponic systems. Besides these vegetables, eggplant, peppers, melons, strawberries and herbs are grown commercially under hydroponic systems in Europe and Japan (Radha & Igathinathane, 2007).

The limited volume of medium and water availability generally causes rapid changes in the status of water and nutrients. Changes in the medium solution such as pH, electrical conductivity (EC) and nutrients level, should be monitored for the efficient use of water and nutrients. Failures in the careful supervision of fertilization and/or the accuracy of irrigation are likely to result in severe plant damage and reduced yields. Hydroponics, however, offers several major advantages in the management of both plant nutrition and plant protection, if the right tools are applied and careful management is carried out (Ronen, 2013).

In agricultural activities, the pH measurement controls the availability of nutrients to the plant in terms of the root medium acidity or alkalinity essential for plants growth. The average level of availability for all essential plant nutrients exists in the pH range of 6.2–6.6 (Tiwari, 2003). One of the advantages of soilless culture is the ability to control pH in the medium solution. This is achieved by adding acid to the irrigation water to change the ratio between NH_4^+ and NO_3^-,

Table 5.15 Crop Salinity Sensitivity.

Crop	EC Threshold (dS/m)
Tomato	2.5
Cucumber	2.5
Lettuce	1.3
Pepper	1.5

which are the only two forms of nitrogen allowed in this cultivation method. It is a common phenomenon that while passing through the root system, the pH will drop slightly due to root respiration and lack of buffer capacity in the soil-less medium.

On the other hand, the EC measurement is commonly used in hydroponics, aquaculture and freshwater systems to monitor the amount of nutrients, salts or impurities in the water. One of the functions of water supplied to soil-less culture, other than fulfilling the need of transporting nutrients in plants, is to maintain a low level of salts in the medium and to prevent a possible build-up of salts. Commonsense tells us material cannot disappear from the system. Although water evaporates from the system, salts do not; hence they will always remain behind. Every anion and cation not consumed by the plant will accumulate in the medium. If the salt is not outside in the drainage it is inside the medium. Therefore, there is a need to balance the EC specifically for each crop according to a threshold level, above which plant productivity will decrease; this threshold levels are shown in Table 5.15 (Ronen, 2013).

Finally, plants require several materials for adequate nutrition. The principle ones, CO_2 and water, usually are available in adequate amounts from the atmosphere and soil, and the supply is continuously replenished by natural phenomena. All the other essential nutrients are normally available from the soil but they are not replenished by nature after the plants take them into its foliage and fruits. For this case or in the soil-less culture case, application of fertilizer becomes necessary.

Thirteen elements have been identified as essential to plant nutrition: nitrogen, phosphorous, potassium, calcium, magnesium, sulphur, iron, manganese, copper, zinc, boron, molybdenum and chlorine. Nitrogen, phosphorous and potassium are needed by plants in relatively large quantities and therefore are called *macro-nutrients*. Calcium, magnesium and sulphur are normally required in lesser but still in considerable amounts. That's why these are called as *secondary nutrients*. The remaining members of the group are known as *micro-nutrients* because the requirement for plant growth is very small. There other three elements considered as non-essential because they do not correct or prevent deficiencies, these are sodium, silicon and aluminum (Tiwari, 2003). Table 5.16 resume these nutrients as well as its form absorbed and its typical concentration amount.

All these nutrients are equally essentially regardless of the fact that they are required and absorbed in widely different amounts. Depending on the analysis of

Table 5.16 Plant Nutrient Uptake.

	Name	Symbol	Form Absorbed	Typical Concentration
A.	Macro-nutrients			
	Nitrogen	N	NH_4^+, NO_3^-	1–5%
	Phosphorus	P	$H_2PO_4^-, HPO_4^-$	0.1–0.4%
	Potassium	K	K^+	1–5%
	Sulphur	S	SO_4^-	0.1–0.4%
	Calcium	Ca	Ca^{++}	0.2–1%
	Magnesium	Mg	Mg^{++}	0.1–0.4%
B.	Micro-nutrients			
	Boron	B	H_3BO_4	6–20 ppm
	Iron	Fe	$Fe\_\_$	50–250 ppm
	Magnese	Mn	Mn^{++}	50–250 ppm
	Copper	Cu	Cu^+, Cu^{++}	5–20 ppm
	Zinc	Zn	Zu^{++}	25-150 ppm
	Molybdenum	Mo	MoO_4^-	<1 ppm
	Chlorine	Cl	Cl^-	0.2–s2%

Table 5.17 Initial Nutrient Solution.

Crops	Ions (mmol/L)								
	NO_3^-	NH_4^+	$H_2PO_4^-$	K^+	Ca^{++}	Mg^{++}	SO_4^-	Na^+	Cl^-
Tomato	12	0	1.5	6	5	2.5	2	<12	<12
Melon	11	0.5	1.5	6	4.5	2	2	<10	<10
Cucumber	14	0.5	1.6	6	4.5	2.2	2	<6	<6

irrigation water, crop species and climatic conditions the nutrient solution is prepared. Table 5.17 is a guide of initial nutrient solutions in for tomato, melon and cucumber:

The nutrition of the plant must be provided entirely through the nutrient solution, which brings the possibility of precise control of mineral nutrition according to the species, phenological time, climatic conditions, etc. to obtain higher profitability for the crop. Now, being inert substrates lacking buffer capacity, mistakes or failures in the control of mineral nutrition or pH adjustment may cause serious damage to the plantation.

5.7.1 Nutrient film technique

Nutrient film technique (NFT) is a form of hydroponics in which plants are grown in narrow, sloped channels. A thin film of recirculating nutrient solution flows through the roots in the channels; the walls of the channels are flexible, which permit the solution to flow around the base of each plant prohibiting light and preventing evaporation. Nutrient solution is pumped to the higher end of each channel and flows past the plant roots by gravity to catchment pipes and a sump as illustrated in Figure 5.109. The

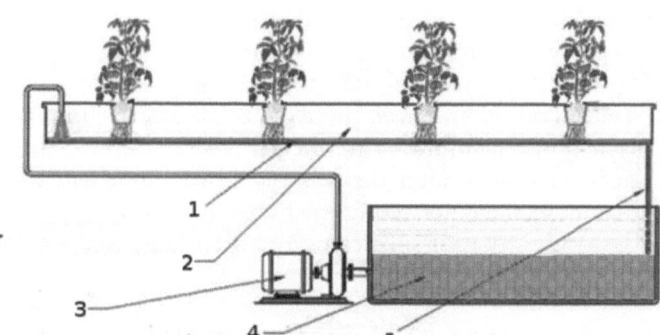

1- Nutrient Solution Film
2- Grow Channel
3- Water Pump
4- Nutrient Solution Reservoir
5- Nutrient Return Tube

Figure 5.109 NFT Hydroponic System.

solution is monitored for replenishment of nutrient salts and water before it is recycled. Capillary material in the channel prevents young plants from drying out, and the roots soon grow into a tangled mat.

A principal advantage of the NFT system in comparison with others is that it requires very less nutrient solution. It is therefore easier to heat the solution during winter months, to obtain optimum temperatures for root growth, and to cool it during hot summers in arid or tropical regions. If it is necessary to treat the nutrient solution for disease control, small volumes are easier to work with.

The channels should not be greater than 15 to 20 m in length. In a level greenhouse, as the recommended slope of the channel is 1 in 50 to 1 in 75, long channels can restrict the height available for plant growth. If the length of greenhouse is more, then with the given slope the elevation difference between the ends of the channels will be so high that the plants at higher elevation will have less head space to grow. If the channel slope is less, it may result in poor aeration of solution. To assure good aeration, the nutrient solution could be introduced into channels at two or three points along the length. The flow of nutrient solution into each channel should be 2 to 3 liter/min, depending on the oxygen content of the solution. The maximum temperature of the nutrient solution should be 30°C. Temperature above 30°C will adversely affect the amount of dissolved oxygen in the solution. The O_2 concentration should be approximately 5 ppm or more, especially in the nutrient solution flowing over the root mat in the channel. Normally channels are made of black plastic coated with white color. NFT system permits economical cooling of plant roots, avoiding the expensive cooling of the entire greenhouse aerial temperature.

The following are the specific advantages of the NFT systems (Radha & Igathinathane, 2007):

• The NFT system eliminates the material and labor costs for steam or methyl bromide pasteurization between crops, as well as the period of 10 to 14 days required for methyl bromide application and aeration.
• NFT has the potential for conserving water and nutrients.

- Recirculation of solution provides an excellent method for reducing nutrient and pesticide effluent from greenhouses.
- NFT systems have the potential for automation.
- Formulation, testing, and adjustment of nutrient solutions can be handled at a central point, and even these operations can be automated.
- The nutrient solutions are mechanically delivered to the crop.
- It is possible to alter the heat level of the nutrient solution by heating or cooling to suit to the plant requirement.
- Use of heavy root substrates and their handling is eliminated.

5.8 GREENHOUSE MODELING AND CONTROL

Manual maintenance of uniform environmental conditions inside the greenhouse is very difficult and cumbersome. A poor maintenance results in less crop production, low quality and low income. For effective control, automatic control systems like microprocessor and computer are used presently to maintain the environment. Automatic control systems sense and measure the environmental parameters, compare it to a standard and, if needed, activate proper device which alters the parameter, to bring the measured parameter to the required level into agreements with the standard (Radha & Igathinathane, 2007).

5.8.1 Greenhouse models

The greenhouse environment is a complex dynamical system. Over the past decades, people have gained a considerable understanding of greenhouse climate dynamics, and many methods describing the dynamic process of greenhouse climate have been proposed. Traditionally, there are two different approaches to describe it (Hu et al., 2010):

- Based on energy and mass flows equations or physical laws describing the process.
- Based on the analysis of input-output data from the process by using a system identification approach.

The main objective about modeling a system relies on the need of simulation for the development of a control system; in other words, to get a first approach of the control system before validating in a physical greenhouse.

Physical based model

The physical based model deals with the inside air temperature and humidity of a greenhouse, and its physical model describes the flow and mass transfers generated by the differences in energy and mass content between the inside and outside air, or by the control or exogenous energy and mass inputs. Most of the analytic models on analysis and control of the environment inside greenhouses have been based on the following state space form:

$$x = f(t, x, u, v) \tag{5.11}$$

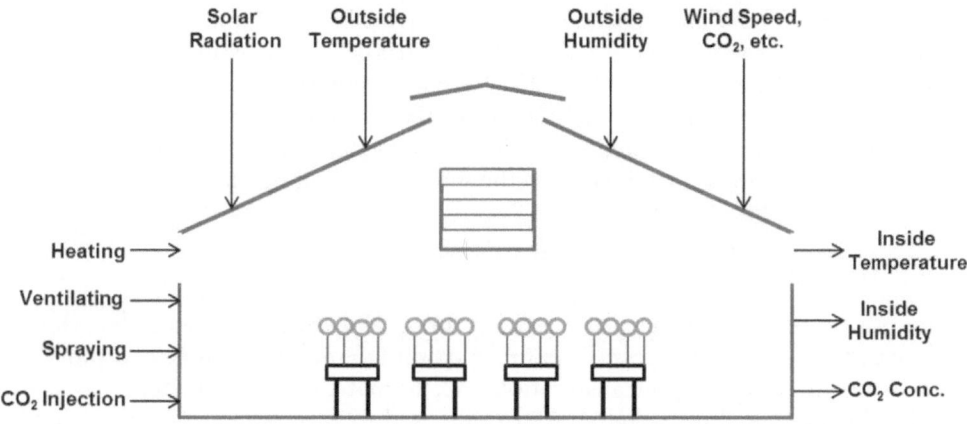

Figure 5.110 Greenhouse Climate Dynamic Model.

where x are states variables like indoor temperature, humidity and carbon dioxide concentration, u are control inputs like energy input by the heating system, fogging systems, ventilation system and CO_2 supply flux, v are external disturbances like solar radiation, outdoor temperature, humidity and wind speed, t denotes time, and $f(\cdot)$ is a nonlinear function.

Considering the related high costs, CO_2 supply systems have not an extensive use, therefore the related variables are not taken into account. To simplify the model, only some primary disturbance variables have been considered, such as solar radiation, outside temperature and humidity. According to the above analysis, the state equations have been formed based on the laws of conservation of enthalpy and matter, and the dynamic behavior of the states is described by using the following differential equations (Albright et al., 2001):

$$\frac{dT_{in}(t)}{dt} = \frac{1}{\varrho C_P V}[Q_{heater}(t) + S_i(t) - \delta Q_{fog}(t)] - \left(\frac{UA}{\varrho C_P V} + \frac{\dot{V}(t)}{V}\right)$$
$$\times [T_{in}(t) - T_{out}(t)] \tag{5.12}$$

$$\frac{dW_{in}(t)}{dt} = \frac{1}{\varrho V}Q_{fog}(t) = \frac{1}{\varrho V}E(S_i(t), W_{in}(t)) - \frac{\dot{V}(t)}{V}[W_{in}(t) - W_{out}(t)] \tag{5.13}$$

where, T_{in} is indoor air temperature (°C), T_{out} is outdoor temperature (°C), V is greenhouse volume (m³), UA is the heat transfer coefficient (W/K), ϱ is air density (1.2 kg/m³), C_p is specific heat of air (1006 J/kgK), Q_{heater} is the heat provided by the greenhouse heater (W), S_i is the intercepted solar radiant energy (W), Q_{fog} is the water capacity of the fog system (gH$_2$O/s), δ is the latent heat of vaporization (2257 J/g), $\dot{V}$ is the ventilation rate (m³/s), W_{in} and W_{out} are the interior and exterior humidity ratios (water vapor mass ratio, gH$_2$O/kg of dry air), respectively, and $E(S_i(t), W_{in}(t))$ is the evapo-transpiration rate of the plants (gH$_2$O/s). It should be noted that the air volume

(V) to be used in the balances is the effective mixing volume. Short circuiting and stagnant zones exist in ventilated spaces, and the effective mixing volume is typically significantly less than the calculated total volume. The effective mixing volume of a ventilated space may easily be as small as 60 to 70% of the geometric volume. This, of course, means indoor air temperature is unlikely to be uniform throughout the air space.

It is also worth noticing that, to a first approximation, the evapo-transpiration rate $E(S_i, W_{in})$ is in most part related to the intercepted solar radiant energy, through the following simplified relation:

$$E(S_i(t), W_{in}(t)) = \frac{\alpha}{\delta} S_i(t) - \beta_T W_{in}(t) \tag{5.14}$$

where, α is an overall coefficient to account for shading and leaf area index, dimensionless and β_T is the overall coefficient to account for thermodynamic constants and other factors affecting evapo-transpiration (i.e., stomata, air motion, etc.).

The climate model provided above can be used in all seasons, and two variables have to be controlled namely the indoor air temperature and the humidity ratio through the processes of heating ($Q_{heater}(t)$), ventilation ($V(t)$) and fogging ($Q_{fog}(t)$). Some works have neglected the action of Q_{heater} for summer seasons, simplifying even more the physical model.

In order to effectively express the state-space form, the inside temperature and absolute humidity are defined as the dynamic state variables, $x_1(t)$ and $x_2(t)$, respectively, the ventilation rate and the water capacity of the fog system as the control variables (neglecting Q_{heater}), $u_1(t)$ and $u_2(t)$, respectively, and the intercepted solar radiant energy, the outside temperature, and the outside absolute humidity as the disturbances, $v_i(t)$, $i = 1, 2, 3$. Hence, the dynamic model can alternatively be written:

$$\dot{x}_1 = \frac{1}{\varrho C_p V} v_1 - \frac{\delta}{\varrho C_p V} u_2 - \frac{UA}{\varrho C_p V} x_1 + \frac{UA}{\varrho C_p V} v_2 - \frac{1}{V} [x_1 - v_2] u_1 \tag{5.15}$$

$$\dot{x}_2 = \frac{1}{\varrho V} u_2 + \frac{\alpha}{\varrho \delta V} v_1 - \frac{\beta_T}{\varrho V} x_2 - \frac{1}{V} [x_2 - v_3] u_1 \tag{5.16}$$

Due to the complexity appearing as the cross-product terms between control and disturbance variables, these equations are obviously coupled nonlinear equations, which cannot be put into the rather familiar form of an affine analytic nonlinear system.

Input-output data based model

Simulation models to describe the dynamic behavior of the air temperature and hygrometry inside the greenhouses based on a restricted number of input-output variables have been published in several studies. Usually, these models ensue of the physical laws have the drawback that are difficult to tune in practice, since they use a larger number of unknown physical parameters (Bennis et al., 2008). Input-output data models are also known as black-box model.

This category is mainly based on soft computing and computational intelligence such as Fuzzy clustering and Neural Networks. Other methods such as least square methods are also applied for system identification. These techniques are applicable when the expert knowledge is not available and only input-output data of the system is available (Trabelsi et al., 2007).

Fuzzy clustering

This procedure is mainly carried out into four steps, when referring to off-line model identification:

- Construction of the regression data.
- Determination of clusters corresponding to a set of local linear sub-models.
- Determination of the antecedent membership function from the cluster parameters.
- Estimation rule's consequence parameter.

The available data samples are collected in a matrix Z formed by concatenating the regression matrix X and the output vector Y:

$$X = \begin{bmatrix} x(1) \\ x(k) \\ x(N-1) \end{bmatrix}, \quad Y = \begin{bmatrix} y(1) \\ y(k+1) \\ y(N) \end{bmatrix}$$

$$Z^T = [X \ Y] \tag{5.17}$$

where: N is the number of data samples.

There are various algorithms construct fuzzy clusters such as: the C-means algorithm, the Gath-Geva algorithm and the Gustafson-Kessel algorithm. Trough clustering, the data set Z is partitioned into N_c clusters. The result is a fuzzy partition matrix $U = [u_{ik}]_{N_c \times N}$, whose element $u_{ik} \in [0, 1]$ represents the degree of membership of the observation in cluster i, a prototype matrix $V = [v_1, \ldots, v_{N_c}]$ and a set of cluster covariance matrices $F = [F_1, \ldots, F_{N_c}]$ (F_i are definite positive matrices). Once the triplet (U, V, F) is determined, the parameters of the rules premises (c_i and σ_i) and the consequent parameters (A_i, B_i and α_i) are computed.

The rules are linear conclusions of the system inputs, for example, for the rule j:

$$\begin{aligned} T_i^j(k+1) = {} & a_{j1T} T_i(k) + a_{j2T} H_i(k) + b_{j1T} O_{uv}(k) + b_{j2T} Rid(k) \\ & + b_{j3T} C_h(k) + b_{j4T} B_{ru}(k) + b_{j5T} Ray(k) + b_{j6T} V_v(k) + b_{j7T} T_e(k) \\ & + b_{j8T} H_e(k) + \alpha_{j1} \end{aligned} \tag{5.18}$$

$$\begin{aligned} H_i^j(k+1) = {} & a_{j1H} T_i(k) + a_{j2H} H_i(k) + b_{j1H} O_{uv}(k) + b_{j2H} Rid(k) \\ & + b_{j3H} C_h(k) + b_{j4H} B_{ru}(k) + b_{j5H} Ray(k) + b_{j6H} V_v(k) + b_{j7H} T_e(k) \\ & + b_{j8H} H_e(k) + \alpha_{j2} \end{aligned} \tag{5.19}$$

With $a_{j1T}, a_{j2T}, b_{j1T}, \ldots, b_{j8T}, \alpha_{j1}$: consequence parameters for the temperature and for the rule j and $a_{j1H}, a_{j2H}, b_{j1H}, \ldots, a_{j8H}, \alpha_{j2}$: consequence parameters for the hygrometry and for the rule j.

Where, T_i and T_e are the internal and external temperature, respectively, H_i and H_e are the internal and external relative humidity, respectively, Ray is the solar radiation,

V_v is the wind speed, C_h is the heating system energy, Ouv is the roofing percentage, Bru is moistening and Rid is the shading percentage.

Neural networks

Artificial Neural Networks (ANN) are collections of mathematical models that reproduce some of the observed properties of biological nervous systems. Their key element is the structure of the information processing system. This system is composed of a large number of highly interconnected processing elements that are analogous to neurons and are coupled together with weighted connections that are analogous to synapses.

There are a broad number of ANNs topologies. Among the most widespread are feed-forward networks. Multi-layer perceptrons (MLP) network with a hyperbolic tangent (tanh) activation function, have proven to be universal approximators. This means that they can approximate any reasonable function f with a subjective accuracy given by:

$$f(u) = \left(\sum_{j=1}^{k} v_{jl} \tau \left(\sum_{i=1}^{n} w_{ij} u_i - \Theta_j \right) - \Theta_l \right), \quad l = 1, \ldots, m \qquad (5.20)$$

where, τ is the tanh function, k is the number of hidden units, v_{jl} and w_{ij} are weights, Θ_i are biases and u the data vector.

The non-linear function f is estimated based on data samples using the Lavenberg-Marquardt optimization technique. The Lavenberg-Marquardt is the standard method for minimization of mean square error criteria, due to its rapid convergence properties and robustness. Neural Networks also have several disadvantages. They require large numbers of data samples due to their large number of degrees of freedom. Problems such as over-fitting and sub-optimal minima may occur more severely than in linear case. Also, this method requires a large computation time for training, i.e. for learning the system behavior, which restricts its application to real time implementations (Boaventura, 2003).

5.9 TYPES OF CONTROLS

A greenhouse environment is an incredibly complex and dynamic environment. The pressures of labor availability and costs, energy costs, and market demands increasingly make efficiency and automation key components for success and profitability. Environment control technology affects all of these critical areas, and many others, so understanding controls and implementing their use is more important than ever. Precise control of the greenhouse environment is critical in achieving the best and most efficient growing environment and efficiency (NGMA, 1998a).

Precise control of different parameters of greenhouse environment is necessary to optimize energy inputs and, thereby, maximize the economic returns. Basically, the objective of environmental control is to maximize the plant growth. The control of greenhouse environment means the control of temperature, light, air composition and nature of the root medium. A greenhouse is essentially meant to permit at least partial control of micro-climate the greenhouse encloses. Obviously, greenhouses with partial environmental control are more common and economical than full fledged systems.

From the origin of greenhouses to the present, there has been a steady evolution of environmental control systems. Five stages in this evolution include:

- Manual controls.
- Thermostats and timers (50–600 USD).
- Step controllers (800–1800 USD).
- Dedicated microprocessors (800–6000 USD).
- Computers (5000–7000 USD).

This chain of evolution has brought about a reduction in control labor and an improvement in the conformity of greenhouse environments to their set points. The benefits achieved from greenhouse environmental uniformity are better timing of crops, higher quality of crops, disease control and conservation of energy (Radha & Igathinathane, 2007).

Manual controls

During the first half of the 20th century, it was common for greenhouse firms to employ a night watch person to regulate temperature. This person made periodic trips through the greenhouses during the night, checking the temperature in each greenhouse and controlling it by opening or closing valves of heating pipes as required. During the day, employees opened or closed ventilators by hand to maintain temperature. Hence the temperatures had to be manually controlled throughout the day during the cropping season. Obviously, there were large deviations on both sides from the desired temperatures, and the success of manual control was mainly based on the skill and experience of the operator (Nelson, 1998).

Thermostats and timers

Thermostat is an automatic device which senses the temperature and activates/ deactivates the attached equipment (fans, heaters, power vents, etc.), with reference to a set temperature. The thermostat may make use of bimetallic strip or thin metal tube filled with liquid or gas as sensor and it produced some physical displacement according to the sensed temperature. These sensors activate a mechanical switch by differential expansion of bimetallic strip or by the movement of the tube due to change in the volume of gas or liquid. Though efficient, the thermostats are not highly accurate and need frequent calibration. For more accurate measurement and control of temperatures, microprocessor and computer based systems using thermocouples or thermistors as sensor are used. These sensors require an electronic circuit to carry the signal to a conventional switch or relay (Radha & Igathinathane, 2007).

These simple devices are low cost and provide limited control. A typical greenhouse zone may require 3 or more individual thermostats to control heating and cooling functions, plus timers for irrigation and lighting control. Additional relays are often necessary to interconnect fans and louvers and other devices that must work together.

A simple zone will generally require 2 or 3 thermostats (1 for heat and 1 or 2 for cooling stages). More complex zones may require 5 or 6 thermostats, including

Figure 5.111 Thermostat (Courtesy of Anjou Automation, www.anjouautomation.com).

multiple stage thermostats for some devices such as vents, plus individual controls or timers for irrigation and lighting.

Beyond the low initial cost there is little if any benefit. They provide very limited control, no coordination between equipment and functions, poor accuracy, and poor energy efficiency. The initial low price is deceiving. The increased energy consumption and the effects of lost production due to poor control devices far exceeds savings from their lower initial cost (NGMA, 1998a).

Step controllers

The primary benefit of these devices is their low initial cost, better equipment coordination and greater accuracy than either single or multiple stage thermostats. These units are not expandable and serve only one zone. They are generally most appropriate for simple greenhouse zones limited to 6–8 total stages of heating and cooling, and in smaller operations not anticipating expansion.

Step controllers bring two benefits to basic temperature control: automatic sequence of operation, and remote sensing. Consequently, a single step controller takes the place of several thermostats. Step controls use a single sensor element to control both heating and cooling functions in a greenhouse zone. That sensor can be located among the plants while the controller can be located more conveniently and safely outside the plant environment. These controllers divide the actions of the greenhouse heating and cooling equipment into steps, or stages, called a sequence of operation. While multiple thermostats with different settings can accomplish the same effect, it is difficult to keep their temperature readings synchronized. As a result, heating and cooling equipment can be on simultaneously (and expensively). A single step can include one or more heat sources or one or more cooling sources. For example, in a zone with three exhaust fans, two unit heaters, three motorized shutters, and an evaporative cooling pad; the equipment might be divided into steps as follows:

Heating Step 1 – First unit heater ON.
Heating Step 2 – Second heater ON (first heater still ON).
Cooling Step 1 – 1 exhaust fan ON, 3 shutters OPEN.
Cooling Step 2 – 2 exhaust fans ON.
Cooling Step 3 – 3 exhaust fans ON, cooling pad pump ON.

As the measured temperature falls below the desired temperature, controller activates Heat Step 1. If the temperature continues to fall, it turns on Heat Step 2. As the zone returns to the desired temperature, the controller first turns off Heat Step 2, then Heat Step 1.

As long as the greenhouse remains near the desired temperature, the controller leaves all the equipment off. When the measured temperature rises above the desired temperature, the controller turns on Cooling Step 1, and if the temperature continues to rise, it activates Cooling Step 2, then Step 3. As zone temperature drops, the controller turns off each cooling step in last-on, first-off order.

Greenhouse step controllers incorporate greenhouse-specific features such as separate day and night temperature settings, gradual equipment start-up after power failures, outdoor temperature influence, provision for humidity control, partial or complete lockout of cooling functions at night, adjustable time delay between steps, and display of the temperature at the remote sensing point (NGMA, 1998a).

Dedicated microprocessors

Dedicated microprocessors can be considered as simple computers. A typical microprocessor will have a keypad and a two or three line liquid crystal display of, sometimes, 80-character length for programming. They have more output connections and can control up to 20 devices. With this number of devices, it is cheaper to use a microprocessor. They can receive signals of several types, such as, temperature, light intensity, rain and wind speed. They permit integration of a diverse range of devices, which is not possible with thermostats.

The accuracy of a microprocessor for temperature control is quite good. Unlike a thermostat, which is limited to a bimetallic strip or metallic tube for temperature sensing and its mechanical displacement for activation, the microprocessor often uses a thermistor. The bimetallic strip sensor has less reproducibility and a greater range between the ON and OFF steps. Microprocessors can be made to operate various devices, for instance, a microprocessor can operate the ventilators based on the information from the sensor for the wind direction and speed. Similarly a rain sensor can also activate the ventilators to prevent the moisture sensitive crop from getting wet. A microprocessor can be set to activate the CO_2 generator when the light intensity exceeds a given set point, a minimum level for photosynthesis (Radha & Igathinathane, 2007).

Computers

Computer systems can provide fully integrated control of temperature, humidity, irrigation and fertilization, CO_2, light and shade levels for virtually any size growing

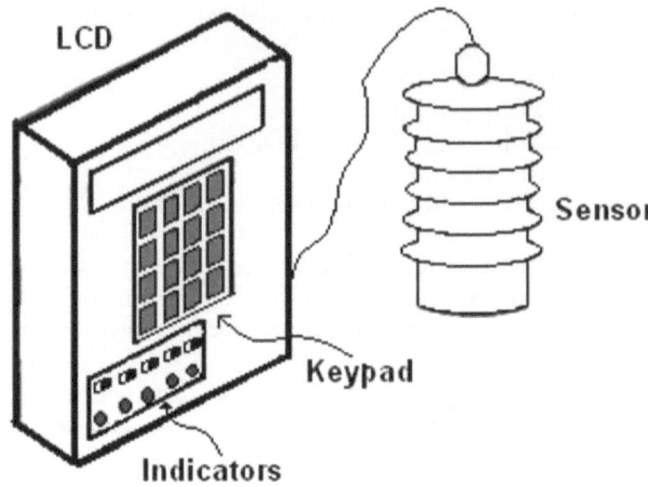

Figure 5.112 Dedicated Microprocessor for Controlling Greenhouse Environment.

facility. Precise control over a growing operation enables growers to realize savings of 15 to 50% in energy, water, chemical and pesticide applications. Computer controls normally help to achieve greater plant consistency, on-schedule production, higher overall plant quality and environmental purity.

A computer can control hundreds of devices (vents, heaters, fans, hot water mixing valves, irrigation valves, curtains and lights) within a greenhouse by utilizing dozens of input parameters, such as outside and inside temperatures, humidity, outside wind direction and velocity, CO_2 levels and even the time of day or night.

Computer systems receive signals from all sensors, evaluate all conditions and send appropriate commands at desired time intervals to each piece of equipment in the greenhouse range thus maintaining ideal conditions in each of the various independent greenhouse zones defined by the grower. Computers collect and record data provided by greenhouse production managers. Such a data acquisition system will enable the grower to gain a comprehensive knowledge of all factors affecting the quality and timeliness of the product.

A computer produces graphs of past and current environmental conditions both inside and outside the greenhouse complex. Using a data printout option, growers can produce reports and summaries of environmental conditions such as temperature, humidity and the CO_2 status for a given day, or over a longer period of time for current or later use. As more environmental factors in the greenhouse are controlled, there comes a stage when individual controls cannot be coordinated to prevent system overlap. An example is the greenhouse thermostat calling for heating while the exhaust fans are still running. With proper software program, which uses the environmental parameters as input from different sensors, can effectively coordinate all the equipment without overlap and precisely control all parameters affecting plant development as desired. Despite the attraction of the computer systems, it should be remembered that the success of any production system is totally dependent on the grower's knowledge of

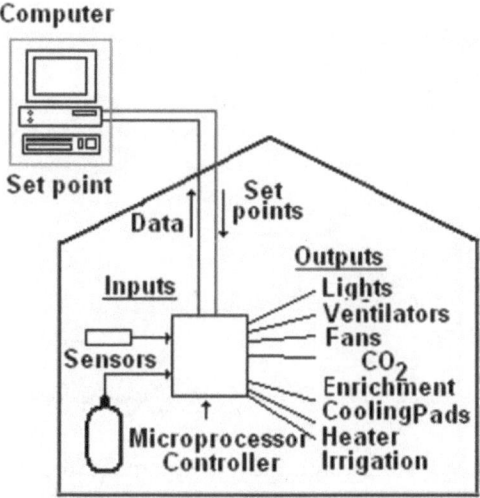

Figure 5.113 Computerized Control Systems in Greenhouse.

the system and the crop management. Computers can only assist by adding precision to the overall greenhouse production practice, and they are only as effective as the software it runs and the efficiency of the operator.

Control techniques

The practice of the greenhouse cultivation consists in setting up suitable techniques and procedures in order to set an accommodation between the internal state of the greenhouse and the plants growing process despite the fluctuations of exterior climate. The recourse of automatic control techniques is currently useful to improve the agricultural production in greenhouses. This can be performed through a good regulation of greenhouse micro-climate.

These last years, the control design of the climatic conditions in the greenhouses has known considerable interest and several teams of applied research have contributed to the development of this area. Numerous strategies and control techniques have been proposed (Bennis et al., 2008):

- Logical control (on/off).
- Linear control – PID.
- Generalized predictive control.
- Optimal control.
- Model predictive control.
- Linear quadratic adaptive control.
- Neural Networks control.
- Fuzzy Logic control.
- Non-linear control.
- Robust control.

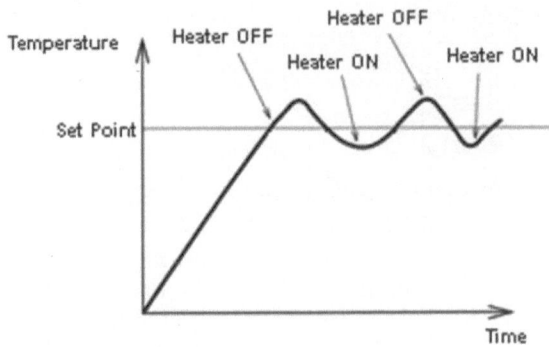

Figure 5.114 On/Off Controller Behavior.

These last techniques are only suitable to implement in dedicated microprocessors or computers due to the necessity to be programmed and some of them require high processing performance. It can be found controllers that combine two or more control techniques.

5.9.1 Control techniques theory

New control strategies for greenhouses are presented, the combine non-linear, Fuzzy Logic and Neural Networks techniques. Specifically, the non-linear methods are based on Sliding Modes and Feedback/Feed-forward Linearization control (Isidori, 1995; Sastry, 1999; Khalil, 2002). Also, control techniques such as On/off and PID are used for comparison objectives. All these techniques are described next.

5.9.2 Bang-bang or on/off control

An on-off controller is the simplest form of control device. The output from the device is either on or off, with no middle state. An on/off controller will switch the output only when the controlled variable crosses the set-point. For heating control, the output is on when the temperature is below the set-point, and off above set-point.

Since the controlled variable crosses the set-point to change the output state, the process variable will be cycling continually, going from below set-point to above, and back below. In cases where this cycling occurs rapidly, and to prevent damage to contactors and valves, an on/off differential, or hysteresis, is added to the controller operations. This differential requires that the temperature exceed set-point by a certain amount before the output will turn off or on again. On-off differential prevents the output from chattering or making fast, continual switches if the cycling above and below the set-point occurs very rapidly.

On/off control is usually used where a precise control is not necessary, in systems which cannot handle having the energy turned on and off frequently, where the mass of the system is so great that the process variable change extremely slowly, or for an alarm.

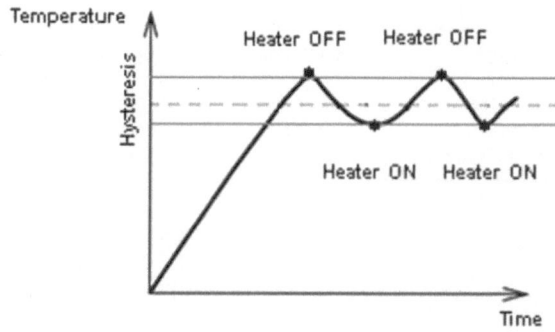

Figure 5.115 On/Off Controller whit Hysteresis Behavior.

5.9.3 PID control

A proportional-integral-derivative controller (PID controller) is a generic control loop feedback mechanism widely used in industrial control systems. A PID controller calculates an "error" value as the difference between a measured process variable and a desired set-point. The controller attempts to minimize the error by adjusting the process control inputs.

The PID controller algorithm involves three separate constant parameters: the proportional, the integral and derivative values, denoted P, I, and D. These values can be interpreted in terms of time: P depends on the present error, I on the accumulation of past errors, and D is a prediction of future errors, based on current rate of change. The weighted sum of these three actions is used to adjust the process via a control element such as a valve, an electric motor or a damper. Defining $u(t)$ as the controller output, the final equation of the PID algorithm is:

$$u(t) = K_p e(t) + K_i \int_0^l e(t)d\tau + K_d \frac{d}{dt} e(t) \tag{5.21}$$

where K_p: Proportional gain, a tuning parameter, K_i: Integral gain, a tuning parameter, K_d: Derivative gain, a tuning parameter, e: Error = Set-point − Process Variable, t: Time or instantaneous time (the present), τ: Variable of integration; takes on values from time 0 to the present t.

In the absence of knowledge of the underlying process, a PID controller has historically been considered to be the best controller. By tuning the three parameters in the PID controller algorithm, the controller can provide control action designed for specific process requirements. The response of the controller can be described in terms of the responsiveness of the controller to an error, the degree to which the controller overshoots the set-point, and the degree of system oscillation. Note that the use of the PID algorithm for control does not guarantee optimal control of the system or system stability.

A proportional controller (K_p) will have the effect of reducing the rise time and will reduce but never eliminate the steady-state error. An integral control (K_i) will have the effect of eliminating the steady-state error for a constant or step input, but it

Table 5.18 Different Greenhouse Orientation by Researchers.

Ctrl Response	Rise Time	Overshoot	Settling Time	Steady-State Error
K_p	Decrease	Increase	Small Change	Decrease
K_i	Decrease	Increase	Increase	Eliminate
K_d	Small Change	Decrease	Decrease	No Change

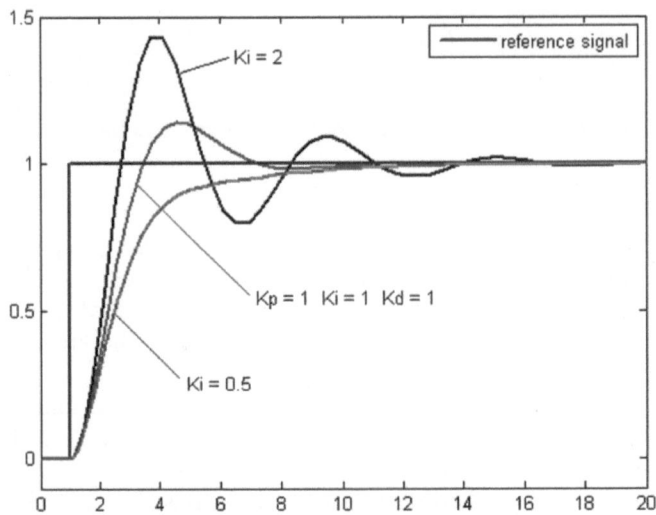

Figure 5.116 PID Controller Response.

may make the transient response slower. A derivative control (K_d) will have the effect of increasing the stability of the system, reducing the overshoot, and improving the transient response (Astrom & Hagglund, 1995).

The effects of each of controller parameters, K_p, K_d, and K_i on a closed-loop system are summarized in the table below.

Note that these correlations may not be exactly accurate, because (K_p), (K_i), and (k_d) are dependent on each other. In fact, changing one of these variables can change the effect of the other two. For this reason, the table should only be used as a reference when you are determining the values for (K_i), (K_p) and (K_d). There are some methods for tuning the PID parameters such as manual tuning, Ziegler-Nichols, root locus method, Cohen-Coon, among others.

Figure 5.116 shows common responses of a PID controller due to set point changes in the controlled variable depending on the tuning values of the PID parameters.

5.9.4   Feedback/Feed-forward linearization

Feedback Linearization is an approach to non-linear control design that has attracted lots of research in recent years. The central idea is to algebraically transform non-linear systems dynamics into (fully or partly) linear ones, so that linear control techniques can be applied.

This differs entirely from conventional (Jacobian) linearization, because Feedback Linearization is achieved by exact state transformation and feedback, rather than by linear approximations of the dynamics.

The basic idea of simplifying the form of a system by choosing a different state representation is not completely unfamiliar; rather it is similar to the choice of reference frames or coordinate systems in mechanics. *Feedback Linearization* equals to some ways of transforming original system models into equivalent models of a simpler form.

Most Feedback Linearization approaches are based on *input-output linearization* or *state-space linearization*. In the input-output linearization approach, the objective is to linearize the map between the transformed inputs (v) and the actual outputs (y). A linear controller is then designed for the linearized input-output model, which can be represented by 5.22 with $r \leq n$ and $w = y$. However, there is an $(n - r)$-dimensional subsystem that typically is not linearized,

$$\dot{\eta} = q(\eta, \xi) \tag{5.22}$$

where η is an $(n - r)$-dimensional vector of transformed state variables and q is a $(n - r)$-dimensional vector of non-linear functions. Input-output linearization techniques are restricted to processes in which these so-called *zero dynamics* are stable.

In the state-space linearization approach, the goal is to linearize the map between the transformed inputs and the entire vector of transformed state variables. This objective is achieved by deriving artificial outputs (w) that yield a feedback linearized model with state dimension $r = n$. A linear controller is then synthesized for the linear input-state model. However, this approach may fail to simplify the controller design task because the map between the transformed inputs and the original outputs (y) generally is non-linear. As a result, input-output linearization is preferable to state-space linearization for most process control applications. For some processes, it is possible to simultaneously linearize the input-state and input-output maps because the original outputs yield a linear model with dimension $r = n$.

Consider, at first, the multi-input multi-output (MIMO) and *square* (systems with as many inputs as outputs) system:

$$\dot{x} = f(x, v) + \sum_{i=1}^{p} g_i(x, v) u_i, \quad y_j = h_j(x), \quad j = 1, \ldots, p \tag{5.23}$$

Here $x \in R^n$ is the state vector, $v \in R^d$ is the external disturbance vector, $u \in R^p$ is the ith control input, $y \in R^p$ is the ith control output, and f, g_i are assumed to be smooth vector fields and h_i to be smooth functions. Supposing f, g_i are analytic vector fields, saying that not only are they truly indefinitely differentiable, they have convergent Taylor Series. The following calculations will be made for $x \in U$, an open subset of R^n. Typically U is an open set containing an equilibrium point x_o of the undriven system, that is a point which $f(x_o, v) = 0$. Differentiating jth of the output y_j respect to time, one obtains

$$\dot{y}_j = \frac{\partial h}{\partial x} f(x, v) + \sum_{i=1}^{p} \frac{\partial h}{\partial x} g_i(x, v) u_i \tag{5.24}$$

$$\dot{y}_j = L_f h_j + \sum_{i=1}^{p} L_{g_i} h_j u_i \tag{5.25}$$

Here $L_f h_j: R^n \to R$ and $L_{g_i} h_j: R^n \to R$ stand for the *Lie derivatives* of h with respect to f and g, respectively. Note that if $L_{g_i} h_j(x,v) \equiv 0$, then the inputs do not appear in the equation. Define the *relative degree* or γ_j to be the smallest integer such that at least one of the inputs appears in $y_j^{\gamma_j}$, that is,

$$y_j^{\gamma_j} = L_f^{\gamma_j} h_j + \sum_{i=1}^{p} L_{g_i}(L_f^{\gamma_j-1} h_j) u_i \tag{5.26}$$

With at least one of the $L_{g_i} h_j(x,v)l = 0$, for some x. Define the $p \times p$ matrix as

$$D(x,v) = \begin{pmatrix} L_{g1}(L_f^{4_1-1} h_1) \cdots L_{gp}(L_f^{4_1-1} h_1) \\ \vdots \qquad \ddots \qquad \vdots \\ L_{g1}(L_f^{4_p-1} h_p) \cdots L_{gp}(L_f^{4_p-1} h_p) \end{pmatrix} \tag{5.27}$$

If a system has well defined vector relative degree, then equation 5.23 may be written as

$$\begin{bmatrix} y_1^{\gamma_1} \\ \vdots \\ y_p^{\gamma_p} \end{bmatrix} = \begin{bmatrix} L_f^{\gamma_1} h_1 \\ \vdots \\ L_f^{\gamma_p} h_p \end{bmatrix} + D(x,v) \begin{bmatrix} u_1 \\ \vdots \\ u_p \end{bmatrix} \tag{5.28}$$

Since $D(x_o,v)$ is non-singular, it follows that $D(x,v) \in R^{p \times p}$ is bounded away from non-singularity for $x \in U$ a neighborhood U of x_o, meaning that $D^{-1}(x,v)$ and has bounded norm on U. Then the state feedback control law

$$\begin{bmatrix} u_1 \\ \vdots \\ u_p \end{bmatrix} = D^{-1}(x,v) \left\{ - \begin{bmatrix} L_f^{\gamma_1} h_1 \\ \vdots \\ L_f^{\gamma_p} h_p \end{bmatrix} + \begin{bmatrix} u^1 \\ \vdots \\ \hat{u}_p \end{bmatrix} \right\} \tag{5.29}$$

Yields the linear closed loop system

$$\begin{bmatrix} y_1^{\gamma_1} \\ \vdots \\ y_p^{\gamma_p} \end{bmatrix} = \begin{bmatrix} u^1 \\ \vdots \\ \hat{u}_p \end{bmatrix} \tag{5.30}$$

Note that this system is, in addition, decouples. Thus, decoupling is a by product of linearization.

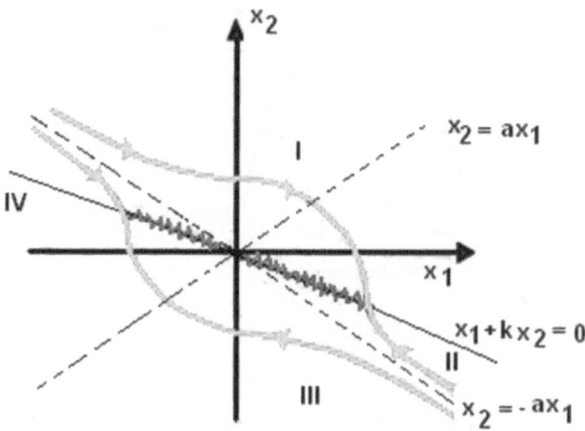

Figure 5.117 Trajectories in the Portrait Phase.

5.9.5 Sliding modes

The term Sliding Modes Control (SMC) first appeared in the context of variable structure systems. Soon sliding modes became the principal operational mode for this class of control systems. Practically, all design methods for variable structure systems are based on SMC which have played and are still playing an exceptional role both in theoretical developments and in practical applications.

In systems with control as a discontinuous state function, the SMC may arise. The control action switches at high frequency, in other words commutates on a specified surface. Systems with sliding modes have proven to be an efficient tool to control complex high order non-linear dynamic plant operating under uncertainty conditions. Also, SMC is very attractive for its excellent performance and easy to implement with simple control algorithm.

In conventional SMC, the control law is used to drive the system errors to a particular hyper plane in the state space, named the sliding surface. When the sliding surface is reached, the system state are kept switching or sliding until stable equilibrium states are achieved (Perruquetti & Barbot, 2002).

Taking as state variables,

$$x_1 = x \tag{5.31}$$

$$x_2 = \dot{x} \tag{5.32}$$

The system can be put in the following state-space representation,

$$\dot{x}_2 = x_2 \tag{5.33}$$

$$\dot{x}_2 = a^2 u \tag{5.34}$$

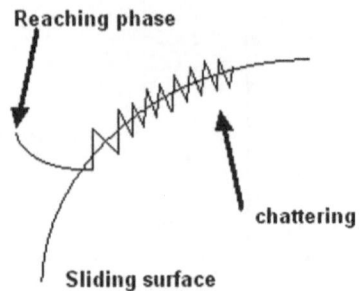

Figure 5.118 The Chattering Phenomenon.

Where the control law is designed under the simplest form,

$$u = -|x|\mathrm{sgn}(x + k\dot{x}) \tag{5.35}$$

$$u = -|x_1|\mathrm{sgn}(x_1 + kx_2) \tag{5.36}$$

In the following, $s = x_1 + kx_2 = 0$, will be called the *switching surface*. As already said switching illustrates the commutation of the control law while crossing the surface $s = 0$. From Figure 5.117 it can be easily seen that:

- The plane is divided into four regions.
- Region I and III trajectories are ellipses given by $a^2x_1{}^2 + x_2{}^2 = constant$.
- Region II and IV trajectories are hyperbolas with asymptotes $x_2 = \pm ax_1$.
- The control law only commutes on the boundary surface $x_1 + kx_2 = 0$.

By a suitable choice of k or the *hitting gain*, all trajectories are directed toward this surface. Consequently, once the surface is reached, a new phenomenon appears: the trajectories are sliding along this surface.

An ideal SMC does not exist since it would imply that the control commutes at an infinite frequency. In the presence of switching time delays and small time constants in the actuators, the discontinuity in the feedback control produces a particular dynamic behavior in the vicinity of the surface, which is commonly known as *chattering* (Perruquetti & Barbot, 2002).

The chattering phenomenon is the main disadvantage of this type of controllers and all efforts are concentrated on decreasing that phenomena by different kind of techniques. This chattering effect is directly related to the hitting gain, a well-chosen k will decrease the chattering effect.

The general SMC law is displayed as,

$$u = u_e + u_d \tag{5.37}$$

where u_e is the equivalent control action which is described by the system behavior and u_d is the hitting control action or the discontinuous part, which ensures a finite time convergence to the chosen surface (Perruquetti & Barbot, 2002).

Now, let us consider the Single-Input-Single-Output (SISO) non-linear first order system:

$$\dot{x} = f_i(x, v) + g_i(x, v)u \tag{5.38}$$

The sliding surface is designed as follows:

$$s = \left(\frac{d}{dt} + \alpha\right)^{n-1} \tilde{x} \tag{5.39}$$

where,

$$n = system\ order = 1$$

$$\tilde{x} = x - x_d \tag{5.40}$$

then,

$$s = \tilde{x} \tag{5.41}$$

$$\dot{s} = \dot{x} - \dot{x}_d \tag{5.42}$$

$$\dot{s} = f_i(x, v) + g_i(x, v)u - \dot{x}_d \tag{5.43}$$

given that,

$$\dot{s} = -k * sgn\ (s) \tag{5.44}$$

then,

$$-k * sgn(s) = f_i(x, v) + g_i(x, v)u - \dot{x}_d \tag{5.45}$$

$$u = g_i^{-1}(x, v)[-f_i(x, v) + \dot{x}_d - k * sgn(s)] \tag{5.46}$$

$$u = g_i^{-1}(x, v)[-f_i(x, v) + \dot{x}_d - k * sgn(x - x_d)] \tag{5.47}$$

where,

$$\hat{u} = \dot{x}_d - k * sgn(x - x_d) \tag{5.48}$$

The variable $\dot{x}_d$ will be only relevant if the controller tracks a desired point which means the system changes frequently the desired point of operation, if the system does not change or slightly changes this derivative is mostly zero. In this case it will be assumed as zero.

$$u = g_i^{-1}(x, v)[-f_i(x, v) + \hat{u}] \tag{5.49}$$

As it can be noticed expressions 5.49 and 5.4 have an equivalent form, following that a Sliding Modes + Feedback/Feed-forward Controller has been developed. In a general form, this controller is extended for a MIMO system and it is presented as:

$$
\begin{bmatrix} u_1 \\ \vdots \\ u_p \end{bmatrix} = D^{-1}(x, v) \left\{ -\begin{bmatrix} L_f^{\gamma_1} h_1 \\ \vdots \\ L_f^{\gamma_p} h_p \end{bmatrix} + \begin{bmatrix} -k_1 * sgn(x_1 - x_{d1}) \\ \vdots \\ -k_p * sgn(x_p - x_{dp}) \end{bmatrix} \right\}
\tag{5.50}
$$

5.9.5.1 Fuzzy logic control

Almost all of the physical dynamical systems in real life cannot be represented by linear differential equations and have a non-linear nature. At the same time, linear control methods rely on the key assumption of small range of operation for the linear model, acquired from linearizing the non-linear system, to be valid. When the required operation range is large, a linear controller is prone to be unstable, because the non-linearities in the plant cannot be properly dealt with. Another assumption of linear control is that the system model is indeed linearizable and the linear model is accurate enough for building up the controller. However, the highly non-linear and discontinuous nature of many, for instance, mechanical and electrical systems does not allow linear approximation. For many non-linear plants i.e. chemical processes, building a mathematical model is very difficult and only the input-output data yielded from running the process is accessible for an estimation. Many control problems involve uncertainties in the model parameters. A controller based on inaccurate or obsolete values of the model parameters may show significant performance degradation or even instability (Mehran, 2008).

Conventional automatic control system design methods involve the construction of mathematical models describing the dynamic system to be controlled and the application of analytical techniques to the model to derive control laws. Although application of Fuzzy Logic to industrial problems has often produced results superior to classical control, the design procedures are limited by the heuristic rules of the system. This implicit assumption limits the application of Fuzzy Logic controller (FLC). Moreover, the majority of FLCs to date have been static and based upon knowledge derived from imprecise heuristic knowledge of experienced operators. The Fuzzy Logic-based approach for solving problems in control has been found to excel in those systems which are very complex, highly non-linear and with parameter uncertainty. We may view a Fuzzy Logic controller as a real time expert system that employs Fuzzy Logic to analyze input to output performance. Indeed, they provide a means of converting a linguistic control strategy derived from expert knowledge into automatic control strategies and give us a means of interrogating the control system evolution and system performance (Tomescu, 2007).

Fuzzy logic Set was presented in 1965 by Lotfi A. Zadeh. After being mostly viewed as a controversial technology for two decades, Fuzzy Logic has finally been accepted as an emerging technology since the late 1980s. This is largely due to a wide array of successful applications ranging from consumer products, to industrial process control, to automotive applications. Fuzzy Logic is closer in spirit to human thinking and natural language than conventional logical systems. The essence of Fuzzy control is to build a model of human expert who is capable of controlling the plant without thinking in terms of a mathematical model. Fuzzy systems are very useful in two general contexts: (1) in situations involving highly complex systems whose behaviors are not well understood, and (2) in situations where an approximate, but fast, solution is warranted (Kaur & Kaur, 2012).

Let X be a universe of discourse. Consider a single-input nth order non-linear system of the following form:

$$\dot{x} = f(x) + b(x)u \tag{5.51}$$

where: $x \in X$, $[x_1, x_2, \ldots, x_n]^T$ is the state vector, $f(x) = [f_1(x), f_2(x), \ldots, f_n(x)]^T$, $b(x) = [b_1(x), b_2(x), \ldots, b_n(x)]^T$ are functions describing the dynamics of the plant and u is the control input of which the value is determined by an FLC.

A Fuzzy controller or model uses Fuzzy rules, which are linguistic if-then statements involving Fuzzy sets, Fuzzy Logic, and Fuzzy inference. Fuzzy rules play a key role in representing expert control/modeling knowledge and experience and in linking the input variables of Fuzzy controllers/models to output variable (or variables). Two major types of Fuzzy rules exist, namely, Mamdani Fuzzy rules and Takagi-Sugeno (T-S) Fuzzy rules. Lets first start with the familiar Mamdani Fuzzy systems. A simple but representative Mamdani fuzzy rule describing the movement of a car is:

If Speed is High and Acceleration is Small then Braking is Modest

where *Speed* and *Acceleration* are input variables and *Braking* is an output variable. "High", "Small" and "Modest" are Fuzzy sets, and the first two are called input Fuzzy sets while the last one is named the output Fuzzy set.

The variables as well as linguistic terms, such as "High", can be represented by mathematical symbols. Thus, a Mamdani Fuzzy rule for a Fuzzy controller can be described as follows:

If x_i is X_{i_1} and X_2 is X_{i_2} and ... and X_n is X_{i_n} then u_i is $Y_i, i = 1, r$

where $x_1, x_2, \ldots, x_n$ are input variables and u_i are output variables. In theory, these variables can be either continuous or discrete; practically speaking, however, they should be discrete because virtually all fuzzy controllers and models are implemented using digital computers. $X_1, X_2, \ldots, X_{i_n}$ are Fuzzy sets which describe the linguistics terms of input variables and Y_i describes the linguistics terms of output variables; "and" are Fuzzy Logic "and" operators. *If x_i is X_{i_1} and X_2 is X_{i_2} and ... and X_n is X_{i_n}* is called the *rule antecedent*, whereas the remaining part is named the *rule consequent*.

Now, let us look at the so-called T-S Fuzzy rules. Unlike Mamdani Fuzzy rules, T-S rules use functions of input variables as the rule consequent. For fuzzy control, a T-S rule corresponding to the Mamdani rule is

If x_i is X_{i_1} and X_2 is X_{i_2} and ... and X_n is X_{i_n} then $u_i = f(x_i), i = 1, r$

where, $f()$ is a real function of any type.

Each fuzzy rule generate an activation degree: $\alpha_i \in [0, 1]$, $i = 1, 2, \ldots, r$, $\alpha_i = (x(t)) = \min(\mu_{i,1}(x_1(t)), \mu_{i,2}(x_2(t)), \ldots, \mu_{i,n}(x_n(t)))$. $f(x_i)$ can be a single value or a function of states vector, $x(t)$. It is assumed that for any $x \in X$ in the input universe

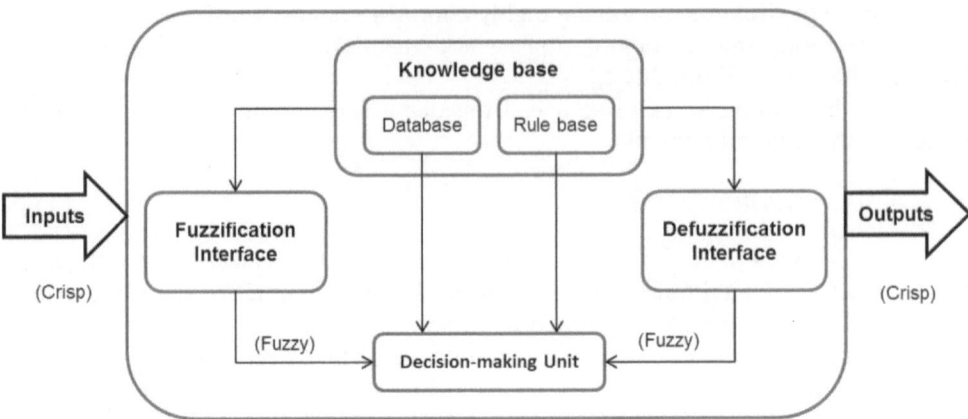

Figure 5.119 Complete fuzzy Inference System.

of discourse X, there exists at least one $\alpha_i \in [0, 1]$, $i = 1, 2, \ldots, r$, among all rules that is not equal to zero. The control signal u_i is a function of α_i and $f(x_i)$. By applying the weighted sum defuzzification method (centroid for Mamdani), the output of the FLC is given by:

$$u = \frac{\sum_{i=1}^{r} \alpha_i u_i}{\sum_{i=1}^{r} \alpha_i} \tag{5.52}$$

where r is the total number of rules.

Fuzzy Inference Systems (FIS) also known as Fuzzy rule-based systems or Fuzzy models are schematically shown in Figure 5.119. They are composed of 5 conventional block: a rule-base containing a number of Fuzzy if-then rules, a *database* which defines the membership functions of the Fuzzy sets used in the Fuzzy rules, a *decision-making unit* which performs the inference operations on the rules, a *fuzzification interface* which transform the crisp inputs into degrees of match with linguistic values, a *defuzzification interface* which transform the Fuzzy results of the inference into a crisp output.

Figure 5.120 utilizes a two-rule two-input Fuzzy Inference System to show different types of Fuzzy system mentioned above. Type 2 is the widely-used Mamdani type Fuzzy system which the output function is determined based on overall Fuzzy output; some of them are centroid of area, min of maxima, maximum of maxima, etc. Type 3 is the Takagi-Sugeno type Fuzzy system. The present focus relies on Fuzzy models that use the T-S rule consequent (Mehran, 2008).

Fuzzy modeling is one of those areas which are often used in control, system identification, classification, decision support systems and fault diagnosis. Fuzzy models can be built by encoding expert knowledge but sometimes the experts are not available or their information is not complete. Therefore, there is the importance and the interest in data driven fuzzy modeling. Different approaches have been developed to obtain Fuzzy models from data. These techniques cover Fuzzy clustering, ANFIS (Adaptive

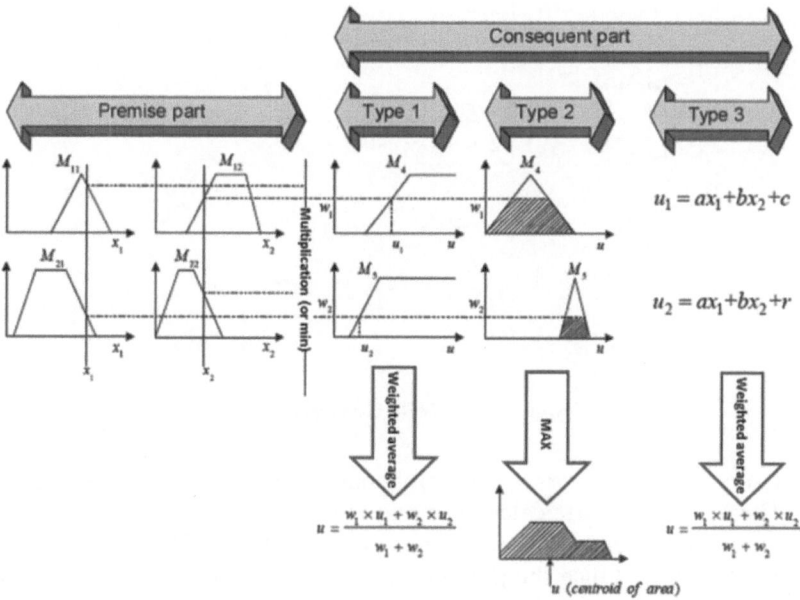

Figure 5.120 Commonly used Fuzzy If-Then Rules and Fuzzy Mechanism.

Neural Fuzzy Inference System), Genetic Algorithms (GAs), statistical information, Kalman filters and combinations of these, among others. The modeling scheme in the present work is ANFIS which will be described next (Cepeda et al., 2012).

5.9.6 Adaptive neural fuzzy inference system – ANFIS

ANFIS is a graphical network representation of T-S fuzzy systems. An ANFIS model combines the artificial neural network's benefits with the fuzzy inference system's profits in a single model. This kind of model has become very popular due to its characteristics: fast and accurate learning, and the capacity of data management. The main objective of ANFIS is to optimize fuzzy system's parameters (to obtain an accurate answer to a problem) through a learning algorithm implementation and a set of inputs and outputs which are responsible of the learning process. Those sets are used to build a fuzzy inference system, from here the membership function parameters are adjusted by a hybrid training algorithm, which combines gradient descent and the least-square method. The least-squares method is actually the major driving force that leads to fast training, while the gradient descent serves to slowly change the underlying membership function that generates the basic functions for the least-squares method. This type of adjustment allows the fuzzy model to learn the data set that is provided.

The adaptive neuro-learning works in a similar form as a neural network. The adaptive neuro-learning model provides a procedure of fuzzy modeled to learn information from a set of data. Figure 5.121 shows the general structure of the ANFIS (Osorio et al., 2011).

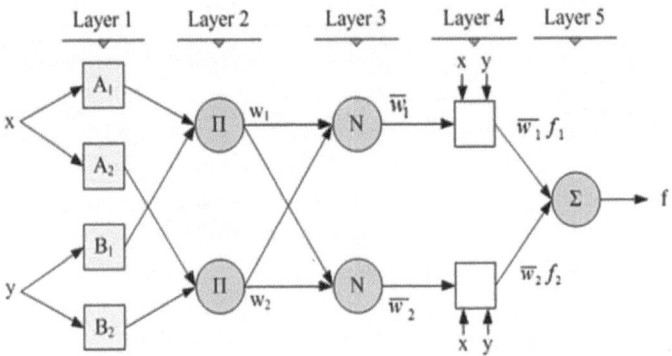

Figure 5.121 ANFIS General Structure.

The associated parameters to the membership functions change during the learning process. These variations are calculated due to a vector called *gradient*. This vector is useful to know how approximate are the results of the ANFIS outputs in relation to the reference outputs. Once the gradient is obtained, many routines of optimization are applied to adjust the parameter and minimize the error. It is important to take into account that two different rules cannot share the same membership function, although the number of rules is the same that the number of membership functions of the output.

An important part of ANFIS is related to the learning algorithms. A hybrid learning algorithm is implemented; this is a combination of least square and back-propagation methods. Considering the least square part and an output y by the parametrized equation:

$$y = \Theta_1 f_1(u) + \Theta_2 f_2(u) + \cdots + \Theta_n f_n(u) \tag{5.53}$$

where u is the model inputs vector, f_n are known functions and Θ are unknown parameters. Substituting each data pair in 5.53 a set of linear equations is obtained.

$$y = A\Theta \tag{5.54}$$

where:

$$A = \begin{pmatrix} f_{1(u_1)} & \cdots & f_{n(u_1)} \\ \vdots & \ddots & \vdots \\ f_{1(u_m)} & \cdots & f_{n(u_m)} \end{pmatrix} \tag{5.55}$$

$$\theta = \begin{pmatrix} \theta_1 \\ \vdots \\ \theta_n \end{pmatrix} \tag{5.56}$$

$$y = \begin{pmatrix} y_1 \\ \vdots \\ y_m \end{pmatrix} \tag{5.57}$$

An error vector is introduced to account for the modeling error,

$$y = A\Theta + e \tag{5.58}$$

$$E(\Theta) = \sum_{i=1}^{m} (y_i - \alpha_i^T \Theta)^2 = e^{Te} \tag{5.59}$$

Equation 5.59 is called the *objective function*. The squared error is minimized when $\Theta = \hat{\Theta}$, called *Least Squares Estimator (LSE)* that satisfies the normal equation,

$$A^T y = A^T A \hat{\Theta} \tag{5.60}$$

If $A^T A$ is non-singular then

$$\hat{\Theta} = (A^T A)^{-1} A^T y \tag{5.61}$$

Moreover, for the back-propagation learning the main part concerns to how to recursively obtain a gradient vector in which each element is defined as the derivative of an error measure with respect to a parameter. Considering the output function of node i in layer l

$$x_{l,i} = f_{l,i}(x_{l-1,1}, \ldots, x_{l-1,N(l-1)}, \alpha, \beta, \gamma, \ldots) \tag{5.62}$$

where $\alpha, \beta, \gamma, \ldots$ are the parameters of this node. Hence, the sum of the squared error defined for a set of P entries, is defined as:

$$E_p = \sum_{k=1}^{N(L)} (d_k - x_{L,k})^2 \tag{5.63}$$

where d_k is the desired output vector and $x_{L,k}$ both for the kth of the pth desired output vector. The basic concept in calculating the gradient vector is to pass from derivative information starting from the output layer and going backward layer by layer until the input layer is reached. The error signal is defined as,

$$\varepsilon_{l,i} = \frac{\partial E_p}{\partial x_{l,i}} \rightarrow \varepsilon_{l,i} = -2(d_i - x_{L,i}) \tag{5.64}$$

If α is a parameter of the ith node at layer l. Thus, it is obtained the derivative of the overall error measure E with respect to α is

$$\frac{\partial E}{\partial \alpha} = \sum_{p=1}^{P} \frac{\partial E}{\partial \alpha} \tag{5.65}$$

The generic parameter α is

$$\Delta\alpha = -\eta \frac{\partial E}{\partial \alpha} \tag{5.66}$$

$$\alpha_{new} = \alpha_{old} - \eta \frac{\partial E}{\partial \alpha} \tag{5.67}$$

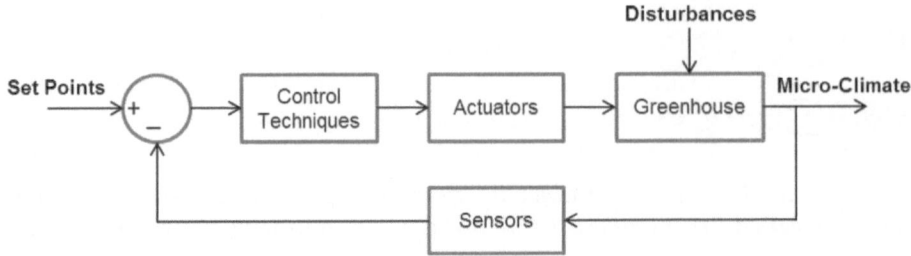

Figure 5.122 Greenhouse Control Loop Schematic.

Where η is the learning rate. For hybrid learning algorithm, each epoch consists of a forward pass and a backward pass. In the forward pass equations 5.67 and 5.61 are implemented in order to calculate the error measure for each training data pair. Then 5.63 is applied to find the derivative of those error measures finding the error signal. In the backward pass, these error signals propagate from the output end towards the input end. The gradient vector is found for each training data entry. At the end of the backward pass for all training data pairs, the input parameters are updated by steepest descent method as given by 5.67 (Osorio et al., 2011).

Finally, before entering next subsection which deals with a controller development, it can be said that all of the necessary elements to build a control system have been described at this point. Sensors, actuators, and control techniques, including greenhouse dynamical models, were already exposed. The control loop for regulating the micro-climate variables inside the greenhouse is illustrated in Figure 5.122.

5.9.7   Green tech control and supervisor

The Tecnológico de Monterrey Campus Ciudad de México has developed a climate control module for greenhouses called *Green Tech Control and Supervisor* (Tecnológico de Monterrey Campus Ciudad de México, 2011). This controller has the task of managing and monitoring the weather conditions inside the greenhouse, temperature and relative humidity, making decisions autonomously (without an operator) of what to do considering the optimal conditions programmed. These decisions are obtained from monitoring the internal and external sensors.

The control system provides a high degree of autonomy to the greenhouse; it is only required that the user, operator or farmer make occasional visits to check the levels of water and nutrients in the storage tanks.

The *Green Tech Control and Supervisor* has four modes of operation which are flexible and adaptable to the requirements of the end user, a manual control (user) and three automated. The autonomous decisions are taken based on three types of intelligent controllers: time control, decision trees and Fuzzy Logic. Once the control decides the action to follow the actuators are activated by a power control module. This stage allows the energy conduction through the single phase or three phase lines as the case of the actuators.

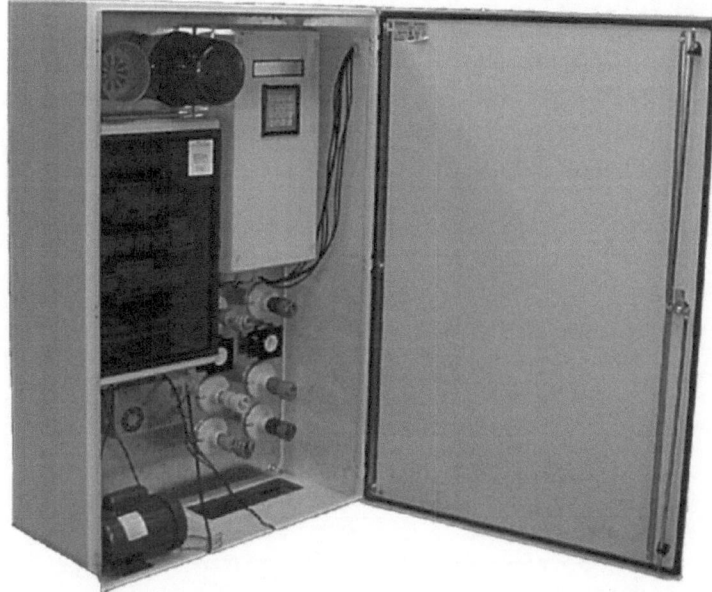

Figure 5.123 Green Tech Control and Supervisor was Patented by Tecnologico de Monterrey Ciudad de Mexico (P. Ponce et al., 2012).

The *Green Tech Control and Supervisor* was developed in stages, a first one, as already mentioned, only took into account the weather conditions regulating variables such as temperature and relative humidity; a second phase of development of the Green Tech Control and Supervisor included an external module for handling irrigation systems, in this case hydroponic.

The automatic control techniques were developed based on the knowledge of skilled operators in greenhouses, a reason why it was not necessary to set the greenhouse model. Linked to the intelligent control strategies that control the climatic conditions within the greenhouse, the *Green Tech Control and Supervisor* also ensures minimum energy consumption and protects the entire greenhouse structure against abrupt climate changes, specifically against rain and wind. No matter what automatic control is selected or running the controller is constantly checking for extreme conditions which may affect the structure, these are (Ponce et al., 2013):

- If the wind speed is equal or above the limit of strong wind, for protection to the physical structure, the controller will close all the vents.
- If the wind speed is strong, the controller checks the wind direction in order to close only the affected vents. This allows cooling of the greenhouse through the non-exposed vents.
- Any finding of rain, the vents will close preventing structural damages due to water accumulations or floods inside the greenhouse.

Table 5.19 Actuators Actions in Manual Control Mode.

Actuators	Keyboard #	Actions
Side Vents	1	Up
	2	Down
	3	Stop
Zenith Vents	1	Up
	2	Down
	3	Stop
Shading Mesh	1	Up
	2	Down
	3	Stop
(De) Humidifier	1	On
	2	Off
Heater	1	On
	2	Off

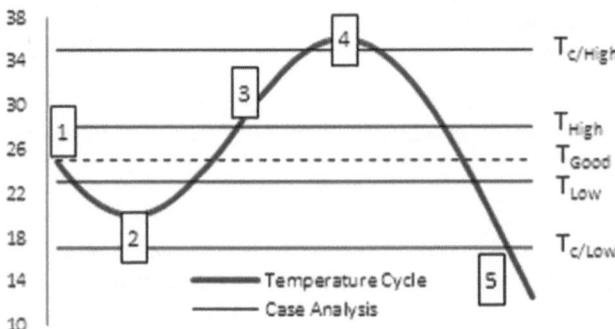

Figure 5.124 Case Analysis for Different Operating Points.

Manual control

This mode is used to control all the actuators movement manually; this means that using the keyboard the user can turn on/off the systems without any parameters or prior programming. Table 5.19 shows the keyboard selection for the different actions of actuators, having the manual control mode (Ponce et al., 2013).

Time control

This control selection is in the automatic menu, which means that the controller will maintain or regulate the temperature reference that you assign. The time control name comes from the fact that the controller waits for a change in the temperature ranges to make a decision of action.

There are three important set points for this control mode: a desired temperature with a certain allowed band, a high critical temperature and low critical temperature as seen in Figure 5.124. This Figure shows five cases in which the actuators will handle different actions in order to meet the needs; these cases are (Ponce et al., 2013):

Case 1 – Temperature in the desired range. The Green Tech Control and Supervisor remains in operating conditions without launching any actuator.

Case 2 – Temperature is out of the band without exceeding the low critical temperature. The Green Tech Control and Supervisor will close the vent system so that the greenhouse begins to heat up.

Case 3 – Temperature is out of the band without exceeding the high critical temperature. The Green Tech Control and Supervisor will open the vent system to cool the greenhouse.

Case 4 – Temperature is out of the band exceeding the high critical temperature. The Green Tech Control and Supervisor will open the vent system and it will activate an alarm system.

Case 5 – Temperature is out of the band exceeding the low critical temperature. The Green Tech Control and Supervisor will close the vent system, it will launch the heater and also the alarm system will be activated.

Decision trees control

Its operation is based on a series of decisions pre-programmed based on the set points assigned by the user.

This control strategy combines the readings received from the sensors and depending on the references consigned the controller makes a decision from a decision network that could trigger the actuators simultaneously. Table 5.20 presents some examples on

Table 5.20 Decision Trees Control Scenarios.

Sensors

Case	Time	T	RH	LI	WS	WD	R
1	Day	High	Yes	x	x	x	Yes
2	Day	High	Yes	Danger	Strong	Left	No
3	Day	Good	No	No Danger	x	x	No
4	Night	Low	Yes	x	x	x	Yes
5	Night	High	No	x	Strong	Right	No

Actuators

Case	LV	RV	LZ	RZ	SM	H	D	A
1	Close	Close	Close	Close	Open	Off	On	On
2	Close	SS	Close	SS	SS	Off	On	On
3	SS	SS	SS	SS	Open	Off	Off	Off
4	Close	Close	Close	Close	SS	Off	Off	Off
5	Open	Close	Open	Close	Open	Off	Off	On

T: Temperature. RH: Relative Humidity. LI: Light Intensity. WS: Wind Speed. WD: Wind Direction. R: Rain. LV: Left Vent. RV: Right Vent. LZ: Left Zenith. RZ: Right Zenith. SM: Shading Mesh. H: Heater. D: Dehumidifier. A: Alarm. SS: Smokestack. x: Not needed.

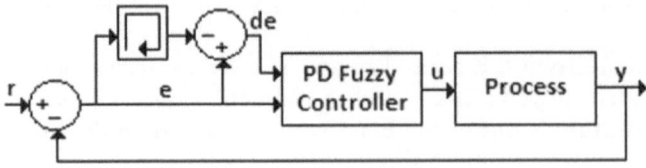

Figure 5.125 PD Fuzzy Control Loop.

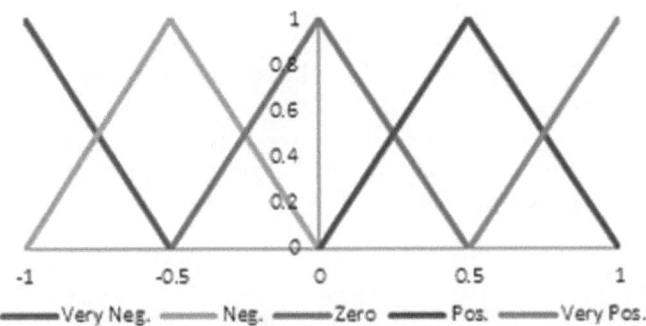

Figure 5.126 Inputs Membership Functions.

how the *Green Tech Control and Supervisor* will perform on different schemes of sensor readings. This table is divided by two components: the sensor readings and the actuators actions, depending on the scenario that is being given (Ponce et al., 2013).

Fuzzy logic control

In this scheme, a PD (Proportional-Derivative) T-S type Fuzzy Control is implemented. The Fuzzy Control is a rule base system that captures the characteristics of how the system should behave to certain inputs by means of if-then rules with fuzzy predicates that establish the relations between the relevant system variables. In particular a PD T-S type Fuzzy Control uses rules like:

If Error is Positive and Error Derivative is Zero then Ventilator is Closed   (3.61)

Where *Positive* and *Zero* represent a fuzzy set of the inputs *Error* and *Error Derivative*, respectively. And *Closed* represents the consequent constant of the output *Ventilator*. The output *Ventilator* is computed as a weighted sum of all rule outputs as shown in 3.45.

The PD T-S type Fuzzy Control keeps the temperature inside the greenhouse at a programmed reference by an intelligent activation of the actuators. The degree of activation depends on how close the current is to the desired temperature. This controller avoids sudden changes with proportional openings. Figure 5.125 displays the PD Fuzzy Control loop implemented for regulating the temperature. For example if the actual temperature barely exceeds the desired point, the vents will rise slightly proportionally or when the temperature is excessive, the opening is total.

The membership functions for the inputs, *Error* and *Error Derivative*, are shown in Figure 5.126, while the output are constants functions or zero-order polynomials. Both the inputs and the outputs of the fuzzy controller are normalized (Cepeda et al., 2013).

Chapter 6

Basic greenhouse control design and implementation

This section describes an implementation of a greenhouse and shows the form of implementing non-conventional control laws. The implementation deals with different parts of the greenhouse as mechanical, electrical and instrumentation. The main goal of this section is to give a complete picture regarding the process of developing a greenhouse. The present chapter is based on low cost structure materials and high control system technology. The structure could use other materials as was shown in previous chapters.

In the development of greenhouses it is essential to make a specifications list of the points of what designers expect at the end areshowcast. The most relevant are described below:

Functionality: Maintain the optimal conditions for crops, ability to operate in different places, rational use of energy, easy mounting and dismantling of the structure.

Dimensions: Optimum height for proper ventilation, wide and long enough for a proper distribution of crops, correct curvature of the top (roof) to prevent condensation and subsequent drip on crops causing damage, enough space for production.

Automation: Correct instrumentation for the greenhouse operation; temperature, relative humidity, light intensity, electrical conductivity and pH control.

Materials: Use of high resistance and non-corrosive materials, implementation of low cost materials, shading mesh with UV protection.

Life-cycle & maintenance: Easy access to the components for maintenance, periodic equipment review to avoid failures, user's manual development.

Cost: Cost reduction without affecting the greenhouse efficiency, use high-tech looking for an optimal relationship between cost and benefit.

To meet the specifications, the design of the intelligent greenhouse integrated and applied different technologies, such as (Cepeda, 2013a):

- Modular structure
- Hydroponic irrigation system
- Automation Systems
- Control strategies

The intelligent greenhouse prototype is located in Mexico City at the Tecnológico de Monterrey Campus Ciudad de México. Due to Mexico at a latitude of 19° 19′ (less than 40°), the greenhouse has a North-South orientation since the angle of the sun is much higher as seen in Figure 6.1

The generation of a micro-climate able to allow full development of the intelligent greenhouse hydroponics, is based on four main inputs: the mechanical, electrical, electronic, and control systems. In this chapter, the development of intelligent greenhouse is described from its construction stage to the completion of the system instrumentation and control. Figure 6.2 depicts the block diagram of the whole cultivation field control system.

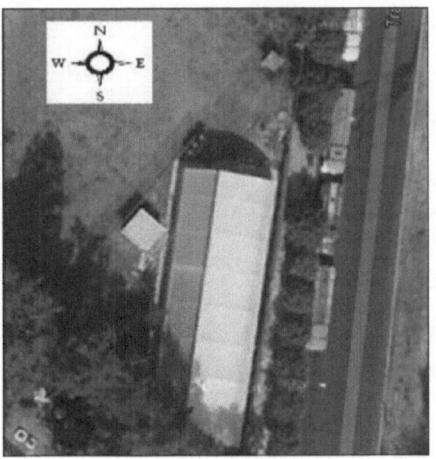

Figure 6.1 Greenhouse Site Selection and Orientation.

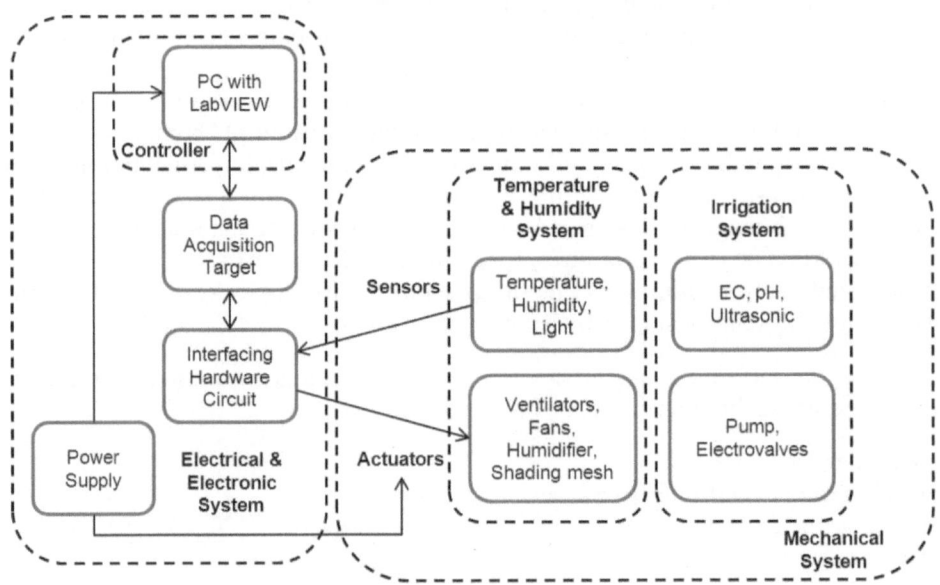

Figure 6.2 Block Diagram of the Whole Cultivation Field Control System.

6.1 MECHANICAL SYSTEMS

The mechanical system contemplates the physical structure and the automation systems, including hydroponics. The automation consists of sensors and actuators which close the control loop for regulating the variables which seeks to provide the adequate micro-climate within the greenhouse.

6.1.1 Structure and materials selection

The physical structure designed, shown in Figure 6.3, is based on a modular architecture which uses columns, beams, arches and connectors; it allows easy assembly/disassembly and scalability with the aim of achieving a quick installation and increase the surface if necessary. According to the Mexican standard NMX-E-255-CNCP-2008, a tunnel design for the structure (arched roof) with a movable zenith was chosen because it prevents accumulation of rainwater and it has great strength that enables greater load capacity supporting strong winds. Intelligent greenhouse dimensions are 10 m long by 10 m wide, which results in a culture area of 100 m^2. The height of the side-walls is 3 m and the height of the upper point is 5.6 m. If the greenhouse to be implemented is located in another country, you must review the local standard regarding mechanical structures.

The selected material for the structure is galvanized steel. This material is widely used in the construction of greenhouses because of its mechanical properties of high resistance to stress (Table 6.1) and therefore guarantees the good performance as a base for a light structure. Galvanized steel can withstand the onslaught of both wet and extremely dry climates. It is noteworthy that such material ensures structural integrity for natural disasters as high winds. Usually greenhouses whose structure is made of steel galvanized coatings using polyethylene or glass.

The cladding chosen is treated plastic (polyethylene) with UV II stabilizer 720 caliber because of its versatility, malleability, substantial mechanical properties and its light weight, thus no heavy loads are generated on the greenhouse structure. According to FAO, polyethylene is used in greenhouses for large agricultural production

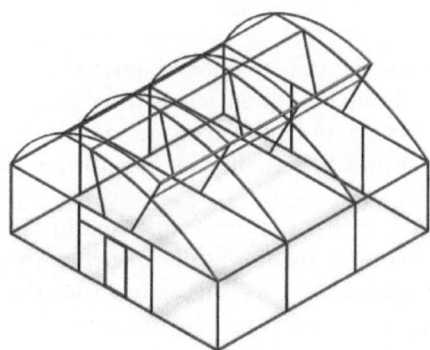

Figure 6.3 Intelligent Greenhouse Structure – CAD.

Table 6.1 Material characteristics.

Galvanized Steel		
General	Mass density	7.85 g/cm³
	Elasticity limit	207 MPa
	Maximum tensile strength	345 MPa
	Young's modulus	200 GPa
Tension	Poisson ratio	0.3 su
	Rupture	76.9231 GPa
	Expansion coefficient	0.000012 su/°C
Thermal Stress	Thermal conductivity	53 W/(m°K)
	Specific heat	450 J/(kg°C)

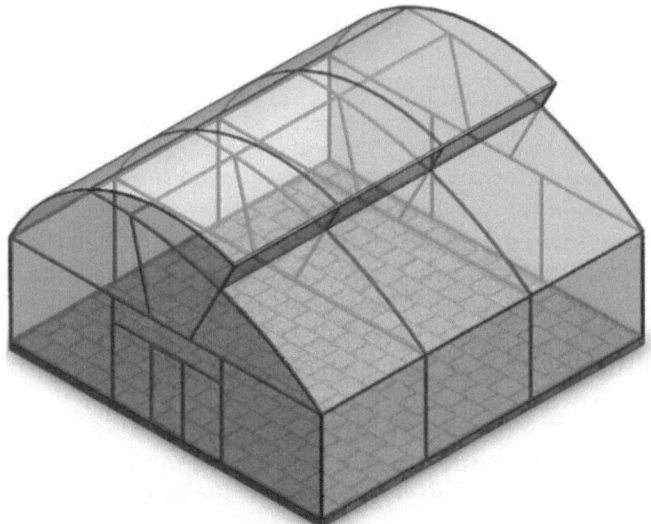

Figure 6.4 Intelligent Greenhouse Structure + Cladding + Floor – CAD.

unlike glass. This is due to the glass monetary cost that it represents. Despite all the advantages, there is a disadvantage in terms of ecological impact; being a polymer within the group of plastics, polyethylene does not degrade easily so its use must be measured. Figure 6.4 shows the intelligent greenhouse design intent adding the covering material and the floor.

Specifically, Table 6.2 describes the materials used in the structure development. The greenhouse floor is made of paving stone in order to provide stability to the structure and not allow the growth of harmful plants inside the greenhouse. The hydroponic and water fogging system, which will be discussed in next sections, are mostly composed by PVC (polyvinyl chloride) because the network of pipes used.

Figure 6.5 displays the intelligent greenhouse prototype in its first stages of construction, where the structure and the covering could be clearly seen.

Table 6.2 Materials Used in the Greenhouse Structure.

Characteristics of the Materials Used

Arches: galvanized pipes 2″ – schedule 30
Posts: galvanized pipes 2″ – schedule 30
Stringers: galvanized pipes 1″ – schedule 30
Treated plastic UV II 720 caliber
Galvanized surrounds OKI-PET
Gateway: angles 1/8 × 1 1/4
Bushings 1 1/4, winch, stringers nipples, poles rods
Cocks N° 10
Screws 3/8 hex head
High strength screws 2″ and 4″
Welding paint
Galvanized gutters 15 × 10 × 15
Expansion anchors for attaching to concrete

Figure 6.5 Intelligent Greenhouse – First Stages of Construction.

6.1.2 Ventilation systems

Facilitating the air renewal and decrease temperature inside the greenhouse natural and forced ventilation systems were designed. The natural ventilation consisting of two side vents and a zenith, and the forced ventilation formed by two exhaust fans; the ventilation systems designed are shown in Figures 6.6 and 6.7, respectively.

To manipulate the vents, a system of pulleys and cables is used. The system basically consists of wrapping/unwrapping the plastic wall on itself to retrieve or deploy as needed. An electric gear-motor rotates and wraps upon itself (Figure 6.8) , a steel wire

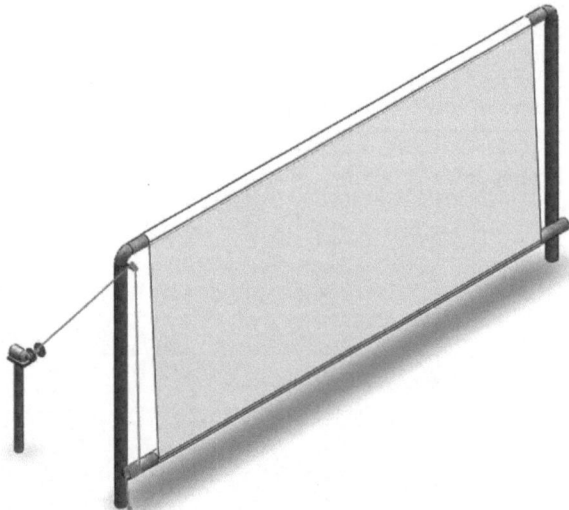

Figure 6.6 Vent System (Side-wall or Zenith) – CAD.

Figure 6.7 Exhaust Fan – CAD.

of 1/8 in which passes over a pulley (Figure 6.8(c)) and then wound on a steel pipe 1 1/4 in (Figure 6.89(d)) with a groove which is fixed on the plastic. This tube serves as the vent bottom bracket (Figure 6.8(b)) and at the same time it rotates and wraps/unwraps the plastic upon itself while going up or down. Finally, the wire ends embedded in the other end.

The data-sheet of the electric gear-motors used for the natural ventilation system is shown in Table 6.3. The same gear-motors are used for the shading system giving a total of five gear-motors in the greenhouse, three vents and two shading meshes.

Lastly, the exhaust fans are located at the front side of the intelligent greenhouse. Each exhaust fan is supported by a structure constructed of PTR (Pipe Thread Reducer) steel and protected against the weather onslaught by micro-tunnels made of the greenhouse plastic as displayed in Figure 6.9.

Table 6.4 presents the data-sheet of the electric motors that comes within the exhaust fans.

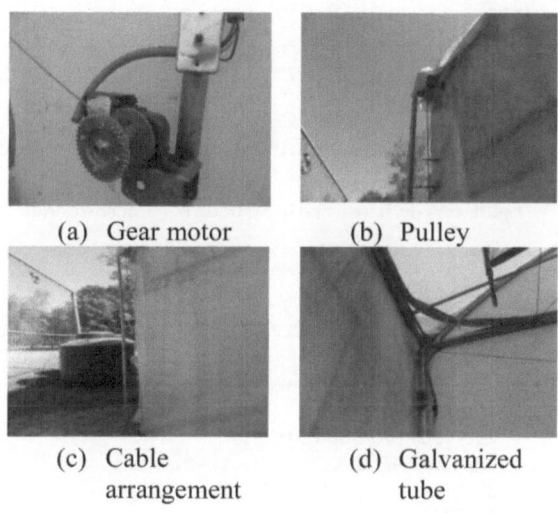

(a) Gear motor   (b) Pulley

(c) Cable   (d) Galvanized
     arrangement   tube

Figure 6.8 Natural Ventilation System.

Table 6.3 Electric Gear-motors Data-sheet.

Dayton AC/DC 2Z798B	
Power	1/15 HP
Voltage	115 Volts
Frequency	50/60 Hz
Speed	6.7 RPM max load – 15 RPM no load
Torsion	28 N mm
Ratio	745:1
Shaft diameter	13 mm
Other	External brushes

Figure 6.9 Exhaust Fan + Structure + Micro-tunnel.

Table 6.4 Electric Motors Data-sheet.

F B370-8T	
Power	1/2–1/3 HP
Voltage	230/460–190/380 Volts
Current	2.6/1.3–2.4/1.2 A
Phase	3
Frequency	60/50 Hz
Speed	825/715 RPM
Insulation Class	B
SF	1
Time rating	Continuous duty
Others	Thermally protected, totally enclosed, lubricated sealed ball bearings

6.1.3 Shading system

The intelligent greenhouse also features a shading system, designed as shown in Figure 6.10 with the intention of protecting the crop against solar radiation regulating the light intensity.

For crop protection, in an area of $100\,\text{m}^2$, a roll of shading mesh was implemented of 50% shade which comes in presentation of $3.7\,\text{m} \, x \, 100\,\text{m}$. This mesh is wrapped in

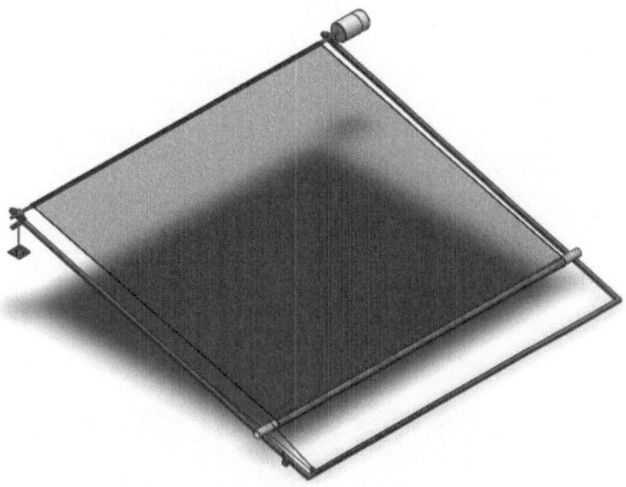

Figure 6.10 Shading System – CAD.

Figure 6.11 Shade Meshes.

aluminum tubes in both the top and bottom and its wrapping/unwrapping function is performed on four aluminum profiles as shown in Figure 6.11.

There is a shading mesh system for both the left and right side of the intelligent greenhouse which consists of independent actuators. In this system, the gear-motors, which are the same as the natural ventilation system, are connected directly to the shaft through bearings, where the mesh is rolled (Figure 6.12). This axis forms the framework on top of the system and, at the other end of the shade mesh that goes up or down, there is an aluminum pipe that serves as a weight-guide and rests on four aluminun rails (Figure 6.12(b)). Attached to this pipe there is a counterweight system

(a) Gear motor + bearing   (b) Aluminum pipe

(c) Pulley   (d) System of shading meshes

Figure 6.12 Shading Mesh System.

consisting of ropes, pulleys and the counterweights themselves (Figures 6.12(a) and b); this system is responsible for tensing the shading meshes with the aim to allow the correct wrapping/unwrapping function of the shading system (Figure 6.12(d)).

6.1.4 Water fogging system

To adjust the relative humidity a water fogging system was selected and designed as exposed in Figure 6.13. This system, besides increasing the humidity, also helps regulating the temperature together with the ventilation system; thermodynamically, increasing relative humidity the temperature decreases and vice versa.

The system comprises a storage tank of 1,100L (Figure 6.14), a domestic water pump (Figure 6.14(a)), a piping network to cover the required area (Figure 6.14 b and c) and 35 foggers (Figure 6.14 (d)). The water pump data-sheet is shown in Table 6.5 and the elements of the piping network are broken down in Table 6.6.

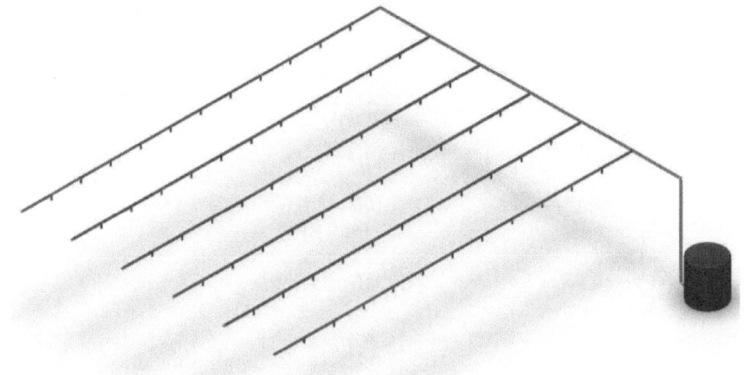

Figure 6.13 Water Fogging System (Humidifier) – CAD.

(a) Storage Tank (b) Water pump

(c) Pipe Network (d) Fogger

Figure 6.14 Shading Mesh System.

Table 6.5 Water Fogging Pump Data-sheet.

Milwaukee MIL-QB60	
Power	1/2 HP
Voltage	110 V
Current	4.2 A
Phase	1
Frequency	60 Hz
Speed	3450 RPM
H.max	35 m
Q.max	40 liter/min
Suction	6 m
Others	IP44
	ISO-9002, Continuous duty

Table 6.6 Elements Used in the Piping Network.

Characteristics of the Materials Used
2GPH foggers
Fogger mini drip valve
Fogger nozzle of 7 mm
Mini valve for PE pipe of 16 mm
PE pipe of 16 mm diameter, caliber 45
Terminal type 8 for pipe of 16 mm
Tee pieces for PE pipe of 16 mm
Elbows for PE pipe of 16 mm

6.1.5 Irrigation system

Furthermore, the irrigation technique is based on a hydroponic technique called NFT (Nutrient Film Technique), which consists in recirculating a nutrient solution through PVC pipes feeding the crop by the roots, while the rest of the plant faces the exterior to continue receiving direct sunlight. Attached to the hydroponic system there is a nutrient supply system. Figure 6.15 presents the whole design of the hydroponic irrigation and nutrients supply system.

The system consists of 10 PVC pipes arranged parallel to each other along the greenhouse as seen in Figure 6.16, with a distance of 84 cm between them. The tubes have a length of 8 m and a diameter of 4 in, and each one has 36 holes every 27 cm approximately, giving a total production of 360 plants. In each hole, a small plastic basket is located whose function is to hold the substrate where the plant grows.

Apart from the main pipes there are also a collector tank (Figure 6.17), a water pump (Figure 6.17(b)) and, distribution and collection pipes (Figures 6.17(c) and 6.17(d), respectively). The collector tank has a capacity of 560 L and is located underground. The water pump used has the specifications shown in the Table 6.7.

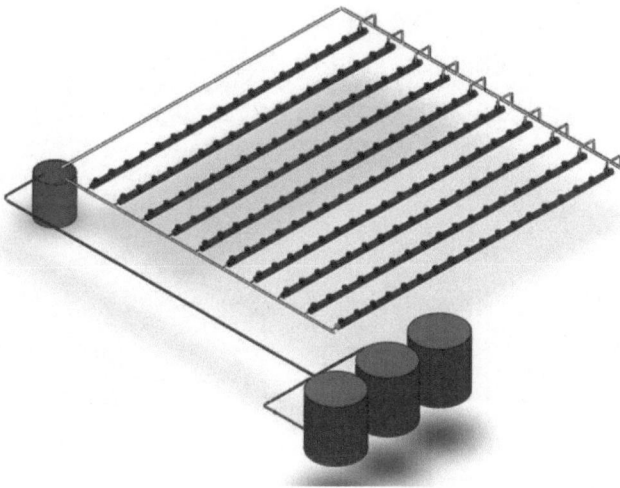

Figure 6.15 Hydroponics and Nutrient Supply System – CAD.

Figure 6.16 PVC Pipe System.

The nutrient supply system consists of three tanks that store the necessary solutions for every stage of crop growth as seen in Figure 6.18. The number of tanks varies according to the requirements of nutrition stages of crop.

Each tank has a volume of 1,200 L and a control valve, as seen in Figure 6.19 and 6.20, at the outlet which allows sending of necessary nutrient solution to the collector tank depending on the stage in which the crop is. The electro-valves are solenoid type 120 V 50/60 Hz (on/off) with a diameter of 1 in.

Moreover, the composition of nutrient solutions for dealing with the growth process of tomatoes is shown in Table 6.8. Also, Table 6.9 shows an irrigation program of 27 weeks depending on the stage of growth of the tomato plant.

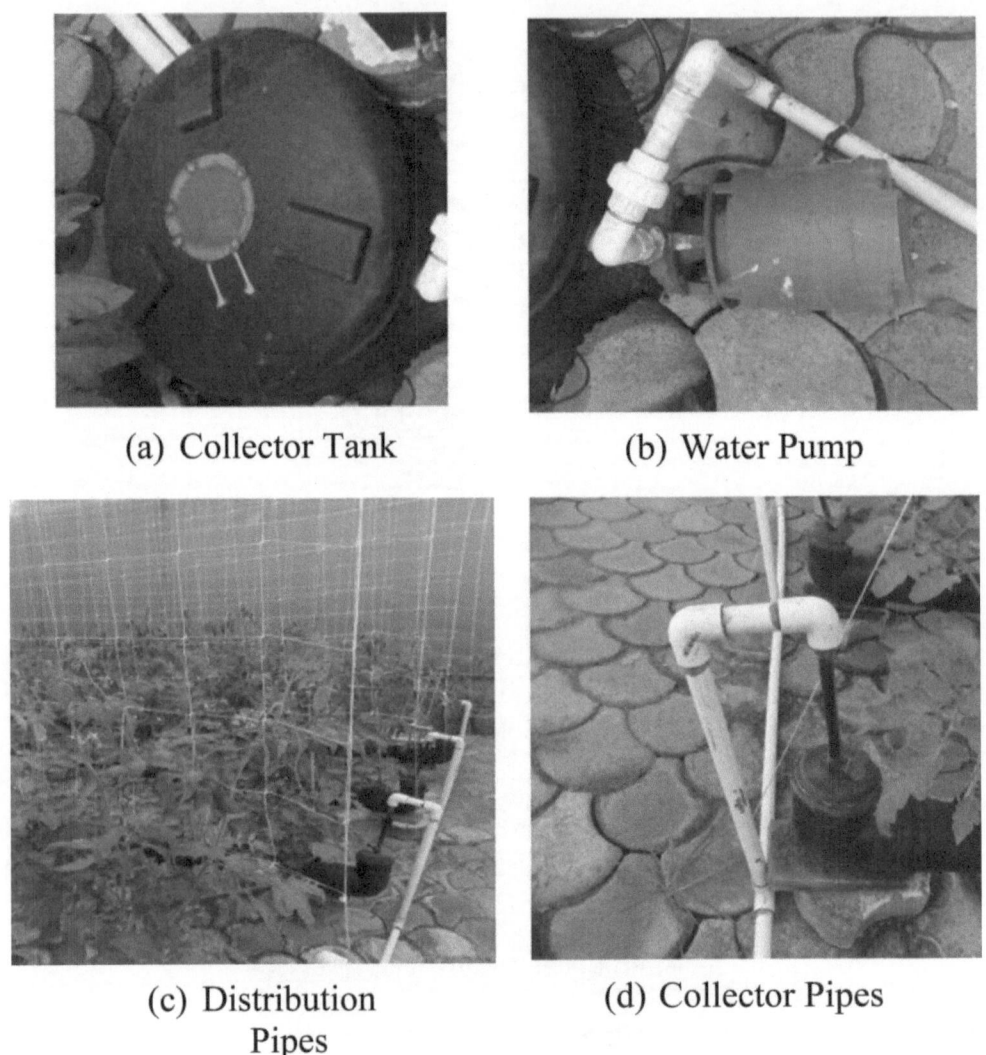

(a) Collector Tank   (b) Water Pump

(c) Distribution   (d) Collector Pipes
Pipes

Figure 6.17 Hydroponic System Components.

The complete design of the intelligent greenhouse is presented in Figure 6.21; this design is the assembly that meets all automation systems described previously. Figure 6.22 shows a quick summary of the intelligent greenhouse construction.

6.1.6 Sensors

The intelligent greenhouse measures temperature, relative humidity, light intensity, electrical conductivity and pH of the nutrient solution, and the level of solution in the collector and storage tanks. For this reason, the intelligent greenhouse is provided

Table 6.7 Hydroponic Pump Data-sheet.

Siemens 1RF3-252-2YC34	
Power	1/4 HP
Voltage	127 V
Current	4.5 A
Phase	1
Frequency	60 Hz
Speed	3450 RPM
Insulation Class	B
SF	1.8
Nominal Efficiency	55%
Others	NEMA 5 thermally protected, drip proof

Figure 6.18 Nutrient Storage Tanks.

Figure 6.19 Electro-valve System.

Figure 6.20 ASCO Electro-valve.

Table 6.8 Nutritive Solution Formulas for Tomatoes.

FORMULA I 12-11-35 + *Micro-nutrients*		FORMULA II 5-11-26 + *Micro-nutrients*	
Total Nitrogen	12%	Total Nitrogen	5%
Phosphorous (P_2O_5)	11%	Phosphorous (P_2O_5)	11%
Potassium (K_2O)	35%	Potassium (K_2O)	26%
Sulphur (S)	1700 ppm	Sulphur (S)	1700 ppm
Magnesium (Mg)	900 ppm	Magnesium (Mg)	900 ppm
Manganese (Mn)	170 ppm	Manganese (Mn)	170 ppm
Zinc (Zn)	540 ppm	Zinc (Zn)	540 ppm
Copper (Cu)	10 ppm	Copper (Cu)	10 ppm
Boron (B)	140 ppm	Boron (B)	140 ppm
Iron (Fe)	30 ppm	Iron (Fe)	30 ppm
Molybdenum (Mo)	0.2 ppm	Molybdenum (Mo)	0.2 ppm
Cobalt (Co)	0.15 ppm	Cobalt (Co)	0.15 ppm
Fulvic acids	1%	Fulvic acids	1%

Soluble powder – Polypropylene bags of 25 Kg

with sensors related to these variables, which are the following. Before proceeding, Figure 6.23 indicates the location where the sensors are placed in the greenhouse.

Temperature and humidity sensors – A temperature/humidity sensor PH1125 (ratiometric) measures temperature in the range of −40°C to 100°C and relative humidity from 10% to 95%. There are two of these sensors, one measuring inside the intelligent greenhouse and the other outside. The internal sensor is located at a height of approximately 10 cm above the plants, in order to measure temperature and relative humidity close to the area where the plants are. Table 6.10 shows the technical information of the temperature/humidity sensor. Figure 6.24 and 6.25 depict the real sensor and a shield (box), respectively, this last built with the aim of protecting them against severe environment.

Table 6.9 Irrigation Plan for Tomatoes.

Week	Nutrients (Formula)	Week	Nutrients (Formula)
I	I	15	II + Ca
2	I	16	II + Ca
3	I + Ca	17	II + Ca
4	I + Ca	18	II + Ca
5	I	19	II + Ca
6	I	20	II + Ca
7	I	21	II + Ca
8	I	22	II + Ca
9	II + Ca	23	II + Ca
10	II + Ca	24	II + Ca
11	II + Ca	25	II + Ca
12	II + Ca	26	II + Ca
13	II + Ca	27	II + Ca
14	II + Ca		

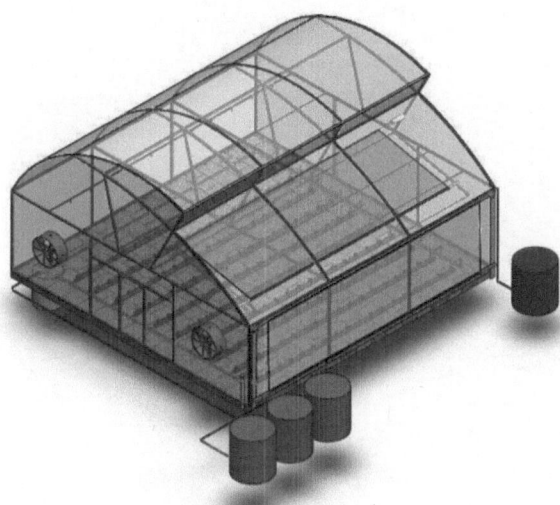

Figure 6.21 Intelligent Greenhouse – CAD.

The sensor output is a voltage ranging from 0 to 5 V, which is translated to temperature and relative humidity (experimental characterization) as follows:

$$T(°C) = 43.454 \cdot V - 58.27 \tag{6.1}$$

$$HR(\%) = 38.12 \cdot V - 40.2 \tag{6.2}$$

Light intensity sensors – In the greenhouse are two light intensity sensors placed under the shading system at a height of approximately 30 cm above the plants and on each side of the greenhouse. The light intensity sensors used are two LDRs (Light

Figure 6.22 Intelligent Greenhouse Summary.

Dependent Resistors) of 2 MΩ. As this works as a variable resistor, the greater the quantity of received light, the lower its resistance. Taking advantage of this feature a voltage divider is used where the LDR is the top resistor as can be seen in Figure 6.26 and 6.27, so giving the maximum voltage during full light.

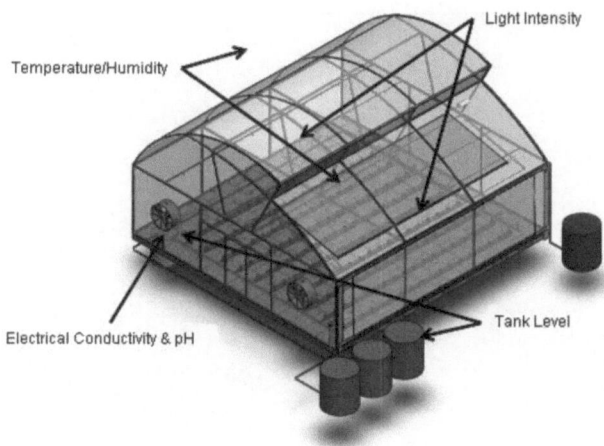

Figure 6.23 Sensor Location.

Table 6.10 Temperature/Humidity Sensor Technical Information.

Phidgets-PH1125

Temperature Sensor	
Current consumption	300 μA
Output impedance	1 kOhm
Range of operation	−40°C to 100°C
Accuracy	1°C
Min/max voltage	4.75–5.25 VDC
Humidity Sensor	
Current consumption	3.6 mA
Output impedance	1 kOhm
Range of operation	10% to 95%
Accuracy @ 55% RH	±2% RH
Accuracy over 10% to 95%	±3% RH Typical
Reaction time for humidity	10 seconds
Min/max voltage	4.75–5.25 VDC

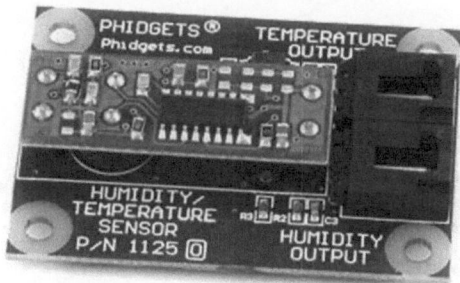

Figure 6.24 Temperature/Humidity Sensor – PH1125.

Figure 6.25 Temperature/Humidity Sensor Shield.

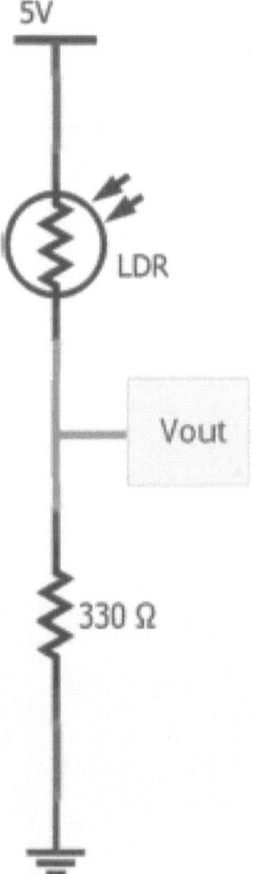

Figure 6.26 Connection Mode of a LDR.

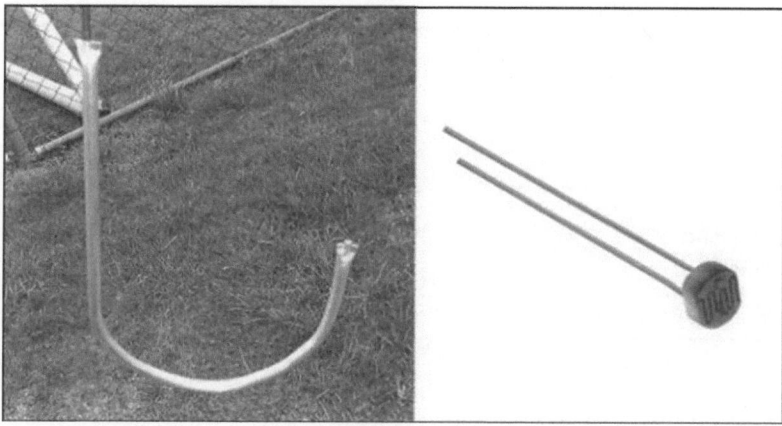

Figure 6.27 LDR.

Table 6.11 pH Probe Specifications.

pH Probe Hanna Instruments – HI 1230	
Reference	Double, Ag/AgCl
Junction/flow Rate	Ceramic, single
Electrolyte	Gel
Max pressure	2 bar
Range	0 to 13 pH; 0 to 80°C (68 to 104°F)
Temperature sensor	No
Amplifier	No
Body material	PEI (PolyEther-Imide)
Cable/connection	Coaxial; 1 m (3.3′)/BNC

Figure 6.27 shows the method used to guarantee that the LDR is always facing up; it consists of an aluminum U-shaped tube. The sensor output is a voltage ranging from 0 to 5 V, which is translated to light intensity (experimental characterization) as follows:

$$LI(kLux) = -2.813 \cdot V^3 + 18.255 \cdot V^2 - 3.8998 \cdot V + 12.981 \tag{6.3}$$

Electrical conductivity and pH sensors – For managing hydroponics and the nutrient supply system, two sensors are required for reading the pH and the electrical conductivity (EC) of the irrigation solution. Both sensors are composed of a probe/electrode and an electrical circuit in charge of signal conditioning and amplification. On one hand, the pH probe specifications are shown in Table 6.11 while, on the other, the electrical conductivity probe was built by two stainless steel screws spaced 1 cm, a

Table 6.12 Ultrasonic Sensor Details.

PING))) Ultrasonic Distance Sensor (#28015)

Perception	Narrow acceptance angle
Range	1 inch to 10 feet (2 cm to 3 m)
Input trigger	+TTL pulse, 2 μs min, 5 μs typ
Echo pulse	+TTL pulse, 115 μs min to 18.5 ms max
Package	3-pin SIP, 0.1″ spacing (ground, power, signal)
Power requirements	+5 VDC; 35 mA active
Communication	positive TTL pulse
Dimensions	0.81 × 1.8 × 0.6 in (22 × 46 × 16 mm)
Operating temperature range	+32 to +158°F (0 to +70°C)

Table 6.13 Atmel 8-bit ATMEGA8535L Microcontroller Specifications.

ATMEGA8535L Microcontroller

Max operating frequency	8 MHz
CPU	8-bit AVR
ISP flash	8 Kb
General purpose I/O	32 lines
Timer/counters	3
Ext interrupts	3
Pin count	40
# of touch panels	16
A/D converter	8-channel 10-bit
Supply voltage	2.7–5.5 VDC

coaxial cable 1 m, a PVC pipe section and polyester resin. Figure 6.28 shows both probes, pH and EC, trapped together.

Sensor probes need an amplification stage with the aim of raising the voltage obtained from mV to V. The circuit schematics are presented in Figures 6.29 and 6.30 where the electronic components used can be easily detailed. Moreover, Figure 6.31 shows a case (box) where the real printed circuits are located, including the power source that feeds them; the power supply needed is +12 VDC and −12 VDC, extracted from a common computer power system.

The sensors' output voltage is calibrated in such a manner to translate the voltage directly into the measured variable. This means that getting 6 V from the pH sensor will indicate directly a 6 in the pH scale and, in the same sense for the EC sensor, if the output voltage is 2 V the EC will be 2 dS/m.

Tank level sensors – For controlling the nutrients supply, in addition to the pH and EC sensors, there is a necessity to know the level of the tanks involved in the system for sending the correct amount of nutrient solution and, moreover, for protecting the system from overflows in the collector tank. Two ultrasonic sensors are used for this purpose as level detectors, which specifications are given in Table 6.12.

The PING))) sensor is interfaced to an ATMEGA8535L microcontroller whose general specifications are given in Table 6.13. A single I/O pin is used to trigger an

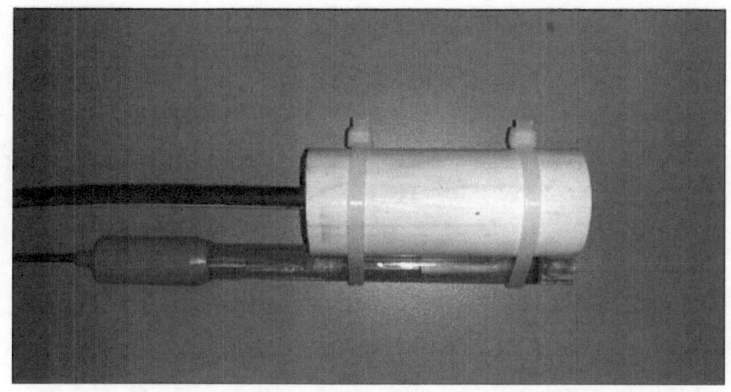

Figure 6.28   pH and EC Probes.

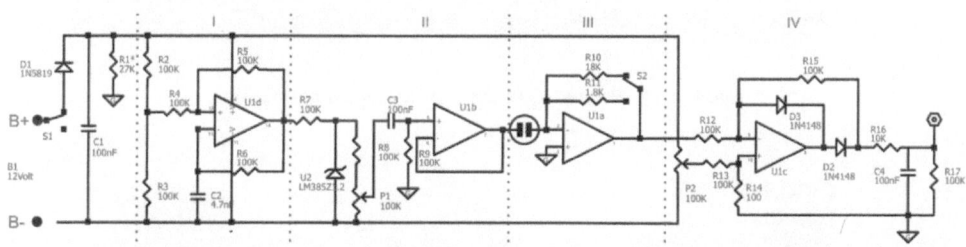

Figure 6.29   Electrical Conductivity Circuit.

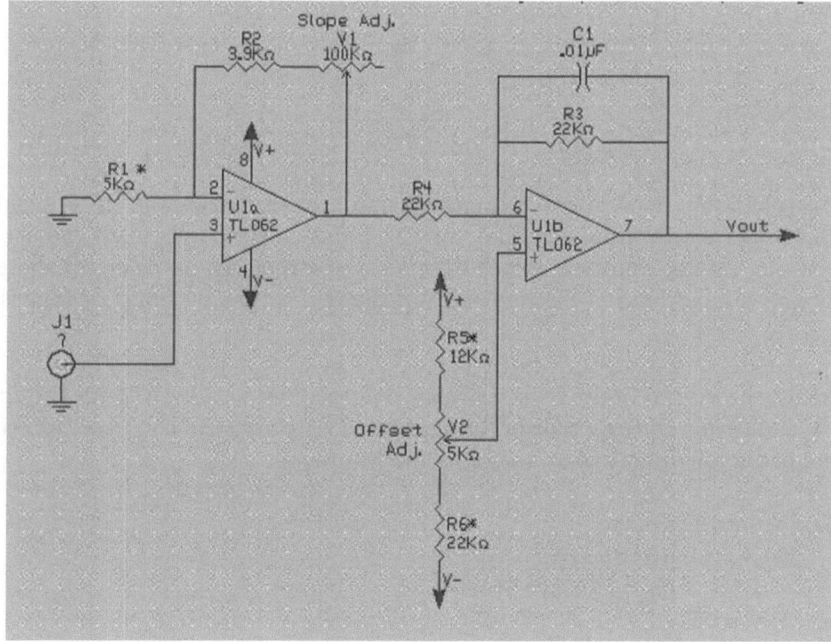

Figure 6.30   pH Circuit.

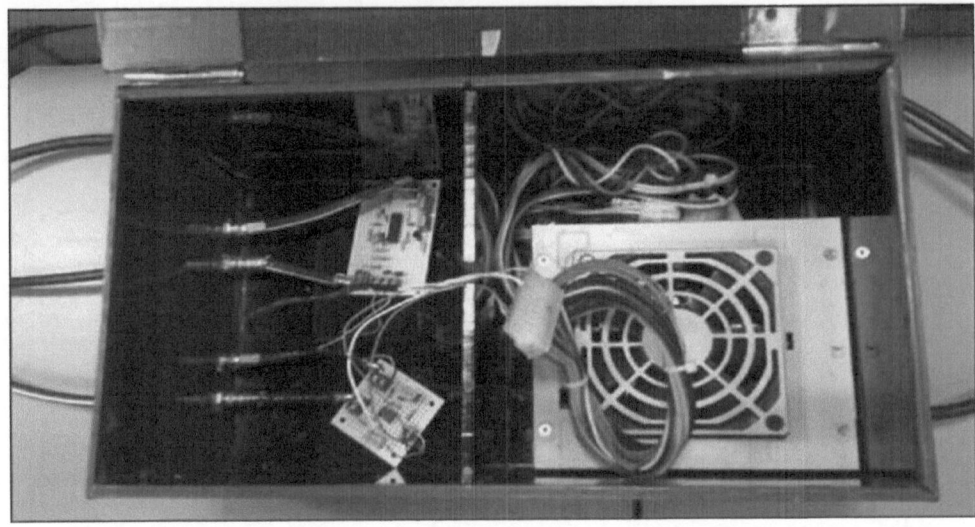

Figure 6.31 Printed Circuits, Power Supply and Casing.

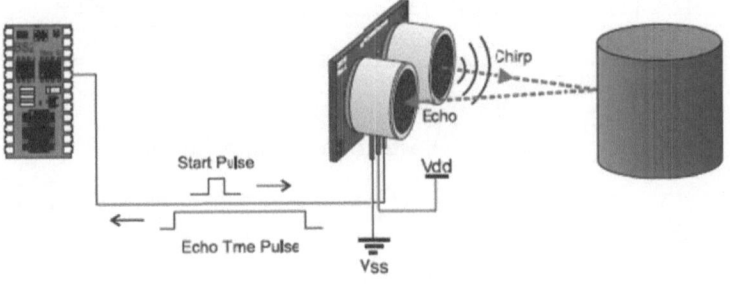

Figure 6.32 PING))) Ultrasonic Sensor Principle.

ultrasonic burst (well above human hearing) and providing an output pulse that corresponds to the time required for the burst echo to return to the sensor. By measuring the echo pulse width, the distance to target can easily be calculated. This principle is illustrated in Figure 6.32 and the real sensor with its package is shown in Figure 6.33.

Finally, the output of the sensor system (microcontroller) is a voltage ranging from 0 to 5 V, indicating linearly a distance between 2 cm and 3 m. The code implemented, in C language, for the microcontroller is the following:

```
#include <avr/io.h>
#include <avr/delay.h>

int main(void)
```

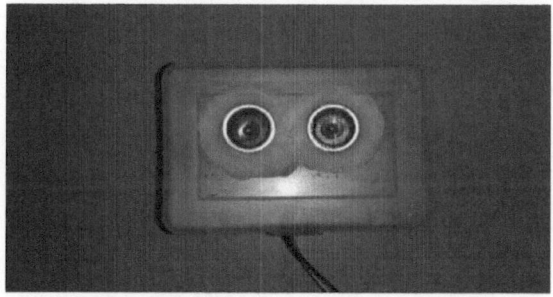

Figure 6.33 PING))) Ultrasonic Sensor.

```
{
    double duration;
    TCCR2 |= (1 << COM21; //Set PWM with timer2
    TCCR2 |= (1 << WGM21) | (1 << WGM20);
    TCCR2 |= (1 << CS21);
    DDRD |= (1 << PD7); //Set port D7 to OUTPUT PWM

    while(1)
    {
        DDRB = 0x01; //Set port B0 as OUTPUT
        PORTB = 0x00; //Set port B0 as LOW
        _delay_us(2); //Wait for 2 uS
        PORTB = 0x01; //Set port B0 as HIGH (Trigger)
        _delay_us(5); //Wait for 5 uS
        PORTB = 0x00; //Set port B0 as LOW
        DDRB = 0x00; //Set port B0 as INPUT

        for (int i=0;i<255;i++)
        {

            for (int j=0;j<36;j++)
            {
                if (PINB & (1<<PB0)) //Wait port B0 to be HIGH
                {
                    duration=i; //Echo pulse width
                    break;
                }
            }
        }
        OCR2 = duration; //PWM proportional to distance
        _delay_ms(100);
    }
}
```

Table 6.14 NI CompactDAQ-9174 Overview.

cDAQ-9174	
Product type	Chassis, measurement device
USB power	External-powered
Power requirements	15W
Number of slots	4
Input voltage range	9V–30V
Counter/Timers	4
Counter/timer resolution	32 bits
Random operating frequency	5 Hz–500 Hz
USB specification	USB 2.0 Hi-Speed

Table 6.15 Relay Module NI-9481 Summary.

NI 9481	
Number of channels	4 electromechanical relay
Relay type	Single pole single throw (SPST)
Switching voltage/current	30 VDC (2 A), 60 VDC (1 A), 250 VAC (2A)
Switching rate	1 operation per second
Life expectancy (operations)	20,000,000 (mechanical) – 100,000 (electrical)

6.2 ELECTRICAL AND ELECTRONIC SYSTEM

Electric and electronic system uses a circuit for interfacing hardware which is coupled to a computer that will store the intelligent control system. Moreover, the system has a data acquisition target for measuring and generating signals.

On one hand, an *IBM Thinkpad* computer with Windows 7 Professional Edition, Pentium IV processor at 2.8 GHz and 2GB of RAM is used. This computer has installed the LabVIEW™ and the Intelligent Control Toolkit for LabVIEW™ programs.

On the other hand, NI CompactDAQ-9174 and NI USB-6009 data acquisition targets are used for acquiring and generating signals. The cDAQ-9174 is a chassis with four slots for handling four different modules, in this case one module of analog inputs NI-9221 and three modules of relays NI-9481 are used. In general, these optimized devices give good accuracy at fast sampling rates. Tables 6.14, 6.15 and 6.16 give a quick overview of the data acquisition targets and Figure 6.34 illustrates them.

In order to control the ventilation system motion, a set of relays and variable speed drives (or variable frequency drives) are implemented. The relays, which specifications are shown in Table 6.17, are used for controlling the gear-motors of the natural ventilation system and the speed drives, given in Table 6.18, for managing the three-phase motors of the exhaust fans.

Table 6.16 NI USB-6009 DAQ and Analog Module NI-9221 Specifications.

	USB-6009	NI-9201
Measurement type	Voltage	Voltage
Analog inputs	8 SE or 4 Diff	8 SE
Analog outputs	2	0
Digital channels	12 I/O	0
Sample rate	48 kS/s	500 kS/s
Analog resolution	14 bits	12 bits
Voltage range	−10V–10V	−10V–10V
Digital logic levels	0V–5V	−
Current (channel/total)	8.5 mA/102 mA	−
Counter/timer	1	0
Counter/timer resolution	32 bits	−
Max counter/timer frequency	5 MHz	−
Pulse generation	No	−

Figure 6.34 National Instruments Hardware – DAQ Systems.

Table 6.17 Power Relays Specifications.

SUN HOLD RHL-2-12D-10H	
Relay type	2 poles/2 throws (DPDT)
Nominal coil voltage	12 VDC
Min/max voltage	9 VDC–15 VDC
Coil power consumption	0.9 W–1.5 W
Contact material	Silver alloy
Max contacts current	12 A (10 A continuous)
Max contacts voltage	28 VDC–250 VAC
Operation time	20 ms Max
Life expectancy (operations)	5,000,000 (mechanical)–500,000 (electrical)

Table 6.18 Variable Speed Drive Altivar 12 Details.

ATV12H037F1	
Product destination	Asynchronous motors
Product specific application	Simple machine
Network number of phases	Single-phase
Rated supply voltage	100–120 V (−15–10%)
Motor power	0.55 hp
Line current	9.3 A at 120 V, 11.4 A at 100 V
Supply frequency	50/60 Hz (+/−5%)
Asynchronous motor control profile	Quadratic voltage/frequency ratio
	Sensorless flux vector control
	Voltage/frequency ratio (V/f)
IP degree of protection	IP20
Speed drive output frequency	0.5–400 Hz
Electrical connection	L1, L2, L3, U, V, W, PA, PC
Analogue input number	1
Analogue output number	1
Discrete input number	4
Discrete output number	2
Acceleration/deceleration ramps	Linear from 0 to 999.9 s, S and U

To generate the gear-motors direction and prevent collisions in the windows and meshes, switch sensors were placed at the end of the run of each system. Figure 6.35 shows the block diagram of the overall connections, while Table 6.19 summarizes the mnemonics.

The computer program LabVIEW™ 2012 is capable of interpreting the measurements made by the sensors and to generate signals through the relay modules and DAQs.

Finally, as a result of putting together the circuits for interfacing hardware, the data acquisition targets and the control computer, the real implementation is achieved as can be seen in Figures 6.36 (a and b) and 6.37. The real implementation consists in two electrical cabinets where the relay arrangement and the variable speed drivers are localized, and also, the computer responsible for making decisions according to the control strategy programmed.

6.3   CONTROL SYSTEM FOR A GREENHOUSES

Two control systems will be presented: a pure on/off and an intelligent control system. This latter combines non-linear (Sliding Modes, Feedback/Feed-forward Linearization) and artificial intelligence control techniques (Fuzzy Logic, ANFIS).

Those techniques are presented in order to show how the non-conventional controllers could improve the greenhouse performance.

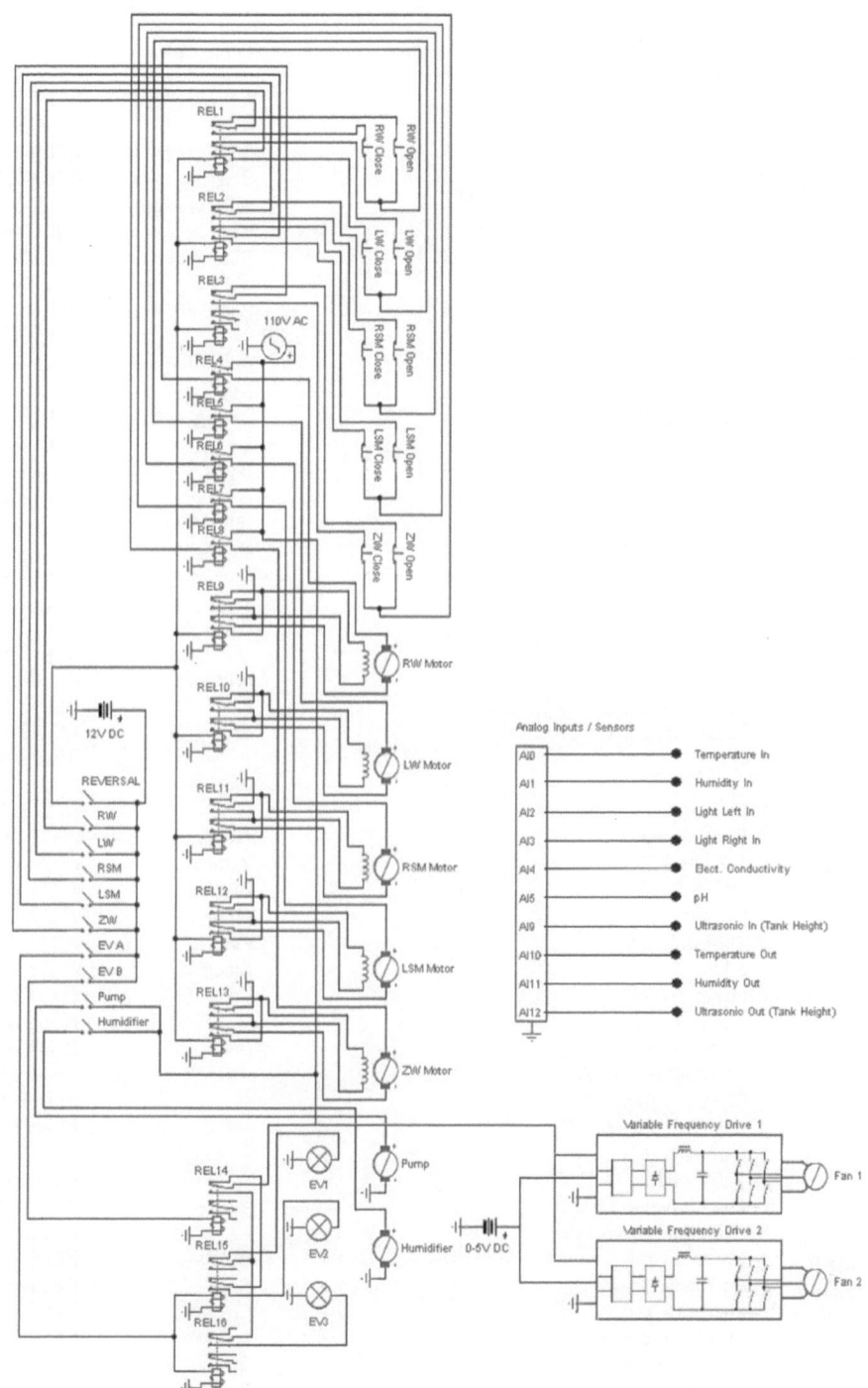

Figure 6.35 Electric/Electronic System Diagram.

Table 6.19 Mnemonics of Electrical/Electronic System.

Name	Description	Connections	Pin	Cable Color	Connected to:
RW Motor		M1-1	–	Yellow/white	REL4-3
	Motor 1/15 HP, 115V,	M1-2	–	Red	REL9-6
	50/60 Hz, 1.3A	M1-3	–	Blue/green	REL9-1, REL9-4
		M1-4	–	Black	REL9-2, REL9-3
LW Motor		M2-1	–	Yellow/white	REL5-3
	Motor 1/15 HP, 115V,	M2-2	–	Red	REL10-6
	50/60 Hz, 1.3A	M2-3	–	Blue/green	REL10-1, REL10-4
		M2-4	–	Black	REL10-2, REL10-3
RSS Motor		M3-1	–	Yellow/white	REL6-3
	Motor 1/15 HP, 115V,	M3-2	–	Red	REL11-6
	50/60 Hz, 1.3A	M3-3	–	Blue/green	REL11-1, REL11-4
		M3-4	–	Black	REL11-2, REL11-3
LSS Motor		M4-1	–	Yellow/white	REL7-3
	Motor 1/15 HP, 115V,	M4-2	–	Red	REL12-6
	50/60 Hz, 1.3A	M4-3	–	Blue/green	REL12-1, REL12-4
		M4-4	–	Black	REL12-2, REL12-3
ZW Motor		M5-1	–	Yellow/white	REL8-3
	Motor 1/15 HP, 115V,	M5-2	–	Red	REL13-6
	50/60 Hz, 1.3A	M5-3	–	Blue/green	REL13-1, REL13-4
		M5-4	–	Black	REL13-2, REL13-3
RS Fan	Exhaust Fan	RSF L1	–	Green	VFD1 U
	1/2 HP, 230V,	RSF L2	–	White	VFD1 V
	60 Hz, 2.6A	RSF L3	–	Black	VFD1 W
LS Fan	Exhaust Fan	LSF L1	–	Green	VFD2 U
	1/2 HP, 230V,	LSF L2	–	White	VFD2 V
	60 Hz, 2.6A	LSF L3	–	Black	VFD2 W
Pump	Pump 1/4 HP,	P	–	–	SLOT2-CH2b
	127V, 60 Hz, 4.5A	N	–	–	Neutro AC
Fog System	Humidifier 1/2 HP,	H	–	–	SLOT2-CH3b
	110V, 60 Hz, 4.2 A	N	–	–	Neutro AC
EV1	Electro-valve D 1″,	E1	–	–	REL15-1
	120V, 60 Hz, 16.1W	N	–	–	Neutro AC
EV2	Electro-valve D 1″,	E2	–	–	REL15-4
	120V, 60 Hz, 16.1W	N	–	–	Neutro AC
EV3	Electro-valve D 1″,	E3	–	–	REL16-3
	120V, 60 Hz, 16.1W	N	–	–	Neutro AC
RW Open	Micro switch	M11-1	–	Black	REL1-2
	normally closed	M11-2	–	Red	REL4-7
RW Close	Micro switch	M12-1	–	Green	REL1-4
	normally closed	M12-2	–	Yellow	REL4-7
LW Open	Micro switch	M21-1	–	Black	REL1-1
	normally closed	M21-2	–	Red	REL5-7
LW Close	Micro switch	M22-1	–	Green	REL1-3
	normally closed	M22-2	–	Yellow	REL5-7

Table 6.19   (Continued)

Name	Description	Connections	Pin	Cable Color	Connected to:
RSM Open	Micro switch	M31-1	–	Black	REL2-2
	normally closed	M31-2	–	Red	REL6-7
RSM Close	Micro switch	M32-1	–	Green	REL2-4
	normally closed	M32-2	–	Yellow	REL6-7
LSM Open	Micro switch	M41-1	–	Black	REL2-1
	normally closed	M41-2	–	Red	REL7-7
LSM Close	Micro switch	M42-1	–	Green	REL2-3
	normally closed	M42-2	–	Yellow	REL7-7
ZW Open	Micro switch	M51-1	–	Black	REL3-2
	normally closed	M51-2	–	Red	REL8-7
ZW Close	Micro switch	M52-1	–	Green	REL3-4
	normally closed	M52-2	–	Yellow	REL8-7
REL1		M21-1	1	–	LW Open
		M11-1	2	–	RW Open
		M22-1	3	–	LW Closed
	Relay Sun-Hold	M12-1	4	–	RW Closed
	RHL-2-12D-10-H	LW	5	–	SLOT1-CH3b
		RW	6	–	SLOT2-CH0b
		R	7	–	SLOT1-CH0b
		0V	8	–	0V DC
REL2		M41-1	1	–	LSM Open
		M31-1	2	–	RSM Open
		M42-1	3	–	LSM Closed
	Relay Sun-Hold	M32-1	4	–	RSM Closed
	RHL-2-12D-10-H	LSM	5	–	SLOT1-CH1b
		RSM	6	–	SLOT1-CH2b
		R	7	–	SLOT1-CH0b
		0V	8	–	0V DC
REL3		–	1	–	–
		M51-1	2	–	ZW Open
		–	3	–	–
	Relay Sun-Hold	M52-1	4	–	ZW Closed
	RHL-2-12D-10-H	–	5	–	–
		ZW	6	–	SLOT2-CH1b
		R	7	–	SLOT1-CH0b
		0V	8	–	0V DC
REL4		–	1	–	–
		–	2	–	–
		M1-1	3	–	RW Motor
	Relay Sun-Hold	–	4	–	–
	RHL-2-12D-10-H	110V	5	–	110V AC
		–	6	–	–
		M11-2, M12-2	7	–	RW Open/Closed
		0V	8	–	0V DC
REL5		–	1	–	–
		–	2	–	–
		M2-1	3	–	LW Motor
	Relay Sun-Hold	–	4	–	–
	RHL-2-12D-10-H	110V	5	–	110V AC
		–	6	–	–
		M21-2; M22-2	7	–	LW Open/Closed
		0V	8	–	0V DC

Table 6.19   (*Continued*)

Name	Description	Connections	Pin	Cable Color	Connected to:
REL6		—	1	—	—
		—	2	—	—
		M3-1	3	—	RSM Motor
	Relay Sun-Hold	—	4	—	—
	RHL-2-12D-10-H	110V	5	—	110V AC
		—	6	—	—
		M31-2; M32-2	7	—	RSM Open/ Closed
		0V	8	—	0V DC
REL7		—	1	—	—
		—	2	—	—
		M4-1	3	—	LSM Motor
	Relay Sun-Hold	—	4	—	—
	RHL-2-12D-10-H	110V	5	—	110V AC
		—	6	—	—
		M41-2; M42-2	7	—	LSM Open/ Closed
		0V	8	—	0V DC
REL8		—	1	—	—
		—	2	—	—
		M5-1	3	—	ZW Motor
	Relay Sun-Hold	—	4	—	—
	RHL-2-12D-10-H	110V	5	—	110V AC
		—	6	—	—
		M51-2; M52-2	7	—	ZW Open/ Closed
		0V	8	—	0V DC
REL9		M1-3	1	—	RW Motor
		M1-4	2	—	RW Motor
		M1-4	3	—	RW Motor
	Relay Sun-Hold	M1-3	4	—	RW Motor
	RHL-2-12D-10-H	N	5	—	Neutro AC
		M1-2	6	—	RW Motor
		R	7	—	SLOT1-CH0b
		0V	8	—	0V DC
REL10		M2-3	1	—	LW Motor
		M2-4	2	—	LW Motor
		M2-4	3	—	LW Motor
	Relay Sun-Hold	M2-3	4	—	LW Motor
	RHL-2-12D-10-H	N	5	—	Neutro AC
		M2-2	6	—	LW Motor
		R	7	—	SLOT1-CH0b
		0V	8	—	0V DC
REL11		M3-3	1	—	RSM Motor
		M3-4	2	—	RSM Motor
		M3-4	3	—	RSM Motor
	Relay Sun-Hold	M3-3	4	—	RSM Motor
	RHL-2-12D-10-H	N	5	—	Neutro AC
		M3-2	6	—	RSM Motor
		R	7	—	SLOT1-CH0b
		0V	8	—	0V DC

Table 6.19   (Continued)

Name	Description	Connections	Pin	Cable Color	Connected to:
REL12		M4-3	1	–	LSM Motor
		M4-4	2	–	LSM Motor
		M4-4	3	–	LSM Motor
	Relay Sun-Hold	M4-3	4	–	LSM Motor
	RHL-2-12D-10-H	N	5	–	Neutro AC
		M4-2	6	–	LSM Motor
		R	7	–	SLOT1-CH0b
		0V	8	–	0V DC
REL13		M5-3	1	–	ZW Motor
		M5-4	2	–	ZW Motor
		M5-4	3	–	ZW Motor
	Relay Sun-Hold	M5-3	4	–	ZW Motor
	RHL-2-12D-10-H	N	5	–	Neutro AC
		M5-2	6	–	ZW Motor
		R	7	–	SLOT1-CH0b
		0V	8	–	0V DC
REL14		–	1	–	–
		E5	2	–	REL15-6
		E4	3	–	REL15-5, REL16-5
	Relay Sun-Hold	–	4	–	–
	RHL-2-12D-10-H	110V	5	–	110V AC
		110V	6	–	110V AC
		EVB	7	–	SLOT3-CH1b
		0V	8	–	0V DC
REL15		E1	1	–	EV1
		–	2	–	–
		–	3	–	–
	Relay Sun-Hold	E2	4	–	EV2
	RHL-2-12D-10-H	E4	5	–	REL14-3
		E5	6	–	REL14-2
		EVA	7	–	SLOT3-CH0b
		0V	8	–	0V DC
REL16		–	1	–	–
		–	2	–	–
		E3	3	–	EV3
	Relay Sun-Hold	–	4	–	–
	RHL-2-12D-10-H	E4	5	–	REL14-3
		–	6	–	–
		EVA	7	–	SLOT3-CH0b
		0V	8	–	0V DC
VFD1		0-5V	AI1	–	USB AO0
	Variable frequency	0V	COM	–	0V DC
	drive 1 HP,	RSF L1	U	–	Right fan
	230V, 1ph to 3ph	RSF L2	V	–	–
		RSF L3	W	–	–
VFD2		0-5V	AI1	–	USB AO1
	Variable frequency	0V	COM	–	0V DC
	drive 1 HP,	LSF L1	U	–	Left fan
	230V, 1ph to 3ph	LSF L2	V	–	–
		LSF L3	W	–	–

Table 6.19 (Continued)

Name	Description	Connections	Pin	Cable Color	Connected to:
SLOT0		TH1-1	AI0	–	THS1-1
		TH2-1	AI1	–	THS2-1
		TH1-4	AI2	–	THS1-4
		TH2-4	AI3	–	THS2-4
	Compact DAQ	LL-1	AI4	–	LLS-1
	SLOT0	LR-1	AI5	–	LRS-1
		EC-1	AI6	–	ECS-1
		PH-1	AI7	–	PHS-1
		–	NC	–	–
		0V	COM	–	0V DC
SLOT1		12V	CH0a	–	12V DC
		R	CH0b	–	REL1,2,3,9,10,11,12,13-7
		18V	CH1a	–	18V DC
		LSS	CH1b	–	REL2-5
	Compact DAQ	18V	CH2a	–	18V DC
	SLOT1	RSS	CH2b	–	REL2-6
		18V	CH3a	–	18V DC
		LW	CH3b	–	REL1-5
		–	NC	–	–
		–	NC	–	–
SLOT2		18V	CH0a	–	18V DC
		RW	CH0b	–	REL1-6
		18V	CH1a	–	18V DC
		ZW	CH1b	–	REL3-6
	Compact DAQ	110V	CH2a	–	110V AC
	SLOT2	P	CH2b	–	Pump
		110V	CH3a	–	110V AC
		H	CH3b	–	Humidifier
		–	NC	–	–
		–	NC	–	–
		12V	CH0a	–	12V DC
		EVA	CH0b	–	REL15-7; REL16-7
		12V	CH1a	–	12V DC
		EVB	CH1b	–	REL14-7
SLOT3	Compact DAQ	–	CH2a	–	–
	SLOT3	–	CH2b	–	–
		–	CH3a	–	–
		–	CH3b	–	–
		–	NC	–	–
		–	NC	–	–
DAQ		LT1-1	AI0	–	LTS1-1
	USB	LT2-1	AI1	–	LTS2-1
	DAQ	–	AO0	–	VFD1 AI1
	6009	–	AO1	–	VFD2 AI1
		0V	COM	–	0V DC
THS1		TH1-1	T0	White	SLOT0-AI0
		5V	–	Red	5V DC
	Temperature/humidity	0V	–	Black	0V DC
	sensor Phidgets 1125	TH1-4	H0	White	SLOT0-AI2
		5V	–	Red	5V DC
		0V	–	Black	0V DC

Table 6.19 (Continued)

Name	Description	Connections	Pin	Cable Color	Connected to:
THS2		TH2-1	TO	White	SLOT0-AI1
		5V	–	Red	5V DC
	Temperature/humidity	0V	–	Black	0V DC
	sensor Phidgets 1125	TH2-4	HO	White	SLOT0-AI3
		5V	–	Red	5V DC
		0V	–	Black	0V DC
LLS	3*Light sensor LDR	LL-1	–	White	SLOT0-AI4
		5V	–	Red	5V DC
		0V	–	Blue	0V DC
LRS	3*Light sensor LDR	LR-1	–	White	SLOT0-AI5
		5V	–	Red	5V DC
		0V	–	Blue	0V DC
ECS		EC-1	–	Copper	SLOT0-AI6
	Electrical Conductivity	+12V	–	–	+12V DC
	Sensor	-12V	–	–	-12V DC
		0V	–	Silver	0V DC
PHS	4*pH Sensor	PH-1	–	Copper	SLOT0-A7
		+12V	–	–	+12V DC
		−12V	–	–	−12V DC
		0V	–	Silver	0V DC
LTS1	Ping))) ultrasonic	LT1-1	–	White	DAQ AI0
	distance sensor	5V	–	Red	5V DC
	#28015	0V	–	Blue	0V DC
LTS2	Ping))) ultrasonic	LT2-1	–	White	DAQ AI1
	distance sensor	5V	–	Red	5V DC
	#28015	0V	–	Blue	0V DC

(a) Relay Arrangement

(b) Variable Speed Drivers

Figure 6.36 Electrical/Electronic System.

6.3.1 Intelligent control system development

The intelligent control system impacts over the climatic conditions in terms of temperature and relative humidity, and the nutrients supply system in terms of electrical conductivity. These controllers are discussed below.

Figure 6.37 Electrical Cabinets and Control Computer.

Sliding Modes + Feedback/Feed-forward Linearization Control

According to the non-linear coupled state-space model of the greenhouse climate, the system could be expressed as:

$$x_1 = \hat{u}_1 = \frac{1}{\varrho C_p V_1^v} - \frac{\delta}{\varrho C_p V_2^u} - \frac{UA}{\varrho C_p V_1^x} + \frac{UA}{\varrho C_p V_2^v} - \frac{1}{V}[x_1 - v_2]u_1 \qquad (6.4)$$

$$x_2 = \hat{u}_2 = \frac{1}{\varrho V_2^u} + \frac{\alpha}{\varrho \delta V_1^v} - \frac{\beta_T}{\varrho V_2^x} + \frac{1}{V}[x_2 - v_3]u_1 \qquad (6.5)$$

where,

$$x = [x_1 x_2]^T = [T_{in} W_{in}]^T \, u = [u_1 u_2]^T = [V Q_{fog}]^T \, v = [v_1 v_2 v_3]^T = [S_i T_{out} W_{out}]^T$$

One can obtain the controller equations in the form of Sliding Modes + Feedback/Feed-forward Linearization, as follows:
From 6.4,

$$\varrho C_p V \hat{u}_1 = v_1 - \delta u_2 - UA x_1 + UA v_2 - \varrho C_p (x_1 - v_2)u_1 \qquad (6.6)$$

$$u_1 = \frac{v_1 - \delta u_2 - UA x_1 + UA v_2 - \varrho C_p V \hat{u}_1}{\varrho C_p (x_1 - v_2)} \qquad (6.7)$$

From 6.5,

$$\varrho V \hat{u}_2 = u_2 + \frac{\alpha}{\delta} v_1 - \delta_T x_2 - \varrho(x_2 - v_3)u_1 \qquad (6.8)$$

$$u_2 = -\frac{\alpha}{\delta} v_1 + \delta_T x_2 + \varrho(x_2 - v_3)u_1 + \varrho V \hat{u}_2 \qquad (6.9)$$

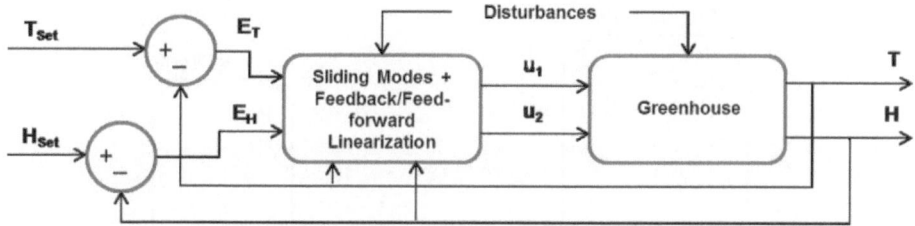

Figure 6.38 Climatic Control Loop.

Replacing 6.9 in 6.7,

$$u_1 = \frac{v_1 - \delta(-\frac{\alpha}{\delta}v_1 + \delta_T x_2 + \varrho V \hat{u}_2) - UAx_1 + UAv_2 + \varrho C_p V \hat{u}_1}{\delta C_p (x_1 - v_2)} \tag{6.10}$$

Replacing 6.7 in 6.9,

$$u_2 = -\frac{\alpha}{\delta}v_1 + \delta_T x_2 + \varrho V (x_2 - v_3)$$

$$\times \left(\frac{v_1 - \delta u_2 - UAx_1 + UAv_2 - \varrho C_p V \hat{u}_1}{\varrho C_p (x_1 - v_2)} \right) + \varrho V \hat{u}_2 \tag{6.11}$$

Finally, rearranging 6.10 and 6.11, the controller equations are derived as:

$$u_1 = \frac{(1+\alpha)v_1 - \delta\beta_T x_2 + UA(v_2 - x_1) - \varrho C_p V \hat{u}_1 - \delta\varrho V \hat{u}_2}{\delta\varrho(x_2 - v_3) + \varrho C_p (x_1 - v_2)} \tag{6.12}$$

$$u_2 = \frac{\varrho C_p (x_1 - v_2)(-\frac{\alpha}{\delta}v_1 + \delta_T x_2 + \varrho V \hat{u}_2)}{+\varrho(x_2 - v_3)(v_1 + UA(v_2 - x_1) - \varrho C_p V \hat{u}_1)}{\delta\varrho(x_2 - v_3) + \varrho C_p (x_1 - v_2)} \tag{6.13}$$

where,

$$\hat{u}_1 = k_1 sgn(x_1 - x_{d_1}) \tag{6.14}$$

$$\hat{u}_2 = k_2 sgn(x_2 - x_{d_2}) \tag{6.15}$$

The climatic control loop designed is illustrated in Figure 6.38.

Fuzzy Logic Control + ANFIS

In many cases, where an automated irrigation system is implemented within a greenhouse, sometimes there is a necessity of supplying the nutrient solution not manually but automatically. In order to achieve this, the electrical conductivity in a collector tank should be monitored and controlled.

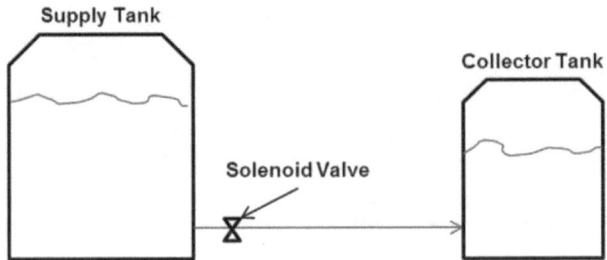

Figure 6.39 Nutrient Supply System.

Figure 6.39 shows a clearer configuration of the nutrient supply system explained and illustrated in previous sections consisting of a supply tank, a collector tank and a solenoid valve; this last allows the flow of the nutrient solution from the supply to the collector tank.

It is necessary to know the dynamic model of the nutrient supply system before continuing. The importance of this lies in the difficulty and complexity of the system to obtain data from experimentation, which means that starting from the nutrient supply model the intelligent controller will be achieved. Both the climate controller as nutrient supply will be validated in the experimental intelligent greenhouse.

The mathematical model for the nutrients supply is derived by doing simple mass balances of the supply tank and collector tank, and a mole balance for the concentration of nutrients in the collector tank as follows:

Valve equation

$$f(t) = C_v \sqrt{\frac{\varrho h_1(t)}{144}} \tag{6.16}$$

Mass balance supply tank

$$\frac{d}{dt} h_1(t) = -\frac{f(t)}{A_1} \tag{6.17}$$

Mass balance collector tank

$$\frac{d}{dt} h_2(t) = \frac{f(t)}{A_2} \tag{6.18}$$

Mass balance collector tank

$$\frac{d}{dt} h_2(t) C_2(t) = \frac{f(t) C_1}{A_2} \tag{6.19}$$

Concentration (ppm) to electrical conductivity (mS/cm)

$$EC = 0.0014C + 0.0318 \tag{6.20}$$

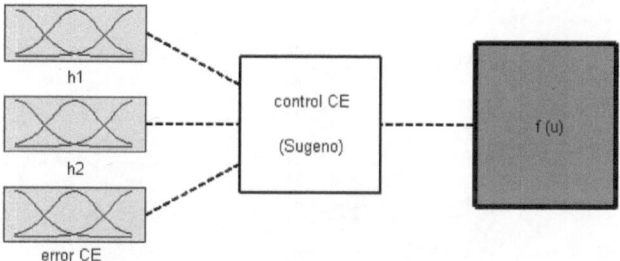

Figure 6.40 Nutrient Supply Fuzzy Model.

The proposed nutrient supply controller is based on Fuzzy Logic, which model is shown in Figure 6.40 and the way to achieve it is by ANFIS (Adaptive Neuro-Fuzzy Inference System). The Fuzzy Logic Controller consists of three inputs: supply tank height h_1(ft), collector tank height h_2(ft), and the electrical conductivity error EC_{error}; and one output which is the time t (min) of activation of the solenoid valve. The Fuzzy Controller is a T-S type controller.

Through the dynamical model a dataset is obtained for training in the ANFIS system with the aim of getting the Fuzzy Logic model. The dataset meets most significant inputs of the system with their respective outputs; 60 different combinations of inputs and outputs were trained. As an example of a small part of the dataset and the ANFIS information it is given:

```
  h1        h2      ECerror     time     ANFIS info:
3.7400    1.4750        0          0     Number of nodes: 78
1.8700    1.4750    0.5000     0.8700    Number of linear parameters: 108
1.8700    2.2125    0.5000     1.3050    Number of non-linear parameters: 27
0.9350    0.7375    1.0000     1.2400    Total number of parameters: 135
0.9350    2.2125    0.2500     0.9300    Number of training data pairs: 60
3.7400    0.7375    0.7500     0.4550    Number of checking data pairs: 0
2.8050    2.2125    1.0000     2.1500    Number of fuzzy rules: 27
```

The Neural Network configuration after the ANFIS training is illustrated in Figure 6.41; the number of training iterations that it took the algorithm to yield an accurate response was 233. It is shown that each input consists of three membership functions and 27 different rules (3^3). Furthermore, Figure 6.42 shows the input membership functions of the nutrient supply controller.

The rules have the form,

If h_1 is High and h_2 is Medium and EC_{error} is Low then $t = f(x)$

The output t consists of 27 linear functions (first order polynomials) of the form:

$$t = (a * h_1) + (b * h_2) + (c * EC_{error}) + d$$

Where a, b, c, d are constants.

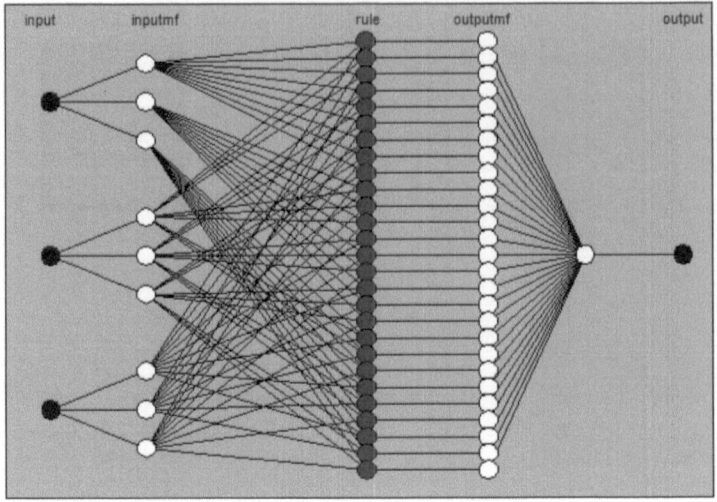

Figure 6.41   Neural Network Configuration.

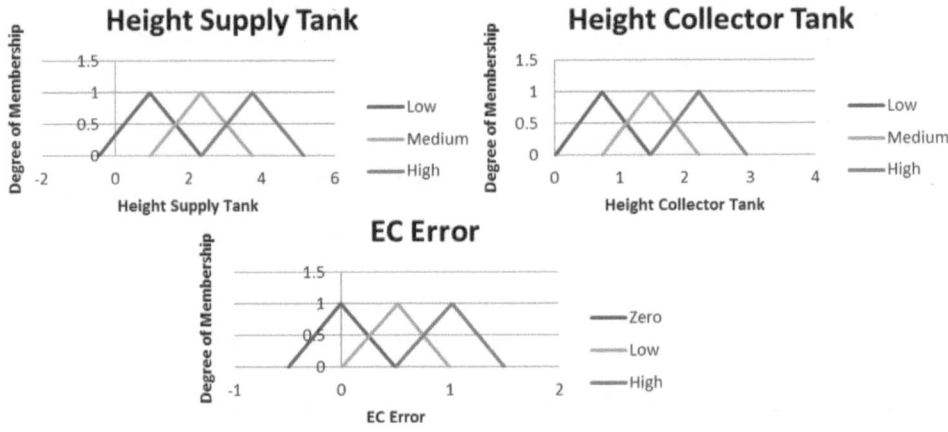

Figure 6.42   Input Membership Functions.

Finally, the nutrient supply control loop designed is summarized as seen in Figure 6.43.

6.3.2   LabVIEW™ controllers programming

LabVIEW™ (Laboratory Virtual Instrument Engineering Workbench) is a system-design platform and development environment for a visual programming language from National Instruments. The programming language used in LabVIEW™, also referred to as G, is a data-flow programming language. Execution is determined by the structure of a graphical block diagram (the LabVIEW-source code) on which the

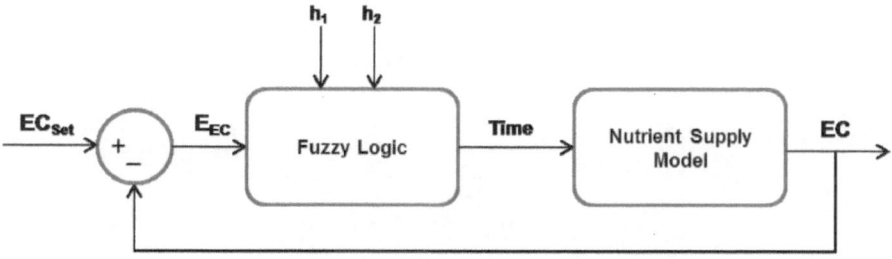

Figure 6.43 Nutrient Supply Control Loop.

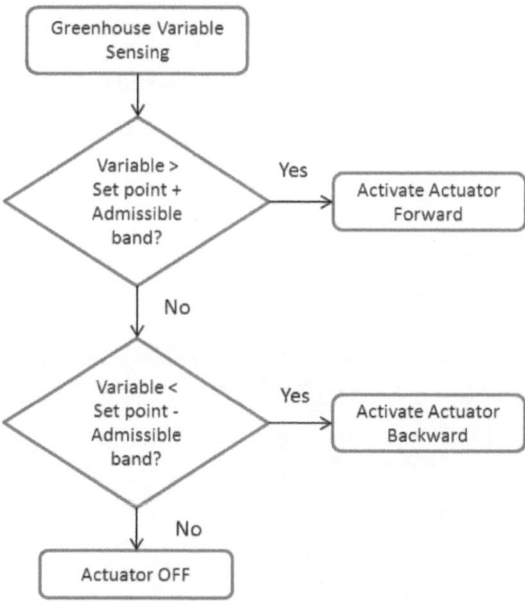

Figure 6.44 Generic On/Off Controller Flow Diagram.

programmer connects different function nodes by drawing wires. These wires propagate variables and any node can execute as soon as all its input data become available. Since this might be the case for multiple nodes simultaneously, G is inherently capable of parallel execution.

The intelligent greenhouse automatic controllers responsible for maintaining the temperature, relative humidity, light and electrical conductivity in the appropriate range to generate higher quality crops as well as higher quantity are programmed in LabVIEW™ 2012. The controllers manage the different actuators to achieve the right performance in terms of desired micro-climate.

In the first case, the on/off conventional controllers are designed for the four variables mentioned above; their performance will be later compared with the non-conventional, non-linear and intelligent controllers. Figure 6.44, which is a flow

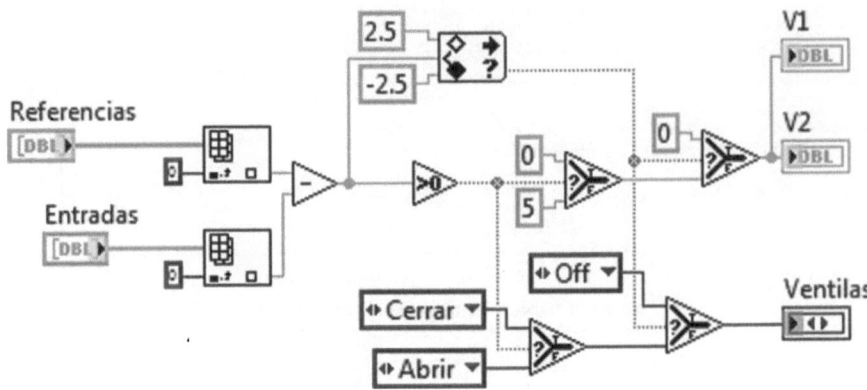

Figure 6.45 Conventional On/Off Temperature Control.

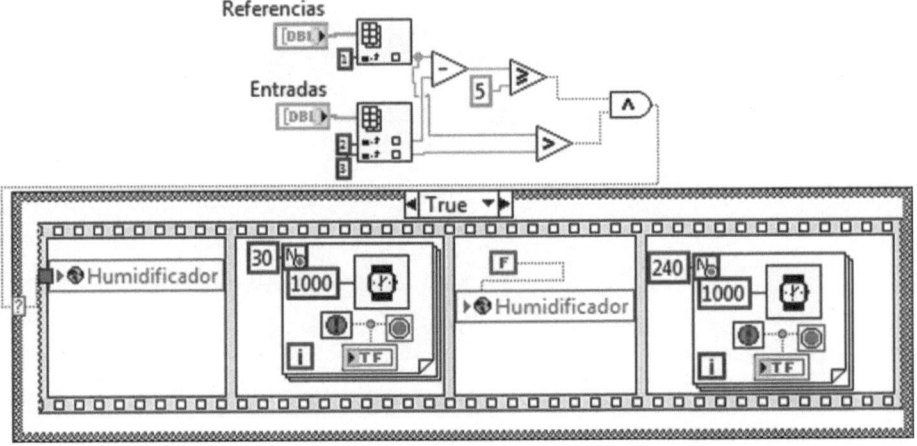

Figure 6.46 Conventional On/Off Relative Humidity Control.

diagram, outlines the generic algorithm for all of the on/off controllers. Then, Figures 6.45, 6.46, 6.47 and 6.48 illustrate the on/off controllers programming in LabVIEW™, one being independent from the other. These controllers basically activate the systems or not depending whether the measured variable is out of a certain range or not, respectively.

On the other hand, Figures 6.49, 6.50 and 6.51 shows the programming for the proposed controllers. These controllers react depending on the control technique used for their design, being Sliding Modes + Feedback/Feed-forward Linearization for the temperature and relative humidity and Fuzzy Logic for the electrical conductivity; the light control remains the same.

Finally, apart from the above controllers there is also an irrigation controller which ensures a cyclic operation of the hydroponics. In other words, the irrigation controller

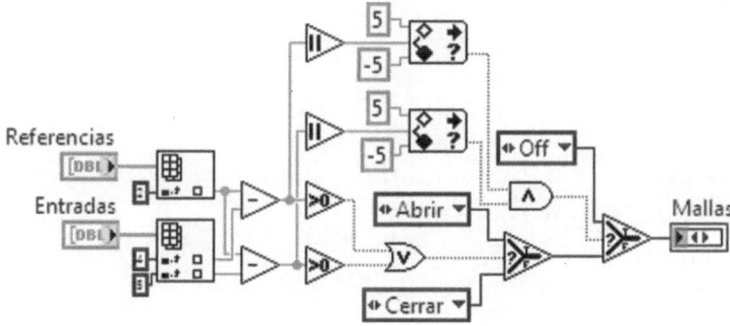

Figure 6.47 Conventional On/Off Light Intensity Control.

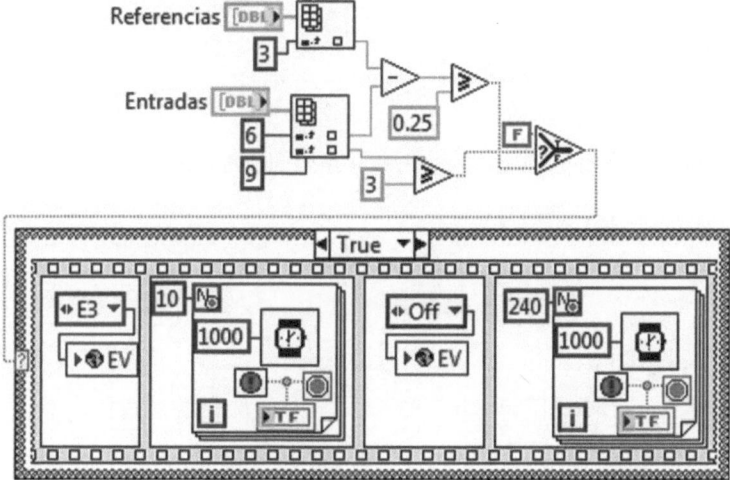

Figure 6.48 Conventional On/Off Electrical Conductivity Control.

is in charge of activating the water pump at certain hours for certain minutes as a timer, ensuring the adequate irrigation of the crops.

How often and long is the irrigation is the user's decision. Figure 6.52 shows the timer control for the irrigation.

6.3.3 Intelligent controllers simulation

A simulation is needed before going to validation because it gives a primary test of the proposed intelligent controllers. The simulation was handled in LabVIEW™ 2012 with help of two toolkits: Intelligent Control Toolkit for LabVIEW™ (ICTL) and, Control Design and Simulation.

Figure 6.53 shows the user interface designed for simulation purposes. This program collects the designed controllers: Sliding Modes + Feedback/Feed-forward

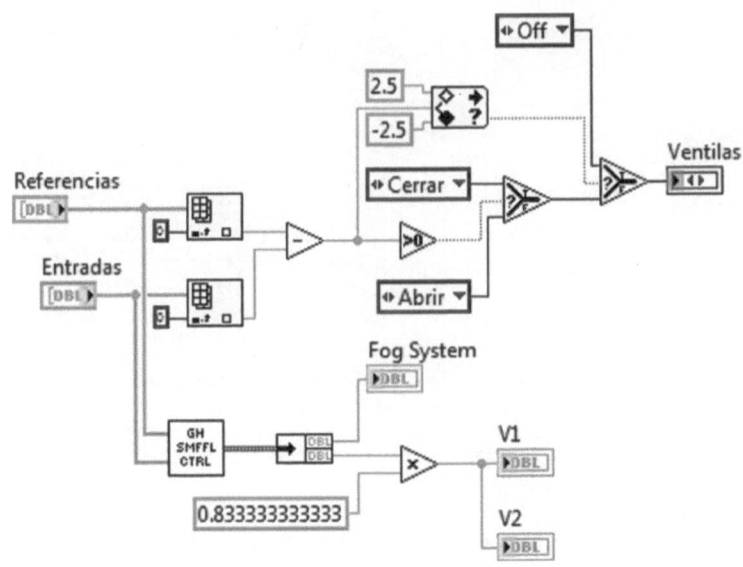

Figure 6.49 Non-linear Temperature and Relative Humidity Control.

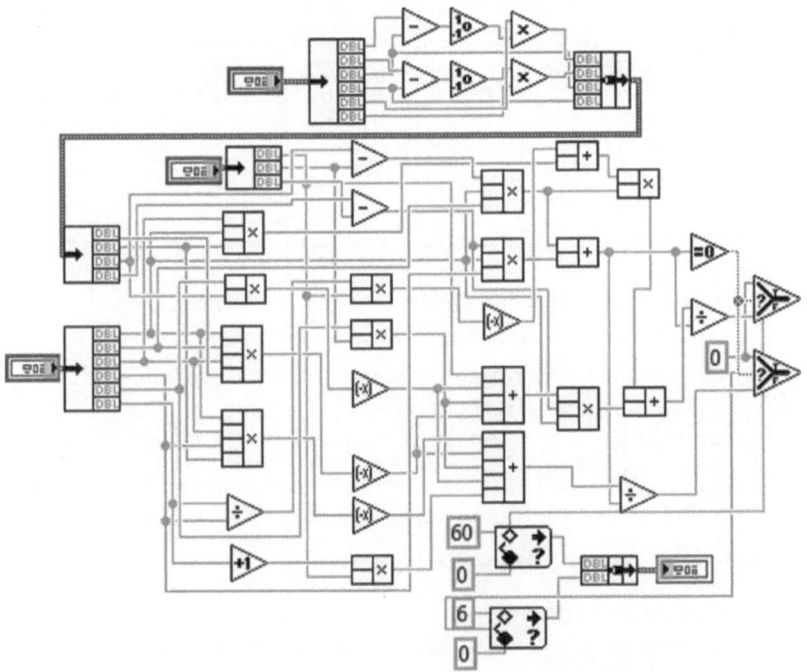

Figure 6.50 Sliding Modes + Feedback/Feed-forward Linearization.

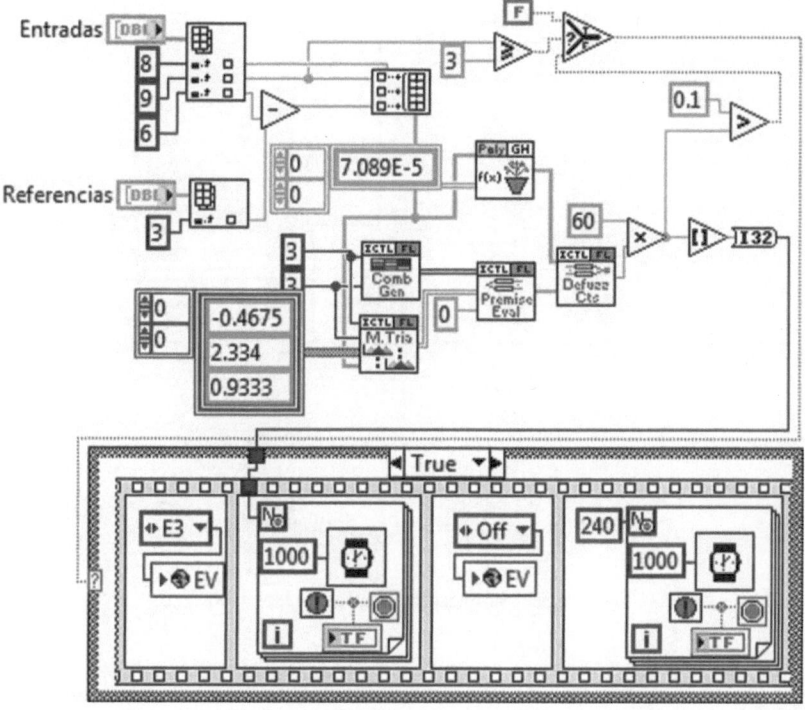

Figure 6.51   Fuzzy Logic Electrical Conductivity Control.

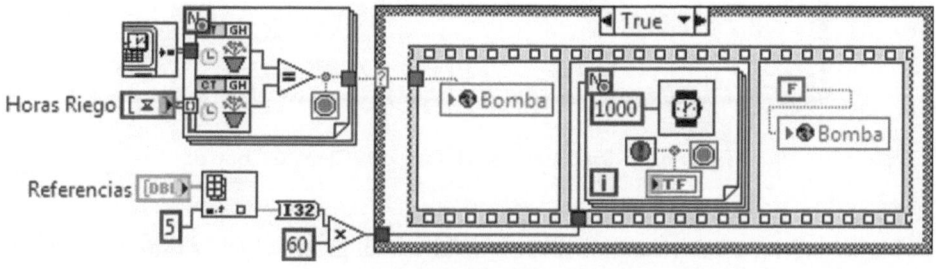

Figure 6.52   Irrigation Timer Control.

Linearitation, and Fuzzy Logic. These controllers are tested through the programmed mathematical models for both climate and nutrient supply, already exposed.

The graphical user interface contains all the necessary controls and indicators for bringing the best user interaction. The controls are basically constants of the dynamical model, variables known as disturbances, initial conditions and controller inputs; on the other hand, as indicators, it can be found the outputs of controllers and controlled variables.

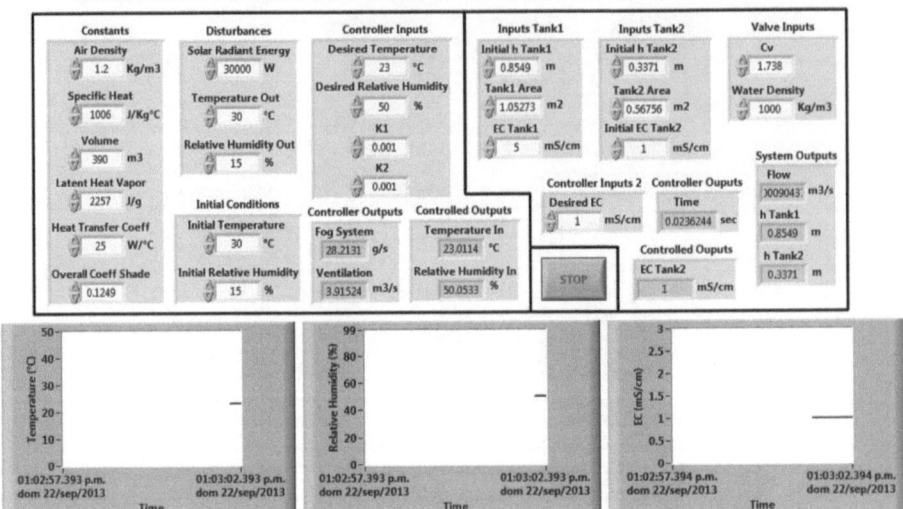

Figure 6.53 Simulation Graphical User Interface.

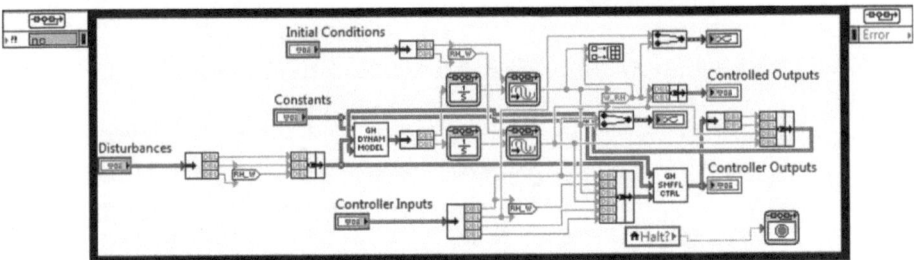

Figure 6.54 Simulation Block Diagram: Climate Control.

Figure 6.54 presents the first part of the block diagram which is the control loop for the climate controller. The main block (subVIs) of this controller is the one that contains the dynamic greenhouse model from equations 6.4 and 6.5, and the other stage contains the Sliding Modes + Feedback/Feed-forward Linearization controller. The programming of this controller was shown in Figure 6.54. The form of the simulation has the generic form of a feedback control loop.

The second part of the simulation block diagram consists in the nutrient supply control system. In Figure 6.55, four blocks (subVIs) can be identified inside the simulation loop, those represents the dynamic model for the nutrient supply system described in equations 6.16 through 6.20. Additional to this, outside the simulation loop can be found the Fuzzy Logic controller block for the nutrient supply system. This controller programming was shown in Figure 6.55. Due to the nature of this controller and the

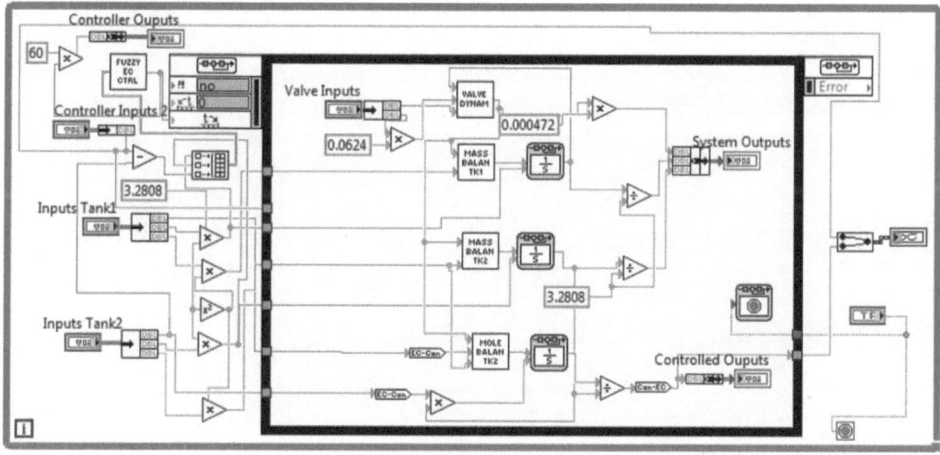

Figure 6.55 Simulation Block Diagram: Nutrient Supply Control.

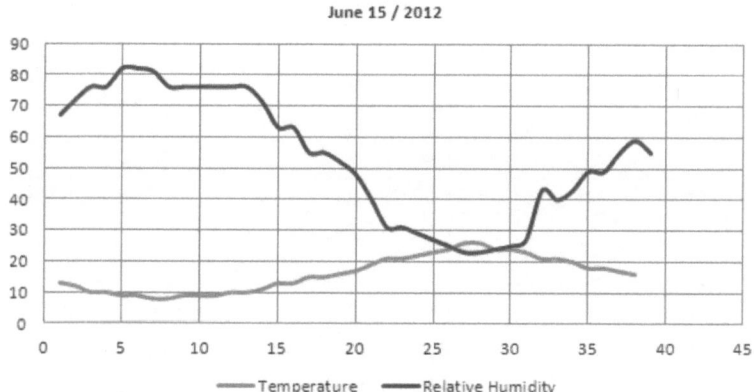

Figure 6.56 Recorded Temperature and Relative Humidity – June 15/2012.

dynamic model used, the controller regulates the simulation time execution emulating the opening of the electro-valve; one reason why it is outside the simulation loop.

Due to the close relationship between the temperature and the relative humidity, there exists some combination of values (pairs) that are valid while others may go against thermodynamic laws. Some values for the desired temperature and relative humidity, simultaneously, can be drawn from Figures 6.56, 6.57 and 6.58; these graphs show real values (thermodynamically valid) of different weather seasons in Mexico City.

6.3.4 Graphical User Interface (GUI) and functions

One of the most important parts of the intelligent greenhouse control system is the graphical user interface (refer to User Manual Appendix). This interface contains all

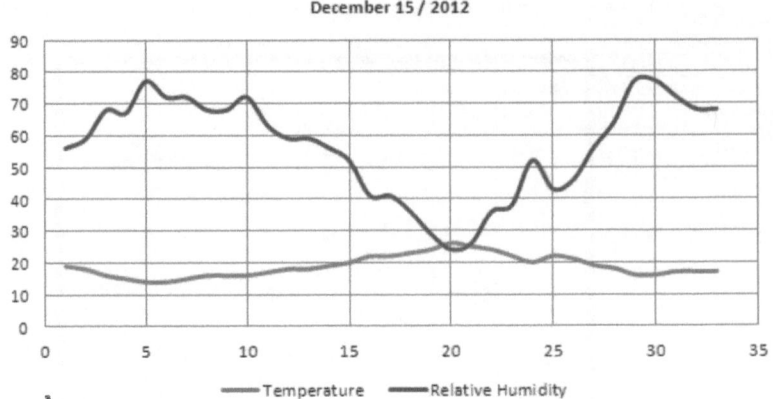

Figure 6.57 Recorded Temperature and Relative Humidity – December 15/2012.

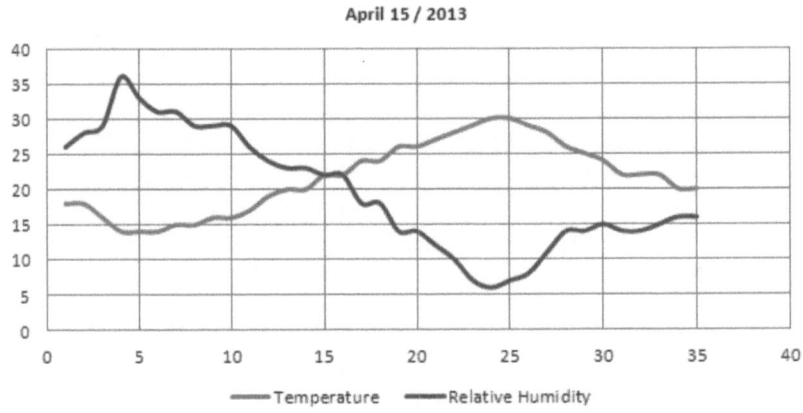

Figure 6.58 Recorded Temperature and Relative Humidity – April 15/2013.

the controls and indicators facilitating the interaction between the user and all the greenhouse aspects. This GUI was developed and designed in LabVIEW™ 2012.

The main GUI objective is to give the user the possibility of programming the crop needs, in terms of micro-climate, depending of its life-cycle stages, which means that the GUI will automatically handle the crop demands programmed by the user.

The first image, Figure 6.59, represents the beginning and welcoming for the user. It allows the user to create a new crop in the system, load it from a database or exit the program. If the user decision was to load an existing crop a window will pop up to choose the text file (*.txt).

Otherwise, if the user decision was to create a new crop, the panel from Figure 6.60 will come out. This is the first step of the crop programming by the user. This panel will need to input the name of the crop and the number of growth stages. Depending

INVERNADERO INTELIGENTE
TECNOLÓGICO DE MONTERREY
CAMPUS CIUDAD DE MÉXICO

Figure 6.59   GUI – Program Start (Salir-Exit, Nuevo Cultivo-New Crop, Cultivo Existente- Crop in the database).

INVERNADERO INTELIGENTE
TECNOLÓGICO DE MONTERREY
CAMPUS CIUDAD DE MÉXICO

Figure 6.60   GUI – Creating a New Crop I (Salir-Exit, Nuevo Cultivo-New Crop, Cultivo Existente-Crop in the data-base).

on this last number the image from Figure 6.61 will appear many times as the number of stages are, so the user can input the micro-climate conditions per stage including the start and finish date.

Consecutively, no matter if the user chose to load an existing crop or created a new one, the panel from Figure 6.62 will appear which is the crop needs summary just entered by the user. This panel has two more functions, to edit or to save; if edit is the decision the program will return the user to the first panel or if saving is the option, the crop summary will be automatically saved as a text file (*.txt) that can be read later by the same program when opening an existing crop.

After continuing with the program process another window as the one seen in Figure 6.63 comes out. This window asks the user to enter the control technique that will be used, conventional on/off or intelligent control.

Finally, the programming phase finishes when accessing the main user interface panel as shown in Figure 6.64. This panel shows all the necessary information about

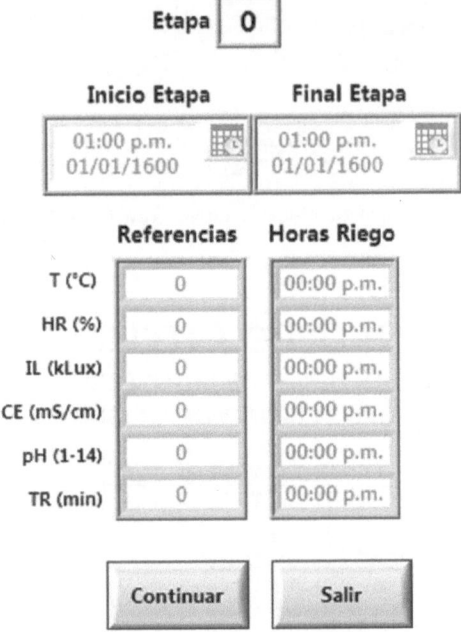

Figure 6.61 GUI – Open an Existing Crop II (Continuar-Continue, Inicio Etapa – Initial Date, Final Etapa- End Date).

the crop control such as name, number of crop stages, stage duration as well as the start and finish date, a clock, the controller chosen, desired micro-climate conditions, activated actuators and measured variables. All this information is updated as the stages go by.

The last function of this main panel is to allow the manual control option no matter what control strategy was chosen. When the manual control is pressed some indicators are hidden and the actuator buttons are shown, as seen in Figure 6.65. The main panel closes after finishing the growing season or being ended by the user, then it returns to the first window.

6.4   CONTROL RESULTS

Here the results of the intelligent greenhouse controllers performance based on simulations and validations will be gone into. Bringing the controllers design to reality could not be possible without the integration of the whole system.

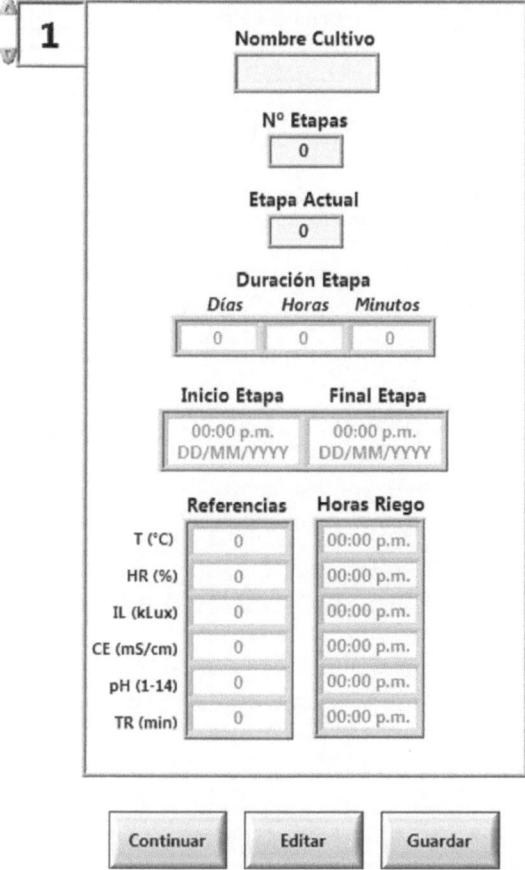

Figure 6.62 GUI – Program Summary (Editar-Edit, Referencias-Reference values, Nombre del Cultivo, Crop Name, Etapas-Periods of time, Etapa actual-Actual Period of time).

INVERNADERO INTELIGENTE
TECNOLÓGICO DE MONTERREY
CAMPUS CIUDAD DE MÉXICO

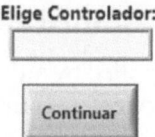

Figure 6.63 GUI – Controller Choice.

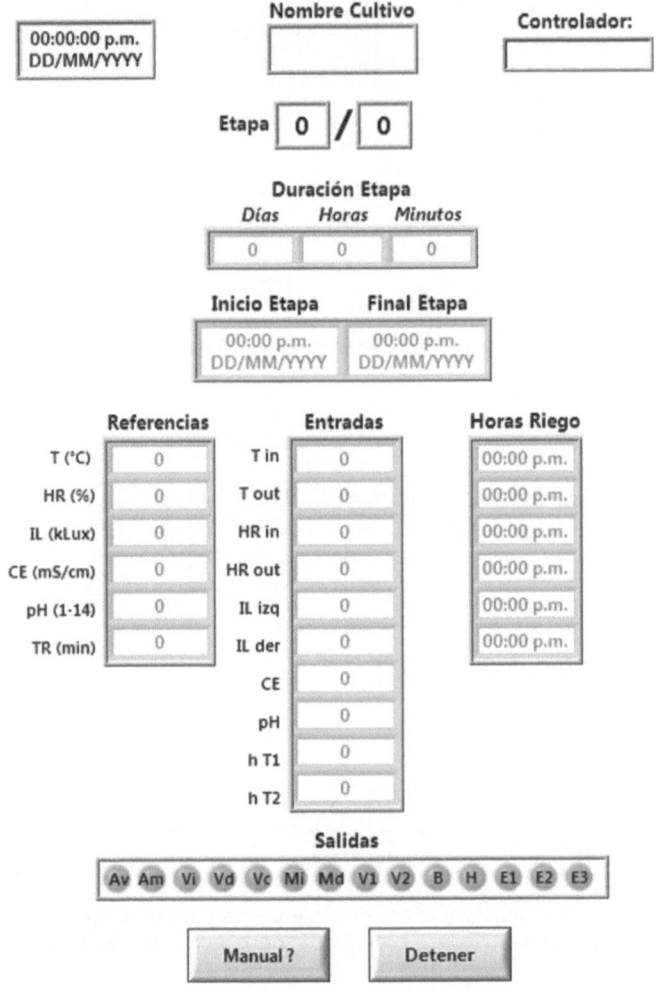

Figure 6.64   GUI – Main User Interface.

6.4.1   Intelligent controllers simultion results

In practice, simulations are broadly applied by industries for generating a first approach of what will be the last product or service. Keeping this in mind, simulations comprise all the theory necessary to carry out the construction of the controls proposed: Sliding Modes + Feedback/Feed-forward Linearization and Fuzzy Logic.

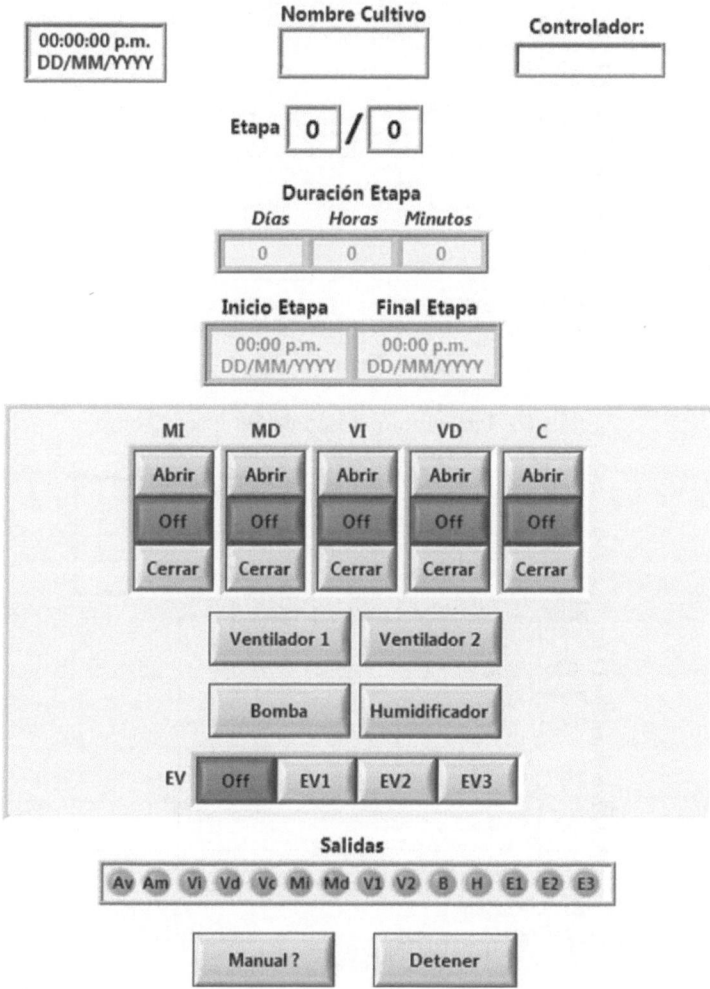

INVERNADERO INTELIGENTE
TECNOLÓGICO DE MONTERREY
CAMPUS CIUDAD DE MÉXICO

Figure 6.65 GUI – Manual Control (Abrir-Open, Ventilador-Fan, Bomba-Electric water pump, Humidificador-Humidificator, Detener-Stop).

As modern engineering is becoming more complex and costlier as is the case with intelligent greenhouses, simulation technologies are becoming crucial to their success. New design tools are needed to support all the software development, testing and validation effort made during the simulation stage; this is crucial because it will cause less time to move into physical prototype and commercial deployment.

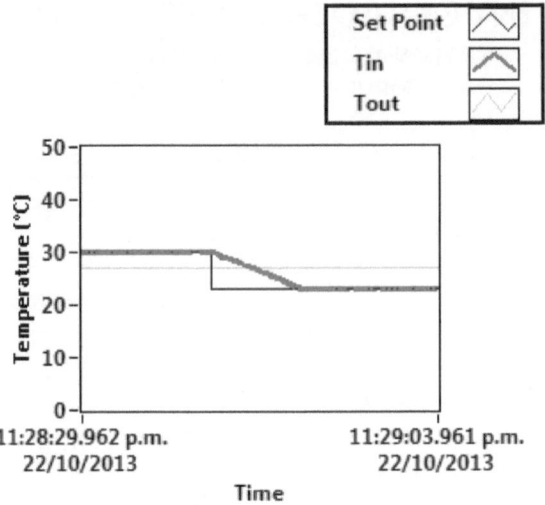

Figure 6.66 Temperature Response Set Point Change.

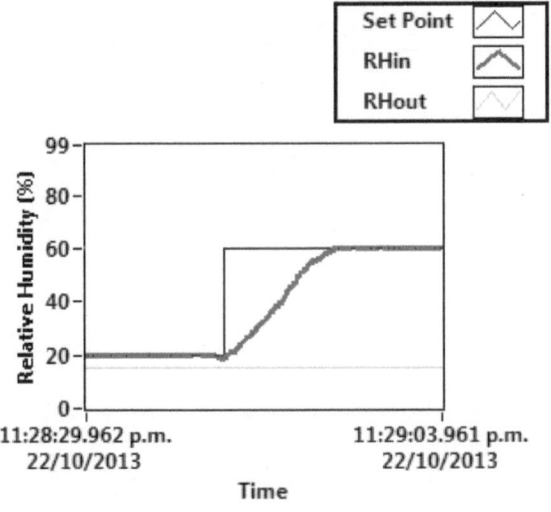

Figure 6.67 Relative Humidity Response Set Point Change.

Moving on, this section handles the intelligent controllers simulation results where the main controlled variables are temperature and relative humidity in terms of the climatic controller and electrical conductivity related to the nutrient supply controller. The simulation was held in LabVIEW™ with help of the Intelligent Control Toolkit for LabVIEW™ (ICTL) and Control Design and Simulation Toolkit.

A first scenario of results is displayed in Figures 6.66 and 6.67 for the climatic controller where a change in the temperature and relative humidity set points is made.

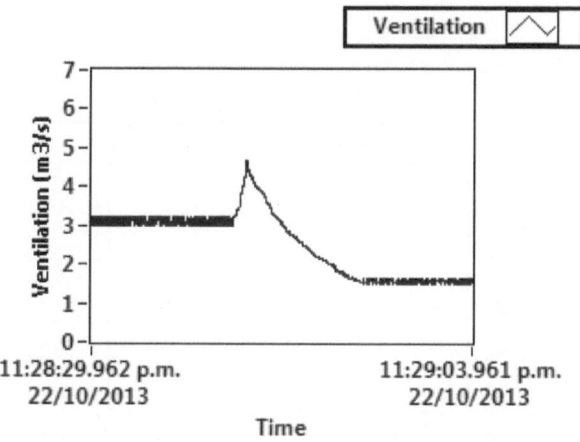

Figure 6.68 Ventilation Response Set Point Change.

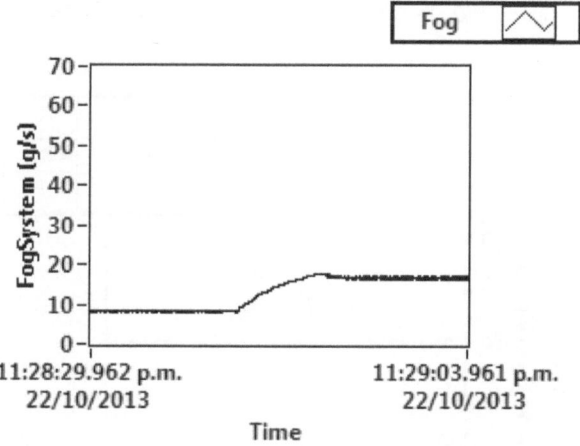

Figure 6.69 Fogging System Response Set Point Change.

The inside conditions were 30°C and 20% for the temperature and relative humidity, respectively. The light intensity was 300 W/m², the outside temperature 27°C and 15% for relative humidity; these last three variables are the main disturbances considered. A sudden change in the desired points was made going from 30° to 23° in temperature and 20% to 60% in relative humidity. The reason why these set points were chosen is because they are the main desired values for tomato cropping, it will be the vegetable grown in the intelligent greenhouse.

The results clearly show the immediate action of the Sliding Modes + Feedback/Feed-forward Controller by reaching both set points.

The controller outputs are linked to the ventilation and water fogging system actions; this combination makes a feasible controller. In this sense, Figures 6.68 and 6.69 illustrate the actuators performance due to the variables set point changes.

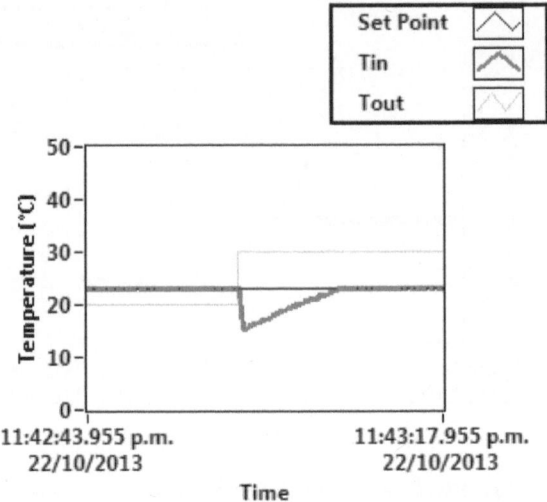

Figure 6.70 Temperature Response Disturbance Change.

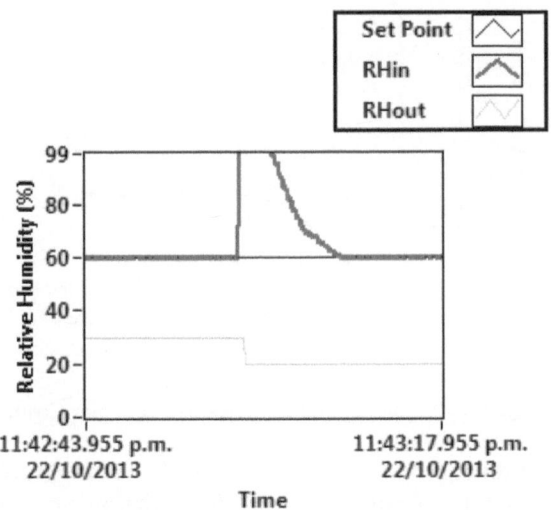

Figure 6.71 Relative Humidity Response Disturbance Change.

The scales used for both controller outputs reach a maximum of 6 m^3/s for ventilation and 60 g/s for water fogging; these values are based on the maximum capacity of the actuators installed inside the real greenhouse prototype.

A second stage of results are shown in Figures 6.70 and 6.71. In this case, changes in the value of disturbances are handled to corroborate that the controller is tolerant to sudden external changes.

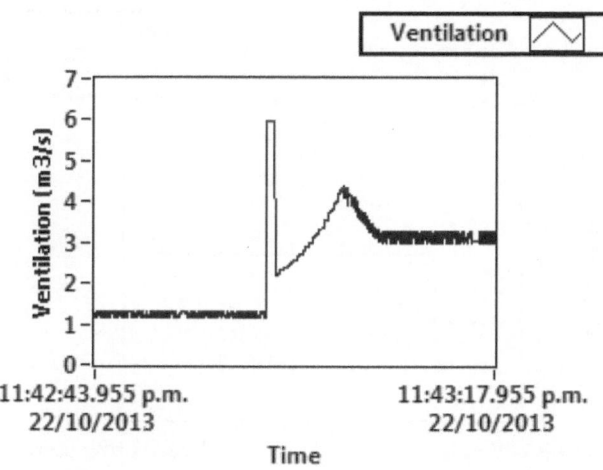

Figure 6.72 Ventilation Response Disturbance Change.

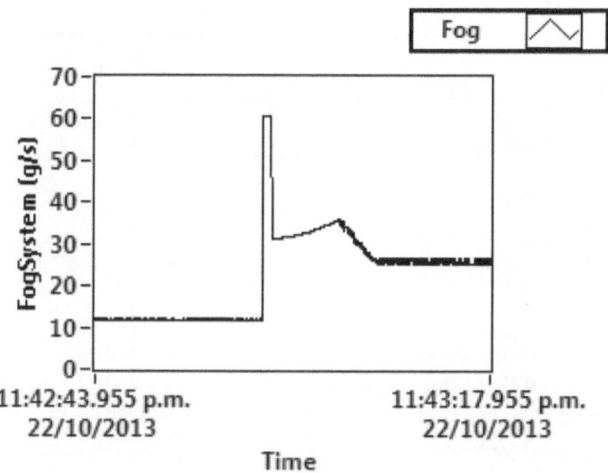

Figure 6.73 Fogging System Response Disturbance Change.

While the desired values for temperature and relative humidity are 23°C and 60%, respectively, the disturbances changed in value from 300 W/m² to 350 W/m² for light intensity, 20°C to 30°C for outside temperature and 30% to 20% for outside relative humidity.

As noticed, the controllers could handle the change in disturbances, the set points were lost for a while but then were reached again thanks to the action of the actuators. The actuators response are shown in Figures 6.72 and 6.73.

Finally, the following test is based on the Fuzzy Logic controller designed through ANFIS (Adaptive Neuro-Fuzzy Inference System) for the nutrient supply system. This

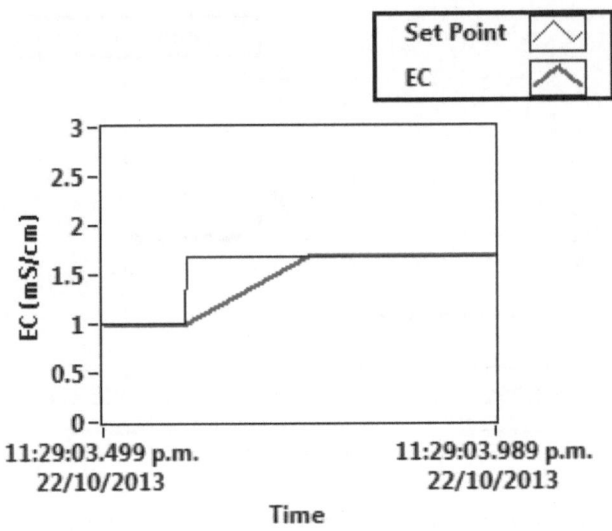

Figure 6.74 Electrical Conductivity Response Set Point Change.

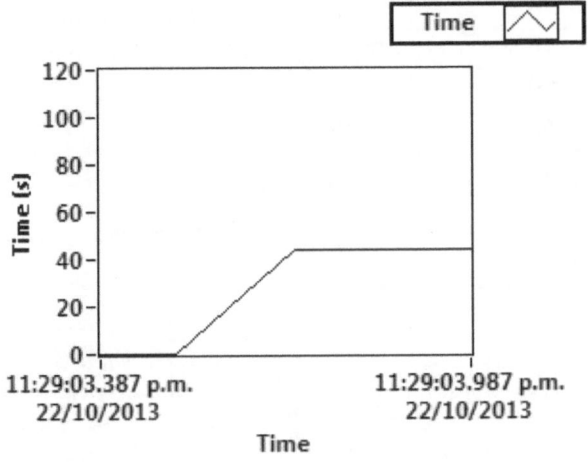

Figure 6.75 Valve Time Response Set Point Change.

system does not contemplate disturbances but set point changes. Figure 6.74 presents the controller response due to a set point change in the electrical conductivity from 1 mS/cm to 1.75 mS/cm.

The simulation results reach the set point by opening an electro-valve for a certain amount of time (Fuzzy Logic control output), 40 seconds approximately. This output response is shown in Figure 6.75 starting from a closed state.

An early conclusion of the first set of results obtained by simulation tell that the Sliding Modes + Feedback/Feed-forward Linearization and Fuzzy Logic controllers

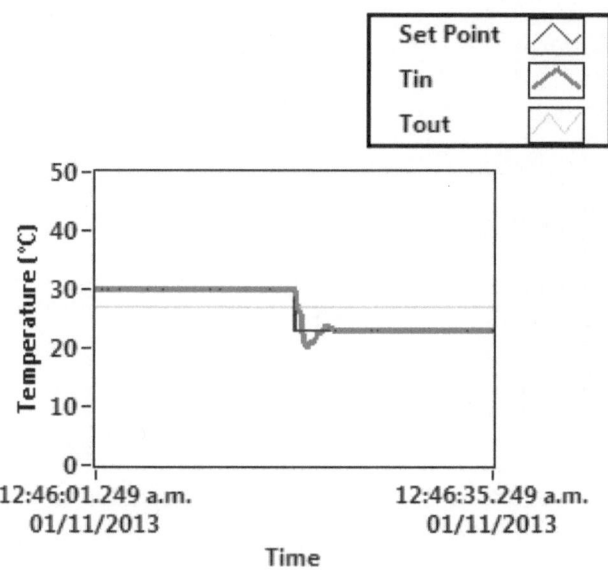

Figure 6.76 Temperature Response Set Point Change.

performed as expected reaching the desired values for the variables involved. No matter if changes in disturbances are presented, the controllers still have a successful action.

6.4.2 PID-type controller simulation results

At this point, the designed controllers can face reality by implementing them in the intelligent greenhouse for validation. In addition, this work also presents an implementation of a simulated PID controller with aim of comparing the controller just presented. This PID is placed maintaining the Feedback/Feed-forward Linearization; a conventional PID could hardly control the greenhouse model due to its non-linearities.

The purpose for presenting this results is to keep the same conditions (temperature, relative humidity, solar radiance) giving a similar scenario for comparison. Figures 6.76 and 6.77 shows the PID controller response making a set point change from 30°C to 23°C and 20% to 60% for temperature and relative humidity, respectively. As disturbances the light intensity was 300 W/m^2, the outside temperature 27°C and 15% for relative humidity.

The controller output related to the PID implemented is shown in Figures 6.78 and 6.79 for ventilation and fogging.

These behaviors compared to the ones seen in the Sliding Modes + Feedback/Feed-forward Linearization controllers output do not have fluctuations; this switching behavior is characteristic of the Sliding Modes.

In the other case, changes in the value of disturbances are made. Figures 6.80 and 6.81 presents the controller response for disturbances changes from 300 W/m^2 to 350 W/m^2 for light intensity, 20°C to 30°C for outside temperature and 30% to 20%

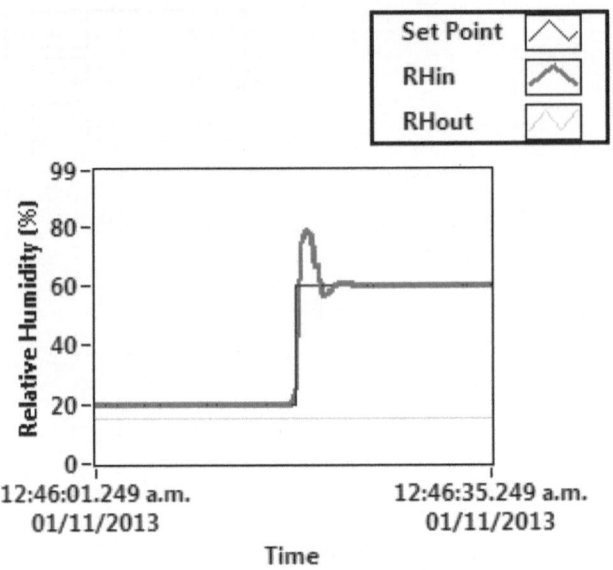

Figure 6.77 Relative Humidity Response Set Point Change.

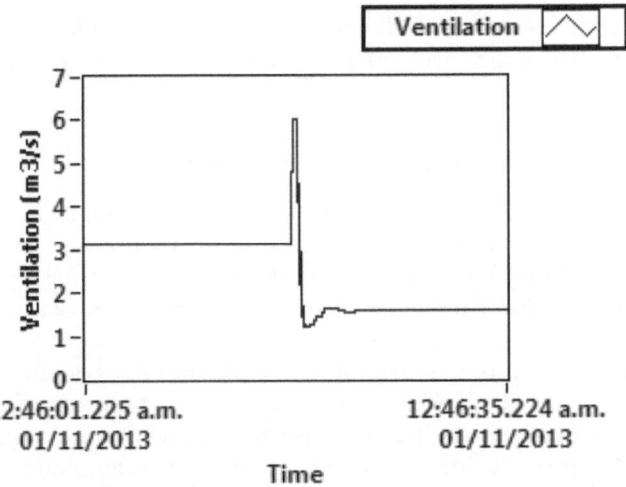

Figure 6.78 Ventilation Response Set Point Change.

for outside relative humidity. The desired values for temperature and relative humidity remained 23°C and 60%, respectively.

Figures 6.82 and 6.83 shows the PID controller response for disturbances changes. It can be noticed that both the Sliding Modes and PID (plus Feedback/Feed-forward Linearization) start and end with an approximate output value; the difference remains on the fluctuations of the Sliding Modes.

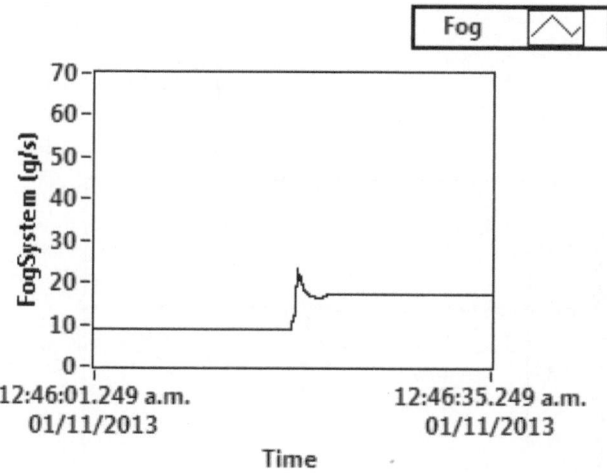

Figure 6.79 Fogging System Response Set Point Change.

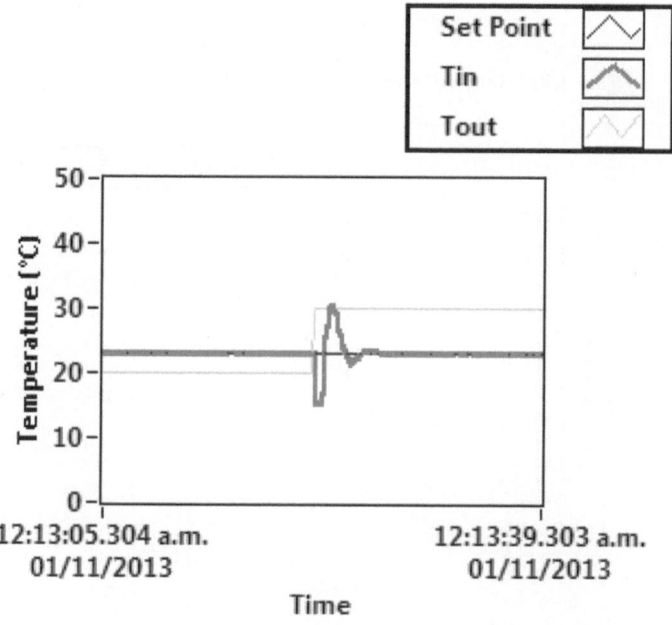

Figure 6.80 Temperature Response Disturbance Change.

It can be seen, that the Sliding Modes + Feedback/Feed-forward Linearization is the most recommended because it has a smooth response reaching the set point. It could be a bit slower than the PID but this last has a non-desirable behavior related to the overshoot presented and its ripples in the response.

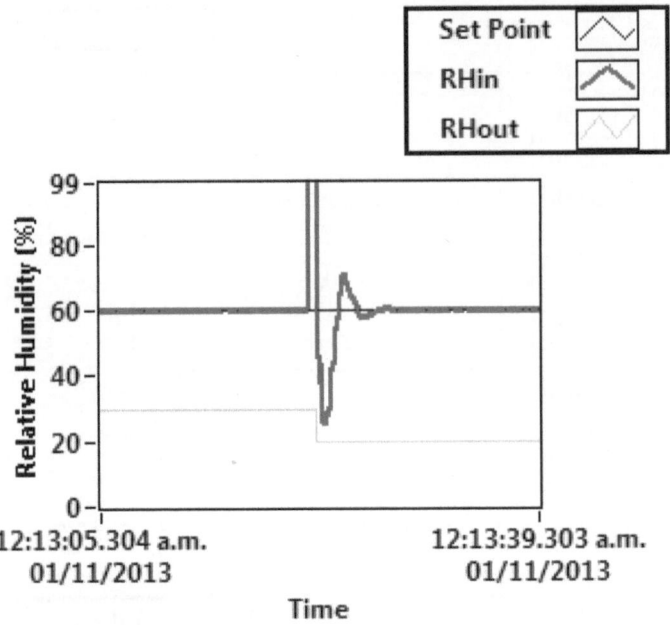

Figure 6.81 Relative Humidity Response Disturbance Change.

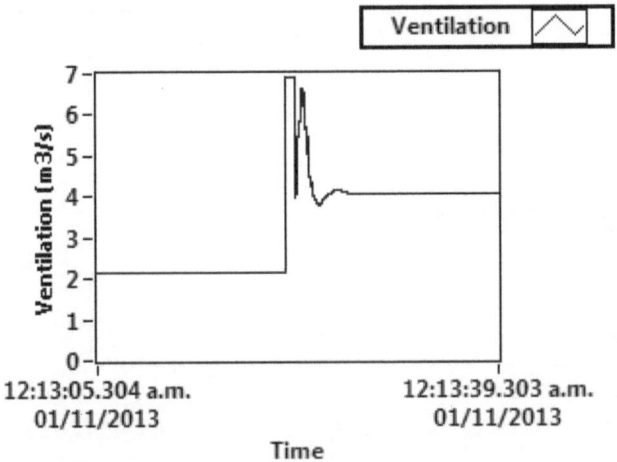

Figure 6.82 Ventilation Response Disturbance Change.

Before introducing the results achieved in validation, other type of controllers results are first presented in order to make a comparison between conventional and non-conventional control. Conventional *bang-bang* or on/off control results are achieved in experimentation and shown in the next section.

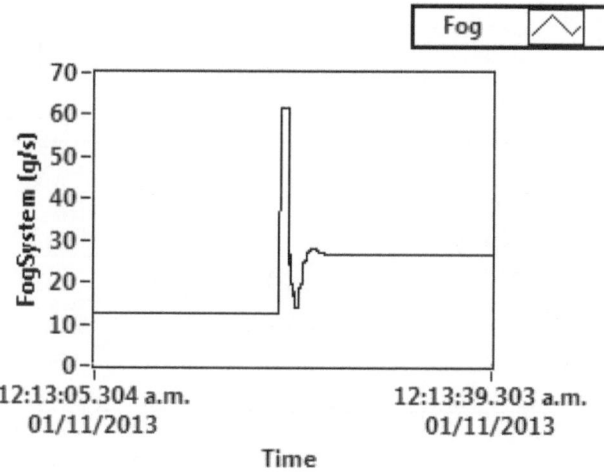

Figure 6.83 Fogging System Response Disturbance Change.

6.4.3 On/off controllers experimental results

This section shows the experimental results for the on/off controllers designed. On one hand, the results are given in a three-day based period (October 19th–October 22nd 2013) for each of the controlled variables involved: temperature, relative humidity, light intensity and electrical conductivity. On the other hand, the controller outputs are also indicated showing when each of the actuators are activated and for how long. The timer control for irrigation system is considered part of the on/off controllers.

Note here that for each controlled variable there is an independent actuator. This can be set as the main disadvantage of this type of controller. Variables such as temperature and relative humidity are closely linked thermodynamically so by controlling one it will affect the other. This issue, by the way, is one of the advantages of the Sliding Modes + Feedback/Feed-forward Linerization Controller just presented in simulation where both variables share the actuators.

The first controlled variable, temperature, is shown in Figure 6.84. The on/off temperature controller faces the drastic climate changes and the main objective is to maintain the temperature at 23° though the actions of the ventilation system.

From Figure 6.85 the inside temperature was close to the set point during the day when the outside temperature was above the desired limit. The reason why it has an offset is because the design rule of the on/off controller allows a band of 2.5° below and above the set point, thus passing the band limits the controller output will consider an action. Another reason is that the greenhouse does not count with a type of heater so at nights when temperature drops the temperature is uncontrollable but crops tolerate (desired, not harmful) temperatures around 15° during dark hours.

Figure 6.86 displays the controller output response by activating/deactivating the vents and the exhaust fans. The vents have three states: open, closed and off (1, −1 and 0 in the graphic), while the fans only go on and off (1 and 0). It can be seen that when the temperature drops at dark hours, below the lower band limit, the vents closes and

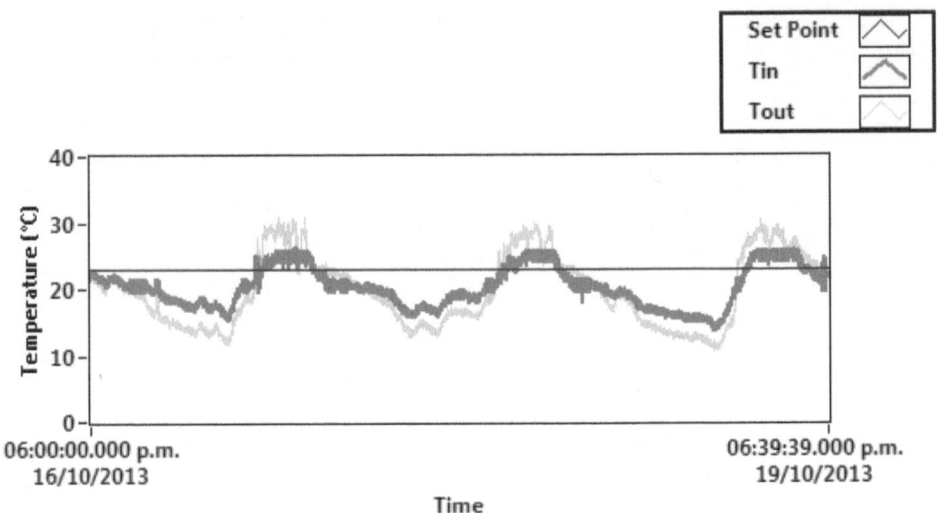

Figure 6.84 Temperature Response On/Off Controller.

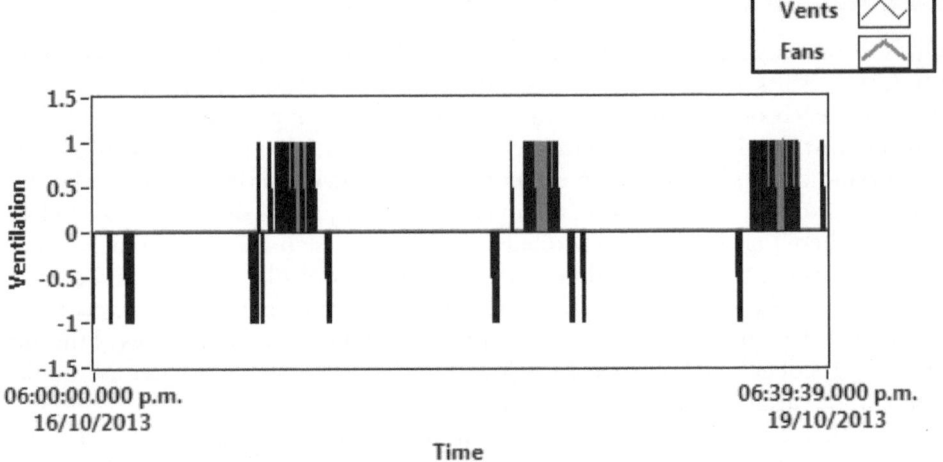

Figure 6.85 Ventilation Response On/Off Controller.

the fans turn off. When the temperature is high during light hours, above the upper band limit the vents open and the fans turn on.

The relative humidity shows a similar behavior of temperature but inverse. In this case the greenhouse can only humidify but not dehumidify, hence the set point can be closely achieved when the outside relative humidity is below the set point.

During night hours and rainy days when the relative humidity rises above the desired point the system is uncontrollable. This rise, unlike that of temperature, could be harmful to crops if it exceeded more than 85%, which is not the case. The reason why relative humidity barely reaches the set point, similar to temperature, is because

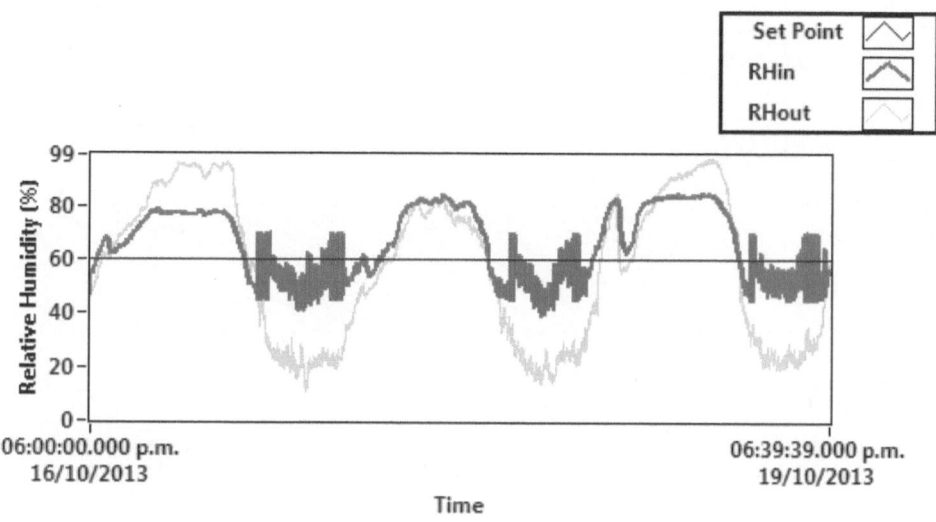

Figure 6.86 Relative Humidity Response On/Off Controller.

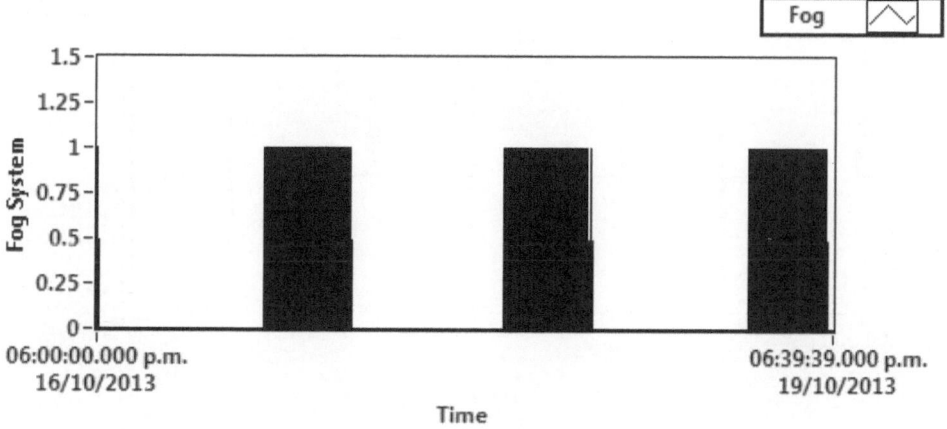

Figure 6.87 Fogging System Response On/Off Controller.

it has an acceptable band of 5% below the reference but still is desirable for crop development.

The way to raise the relative humidity is through the water fogging system which is activated when inside relative humidity is below the lower band limit and deactivated in other cases. Figure 6.87 shows when the fogging system works for raising the inside relative humidity (1) and in other cases it's off (0). If compared to the last figure, the hours when there is a need to raise the relative humidity match with the activation of the fogging system; this is the case mostly during light hours when the outside relative humidity is extremely low.

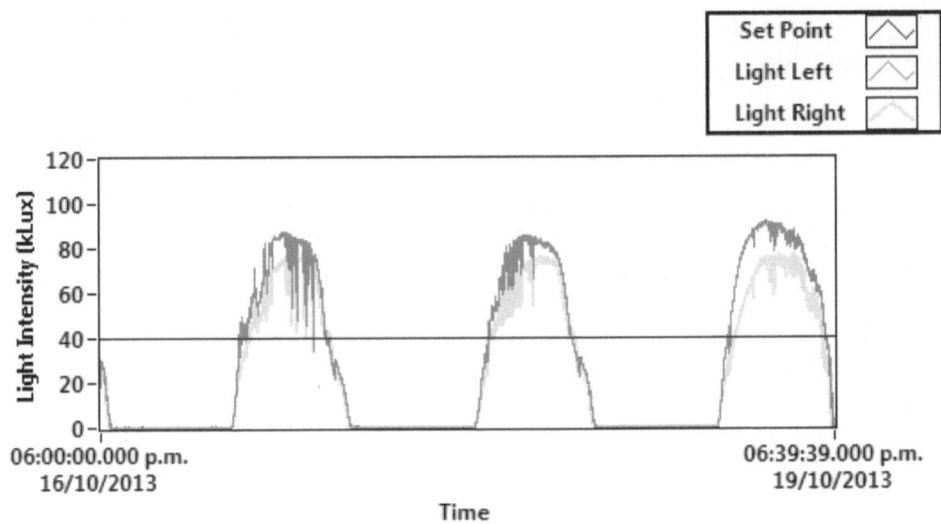

Figure 6.88 Light Intensity Response On/Off Controller.

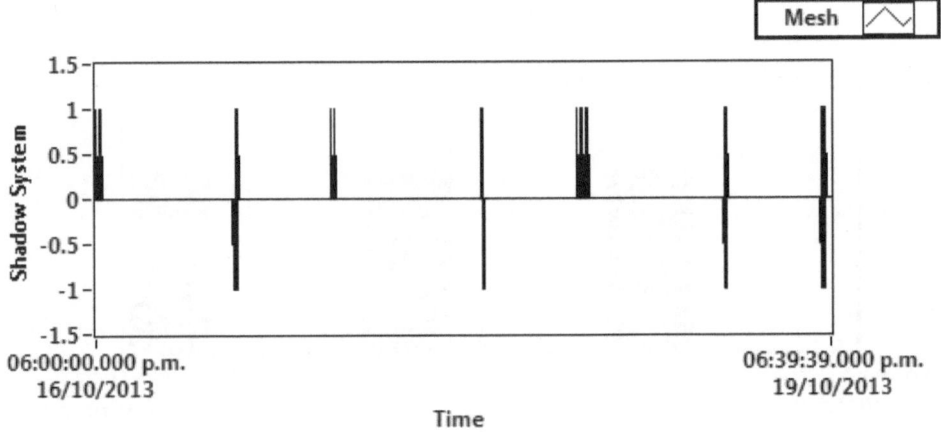

Figure 6.89 Shadow Mesh Response On/Off Controller.

Consecutively, the light intensity is sensed in both sides of the greenhouse. The main purpose of the on/off light intensity controller is to allow the pass of certain amount of sunlight. Figure 6.88 illustrates the behavior of the light inside the greenhouse where the set point is not followed but the performance is desirable. It is known that for the correct crop development, plants must absorb high amounts of light for the photosynthetic process.

As noticed the maximum light intensity is reached around noon and minimum at dark hours.

The actuator in charge of blocking/unblocking sunlight is the shadow mesh system which has three action states like the vent system. Figure 6.89 gives the action

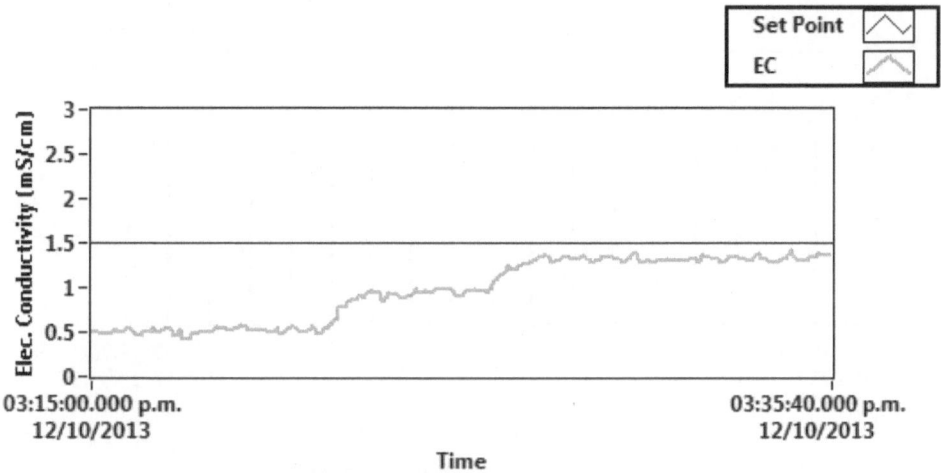

Figure 6.90   Electrical Conductivity Response On/Off Controller.

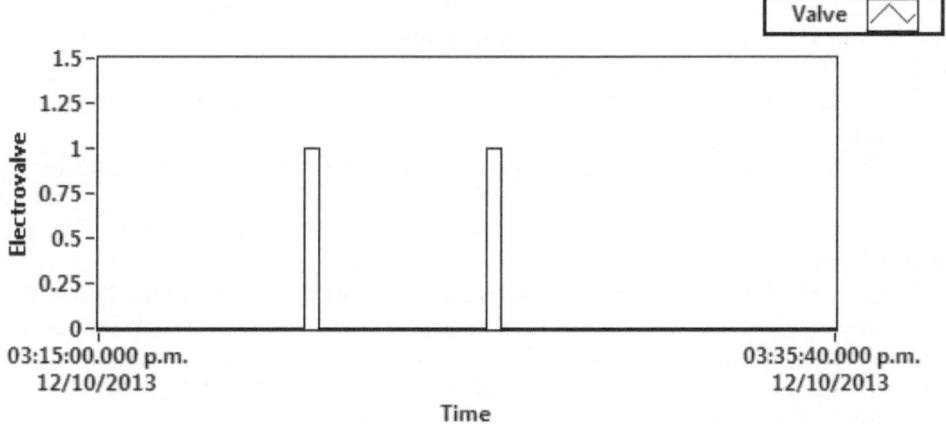

Figure 6.91   Electro-valve Response On/Off Controller.

performance of this actuator where the mesh is opened below the set point and closed above. It only blocks around 50% of the sunlight.

The last variable controlled in this section is the electrical conductivity. As known this variable defines the amount of nutrients contained in a solution reason why it is extremely important to control for the successful crop feeding. Figure 6.90 exhibits the on/off controller performance. The electrical conductivity does not reach the reference because it enters an allowable band. Even this reason, the levels achieved are widely desirable for the right crop growth.

The on/off electrical conductivity controller output which is the electro-valve state open or close (1 or 0) is shown in Figure 6.91. It presents the occurrence of two valve openings because, unlike the above on/off controllers, when the electrical conductivity

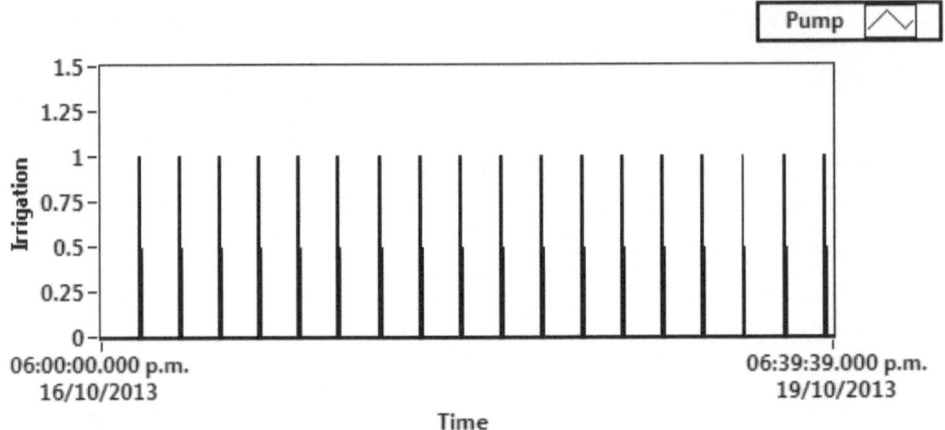

Figure 6.92   Irrigation Pump Response.

is below a band limit of 0.25 mS/cm from the set point, the controller opens the electro-valve for 20 seconds, then waits some minutes till the solution gets even and senses again, if the variable still out of the band limits the controller will take the initial decision until it gets into the desirable band limits.

In the electrical conductivity response from Figure 6.92 it can be observed the electro-valve openings impact.

Finally, the timer control that activates the irrigation pump is presented in Figure 6.92. A cyclic activation every four hours is given with a time length of ten minutes each. The hours were 2:00 a.m, 6:00 a.m, 10:00 a.m, 2:00 p.m, 6:00 p.m and 10:00 p.m.

After seeing the results obtained, one can infer that the on/off control technique has the advantage over other control strategies of been a simple system demanding low computational capacity. Despite of this it can handle non-linear systems such as the greenhouse presenting some level of robustness, reason why is the most common type of controller seen in the market around the world.

6.4.4   Intelligent controllers experimental results

The main problem that should be taken into account in greenhouse control is the complex interaction between the inside and outside parameters. Some conventional control systems such as conventional PID (Proportional-Integral-Derivative) controllers are not suitable for this type of application because a greenhouse model shows non-linear behavior on many points. Hence, it is not possible to achieve an accurate mathematical model reason why this work focused on combining non-conventional techniques that could handle the high non-linearities of the greenhouse.

This section shows the validation of the proposed controllers already tested in simulation: Sliding Modes + Feedback/Feed-forward Linearization and Fuzzy Logic. In other words the intelligent controllers are experimentally validated. Basically, these results are presented also in a three-day based period (October 19th–October 21st) as the previous section shows.

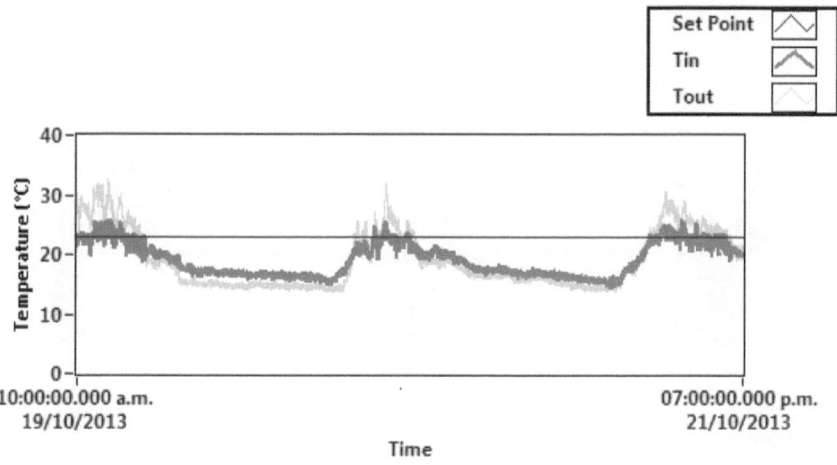

Figure 6.93 Temperature Response Non-Linear Controller.

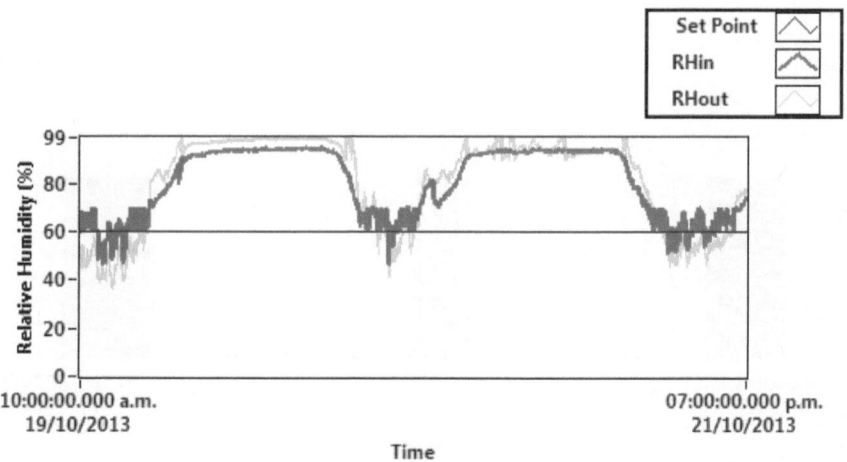

Figure 6.94 Relative Humidity Response Non-Linear Controller.

Figure 6.93 shows the first set of results for the controllers validation. It presents the inside temperature behavior following the temperature reference. Unlike the on/off controllers, the Sliding Modes + Feedback/Feed-forward Linearization does not need band limits, instead it is always trying to reach the desired set point.

On the other hand, similarly to the on/off temperature controller, it can hardly exercise control when outside temperatures are low due to the lack of the heating actuator.

The climatic controller also contemplates the relative humidity the response of which is shown in Figure 6.94. As in the case of the temperature behavior there is no more band limit as that of the on/off controllers. The desired inside relative humidity

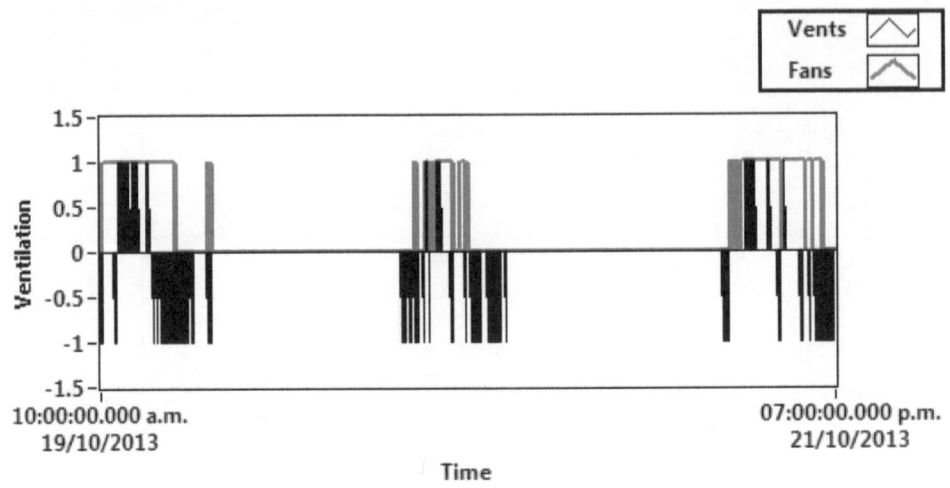

Figure 6.95 Ventilation Response Non-Linear Controller.

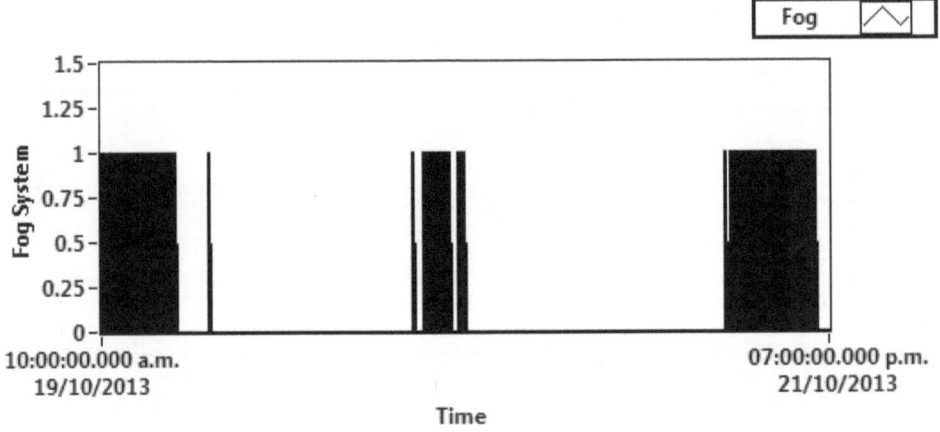

Figure 6.96 Fogging System Response Non-Linear Controller.

also can only be achieved when the outside relative humidity is below the set point. In particular, this set of results were taken from a rainy day, reason why the relative humidity values are higher than normal. Again, if the outside relative humidity is above the reference the system become uncontrollable.

The actuators are shared for both the temperature and relative humidity. The control technique implemented have the decision of the amount of ventilation and fogging is needed for regulating the temperature and relative humidity together.

Despite of the results shown in Figures 6.95 and 6.96 that only illustrates when the activation/deactivation occurs, the real controller outputs will be the range of values presented in the simulation section translated to the real actuators. This means that

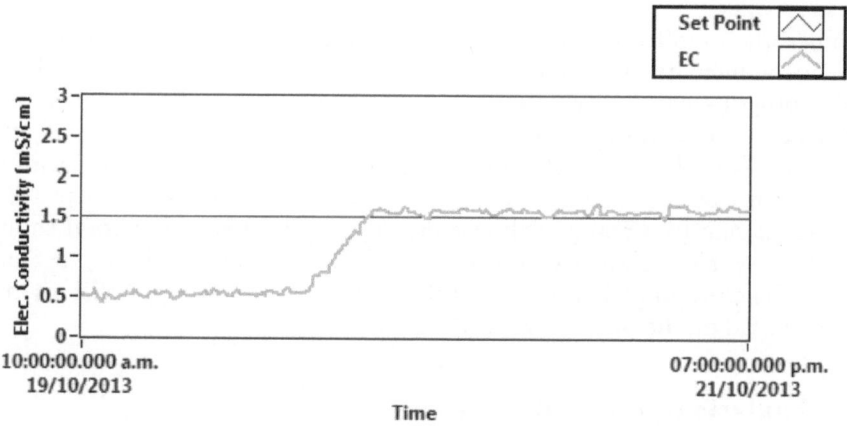

Figure 6.97 Electrical Conductivity Response Fuzzy Controller.

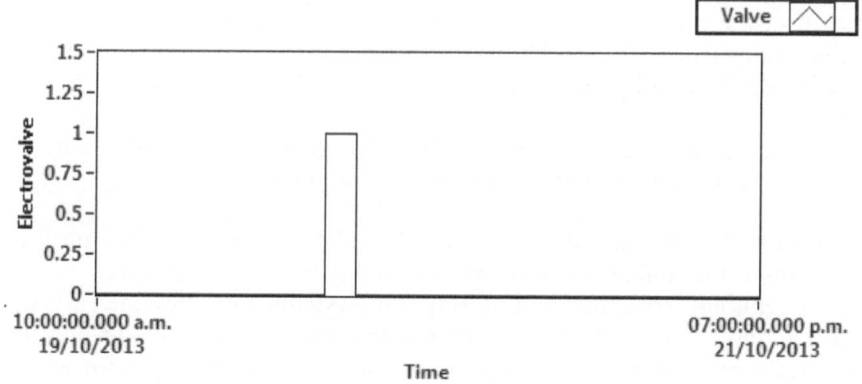

Figure 6.98 Electro-valve Response Fuzzy Controller.

the proposed controller manage the resources depending on the demand not totally on or totally off.

Generally, the actuators are activated when the temperature is above the reference and the relative humidity below its desired point. Note again that this scenario is mostly observed during light hours.

Finally, the nutrients supply Fuzzy Logic control for the electrical conductivity becomes the last set of results. Figure 6.97 gives the response of the controlled variable. This controller easily reaches the desired point with a minimum offset value that is completely allowable.

Figure 6.98 shows the controller output in terms of the amount of time of the electro-valve opening depending on the demand of nutrients required. It can be seen that the electro-valve only performed one opening for almost 46 seconds and it was enough to reach the set point, a clear difference from the on/off electrical conductivity controller that needed more than one electro-valve opening without getting so close to the desired value.

Unlike the on/off controller, the Fuzzy Logic controller not only measures the electrical conductivity error but also the tank levels where the nutrient solutions are stored in order to make an accurate decision for the electro-valve opening time length.

The proposed controllers validation has been carried out, and obviously there are a lot of facts not taken into account during simulation where ideal scenarios prevail. The importance here is that beginning from a simulation it is possible to extrapolate the controllers design to a real greenhouse prototype. The results are good enough and the greenhouse micro-climate could be controlled, even when the lack of actuators are a limiting. The proposed controller is robust and complex where decisions are made smoothly based on the micro-climate demand.

6.4.5   Analysis of control systems

The main objective of this section is to begin a discussion between the different packages of results presented above including the intelligent greenhouse development. The approach consists of analyzing the results of the proposed control both simulation and experimentation while proving the superiority of the proposed control over the on/off, both experimental. In other words, a brief comparison is carried out between the on/off and the intelligent controllers showing the advantages and drawbacks of each strategy.

The next remarks represent the analysis based on the overall results obtained through this work, including all the stages of development:

- A product design methodology was followed from specifications until the construction of the functional prototype of an intelligent greenhouse.
- The greenhouse structure and automation systems were designed based on the needs found in the market demand, systems that were subjected to optimization during the development and construction of the intelligent greenhouse.
- The selection of the greenhouse materials was based on the so-called conventional. These present strong mechanical properties but cost and pollution have led to seek others such as eco-materials.
- An important thing to keep in mind is the strict connection between the structure, automation systems and control; if the structure or automation systems are not properly fitted the control won't give its best performance.
- The level of technology will determine how excellent or poor the production will be; according to the climatic conditions of the greenhouse suitable for crop growth, a better harvest and growth in less time will be achieved.
- The simulations and control theory exposed in the present work were the starting point for getting a robust control design, development and implementation.
- Simulation comparison between Sliding Modes and PID (plus Feedback/Feedforward Linearization) clearly marked a higher degree of confidence with the first. It demonstrated a soft response and accurate while the PID showed ripples with undesirable overshoots.
- The response time of the controlled variables defers from simulation to experimentation because simulations does not contemplate the large death time had in a real

greenhouse system. Moreover, simulations represented an ideal scenario where the greenhouse model did not include other disturbances such as wind speed or CO_2 concentration that also affect the greenhouse micro-climate. Aware of these facts, the proposed control validation was a success.

- The control strategies developed: on/off, Sliding Modes + Feedback/Feed-forward Linearization, PID and Fuzzy Logic, proved that they are capable of controlling the greenhouse at the points where they could. This was observed in the results where all of them tried to track the desired set points.
- Due to the lack of actuators for heating and dehumidifying, the controllers could not regulate when temperatures were low and relative humidities were high but still the micro-climate was desirable and will not harm future crops. Here, it can be inferred that the greenhouse and controllers will have a better performance during summer season were the temperatures are high and relative humidities low.
- The control unit would not be possible without the software support of LabVIEW™ 2012 and National Instruments hardware that handled the controllers, a graphic user interface and the hardware interface as a link to the intelligent greenhouse world.

Advantages/drawbacks of proposed control over conventional on/off (experimentation)

Advantages:

- The proposed controller showed a better response by tracking the set point while the on/off controller remained in the allowable band with an offset.
- As long as the intelligent controller only demands the actuators what is required for regulation, the on/off control makes the actuators going up the maximum capacity or the lowest (two states).
- The on/off controller uses the actuators independently even when the variables such as temperature and relative humidity are thermodynamically linked (one affects the other). On the other hand, the proposed climatic controller makes the decision of action combining the two variables possible.

Disadvantages:

- Complexity over simplicity which requires the proposed controller to place high demands on computational resources.
- Due to last drawback the proposed controller is high cost.
- It needs more time for development.

Chapter 7

Automatic greenhouse operation

One of the most important crops on the planet is the tomato, hence it will form the case study presented in this chapter. However, at the end of the chapter tables are presented covering other crops. The tomato is an herbaceous perennial, but is usually grown as an annual in temperate regions since it is killed by frost. It originally had an indeterminate plant habit, continuously producing three nodes between each inflorescence, but determinate varieties have been bred with a bushlike form in which the plant is allowed to produce side shoots and the plant terminates with fruit clusters. Determinate varieties have fewer than three nodes between inflorescences with the stem terminating in an inflorescence, producing fruits that are easily machine harvested and primarily grown for processing.

Tomato belongs to the genus Lycopersicon, especially *L. esculentum*, that is grown for its edible fruit. The genus *Lycopersicon* of the family Solanaceae is believed to have originated in the coastal strip of western South America, from the equator to about 30° latitude south. The species is native to South America, especially Peru and the Galápagos Islands, being first cultivated in Mexico. In the mid-16th century, the tomato was introduced into Europe, primarily featured in early herbals. It was grown for the beauty of its fruit but was not often eaten, except in Italy and Spain. The tomato was introduced back into America from Europe in the 18th century, although it only became an important fruit in the 20th century. It is believed that the American Indians ate the tomato a long time ago.

Although it is certain that the origin of *Lycopersicon esculentum* was South America, the tomato was probably first cultivated in Mexico. Therefore, the seeds of tomato first taken to Europe came from Mexico after Cortez sacked Tenochtitlan, present day Mexico City in 1519, since it was not until 1533 that Peru and Equador were conquered by the Spaniards.

After introduction of the tomato into the United States, it was grown and brought to the table by Thomas Jefferson. George Washington Carver grew and recommended the tomato in an attempt to introduce the fruit into the diet among the poor in Alabama whose diets were woefully deficient in vitamins.

Breeding of the tomato over the past 50 years has substantially changed the characteristics of the plant and its fruit. Varieties available today for use by both the commercial and home gardener have a wide range of plant characteristics; they are resistant to many of the tomato-affected blight and wilt diseases; and are specifically adapted to a particular set of growing conditions, such as high tropical temperatures, field and greenhouse conditions, and fresh market versus processing tomato fruit. Maturity dates range from about 60 to more than 95 days, although several 45-day determinate varieties have been introduced for production in the very northern latitudes; and fruit size, color, texture, and acidity can be selected by variety, whether adapted to field or greenhouse conditions, or long or short days. Genetic engineering techniques applied to tomato breeding have been used to produce fruit with a long shelf life. The commercial production of tomatoes in the tropics offers an unique challenge in terms of varieties that can withstand high temperatures, and disease and insect pressures.

Botanical Name
Division: Anthophyta
   Class: Dicotyledons
      Family: Solanaceae
         Genus: *Lycopersicon esculentum*   Milltomato
                 Lycopersicon pimpinellifolium   Millcurrant tomato
                 Lycopersicon esculentum var. *cerasiforme*   Cherry tomato

Common Names
The common names for tomato in eight countries are

Country	Name
Danish	Tomat
Dutch	Tomaat
French	Tomate
German	Tomate
Italian	Pomodoro
Portuguese	Tomate
Spanish	Tomate
Swedish	Tomat

7.1   FIELD VERSUS GREENHOUSE GROWN FRUIT

The competition between field and greenhouse-grown fruit continues although the majority of fresh market fruit is and will continue to be field grown. Today, much of the fruit is being produced some distance from the market. The question of quality between field- and greenhouse-grown fruit is of major importance for the future of the greenhouse tomato industry. In general, greenhouse-grown fruit is vine ripened and can be delivered to the local market within a day or two of harvest. Most field-grown fruit is harvested before the fruit is fully ripe and shipped to the market, ripening occurring either naturally during shipment or by ethylene treatment.

Soil field-grown fruit can be coated with soil or dust particles, which although removed by washing prior to placement in the market can affect the self-life. Normally the shelf life of greenhouse fruit is better than that of field-grown fruit, which may be due to some soil residue remaining on the fruit.

Most soil field-grown plants require the use of pesticides and fungicides to keep them pest free, and soils are frequently treated with sterilizing chemicals to eliminate soil borne pests as well as being treated with herbicides to control weeds. Some residues from these applied chemicals can remain on the surface or in the fruit, normally at concentrations well below those considered physiologically significant, and therefore safe for human consumption. However, for some consumers, any presence of applied chemicals on or in the fruit would be considered unacceptable.

Greenhouse production will never be able to match field-grown fruit in terms of volume of supply, but for the quality-demanding customer, greenhouse vine-ripened fruit will remain in demand, if of high quality. How this demand is supplied will be determined by a number of factors. If greenhouse tomato plant production can be done pesticide free, the fruit can be so identified in the marketplace, making such fruit of increased value to many consumers.

The time period in the field is determined by the length of the growing season. In addition, keeping the tomato plant productive over the whole season may not be possible due to climatic (early or late frosts, drought, excess moisture, etc.) or other conditions (insect and disease pressures).

In the greenhouse, the tomato plant can be maintained for periods of 6 to 9 months in duration, or even longer, by training the plant up a vertical supporting twine, removing older leaves as the lower fruit clusters are harvested, and by lowering the main plant stem to keep the whole plant within easy reach of workers. This process can be sustained as long as the plant is actively growing, free from disease and other stresses. In the greenhouse, it is possible to control the environment and those factors than affect the plant's well-being, and thereby keeping the tomato plant productive over a long period of time.

7.2   PLANT FORMS

Cultivated tomato is divided into two types, *indeterminate* and *determinate*, the former being the single vine type usually trained to maintain a single stem with all the side shoots removed, and the latter terminating in a flower cluster with shoot elongation stopping. Determinate cultivars are usually earlier than indeterminant ones and are especially desirable where the growing season is cool or short, or both. With fruit ripening nearly at one time, it makes this plant type suitable for mechanical harvesting. Indeterminate plants are for long-season production because this form of the tomato plant will continuously produce fruit for an extended period of time if properly maintained. The approximate time from planting to market maturity for an early variety is from 50–65 days while for a late variety from 85–95 days.

7.3   PHOTOSYNTHETIC CHARACTERISTICS

When chlorophyll-containing plant tissue is in the presence of light, three of the essential elements, carbon (C), hydrogen (H), and oxygen (O), are combined in the process called *photosynthesis* to form a carbohydrate. Carbon dioxide (CO_2) is from the air, and water (H_2O) is taken up through the roots. A water molecule is split and combined with CO_2 to form a carbohydrate while a molecule of oxygen (O_2) is released. Tomato is a C_3 plant since the first product of photosynthesis is a 3-carbon carbohydrate.

The photosynthetic process occurs primarily in green leaves and not in the other green portions (petioles and stems) of the plant. The rate of photosynthesis is affected by factors external to the plant, such as:

- Air temperature (high and low)
- Level of CO_2 in the air around the plant
- Light intensity and quality

Under most conditions, both in the greenhouse as well as outdoors, the energy level impacting the plant canopy is the factor that influences plant growth; and for tomato, that energy level is frequently exceeded. In any growing system, the ability to control both the total amount of energy received over a period of time and the energy level at any one point in time determines plant performance.

Based on scientific terms for the production of greenhouse tomatoes, control of the light and air environment could be described as "process management" of the growing system in which assimilation, translocation, allocation, and uptake are the factors requiring control to maintain a vigorously growing tomato plant and high fruit yields; these factors are becoming controllable as greenhouses install computer-directed control devices.

7.4 TEMPERATURE REQUIREMENTS

The tomato plant prefers warm weather because air temperatures, 10°C or below, will delay seed germination, inhibit vegetative development, reduce fruit set, and impair fruit ripening. The tomato plant cannot tolerate frost. High air temperature, above 35°C, reduces fruit set and inhibits development of normal fruit color. The optimum range in air temperature best suited for normal plant growth and development and fruit set is between 18.5°C and 26.5°C, with day and nighttime temperature ranges being 21–29.5°C and 18.5–21°C, respectively. The growing degree day base temperature is 10.5°C, a temperature below which growth is negligible; while on the contrary, the best growing temperature is 21–24°C, the minimum being 18.5°C and the maximum being 26.5°C.

Although air temperature is critical for normal vigorous plant growth, the canopy (leaf) temperature may be far more important, a factor that can be controlled in environmental chambers and greenhouses, the optimum range being between 20 and 22°C. The combination of air temperature, relative humidity, and plant transpiration rate will determine the canopy temperature.

A tomato plant exposed to cool air temperature, less than 15.5°C for extended periods of time, will begin to flower profusely with flower clusters appearing at terminals, typical of determinate plants. Two flowers may fuse together forming an unusually large flower. Flowers will remain open on the plant for several weeks without the formation of fruit. If a fruit does appear, it may be ribbed in appearance indicating incomplete pollination, or two or more fruit may fuse together.

7.5 CARBON DIOXIDE

The normal atmosphere contains about 300 mg/L [parts per million (ppm)] carbon dioxide (CO_2); and in a tomato greenhouse canopy, it can be quickly drawn down to 200 ppm. Carbon dioxide level is thought not to be a problem if the normal atmospheric level can be maintained in the plant canopy. However, the tomato plant, being a C_3 plant, is highly responsive to elevated CO_2 in the air surrounding the plant. In a greenhouse, elevating the CO_2 content to 1000 ppm can have a significant effect on the tomato plant growth and yield. Under high light intensity with 1000–1500 ppm CO_2 levels, tomato leaves became thickened, twisted, and purple; and the intensity of deformity increased with increasing CO_2 concentration.

7.6 LIGHT QUALITY AND INTENSITY

Light interception by the plant canopy is influenced by the leaf area exposed to incoming radiation with plant spacing having a significant effect on interception. One of the primary reasons why greenhouse tomato yields far exceed that obtainable for field-grown plants is the greater interception of light energy due to the increased leaf area indices of the greenhouse plants. The value of the lower leaves on the tomato plant is considerable in terms of their contribution to plant growth and fruit yield.

High light intensity is probably as detrimental to tomato fruit production as low light intensity is. With high solar radiation impacting fruit, cracking, sunscald, and green shoulders can be the result. In addition, high light intensity can raise the canopy temperature, resulting in poor plant performance. In southern latitudes and during the summer months in all latitudes, greenhouse shading is essential to maintain production of high quality fruit. Under low light conditions, light supplementation is more effective by extending the hours of light rather than attempting to increase light intensity during the sunlight hours.

Plants respond to both light intensity and quality. When there is excess blue light with very little red light, the growth will be shortened, hard, and dark in color; if there is excess red light over blue light, the growth will become soft with internodes long, resulting in lanky plants. The tomato plants in the glass-covered house were tall and light green in color, while those in the fiberglass-covered house were short and dark green in color; the differences in plant appearance were due in part to wavelength light filtering. However, fruit yields and quality were comparable in both houses; although in the fiberglass-covered house, the cultural requirements were easier to manage with shorter plants.

7.7 WATER REQUIREMENT

The tomato plant needs plenty of water but not an excess because tomato roots will not function under waterlogged (anaerobic) conditions. When the moisture level surrounding the roots is too high, epinasty, poor growth, later flowering, fewer flowers, and lower fruit set occurs; and fruit disorders such a fruit cracking will occur when water availability is inconsistent. The tomato plant responds quickly to fluctuations in radiation, humidity, and temperature, factors that significantly impact the plant.

The ability of the tomato plant to adjust to these conditions determines the rate of plant growth as well as the yield and quality of fruit. Even under moderate water stress, photosynthesis is slowed because the movement of gases through the stomata is restricted when the plant is under moisture stress.

The size of the root system is determined not only by the genetic character of the plant but also by the rooting conditions. The extent of root growth will be determined by soil physical conditions and levels of soil moisture. Under high soil moisture conditions or around a drip emitter, root growth will be less than where there is not an excess of water present. A mature tomato plant will wilt if the plant is not able to draw sufficient water through its roots, a condition that can occur if the rooting medium is cool or the rooting zone is partially anaerobic. Also the size of the root system may be an important factor, but just how large the root system must be to ensure sufficient rooting surface for water absorption is a factor that is not known.

Under low moisture conditions surrounding the roots, there will be fewer flowers per truss, lower fruit will set if at 25% less than that needed, and blossom-end rot (BER) incidence will be high.

Soil moisture control in the field was obtained by maintaining the underlying water table over a raised plastic-covered bed. Today, the use of the plastic culture technique and drip irrigation provides the control needed to maintain the supply of water and essential plant nutrients at optimum levels. The drip irrigation technique is in wide use both in the field and in the greenhouse for supplying water to the plant at precise rates and times.

The water requirement for field tomatoes is 2000–6600 m^3/ha under normal air temperatures. In a greenhouse setting, a tomato plant in full fruit production will consume about 1 L of water per day.

Fruit yield and quality are factors that are affected by the amount of water available to the plant.

7.8   FACTORS AFFECTING GREENHOUSE PRODUCTION

The major factors that affect greenhouse tomato production are

- Light, both intensity and length
- Carbon dioxide (CO_2) level in the greenhouse
- Temperature and humidity control, both low and high
- Disease and insect control
- Nutritional management over the life of the tomato plant
- Varietal plant characteristics
- Management skill required to produce higher plant yields of quality fruit

The following are the optimum ranges for the following factors.

In the past 20 years, there have been a number of very significant developments that have affected the ability to produce high quality fruit in a greenhouse-controlled environment, which include:

- There has been a change from growing in soil to some form of soilless production, such as Nutrient Flow Technique (NFT) hydroponics, or perlite bag or rockwool slab–drip irrigation systems, which provides a degree of nutrient element control not possible in soil and eliminates soil factors that are difficult to control.

Table 7.1 Optimum Ranges.

Factor	Growth Range	Optimum Range (°C)
Air temperature	Germination to seedling stage	24–26
Sunny daytime	Seedling to termination	24–27
Cloudy daytime	Seedling to termination	22–24
Night air temperatura	Seedling to termination	18–20
Root temperature	Germination to early growth	24–27
	Vegetative to termination	20–24
		(%)
Relative humidity	Germination to early growth	75–88
	Seedling stage	70–80
	Vegetative to termination	60–80
pH nutrient solution	Germination to early growth	5.5–6.5
		(dS m^{-1})
Electrical conductivity of nutrient solution	Germination to early growth	1.8–2.0
Sunny day	Seedling to termination	1.5–2.0
Cloudy day	Seedling to termination	2.5–4.0

- Cultivars have been bred specifically for greenhouse conditions and low light situations, having either resistance or tolerance to common tomato plant diseases and insects; and having significantly increased fruit yield potentials, producing fruit with specific fruit characteristics to meet particular consumer preferences.
- Introduction of bumblebees for flower blossom pollination eliminates the need to hand pollinate, a major labor-intensive operation.
- Use of predator insects and other nonchemical techniques can either eliminate or significantly reduce the need for chemicals to control plant-damaging insects and disease as well as integrated pest management (IPM) procedures.
- Computer control of the growing system and greenhouse environment is based on factors being continuously and automatically monitored.

7.9 GREENHOUSE STRUCTURES

The greenhouse grower has a variety of greenhouse structures to choose from in terms of size and covers, the use of plastic-covered greenhouses being increasingly used worldwide. For the single owner–operator, a standard-sized greenhouse would be 30–40 ft wide and 100–140 ft in length, with the cover being a single or double air-separated layer of plastic film. For larger installations, greenhouses are normally gutter-connected with combinations of covers being either totally plastic or glass, or a mix of glass and plastic film or sheets of fiberglass or plastic.

The size and design of the installed heating and cooling systems will vary considerably depending on location (latitude). In northern latitudes, an efficient heating system is the dominate requirement, while in southern latitudes, cooling efficiency is the dominant requirement. The heating and cooling system must be able to maintain an optimum air temperature within the range of 21–24°C, keeping the minimum temperature from dropping below 18.3°C and the maximum temperature from exceeding 29.4°C. Some provision for shading the greenhouse during periods of high light intensity is also required, even in the northern latitudes if fruit production is to continue through the summer months. Floor heating is proving to be very advantageous in keeping the rooting media from dropping below the optimum rooting temperature of 21°C.

Air movement within the greenhouse is important, with warm dry air introduced at the bottom of the plant canopy so that air flow is from the base of the plant up through the canopy into the greenhouse gable. The objective is to keep the plant canopy as dry as possible, which prevents the development of diseases and a potential habitat for insects in the older, lower foliage. The ideal relative humidity is 50%, with the acceptable range being from 40 to 70%.

7.10   PEST MANAGEMENT

It is essential that a pest management program be developed and carefully followed to ensure that disease and insect infestations do not occur. Protective and treatment procedures should be timely because after-the-fact treatment or treatments may not be able to control an established disease or insect infestation.

The insect population in the greenhouse is best monitored by placing yellow or blue colored sticky boards at intervals within the plant canopy, and by examining daily the boards to determine the number and species of insects on the board. As insect numbers accumulate on the board, procedures for insect population control can be instituted by using chemical procedures or by introducing predator insects into the greenhouse.

Before any chemical or biological treatments are applied, the disease or insect species should be confirmed by a plant pathologist or entomologist, respectively, and any treatment should be applied based on procedures that conform to current chemical regulations.

7.11   CULTURE SYSTEMS

The demands of the marketplace, the growing environment (such as light intensity and duration) and outside air temperatures will dictate to a considerable degree which tomato plant culture system can be efficiently employed. A single initial planting and fruit harvesting over a long period of time is one system; also there are several versions of multi-cropping in which the tomato plant is allowed to develop to a certain point, is topped allowing already set fruit to mature, and then is removed from the greenhouse.

Keeping a tomato plant in profitable production over an extended period of time requires considerable management skill, while multi-cropping systems must be carefully managed to efficiently use the greenhouse space and maintain a constant supply of fruit to satisfy market requirements. Success depends on maximizing the greenhouse growing space for high yield fruit production, minimizing costs of production, and sustaining a flow of high quality fruit to the marketplace, thereby making the growing system, whichever is selected, conforms not only to environmental conditions but also to market demands.

The rapid expansion of the greenhouse tomato industry suggests that the various systems of growing currently being employed, particularly hydroponics are both profitable and able to meet consumer demands for high quality fruit.

There are basically three hydroponic growing systems that have been or are being used to grow tomatoes commercially. Initially, the ebb-and-flow method (or modifications of the concept) was the method in wide use from the late 1930s into the 1950s. In the mid-1970s, Allan Cooper introduced his nutrient film technique (NFT), which substantially changed the basic concept of hydroponic growing; this system is relatively inexpensive to install and maintain, and is quite precise in its control of the nutrient–root environment.

With the introduction of drip irrigation combined with fertilizer injector systems, placement of water or a nutrient solution at the base of the tomato plant on a regulated basis became possible. With this type of water–nutrient solution delivery system, the use of rockwool slabs and perlite bags as the major growing media came into wide use.

7.12 PLANT NUTRITION

There are 16 elements that have been identified as being essential for the normal growth and development of all plants. The form for uptake and general functions for the 16 essential elements in plants are given in Table 7.2. Of the 16 essential elements required by plants, 13—nitrogen (N), phosphorus (P), potassium (K), calcium (Ca), magnesium (Mg), and sulfur (S), known as the *major elements;* and boron (B), chlorine (Cl), copper (Cu), iron (Fe), manganese (Mn), molybdenum (Mo), and zinc (Zn), known as the *micronutrients*—must be present at specific concentrations in the nutrient solution supplied to the plant to sustain normal growth.

The success of any growing system is based on the ability of the grower to maintain the nutrient element status of the plant without incurring insufficiencies; this is not an easy task, particularly when growing in soil in the field. Even successfully growing plants hydroponically in the greenhouse can be a formidable task.

The level of an essential element in the tomato plant determines the plant's nutritional status and affects the plant quality. The tomato plant has been established based

Table 7.2 The Essential Elements, Their Form for Uptake, and Functions in the Plant.

Essential Element	Form for Uptake	Functions in the Plant
C, H, O, N, S	Ions in solution (HCO_3^-, NO_3^-, NH_4^+, SO_4^{2-}), or gases in the atmosphere (O_2, N_2, SO_2)	Major constituents of organic substances
P, B	Ions in solution (PO_4^{3-}, BO_3^{3-})	Energy transfer reactions and carbohydrate movement
K, Mg, Ca, Cl	Ions in solution (K^+, Mg^{2+}, Ca^{2+}, Cl^-)	Nonspecific functions, or specific components of organic compounds, or maintaining ionic balance
Cu, Fe, Mn, Mo, Zn	Ions or chelates in solution (Cu^{2+}, Fe^{2+}, Mn^{2+}, MoO^-, Zn^{2+})	Enable electron transport and catalysts for enzymes

Table 7.3 Essential Elements Normal and Deficient Levels.

Element	Normal Range (%)	Deficient (%)
Major Elements		
Nitrogen (N)	2.8–6.0	<2.0
Phosphorus (P)	0.3–0.9a	<0.2
Potassium (K)	2.5–6.0b	<1.5 vegetative (<2.5 fruiting)
Calcium (Ca)	0.9–7.2c	<1.0
Magnesium (Mg)	0.4–1.3	<0.3
Sulfur (S)	0.3–4.2	–
	[mg kg^{-1} (ppm)]	[mg kg^{-1} (ppm)]
Micronutrients		
Boron (B)	25–100	<20
Chlorine (Cl)	Not known	–
Copper (Cu)	5–20	<4
Iron (Fe)	40–300	<40 (<50 may be deficient)
Manganese (Mn)	40–500	<30
Molybdenum (Mo)	0.9–10.0	Not known
Zinc (Zn)	20–100	<16

a Levels in excess of 1.00% can be detrimental to the plant.
b Relationship between K and Ca may be more important than either element alone.
c Levels less than 1.50% may result in significant BER in fruit.

on the optimum concentration level of the essential elements in the tomato plants. The average normal range and deficient level for the essential elements found in tomato are shown in Table 7.3.

Photographs of visual deficiency or excess (toxicity) symptoms of the essential elements in tomato are scattered among various. Photographs of good quality have appeared on the Internet, which may be the best means of obtaining visual identification of insufficiencies associated with the essential elements.

There are two elements—silicon (Si) and nickel (Ni) that have been suggested as being essential for plants. The major role of Si has been found to be in the strengthening of the stem of rice, as well as other grain crops, plus the possibility that Si may also contribute to the stem strength of the tomato plant. Silicon has also been found to be a factor in preventing the penetration of fungus hypha (disease resistance) into plant leaf cells, therefore making the plant more resistant to fungus attack. Since this disease resistance aspect of Si could be of major benefit for tomato plants being grown hydroponically in the greenhouse, several soluble formulations of Si are available for addition to the nutrient solution.

There are three other elements, sodium (Na), vanadium (V), and cobalt (Co), which fall into the category as being beneficial, because Na can partially substitute for potassium (K) and V for molybdenum (Mo), while Co is required by the nitrogen (N_2) fixing bacteria in leguminous plants. However, none of these three elements have been found to be beneficial to the tomato plant, except for Na that might be a factor in enhancing the flavor of tomato fruit.

Table 7.4 Essential Element Concentrations.

Element	Concentration [mg L⁻¹ ppm]
Major Elements	
Nitrogen (N)	200
Phosphorus (P)	60
Potassium (K)	300
Calcium (Ca)	170
Magnesium (Mg)	50
Micronutrients	
Boron (B)	0.3
Copper (Cu)	0.1
Iron (Fe)	12.0
Manganese (Mn)	2.0
Molybdenum (Mo)	0.2
Zinc (Zn)	0.1

The challenge for the hydroponic grower is to maintain the nutrient element status of the tomato plant to keep it productive over an extended period of time. The initial composition of the nutrient solution, its rate of delivery, and adjustment in composition with both the changing status of the plant and environmental conditions are significant factors.

The theoretically ideal nutrient solution formula for the NFT system given by Cooper (1996) is shown in Table 7.4. This formula gives the following essential element concentrations in the "starter" solution.

With the NFT system, one method for determining when to make a nutrient solution irrigation is that for every $0.3\,MJ/m^2$ of radiation received, an irrigation is scheduled. Similar procedures are being used for other hydroponic systems, regulating the frequency and amount of nutrient solution applied based on solar radiation received.

The influence of stage of plant growth is also a factor in determining what the elemental concentration ranges should be for the NFT and rockwool techniques. As the stage of growth advances, there is an increase in the N, K, and Mg concentrations, while the other elements remain at constant concentration (Benton, 1999).

7.13 GENERAL GREENHOUSE REFERENCE CONDITIONS

To use the automatic greenhouse control for different crops it is necessary to know the optimal reference conditions for each crop. Those conditions are defined by the user as initial conditions and the greenhouse controller will follow those references according to the crop stage.

It is important to identify the main conditions of nutrients that allow growing crops in optimal conditions. When the environmental conditions and nutrient conditions are well-defined the complete cycle of the cultivation process can be completed by the

Table 7.5 General Reference Conditions.

Crop	Day Temperature [°C]	Night Temperature [°C]	Relative Humidity Range [%RH]	Light Intensity Range [kLux]	pH
Tomato general conditions (Cherry)	21–24	14–17	60–70	20–30	6.5–7.5
Lettuce	16–21	12–13	60–80	5–20	6.7–7.4

Table 7.6 General vegetable nutrient formulas for 3 growing steps (these quantities are valid for nutrients dissolved in 100 liters of water and are not the optimal values).

		Quantity [g]		
Name	Formula	Premature Germination	Flowering	Increasing Flowering
Calcium Nitrate	$Ca(NO_3)_2$	90.0	0.0	0.0
Magnesium Sulfate	$MgSO_4$	30.0	0.0	50.0
Monopotassium Phosphate	KH_2PO_4	20.0	0.0	0.0
Potassium Nitrate	KNO_3	35.0	110.0	75.0
Potassium Sulfate	K_2NO_4	15.0	0.0	0.0
Iron Sulfate	$FeSO_47H_2O$	10.0	0.0	0.0
Manganese Sulfate	$MnSO_44H_2O$	1.0	50.0	5.0
Boric Acid	H_2BO_3	0.5	0.0	0.0
Zinc Sulfate	$ZnSO_47H_2O$	0.5	0.0	1.0
Copper Sulfate	$CuSO_45H_2O$	0.5	0.0	1.0
Calcium Chloride	$CaCl_26H_2O$	0.0	70.0	150.0
Sodium Nitrate	$Na(NO_3)$	0.0	100.0	100.0
Ammonium Chloride	$(NH_4)Cl$	0.0	15.0	20.0
Calcium Phosphate	$Ca(PO_3)$	0.0	30.0	7.0
Ferrous Sulfate	$FeSO_47H_2O$	0.0	15.0	15.0

automatic controller. Tables 7.5 and 7.6 list some environmental and nutrient basic reference conditions which are needed in the automatic controller. Since there are a lot of tomato variations, it is recommended to set the reference information according to the specific tomato classification. However, general conditions could be used for different tomatoes and the results will be good enough. For increasing the quality of the cultivation process general conditions are not used in the automatic controller. It is observed in Table 7.5 that the general references values of tomato and lettuce are close to each other.

References

Alabama , A. & Auburn Universities (2013) Hobby Greenhouse. Construction. *www.aces.edu.*

Albright, L., Arvanitis, K., & Drysdale, A. (2001). Environmental Control for Plants on Earth and in Space. IEEE Control Systems, 21 (5), 28–47.

ALECO (2009) Plásticos orgánicos. ALECO consult Internacional. España ALECO.

AMCI. (2008). Norma Mexicana para el Diseno y Construccion de Invernaderos (NMXE-255-CNCP-2008).

ASERCA. (2012). Un Horizonte ASERCA del Mercado Agropecuario. Claridades Agropecuarias, 1–52.

Ashby, M. F. (1999) Materials Selection in Mechanical Design. *Butterworth-Heinemann, Oxford, UK.*

Astrom, K., & Hagglund, T. (1995). PID Controllers: Theory, Design and Tuning. Instrument Society of America.

Balas, M., & Balas, V. (2008). Modeling Passive Greenhouses. The Sun's Influence. In 2008 international conference on intelligent engineering systems (ines) (pp. 71–75).

Bartzanas, T., Boulard, T. & Kittas, C. (2002) Numerical simulation of the airflow and temperature distribution in a tunnel greenhouse equipped with insect-proof screen in the openings. *Computers and Electronics in Agriculture,* 34, 207–221.

Bastida T., A. (2011). Los Invernaderos y la Agricultura Protegida en Mexico. Chapingo, Mexico: Serie de Publicaciones Agribot.

Beer, F. P. (2004) *Mecánica de materiales,* México, Mc Graw Hill.

Bellows, B. & Adam, K. (2008) Solar Greenhouse Resources. Horticulture Resource List. NCAT Agriculture Specialists. U.S.A., ATTRA National Sustainable Agriculture Information Service.

Bennis, N., Duplaix, J., Enea, G., Haloua, M., & Youlal, H. (2008). Greenhouse Climate Modelling and Robust Control. Computers and Electronics in Agriculture, 61, 97–107.

Benton, J. (1999). Tomato Plant Culture: In the Field, Greenhouses, and Home Garden. CRC Press LLC.

Boaventura, J. (2003). Greenhouse Climate Models: An Overview. In Efita 2003 conference (pp. 823–829).

Both, A. (2008). Greenhouse Temperature Management. New Jersey Agricultural Experiment Station.

Bouzo, C., & Gariglio, N. (2009). Tipos de Invernaderos. Retrieved from http://www.ecofisiohort.com.ar/wp-content/uploads/2009/10/Tipos-de-Invernaderos.pdf.

Bowman, G. E., & Weaving, G. (1970). A Light-modulated Greenhouse Control System. Journal of Agricultural Engineering Research, 3 (15), 255–264.

Bucklin, R. A. (2012). Florida Greenhouse Design.

Building and Construction authority (2008) The code for Environmental Sustainability of Buildings. Version 1.0 Building and Construction Authority, http://www.bca.gov.sg/Envsus legislation/others/Env_Sus_Code.pdf.

Caponetto, R., Fortuna, L., Nunnari, G., & Occhipinti, L. (1998). A Fuzzy Approach to Greenhouse Climate Control. In Proceedings of the 1998 American control conference (pp. 1866–1870).

Caponetto, R., Fortuna, L., Nunnari, G., Occhipinti, L., & Xibilia, M. (2000). Soft Computing for Greenhouse Climate Control. IEEE Transactions on Fuzzy Systems, 8 (6), 753–760.

Castilla, N. & Hernandez, J. (2007) Greenhouse technological packages for high quality production. Acta Hortic, 761, 285–297.

Cedillo, E., & Calzada, M. (2012). La Horticultura Protegida en Mexico Situacion Actual y Perspectivas. Encuentros UNAM, 1–10.

Cepeda, P., Ponce, P., & Molina, A. (2012). A Novel Speed Control for DC Motors: Sliding Mode Control, Fuzzy Inference System, Neural Networks and Genetic Algorithms. In 2012 11th Mexican international conference on artificial intelligence (MICAI) (pp. 116–121).

Cepeda, P., Ponce, P., Molina, A., & Lugo, E. (2013). Towards Sustainability of Protected Agriculture: Automatic Control and Structural Technologies Integration of an Intelligent Greenhouse. In 2013 11th IFAC workshop on intelligent manufacturing systems (IMS).

Cepeda, P. (2013a). Design, Automation and Control of a Sustainable Intelligent Greenhouse. Tecnológico de Monterrey Campus Ciudad de México.

Cepeda, P., Ponce, P., Romero, D., & Molina, A. (2011). Fuzzy C-Means Clustering Technique Applied for Modeling Parameters of an Intelligent Greenhouse Open Control System. In 2011 IEEE electronics, robotics and automotive mechanics conference (CERMA) (pp. 283–288).

Cepeda, P., Rocha, R., Ponce, H., Garcia-Ravize, A., Romero, D., Ponce, P., & Molina, A. (2010). Invernadero Inteligente basado en un Enfoque Sustentable para la Agricultura Mexicana. In 2010 viii congreso internacional sobre innovacion y desarrollo tecnologico (ciindet) (pp. 1–8).

Cepla & Cipa (2006) Búsqueda y desarrollo. La ventilación en invernaderos. Filclair. Greenhouses technology, 124, http://www.filclair.fr/filclair-infos/busqueda-y-desarrollo-3.html.

Chao-gang, Y., Yi-bin, Y., Jiang-ping, W., & Jamal, N. (2005). Determining Heating Pipe Temperature in Greenhouse using Proportional Integral plus Feedforward Control and Radial Basic Function Neural-Networks. Journal of Zhejiang University SCIENCE (JZUS), 6A(4), 265–269.

Chouchaine, A., Feki, E., & Mami, A. (2012). A Feedback Linearization Control Based on a Takagi-Sugeno System to Control the Inside Temperature of a Greenhouse. In 2012 16th IEEE mediterranean electrotechnical conference (melecon) (pp. 693–698).

Chunfeng, Z., & Yonghui, C. (2011). Applications of DMC-PID Algorithm in the Measurement and Control System for the Greenhouse Environmental Factors. In 2011 chinese control and decision conference (ccdc) (pp. 483–485).

Clements, R., Haggar, J., Quezada, A., & Torres, J. (2011). Technologies for Climate Change Adaptation - Agriculture Sector (X. Zhu, Ed.). Roskilde: UNEP RisoCentre.

Cline, W. (2007). Global Warming and Agriculture: Impact Estimates by Country. In Center for global development. Washington DC, USA.

CONEVAL. (2011a). Informe de Evaluacion en Materia de Acceso a la Alimentacion.

CONEVAL. (2011b). Pobreza en Mexico y en las Entidades Federativas 2008–2010.

Cook, R., & Calvin, L. (2005). Greenhouse Tomatoes Change the Dynamics of the North American Fresh Tomato Industry. In Err-2, USDA/ERS.

Cooper, P. I. & Fuller, R. J. (1983) A Transient Model of the Interaction between Crop, Environment and Greenhouse Structure for Predicting Crop Yield and Energy Consumption. J. agric. Engng Res., 28, 401-417.

Cox, S. (1987). Electronics in UK Agriculture and Horticulture. 1987 IEE Proceedings on Physical Science, Measurement and Instrumentation, Management and Education – Reviews, 134 (6), 466–492.

Cornell University (2010) Greener Greenhouses and Growth Chambers. *Cornell University Agricultural Experiment Station* EUA, http://cuaes.cornell.edu/cals/cuaes/sustainability/sustainability.cfm.

CSIC (2013) Nuevo plástico biodegradable de origen natural OTRI. Universidad de Malaga. España, UMAPATENT.

Couto, N., Rouboa, A., Monteiro, E. & Viera, J. (2012) Computational Fluid Dynamics Analysis of Greenhouses with Artificial Heat Tube. *World Journal of Mechanics*, 2, 181–187.

Cuiping, H. & Chengwei, M. (2008) Boundary Setting in Simulating Greenhouse Ventilation by Fluent Software. *Computer And Computing Technologies In Agriculture, Volume II The International Federation for Information Processing* 259, 1391–1395.

Davis, P. F., & Hooper, A. W. (1991). Improvement of Greenhouse Heating control. 1991 IEE Proceedings Control Theory and Applications, 138 (3), 249–255.

Dieck, G. (2000). Instrumentacion, Acondicionamiento Electronico y Adquisicion de Datos. Mexico: Trillas ITESM.

Elsner, B. V., Briassoulis, D., Waaijenberg, D., Mistriotis, A., Zabeltitz, C. V., Gratraud, J., Russo, G. & Suay-Cortes, R. (2000), a, Review of Structural and Functional Characteristics of Greenhouses in European Union Countries: Part I, Design Requirements. *J. agric. Engng Res*, http://www.idealibrary.com.

Elsner, B. V., Briassoulis, D., Waaijenberg, D., Mistriotis, A., Zabeltitz, C. V., Gratraud, J., Russo, G. & Suay-Cortes, R. (2000), b, Review of Structural and Functional Characteristics of Greenhouses in European Union Countries, Part II: Typical Designs. *J. agric. Engng Res.*, http://www.idealibrary.com.

Emerson, R. W. (2009) Truly greenhouses. Our Farm by Earth Flora Inc, flowerdepot.

Enriquez, G. (1994). El ABC de las Instalaciones Electricas Industriales (Grupo Noriega, Ed.). Mexico, D.F.: Editorial Limusa.

Enriquez, G. (1999). Manual de Instalaciones Electricas Residenciales e Industriales (2nd ed.; Grupo Noriega, Ed.). Mexico, D.F.: Editorial Limusa.

Everhart, E. Hansen, R., Lewis, D., Naeve, L., Taber, H. & Huntrods, D. (2010) Lowa High Tunnel Fruit and Vegetable Production Manual. Iowa State University, Copyright© 2009, Leopold Center for Sustainable Agriculture, Iowa State University of Science and Technology.

Explorando Mexico. (2012). El Campo Mexicano. Retrieved from http://www.explorandomexico.com.mx/about-mexico/6/43/

Fang, X., Jialiao, C., Libin, Z., Hongwu, Z. (2006). Self-Tuning Fuzzy Logic Control of Greenhouse Temperature Using Real-Coded Genetic Algorithm. Proceedings of IEEE International Conference on Control, Automation, Robotics, and Vision, pp. 1–6.

Fang, X., Junqiang, S., & Jiaoliao, C. (2006). Rough Sets Based Fuzzy Logic Control for Greenhouse Temperature. In 2006 proceedings of the 2nd IEEE/ASME international conference on mechatronic and embedded systems and applications (mesa) (pp. 1–5).

Fao (2002) Dirección de Producción y Protección Vegetal 90. El cultivo protegido en clima mediterráneo. Manual preparado por el grupo de cultivos hortícolas. Dirección de producción y protección vegetal organización de las naciones unidas para la agricultura y la alimentación roma.

FAO. (2012a). El Estado Mundial de la Agricultura y la Alimentacion (Tech. Rep.). Rome, FAO.

FAO. (2012b). Statistical Yearbook: World Food and Agriculture (Tech. Rep.). Rome, FAO.

FAO, WFP, & IFAD. (2012). The State of Food Insecurity in the World 2012. Economic growth is Necessary but not sufficient to accelerate reduction of hunger and malnutrition. (Tech. Rep.). Rome, FAO.

FAOSTAT. (2013). FAOSTAT Agriculture Data. Retrieved from http://faostat3.fao.org/home/index.html#VISUALIZE

Fitzgerald (2007) *Mecánica de materiales*, México, Alfaomega.Flores, D., & Ford, M. (2010). Greenhouse and Shade House Production to Continue Increasing (Tech. Rep.).

Flores-Velázquez, J., De La Torre-Gea, G., Rico-García, E., López-Cruz, I. L. & Rojano-Aguilar, A. (2012) Advances in Computational Fluid Dynamics Applied to the Greenhouse Environment IN OH, H. W. (Ed.) *Applied Computational Fluid Dynamics*. InTech.

Flores-Velázquez, J., Mejía-Saenz, E., Montero-Camacho, J. I. & Rojano, A. (2011) Numerical analysis of the inner climate in a mechanically ventilated greenhouse with three spans. *Agrociencia*, 45 no. 5, 545–560.

Fourati, F., & Chtourou, M. (2007). A Greenhouse Control with Feed-Forward and Recurrent Neural Networks. 2007 Simulation Modelling Practice and Theory, 15, 1016–1028.

Gaceta Oficial del Distrito Federal (2004) Normas técnicas complementarias sobre criterios y acciones para el diseño estructural de las edificaciones. *Gaceta oficila del Distrito Federal*. México, Administración pública del Distrito Federal. Jefatura de Gobierno.

Gassó-Busquets, F. & Solomando-Valderrabano, S. (2011) Estructura e instalaciones de un in invernadero *Mecánica*. Barcelona España, Escola Universitária d'Enginyeria Técnica Industrial de Barcelona.

Giacomelli, G. A. (2002) Introduction to Greenhouse Glazing. *Department of Agriculture & Biosystems Engineering. The University of Arizona.*

Gomez-Melendez, D., Lopez-Lambrano, A., Herrera-Ruiz, G., Fuentes, C., Rico-Garcia, E., Olvera-Olvera, C., Verlinden, S. (2011). Fuzzy Irrigation Greenhouse Control System Based on a Field Programmable Gate Array. African Journal of Agricultural Research, 6 (13), 3117–3130.

Greenhouses. (2011). Greenhouses. Retrieved from http://0-www.britannica.com.millenium.itesm.mx/EBchecked/topic/245223/greenhouse.

Gurban, E., & Andreescu, G. (2012). Comparison Study of PID Controller Tuning for Greenhouse Climate with Feedback-Feedforward Linearization and Decoupling. In 2012 16th international conference on system theory, control and computing (icstcc) (pp. 1–6).

Guy, B. (2012). Manejo de cultivos en altas temperaturas. In Curso de agricultura protegida - fundacion produce sinaloa.

Heisler, G. M. & Dewalle, D. R. (1988) Effects of windbreak structure on wind flow. *Agric Ecosyst Environ* 22(23), 41–69.

Hiram Ponce, graduate thesis ITESM CCM.

Hu, H., Xu, L., & Wei, R. (2010). Nonlinear Adaptive Neuro-PID Controller Design for Greenhouse Environment Based on RBF Network. In The 2010 international joint conference on neural networks (IJCNN) (pp. 1–7).

IAASTD. (2009). Latin America and the Caribbean (LAC) Report. Beverly D. McIntyre.

ICAMEX. (2011). Agricultura Protegida. Retrieved from http://portal2.edomex.gob.mx/icamex/investigacionpublicaciones/otras alternativas/agricultura protegida/index.htm.

InfoAgro. (2006). Control Climático en Invernaderos. Retrieved from http://www.infoagro.com/industriaauxiliar/controlclimatico.htm.

InfoAgro. (2010). Conductividad Electrica. Retrieved from http://www.infoagro.com/instrumentos medida/doc conductividad electrica.asp?k=53.

Iribarne, L., Torres, J. A. & Peña, A. (2007) Using computer modeling techniques to design tunnel greenhouse structures. *Sciencedirect. Computers in industry*, 58.

Isidori, A. (1995). Nonlinear Control Systems. Berlin; New York: Springer.

Javadikia, P., Tabatabaeefar, A., Omid, M., Alimardani, R., & Fathi, M. (2009). Evaluation of Intelligent Greenhouse Climate Control System Based Fuzzy Logic in Relation to Conventional Systems. In 2009 international conference on artificial intelligence and computational intelligence (AICI) (pp. 146–150).

Javadi-Kia, P., Tabataee-Far, A., Omid, M., Alimardani, R., & Naderloo, L. (2009). Intelligent Control Based Fuzzy Logic for Automation of Greenhouse Irrigation System and Evaluation in Relation to Conventional Systems. World Applied Sciences Journal, 6 (1), 16–23.

Jenkins, B. M. (1985). Alternative Greenhouse Heating Systems. California Agriculture, 5–7.

Johnson, H. (2013). Soilless Culture of Greenhouse Vegetables. UC Davis, Vegetable Research and Information Center.

Juárez-López, P., et al. (2011) Estructuras utilizadas en la agricultura protegida. Revista Fuente. 3:8(2007 - 0713): p. 21–27.

Juarez, P., Bugarin, R., Castro, R., Sanchez, A., Cruz, E., Juarez, C., Balois, R. (2011). Estructuras utilizadas en la Agricultura Protegida. Revista Fuente, 21–27.

Kacira, M. (2012). Greenhouse Environmental Control. 2012 UA-CEAC Greenhouse crop production & engineering design short course.

Kacira, M. (2013). Choose The Right Greenhouse Style. Retrieved from http://www.growing produce.com/article/32536/choose-the-right-greenhouse-style.

Kaur, A., & Kaur, A. (2012). Comparison of Mamdani-Type and Sugeno-Type Fuzzy Inference Systems for Air Conditioning System. International Journal of Soft Computing and Engineering (IJSCE), 2 (2), 323–325.

KAVEH, A. (2006) Optimal Displacement Method of Structural Analysis, France, Wiley.

Kendirli, B. (2006) Structural analysis of greenhouses: A case study in Turkey. Building and Environment, 41, 864–871.

Khalil, H. K. (2002). Nonlinear Systems. Upper Saddle River, N.J.; Mexico: Prentice Hall.

Kiselev, M. F., Safonov, V. S. & Chubarova, N. P. (1975) Construction of Agricultural Structures (Greenhouses) on Heaving Soils. Scientific-Research Institute of Bases and Underground Structures. Translated from Osnovaniya, Fundamenty i Mekhanika Gruntov, 3, 11–13.

Kolokotsa, D., Saridakis, G., Dalamagkidis, K., Dolianitis, S., & Kaliakatsos, I. (2010). Development of an Intelligent Indoor Environment and Energy Management System for the Greenhouses. 2010 Energy Conversion and Management, 51 (1), 155–168.

Kumar Jaypuria, S. (2008) Heat treatment of low carbon steel. Project report. Department of Mechanical Engineering, National Institute of Technology, Rourkela-769008., Session: 2008-09.

Lafont, F., & Balmat, J. (2002). Optimized Fuzzy Control of a Greenhouse. Fuzzy Sets and Systems, 128, 47–59.

Leal, H. (2011) Selección de materiales. Universidad Politécnica de puerto Cabello. Ingeniería en materiales industriales., sección 11–20.

Lucero, J., & Sanchez, C. (2012). Inteligencia de Mercado de Pimiento Morron Verde. Centro de Investigaciones Biologicas del Noroeste, 1–83.

Mallick, P. K. (1993) Fiber Reinforced Composites. Marcel Dekker Inc., New York.

Matt, S. (2010) State of the Industry report 2010 National Greenhouse Manufacturers Association Spotlight (NGMA). U.S.A., http://www.ngma.com/spotlight/NGMA_Supplement_2010.pdf.

May Nava, Fernando Díaz, and Efrén Ortiz undergraduate Thesis ITESM CCM (2009).

Mehran, K. (2008). Takagi-Sugeno Fuzzy Modeling for Process Control. Industrial Automation, Robotics and Artificial Intelligence (EEE8005).

Mississippi State University (2012) High Tunnel Crop Production Project. msucares,com. U.S.A., Mississippi State University.

Moreno, A., Aguilar, J., & Luevano, A. (2011). Caracteristicas de la Agricultura Protegida y su Entorno en Mexico. Revista Mexicana de Agronegocios, 763–774.

Morgan, L. (2012). Build It Right – Determining Greenhouse Design by Climate. Retrieved from http://www.maximumyield.com/inside-my-com/asktheexperts/item/120-build-it-right%E2%80%94determining-greenhouse-design-by-climate.

Morgan, L. (2012) Build it Right – Determining Greenhouse Design by Climate. *Maximum yield*. USA, http://www.maximumyield.com/inside-my-com/asktheexperts/item/120-build-it-right%E2%80%94determining-greenhouse-design-by-climate.

Morimoto, T., Takeuchi, T., Hashimoto, Y. (1993). *Growth Optimization of Plant by Means of the Hybrid System of Genetic Algorithm and Neural Network*. International Joint Conference on Neural Networks, Japan, pp. 2979–298.

Muñoz, P.; Antón, A.; Montero, J. (2002). Tendencias en la Construcción de Invernaderos: normas CEN y UNE. Sistemas de Control Ambiental y Posibilidades de Mecanización de Operaciones de Cultivo y Riego.

Muñoz, P., Anton, A., & Montero, J. (2013). Tendencias en la Construccion de Invernaderos: normas CEN y UNE. Sistemas de Control Ambiental y Posibilidades de Mecanizacion de Operaciones de Cultivo y Riego.

Nachidi, M., Benzaouia, A., & Tadeo, F. (2006). Temperature and Humidity Control in Greenhouses using the Takagi-Sugeno Fuzzy Model. In 2006 IEEE computer aided control system design (cacsd), 2006 IEEE international conference on control applications (cca), 2006 IEEE international symposium on intelligent control (isic) (pp. 2150–2154).

Nachidi, M., Rodriguez, F., Tadeo, F., & Guzman, J. (2011). Takagi-Sugeno Control of Nocturnal Temperature in Greenhouses using Air Heating. ISA Transactions, 50, 315–320.

National Instruments. (2013). NI Products. Retrieved from http://www.ni.com/products/esa/

Nelson, P. (1998). Greenhouse Operation and Management (5th ed.). Upper Saddle River, United States: Prentice Hall.

Newman, S. E. (2013). Sustainable Commercial Greenhouse Production.

NGMA. (1996). Standards for Design Loads in Greenhouse Structures.

NGMA. (1998). Greenhouse Heating Efficiency Design Considerations.

NGMA. (1998a). Greenhouse Environment Control System Considerations.

Nowosielsky, R., Kania, A. & Spilka, M. (2007) Development of ecomaterials and materials technologies, Journal of Achievements in materials and manufacturing engineering, 21, 27–30

Occhipinti, L., & Nunnari, G. (1996). Synthesis of a Greenhouse Climate Controller using AI-based Techniques. In 1996 8th mediterranean electrotechnical conference (melecon) (pp. 230–233).

Oduk, M., & Allahverdi, N. (2011). The Advantages of Fuzzy Control Over Traditional Control System in Greenhouse Automation. 2011 International Journal on Artificial Intelligence and Machine Learning (AIML), 91–97.

OECD. (2012). Mexico. In Agricultural policy monitoring and evaluation 2012 – OECD countries. OECD Publishing.

Ortiz-Vértiz, G. (2013) Agricultura Protegida. *Subsecretario SEDARH*.

Osorio, J., Molina, A., Ponce, P., & Romero, D. (2011). A Supervised Adaptive Neuro-Fuzzy Inference System Controller for a Hybrid Electric Vehicle's Power Train System. In 2011 9th IEEE international conference on control and automation (icca) (pp. 404–409).

Pasgianos, G., Arvanitis, K., Polycarpou, P., & Sigrimis, N. (2003). A Nonlinear Feedback Technique for Greenhouse Environmental Control. Computers and Electronics in Agriculture, 40 (1–3), 153–177.

Pazos, A., Sierra, A., Buceta, W. (2009). *Advancing Artificial Intelligence Through Biological Process Applications*. Medical Information Science Reference, United States of America.

Perdigones, A., Peralta, I., Nolasco, J., Munoz, M., & Pascual, V. (2004). Sensores para el Control Climatico en Invernaderos. Revista Horticultura, 44–49.

Perez, M., & Sanchez, J. (2012). Energias Renovables en los Invernaderos. Cuadernos de Estudios Agroalimentarios (CEA) - Universidad de Almeria.

Perruquetti, W., & Barbot, J. (2002). Sliding Mode Control in Engineering. New York: Marcel Dekker.

pH Info. (2005). pH-Meter. Retrieved from http://www.ph-meter.info/

Pinon, S., Camacho, E., Kuchen, B., & Pena, M. (2005). Constrained Predictive Control of a Greenhouse. Computers and Electronics in Agriculture, 49, 317–329.

Ponce, H. (2009). Intelligent Control System for a Sustainable Portable Greenhouse using LabVIEW. Tecnologico de Monterrey Campus Ciudad de Mexico.

Ponce P., Molina A Advanced Controller for Greenhouses, DYCAF 2012.

P. Ponce A. Molina, H. Ponce, M. Nava, F. Díaz, E.Ortiz A Novel Intelligent Greenhouse using LabVIEW internal Report (2009).

Ponce-Cruz, Pedro, Ramírez-Figueroa, Fernando, Intelligent Control Systems with LabVIEW™ 2010.

Ponce, P. (2011). Panorama Mexicano: Revision de los Datos de la Industria de los Invernaderos en Mexico. Retrieved from http://www.hortalizas.com/noticias/?storyid=2721.

Pucheta, J., Patino, H., Fullana, R., Schugurensky, C., & Kuchen, B. (2006). A Neuro-Dynamic Programming-Based Optimal Controller for Tomato Seedling Growth in Greenhouse Systems. Neural Processing Letters, 25, 241–260.

Pucheta, J., Schugurensky, C., Fullana, R., Patino, H., & Kuchen, B. (2006). Optimal greenhouse control of tomato-seedling crops. Computers and Electronics in Agriculture(50), 70–82.

Putter, E., & Gouws, J. (1996). An Automatic Controller for a Greenhouse using a Supervisory Expert System. In 1996 8th mediterranean electrotechnical conference (melecon) (pp. 1160–1163).

Quan, V., Gupta, G., & Mukhopadhyay, S. (2011). Review of Sensors for Greenhouse Climate Monitoring. In 2011 IEEE sensors applications symposium (SAS) (pp. 112–118).

Quiminet.COM (2011) Consejos para alargar la vida de los materiales en un invernadero QuimiNet.com, http://www.quiminet.com/articulos/consejos-para-alargar-la-vida-de-los-materiales-de-un-invernadero-50777.htm.

Radha, K., & Igathinathane, C. (2007). Greenhouse Technology and Management. Hyderabad, IND: Global Media.

Rajaoarisoa, L., M'Sirdi, N., & Balmat, J. (2012). Micro-Climate Optimal Control for an Experimental Greenhouse Automation. In 2012 2nd international conference on communications, computing and control applications (CCCA) (pp. 1{6).

Ronen, E. (2013). Managing the Root Zone in Soilless Culture.

Rouboa, A. & Monteiro, E. (2007) Computational fluid dynamics analysis of greenhouse microclimates by heated underground tubes. *Journal of Mechanical Science and Technology*, 21, 2196–2204.

Ruiz-Funes, M., & Smith, K. (2012). Food Security: A G20 Priority The Input Of Mexico's Experiences. Mexico: Secretaria de Relaciones Exteriores, Instituto Matias Romero.

Rutledge, A. D. (1998) Commercial greenhouse tomato production. Agricultural extension service. The University of Tennessee.

SAGARPA. (2008). Delegaciones Estatales. Retrieved from http://www.amhpac.org/contenido/plan nacional de agricultura protegida 2009.pdf.

SAGARPA. (2012). Agricultura Protegida 2012. Retrieved from http://www.sagarpa.gob.mx/agricultura/Paginas/Agricultura-Protegida2012.aspx.

Salazar, R., López, I., Rojano, A. (2008). *A Neural Network Model to Control Greenhouse Environment*. Sixth Mexican International Conference on Artificial Intelligence, pp. 311–318.

Samperio, G. (2009). *Hidroponía Básica,* Diana, Mexico.

Sastry, S. (1999). Nonlinear System : Analysis, Stability, and Control. New York: Springer.

Schnelle, M. A., Dobbs, S. H., Needham, D. C. & Dole, J. M. (1990) The hobby greenhouse. *Division of Agricultural Sciences and Natural Resources, Oklahoma State University*, http://pods.dasnr.okstate.edu/docushare/dsweb/Get/Document-2271/HLA-6705web.pdf.

Service, W. V. (2012) Planning and Building a Greenhouse. Recuperado el 22 de Abril de 2013 http://www.wvu.edu/~agexten/hortcult/greenhou/building.htm#Structural Materials.

Sethi, V. (2009). On the Selection of Shape and Orientation of a Greenhouse: Thermal Modeling and Experimental Validation. Solar Energy, 83, 21–38.

Shihua, L., Shiyan, L., & Limei, J. (2008). Application of Adaptive Fuzzy Controller in Intelligent Greenhouse Control System. In 2008 IEEE international conference on automation and logistics (ICAL) (pp. 1708–1712).

S.L. Speetjens, J.D. Stigter*, G. van Straten Smith, S. (1992) *Greenhouse gardener's companion*, USA, Fulcrum Publishing.

Smith, S. (2002) *Greenhouse gardener's companion*, Colorado, USA, Fulcrum Publishing.

Tadj, N., Fezzioui, N. & Draoui, B. (2013) Influence of vents arrangement on greenhouse thermal driven ventilation *16 èmes Journées Internationales de Thermique (JITH 2013) Marrakech (Maroc)*.

Tecnologico de Monterrey Campus Ciudad de Mexico. (2011). Controlador Climatico para Invernaderos con Controles Adaptables y Modulo de Potencia Externa. Exp: MX/a/2011/009101, Folio: MX/e/2011/060799.

Teorema Ambiental (2002) Invernaderos, alta rentabilidad. Teorema ambiental. Sección Ciencia y tecnología. Revista Técnico Ambiental. 3W México.

Teorema Ambiental (2002) Invernaderos, alta rentabilidad. *Teorema ambiental. Sección Ciencia y tecnología.* . Revista Técnico Ambiental. 3W México.

The Eco-efficiency Centre (2008) Fact Sheet: Eco-Efficiency in the Greenhouse Industry Eco-Efficiency Centre Committed to Excellence and Efficiency, http://eco-efficiency.management.dal.ca/Files/Business_Fact_Sheets/greenhouse_fs.pdf.

Tiwari, G. (2003). Greenhouse Technology for Controlled Environment. Pangbourne, England: Alpha Science International Ltd.

Tomescu, M. L. (2007). Stability Analysis Method for Fuzzy Control Systems Dedicated Controlling Nonlinear Processes. Acta Polytechnica Hungarica, 4 (3), 127–141.

Towards an adaptive model for greenhouse control Trabelsi, A., Lafont, F., Kamoun, M., & Enea, G. (2007). Fuzzy Identification of a Greenhouse. 2007 Applied Soft Computing, 7 (3), 1092–1101.

Tribal Engineering LLC. (2010). Intelligent Control Toolkit for LabVIEW (ICTL). Retrieved from http://www.tribalengineering.com/technology/ictl-ictl.aspx.

UNE EN-13031-1. (2002). Invernaderos, Proyecto y Construccion. In Parte 1: Invernaderos para produccion comercial. Madrid, Spain: AENOR.

University of Arizona, & CEAC. (2012). Total Areas in Major Greenhouse Production Countries. Agricultural & Biosystems Engineering.

Vaisala. (2011). Greenhouse Climate Measurements Ensure Optimal Plant Growth (Tech. Rep.). *Wageningen University, Systems and Control Group, Bornsesteeg 59, 6708 PS Wageningen, The Netherlands.*

Yingchun, K., & Yue, S. (2010). A Greenhouse Temperature and Humidity Controller Based on MIMO Fuzzy System. In 2010 international conference on intelligent system design and engineering application (ISDEA) (pp. 35–39).

Yousefi, M., Hasanzadeh, S., Mirinejad, H., & Ghasemian, M. (2010). A Hybrid Neuro-Fuzzy Approach for Greenhouse Climate Modeling. In 2010 5th IEEE international conference intelligent systems (is) (pp. 212–217).

Yuste Pérez, M. P. (2008) Biblioteca de la agricultura, España, Lexus.

Zabeltitz, C. (1990) Appropriate Greenhouse Constructions for Mild Climates. *Der TRopenlandwirt, Zeitschrift für die Landwirtschaft in den Tropen und Subtropen.* Jahrgang.

Zhou, X., Wang, C., & Lan, H. (2009). The Research and PLC Application of Fuzzy Control in Greenhouse Environment. In 2009 sixth international conference on fuzzy systems and knowledge discovery (fskd) (pp. 340–344).

Subject index

GLOBALISING WORLDS AND
NEW ECONOMIC CONFIGURATIONS

Globalising Worlds and New Economic Configurations

Edited by
CHRISTINE TAMASY
University of Auckland, New Zealand
and
MIKE TAYLOR
University of Birmingham, UK

Routledge
Taylor & Francis Group

LONDON AND NEW YORK

Contents

List of Tables

List of Figures

List of Contributors

Dr John Bryson is Professor of Enterprise and Economic Geography at the School of Geography, Earth and Environmental Sciences, University of Birmingham, UK.

Dr Simon Burke is Senior Lecturer and Head of Department of Economics, University of Reading, UK.

Dr Frank Calzonetti is Professor of Geography and Vice President for Research Development, The University of Toledo, USA.

Dr Michael C. Carroll is Associate Professor of Economics and Director of the Center for Regional Development, Bowling Green State University, USA.

Dr Julie Cidell is Assistant Professor of Geography at the University of Illinois at Urbana-Champaign, USA.

Dimitra Dimitropolou is a Ph.D. Candidate in the Department of Economics, University of Reading, UK.

Dr Sabine Dörry is lecturer in Geography at the Department of Human Geography, Goethe-University of Frankfurt am Main, Germany.

Dr Bolesław Domański is Professor of Geography and Director of the Institute of Geography and Spatial Management, Jagiellonian University in Cracow, Poland.

Dr Arnt Fløysand is Professor at the Department of Geography, University of Bergen, Norway.

Dr Joseph P. Frizado is Associate Professor at the Department of Geology, Bowling Green State University, USA.

Dr Hallgeir Gammelsæter is Professor at the Molde University College, Norway.

Dr Robert Guzik is Economic Geographer at the Department of Regional Development, Jagiellonian University, Krakow, Poland.

Dr Krzysztof Gwosdz is Economic Geographer at the Department of Regional Development, Jagiellonian University, Krakow, Poland.

Håvard Haarstad is a Ph.D. Candidate in Geography at the Department of Geography, University of Bergen, Norway.

Dr James W. Harrington is Professor of Geography, University of Washington, Seattle, USA.

Dr Roger Hayter is Professor of Geography, Simon Fraser University, British Columbia, Canada.

Dr David Hayward is Senior Lecturer at the School of Geography, Geology and Environmental Science, The University of Auckland, New Zealand.

Dr Stig-Erik Jakobsen is Associate Professor in Human Geography at the Department of Sociology and Human Geography, University of Oslo, Norway.

Dr Hege Merete Knutsen is Professor in Human Geography at the Department of Sociology and Human Geography, University of Oslo, Norway.

Dr Jeong Hyop Lee is Research Fellow at the Science and Technology Policy Institute (STEPI), Seoul, Korea.

Dr Richard Le Heron is Professor of Geography at the School of Geography, Geology and Environmental Science, The University of Auckland, New Zealand.

Dr John R. Lombard is Assistant Professor of Urban Studies & Public Administration and the Director of the Center for Real Estate and Economic Development, Old Dominion University, USA.

Tomasz Majek is a Ph.D. Candidate in Geography at the University of Luxembourg.

Dr Juliana Mansvelt is Senior Lecturer in Geography at Massey University, Palmerston North, New Zealand.

Dr Philip McCann is Professor of Economics in the Department of Economics, University of Waikato, New Zealand, and Professor of Urban and Regional Economics, Department of Economics, University of Reading, UK.

Dr Philip McDermott was formerly Professor of Planning at Massey University, Palmerston North, New Zealand, and has been a director of several New Zealand economic development consultancies during the past 25 years. He is currently a Principal in CityScopeConsultants, Auckland, New Zealand.

Dr Abd Rahim Md Nor is Professor of Human Geography and Head, Postgraduate Environmental Management Program, Faculty of Social Science and Humanities, Universiti Kebangsaan Malaysia (National University of Malaysia).

Dr Nor Ghani Md Nor is Professor and Chairman in the Centre for Economic Studies, Faculty of Economics and Business at Universiti Kebangsaan Malaysia (National University of Malaysia).

Dr Caroline Miller is Senior Lecturer in the Planning Programme in the School of People, Environment and Planning at Massey University, Palmerston North, New Zealand.

Dr Ivo Mossig is Professor of Human Geography at the Department of Geography, University of Bremen, Germany.

Dr Laurence Murphy is Professor and Head of the Department of Property, The University of Auckland, New Zealand.

Sabine Panzer is a Ph.D. Candidate in Geography at the University of Jena, Germany.

Dr Bruce W. Smith is Professor of Geography and Research Fellow in the Center for Regional Development at Bowling Green State University, USA.

Dr Christine Tamásy is Heisenberg Research Fellow at the School of Geography, Geology and Environmental Science, The University of Auckland, New Zealand, and 'Privatdozentin' at the Faculty of Economics and Social Sciences, University of Cologne, Germany.

Dr Michael Taylor is Professor of Human Geography at the School of Geography, Earth and Environmental Sciences, University of Birmingham, UK.

Nicholas Velluzzi is a Ph.D. Candidate in Geography at the University of Washington, Seattle, USA.

Dr Steffen Wetzstein is a Lecturer at the School of Geography, Environment and Earth Sciences, Victoria University of Wellington, New Zealand.

Dr Johannes Winter is Research Assistant at the Department of Economic and Social Geography, University of Cologne, Germany, and political consultant in Munich.

Chapter 1
Researching New Economic Configurations: Theory and Context

Christine Tamásy and Mike Taylor

In recent decades, processes of economic globalisations have stretched, modified, reconfigured and transformed the circuits of capital that bind together the evolving elements of the world economy (Murray 2006; Griffin 2003). Structures of production, distribution, consumption, investment and finance have evolved rapidly to create complex and hybrid forms of organisation to capitalise on emerging opportunities and to incorporate new players (Yeung 2002; Sassen 2002). At the same time, these new structures have been recognised as being vulnerable and fragile as is evident in the unequal power relationships within value chains (Gereffi et al. 2005), the potential for globalisation to promote labour exploitation, the pressure it puts on global resources, and the fragility of the global financial system (Helleiner 1995), that is so fundamental to its operation, that became evident with the onset of the 'credit crunch' in 2007. Understanding these new economic configurations and their geographic patterns involves a number of tasks. These include, building and developing new theoretical arguments drawing on, extending and integrating, for example, economists' thinking on endogenous growth (Romer 1994) and the socially constructed institutionalist thinking of geography's 'new regionalism' (Taylor and Plummer 2003; Plummer and Taylor 2003; McCann 2007), especially ideas on value chains and networks. This is a major task that has significant methodological as well as theoretical dimensions, especially issues of measurement, the use of qualitative data, and the analytical potential of multi-methodology research.

Globalising Worlds seeks to contribute to this project of developing new understandings by bringing together theoretically-informed empirical analyses from the Asia/Pacific region, Europe and North America to illustrate different ways in which new economic configurations have evolved in different contexts, and better to understand individual, local and regional responses to a variety of global challenges and opportunities. The collection of chapters is derived from the 2006 meeting of the International Geographical Union's Commission on *The Dynamics of Economic Spaces* held in Auckland (New Zealand). Based on their specific contributions to the conference theme, *Globalising Worlds: Geographical Perspectives on Old and New Value Chains, Commodity Chains, Supply Chains*, selected delegates were invited to write chapters for the book that were, subsequently, internationally refereed.

Globalising Worlds is not a conference proceedings. All 24 of the volume's chapters have direct and indirect linkages and move discussion on aspects of globalisation into different geographical contexts, adding nuances and synergies to the debate. The 38 authors are scholars from around the world and their contributions reflect their distinct regional perspectives on key issues. Additionally, the authors represent a broad range of disciplines including economics, geography, political science and management. The book offers a new exploration of the economic impacts of globalisation, and the distinctive contribution of human geography, and especially economic geography, to the debate in this field. It critically appraises new economic configurations from a geographical perspective, illustrates how network and value chain theories lead to a better understanding of the globalisation phenomenon and examines impacts of these transformations 'on the ground' using examples from the Asia/Pacific region, Europe and North America. It focuses on the underlying processes of globalisation within which new economic configurations will be better understood by using geographical perspectives.

Globalising Worlds is organised into seven closely related, but distinct parts. The chapters do not attempt to be comprehensive, aiming instead to provide detailed empirically-based insights into the impact of global processes on places. Six parts of the volume explore distinctive geographical aspects of globalisation processes. These processes include:

1. *cross-border business relations*, involving trust-based relationships and enforceable contracts, the operations of market makers and trade regulation and protectionism;
2. *international investment flows*, with a particular emphasis on foreign direct investment and its impact at different spatial scales, but also the impact of investment trusts as an example of international financial services;
3. *production chains* and their global configuration, explored in the context of automobile production;
4. *the dynamics and reconfiguration of enterprise clusters*, including issues of technological change;
5. *the dynamics of human capital resources*, exploring issues of labour recruitment, knowledge transfer and entrepreneurship; and
6. *issues of business vulnerability, competition and persistence.*

To round out this geographical discussion, the final part explores a number of issues in the specific context of New Zealand as it attempts to compete from the edge of the global economy. These issues include primary commodity trading, local governance in a neo-liberal environment, and the shifting environment of consumption.

The three chapters of Part 1 deal with the cross-border business relationships that are at the heart of the global economy, exploring the roles of trust-based relationships versus enforceable contracts, the role of 'market makers' who broker cluster-based production into international markets, and the impact of trade regulation and protectionism. All three chapters demonstrate the very different consequences of large versus small firms in these transactions, reflecting the different levels of power they are able to bring to bear. *Hege Merete Knutsen* explores the consequences of trade protection for cross-border flows. She discusses how the phase-out and abolition of the Multifibre Arrangement led to new economic configurations in the textile and garment industry. She explores the extent to which new regional trade agreements serve as alternative protectionist measures imposed by the countries of the EU and the US in place of quota regulations in the textile and garment industries, and the extent to which such measures are challenged by low-cost manufacturers in Asia. This backdrop of regulation though regional trade arrangements is used to explain industrial development in countries that are latecomers to international trade in the textile and garment industries.

Sabine Dörry investigates cross border relationships between businesses, and relational governance in the international package tourist industry that links Germany and Jordan. She argues that international package tours are characterised by uncertainty about the character and quality of the product to be exchanged between in-country providers and package tour operators. To avoid opportunistic behaviour on the part of one party or the other, business relationships are framed in terms of both formal and informal arrangements, although formal contractual arrangements are unenforceable across national boundaries. The analysis in this chapter shows that a mix of trust-enhancing and controlling activities are used by German firms to shape their business relationships, to bridge the 'formal distance' that separates them and, at the same time, to ensure the experience of the tourist and maintain the commercial transactions between the German and Jordanian businesses involved.

In the context of the US motion picture industry and the clustering of this industry in the Los Angeles/Hollywood district of California, *Ivo Mossig* discusses the major role of 'market makers' in sustaining this type of clustered production through the cross-border business relationships they both foster and exploit. In terms of the links between Los Angeles/Hollywood and the German market, he discusses the consequences of a powerful global distribution system that guarantees the revenues to produce movies with above-average production budgets. While the 'majors' have their own internal distribution systems, smaller 'independent' producers must use specialist distribution companies. The international revenues generated world-wide through these market makers enable the industry to engage the most popular actors and to deploy extremely high marketing and promotional budgets. Indeed, it is argued that these cross-border connections are decisive for maintaining the motion picture cluster in Los Angeles/Hollywood.

The three chapters of Part 2 present geographical perspectives on international investment flows as they impact at the international, national and regional scales. Two chapters are concerned with foreign direct investments (FDI) while the third is concerned with real estate investment trusts. What they demonstrate very clearly is the strong geographically-differentiated patterns that investment flows are creating and reinforcing within the global economy. *Arnt Fløysand* and *Håvard*

Haarstad focus on FDI in development strategies. They argue that global flows of FDI are concentrated heavily in some parts of the world, especially the developed countries, and largely bypass others. Yet, FDI is seen as an important tool for poverty reduction in the bypassed regions. Through a case study of Norwegian investment flows, which mirror the global pattern of FDI flows, the authors demonstrate a strong tendency towards agglomeration. Their analysis indicates that market mechanisms create and reinforce this skewed distribution of FDI, and that FDI inflows to regions such as Sub-Saharan Africa do not appear to be on the horizon.

Dimitra Dimitropoulou, Simon Burke and *Philip McCann* analyse FDI flows at the finer, regional scale of inflows into the UK. Their investigation shows that there are clear differences in the location behaviour of different types of inward FDI, with the regions in the south and east of the UK exhibiting high proportions of investment in research and development and headquarter activities, while investment flows into manufacturing and distribution are more evident in the country's geographically more peripheral regions. In addition, however, having controlled for these various different characteristics, their results suggest that existing regional specialisation is the single most important determinant of where inward FDI will locate.

Laurence Murphy examines the genesis and development of Listed Property Trusts (LPTs) in New Zealand, paying particular attention to growth strategies, investment portfolios, issues of risk and governance, and the role of foreign (especially Australian) ownership within them. He begins by arguing that global investment in Real Estate Investment Trust (REIT) structures has risen dramatically since the 1990s. As the author discusses, the growth of securitised property markets has significant implications for the commercial property sector and transnational flows of investment.

Part 3 of the volume brings together four chapters that focus on the automobile industry and the globalising networks within which it is embedded. Three chapters explore the developments in this key and emblematic sector of global production in the empirical context of Poland, where investment and production has expanded significantly in recent years. The fourth chapter takes a very different perspective, viewing the industry from the Korean perspective of a de-territorialising automotive production cluster. *Boleslaw Domański, Robert Guzik* and *Krzysztof Gwosdz* discuss the new international division of labour and the changing role of the periphery through the case study of the Polish automotive industry. The authors' analysis shows that the recent expansion of Poland's automotive production and exports are built not so much on low costs, as was the case in the past, but now on high quality, reliability, adaptability, and fast response. They interpret this as created localised capabilities. It accords with a corporate strategy of complementary specialisation, in which the fundamental reorganisation of tasks and functions within a value chain provides the basis for improved efficiency. Currently, it brings complex, high-skill and capital-intensive processes to Central Europe and, potentially, engineering functions may follow.

Tomasz Majek and *Roger Hayter* critically examine the concept of the hybrid branch plant in the context of a Toyota plant in Poland. As an example of hybrid lean production, the Toyota branch plant matches the performance of its parent firm and yet its practices have been modified to reflect Polish circumstances. The authors discuss why the adaptation of lean production is not necessarily detrimental to performance, and explore the implications of this adaptation as a way of better understanding processes of industrial learning and regional competitiveness.

Johannes Winter argues that the traditional division of competencies in the automotive industry, characterised by the localisation of knowledge-intensive capacities of the value chain in core regions and labour-intensive activities in semi-peripheral regions, has lost its universal validity. Instead, a new international allocation of corporate competencies has become apparent. Now, owing to the transmission and self-acquisition of corporate competencies, semi-peripheral locations can go through different forms of upgrading. A conceptual framework for the subdivision of corporate competencies is developed in the chapter and applied in the empirical context of the regional development of the Polish automotive industry.

Jeong Hyop Lee analyses the Ulsan automobile cluster in Korea by using a multi-dimensional spatial framework. The Ulsan automobile cluster has evolved as the home town of Hyundai, and is characterised as an agglomeration of assembly lines and suppliers which is, however, threatened with the possibility of de-territorialisation of the local production system. In short, Korea is one of the countries losing the production capacity in this industry that countries like Poland are gaining. The author argues that the engineering networks, which are the source of the competitiveness of Hyundai, need to be focused in the co-called 'Auto Valley' project, recently initiated by the Ulsan government.

Part 4 of the book explores cluster developments in a globalising world from three perspectives: the definitional perspective of cluster identification; the reconfiguration of production in a cluster facing intense global competition; and the emergence of new and innovative capacities in a cluster challenged by globalisation. *Michael Carroll, Joseph Frizado* and *Bruce Smith* identify potential cluster areas by using local indexes of spatial autocorrelation (LISA). Because geographic proximity (co-location) is a necessary but not a sufficient condition for potential clustering activity, identifying industry location and density becomes the first phase in the development of any cluster-driven development policy. The chapter explores the benefits and limitations of the LISA approach and addresses alternative methods of creating the spatial weights matrix integral to LISA methodologies. The US greenhouse industry is used to illustrate the approach.

Michael Taylor and *John Bryson* analyse the restructuring of metal manufacturing in the West Midlands region in the UK. This region is the UK's industrial heartland, where manufacturing is more important within the economic structure than in any other region in the country. It has long been recognised as having a distinctive structure of production based on the local linkage and clustering of essentially small metal manufacturers – a 'locational integration' that has much in common with current policy and academic debate regarding the clustering of economic activity. In recent years, however, the role of manufacturing, regionally as nationally, has been in decline in terms of both employment and gross value added. Associated with this decline has been a reconfiguring of metal manufacturing production within the West Midlands region. The chapter explores the processes of polycentric development that are reshaping this region's metal manufacturing activities.

Frank J. Calzonetti provides a very clear analysis of cluster dynamics in the face of global economic pressures related to an element of the automotive industry in northwest Ohio. The analysis of this chapter shows how Toledo is developing an emerging cluster in thin-film photovoltaics that has its roots in the antecedent 'glass cluster' of this part of northwest Ohio. After a long gestation period, the alternative energy cluster is now a 'visible' but still small contributor to the region's economy. The chapter draws together the evidence of progress towards the creation of a new technology cluster in alternative energy that is indebted to an earlier cluster in glass manufacturing and innovation, showing that Toledo may be completing a circle of development that started in the first half of the twentieth century with national leadership in the technology of glass, followed by a period of decline, and then the rebirth in a new industry that benefited from the knowledge and culture from the antecedent cluster.

Part 5 of the book draws together four chapters on labour, knowledge and entrepreneurship, that lie at the heart of human capital resources, and the way they are impacted locally by global pressures and processes. It explores labour market intermediation and skilled labour migration as they impact commodity chains and knowledge transfer. It also explores, through detailed case studies in Germany and the US, the relationships between banks and their business clients, and the deployment of government funds to offset the local economic stress that economic change and globalisation can create. *James W. Harrington* and *Nick Velluzzi* explore labour market intermediation (LMI) in the context of the changing world of work in the face of global economic pressures and changing patterns of production. They conceptualise LMI by applying theories of intermediation and commodity chains. Because labour market intermediaries (e.g. temporary services firms, job-placement agencies, and workforce development networks) can be based in the profit-making, government, or non-governmental organisation (NGO) sectors, the discussion uses examples and outcomes to illustrate the way in which the governance and motivation of commodity chain intermediaries affects their uncertainty reduction and risk-shifting.

Christine Tamásy discusses the relationship between globalisation and skilled labour migration processes. She cautions against simplistic interpretations based on fuzzy concepts or wobbly data and re-conceptualises skilled migration as one element of the international mobility of knowledge that is central to globalising economic strategies. It is argued that the mobility of workers enables the spread of tacit knowledge in intra- and inter-firm networks, because organisational and relational proximity is integral to the transmission of knowledge that is difficult to communicate.

Sabine Panzer analyses the relationships between banks and businesses in Thuringia, Germany. The German banking sector has undergone remarkable spatial consolidation of its branch office networks during the last decade. For many business enterprises the distance between them and their major banks has significantly increased. However, trust-based relationships between businesses and their banks remain undiminished as banks provide important financial advice based on their knowlegde of capital and financial markets. In this role, banks serve to reduce the insecurity in processes of economic exchange. The chapter explores whether trust-based relationships break down with greater spatial distances between banks and businesses, or whether the necessary social nearness can be built and sustained over greater geographical distances.

John Lombard investigates the spatial aspects of the distribution monies from the Governor's Development Opportunity Fund (GOF) of the Commonwealth of Virginia which is a cash-based discretionary economic development incentive fund controlled by the executive branch designed to act as a 'deal closer' for economic development projects. Through an analysis of the GOF across Virginia at the Census tract level, the chapter explores the extent to which Virginia communities suffering high socio-economic stress coincide with those communities receiving discretionary GOF awards. Using GIS and data for 1996 to 2003, the study shows that those communities with the highest need for economic development are not those areas benefiting from fund allocations.

In Part 6, three chapters explore issues of globalisation associated with business vulnerability, including new areas of competition, and the countervailing persistence of local business flavour. *Abd Rahim Md Nor* and *Nor Ghani Md Nor* examine the emergence of a low-cost carrier in a globalising Malaysia and its expansion in the country's aviation industry domestically and regionally. The chapter explores the factors that have led to the success the country's first low cost carrier. It examines the impact of the airline on the mobility of the people in Malaysia in particular, and in the Southeast Asian region in general. It also explores the impact of the airline on the existing national carrier, MAS (Malaysian Airline System), which operates full-service domestic and regional services.

Stig-Erik Jakobsen, Arnt Fløysand and *Hallgeir Gammelsæter* examine the relationship between globalisation and local flavour in business organisations by using Norwegian elite football clubs as a case study. They argue that over the last century, football has constituted itself as a global business that establishes, maintains and transforms the rules of the business across nations and boundaries. Similar to professional football in other countries, Norwegian top football has experienced increased commercialisation, characterised by a growth in turnover, development of larger and more complex organisation and a more prominent position for economical institutions and market transactions. The conventional view is that this has resulted in a de-coupling of clubs from local contexts. The chapter tests this view by analysing the spatiality of clubs focusing on, amongst other things, their environment linkages, the inter-dependency between football clubs and their institutional contexts.

Julie Cidell identifies some of the ways in which transportation is a fundamental part of commodity chains. She does so by examining two particular natural disasters, the Great Hanjin Earthquake of 1995 and Hurricane Katrina of 2005, to see how the damage and disruptions which those events caused reverberated throughout the links of different commodity chains. The vulnerability of many of the world's transportation nodes to natural disasters means that the commodity chains of which they are a part are vulnerable along their lengths, whether or not the actual sites of production are vulnerable themselves.

Part 7 of the book comprises four chapters that analyse new economic configurations in trade-exposed New Zealand, where economic actors compete from the edge of the global economy. They explore in this context the local economic, governance and social adjustments that have accompanied neo-liberal responses and reactions to the opportunities and problems associated with globalisation. *David Hayward* shows that primary sector-based exports continue to be the mainstay of New Zealand's international trade. In public policy discourse, this is held to be undesirable and both a cause of the country's poor comparative trade performance as well as an inhibitor to its improvement. Prevailing thinking holds that international economic engagement is vitally important for economic prosperity, and especially so for a small open economy. The chapter reconsiders and reinterprets the evidence for this view based on comparative ratio measures of trade and GDP, and rejects the simplistic notion of commodities that is generally employed to support it. Instead, it is argued that attention needs to be shifted away from product types and placed on the value embodied within traded commodities without prejudice to their specific forms, which provides a different interpretation of the country's primary sector-based exports.

Steffen Wetzstein argues that in a largely market-driven governance environment and under globalising conditions, the regulation of Auckland's economy is primarily discursive rather than regulatory and interventionist. Important emerging discursive practices of governance are story-telling, benchmarking and indicatorisation, through which new actor imaginaries and capacities are constituted. These promise to be better suited to link Auckland actors and activities into globalising value chains. However, while processes and relationships of the state-regulatory apparatus are re-worked and aligned, the effects of contemporary governance on private investor behaviours are not, as yet, known. Thus, influencing Auckland's global economic participation remains a difficult political and policy task.

Richard Le Heron and *Phil McDermott* outline the emergence of the ambitious Metro Project that is focused on the economic transformation of Auckland. At the heart of the project is the question, 'how might the international competitiveness of companies, governments and individuals be fostered under neo-liberalising conditions?' It is argued in the chapter that the project must be understood on a number of levels: as the culmination of a two year process aimed at developing an action plan to ensure Auckland becomes a world-class city-region; as an extension of a central government initiative aimed at stimulating growth and innovation; as a new generation economic partnership between local interests and central government; and, as a contest over distributional issues in the Auckland region.

Juliana Mansvelt and *Caroline Miller* explore the shifing environment of consumption in a globalising, neo-liberal environment in the context of catalogue shopping. They begin by establishing the conceptual basis for using catalogue shopping as a means of conducting a situated investigation of the times and spaces in which people shape and encounter commodities and the significance of commodities and things in the assembling of such networks. They examine the historical emergence of catalogue shopping in New Zealand and Britain, exploring changes in retailing and consumer practices. Using a case study of company 'Christmas catalogues', they explore how Christmas shopping 'at home' is a practice which shapes, and is shaped by, wider subject positions, socialities and spatialities.

The chapters of this volume combine to suggest an agenda for future research on the geographical dimensions of globalisation that is economically, socially and politically engaged. They provide examples of the depth of empirical research that is needed to come to grips with these processes. It is only on such a foundation that we can hope adequately to theorise and understand the processes of globalisation shaping the world today.

Acknowledgments

We would like to thank Igor Drecki, Geographics Unit Manager at the School of Geography, Geology and Environmental Science (SGGES), who finalised the cartographic work for this book. We also would like to thank Beryl Jack for the secretarial support. Finally, we are very grateful to the international reviewers who refereed individual chapters. The refereeing process consisted of two parts: an assessment by a referee selected from conference participants and an international referee. The reviewing process was anonymous.

References

Gereffi, G., Humphrey, J. and Sturgeon, T. (2005), 'The Governance of Global Value Chains', *Review of International Political Economy* 12:1, 78-104.

Griffin, K. (2003), 'Economic Globalization and Institutions of Global Governance', *Development and Change* 34:5, 789-808.

Helleiner, E. (1995), 'Explaining the Globalization of Financial Markets: Bringing States Back In', *Review of International Political Economy* 2:2, 315-341.

McCann, P. (2007), 'Observational Equivalence? Regional Studies and Regional Science', *Regional Studies* 41:9, 1209-1221.

Murray, W.E. (2006), *Geographies of Globalization* (Routledge: London).

Plummer, P. and Taylor, M. (2003), 'Theory and Praxis in Economic Geography: "Entrepreneurship" and Local Growth in a Global Economy', *Environment and Planning C* 21, 633-649.

Romer, P.M. (1994), 'The Origins of Endogenous Growth', *Journal of Economic Perspectives* 8, 3-22.

Sassen, S. (2002), 'The Urban Impact of Economic Globalization', in Lin, J. and Mele, C. (eds), *The Urban Sociology Reader* (Routledge: London), 230-240.

Taylor, M. and Plummer, P. (2003), 'Drivers of Local Growth: Ideologies, Ambiguities and Policies', *The Australasian Journal of Regional Studies* 9:3, 239-257.

Yeung, H.W. (2002), 'The Limits of Globalization Theory: A Geographic Perspective on Global Economic Change', *Economic Geography* 78:3, 285-305.

Chapter 2

Impeding Industrial Development? Regional Trade Arrangements as Response to Quota Abolition in the Textile and Garment Industry

Hege Merete Knutsen

Introduction

After a 10 year phase-out programme the export quota trade of the Multi-Fibre Arrangement was abolished for WTO countries on 1 January 2005. This chapter explores how and why the phase-out and abolition of this quota system led to new economic configurations in the textile and garment industry. Two topics are discussed. The first concerns the extent to which regional trade arrangements serve EU countries and the US as protectionist measures that are alternative to quota regulations. The second explores the extent to which such measures are challenged by low-cost manufacturers in Asia. Against this backdrop, the chapter explores the consequences of these regional trade arrangements for industrial development in countries that are latecomers to international trade in the textile and garment industry.

The export quota trade of the Multi-Fibre Arrangement was a measure used by the countries of Western Europe and the US to protect their domestic textile and garment industries from low-cost imports. Access to quotas also protected a number of manufacturers in developing countries who would not have been among the most competitive exporters. However, abolition of the quota system did not guarantee a level competitive playing field. Some developing countries, especially in the South, are heavily dependent on the textile and garment industry, and industry representatives in those countries fear that the quota system will be replaced by more subtle protectionist measures that will continue to restrict their access to markets in the North.

Criticism of conventional economic approaches to trading blocs (Michalak and Gibb 1997; Pomfret 2003) suggests that the objectives and outcomes of regional trade arrangements will vary with time, place and branch-specific conditions. This is the point of departure of the present analysis. More specifically, the case of the textile and garment industry sheds light on the outcomes of regional trade arrangements in a technologically mature and labour-intensive industry. It is part of regional and global buyer-driven production networks that have strong asymmetric power relations that favour retailers, designers and trading companies.

The analysis is qualitative and based on a combination of primary and secondary data including semi-structured interviews and informal personal communication with national and foreign industry representatives in Vietnam, Sri Lanka, Turkey, Brussels, Norway and Sweden in the period 1999–2003, and international statistics and reports. The main focus is on the European trading bloc, which is contrasted with the American trading bloc using secondary data from the US.

Trading Blocs and Protectionism in a Network Economy

Vinerian customs union theory addresses whether trading blocs have a trade creation or a trade diversion effect. According to Krugman (1991), regional trading blocs may in theory divert trade from low-cost to high-cost suppliers. He distinguishes between increasing protectionism and 'beggar thy neighbour' effects of trading blocs. Increasing protectionism refers to the adoption of higher trade barriers against outsiders such as increasing tariffs and more quota regulation. 'Beggar thy neighbour' effects occur even without this. It refers to a decline in the relative prices that non-members get for their products. Prices decline because the demand for their products decline when trade inside the blocs increases. However, in practice, trading blocs create more trade than they divert. This is because trade between member countries will replace domestic production. Greater market size is not the factor that stimulates trade within trading blocs. Neighbours tend to trade more with each other than with distant partners because of factors such as trust and lower costs of transport and services. This is also why regional trading blocs are sometimes referred to as 'natural trading blocs'.

Bergsten (1991) disagrees and emphasises that, in practice, trade diversion is a goal of trading blocs. Blocs in the North will divert trade from low-cost producers in the South to high-cost producers in the North. Moreover, the notion of natural trading blocs is weakened because technological innovations have reduced the costs of transport and communications, and trading blocs tend to stretch over large geographical areas. Although trade may be shifted to low-cost countries in the trading bloc, these may not be as efficient as low-cost countries outside the bloc.

Michalak and Gibb (1997) consider regionalism and the formation of trading blocs since the 1980s as a restructuring response to the erosion of the Fordist mode of production. However, the goals of restructuring are contradictory. They embrace both the pursuit of flexibility at an international scale in terms of labour and capital markets, and also the protection of domestic markets from outside competition. This raises two issues of relevance to the analysis. Firstly, insiders within a trading bloc may experience protectionist measures in the form of conditional market access. This may impede industrial development in the longer run. Secondly, in order to ensure flexibility, trading blocs may be fairly open and practise liberal external trade policies. This facilitates market access for outsiders into the blocs (Pomfret 2003), and contrasts with established thinking that insiders benefit while outsiders lose.

Most of the lead firms in textile and garment networks are from the US and Western Europe. Trade policy affects power relations in these buyer-driven networks and who appropriates the bulk of the surplus value. The quota system brought suppliers from more countries into the buyer-driven networks, and manufacturers with access to quotas in high demand had a bargaining chip in price negotiations with their buyers. In respect of trading blocs, however, duty and tax concessions within a bloc open the opportunity for buyers to earn a surplus. Trading blocs may also represent an opportunity to cut lead-times due to the shorter distances involved. Small inventories keep costs down, and the ability to adjust to sudden changes in fashion makes it possible to attain premium prices in the market. The analysis of this chapter addresses whether such concessions and conditions are sufficient to attract buyers to source in the trading blocs when the quota system expires and price competition increases further.

Regionalism, Regionalisation and Counter-processes

In the EU, regional trade arrangements take place within the Pan-Euromed zone, which refers to Western Europe including Turkey, Eastern Europe, the Commonwealth of Independent States of the former Soviet Union (CIS), North Africa, Israel, Libya, Syria and the occupied territories. It is not one formalised free-trade zone, but consists of countries with different types of trade arrangements with the EU. The US has practised conditional quota-free imports of garments from countries in the Caribbean Basin and Mexico since the late 1980s. From the inception of the North American Free Trade Agreement NAFTA in 1995, Mexico has enjoyed quota-free exports to the US of garments that are manufactured from textiles, i.e. from the yarn forward, in one of the NAFTA countries. Full quota freedom was obtained from 2004. Likewise, tariffs on exports from Mexico to the US on garments originating in the NAFTA countries were eliminated by 1999 (ECLAC 2000). The Caribbean Basin Trade Partnership Act and the African Growth and Opportunity Act were signed as a part of the US Trade and Development Act in 2000. Countries included in the two arrangements obtain quota-free and tariff-free exports of garments when the fabric or yarn is imported from the US or are in short supply in the US. The objective of the agreements has been to protect jobs in the US, with the argument that Asian suppliers are unlikely to use US textiles.

The WTO Agreement of Textile and Clothing prevented stricter quota regulations on latecomers and outsiders to the trading blocs that were members of the organisation. However, preferential treatment of insiders to a trading bloc such as favourable tariff levels and favourable quota regulations that may shut other suppliers out, can also be considered a form of protectionism. When Sweden phased out the quota system in the early 1990s, China gained shares from Portugal among others and became the largest single exporter of garments to Sweden. When Sweden joined the EU in 1995, however, it had to re-impose import controls. The share of imports from China then declined by six percentage points from 1994 to 2000 (data sheets obtained from

Weyler at the Swedish Association of Textile Importers, 2002). In line with Bergsten (1991), this illustrates how trade diversion from one of the lowest-cost countries to higher-cost countries within the EU-dominated bloc has taken place.

Outward-processing traffic in the EU-dominated bloc and production sharing in the US-dominated bloc occur when textiles from the EU countries and the US are assembled into garments elsewhere in the blocs and re-exported. When the arrangements are governed by a regulatory mechanism they function as protectionist measures. The EU countries and the US obtain an outlet for their textiles, while the other members of the blocs attain market access by the means of quota concessions and lower import duties. This form of protection of the textile industry should be understood in light of the fact that both the EU and the US lost about 50 per cent of the employment in the textile and garment industry in the 1980s and 1990s. As the textile sector is more technologically advanced, the policy is to retain high quality production in this sector (Euratex 2002; US Census Bureau – undated). In the early 1990s, outward-processing traffic linked to quota concessions was the engine for the transfer of garment production to lower-cost locations in the Pan-Euromed zone (Smith et al. 2002). However, in 1995 and 1998, imports from Turkey and countries in Central and Eastern Europe, were fully liberalised. As in the case of the EU, production sharing is still practised between the US and Mexico and the US and the Caribbean countries, but the legal mechanism only applies in the latter case.

Table 2.1   Share of imports of textiles and garments to the EU 1994–2005

	Textiles (%)				Garments (%)			
	1994	2000	2003	2005	1994	2000	2003	2005
	EU15	EU15	EU15	EU25	EU15	EU15	EU15	EU25
Intra-EU	68	63	62	68	43	39	40	45
C/E Europe, Baltic States, CIS[1]	3	5	8	1	8	10	11	1
Other Europe	7	7	8	9	8	9	10	14
of which Turkey	(2)	(4)	(5)	(5)	(6)	(6)	(8)	(8)
Asia	15	18	17	19	30	32	30	34
of which China	(3)	(4)	(5)	(8)	(8)	(9)	(12)	(18)
Africa	2	2	2	1	7	8	7	6
Middle East	2	2	1	1	1	1	1	0
North America	3	3	2	2	1	1	0	1
Other America	1	0	0	0	1	0	0	0
Total %	101	100	100	101	99	100	99	101
US$ mill.	48824	49774	52534	65825	68331	85909	101294	128702

Note 1: With the expansion to EU25, most of the imports from Central and Eastern Europe and the Baltic States are recorded as intra-EU imports.

Source: Based on WTO (1996, 2001, 2004, 2006).

Trade in textiles and garments became increasingly more regional in the 1990s (Mortimore 1999). Within the Pan-Euromed zone there has been a shift to lower-cost suppliers such as Central and Eastern Europe, the Baltic States, CIS and Turkey. The shares of imports of both textiles and garments to the EU from the US have been low since 1994, while imports of textiles and garments from Asia have grown at a slow pace. Imports from China in 2005 cannot be seen to undermine intra-EU exports at the aggregate level (Table 2.1).

In textiles, the consolidation of the US-dominated bloc happened mainly at the expense of imports from Europe (Table 2.2). In spite of regionalism, there has been a large increase in imports of Asian textiles into the US since the mid-1990s. In garments, however, imports from Asia into the US were stalled after 1995, but picked up again after China joined the WTO in 2001. By 2003, China became the single largest exporter of both textiles and garments to the US, surpassing Western Europe in the case of textiles and Mexico in the case of garments. This reveals a counter-tendency to former processes of regionalisation in the US-dominated bloc. The differences between the EU-dominated bloc and US-dominated bloc call for a discussion of the role of geographical proximity versus price competition in the explanation of regionalisation.

Table 2.2   Share of imports of textiles and garments to the US 1995–2005

	Textiles (%)				Garments (%)			
	1995	2000	2003	2005	1995	2000	2003	2005
North America	17	23	19	17	9	17	13	10
of which Mexico	(7)	(10)	(9)	(2)	(7)	(14)	(10)	(8)
Other America	4	3	3	3	15	17	16	15
Asia	49	49	54	58	64	56	59	65
of which China	(12)	(12)	(20)	(26)	(15)	(13)	(17)	(26)
Western Europe	26	21	19	18	7	6	6	5
C/E Europe, Baltic States, CIS	2	1	1	0	1	1	2	0
Middle East	2	3	3	3	2	2	3	2
Africa	1	1	1	1	2	2	3	3
Total %	101	101	100	100	100	101	102	100
US$ mill.	10441	15709	18289	22538	41367	66392	71276	80071

Source: Based on WTO (2001, 2004, 2005, 2006).

Buyers' Preferences: Proximity or Price?

The representative of the Swedish Association of Textile Importers holds that price is more important than lead-time, and thus that quota abolition, if it actually happens the way it is intended, will impede regionalisation (interview 2002). Likewise, US buyers, i.e. those without factories, plan to source 70–80 per cent of their products from China from 2005 (Todaro 2003). Geographical proximity is not an issue any longer in sourcing decisions. One supplier in Hong Kong may replenish as many as 1000 outlets in the US per day. Moreover, the buyers plan to reduce the number of suppliers and nations that they source from. The reason for this is that large US buyers prefer to deal with 'equally' large and well-organised suppliers who only require 'a sketch and a check' to deliver large quantities fast (Todaro 2003, 1; interviews Vietnam 2002; personal communication Norway 2003). This case weakens the emphasis that Krugman (1991) places on geographical proximity in trading blocs as an engine of trade creation. Trade within the US-dominated bloc is not more attractive than it being shifted to China during the last few years of the quota phase-out (Table 2.2).

In the EU-dominated bloc, geographical proximity enables response to fast changes in demand and easier control over sub-contractors. Another advantage is high quality standards (Stengg 2001). Although Norway abolished quota regulations on 1 January 1999, this is confirmed by two Norwegian buyers with broad experience in the garment sector. They argue that sourcing in the Pan-Euromed zone has a role to play when chain stores require fast replenishment, especially of finer or superior quality products (personal communication 2003).

The lock-in effect of the industrial networks that were established in the early 1990s explains the continuation of the outward-processing traffic after trade with Turkey and Eastern and Central Europe was fully liberalised (Smith et al. 2002). However, this cannot be seen in isolation from the appropriate combination of quality of production and lead-time offered by the latecomers in the Pan-Euromed zone. A similar lock-in effect is not expected to last in the case of Mexican exports to the US. In spite of geographical proximity, lead-times are too long. The bulk of exports from Mexico and the Caribbean countries are basic garments, which meet strong price competition from China. Moreover, the material that the Caribbean countries import from the US in order to obtain preferential market access is more costly than imports of similar material from Asia (Department of State Washington Files 2004; Textiles Intelligence 2004).

The share of imports by retailers and branded marketers is almost the same in the EU and the US (Gereffi 2002). Hence, the higher inclination to source regionally in the EU-dominated bloc is not because buyers in the EU are to a larger extent manufacturers who depend on market access for their own textiles. Differences in the structure of demand explain why price competition in the textile and garment industry is even keener in the US, and thus why regionalisation seems to be more embedded in the EU. In the US, requirements for large quantities of each item strengthen price competition. In the EU, the pattern of demand is more varied and buyers are generally smaller. There is more room for economies of differentiation and higher prices in the market, but this also implies that reduction of lead times and geographical proximity become more important. Another explanation is that the textile sector is more technologically advanced and thus less subject to competition in the first-comer EU countries than in the US.

Consequences for Industrial Development

Insiders

About 80 per cent of the exports of textiles and garments from Turkey go to the EU market. Moreover, Turkey is the Pan-Euromed country that has experienced the greatest success in penetrating the EU market. A vertically integrated industry from production of cotton fibres to the manufacture of textiles and garments facilitates short lead-times. There is flexibility to take smaller orders, quality is good, and the industry has already developed some internationally renowned brand names. Outward-processing traffic has not been an issue in Turkey-EU relations. Together with full quota freedom and zero tariffs, this explains how the country has been in a position to further develop integration of its textile and garment sectors. Turkish manufacturers also enjoy a cost advantage in comparison with Poland, the Czech Republic and Hungary, and higher productivity compared with Romania and Bulgaria. It takes three days to transport products by truck to Germany. In comparison, transport from China to Europe by ship takes about 27 days. Nevertheless, industry representatives in Turkey fear that quota abolition and competition from lower cost production in China will lead to a shake-out of small and medium-sized manufacturers (interviews in Turkey 2003). Unlike Turkey, as much as 80 per cent of exports of garments from Poland take place as outward-processing traffic (Pluta 1998). Other countries that are strongly involved in this are Romania, Slovenia and Hungary (Morris 2002). The processing of imported fabrics in the latecomer countries results in little value-added and limited scope for industrial upgrading (Smith et al. 2002).

NAFTA boosted production of both textiles and garments in Mexico. However, much of the textile industry is technologically backward and inefficient, and the goods are of poor quality. Production sharing still dominates exports of garments from Mexico to the US and impedes vertical integration, although vertical integration was attained in the production of blue jeans (Gereffi 2002; Textiles Intelligence 2004). Unlike Mexico, the Caribbean countries are still subject to the legal mechanism of production sharing. Because value-added only is subject to duty upon re-entry into the US, there is no incentive to upgrade production technologically. With a longer distance to the market, the countries that are part of the African Growth and Opportunity Act are even worse off.

Outsiders

When Sri Lanka attained quota-free exports to the EU from 1 March 2001, it was on the condition that Sri Lanka reduces tariffs on imports of textiles and garments from the EU and does not apply non-tariff barriers on such items (European Commission 2001). In Vietnam the EU offered quota enlargement as part of an outward-processing arrangement. Further quota enlargement was offered in 2003 subject to reduction of import duties on textiles and garments from the EU.

Vietnam also had to liberalise and open other economic sectors for investments from the EU (DG Trade 2003). Quota enlargement and quota abolition were used as a bargaining chip for tariff reductions and better market access.

Arrangements that encourage imports of textiles discourage the creation of backward linkages in the garment industry. Such linkages increase local value-added which is important in order to sustain the development of the industry in the longer run. The garment industries in Sri Lanka and Vietnam do not have any backward linkages to speak of. In Sri Lanka net export earnings in the textile and garment industry are 30–40 per cent less than the value of gross exports. In Vietnam, the textile and garment industry together imported more than the total value of exports in 1999-2000 (Knutsen 2004).

China and India, the two developing countries that are assumed to become the main beneficiaries of quota abolition, have the advantage of local fibre production and an integrated textile and garment industry. The same is true of Turkey, which is the largest Pan-Euromed exporter of garments to the EU. It may be difficult for small and poor countries with a limited domestic market, such as Sri Lanka, to develop competitive backward linkages. Vietnam, however, with 80 million inhabitants, is about the same size as Turkey, and has the advantage of cotton cultivation.

In sum, the contradictory objectives of regional trade arrangements that Michalak and Gibb (1997) are concerned with are evident in the textile and garment industry. The fact that outward-processing traffic and production-sharing agreements are applied to manufacturers in countries both within and outside the respective blocs, illustrates this very well. On the one hand, they are meant to protect the textile industry and employment in the first-comer EU countries and the US. On the other hand, buyers from these countries depend on the use of cheap labour and capital in low-cost countries in order to survive and grow in the market. The explanation is that low margins and large sales are the fundamental source of profits in the textile and garment industry. This is a result of mature technology and saturated markets. Open regionalism is an advantage in this context because more countries can then be tied into trade arrangements.

'Beggar thy Neighbour'?

From 2000 to 2001, the overall import price of garments to the EU fell by 4.8 per cent, whereas the import price of garments to the EU from Asia, Latin-America and the ACP-countries declined by 9–10 per cent (Euratex 2002). This indicates that manufacturing suppliers outside the Pan-Euromed zone are subject to stronger pressure to reduce prices in line with the 'beggar thy neighbour' effect that Krugman (1991) has described. However, the problem of the outsiders is exacerbated by China's accession to the WTO. Exports from Sri Lanka to the EU started to decline in 2001. At the same time, exports to the EU from the largest suppliers, i.e. China, Turkey and Romania experienced high annual growth rates (statistics obtained at DG Trade 2002; WTO 2004). Similarly, when quota restrictions were lifted on China's exports of 29 categories of garments to the US in 2002, prices of all suppliers dropped, but the average Chinese price dropped to a lower level than the prices of the others (ATMI 2003).

The race to the bottom price-wise and in combination with good quality and accountability cannot, however, be seen in isolation from the practice of buyers who control and co-ordinate the networks both inside and outside the trading blocs. Shortage of demand and crises in retail mean that buyers cut prices to sell more. Buyers apply pressure for 'Chinese prices', while manufacturing suppliers in Vietnam and Sri Lanka experience a decline in their profit margins (Knutsen 2004). In Turkey, too, margins are becoming slimmer and some are selling 'for their costs only' (interview Turkey 2003).

There is a clear conflict of interest between textile and garment manufacturers in the US and the EU who are worried about competition from China, and buyers without factories in the US and the EU who are eager to source more from China. Eventually, the US decided to employ safeguards against imports of some categories of garments in late 2003, and in 2005 both the EU and the US negotiated new agreements with China. Ten categories of textiles and garments became subject to quota regulations in the EU and thirty-four in the US. The agreements will last until the end of 2007 in the case of the EU and until the end of 2008 in the case of the US (Economist Intelligence 2005; EU 2005). This is what buyers who are not manufacturers fear could develop into a new multilateral quota system.

Summary and Conclusions

The processes of regionalisation in both the EU-dominated and US-dominated trading blocs have been fuelled by regionalism in an attempt to secure markets for textiles manufactured in the first-comer countries in the North. This is reflected in the former outward-processing mechanism in the EU-dominated bloc and the production-sharing mechanism that the US offers in the US-dominated bloc. The fact that the EU offers outward-processing mechanisms to latecomers outside the trading bloc corroborates the urge to secure markets and is a sign of open regionalism in line with Michalak and Gibb (1997) and Pomfret (2003). In the case of the US, preferential treatment in the trading bloc is no longer sufficient to act as a replacement for the quota system. Regionalisation in the EU-dominated bloc seems to be more embedded than in the US-dominated bloc. This cannot be explained by the nature of preferential treatment only. The market structure in the EU makes greater demands on geographical proximity. A technologically advanced textile sector in the first-comer EU countries also promotes intra-regional trade. In sum, regional trade arrangements do not fully replace the quota system as a protectionist measure, but the new quota agreements with China show that first-comer countries in the EU and the US still use leverage to protect their textile industries.

The advantage of open regionalism to outsiders is market access, but this alone does not ensure industrial development. Regional trade arrangements serve to impede industrial development both for outsiders and insiders in relation to trading blocs. For outsiders this is when insiders squeeze them out of the market and when their exports to the trading blocs are subject to conditions that discourage local linkages in the industry. For insiders too, outward-processing traffic and production-sharing mechanisms discourage an integrated textile and garment industry. Turkey has benefited from inclusion in the EU trading bloc, but the country had attained linkages in production before inclusion and was also not subject to the mechanism of outward-processing traffic. As in the case of Eastern Europe and Mexico, it is difficult to graduate from these types of production relationships where little is done to upgrade the manufacturers even when the legal mechanism is abolished and trade becomes freer.

New economic configurations arise. It is likely that the bulk of trade will continue to take place in the form of exports to the North from the lowest-cost countries, predominantly China, but more so in the US than the EU. Most of the remaining exports will come from flexible production within the trading blocs. This is explained by the intense price competition in a saturated market, and the fact that that the networks are dominated by lead firms that are not manufacturers themselves but shop around for the best deals. Low-cost, developing countries that do not belong to the lowest-cost countries anymore and low-cost countries that are not part of a trading bloc gradually get squeezed out of international trade.

It is difficult for latecomer countries to benefit from inclusion in regional trade arrangements in a technologically mature, labour-intensive and price-sensitive industry. First-comer countries resort to regional trade arrangements in order to reduce loss of employment in their home countries. Due to intensity of the fight for market access in first-comer countries, latecomer countries accept conditional access. Meanwhile, lead firms in networks put pressure on suppliers for lower prices and shorter lead-times to be able to retain their position and grow in the market. This amounts to a double squeeze on latecomers.

References

ATMI (2003), *The China Threat to World Textile and Apparel Trade* <http://www.atmi.org>.

Bergsten, C.F. (1991), *Commentary: The Move toward Free Trade Zones*. Paper for the Symposium of the Federal Reserve Bank of Kansas City 1991.

Department of State Washington Files (2004), *China Expected to become Dominant Textile, Apparel Supplier* <http://usembassy-australia.gov/hyper/2004/0210/epf205htm>.

DG Trade (2003), *EU and Vietnam Agreement on Mutually Beneficial Trade Liberalisation* <http://europa.eu.int/comm/trade/miti/devel/pr.170203_en.htm>.

ECLAC (2000), *Foreign Investment in Latin-America and the Caribbean, 1999*. Economic Commission on Latin-America and the Caribbean (UN, Santiago: Chile).

Economist Intelligence (2005), *Industry Briefing*. 10 November <http://www. viewswire.com/ index.asp?layout=ArticleVW3&article_id=107957469>.

EU (2005), *EU-China Textile Agreement 10 June 2005*. Memo-05-201 Brussels 12 June <http://europa.eu.int/comm/external_relations/china/intro/memo05_201. htm>.

Euratex (2002), 'General Economic Situation in 2001'. *Short Term Perspectives. Euratex Bulletin* 2002/2.

European Commission (2001), *EU Removes Textiles Quotas for Sri Lanka in Market Access Deal*. Press release 1 March <http:/europa.eu.int/comm/trade/ goods/textile>.

Gereffi, G. (2002), *Outsourcing and Changing Patterns of International Competition in the Apparel Commodity Chain*. Paper Presented at the Conference 'Responding to Globalisation: Societies: Groups and Individuals', Boulder, Colorado. 4-7 April.

Knutsen, H.M. (2004), 'Industrial Development in Buyer-driven Networks: The Garment Industry in Vietnam and Sri Lanka', *Journal of Economic Geography* 4: 5, 545-564.

Krugman, P. (1991), *The Move Toward FreeTtrade Zones*. Paper for the Symposium of the Federal Reserve Bank of Kansas City 1991.

Michalak, W. and Gibb, R. (1997), 'Trading Blocs and Multilateralism in the World Economy', *Annals of the Association of American Geographers* 87: 2, 264-279.

Morris, D. (2002), *The Multilateral Trading System and the New Political Economy for Trade in Textiles and Clothing. An Introduction for Least Developed Countries*. International Trade Centre UNCTAD/WTO (ITC). Technical Paper (Geneva: UN Publications Unit).

Mortimore, M. (1999), 'Apparel-based Industrialisation in the Caribbean Basin: A Threadbare Garment?' *CEPAL Review* 67, 119-136.

Pluta, A. (1998), *Made in Eastern Europe. Summary of the Clean Clothes Campaign's Report on the East European Garment Industry* <http:/www. cleanclothes.org/ publications/easteusummary.htm>.

Pomfret, R. (2003), 'Introduction', in Pomfret, R. (ed.), *Economic Analysis of Regional Trading Arrangements*. The International Library of Critical Writings in Economics (Cheltenham: Edward Elgar), xi-xvii.

Smith, A., Rainnie, A., Dunford, M., Hardy, J., Hudson, R., and Sadler, D. (2002), 'Networks of Value, Commodities and Regions: Reworking Divisions of Labour in Macro-regional Economies', *Progress in Human Geography* 26: 1, 41-63.

Stengg, W. (2001), *The Textile and Clothing Industry in the EU*. A Survey. Enterprise Papers, No. 2 (Brussels: Enterprise Directorate-General, European Commission).

Textiles Intelligence (2004), *Report Summary: The Rise and Fall of the Garment Industries in Mexico and the Caribbean Basin*, Issue 112 <http://textilesintelligence.com>.

Todaro, M. (2003), *The Land of Not-China*. AAPNetwork <www.aapnetwork.net/main.html?page=notchina>.

U.S. Census Bureau (2003), *Statistical Abstract of the United States: 2003. The National Data Book*. US Department of Commerce, Economics and Statistical Administration.

U.S. Census Bureau (undated), *Manufactures- Summaries 1982-1995* <http://www.census.gov/statab/www>.

WTO (1996), *International Trade Statistics 1997* (Geneva: WTO).

WTO (2001, 2004, 2005, 2006), *International Trade Statistics* <http://www.wto.org>.

Chapter 3

Crossing Juridical Borders: Relational Governance in International Package Tourism from Germany to Jordan

Sabine Dörry

Introduction

The purpose of this chapter is to explore the under-examined aspects of informal institutions and small and medium-sized enterprises (SMEs) in the tourism industry in the context of global value chains (GVCs). The focus of the empirical elements of the study is the operation of the package tourism industry between Germany and Jordan. The package tourism business is surrounded by uncertainty. Using the GVC concept as an heuristic allows the problems inherent in 'transactions' between firms separated by distance that are involved in the production of package tourism to be demonstrated. International package tours are 'experience goods' that involve a considerable element of uncertainty prior to the transaction in regard to the character and quality of the product to be exchanged. A transaction that largely avoids a so-called 'opportunistic' action by one of the business partners must – and generally will – be 'framed' by formal institutions. However, these formal structures only apply to national territories where they are generally implemented by the nation state. In the international package tourism business, national borders exist only on maps and therefore no common international law exists to enforce private contracts. Hence, there is a potential conflict that could prevent a transaction from taking place. In tourism, substituting destinations is common practice among tour operators when a destination fails to fulfil economic expectations or is hit by external shocks, and this remains a major challenge to the industry.

The GVC approach[1] draws together the new economic relations and divisions of labour that characterise the increasing global integration of the world economy. Within the dynamic GVC concept, the spatial dimensions of events at the macro level, and the organisational structures of firms at a micro scale, are elaborated. The concept builds on a perspective of dependency between industrialised and

---

1 Bair (2005) provides a comprehensive discussion on the background and development of the GVC research 'school'; its intentions and approaches including the concept of global production networks.

developing economies. It is assumed that the local production of goods (and services) in developing countries is highly dependent on firms being embedded into GVCs, which are 'governed' by powerful firms in industrialised countries. These firms are able to organise the production chain for their benefit because of their tremendous purchasing power and their exclusive access to final consumer markets. Therefore, the ability of suppliers in developing countries to develop and to foster regional development is controlled by the interest of the purchasing firm. If the interests of that lead purchasing firm are not met, development is likely to be blocked. Such interrelationships of industrial and trading activities at different geographical scales shape and are shaped by globalisation.

Gereffi (1995) outlines different analytical levels, combining activities of organisation and co-ordination from the most influential firm within a product-specific GVC. These chains combine the *input-output* relationships that link value-adding activity involving products, services, and resources in a range of industries within a particular *geographic localisation*. At the same time, the chain's operations are co-ordinated by the *governance structure* built on the authority and power relationships between firms and the *institutional framework*. So far, researchers have focused on governance issues surrounding the activities within GVCs of powerful transnational corporations (TNC). These TNCs dominate their supplier networks by enforcing standards and quality requirements by undertaking for the chain the highly profitable functions such as research and development, product design, sales and marketing, or market research.

However, this characterisation of GVCs has a number of theoretical shortcomings of which two are highlighted in this chapter. Firstly, the neglected impact of external events on GVCs is problematic (Bair 2005). Certainly, the divorce of issues of 'intra-chain' governance from insufficiently examined 'external' institutional processes is of concern. Hess and Yeung[2] (2006, 7), for example, acknowledge that in the recent past, different authors have incorporated into analyses '... the role of the state and other non-firm institutions as important agents'. Although this is a valuable addition, it is an incomplete definition of a chain's co-ordination structure. Institutions in general serve to control and direct individual behaviour that tends to reduce uncertainty in socio-economic relations (North 1994). The analysis of this chapter highlights through empirical evidence the importance of the broader, encompassing and dynamic interplay between co-ordination within value chains (*governance*) and the external influences on the chain (*institutional framework*). I argue that informal institutions, including conventions, customs, norms, or social routines in a socio-economic environment (see for example, North 1984) might therefore be a specific aspect of both the institutional framework *and* governance characteristics which cannot be analysed separately.

---

2 Hess and Yeung advocate the 'Manchester School' of Global Production Networks (GNP) focusing on similar themes to GVC research (2006).

The second theoretical shortcoming is that this broader institutional and governance framework applies particularly to the operations of small and medium-sized enterprises (SMEs) that have so far been excluded from the GVC debate. This limitation is significant because the economic 'Mittelstand'[3] is typically the metaphorical spine of an economy. Besides, SMEs must defend or extend their own economic positions within a coalescing world economy as well as co-ordinate the actions of their production chains. The main focus of the chapter lies at this micro level of empirically-examined socio-economic co-ordination mechanisms among fragmented production activities.

Jordan only affords limited access to typical '3S' (sun, sand and sea) tourism activities but enjoys broad interest among culturally and religiously motivated German tourists. Round trips offered by tour operators are the predominant type of travel to Jordan. Small and medium-sized specialist tour operators dominate this area of business. Only recently have TNCs, like TUI (http://www.tui-group.com/en/), discovered the potential of Jordan for their product portfolio.

Two types of central agent usually handle tourist flows between Germany and Jordan in the package tourism sector; (1) tour operators in the source market (Germany), and (2) inbound agencies at the destination (Jordan). Tour operators, acting as independent companies, offer services under one name while carrying the entire product liability. Furthermore, these entities hold exclusive rights as gatekeepers to final tourist markets. Increasingly, product creation is designed to meet customer demand in the country of origin. Consequently, tour operators need to co-ordinate and control their supply chain efficiently in order to gain follow-up business and ensure customer satisfaction. Compared to most classical '3S' holiday trips, the co-ordination of a mix of activities that makes up a round trip (e.g. local and regional transfers, different hotels at different places, daily sightseeing tours and events, individual extension packages) is a complex task requiring destination-specific knowledge. Inbound agents are typically responsible for handling tourist activities at the destination. They possess specific local knowledge and book hotel rooms and handle operational tasks during the tourists' stay. They are able to realise economies of scale by combining demand from a range of different tour operators, thus achieving lower prices in local facilities. Because the agents are embedded in the same culture as the local suppliers they use, arrangements are made easier for foreign tour operators, reducing both risk and transaction costs.

---

3 The term 'Mittelstand' covers the broad SME sector in Germany which covers 99.7 per cent of German companies and has huge political and economic clout.

The Global Connections of Small Tourism Businesses

Small service firms operating within GVCs differ greatly from those in the primary and secondary sectors dealing in tangible products (see Dörry and Mosedale 2006). The service sector is an eclectic industry offering intangible services that can not be adequately accommodated within the classical input-output framework of the GVC concept.

One such attribute of the tourist service sector is the need for tour operators to recognise both the tangible and intangible requirements of customers. What the consumer is buying is the experience, which cannot be portrayed by catalogues (Britton 1991). This disconnection between sales hyperbole and consumer desire is problematic because the judgement from the traveller is not made until after the journey has been completed and the 'experience' has been consumed.

As a consequence of the simultaneity of execution of service production and consumption (uno-actu), the customer is not only consumer but also producer of the tourist experience – appropriately termed the 'moment of truth' in the managerial literature. Tour operators need to ensure positive experiences for the consumer from beginning to end since the experience heavily relies on all aspects of the travel package.

Finally, the 'uno-actu' principle makes it difficult for tour operators to intervene directly in the customers' production process. The challenge for the tour operator is a clear spatial dilemma. The operator must oversee and ensure the suppliers' activities at a distance to guarantee that the inbound agent meets the agreed service quality. There are few examples of analyses of this type of consumption-oriented value chain (see Clancy 1998; Mosedale 2006). Thus, interactions within the tourism sub-sectors on a product-specific base, like a package tour, have still not been analysed from a theoretical perspective such as that suggested by Gereffi (1995), for example.

Relationships on the production side of the chain suggest a multi-level principal-agent (PA) problem. Three levels are apparent; (1) the fact that tourists are uncertain about the quality of the trip, (2) that tour operators are uncertain about the quality of the arrangements made by the inbound agent, and (3) that inbound agents are uncertain about the performance of their local suppliers (Figure 3.1). The first and third sets of relationship can be described as 'conventional' PA constellations which are embedded in the same national context. These relationships are outlined in this section. The very different relationships between tour operators and inbound agents are delineated in the section that follows.

Business relationships are characterised by information asymmetries. In classical-economic reality, individuals engage in business for profit rather than altruism. Both parties expect to gain positively from their co-operation, but neither can eliminate the risk of each other's opportunism. The inbound agent (the principal) knows less than the local suppliers (the agents) and cannot reasonably monitor every aspect of their performance. It is assumed that the agent does not necessarily work in principal's best interest (Arrow 1985). The principal's information deficit encourages agents' opportunism necessitating efficient control mechanisms on the part of the principal. According to game theorists, this could lead to the establishment of informal rules, such as trust or reciprocity (Young 1998). In Jordan, the entire ground handling process lies in the inbound agents' realm of responsibility and introduces the first PA constellation between inbound agents and local suppliers legally and culturally embedded in the same national context. The second PA constellation relies on the relationship between tourists and tour operators. Both are embedded in an institutional environment characterised by strict German law (Deutsches Reiserecht) which offers considerable protection to the consumer.

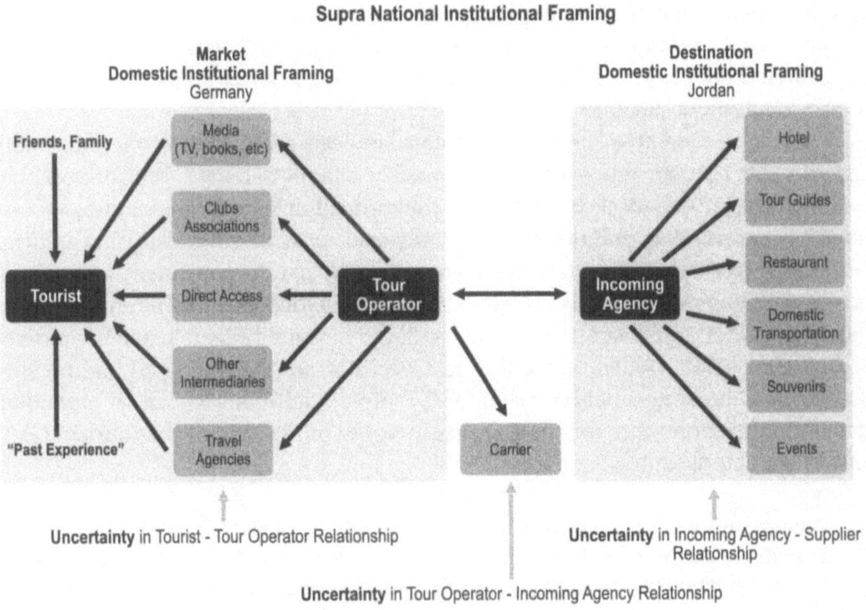

Figure 3.1   GVC and its 'assembly nodes' in tourism

Source: Based on Schamp (2007)

Relational Governance

While the PA relations discussed thus far operate in unique legal environments, tour operators and inbound agents need not only to bridge geographical distance but also to address geopolitics, customs and established networks to do business without legal enforcement tools for private contracts. Looking at business relationships across different national legal systems, German economists, Schmidt-Trenz and Schmidtchen (1991) and Egbert (2006) have contributed fruitful insights to explain this phenomenon.

Core firms of GVCs need to ensure that each business in the supply process 'comes together as an integrated whole' to determine 'how financial, material, and human resources are allocated and flow within a chain' (Gereffi 1995, 116). SMEs have difficulties performing this way since they often have limited buying power, internal resources and capacities at their command. In fact, due to the functional 'competence splitting' between tour operators and inbound agents, the strong reliance on the inbound agents' destination-specific knowledge and skills to smoothly handle all local suppliers weakens the tour operators' positions. Nevertheless, in tourism production systems, tour operators still hold the strategic position in the final consumer market, re-establishing PA relationships. The chapter seeks to demonstrate the alternative actions of firms embedded in different national legal systems when they cannot count on established legal institutions to enforce their 'rights', a situation one interviewee referred to as the 'law of the jungle' (Interview, 30 June 2004).

Criticising Gereffi's (1994) dichotomy between producer-driven commodity chains (PDCC) and buyer-driven commodity chains (BDCC), McCormick and Schmitz (2002, 43) argue for a more differentiated debate about socio-economic power relations. They refer to so-called 'balanced networks' declaring that all firms belonging to a network tend to have equal power and do not inherit a dominant position. Likewise, a variety of governance forms occurs in different GVCs of the same sector at the same time overcoming the rather holistic categorisation of PDCC and BDCC which, according to Gereffi et al. (2005, 84-85) can be sub-divided into market, modular, relational, captive, and hierarchical. In particular, the relational form of governance seems to apply in the case of the tourism GVC and is characterised by:

> complex interactions between buyers and sellers, which often creates mutual dependence and high level of asset specificity [...] the power balance between the firms is more symmetrical, given that both contribute key competence. There is a great deal of explicit co-ordination in relational global value chains, but it is achieved through a close dialogue between more or less equal partners. (Gereffi et al. 2005, 84-85)

To examine the co-ordination of this 'competence splitting', in-depth personal interviews with 37 German tour operators were carried out between May 2004 and December 2005. Interview partners included some brand named companies, but most were independent SMEs with regular packages to Jordan, as well as occasional providers. The interviews were transcribed for the purpose of qualitative analysis and interpretation. To avoid a one-sided tour operator's perspective, six central inbound agents in Jordan dealing among others with the German market were also interviewed. Some of the most interesting empirical results, which show how small and medium-sized travel firms reduce their risk of opportunism and deal with the existing information gap against their incoming agents, are elaborated in the next section.

Principals, Agents, and the Challenge of Mastering Opportunism

Large integrated travel companies within this analysis, such as TUI, do exert restrictive control over their partners in Jordan. That is mainly due to their immense human resources, large legal departments, and extensive experience with legal cases. Indeed, they bypass the lack of a common juridical system by obliging their business partners to take out liability insurance. Since German insurance companies usually provide this insurance, TNCs can easily bridge the 'lawless' space for their own benefit. Such insurance structures the common legal ground for co-operation on a one-year basis, and the large companies are able to sue their inbound agent partners in cases of a breach of contract. Strictly speaking, the 'commanding' firm transfers the legal obligation to its own national legal system to obtain enforceable legal coverage in the event of damage or loss to its business. At the same time, they insure their partners against consequences such as bankruptcy to save their investments and to ensure that the business relationship can continue.

A very different scenario emerges in the context of SME tour operators. Their strategic position in the value chain identifies them as principals, albeit weaker than the integrated travel corporations such as TUI. Because of their lower resource bases, they struggle to operate like the integrated firms. To avoid opportunistic behaviour on the part of inbound agents, SME tour operators need to establish alternative effective control mechanisms to solve the challenge more subtly. Formal written contracts between SME tour operators and inbound agents generally do not exist. In fact, this is not only due to high set-up costs but also because there is no supra-national formal institution that could ensure the legal enforcement of such private contracts.

Heterogeneous Ways of Co-operation and Control[4]

Tour operators are fully liable to customers for shortcomings in services provided by agents at the destination. Since success depends upon comparatively low fault tolerance, operators have developed particular management practices that range from up-front modification of package production to mechanisms of control implemented upon completion of the traveller's journey. Overall, heterogeneous and fragmented structures of co-operation were found in this empirical context. Despite Jordan's relatively minor importance as a holiday destination, at least according to the average number of tourist arrivals, co-operation between tour operators and inbound agents seems to be determined by a mix of factors of economic strength, indicating the realised numbers of guests, and informal, rather socially established institutions such as trust and reciprocity resulting from frequency and quality of performed trips.

---

4 For a fuller empirically-based discussion of the stability of economic relations within an unstable political environment see Dörry (2008).

Not only management practices but also the number of local inbound agents that an SME tour operator works with is considered an essential control instrument. Some tour operators maintain relatively stable relationships with more than one inbound agent business partner on the grounds of strategy and competition. Strategically, having more than one inbound agent allows tour operators to deal with bottlenecks at seasonal peaks in, for example, hotel provision. Competitively, having multiple inbound agents encourages competition in terms of quality and price that gives tour operators a competitive edge in the aggressive markets in which they deal.

In terms of service procurement, evidence suggests that powerful tour operators negotiate prices with local partners only partly upfront. Instead, they fix their own prices and re-negotiate contracts in their favour. Incoming agents have clear reasons to accept such dictated conditions. Brand named tour operators have images built on reputation which can, in turn, help inbound agents to acquire new business from other tour operators. However, a large number of 'weaker' tour operators in Jordan were less prescriptive in the way they bought services from their local business partners.

The product managers of tour operators use regular travel to tourist destinations to exercise and maintain control over suppliers even though they have no legal power. They become acquainted with 'their' products while checking services such as hotel quality. Few tour operators were unable, for financial reasons, to make regular inspection visits. Without the specific knowledge and experience of these visits, power over the travel product shifts to the inbound agent.

Just as co-operation between players in the supply chain varies between players, so do payment modes. Some tour operators pay for packages fully or partly in advance, while others are invoiced for a travel package after the holiday has been completed. If there are justifiable complaints against an inbound agent, the chances of a tour operator realising financial compensation from the agent are mixed. In long-term relations, ex-post payments were far more usual than in short-term or only recently set up relationships. It is clear that trust, based on the tour operators' economic strength and business reputation in Jordan, and self interest on the part of tour operators and inbound agents create stable business relations even in the absence of legal sanctions (Sayer 2000).

Final Option: Exit?

Interview data suggest that there is no simple regulatory procedure that links partners in the tourism supply chain examined here. Verbal contracts and email requests, especially amongst independent SMEs, enable both business partners in a relationship to save costs by cutting bureaucracy and flexibly changing business partners when performance is low. Under such circumstances, the exit of one of the business partners can be simple, quick and cheap because there are no legal obstacles. Serious service failings, non-competitive prices, or dishonesty are the most common reasons for tour operators breaking trading relationships. It can be argued that a tour operators' 'voice' (Hirschman 1970) is backed up by not only the tour operator's high sunk costs in search for local partners but also by their threat of exit as it affects inbound agents. This threat can be used to develop long-term relationships. In most cases the 'voice' option is chosen over the exit strategy, but each firm has to find its own optimal mix of exit and 'voice' (Sayer 2000). Firms interested in using Jordan as a long-term destination invest time to ensure inbound agents understand their service expectations. Exit is, therefore, a last resort.

However, as the responsibility of inbound agents, supply or product failings can only be rectified through negotiation because tour operators have little power as they do not control huge demand (Schamp 2007). This situation strengthens the position of the inbound agents because of their embeddedness in and familiarity with a particular cultural environment and their negotiating tactics with their subordinate suppliers in Jordan that the tour operators rely on.

Dynamic Interplay: A More Holistic View of GVC Co-ordination

The purpose of this chapter has been to explore the impact of economic and social processes on GVC governance. In the context of the tourism GVC it has been shown that some firms are unable fully to exploit the potential of such socio-economic integration within global production and trading relationships. Hence, the aim this study has been to explore *how* and *why* successful firms of different sizes and with different resources are able to exercise the power of a principal in a principal-agent relationship in the face of uncertainty and the absence of legal sanction. These firms can be successful while embedded in various national legal systems, as the example of the tourism service sector shows.

In the GVC debate, the legal enforcement of contracts by the core firm in a chain is assumed. For TNCs this power has been obvious because of their strong governance position in producer-driven and buyer-driven commodity chains. The heuristics of the GVC are also useful for analysing SMEs. However, some of the established terms and definitions do not apply because of the slightly different circumstances these smaller firms operate within. In the SME context, more emphasis needs to be placed on informal rather than formal business arrangements, and the adaptation that occurs as a result of repeat business that facilitates smoother inter-firm relations.

Returning to the GVC debate, a sound understanding of institutions is dependent upon a more precise understanding of enforcement mechanisms (Wiggins 1991), especially in the tourism sector. New governmental regulations can be enforced at very short notice, as has happened in Jordan on a regular basis not least for geopolitical reasons. The analysis reported here shows that it is hard to draw strict distinction between governance and institutional frameworks as they impact on the tourism GVC.

It has been shown that myriad informal mechanisms and formal institutions interact. Not every company uses the same amount of internal firm resources. Large firms usually have access to more information and also possess greater capacity for information 'digestion' and complex problem-solving. Thus, whilst some tour operators accentuate formal regulations to configure their supplier relationship, others (mostly SMEs) concentrate on more informal institutions. It is vital to emphasise that each firm acts in a unique business environment comprising a unique set of cumulative formal and informal institutions.

Acknowledgements

Eike W. Schamp provided critical and helpful comments on various drafts of this chapter. His perceptive suggestions and questioning are greatly appreciated. I also thank two anonymous reviewers for their valuable remarks. This chapter has benefited from a three year research project funded by the German Research Council (DFG).

References

Arrow, K.J. (1985), 'The Economics of Agency', in Pratt, J.W. and Zeckhauser, R.J. (eds), *Principals and Agents: The Structure of Business* (Boston: Harvard Business School Press), 37-51.

Bair, J. (2005), 'Global Capitalism and Commodity Chains: Looking Back, Going Forward', *Competition and Change* 9, 153-180.

Britton, S. (1991), 'Tourism, Capital, and Place: Towards a Critical Geography of Tourism', *Environment and Planning D: Society and Space* 9, 451-478.

Clancy, M. (1998), 'Commodity Chains, Services and Development: Theory and Preliminary Evidence from the Tourism Industry', *Review of International Political Economy* 5, 122-148.

Dörry, S. (2008), 'Business Relations in the Design of Package Tours in a Changing Environment: The Case of Tourism from Germany to Jordan', in Burns, P. and Novelli, M. (eds), *Tourism and Mobilities: Local-Global Connections* (Oxfordshire: CABI), 204-218.

Dörry, S. and Mosedale, J. (2006), *Commodity Chain Analysis and Tourism: A Progressive Synthesis?* Paper presented to Cutting Edge Research in Tourism: New Directions, Challenges and Applications. Surrey.

Egbert, H. (2006), 'Cross-border Small-scale Trading in South-Eastern Europe: Do Embeddedness and Social Capital Explain Enough?', *International Journal of Urban and Regional Research* 30, 346-361.

Gereffi, G. (1994), 'The International Economy and Economic Development', in Smelser, N.J. and Swedberg, R. (eds), *Handbook of Economic Sociology* (Princeton: University Press), 206-233.

Gereffi, G. (1995), 'Global Production Systems and Third World Development', in Stallings, B. (ed.) *Global Change, Regional Response. The New International Context of Development* (Cambridge: Cambridge University Press), 100-142.

Gereffi, G., Humphrey, J. and Sturgeon, T. J. (2005), 'The Governance of Global Value Chains', *Review of International Political Economy* 12, 78-104.

Hess, M. and Yeung, H. (2006), 'Whither Global Production Networks in Economic Geography? Past, Present, Future', *Environment and Planning A* 38, 1193-1204.

Hirschman, A.O. (1970), *Exit, Voice, and Loyalty: Responses to Decline in Firms, Organisations, and States* (Cambridge, Mass. [et al.]: Harvard Univ. Press).

McCormick, D. and Schmitz, H. (2002), *Manual for Value Chain Research on Homeworkers in the Garment Industry* (Brighton: Institute of Development Studies).

Mosedale, J. (2006), 'Tourism Commodity Chains: Market Entry and its Effects on St Lucia', *Current Issues in Tourism* 9, 436-458.

MoTA (2006), *Tourism Statistics.* (Amman/Jordan: Ministry of Tourism and Antiquities, Information Statistics Department).

North, D.C. (1984), 'Transaction Costs, Institutions, and Economic History'. *Journal of Institutional and Theoretical Economics* 140, 7-17.

North, D.C. (1994), 'Economic Performance through Time', *American Economic Review* 84, 359-368.

Sayer, A. (2000), *Markets, Embeddedness and Trust: Problems of Polysemy an Idealism.* Lancaster, Online Papers Sociology Department, Lancaster University, UK.

Schamp, E.W. (2007), 'Wertschöpfungsketten in Pauschalreisen des Ferntourismus – zum Problem ihrer governance', *Erdkunde* 61, 147-160.

Schmidt-Trenz, H.-J. and Schmidtchen, D. (1991), 'Private International Trade in the Shadow of the Territoriality of Law: Why Does It Work', *Southern Economic Journal* 58, 329-338.

Wiggins, S.N. (1991), 'The Economics of the Firm and Contracts: A Selective Survey', *Journal of Institutional and Theoretical Economics* 147, 603-661.

Young, H.P. (1998), *Individual Strategy and Social Structure. An Evolutionary Theory of Institutions* (Princeton: Princeton University Press).

Chapter 4

Global Distribution and Cluster Development: Hollywood and the German Connection

Ivo Mossig

Introduction

Few industries are as spatially highly concentrated worldwide as the motion picture industry in the Los Angeles and Hollywood area of the US west coast (referred to in this chapter as 'Hollywood') and the reasons for such concentration are a topic of longstanding interest in economic geography. In the early 1990s, Porter (1990) referred to these spatial concentrations of economic activities as clusters. Since then numerous studies and much empirical work has been published emphasising the advantages of spatial proximity for co-localised enterprises within a cluster. But these advantages do not offer sufficient explanation for the continued existence and growth of a cluster. Linkages across cluster boundaries are also significant for the success or stagnation of a cluster. Here, distribution structures to external markets are of central relevance because they generate the revenue stream that maintains economic existence of the enterprises within a cluster. In other words, a cluster's success depends on the effective distribution of its products on external markets.

This chapter deals with the example of the motion picture industry. It demonstrates the major significance of a powerful global distribution system for the development and continued existence of the world's most important movie cluster in Hollywood. The dominance of movies from Hollywood on the world market is demonstrated, and the connections to the German market are analysed as a case study providing an empirical perspective on the global commodity chain of motion picture distribution. A key question that is discussed is how the advantages of spatial proximity within a cluster can be augmented by global distribution linkages. The powerful oligopoly of the world's largest media companies, together with their international subsidiaries, dominate the world market, and it is within the organisational structures of these large media corporations that the distribution of movies, so important for cluster development in Hollywood, takes place. It is argued that cross-boundary linkages, as well as internal cluster networks, are major forces that promote and maintain economic clusters and deserve greater research attention.

The Motion Pictures and the Cultural Industries

In addition to the so-called 'fine arts' (e.g. theatre, music, painting, literature), the entertainment and media industry, publishing and printing, as well as advertising, are branches of the cultural industries that have experienced above-average economic growth and have generated increasing numbers of jobs (Scott 2000; Power 2003; Mossig 2006). The motion picture industry is a central part of these cultural industries. In Los Angeles County, employee numbers in this industry increased from 50,264 to 159,000 between 1980 and 2000, and more than 61 per cent of US motion picture employment is in Hollywood (Scott 2005).

Movies have distinctive characteristics relevant to their distribution arrangements. They commercialise immaterial content such as fictional stories, information and emotions, that stand in contrast to material products and consumer goods. They are a medium for this content and can be transported extremely cheaply. Today, in the B2B sector and increasingly in the B2C sector as well, the internet can be used practically free of charge for the distribution of this material, so that transportation costs, transportation sensitivities and long distances need no longer be considered barriers for the distribution of movies as a commodity.

In this respect, every movie is unique. Only through sequels and links with well-known artists can a new movie be connected with a former success. Also, movies are consumed much more on impulse and momentary preference than are material goods. Word-of mouth recommendation is vital to their success.

The costs of producing the first copy of a movie are extremely high while duplication costs are relatively minor. There is no limit to the numbers who can consume the commodity – the content of the movie – because it does not pass into the ownership of the paying spectator. As a consequence, there are strong incentives for media enterprises to draw in as large an audience as possible (Beck 2002) by spending heavily on advertising and, as a cultural product, it can be distributed to a global audience.

Local Nodes and Global Commodity Chains

During the last 10 years clusters have become an important area of research in economic geography. Porter (2000, 254) has defined a cluster as, '... a geographically proximate group of interconnected companies and associated institutions in a particular field, linked by commonalities and complementarities'. Most empirical studies have dealt with the internal structures of clusters, and the vertical and horizontal, formal and informal relationships between firms within networks that have been interpreted as creating local competitive advantage (Maskell 2001; Storper and Venables 2004; Bathelt et al. 2004). There are also advantages from labour market pooling and the local availability of intermediate inputs (Krugman 1991), while the spatial proximity and face-to-face contacts in clusters facilitate the transfer of tacit knowledge.

Up to now, few studies have dealt with the role of linkages and network relations across cluster boundaries. Bathelt et al. (2004) have argued that these external linkages and relations are necessary for knowledge creation with a successful cluster depending on strong pipelines to build and co-ordinate these external linkages. Essential to this process is the ability of local actors to absorb and to convert this knowledge into profits. At the same time, outside knowledge is needed successfully to deliver a cluster's output to the market beyond its boundaries. The motion picture industry shows that powerful distribution structures are a decisive force in maintaining the dominant position of Hollywood at the global scale over the last decades (Aksoy and Robins 1992).

In economic geography, Gereffi's (1994, 1999) concept of global commodity chains has established itself as a valuable framework for analysing global network relations. According to this concept, a global commodity chain has four characteristics. Firstly, it has a clear input-output structure involving products and services linked in sequence as value-adding economic activity. Secondly, it has a distinctive spatiality or territoriality that can be either spatially concentrated or spatially dispersed. Thirdly, it has a distinctive governance structure built on authority and power relations which determine how financial, material and human resources are linked into the flow of the chain. This governance structure can assume two basic forms being either 'buyer-driven' or 'supplier-driven'. Fourthly, a chain has an institutional framework that identifies how local, national, and international conditions and policies shape the globalisation process at each stage in the chain (Gereffi 1994, 1999; Hassler 2004).

The particular firm, subsidiary or agent whose job it is to link the internal production capabilities of a cluster and external markets holds an important position (Schamp 2000). The 'impannatore' in the Third Italy, for example, has been described as a typical figure connecting the industrial district of Prato into global commodity chains (Becattini 1991). The empirical analysis of the global commodity chains for motion pictures show these 'market makers' in the context of the motion picture industry in Hollywood.

The Development Path of the Motion Picture Cluster in Hollywood

The development path of the motion picture industry in Hollywood (Mossig 2006; Scott 2005; Wilson 1998) started in 1907. The shooting of the movie 'The Count of Monte Cristo' was shifted to Santa Monica near Los Angeles because of bad weather at the original location, Chicago. Producers soon recognised the advantages of the prolonged periods of dry and sunny weather and in the subsequent years more and more film productions moved to Hollywood. Producers on the West Coast of the US soon won the favour of the public by emphasising that the actors were 'stars' while their competitors on the East Coast continued to work with mediocre actors in order to avoid the higher fees. At this time in the early twentieth century, the major studios were founded that continue to dominate the international motion picture industry today: Universal (established in 1912), Warner Bros. (1913), Paramount (1914), 20th Century Fox (1917), Columbia Tristar (1920), Walt Disney Co. (1923) and Metro-Goldwyn-Meyer (1924). Besides the seven major studios, a new group of so-called independent producers, or 'independents', was formed in the early 1980s. Because the seven large studios called themselves majors, every movie that was not produced by a major studio was considered to be a production of the independents. The independents are, therefore, a residual category. The dominance of the major studios, however, continues despite the emergence of the independents. About 75 per cent of the domestic box office revenues in the US are generated by major studio productions.

In the early 1980s, the current trend of modern blockbuster cinema arose. Such movies are produced with increasingly expensive special effects and, consequently, production budgets have risen sharply. Figure 4.1 shows that box office revenues have not been able to compensate for the increasing production budgets of a movie. As production costs could not been covered by revenues gained on the US domestic cinema market, a growing need for a consistent commercialisation on a global scale has arisen.

The unique position of Hollywood as the most important location for motion picture production worldwide can be demonstrated in terms of total production budgets available to the producers at particular locations. In 2003, US$22.46 billion were spent worldwide on the production of movies, with 65 per cent of that amount (US$14.6 billion) being spent by producers in the US. The lion's share went to producers in Hollywood because the seven large major studios and most

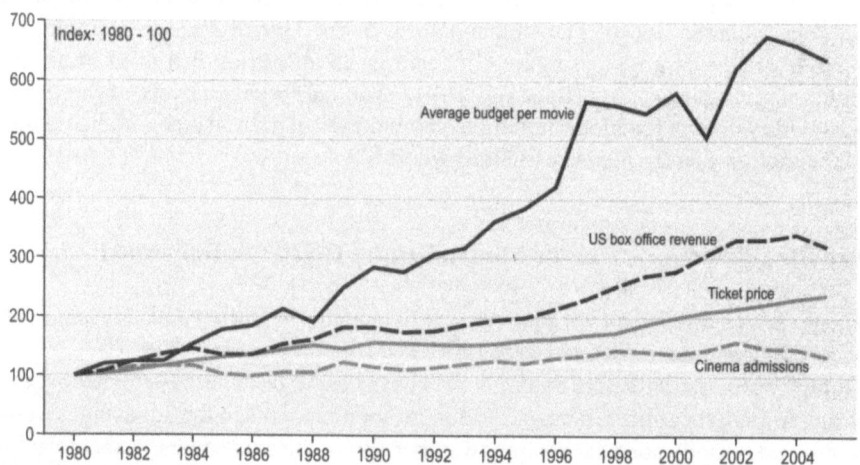

Figure 4.1   Development of production costs per movie at a major studio compared to the cinema admission, average ticket prices and box office revenues in the US 1980–2005

Source: Mossig 2006

of the US independents are located there. The oligopoly of the seven major studios in Hollywood controls a production budget of US$12.4 billion and thus they have more than half of the total budgets worldwide (55 per cent) at their disposal. The sums and shares of the other countries are very modest compared to the US (see Table 4.1).

Table 4.1 Number of movies produced and total budgets of motion picture productions in selected countries, 2003

Country	Number of movies	Total budgets (billion US $)	Average budgets (million US $)
US - major studios	194	12.38	63.8
US - independents	399	2.21	5.5
US total	593	14.59	24.6
Great Britain	175	1.90	10.83
Japan	287	1.34	4.66
France	212	1.30	6.15
Germany	107	0.85	7.96
World	4087	22.46	5.50

Source: Own calculation based on Mossig (2006), MPAA (2006).

Hollywood producers find that the additional revenues gained through successful global distribution are necessary to pay back the large capital investments that are made in high-budget movies. Over a number of decades the motion picture industry has collected increasingly more venture capital from national and international investors. According to some estimates, more than 30 per cent of the costs of film production are covered by institutional investors (*The Economist* 2007). Until recently, the major studios had not allowed financiers to invest in a single money-making movie project like the next sequel of *Harry Potter*. Instead, they have offered them packages of films that would be produced and released over the following months or years. The investors found these packages less risky than individual films. Independent producers cannot offer such packages and have to find private investors for their single film projects. They attract money for their movies, by promoting the records and reputations of the creative people they have engaged; especially the reputations of producers, directors and popular actors. But the best way for both major studios and independents to attract investors is to be high profile, successful movie makers. Reputation and revenues are the key to future financing.

The motion picture industry in Los Angeles/ Hollywood is the only place worldwide where capital investment is transferred on a large scale into the movie industry. In Germany, for example, large amounts of private capital (so called 'stupid German money') have been funnelled into specialised media funds because of former tax advantages. Experts have estimated that between US$1.5 to US$2.0 billion of capital German capital is invested each year in the Hollywood film industry. The magnitude of this capital outflow from Germany and the attractiveness of Hollywood is demonstrated by the fact that less than US$900 million is spent annually on motion picture production in Germany (Mossig 2006; Screen International 2004).

Big budgets allow new film projects to be developed and realised to meet international demand. They make it possible to engage the most popular and successful people worldwide both in front of and behind the camera (actors, directors, cameramen, producers). As a consequence, successful international distribution is much more likely, further raising the attractiveness of projects for capital investors. This is a powerful circle of capital accumulation that even flops or occasional successful foreign movies can not break.

Table 4.1 shows the differences in the average production budgets per movie in 2003. The dominance of the seven major studios and their superior financial opportunities for producing movies are obvious. But, average budgets must be interpreted with care as only a small selection of independent movies were released at the cinemas. Table 4.2 shows that only 192 movies entered the US cinemas to any significant extent[1] in the 12 months between 1 July 2003 and 30 June 2004 (Mossig 2006). There were 117 movies produced by major studios and US independents produced 55. Using box office revenues as an indicator, it is clear that the 20 movies from foreign countries outside the US did not play any role on the US cinema market (see Table 4.2).

Table 4.2   Number of movies produced, total box office revenues and box office revenues per movie made in the US, by movies released between 1 July 2003 and 30 June 2004

Origin	Number of movies	Box office revenues (US $ million)	Revenues per movie (US $ million)
Major studios	117 (60.9%)	7,213.0 (73.1%)	61.6
US-independents	55 (28.6%)	2,567.2 (26.0%)	46.7
Foreign countries	20 (10.4%)	93.2 (0.9%)	4.7
US market	192 (100%)	9,873.4 (100%)	51.4

Source: Own calculation based on Mossig (2006)

A comparison of the data in Table 4.1 and Table 4.2 shows that, especially in the independent sector, considerably more movies are produced in a year than are shown in the cinemas. Thus, the average production cost per movie listed in Table 4.1 is only significant in the case of the major studios. The budgets of independent movies are understated when expressed as a mean value because that mean also includes all the small movies with extremely low budgets that have no cinema release. A special analysis of only independent movies released in theatres calculates the average production costs of a US-independent movie at US$35.5 million (Mossig 2006). Thus, the average budgets of independent producers in

---

1 A movie had to be listed for at least one week in the TOP 30 most-watched movies in the US. In the case of the German cinema market (see Table 4.3) only the TOP 10 movies could be counted each week owing to the limitations of the published data.

the US are much lower than those of the major studios, but they have significantly more financial resources at their disposal than their international competitors in other countries.

The dominance of US motion pictures on German cinema screens is indicated in Table 4.3 by the number of movies and even more clearly when box office revenues are used as an indicator. Again, movies from foreign countries except the US do not play an important role, with a market share of only 1.6 per cent. Among the US movies, those of the seven major studios in Hollywood are dominant in Germany too.

Table 4.3   Box office revenues in Germany from movies released between 1 July 2003 and 30 June 2004, by country of origin

County of origin	Number of movies released		Box office revenues (US $ million)		Revenues per movie (US $ million)
Germany	22	(16.2%)	73.4	(8.4%)	3.3
German co-productions	4	(2.9%)	29.0	(3.3%)	7.3
Germany	26	(19.1%)	102.4	(11.7%)	3.9
US-Major Studios	71	(52.2%)	589.2	(67.6%)	8.3
US-Independents	26	(19.1%)	165.3	(19.0%)	6.4
US	97	(71.4%)	754.5	(84.6%)	7.8
Other foreign movies	13	(9.5%)	14.1	(1.6%)	1.1
German market	136	(100%)	871.0	(100%)	6.4

Source: Own calculation based on Mossig (2006)

But cinemagoers in other western industrial nations, as well as in Germany, contribute to financing of the high production and marketing costs of Hollywood movies. Cinema audiences consume either domestic productions or US movies, especially those which are expensively produced, cast with international stars and widely advertised as blockbusters by the seven major studios. In comparison to the very small domestic share in Germany, the market share of domestic movies in the other countries varies. In France, the average domestic share between 1998 and 2002 was relatively high at 32.8 per cent. In Japan it was 32.6 per cent. In Great Britain (which, along with Germany, Japan and France is one of the four most important markets outside the US) the share amounted to 17.1 per cent (Mossig 2006). As a whole, this shows how powerful international distribution permanently supports the motion picture cluster in Hollywood by earning major revenues on the international markets it dominates. Besides the investment capital that supports the superior production budgets, production opportunities and distributional power of the Hollywood movie picture industry, there are other external linkages that support this motion picture cluster. Firstly, it has an enormous appeal to international talent both in front of and behind the camera. They are attracted by the highest earnings and best working conditions and their creative potential is withdrawn from their home countries. Secondly, the powerful investment circle described above has a strong basis in the large US domestic market. Global box office revenues amounted to US$25.2 billion in 2004 with US$9.53 billion (37.8 per cent) in the US alone. This share was even higher in the preceding years: 46.7 per cent in 2003 and 48.1 per cent in 2002 (MPAA 2006). Except for a few isolated successes, motion pictures produced in a language other than English have a considerable disadvantage in recouping their production costs through box offices revenues, and thus for further commercialisation, because of the lack of a comparably large and financially potent domestic market. Available production budgets are as a rule significantly smaller.

Co-ordination of Distribution Structures: The Global Commodity Chain of the Motion Picture Industry in Hollywood

To explore how the advantages of spatial proximity within a cluster are tied in with global distribution linkages, the global commodity chain of the motion picture industry and its connections to the German market have been analysed as a case study. In its simplest form, the global commodity chain for movie distribution can be subdivided into the following sequence of relationships: from (1) the movie production company, to (2) the global distributor, to (3) the local distributor, to (4) cinemas, DVD/video/TV. The data for this analysis have been gathered through 27 guided interviews with experts working at the interfaces between Hollywood and the German market, 22 in Los Angeles, and five in Munich and Berlin in Germany.

In the case of the seven major studios the first three steps of the commodity chain are transacted in their entirety by subsidiaries of the respective media enterprise. Therefore, the important market makers whose job it is to make and have connections to external markets are part of the major studios. Thus, the interaction processes needed to overcome the spatial, social and cultural distances take place within the directive structures of these firms. Processes of negotiation and interaction beyond the boundaries of the enterprise (which requires face-to-face contact as well as relations of trust and reciprocity between the actors engaged and, according to the arguments of the recent cluster debate, are supported by spatial proximity) occur only at the interface between the local distribution subsidiary of the major studio and the cinema groups in the respective countries. The major studios benefit from their integration into the largest media enterprises worldwide by exploiting the further opportunities for realising a cross-media commercialisation of the movie content through other subsidiaries of the media enterprise.

Every distribution unit of the major studios has an average of 15-20 movies each year at German cinema screens including the eagerly-awaited blockbuster movies which are extensively advertised before their release, films that are sequels of former successes, or filmings of internationally successful novels. Furnished with this attractive basket of goods, the local distribution units of the major studios are powerful actors in negotiations with local cinema groups or TV networks because they are in possession of the goods the cinema groups and TV networks need for their continued commercial existence.

In contrast, the global commodity chain for the distribution of independent movies is connected in a quite different way because the independent producers do not have their own global distribution networks with subsidiaries in many countries. For these movie makers, every link of the global commodity chain consists of specialised firms and for each movie the stages of movie production, global distribution and local distribution have to be connected anew in order to sell an independent movie from Hollywood into foreign markets. Owing to the immense surplus of independent movies, the first hurdle for an independent movie producer is to find a global distributor to represent the movie professionally through its contacts on the global market. These global distributors are the most important market makers for producers of independent movies. Good global distributors know the local distributors in different countries and regions and the preferences of the local distributors and cinemagoers. They have the specialist distribution knowledge that independent movie producers do not have.

To develop this knowledge, interviewees suggested that face-to-face contacts and personal communication and consultation are decisive, just as they are within clusters. In contrast to the clusters, however, co-ordination is more complicated because of the absence of spatial proximity as well as the lack of opportunities for 'chance encounters'. Instead, personal meetings and important formal and informal exchanges of information take place at film markets and film festivals, especially the two-week Film Festival and Film Market in Cannes which takes place annually in May, and the American Film Market (AFM) in Beverly Hills/ Los Angeles each November. All those active in the international motion picture business gather at these events and transform Cannes or Beverly Hills into a temporary one- or two-week long international motion picture cluster. There are other festivals and film markets in addition to these which act as further meeting points, such as the 'Berlinale' in Berlin in February and the festivals in Toronto and Venice, both in September.

Interviewees emphasised that these well-attended meetings are extremely important for them. The programme of events at the festivals and markets are scheduled very tightly compared to the day-to-day meetings in a cluster. Some of those interviewed have reported that during festival weeks they have meetings every twenty to thirty minutes from 9.00 am until late evening, daily, in order to meet as many business partners as possible. These film festivals and the associated film markets overcome the problems of distance and act as dense pools of information.

Conclusion

The example of the motion picture industry explored in this chapter has shown that the existence of a cluster depends not only on the internal structures, but is also decisively supported by a powerful global distribution system. It has been shown that the seven major studios with their headquarters in Hollywood dominate the financially strong international markets for motion pictures. The major studios are part of the largest media enterprises worldwide and co-ordinate the distribution of their movies to different countries through wholly-owned subsidiaries. Thus, the important market makers, the distributors, are parts of these firms. Here the established firms have an advantage by having global networks and legions of marketing men. Outside this dominant oligopoly, a huge number of independent producers compete for the remaining market share. The individual links of the global commodity chain for the distribution of independent movies are outsourced to specialised firms. The links of the chain have to be newly connected for every movie. Due to their experience, knowledge and contacts with local counterparts in their respective countries, global distributor companies are the crucial market makers for independent movies. They possess the decisive position for overcoming the distances between the independent producers located in Hollywood and the foreign markets. Personal contacts and informal information are very important for the co-ordination of such decentralised distribution systems. But in contrast to a cluster, the economic actors do not have the advantage of spatial proximity which supports the important communication processes. International film festivals and film markets have the important function of being meeting points for the building up and cultivating of personal contacts. They are temporary clusters of the international motion picture industry.

References

Aksoy, A. and Robins, K. (1992), 'Hollywood for the 21st Century: Global Competition for Critical Mass in Image Markets', *Cambridge Journal of Economics* 16, 1-22.

Bathelt, H., Malmberg, A. and Maskell, P. (2004), 'Clusters and Knowledge: Local Buzz, Global Pipelines and the Process of Knowledge Creation', *Progress in Human Geography* 28:1, 31-56.

Becattini, G. (1991), 'The Industrial District as a Creative Milieu', in Benko, G. and Dunford, M. (eds), *Industrial Change and Regional Development: The Transformation of New Industrial Spaces* (London: Belhaven), 102-114.

Beck, H. (2002), *Medienökonomie. Print, Fernsehen und Multimedia* (Berlin, Heidelberg, New York: Springer).

Gereffi, G. (1994), 'The Organisation of Buyer-Driven Global Commodity Chains: How U.S. Retailers Shape Overseas Production Networks', in Gereffi, G. and

Korzeniewicz, M. (eds), *Commodity Chains and Global Capitalism* (Westport: Greenwood*)*, 95-122.

Gereffi, G. (1999), 'International Trade and Industrial Upgrading in the Apparel Commodity Chain', *Journal of International Economics* 48, 37-70.

Hassler, M. (2004), 'Raw Material Procurement, Industrial Upgrading and Labour Recruitment: Intermediaries in Indonesia's Clothing Industry', *Geoforum* 35 441-451.

Krugman, P. (1991), *Geography and Trade* (Leuven: Leuven University Press and Cambridge, Mass., London: The MIT Press).

Maskell, P. (2001), 'Towards a Knowledge-based Theory of the Geographical Cluster', *Industrial and Corporate Change* 10 921-943.

Mossig, I. (2006), *Netzwerke der Kulturökonomie. Lokale Knoten und globale Verflechtungen der Film- und Fernsehindustrie in Deutschland und den USA* (Bielefeld: transcript).

MPAA (2006), *U.S. Theatrical Market: 2005 Statistics* <http://www.mpaa.org>

Porter, M.E. (1990), *The Competitive Advantage of Nations* (New York: Free Press).

Porter, M.E. (2000), 'Locations, Clusters, and Company Strategy', in Clark, G., Feldman, M.P., Gertler, M.S. (eds), *The Oxford Handbook of Economic Geography* (Oxford: Oxford University Press), 253-274.

Power, D. (2003), 'The Nordic "Cultural Industries": A Cross-national Assessment of the Cultural Industries in Denmark, Finland, Norway and Sweden', *Geografiska Annaler B* 85, 167-180.

Schamp, E.W. (2000), *Vernetzte Produktion. Industriegeographie aus institutioneller Perspektive* (Darmstadt: Wissenschaftliche Buchgesellschaft).

Scott, A.J. (2000), *The Cultural Economy of Cities. Essays on the Geography of Image-Producing Industries* (London, Thousand Oaks, New Delhi: Sage).

Scott, A.J. (2005), *On Hollywood. The Place, the Industry* (Princeton, Oxford: Princeton University Press).

Screen International (2004): *Hollywood's Indie Matchmakers*, 25-28 (7 May 2004).

Storper, M. and Venables, A.J. (2004), 'Buzz: Face-to-face Contact and the Urban Economy', *Journal of Economic Geography* 4, 351-370.

The Economist (2007), *Hollywood's New Model*, 81-82 (15 March 2007).

Wilson, J.M. (1998), *Inside Hollywood. A Writer's Guide to Researching the World of Movies and TV* (Cincinnati: Writer's Digest Books).

Chapter 5

Foreign Direct Investments in Development Strategies: Norwegian FDI and the Tendency for Agglomeration

Arnt Fløysand and Håvard Haarstad

Introduction

With globalisation and increased flows of private capital across borders, major development institutions argue that foreign direct investment (FDI) can be a significant tool for poverty reduction. Their recommendation is that poor countries should restrict government intervention to providing the basic governance to improve the 'investment climate', and in turn gain spillovers and other benefits for their development. However, experience with the uneven patterns of FDI indicate that investments flow in relatively resilient paths and networks, and that the relationships between good investment climates and investment flows are probably more complex than development strategists are willing to admit. Governments are advised to avoid policies that interfere with the interests of investors, which tie their hands and prevent them from pursuing alternative policies even though the potential for most poor countries to attract significant amounts of FDI is unproven. This study interviewed in-depth Norwegian investment decision-makers with FDI in developing countries to uncover the motivations behind these investments. When looking at the motivations behind investment decisions, it does not appear likely that the poorest countries in the world can attract significant amounts of FDI.

FDI as a Tool for Poverty Reduction

The main message of the World Bank's annual *World Development Report* for 2005 is that the investment climate in a country is central to growth and poverty reduction. The fundamental role of governments is to 'deliver the basics', such as securing property rights, contract enforcement, fostering a skilled workforce and cushioning workers against economic change. Government interventions 'beyond the basics' are to be treated with caution, 'and should not be viewed as a substitute for broader investment climate improvements'. The Bank's advice, then, is that governments refrain from policies that interfere with the market and instead remove trade protection and related 'distortions' (World Bank 2004, 9-14). The investment climate perspective puts the actors making the investments, the firm, at the centre of discussion and policy recommendations, while governments limit their role to providing the conditions that may attract investment decision-makers to invest in their economy.

Similarly, the 'Monterrey Consensus' arrived at during the United Nations Conference on Financing for Development in 2002, concluded that creating a good investment climate is a critical role for governments (UN 2002). To achieve increased capital flows, the Consensus prescribes a transparent, stable and predictable investment climate, with proper contract enforcement and respect for property rights. UNCTAD, in the *World Investment Report 2003* on 'FDI policies for Development: National and International Perspectives' stated that national policies are key for attracting FDI, increasing benefits from it and assuaging concerns about it (UNCTAD 2003, 86). Governments should provide a basic regulatory framework, basic social provisions (such as infrastructure) and an educated labour force. They should then refrain from further active policy making and providing welfare to their citizens.

There seems to be little dispute that FDI can influence significant poverty reduction in the economies that receive it. This does not automatically mean that creating a favourable investment climate should be the primary concern for governments of poor countries. There is reason to question whether reliance upon market mechanisms will attract FDI flows to under-financed economies to the degree necessary to justify the central position of FDI in poverty reduction strategies. As Schultz (2001, 107) notes, 'if good governance, democracy and transparency were prime determinants of private capital flows, China would certainly not rank in first place'.

The Geography of FDI: Agglomeration and 'the Triad'

A look at current FDI patterns reveals that investments are heavily concentrated in developed and high-growth regions. Aggregate data show enormous changes in the amounts and the pattern of capital flows from industrial countries to emerging economies in the 1980s and 1990s. UNCTAD figures indicate that the inflows of FDI to all developing countries increased from US$3.8 billion in 1970 to more than US$77 billion in 1993, and to just over US$103 billion in 1994. Between 1995 and 2000 these amounts took a quantum leap to US$254 billion, and in 2005 inflows to developing countries totalled US$320 billion.

However, the tendency over the past decades is for the poorest countries under the 'developing country' category to receive less, the richest to receive more, relative to total world FDI. While all regions have seen growth, this growth is widely asymmetrical, not only between 'developed' and 'developing' countries, but

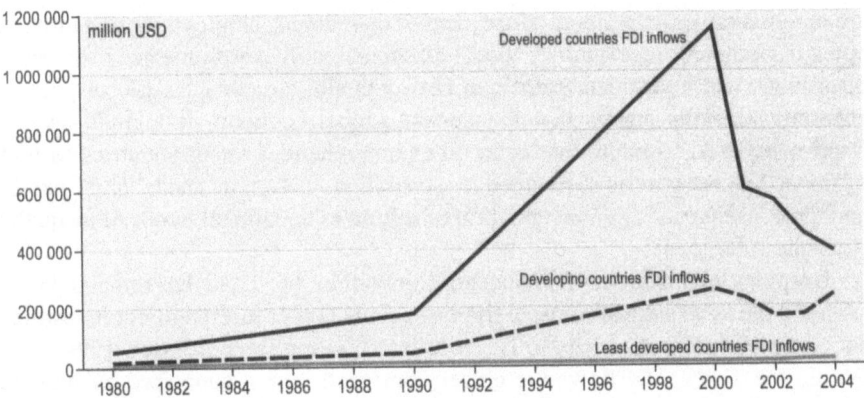

Figure 5.1   FDI inflows to developed, developing and least developed countries between 1980 and 2004

Source: UNCTAD

also within the category of 'developing countries'. In 2005, developing countries representing four fifths of the world's population received 35 per cent of total FDI flows (UNCTAD Statistics Database). Figure 5.1 shows FDI inflows to developed, developing and least developed countries between 1980 and 2004.

By 'deconstructing' the developing country category, imbalances between regions of the developing world emerge. In 2000, fifteen countries accounted for over 80 per cent of FDI to developing countries, while the 49 least developed countries received only 0.3 per cent (UNCTAD 2001, xiii). In 2005, Africa's inflows had shot up to US$31 billion, but the region's share of global flows remained at merely 3 per cent (UNCTAD 2006, xix). Sub-Saharan Africa, with 10 per cent of the world's population, according to the UN Population Division, received only 2.2 per cent of total FDI inflows in 2004. Though Africa's share of global inflows has risen somewhat in recent years, the continent's share remains much lower than it was in the 1970s and 1980s, when the Sub-Saharan portion of FDI to total world inflows peaked with 5.5 per cent in 1974. It can be argued that relative figures hide important development trends since absolute increases in investments can be helpful for poor countries regardless of the broader picture. The purpose here is to show that the universal FDI trends should not be assumed to be reproduced in poorer countries.

The explosion of FDI is seen as a symptom of globalisation, an indication that the world is now borderless and that flows of capital and goods are no longer geographically constrained. But there are structural asymmetries in the global economy apparent in the flows of FDI. As Amin argues, rather than melting everything that is solid into air, globalisation signals the rise of new spatio-temporalities instead of the eradication of existing spatio-temporalities. This affects what goes on in places and redraws rather than erases institutional boundaries (Amin 2002). Capital is still embedded in space and most of the largest multinational corporations are regional rather than global, since most of their assets, employment and turnover come from their home countries. Local differences still exert influence over capital operations, and investment paths and networks still structure the global political economy. Castells argues that the current global economy is a multi-layered, multi-networked structure that cannot be comprehended using countries as units of trade. The network is segmented by separations between 'inside' and 'outside' and the consequence is a concentration of resources and global trends of inequality (Castells 2000, 134).

Despite globalisation, FDI concentration within 'the Triad' has remained high. 'Among the key features of these triadic regions is that foreign direct investment and international trade tends to be "localised" within them' (Jessop 1994, 270). Within the Triad, FDI flows are closely mirrored in and supported by bilateral trade agreements and double taxation treaties. Triad members form bilateral trade agreements with their most important partners to protect their investments, which further strengthens the Triad investment pattern. And, even though investments between developing countries (South – South) have been increasing, particularly from developing countries in Asia to Africa (UNCTAD 2006, 43), South – South investments have also tended to be concentrated in the region of origin (World Bank 2004).

The paradigmatic development strategy of attracting FDI assumes that these structural asymmetries can be significantly redistributed by the creation of a favourable investment climate. This, in turn, assumes that market forces will significantly attract FDI flows to these countries should a favourable investment climate be achieved. An actor-centred focus on investment motivations can shed light on the actions that make up patterns on the macro scale, tell us something about why investments flow where they flow, and in turn whether FDI can potentially be attracted to poor countries. As the following section shows, theoretical insights into investment motivations are more complex than dominant development strategies take into account.

Explaining FDI Patterns: What Motivates Investors?

FDI Motivations in Theory

Traditionally, theory on FDI has seen foreign engagements of companies in terms of motivation to gain strategic positioning and access to resources. Hymer (1976) viewed the internalisation of a firm's activities as a means of increasing its market power. Inspired by transaction cost theory, other studies have explained FDI as a consequence of multinational firms preferring internalisation of transactions within the firm instead of trade or licensing (Hennart 1982). Dunning (1981, 1988) combined the market power and the transaction cost approach when explaining international production. In Dunning's model, three conditions must be met for FDI to occur: ownership advantages; FDI must be preferred over trade or licensing; and there must be location advantages. Lim (2001), in his summary of recent literature, lists incentives that may be important in affecting FDI inflows; economic distance/transaction costs, size of host market, agglomeration effects, factor cost, fiscal incentives, business/investment climate, and openness to trade. Other factors important in attracting FDI were agglomeration/clustering, low labour costs, market size, and FDI policies (Lim 2001).

These general conditions help us to explain why foreign investments occur. But we are particularly interested in why FDI occurs at one site and not at another, and how these actions create resilient paths and networks. Some studies conclude that market size, liberalised FDI, infrastructure or human capital are determinants of FDI (Noorbakash et al. 2001; Zhang 2001). Actor-centred perspectives on motivations behind FDI separate these into three main categories; market seeking, resource seeking and efficiency seeking. Organisational theory introduces a fourth objective: strategic interaction in relation to rivals. The decision of a company to locate somewhere will likely be a compound of these factors.

Most of the empirical analyses of the determinants of FDI use cross-country regressions to identify the characteristics of the countries attracting FDI (Aseidu 2002), but correlations between isolated factors do not reflect the real dynamics and causalities behind FDI flows. There is a need to supplement this with qualitative studies that aim to uncover perspectives missed by statistical analyses. In order to understand market driven patterns of FDI, exploring the motivations of individual investment decision-makers can be helpful. Findings from the actor-centred approach should be expected to forecast macroeconomic patterns, since these countless investment decisions make up the global distribution of FDI. This will be illustrated by examining Norwegian FDI and the pattern it forms.

Patterns of Norwegian FDI

Norwegian investments represent a small portion of global FDI flows. Although they have increased sharply in the short term, they represented 0.7 per cent of world total stock in 2004. The observed pattern demonstrates that Norwegian FDI is concentrated in a few sectors and regions and, as with most FDI, almost bypasses Sub-Saharan Africa. Developed countries received 76 per cent of total Norwegian FDI inflows in 2004 (Statistics Norway – most recent available data). Despite characteristics of the Norwegian economy that increases FDI to South and Central America, the main investment patterns of the global FDI flows are reflected in the Norwegian patterns. Investments are heavily concentrated in some regions, while others receive almost no investments. FDI to Sub-Saharan Africa has typically not exceeded one per cent, but reached almost five per cent in 2004 thanks to investments in Angola and Liberia. Like global flows, Norwegian patterns show a significant increase in FDI to South, East and Southeast Asia (Statistics Norway). The stock of Norwegian FDI in some selected regions is illustrated in Figure 5.2.

Divergence in Norwegian investment patterns from global patterns can largely be explained in terms of the specific context of the Norwegian economy. Firstly, there is a heavy concentration of investment in the petroleum industry which reflects offshore competencies. Companies such as Statoil have developed competencies in deep sea drilling as a result of their experience off the Norwegian coast. This developed in partnerships with foreign firms making FDI in Norway, and the Norwegian companies now apply these competencies in FDI projects off the coast of Brazil, Nigeria and Azerbaijan. Secondly, Norwegian companies have particular competencies in fish farming because of the favourable fish farming conditions in the fjords. This has resulted in an increasing presence in places with similar conditions, for example Chile. Third, statistics show heavy 'investments' in tax havens such as Cayman Islands and Liberia, which are likely to originate in the historically strong Norwegian shipping industry. Since Norwegian investments largely mirror global patterns, and the divergences can be explained by the specific characteristics of the Norwegian economy, there is little reason to believe that the motivations of Norwegian investors should be much different from those of other countries.

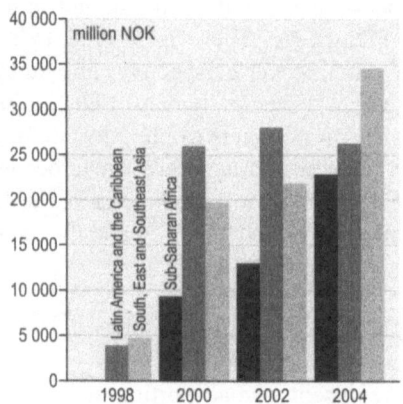

Figure 5.2   Norwegian FDI stock in selected regions, 1998–2004
Source: Statistics Norway

Motives behind Norwegian FDI

In order to survey the motivations behind Norwegian investments in developing countries, we conducted a series of in-depth interviews with Norwegian corporate leaders. For the purpose of controlling for the representativeness of informants, we gathered a list of 91 companies that had at least one investment project in a developing country. These were divided into regions and sectors in order to create a sample covering all regions and sectors. The survey consists of 25 qualitative in-depth interviews with corporate leaders from companies with green field and mergers and acquisitions (M&A) FDI distributed through South America, Africa, and Asia in a range of sectors.

The main motivating factors for FDI are usually separated into access to resources, efficiency seeking, positioning in growing markets and others. While simple categorisation is somewhat reductive, it gives a significant indication of the investment motivations of the interviewees. Market-seeking was the dominant factor, as 13 (more than half of the interviewees) can be categorised as such. Efficiency-seeking describes the predominant motivation for four of these, resource-seeking for three and 'other' for five. However, the motivations behind an investment were usually compounds of these categorisations. For the companies that were not attracted to a location because of access to resources, strategic positioning in relation to competitors and perceived future market growth was an important factor for nearly all interviewees. One interviewee from a communications technology company investing in India said:

> There were several [reasons]. First of all the market in India was attractive, and it was interesting for us to get access there. There are several Asian countries that we are trying to get access to, and it seemed that India was a good place to start. So we had decided to try to gain a position in Asia, and India was interesting for that reason. Then there was the access to production and engineering resources at a low costs that was interesting. But the market was the main reason.

The company entered by forming a joint-venture where they owned 51 per cent and Indians owned 49 per cent, but co-operation failed and the Norwegian company acquired the entire business. Another company, investing in a paint factory in China, had previous experience of the region through exports to Thailand and Singapore, but decided to invest in a factory in China to gain a position selling paint to the growing ship building market:

> China is important in shipbuilding, there is huge activity. To grow we had to get established there. Today they are number three in the world in shipbuilding, in a few years they will be number two, and eventually number one. It was the market that attracted us there. It will be a huge factory.

A company investing in India said labour costs were favourable, but the main reason for investing was the growing Indian market:

> We wanted to access the best market in the world. It is the second largest country in the world. The relocation is done through a subsidiary in England. They had some experience from having India as a colony, I guess. No one in [headquarters] had any experience, but we got in. The growth in India came a bit later than anticipated, but it is good now. We also found a partner in India; without that we would not have entered on our own.

Motivations to access growing markets were in several cases compounded with the locations of resources, particularly within resources where Norwegian companies have special experience from the Norwegian context. Gaining access to resources in developing countries can be a strategic positioning in relation to other Norwegian firms:

> Chile has a dominant position in fish farming, we wanted to be a part of that. There are four fish farming regions in the world, and Chile is about to become really huge. We knew Chile from our experience within the sector.

It was apparent from the interviews that Norwegian firms investing in developing countries pay close attention to growing markets and projected growth, and that perceptions of projected growth have significant influence over investment decisions. The terms 'getting in on' and 'being a part of' were frequently used to describe investment motivations. As the fish farming companies investing in Chile illustrate, even within resource extracting industries, strategic location and expected market growth are important factors behind investment decisions. Norwegian resource-based investments tend to follow particular competence and experience gained from the Norwegian context, such as fish farming, into global strategies towards both Norwegian and international firms.

Efficiency-seeking investments are slightly different, in that investment strategies are not necessarily directed at competitors and growing markets, but at cost saving measures through reducing transport and labour costs or getting behind tariff barriers. This has traditionally been seen as an important factor when foreign firms invest in less developed countries by setting up production facilities for exports to other markets. A cleaning product company had a sales office in Indonesia and sold products to Malaysia for five years before deciding to establish a subsidiary in Malaysia to avoid crossing tariff barriers. The Indonesian subsidiary was closed and, instead, the Malaysian subsidiary distributes to other countries in the region. Low labour costs were described as a dominant reason for investing by only one company that operates in China.

Personal connections and personal knowledge of an area, often gained through other activities, were important motivating factors. Several interviewees reported that they had being contacted directly by a representative of a country and invited to invest. There are several examples of coincidental familiarity of an area from employees who had previously worked for other companies in the area, or friends that recommend a deal which eventually resulted in an investment project. These factors are often ignored by economic theory, particularly by theory on FDI, since it generally deals with large-scale strategies and big firms. In the Norwegian case, and with the smaller companies that were included in our survey, these factors are important for investments in developing countries. In the interviews we sought with firms investing in Sub-Saharan Africa, the reason that they had invested there were either personal (one wanted to 'do something for the Third World'), resulting from contracting through Norwegian aid projects or resource seeking. The aid related investments had been contracted to build infrastructure by the Norwegian aid organisation NORAD and continued their presence because of the experience they had gained in NORAD operations. In other words, the investments that did end up in Sub-Saharan Africa were not motivated purely by economic concerns.

FDI and the Tendency for Geographical Agglomeration

The interviews with Norwegian investment decision-makers showed that *strategic location* in growing markets is the dominant motivating factor behind Norwegian FDI. This involved both strategic interaction with competitors and the desire to be a part of growing markets – often two sides of the same coin. Our study indicates that the actual investment decisions of internationalising firms are mainly strategic, as the classical FDI literature asserts. Actor-centred motivations forecast patterns of investments at the macro scale. Hence, the heavy concentration of FDI in some regions and investment paths along the paths of the Triad structure of trade and trade agreements within the world economy is not surprising. While factors such as human capital can certainly make a difference in a company's decision to invest in one country or another, investment decision-makers seem to be mainly looking for high growth economies. In other words, FDI is more likely to flow to regions where there is already a concentration of investments. Due to the importance of investment decisions relating to strategic positioning within growing markets, and in line with what classical theory calls 'strategic interaction', we conclude that FDI has an inherent *tendency for agglomeration*. FDI is likely to agglomerate despite improved 'investment climates' in poor countries, because of the importance of strategic location.

The importance of strategic location for internationalising firms, and in turn the tendency for agglomeration, is firmly embedded in the theory of FDI, but seems conveniently forgotten when broader development strategies are shaped. The development strategies of major multilateral organisations, such as the World Bank, assume that through creating a favourable investment climate, and limited government delivering basic services, developing countries can attract FDI and affect poverty reduction. FDI can certainly provide important resources for development. The problem is that 'basic governance' limits the ability of governments of poor countries to attempt more active interventions. If governments should stick to 'providing basic services', this ties their hands in attempting other solutions to provide welfare for their citizens, while their potential to attract significant amounts of FDI seems low.

Acknowledgements

This study was financially supported by the Meltzer Foundation and the Research Council of Norway.

References

Amin, A. (2002), 'Spatialities of Globalisation', *Environment and Planning A* 34, 385-399.

Aseidu, E. (2002), 'On the Determinants of Foreign Direct Investments to Developing Countries: Is Africa Different?' *World Development* 30:1, 107-119.

Castells, M. (2000), *The Rise of the Network Society* (Oxford: Blackwell).

Dunning, J.H. (1981), International Production and the Multinational Enterprise (London: Allen & Unwin).

Dunning, J.H. (1988), 'The Eclectic Paradigm of International Production. A Restatement and Some Possible Extensions', *Journal of International Business* 19, 1-31.

Hennart, J.F. (1982), *A Theory of Multinational Enterprise* (Ann Arbor: University of Michigan Press).

Hymer, S. (1976), *International Operation of National Firms: A Study of Foreign Direct Investment* (Cambridge: MIT Press).

Jessop, B. (1994). 'Post-Fordism and the State', in A. Amin (ed.) *Post-Fordism. A Reader* (Oxford: Blackwell), 251-279.

Lim, E.G. (2001). *Determinants of, and the Relation Between, Foreign Direct Investment and Growth: A Summary of the Recent Literature*. IMF Working Paper, WP/01/175, International Monetary Fund.

Noorbakash, F., Paloni, A. et al. (2001), 'Human Capital and FDI inflows to Developing Countries: New Empirical Evidence', *World Development* 29:9, 1593-1610.

Schultz, B. (2001), 'Poverty and Development in the Age of Globalisation: The Role of Foreign Aid', in F. Wilson, N. Kanji and E. Braathen (eds), *Poverty Reduction: What Role for the State in Today's Globalised Economy?* (London: Zed Books/CROP), 95-109.

UN (2002), *Report of the International Conference on Financing for Development. Monterrey, Mexico, 18-22 March 2002* (New York: United Nations).

UNCTAD (2001), *World Investment Report 2001. Promoting Linkages* (New York and Geneva: United Nations Conference on Trade and Development).

UNCTAD (2003), *World Investment Report 2003. FDI Policies for Development: National and International Perspectives* (New York and Geneva: United Nations Conference on Trade and Development).

UNCTAD (2006), *World Investment Report 2006* (New York and Geneva: United Nations Conference on Trade and Development).

World Bank (2004), *World Development Report 2005* (New York, The World Bank: Oxford University Press).

Zhang, K.H. (2001), 'What Attracts Foreign Multinational Corporations to China?' *Contemporary Economic Policy* 19:3, 336-346.

Chapter 6

Multinational Investment
in UK Regions

Dimitra Dimitropoulou, Simon Burke and Philip McCann

Introduction

The aim of this chapter is to examine the factors that determine the location of foreign direct investment (FDI) projects in UK regions. The chapter attempts to throw light on these issues by employing a dataset on inward FDI, disaggregated to the level of both the individual firm and region, and also disaggregated by manufacturing and service FDI. The motivation for this work is that a better empirical knowledge of the forces which drive the location behaviour of FDI will help understanding of the links between FDI and regional development. Unfortunately, however, previous empirical analyses of UK inward FDI have tended to be based on aggregated data (Hill and Munday 1994). This has limited our ability to understand the micro-determinants of the regional geography of inward FDI. At the same time, where more disaggregated data has been available, the analyses have tended to be based primarily on observations of manufacturing industry (Jones and Wren 2006). Location theory, however, suggests that manufacturing and service industries will exhibit quite different types of location behaviour. The result of this is that our ability to make more general statements regarding the determinants of the location behaviour of FDI on the basis of manufacturing observations alone has previously been rather limited. This chapter aims to improve on the current situation by providing evidence on the links between FDI location behaviour and the characteristics of regions across all types of industries.

The chapter is organised in six sections. Following the introduction, Section 2 outlines the recent patterns of inward FDI in UK regions, demonstrating that there has been a relative shift in favour of London and the southern and eastern regions of the UK. Section 3 discusses insights from location theory and concludes that the most appropriate approach, wherever possible, is to analyse the characteristics of both the firm and the region within the same analytical and empirical framework. Section 4 discusses the dataset and methodology employed and section 5 presents the results of the analyses. Section 6 provides a discussion of these findings and draws conclusions.

FDI and UK Regions

The global stocks of both inward and outward foreign direct investment (FDI) activity have continued to rise annually (OECD 2006) and these trends have also been mirrored in terms of inward FDI stocks amongst UK regions. Indeed, for the previous two decades up until 2002, when Germany first outperformed the UK as a location for FDI, the UK had consistently exhibited the largest stocks of inward FDI in the EU, and the UK still has by far the largest FDI inflows relative to the size of the economy for any of the major European economies (OECD 2006). In terms of flows of FDI, the UK has been annually recording between 800 and 1000 inward investment projects in each of the years 2000 to 2004 (UKTI 2004), and since 2004 UK inward FDI flows have once again emerged as being by far the largest in Europe. The evidence also suggests that these investments have had significant impacts on the performance of the UK's regions, and that immigrant manufacturing FDI in particular, has been very important in the regions outside the South East (Hill and Munday 1994).

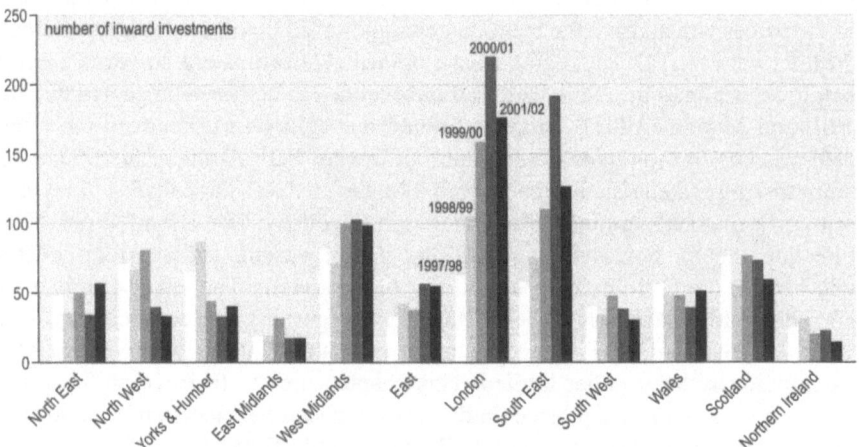

Figure 6.1   Total inward investment projects in UK regions

Figure 6.1 shows that London and the South East are not only the two regions which have consistently received the largest number of inward investment projects, but also these regions have recently generally experienced increasing inward FDI. This has also been the case in the East region and the West Midlands, although the total number of projects is very much lower in these regions than in either London or the South East. As such, the trajectory of the bar columns in Figure 6.1 suggest many of the regions which are more geographically peripheral with respect to London have recently been experiencing falling inward FDI relative to the more central locations. Because of these scale differences, the implication is therefore that both London and the South East have become relatively more important as locations for inward FDI investments, in comparison with the regions outside the South East.

A more careful examination of the data, however, suggests that there have been some subtle shifts in both the nature and spatial patterns of inward investment in UK regions. When, as in Figure 6.2 and Figure 6.3, inward investment is divided into two broad categories – manufacturing and non-manufacturing investments – it is clear that the number of manufacturing FDI projects is very low in London, the East and the East Midlands. On the other hand, some of the regions which have previously experienced large numbers of inward FDI projects, such as Scotland, the North West, and Yorkshire and Humberside, have recently seen significant reductions in the numbers of such projects. At the same time, while the number of inward FDI projects in the West Midlands, East Midlands and Wales, have remained largely stable, in the southern and eastern regions, the number of manufacturing investment projects has remained either largely stable, or has even increased. As such, the relative falls in inward FDI manufacturing projects have tended to be most noticeable in many of the regions which are geographically peripheral from London and the South East. Therefore, even though the actual numbers of inward manufacturing investment projects in London and the South East are lower than in many of the other more peripheral regions, the trend in London and the South East is rather different from that in many of the other regions. The result of this is that over the last few years, the southern and eastern regions have tended to figure relatively more prominently in the UK as bases for inward manufacturing FDI.

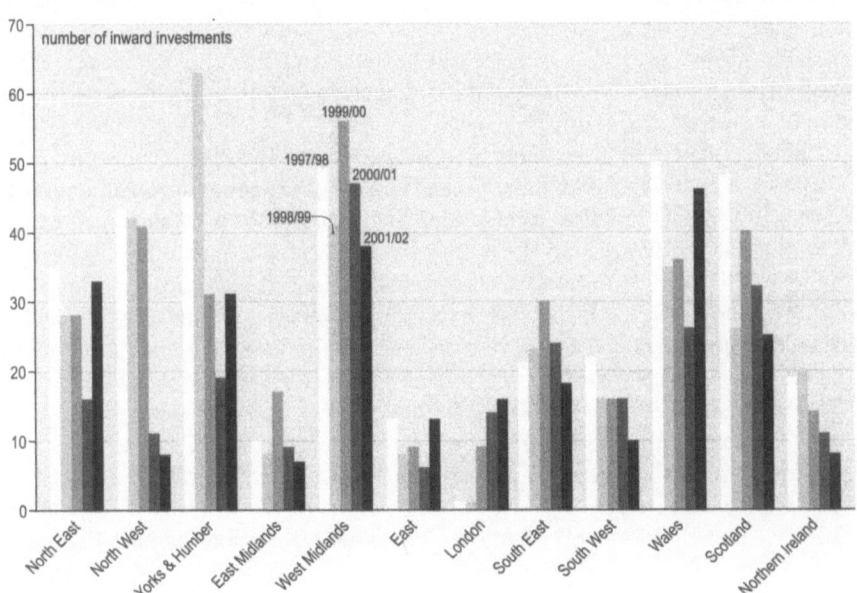

Figure 6.2   Manufacturing FDI projects in UK regions

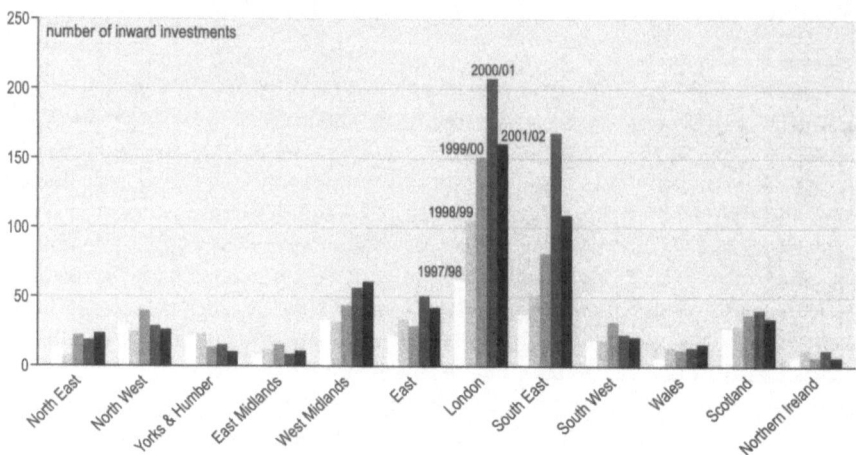

Figure 6.3   Non-manufacturing FDI projects in UK regions

A rather different pattern emerges for inflows of service FDI in UK regions. As is shown in Figure 6.3, in the non-manufacturing service industries there has been a steady increase in both the total number of inward investments in the UK[1] as well as in the relative dominance of the southern and eastern regions. This general trend towards an increasing share of service industry activity amongst inward FDI projects in many ways reflects the general domestic structural changes taking place within the UK economy as a whole (DTI 2001). However, what is particularly noticeable from Figure 6.3, is the relative strength of these restructuring effects regarding inward FDI. There is a generally increasing relative bias towards London and the South East as locations for service inward FDI. Not only are these regions experiencing growth in these types of projects, but the absolute numbers involved are very large relative to all other regions. While the number of non-manufacturing inward FDI projects has also increased in the East, the West Midlands, Scotland, the North East and Wales, and has remained largely stable in all other regions, the absolute numbers of projects in each of these regions is in comparison relatively very low. These trends in the UK's regional patterns of inward service FDI appear, therefore, to be even more marked than the general trends in structural transformation which are taking place in all UK regions (McCann 2005).

Taken together, the total effect of these inward FDI changes appears to be a significant and continuous fall in the relative importance of the other English regions as bases for inward FDI projects, and a significant increase in the relative importance of the south eastern regions as locations for inward FDI. If the apparent relative weakening of inward FDI flows into other UK regions persists in the long

---

1 Except for the most recent year for which data is available (2000/01), which is when both the dot-com collapse and the 9/11 disaster occurred. However, the total value of inward FDI in the UK increased during this period.

run, this will create significant problems for policy-makers. For many years, the encouragement of inward FDI has been a major policy for countering the relative under-performance of many of the non southern and eastern UK regions. However, it may be that the very reasons why some regions are lagging behind others, are themselves part of the explanation as to why many multinational enterprises now prefer to invest in London and the South East. Therefore, if the lagging regions, which for a variety of reasons appear unable to generate sufficient levels of domestic economic activity, are also the same regions which are increasingly unable to generate sufficient levels of compensating inward FDI flows, then the problems of uneven regional development in the UK may be further exacerbated.[2]

While raw FDI data show the emergence of new patterns of investment in UK regions, they do not explain them. However, the various strands of FDI location theory offer insights into the processes involved. This is the topic of the next section.

The Geography of Foreign Direct Investment

The analysis of the relationship between the competitive behaviour of multinational enterprises (MNEs) and the resulting spatial patterns of their FDI, has previously been undertaken in three quite distinct fields; international business management, new international trade theory, and traditional economic geography.

In international business literature (Vernon 1966; Johanson and Vahlne 1977), the location of FDI has traditionally been analysed within the framework of Dunning's (1977) 'eclectic' paradigm, in which multinational enterprise is viewed in relation to firm organisation, location and the internalisation of knowledge. It is only relatively recently that the MNE has begun to be incorporated into theoretical treatments of the location of FDI, which has developed as new trade theory. In new trade theory, economies of scale and product differentiation provide a rationale for the existence of MNEs. However, although there have been rapid advances in this field regarding issues such as the internalisation and pricing of knowledge assets (Markusen 2002), the advantages of horizontal (Markusen and Venables 2000) or vertical integration, and the advantages of subcontracting and licensing versus FDI (Horstmann and Markusen 1987), in this schema each individually

---

2 Anecdotal evidence suggests that the size of many of the individual inward service sector FDI investments are somewhat smaller than many of the inward investment manufacturing projects located in the more geographically peripheral regions. Therefore, the raw number of inward projects is likely to be only an imperfect indicator of the relative employment contribution made by inward FDI to each region. However, given the fact that over the last fifteen years, service sector FDI now accounts for two-thirds of total inward FDI in OECD countries, it is unlikely that the different average sizes between these individual investment types across UK regions would be sufficient to counterbalance the general changes in the regional patterns of inward investment outlined in this section.

differentiated product is still identified with a single firm at a single location. As such, there is still no multi-plant or multinational production, at least in the sense that the international business literature would recognise. Nor are there any explicitly geographical elements in either the international business or the new international trade theory traditions, at least of a form that economic geographers would recognise. In contrast, it can be argued that the literature on the location of FDI, which is the traditional economic geography literature on MNE locational behaviour, is far too general to act as an explanatory framework. While there have been some attempts to incorporate organisational issues into this approach (Hayter 1997), the traditional economic geography approach often adopts as its analytical framework over-stylised geographical versions of the product cycle model, many of which are based on a range of assumptions which may no longer be tenable (McCann and Mudambi 2004, 2005). In particular, the modern organisational structure, logic and behaviour of the MNE appear to have changed significantly over the last three decades since the product cycle model was first developed. MNEs are nowadays acknowledged to adopt a much more sophisticated approach to multinational organisation and parent-subsidiary relationships than the simple hierarchical model implied by the product cycle theory, although the exact features of this are still open to question (Birkinshaw 1996). These issues are further complicated by the fact that in many host regions and countries, the majority of FDI investments are actually in service industries. Meanwhile, the theoretical treatment of the FDI behaviour of service industries is still in its infancy (McCann and Mudambi 2004, 2005).

Over recent years, an increasing awareness of the different analytical limitations faced by each of these three diverse literatures has led to an emerging consensus, and a gradual convergence of these three literatures (McCann and Mudambi 2004, 2005). Specifically, it has become increasingly apparent that not only a more explicitly spatial focus on the MNE is required, but also that the analytical treatment of FDI must have much more of a micro-economic flavour to it. As a result, in recent years, the locational analysis of the MNE at the sub-national regional level has come to be regarded as an ever more important issue by many analysts and policy-makers in the international business field (Mucchieli and Mayer 2004) as well as by regional economists and economic geographers. As such, identifying the conditions under which MNEs will locate FDI not only in individual countries, but more specifically in large or small urban centres, in central or peripheral locations, or in sectorally specialised or diversified areas, is now regarded as essential for understanding the nature of the FDI, and also for designing appropriate public policies. Issues such as agglomeration, industrial clustering, local and regional knowledge spillovers, or technological assets and local network embeddedness (McCann and Mudambi 2004, 2005), have therefore come to be perceived as being critical to our understanding the location of FDI. At this more disaggregated firm and spatial scale, it is therefore argued that the location of FDI can only be analysed by discussing more explicitly spatial and organisational issues at the level of the individual MNE establishment, while also taking account of the characteristics of the region itself (Phelps 1997). As such, the theoretical treatment of the location behaviour of MNEs is generally moving in the direction of increasing disaggregation, both in term of firm organisation and also in terms of geography.

Unfortunately, however, a lack of appropriate FDI data, disaggregated by both location and sector, has meant that empirical work on FDI location is generally failing to keep up with theoretical developments. This chapter attempts to partially fill this gap by employing a unique dataset of inward FDI investments in UK regions, broken down according to the firm, the sector and the location. The next sections describe both the dataset and also the econometric methodology used in this study to identify the determinants of the location of inward FDI in UK regions.

The Dataset and Methodology

The analysis in this chapter proceeds in the manner suggested in the FDI location literature by focusing on the firm level and the host region level (McCann and Mudambi 2004, 2005) using a dataset in which FDI investments, including both manufacturing and service activities, can be identified at the level of the individual establishment and the individual region.

UK Trade & Investment (UKTI), which is part of the UK Department of Trade and Industry (DTI), is the overarching official government body responsible for support of overseas businesses wishing to invest in the UK. UKTI offers a subsidised consultancy service for inward investing firms and, on the basis of the many firms which have taken advantage of this service, UKTI has built up a detailed dataset containing information on foreign inward investment projects which have been realised in the UK. The dataset contains information on the name and nationality of the investing firm, the industrial sector of the firm, the exact region where the project is being located, the type of project undertaken, and the scale of the investment. The complete dataset for which all variables are available contains 2265 inward FDI projects which were located in the UK between 1997 and 2003. For the purposes of econometric analysis, the UKTI firm-specific and project-specific FDI data are merged with publicly available UK regional economic data.

To use available data, inward FDI projects are grouped into three distinct categories in terms of the parent nationalities: North America, Europe and Asia Pacific. In addition, the year each project commenced is known. The FDI project type is known in terms of whether it is a New Investment, Acquisition, Merger, Joint Venture or an Expansion. The operational nature of the project is also known, not only in terms of being a manufacturing or service operation, but also in terms of it being an assembly, distribution, sales, e-commerce, call centre, R&D, or headquarters activity. The data are available at the four-digit sectoral level. Finally, the data are broken down into nine English regions plus the three additional country-regions of Wales, Scotland and Northern Ireland. These twelve regions correspond to the NUTS 1 European classification of regions. To analyse the characteristics and determinants of FDI investments in UK regions a logit model is used (see Table 6A.1, Table 6A.2 and the model description in the Appendix).

Results

The model results reported here have good explanatory power. There are two reasons for this. Firstly, the conditional logit model passes the Haussman test, thereby ruling out any problems associated with independence-of-irrelevant alternatives. Secondly, the goodness of fit level of the model, indicated by a pseudo-R squared value of 0.1844, is good for a logit model (McFadden 1979; Louviere et al. 2000), for which a value of 0.2 would be equivalent to an orthodox R squared value of 0.8. This indicates that the overall model estimated here captures the FDI location behaviour in general very well.

In terms of the behaviour of the individual explanatory variables, we can first consider the variables which simply reflect regional characteristics. Here, the size of the regional economy *logrgdp*, the existing size of the regional R&D base *logrdgt*, and the location quotient *loglq*, are all positive and significant. However, neither the level of regional GDP per capita *logrgcap*, the regional wage *logrwe*, nor the existing regional stock of FDI *logfditot* have any statistical significance, while the effect of regional unemployment *logunm* is only slightly significant.

In the hybrid conditional logit model there are no pure firm-specific variables, but rather the effect of firm specific attributes are captured by the interaction terms which combine project and firm-specific characteristics with regional dummies. In terms of these variables we see that the location of headquarters activities *hqs* is not statistically different in any region from London, except for the case of the South West *sw*, which is more likely to receive such investments than London. Meanwhile, North American investments *nam* are by far the largest single category of UK inward FDI. We therefore estimate the location of these investment projects relative to the *eu* and *as* combined as the baseline case. Here the evidence suggests that the South East *se* and East *e* regions are statistically more likely to attract North American investment whereas the North East *ne* and Wales *wales* are statistically less likely to attract North American investments relative to London. In a similar manner we estimate the case of new investments *new* relative to the other relevant project type alternatives (*acq ven mer*) combined as the baseline case and with respect to London. On this point, the evidence implies that all regions, except for Scotland and East Midlands, are statistically less likely to attract *new* investments in comparison with London. In terms of the split between tangibles, *tan,* such as manufacturing and distribution and intangibles, *intan,* we see that the South West, West Midlands, Yorkshire & Humberside, and Wales, are all statistically more likely than London to receive investments in sectors dealing in tangibles rather than intangibles. This is unsurprising and confirms prior expectations (Hill and

Munday 1994). Meanwhile, for R&D related investments *rd*, we see that the South East *se*, South West *sw*, East *e* and Scotland *scot* are all statistically more likely to receive R&D related investments *rd* than London, while West Midlands is less likely to do so.

Discussion

This analysis of FDI location behaviour has been conducted using a dataset with a unique level of both coverage and disaggregation. No previous large-scale analyses of FDI location have been able to separate and disaggregate manufacturing FDI from non-manufacturing FDI at the level of detail of an individual FDI project. The unique dataset has allowed us to employ a conditional logit model of location behaviour, which is a well-known but difficult technique successfully to employ.

The first point to notice from the estimated results for the conditional model, is that the single most important explanatory variable in terms of the *z* score is the location quotient *log lq*. Moreover, this is also the most important individual variable in terms of its effect on the overall goodness of fit, with the model fit increasing from 0.1594 to 0.1844 with the inclusion of this individual variable. Meanwhile, regional industrial diversity, which is calculated on the basis of a Herfindahl index at a two-digit level, is very highly correlated with the regional dummies and cannot be used independently within the conditional model. However, it is sufficient to note that London, which is the leader in terms of inward FDI projects, is by far the most specialised, and least diversified, region in the UK. As such, having controlled for all of the various firm and regional characteristics we see that in terms of the probability of a region receiving inward FDI projects, the existing pattern of regional specialisation is, therefore, a crucial indicator of the likelihood of inward FDI taking place. While the knowledge and research base of a UK region also appears to be important for multinational FDI, an observation consistent with other countries (Cantwell and Iammarino 2003), it cannot, however, be an explanation of the primary role of London for inward FDI. This is because the South East *se*, South West *sw*, East *e* and Scotland *scot* are all statistically more likely to receive R&D related investments *rd* than London, mirroring results reported elsewhere (Gordon and McCann 2005). Moreover, the location of headquarters' activities appears to be much less important than is often supposed. As such, London appears to benefit primarily in terms of receiving very large numbers of new investments, even after controlling for all firm and region specific characteristics.

These results, therefore, suggest that much of the existing knowledge-agglomeration literature on FDI, which assumes that relatively buoyant and diverse regions dominated by headquarter operations are key locations for multinational investments, may actually be far too general in its analysis. On the contrary, patterns of existing regional specialisation would appear to be crucial in determining both the likelihood and spatial distribution of inward FDI. Moreover, in terms of the issues raised by endogenous growth theory, London appears to benefit not only from being the recipient of the largest number of inward FDI investment, but also the fact that these are systematically newer investments than in most other regions. Both the quality as well as the quantity of inward FDI investments would appear to be a major advantage of the London economy, ensuring that it is always at the production possibility frontier.

The analysis undertaken in this chapter is unique in that no empirical analyses of FDI location behaviour has previously been based on such disaggregated and comprehensive inward FDI data. This level of disaggregation allows for the use of a conditional logit model analysis of location behaviour at the level of an individual FDI project, and the resulting goodness of fit of the model allows for confidence in the results. However, as well as providing for an original application of the conditional logit model technique, the empirical findings regarding the dominant role of regional specialisation in attracting inward FDI are also entirely new. In the case of UK regions, having controlled for all firm and regional specific characteristics, the dominant role played by regional specialisation has never been fully observed before. Moreover, these findings largely go against prevailing wisdom, which tend to favour locational explanations based on regional sectoral diversity.

References

Birkinshaw, J. (1996), 'How Multinational Subsidiary Mandates are Gained and Lost', *Journal of International Business Studies* 27:3, 467-498.

Cantwell, J.A., and Iammarino, S. (2003), *Multinational Corporations and European Regional Systems of Innovation* (London: Routledge).

DTI (2001), *Productivity in the UK: The Regional Dimension* (London: Department of Trade and Industry).

Dunning, J.H. (1977), 'Trade, Location of Economic Activity and the Multinational Enterprise: A Search for an Eclectic Approach', in Ohlin, B., Hesselborn P.O. and Wijkman, P.M. (eds), *The International Allocation of Economic Activity* (London: Macmillan), 395-418.

Gordon, I.R. and McCann, P. (2005), 'Innovation, Agglomeration and Regional Development', *Journal of Economic Geography* 5:5, 523-543.

Greene, W.H. (2003), *Econometric Analysis* (New York: Prentice Hall).

Hayter, R. (1997), *The Dynamics of Industrial Location: The Factory, The Firm and the Production System* (Chichester: John Wiley & Sons).

Hill, S. and Munday, M. (1994), *The Regional Distribution of Foreign Manufacturing Investment in the UK* (London: Macmillan).

Horstmann, I.J. and Markusen, J.R. (1987), 'Strategic Investments and the Development of Multinationals', *International Economic Review* 28, 109-121.

Johanson, J., and Vahlne, J. (1977), 'The Internationalisation Process of the Firm: A Model of Knowledge Development and Increasing Foreign Market Commitments', *Journal of International Business Studies* 7, 22-32.

Jones, J., and Wren, C. (2006), *Foreign Direct Investment and the Regional Economy* (Aldershot: Ashgate).

Louviere, J.J., Hensher, D.A. and Swait J.D. (2000), *Stated Choice Methods* (Cambridge, UK: Cambridge University Press).

Markusen, J. (2002), *Multinational Firms and the Theory of International Trade* (Cambridge, MA: MIT Press).

Markusen, J.R. and Venables, A.J. (2000), 'The Theory of Endowment, Intra-Industry and Multinational Trade', *Journal of International Economics* 52, 209-235.

McCann P. and Mudambi, M. (2004), 'The Location Decision of the Multinational Enterprise: Some Theoretical and Empirical Issues', *Growth & Change* 35:4, 491-524.

McCann P. and Mudambi, M. (2005), 'Analytical Differences in the Economics of Geography: The Case of the Multinational Firm', *Environment and Planning A* 37:10, 1857-1876.

McCann, P. (2005), *Cities, Regions and Competitiveness: The Location Decisions of High Value Inward Investors*, Report to the UK Office of the Deputy Prime Minister, London.

McFadden, D. (1979), 'Quantitative Methods for Analysing Travel Behaviour of Individuals', in Hensher, D.A. and Storper, P.R. (eds), *Behavioural Travel Modelling* (London: Croom Helm), 279–318.

Mucchieli, J.L. and Mayer, T. (eds) (2004), *Multinational Firms' Location and the New Economic Geography* (Cheltenham: Edward Elgar).

OECD (2006), OECD *Factbook 2006: Economic, Environmental and Social Statistics* (Paris: OECD).

Phelps, N.A. (1997), *Multinationals and European Integration: Trade, Investment and Regional Development* (London: Jessica Kingsley).

Powers, D., and Xie, Y. (2000), *Statistical Methods for Categorical Data Analysis* (San Diego: Academic Press).

Train, K.E. (2001), *Discrete Choice Methods with Simulation* (Cambridge, UK: Cambridge University Press).

UKTI (2004), UK *Inward Investment 2003-2004: Report by UK Trade and Investment* (London: UK Trade and Industry).

Vernon, R. (1966), 'International Investment and International Trade in the Product Cycle', *Quarterly Journal of Economics* 80, 190-207.

Wooldridge, J.M. (2002), *Econometric Analysis of Cross Section and Panel Data* (Cambridge, MA: MIT Press).

Appendix

Table 6A.1   Description of the full data-set variables

Variable	Description
New	new investment
Acq	Acquisition
Mer	Merger
Ven	joint venture
Exp	Expansion
Tan	tangibles (manufacturing or distribution)
Intan	intangibles (services)
Rd	Research & Development
Hqs	Headquarters
Se	South East
Sw	South West
E	East
Wm	West Midlands
Em	East Midlands
Y	Yorkshire & Humberside
Nw	North West
Ne	North East
Wales	Wales
Scot	Scotland
Ni	Northern Ireland
Nam	North American firm
Eu	European firm
As	Asia-Pacific firm
Logrgdp	Log regional GDP
Logrgcap	Log regional GDP per capita
Logrwe	Log regional wage
Logunm	Log regional unemployment
Logfditot	Log of total regional FDI
Logrdgt	Log total regional R&D
Loglq	Log of the location quotient

Table 6A.2   Conditional logit model

Variable	Coef.	Std. Err.	Z	P>|z|
Se	1.6221*	0.8902	1.82	0.068
Sw	11.1152***	3.4359	3.24	0.001
E	8.8025***	2.4479	3.60	0.000

Variable	Coef.	Std. Err.	Z	P>|z|
Wm	10.1762***	2.9354	3.47	0.001
Em	13.5595***	4.0077	3.38	0.001
Y	10.4091***	3.2725	3.18	0.001
Nw	6.8840***	2.0221	3.40	0.001
Ne	23.4003***	7.0228	3.33	0.001
Wales	20.2925***	6.2768	3.23	0.001
Scot	8.7808***	2.8866	3.04	0.002
Logrgdp	16.8663***	5.5418	3.04	0.002
Logrgcap	-4.2942	5.2081	-0.82	0.410
Logrwe	-0.6152	1.0687	-0.58	0.565
Logunm	1.0661*	0.5975	1.78	0.074
Logfditot	0.1981	0.1393	1.42	0.155
logrdg_t	-1.6456***	0.5380	-3.06	0.002
Loglq	0.8728***	0.0672	12.98	0.000
nam_se	0.3764**	0.1647	2.29	0.022
nam_sw	0.4104	0.2866	1.43	0.152
nam_e	0.3780*	0.2194	1.72	0.085
nam_wm	0.0255	0.1844	0.14	0.890
nam_em	-0.1759	0.3096	-0.57	0.570
nam_y	-0.0628	0.2250	-0.28	0.780
nam_nw	0.0908	0.2126	0.43	0.669
nam_ne	-0.5895***	0.2390	-2.47	0.014
nam_wales	-0.6665**	0.2817	-2.37	0.018
nam_scot	0.2270	0.2658	0.85	0.393
new_se	-0.9349***	0.1723	-5.43	0.000
new_sw	-1.2985***	0.3193	-4.07	0.000
new_e	-0.9428***	0.2307	-4.09	0.000
new_wm	-1.4853***	0.2052	-7.24	0.000
new_em	-0.4787	0.3230	-1.48	0.138
new_y	-0.9349***	0.2467	-3.79	0.000
new_nw	-1.8907***	0.2522	-7.50	0.000
new_ne	-0.8779***	0.2506	-3.50	0.000
new_wales	-0.6559**	0.2891	-2.27	0.023
new_scot	-0.3412	0.2753	-1.24	0.215
tan_se	0.3644*	0.2116	1.72	0.085
tan_sw	0.5409*	0.3272	1.65	0.098
tan_e	0.0609	0.2650	0.23	0.818
tan_wm	0.7499***	0.2549	2.94	0.003
tan_em	-0.0364	0.3603	-0.10	0.919
tan_y	0.9486***	0.2893	3.28	0.001

Variable	Coef.	Std. Err.	Z	P>|z|
tan_nw	-0.1398	0.2647	-0.53	0.597
tan_ne	0.4639	0.2968	1.56	0.118
tan_wales	0.8375**	0.3579	2.34	0.019
tan_scot	0.3256	0.3155	1.03	0.302
rd_se	0.4756*	0.2597	1.83	0.067
rd_sw	1.0981***	0.3784	2.90	0.004
rd_e	1.3876***	0.2864	4.84	0.000
rd_wm	-0.6941*	0.3892	-1.78	0.075
rd_em	-0.2354	0.6323	-0.37	0.710
rd_y	-0.3388	0.4794	-0.71	0.480
rd_nw	0.0911	0.3667	0.25	0.804
rd_ne	-0.8818	0.5588	-1.58	0.115
rd_wales	-1.1446	0.7570	-1.51	0.131
rd_scot	0.8314**	0.3781	2.20	0.028
hqs_se	0.1797	0.4346	0.41	0.679
hqs_sw	1.2965*	0.6925	1.87	0.061
hqs_e	0.2355	0.6597	0.36	0.721
hqs_wm	-14.5822	871.0942	-0.02	0.987
hqs_em	-14.4210	1286.048	-0.01	0.991
hqs_y	-0.3571	1.0683	-0.33	0.738
hqs_nw	-14.7294	1157.036	-0.01	0.990
hqs_ne	-14.5936	1100.891	-0.01	0.989
hqs_wales	-14.1270	1076.25	-0.01	0.990
hqs_scot	-0.7306	1.1158	-0.65	0.513
Number of observations: 18942				
LR chi2(67): 1523.08				
Prob > chi2: 0.0000				
Log likelihood: -3367.637				
Pseudo R2: 0.1844				

Model Description

To demonstrate how the location of FDI investment and firm preferences are related, suppose that there are J potential locations ($j = 1, ..., J$) in which each individual MNE i ($i = 1, ..., I$) can invest. We can write the deterministic (observable) part of the utility function of the individual MNE firm i investing in location L to location j as:

$$V_{ij} = V(\mathbf{X}_i, \mathbf{Y}_j) \tag{1}$$

where $\mathbf{X}_i$ is a vector of firm-specific characteristics of MNE firm i, and $\mathbf{Y}_j$ is a vector of characteristics of the region j. Utility in this sense represents the expected returns to the MNE of undertaking a particular investment at a particular location. Introducing a random error of unexplained firm and location-specific variables given as e_j, the FDI location-utility function becomes:

$$U_{ij} = V(\mathbf{X}_i, \mathbf{Y}_j) + e_{ij} \tag{2}$$

Utility is now composed of a deterministic portion of observable firm and location characteristics and a random portion containing the unobservable attributes of both the firms and the location alternatives.

By this argument, the probability $P(m_{ij})$ that an individual MNE firm i will invest in a particular location j rather than in any other alternative location j', is the probability that the individual will maximise their potential returns to investment by locating at that particular alternative region j rather than in any other region j'. Formally, we can write this as:

$$P(m_{ij}) = prob\,[U_{ij} = V(\mathbf{X}_i, \mathbf{Y}_j) + e_{ij} > U_{ij'} = V(\mathbf{X}_i, \mathbf{Y}_{j'}) + e_{ij'}] \tag{3}$$

$$j \neq j'$$

$$j, j' \in J$$

Assuming that utility has a deterministic portion $[V(\mathbf{X}_i, \mathbf{Y}_j)]$ which is linear in its parameters and an error term e_{ij} which has a Gumbel distribution, the logit family of models can be used to study the sequential migration process.

With only two possible location choices available, then estimating the probability that an individual will choose location $y = 1$ as against the alternative location $y = 0$, can be modelled simply by using a dichotomous logit model, the structure of which is given as (Train 2001; Wooldridge 2002):

$$P_i(y = 1) = \frac{1}{1 + e^{-\mathbf{N}_{ij}\beta}} \tag{4}$$

where $\mathbf{N}_{ij}$ includes both the firm-specific and the location-specific characteristics, and β represents the vector of parameters to be estimated. On the other hand, in the case of more than two choices, we must distinguish explicitly between the characteristics of the chooser from the characteristics of the choices. Some researchers (Greene 2003) perceive the MNL model to be simply an extension of the dichotomous logit model in which the dependent variable has more than two categories. Other researchers (Powers and Xie 2000) make an explicit distinction between the pure MNL model, which contains information only about the choosers, and the pure conditional model, which contains information only about the choices. Therefore, in the case of more than two alternative location choices ($k = 1, ...,K$) in

which we focus only on the characteristics of the chooser, the structure of the pure multinomial (MNL) logit model is (Train 2001; Wooldridge 2002):
where $\mathbf{X}_i$ is the vector of firm-specific characteristics, i.e. the characteristics of the

$$P_i(k) = \frac{e^{\beta_k \mathbf{X}_i}}{\sum_h e^{\beta_h \mathbf{X}_i}} \tag{5}$$

chooser only and not the choices, and β_k represents the vector of parameters to be estimated.

A pure conditional logit model would exhibit a similar structure to equation (5), except that vector $\mathbf{X}_i$ is replaced by vector $\mathbf{Y}_j$ which represents the characteristics of the choices only and not the chooser.

In order to estimate the likelihood of locating in alternative regions as a function of both the firm-specific and the location-specific characteristics, it is therefore necessary to construct a hybrid conditional logit model which combines elements of both a pure multinomial model and a pure conditional model. The structure of this is given as:

$$P_i(k) = \frac{e^{\beta_k \mathbf{W}_i + \gamma\, \mathbf{z}_{ik}}}{\sum_h e^{\beta_h \mathbf{W}_i + \gamma\, \mathbf{z}_{ih}}} \tag{6}$$

where $\mathbf{W}_k$ is a vector of firm-specific variables, $\mathbf{Z}_{ik}$ includes both location-specific and firm-specific characteristics, and β_k and γ are the parameters to be estimated.

Chapter 7

Globalising Commercial Property Markets: The Development and Evolution of the Listed Property Trust Sector in New Zealand

Laurence Murphy

Introduction

In 2005 it was estimated that global direct investment in commercial real estate amounted to US$15 trillion dollars. However, notwithstanding the size of this market, the transaction costs and information-rich character of direct investment means that there is a 'home bias' in the investment strategies of large-scale investors. In response to investors' needs for greater liquidity, indirect property investment vehicles such as Real Estate Investment Trusts (REITs) in the US and Listed Property Trusts (LPTs) in Australia and New Zealand have emerged since the 1970s. As of 2006, the global REITs sector had a market capitalisation of US$608 billion and property assets in excess of US$890 billion (Ernst & Young 2006).

To date, the development of REITs has been characterised by significant geographical unevenness at a global level. REITs first developed in the US in the 1960s and are significant in the Asia-Pacific Region, but less important in Europe. However, the recent introduction of REIT legislation in the UK (January 2007), and the likely development of REIT structures in Germany, is expected to generate a European REIT market with a potential market capitalisation of US1.5 trillion by 2011 (Hughes 2006).

Academic and industry discourses on the growth of REITs emphasise their role in the globalisation of international property investment (Gotham 2006). These accounts centre on the role of the US as a leader in REIT development and the potential impacts of an emerging REIT market in Europe, and especially the UK. The focus on large property markets reflects both the global geography of the commercial property investment market (five countries account for 68 per cent of global investment grade commercial property, Hughes and Arissen 2005, 2) and wider understandings of the centralising tendencies of geographies of finance (Dicken 2007). However, REITs are entities that are necessarily involved in the ownership, management and, at times, development of locally embedded property assets. While the growth of REITs has implications for the *flow* of investment funds into property at a global level, these entities are required to participate in local property markets consisting of distinct institutional structures (Seabrooke et al. 2004).

The recent emergence of REITs in the UK and other large markets in Europe stands in marked contrast to the long-standing experience of REIT development in Australia and New Zealand. Australia had the second largest REIT market in 2006 (market capitalisation US$77 billion) and New Zealand LPTs produced the highest total returns of international REITs in the year to June 2006 (24.6 per cent) (Ernst & Young 2006). In this chapter I argue that the development of LPTs in New Zealand, and their ongoing incorporation into the Australian market, provide an important case study for exploring the impact of REITs on a local, regionally significant but globally 'peripheral', market. The analysis offers an alternative to accounts of globalisation that centre on global cores (e.g. global cities) and highlights the significance of local economic and institutional conditions in shaping the dynamics of urban property markets under conditions of globalisation. The New Zealand case study provides information on the impacts of REITs on a 'peripheral' (regional) property market and thus has the potential to offer insights into the possible effects of REIT development on regional property markets in the UK (Bryson 1997) and mainland Europe.

The remainder of the chapter is in four parts. First, key literatures relating to geographical dimensions of property investment are reviewed. These literatures address the tensions between the local embeddedness of property assets and the increasing demands of finance capital for greater flexibility. Next, key trends in indirect investment at a global level are reviewed, highlighting the benefits of indirect investment for investors. The third section provides an account of the rise of the New Zealand LPTs, examines the nature and effects of the growth of LPTs on the local commercial property market and outlines the implications of increased foreign ownership. In the concluding section, it is argued that LPTs/REITs alter the nature and operation of local commercial property markets and simultaneously integrate local markets into global flows of capital.

Investment in the Built Environment

Commercial property is both a physical and financial asset. As a physical entity a building provides a use value and is demanded by economic interests to satisfy their space needs. As a financial asset, property ownership provides access to a stream of income (rent) that is capitalised in the price of a property or its exchange value. Since property can function as a financial asset, its price is exposed to speculative tendencies and this is especially evident during the boom phase of a property cycle when the exchange value of property drives occupier and investor behaviour (Coakley 1994; Moricz and Murphy 1997). As the volume of investment into property markets increases, property prices are bid up and developers, responding to investment demand, have a tendency to overproduce new space compared to underlying user demand (MacLaran 2003).

Since the 1980s it has been argued that there has been a 'deepening of financial links between real estate ... and the financial markets' (Coakley 1994, 705). Moreover, Warf (1994, 325) has argued that 'finance capital is not a passive actor in the construction of landscapes, but is an active participant with a logic of its own'. Attempts to uncover this 'logic' have tended to focus on the dynamic interrelationship of investment capital and property development processes in global cities (Fainstein 1994; Pryke 1994). However, Wood (2004) cautions against this focus on global cities and argues that the experiences of these places cannot be extrapolated to other types of cities. Paying particular attention to development processes he contends that, notwithstanding the increased linkages between global finance capital and property development, the property development industry tends to be dominated by 'local firms embedded within particular metropolitan markets' (Wood 2004, 137).

Wood's (2004) critique of general accounts of globalisation tendencies within commercial property markets rightly challenges the emphasis placed on global cities, since these are, in many ways, exceptional places. Clearly there is a need to examine how property markets are being transformed in 'other'/peripheral cities to uncover the extent of globalisation (Charney 2003a). However, in focusing on the development industry, Wood examines what is arguably the most locally embedded dimension of property markets. In terms of investment practices, there is a need to examine the ownership and ongoing management of properties and the evolution of direct and indirect property investment practices.

In contrast to 'pure' financial instruments, commercial property markets are constrained by market size, limited turnover and local planning, tax and legal structures that affect property rights. Property markets are also characterised by heterogeneity (construction type, use, lease structures) and large lot sizes which means that direct investment requires substantial investment on the part of investors and involves ongoing management costs. In addition, trading in property involves considerable transaction costs and requires access to considerable local knowledge (see Ball et al. 1998). These characteristics impede the flow of funds to the direct property market and act as barriers to the globalisation of property markets (Clark and Lund 2000; Gotham 2006; Seabrooke et al. 2004).

The increasing securitisation of real estate assets reflects a response to property investors' demands for greater liquidity. The rise of REITs, where investors invest in trusts that are legally restricted to manage a property portfolio and are required to pay out the majority of their income as dividends, represents an important response to investors needs. Clark and Lund (2000, 469) regard the rise of REIT structures as revolutionary, as these structures offer a variety of benefits to investors including: divisibility of investment, spread of risk over a portfolio, 'separation of investment from the function of procuring local knowledge necessary for risk assessment' and increased ease of entry and exit (see also Gotham 2006). In this context, REITs offer the opportunity for the increased globalisation of property investment.

While REIT units (shares) can be traded globally, the underlying value of a REIT is determined by specific buildings that make up the property portfolio. Since buildings are embedded in specific local economies they are susceptible to place-based devaluation. In this context Weber (2002, 521) argues that '... spatialised capital, unlike derivatives or corporate equities, has the unique (dis)advantage of having its value held hostage to the vagaries of proximity and its relationship to other properties'. Yet, notwithstanding the continued importance of local market conditions, Clark and Lund (2000, 470) maintain that:

> REITs seem rather to involve a radical step towards the translucency of investments in property capital, breaking the local barrier. The globalisation of commercial property market investment is intrinsically tied to transformations which increase the liquidity of this form of fixed capital, and in this context, REITs play an important role.

Commercial property investment involves ongoing interactions between highly localised development processes embedded in localised knowledge structures, that are increasingly exposed to transnational and global processes centred on investment flows. REITs represent an important conduit for these interactions since, as Gotham (2006, 266) argues:

> The significance of REITs is that they are a financial vehicle that links diverse actors and organised interests within real estate to one another even though these actors and interests may be geographically distant and, at the same time, disengaged from local settings.

While REITs participate in the process of delocalisation ('the conversion of real estate into a liquid financial asset' Gotham 2006, 238) they also engage in managing and developing property assets in specific locales. REITs channel the flow of investment into property *but* also adopt strategies and management practices that shape local property markets. In this context I would argue that REITs are not just investment vehicles but that they are active agents involved in rent setting and tenant management. Moreover, as REITs attract increasing volumes of investments they can, especially in smaller markets, assume a level of dominance in the market whereby they determine exchange values.

This chapter examines the role of REITs in the globalisation of a commercial property market outside of a 'global cities' context. By focusing on the experience of New Zealand, it is possible to examine the role of REITs as conduits of globalisation beyond the core and to reflect on the material effects of REITs in transforming local property markets.

The Rise of Indirect Property Investment

The securitisation of property involves financing a pool of similar but unrelated assets by issuing to investors a security (a unit or share) that represents a claim on the cash flows generated by the underlying property asset. REITs are publicly listed property investment companies that own property assets. REITs offer investors considerable liquidity over direct property investment and provide significant tax advantages (Ball et al. 1998).

REITs developed in the US in the 1960s, but it was not until changes in tax legislation in the 1990s that the growth in REITs took off. In 1972, the market capitalisation of the US REIT industry was US$1.5 billion, but this rose to $195 billion in 2003 (Gotham 2006) and US$395 billion in 2006 (Ernst & Young 2006). US REITs operate throughout the major US metropolitan regions and have substantial investments in Europe and Japan. Gotham (2006, 265) cites several examples of major investments by US REITs in Europe, including GE Capital Real Estate's acquisition of MEPC, Britain's 'fourth-largest publicly traded real estate operating firm, and owner operator of a dozen business parks'.

The success of the US REIT sector has had a significant demonstration effect on other countries. REIT legislation was enacted in the Netherlands in 1969 and Australia in 1971 and since 2000 a number of countries have followed suit including Japan (2001), South Korea (2001), Singapore (2002), Hong Kong (2003) France (2003) and most recently the UK (2007), which has the largest publicly listed property sector in Europe.

Outside of the US, Australia's Listed Property Trust (LPT) sector is the most developed REIT market. In 2004, Australian LPTs had total assets in excess of A$100 billion and accounted for over A$73 million in market capitalisation 'representing over 8 per cent of the total Australian stockmarket capitalisation' (Newell 2005, 211). The rapid growth of Australian LPTs has entrenched their position in the commercial property market and in 2004, LPTs held '49 per cent of all institutional-grade property in Australia' (Newell 2005, 211). As LPTs have assumed a position of dominance within the Australian market, their opportunities for growth have been constrained by the availability of investment-grade properties. Consequently, to overcome local supply restrictions, Australian LPTs have invested offshore. Westfield, Multiplex, Lend Lease and General Property Trust have all engaged in property investment in the US and UK, and Australian LPTs are actively engaged in European property markets.

In contrast to institutional investors that allocate a proportion of their funds to property, REITs are required to invest the majority of their funds in property markets. Thus, from a property perspective, the emergence of REITs/LPTs has at least three implications. Firstly, indirect property investment adds to the weight of capital flowing into property markets and as the flow of funds increases, yield compression occurs and capital values increase. In effect, the increased volume of funds chasing investment-grade property pushes up prices. Secondly, as they come to dominate their home markets, REITs need to seek out overseas markets and thus contribute to the increasing globalisation of property investment. Australian LPTs and US REITs are active in global markets and participate in the development of transnational ownership and management processes. Thirdly, as REITs/LPTs enter new markets, they alter market structures and practices. For example, in contrast to the practices of pension funds that have traditionally adopted arms length property management practices and relied on long leases to secure their returns, Gibson and Lizeri (1999) argue that US REITs operating in the UK offer the potential for more flexibility in lease structures. Since REITs/LPTs actively manage their property portfolios, as they enter new markets they have the potential to significantly alter the operation of local /national property markets.

New Zealand's Listed Property Trust Sector

Following significant deregulation of New Zealand's financial system in 1984, commercial property development (especially office development) experienced an unprecedented boom. At the forefront of this development were listed property companies that were heavily engaged in speculative development processes. The 1987 share market crash had a profound impact on the property sector. The office vacancy rate rose from 5 per cent in 1987 to 25 per cent in 1989 and rents dropped by 30 per cent. Faced with declining rental income and capital values, many companies went bankrupt. 'Between 1987 and 1989, the total market capitalisation of property companies ... declined by 78 per cent from NZ$5782 million to NZ$1278 million' and 'of the top 20 property companies operating in 1987, 11 where in receivership, defunct or subject to takeover by 1989' (Moricz and Murphy 1997, 177). The property crash resulted in a significant oversupply of office space and no new development occurred in Auckland for eight years (Murphy 2003).

The 1990s marked a new phase in the development of the commercial property market. High vacancy levels (30 per cent in 1991) and depressed rentals (prime rents in Auckland during the 1990s were 50 per cent lower than rents in 1989) provided little opportunity for speculative development. However, prime commercial space was available and opportunities existed for investors to construct portfolios of well-managed properties. It was during this period that LPTs emerged in New Zealand and have since assumed considerable significance in the market.

In 1993, Kiwi Income Property Trust became the first LPT in New Zealand. Between 1993 and 2006, the sector grew to a market capitalisation of $NZ3.6 billion, approximately 4.9 per cent of the New Zealand share market capitalisation, and owns assets in excess of NZ$4.5 billion. In 2005, there were eight LPTs and the top three trusts accounted for 61 per cent of all assets held by trusts. In order to examine the implications of the growth of this sector, it is important to understand the structural character of the sector and the ways in which the sector has evolved.

Structural Characteristics of the LPT Sector

With assets in excess of NZ$1.3 billion, Kiwi Income Property Trust is the largest property trust, followed by AMP New Zealand Office Trust (NZ$842 million) and ING Property Trust (NZ$810 million) (see Figure 7.1). The relative size of LPTs is important since as LPTs grow they derive greater market power and their rent setting strategies have a greater bearing on the development of market trends. LPTs in New Zealand are primarily externally managed which means that the management of the portfolio is undertaken by a management group that are paid in relation to their performance and to the growth in assets under management. In these circumstances, there is a management imperative to increase the size of the portfolio.

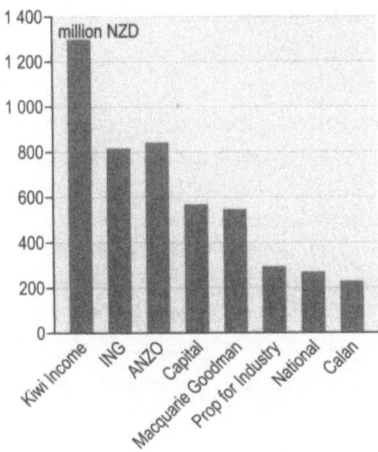

Figure 7.1   Listed property trusts by assets, 2005

Source: Published annual reports of all LPTs

Only two LPTs, ANZO, which focuses on office space, and Calan, which focuses on healthcare facilities, specialise in a single property sector (see Figure 7.2). The remaining LPTs have mixed property portfolios that include office, retail, industrial and development assets. While a mixed sector property portfolio offers diversification benefits, it is likely that the absolute size of the New Zealand property investment market requires LPTs to pursue diversification strategies. All of the major LPTs have exposure to commercial office space but the relative weightings of retail and industrial space varies. Of the 'big three' LPTs, Kiwi Income Property Trust has the largest retail profile worth NZ$523 million (43 per cent of its assets) and a relatively minor engagement in industrial space (NZ$7 million, 0.6 per cent of assets).

LPT property portfolios exhibit distinct geographies (Figure 7.3) that reflect the underlying commercial geography of the country. In New Zealand the absolute

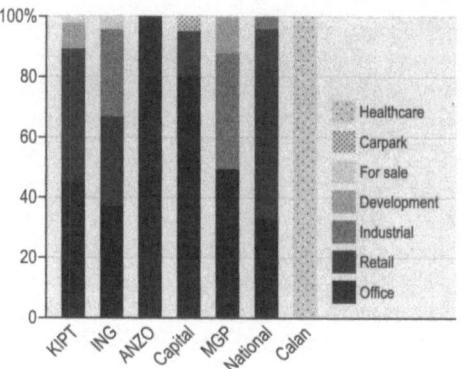

Figure 7.2   LPT assets by sector, 2005

Source: Published annual reports of all LPTs

size and location of investment-grade office constrains the operation of LPTs
but the differences in the two largest urban markets (Auckland is dominated by
large national and international corporate tenants and Wellington is dominated by
government agencies) offers diversification benefits to investors.

Notwithstanding the spread of investment through the urban system, LPTs
demonstrate a strong bias toward investing in Auckland and this is particularly
the case for the larger LPTs. This geographical bias reflects the underlying logic
and rationale of LPTs. Given the significance of rental income in determining the
value and success of LPTs there is a compelling logic to invest in larger centres

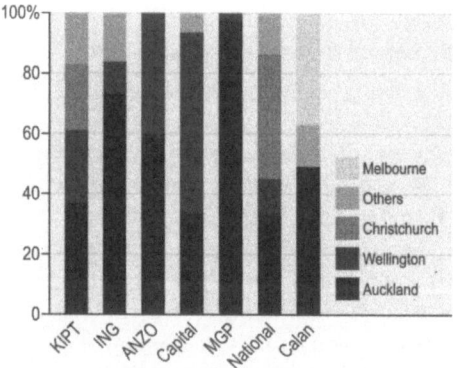

Figure 7.3   LPT assets by geographical area, 2005

Source: Published annual reports of all LPTs

(c.f. Charney 2003b) that are less vulnerable to the economic fortunes of one or two key properties or tenants.

Overseas Ownership

The sheer size of Australian LPTs and their entrenched ownership position of investment grade commercial space in Australia, has meant that Australian LPTs have increasingly looked for international investment opportunities. In search of geographical diversification, Australian investors have taken over New Zealand LPTs and as of 2005, owned all of the largest New Zealand LPTs (see Table 7.1).

From an Australian perspective, New Zealand LPTs provide a relatively easy opportunity for internationalisation. The economies are closely interlinked. Australian investors are familiar with the legal structures surrounding commercial property investment in New Zealand and would have ready access to the local networks required to undertake investment in property markets in New Zealand. In addition, New Zealand LPTs have tended to be more profitable than Australian LPTs (BDO 2004; Ernst & Young 2006), and this would have attracted Australian attention.

Table 7.1   Ownership of LPTs

Kiwi Income Property Trust
1998 Lend Lease (Aus) 50%
2002 Colonial First State Property (CFSP) acquire the management company
CFSP Parent - Commonwealth Bank of Australia

ING Property Trust
Manager 50% ING(NZ) (part of ING Group)
50% Symphony Investments Ltd (NZ)

ANZO
1997-2003 AMP Capital Investors (NZ)
2004 AMP Ronin Management Ltd (Onyx Property, Australia) Onyx purchased by Multiplex
AMP Multiplex Management Ltd (Aus)

Macquarie Goodman Property Trust
2003 MG Management Ltd (Aus) acquired management rights and 20% of MGP
2005 MGM shares stapled to Macquarie Goodman Investment (ASX) units

The exposure of New Zealand's LPT sector to foreign ownership has had several implications. First, Australian investment has increased the weight of capital in the market and, as a consequence, has led to a decline in yields by 200–300 basis points since 2000 (Jones Lang La Salle 2005), and a rise in capital values. In this context, it can be argued that the rise in commercial property values in New

Zealand has been increasingly affected by the investment decisions of Australian interests. Secondly, foreign ownership has resulted in New Zealand LPTs increasingly conforming to the performance metric of the Australian LPT sector. Historical differences in the performance of New Zealand and Australian LPTs are declining as New Zealand LPTs are incorporated into the larger Australian market. Since 2000, the pricing techniques of REITs and other listed property companies has resulted in a 'substantial reduction of the return premium' and the Auckland and Wellington markets have become aligned with the Sydney market (Jones Lang LaSalle 2005, 5). Thirdly, the nature of New Zealand LPTs is being transformed as Australian LPTs change their management structures. While New Zealand LPTs have traditionally operated under external management structures, Australian LPTs are increasingly adopting so-called 'stapled structures'. Under these structures, when a unit holder buys a unit, they also buy a share in a management company which is often involved in property development processes. The evolution of stapled structures has significant implications for the risk profiles of LPTs.

As LPTs have evolved to assume a significant position across all of the major property classes and urban centres in New Zealand, they have become embedded in management structures that increasingly incorporate interests that are distant from the local market. Moreover, the fortunes of New Zealand LPTs are increasingly intertwined with the strategic development of Australian LPTs and their performance in Australian, US and European commercial property markets.

Conclusions

Apart from large global cities, commercial property markets have tended to be dominated by 'local' development and investment interests. However, the rise of indirect property investment and in particular the growth of REITs globally, has had significant implications for local commercial property markets. REITs offer the potential for diversification benefits for investors and offer enhanced opportunities for transnational investment. In addition, it has been argued that REITs provide increased liquidity for investors and are consequently significant players in the transformation of local property markets and the globalisation of these markets (Clark and Lund 2000).

This chapter has shown that since the early 1990s New Zealand LPTs have experienced significant growth and now manage in excess of NZ$4 billion of commercial property. LPTs are now significant owners and managers of office, retail and industrial space throughout New Zealand and particularly in the major urban centres of Auckland and Wellington. Their investment strategies provide not only returns to investors, but have significant implications for existing and future occupiers of the spaces they own. In managing their property portfolios, these LPTs influence the everyday operations of service, retail and industrial users. Significantly, New Zealand's largest LPTs are increasingly overseas owned and, as a consequence, are correspondingly subject to management processes determined at a distance from the 'local' New Zealand market.

In contrast to previous rounds of foreign investment in New Zealand's built environment, and especially the flow of foreign lending to property companies in the 1980s (Murphy 2003), the recent development of the LPT sector has resulted in a structural incorporation of New Zealand property into Australian operations. This process has implications for 'local' markets. As Gotham (2006, 269) argues, 'the enmeshment of real estate into global financial markets and flows suggests that the local impact of global-local developments is amplified, while ostensibly remote and localised events can have prodigious global consequences'. In this context, the future management and 'value' enhancement of office, retail and industrial space in Auckland, Wellington, Christchurch and smaller centres is increasingly being aligned with transnational interests and experiences.

At a wider level, the New Zealand case study illustrates the extent to which REITs can come to dominate a regional property market and align the performance of this market with international financial trends. Significantly this 'aligning' is occurring outside a 'global city' context. The New Zealand experience suggests that current attempts to create a REIT market in Europe may have significant implications for regional property markets.

References

Ball, M. Lizieri, C., and MacGregor, B.D. (1998), *The Economics of Commercial Property Markets* (London: Routledge).

BDO (2004), *BDO Listed Property Trust Survey, 2004 Edition*, BDO, Australia.

Bryson, J. (1997), 'Obsolescence and the Process of Creative Reconstruction', *Urban Studies* 34:9, 1439-1458.

Charney, I. (2003a), 'Local Dimensions of Global Investment: Isreali Property Firms in Central Europe', *International Journal of Urban and Regional Research* 27: 2, 441-47.

Charney, I. (2003b), 'Spatial Fix and Spatial Substitutability Practices among Canada's Largest Office Development Firms', *Urban Geography* 24:5, 386-409.

Clark, E. and Lund A. (2000), 'Globalisation of a Commercial Property Market: The Case of Copenhagen', *Geoforum* 31, 467-475.

Coakley, J. (1994), 'The Integration of Property and Financial Markets', *Environment and Planning A* 26, 697-713.

Dicken, P (2007), *Global Shift*, 5th Edition (London: Sage Publications).

Ernst & Young (2006) *Global REIT Report 2006*, Ernst & Young, Australia.

Fainstein, S. (1994), *The City Builders* (Oxford: Blackwell).

Gibson, V. and Lizieri, C. (1999), 'New Business Practices and the Corporate Property Portfolio: How Responsive is the UK Property Market', *Journal of Property Research* 16:3, 201-218.

Gotham, K. (2006), 'The Secondary Circuit of Capital Reconsidered: Globalisation and the US Real Estate Sector', *American Journal of Sociology* 112:1, 231-275.

Hughes, F. (2006), 'Viewing UK-REITs in a Global Context, in Page, N. (ed.), *A New Route up the Property Ladder* (London: London Stock Exchange and White Page Ltd.), 56-62.

Hughes, F. and Arissen, J. (2005), *Global Real Estate Securities - Where Do They Fit in the Broader Market* < http://www.epra.com/media/Size_of_the_Total_ Real_Estate _Markets.pdf> Accessed 26 March 2007.

Jones Lang LaSalle (2005), *The Yield Puzzle* (Auckland: Jones Lang La Salle).

MacLaran, A. (2003), 'Masters of Space: The Property Development Sector', in MacLaran. A. (ed.), *Making Space: Property Development and Urban Planning* (London: Edward Arnold), 7-62.

Moricz, Z. and Murphy, L. (1997), 'Space Traders: Reregulation, Property Companies and Auckland's Office Market, 1975-1994', *International Journal of Urban and Regional Research* 21:2, 165-179.

Murphy, L. (2003), 'Remaking the City: Property Processes, Planning and the Local Entrepreneurial State', in MacLaran, A. (ed.), *Making Space: Property Development and Urban Planning* (London: Edward Arnold), 172-193.

Newell, G. (2005), 'Factors Influencing the Performance of Listed Property Trusts', *Pacific Rim Property Research Journal* 11:2, 211-227.

Pryke, M. (1994), 'Looking Back on the Space of a Boom: (Re)developing Spatial Matrices in the City of London', *Environment and Planning A* 26, 235-264.

Seabrooke, W., Kent, P. and How, H. (2004), *International Real Estate: An Institutional Approach* (Oxford: Blackwell).

Warf, B. (1994), 'Vicious Circle: Financial Markets and Commercial Real Estate in the United States', in Corbridge, S., Martin, R. and Thrift, N. (eds), *Money, Power and Space* (Oxford: Blackwell), 309-326.

Weber, R. (2002), 'Extracting Value from the City: Neoliberalism and Urban Redevelopment', *Antipode* 34:3, 519-540.

Wood, A. (2004), 'The Scalar Transformation of the US Commercial Property-development Industry: A Cautionary Note on the Limitations of Globalisation', *Economic Geography* 80:2, 119-140.

Chapter 8

The New International Division of Labour and the Changing Role of the Periphery: The Case of the Polish Automotive Industry

Bolesław Domański, Robert Guzik and Krzysztof Gwosdz

Introduction

After the fall of communism, Central and Eastern Europe emerged as a new periphery of the capitalist economy. It was characterised by low incomes, backward technology and weak linkages with the West European core. Consequently, it was often perceived as a place for simple low-cost and/or local market-oriented production. In recent years, the immense changes that have taken place have resulted in the strong integration of Central European manufacturing within European production networks. The rate of growth in output has been higher than that in Western Europe and this growth has been accompanied by thriving exports. Underpinned by substantial foreign investment, Central European economies have achieved an increasing trading surplus in manufactured goods with the UK and Germany. None of the industries reflects these trends better than the automotive sector. The industry is relatively intensive in both technology and human capital, thus representing a traditional manufacturing stronghold of the advanced countries. From the processes of change that can be observed it can be suggested that a new international division of labour is underway.

A number of authors have explored the new processes operating in the post-communist automobile industry, for example, Richet and Bourassa (2000), Pavlinek (2002), Worrall et al. (2003), Tulder (2004), Pavlinek et al. (2009). Sadler (1999, 110) asked whether, 'a process of hollowing-out might be taking place in the European automobile industry, involving the effective relocation of out-sourced production [of components] away from Europe, leading to an industry with a hollow core of final assembly tasks'. It can be argued that the increasing involvement of major auto component manufacturers in Central Europe is a new alternative to the sourcing of the components outside Europe advanced in the 'hollowing-out' thesis. This may be related to broader tendencies towards 'Europeanisation' of manufacturing, i.e. creating company production systems oriented to Europe as a whole (Hudson and Schamp 1995; Sadler 1999; Freyssenet et al. 2003). Bordenave and Lung (1996) anticipated the scenario of continental integration and the dispersal of component production with the gradual emergence of a pan-European vertical division of labour. In 2000, intraregional trade had reached 70 per cent of the EU countries' automobile exports in comparison to 58 per cent in 1980 (Layan and Lung 2004).

While, on the one hand, the emergence of regional production systems creates opportunities for industrial upgrading in developing countries, on the other hand, a growing preference for the use of the same suppliers in different locations and the transfer of design responsibilities from the assemblers to the suppliers limit the opportunities for local producers in such countries to supply components (Humphrey and Memedovic 2003). Lung (2000) pointed out the serious uncertainties – in the form of volatile demand, exchange rate fluctuations and politics – faced by transnational corporations (TNCs) operating in emerging countries as automobile producers. Humphrey et al. (2000) placed Poland in the 'integrated peripheral markets' category, which indicates its integration into the motor industry of the EU with comparative advantages in the assembly and labour-intensive manufacture of components.

This chapter addresses the question of the extent to which recent changes in the Polish automotive sector may indicate a significant shift in core-periphery relationships. Is it justified to argue for an upgraded role of Poland's manufacturing in the European division of labour so that it may no longer be treated as peripheral? Or is Poland primarily locked in a vicious circle of hosting the activities at the lower end of the value-added chain? In other words, the question is whether foreign investment in the Polish auto industry mainly conforms to given location conditions, and thus reinforces a peripheral model of production, or whether it challenges and alters those conditions.

The chapter begins by discussing views on the nature of the periphery from the perspective of various theoretical concepts of production, transnational corporations and regional development. The automotive industry in Poland is explored on the basis of in-depth interviews at 75 large and medium-sized companies and a statistical analysis of some 800 business entities. This allows a comparison of the characteristics revealed and the trends identified among automotive manufacturers located in Poland and features of the periphery derived from theoretical considerations. A more detailed analysis of territorial embeddedness of automotive production and localised capabilities can be found elsewhere (Gwosdz and Micek 2009).

The Economic Periphery in Theoretical Perspective

The classic concept of a product life cycle suggests a movement of the manufacture of mature and standardised products from the economic core to the economic periphery (Vernon 1996). The movement stems from the producer's search for low-cost semi-skilled and unskilled labour and results in exports from less developed countries to the home market. At the same time, new products are introduced in the latter. The theory, at least implicitly, rests on the assumption of a stable core and periphery.

Various factors can account for spatial inertia in manufacturing activities. Core countries and regions enjoy the advantages of accumulated capital and knowledge, large markets and decision-making powers. These factors lie behind the processes of cumulative, self-reinforcing development, which reproduce and reinforce the dominance of the core over the periphery; the position of the latter being dependent on the core in terms of decisions, capital and innovation. The different innovation capabilities of various territories reflect a highly localised process of creation and accumulation of knowledge and find expression in the notions of innovative milieus and learning regions. A geographical agglomeration of economic activities creates traded interdependencies, as well as untraded interdependencies, including easier social interaction and face-to-face contacts.

Porter's (2000) popular notion of a cluster provides another perspective on the perpetuation of the developed countries and regions, and the inferior situation of those lagging behind. Localised geographical clustering of companies fosters the impact of the four interconnected determinants: factor conditions, demand conditions, related and supporting industries, and finally the strategy and structure of firms in relation to domestic competition.

The concept of a new international division of labour explicitly demonstrates a change in global relations between certain parts of the periphery and the core. It is manifested in a faster growth of output and especially of exports in peripheral economies than in the core. Whereas initially products manufactured in these countries often had features compatible with the product life cycle theory, at later stages their production growth could no longer be accounted for in these terms. An important role in this growth has been attributed to TNCs, but a successful expansion of indigenous producers cannot be neglected.

There are various theoretical explanations of foreign direct investment. Hymer (1976) put forward the notion of the ownership advantages possessed by foreign investors over domestic competitors. These could stem from know-how, human skills, brand names, capital capabilities, etc., which may be lacking in the periphery. The latter may also be disadvantaged in terms of market characteristics, labour attributes, and other location-specific factors, which are emphasised in Dunning's (2000) eclectic model of FDI along with the conditions affecting a firm's decision to internalise manufacturing activity.

The dynamic capability approach contends that the competitive advantage of a firm is built on its access to and control over tangible or intangible assets and the ability to combine them in building distinctive competences. Maskell (2001) argues that a firm's competences can be built on 'created localised capabilities'. Peripheral countries differ in these capabilities, which may find expression in the attributes of production located by TNCs.

Finally, one of the major hypotheses of the global commodity chains framework is that development requires a link-up with the major 'lead firms' in an industry

(Gereffi 1999). The lead firms control access to major resources, such as product design, new technologies, or brand names.

Theoretical concept Groups of features	Product life cycle	Uneven development	Innovative milieux and learning regions	Porter's clusters	New international division of labour	Dunning's eclectic paradigm	Dynamic capabilities	Global commodity chains
product	●	◐	●	◐	●	◐	◐	◐
production process	●	◐	●	○	●	●	◐	◐
labour	●	◐	●	●	●	◐	●	◐
market	◐	●	○	●	◐	◐	○	●
suppliers	○	●	◐	●	●	◐	●	●
non-production competences	●	●	○	◐	●	●	●	●
others*	○	●	○	●	◐	●	●	●

* - sunk costs, embeddedness, and other features highlighted in the above concepts which do not fit into the distinguished groups of characteristics; for complete list of features see figure 8.2

● strongly related to the concept
◐ related to the concept
○ weakly related to the concept

Figure 8.1   Theoretical concepts

All in all, the division of economies into core and peripheral (including emerging markets) can be identified in several theoretical concepts. The conceptual discussion allows one to distinguish a list of features attributed to production in the periphery grouped into seven main categories: the product, the production process, labour, the market, suppliers, non-production competencies, and others (Figure 8.1). They can be used as a template for assessing whether the automotive industry in Poland conforms to or departs from the general categories of periphery or semi-periphery viewed as ideal types. For example, in terms of production processes the periphery should be characterised by less advanced technology, limited automation, low capital-intensiveness, and more traditional organisation of production. In addition, non-production competencies, such as R&D, marketing, and purchasing would be limited.

Automotive Production in Poland: Features of the Periphery?

In this analysis, the contemporary characteristics of the automotive sector in Poland are explored primarily at the microeconomic level. The picture that emerges of an enormously varied set of foreign and indigenous companies is inevitably complex. Managers from 75 large and medium-sized companies were asked to compare the features of automotive companies and subsidiaries operating in Poland with their counterparts in Western Europe. The responses were expressed on a simple scale from one (significantly lagging behind Western Europe) to five (significantly better than in Western Europe), with three meaning similar to Western Europe.

Some characteristics needed to be assessed at a macroeconomic level of an entire sector/country (for example, impact on the local economy, dependence on one sector, TNCs, or a single firm), and this was undertaken using available statistical data.

In total, 43 characteristics of the Polish automotive sector have been assessed, and the median values of these assessments are presented in Figure 8.2. In this chapter, however, owing to space, only the most important and interesting characteristics and issues are explored that relate in particular to products, production processes, labour, markets, suppliers, and non-production competencies. These issues are dealt with in the remaining sections of this chapter.

Products

In the 1990s, major automobile companies saw Poland basically as a market and a source for lower-end cars and components. Consequently, simple labour-intensive products and products in a mature stage of their life cycle were manufactured, representing relatively low value-added, e.g. wiring harnesses and upholstery. At the same time, products with a high level of technological complexity that mobilised specialised competencies continued to be made in the traditional automobile heartlands of Western Europe. This pattern of production conformed with ideas of the product-cycle theory (Vernon 1996).

Since the end of the last decade, the picture has changed profoundly. Between 1990 and 2005, the value of foreign direct investment in the Polish motor vehicle industry exceeded US$10 billion. The output of the narrowly-defined automotive sector (NACE 34) increased nine-fold in Poland between 1991 and 2006 (at constant prices). During the last decade, the output of component manufacturers was growing more than twice as fast as the production of complete vehicles. Poland's exports of motor vehicles and components reached US$19.3 billion in 2006 (in comparison to US$0.6 billion in 1992) and exceeded imports by more than US$7 billion.

The export of complex and relatively high value-added products, e.g. engines, braking systems, and steering systems, has gained in significance (Figure 8.3). Manufacturers have been increasingly selecting their Polish plants for the production of new state-of-the-art products, such as engines, intended for the Europe-wide market. These products include both finished and intermediate goods, high-volume standardised products and low-volume niche products. At the same time, the production of certain simple labour-intensive components has remained in Poland, although it is beginning to be moved to cheaper locations such as Romania and Ukraine.

There is a clear trend towards greater economies of scale which finds its expression in a continuous expansion of the Polish production plants. This facilitates profitability and is part of a broader strategy of supra-national specialisation of factories and their integration into the European production network. In some cases, this entails delocalisation (reduction or closure) of production in Western

Europe. Whereas the economies of scale achieved in Polish plants are now often similar to those in the core countries, there is still some distance in the economies of scope *vis-à-vis* Western Europe. However, the tendency to broaden the range of

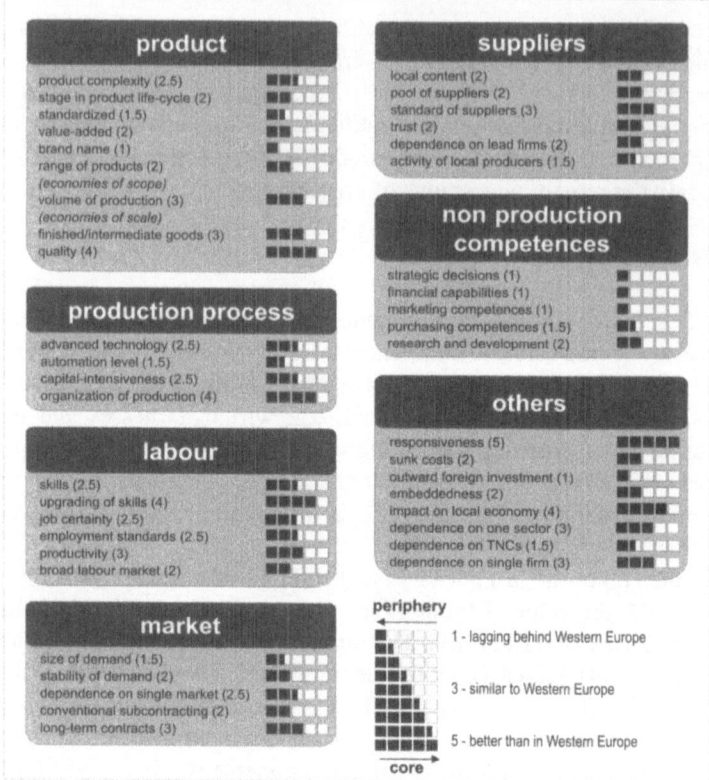

Figure 8.2   Automotive industry in Poland compared to Western Europe

manufactured products is evident in all large corporations operating in Poland.

One of the surprising findings of the study has been a widely-expressed opinion that by and large, Polish plants achieve higher quality standards than their counterparts in the automobile heartlands of Germany, France and the UK. This finds expression in their increasing export capabilities and is supported by quality indicators in intra-corporation assessments. Many of the plants investigated rank number one in Europe in terms of their ppm indicators (i.e. the lowest number of defective parts per million units produced). This shows the major progress achieved since 2000, when a study on foreign-owned companies in the automotive, electronics and electrical engineering industries in Poland revealed that just a few of the best plants approached West European quality standards (Domański 2001). What remains a classic peripheral economy feature in Poland is the lack of local

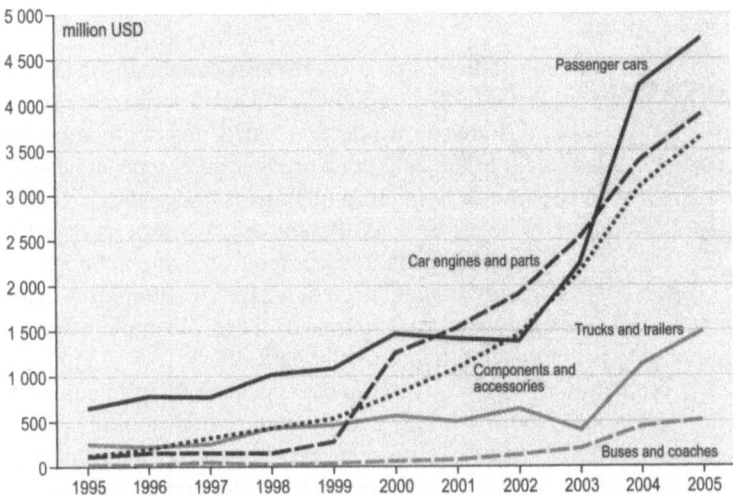

Figure 8.3   Exports of major automotive products from Poland, 1995–2005

brand name products – an effect of the predominance of large foreign corporations. The exceptions are Polish-owned OEMs in bus manufacturing.

Production Process

Since 1990, trends in production processes, including technology, have generally been similar to the changes in product characteristics. Assembly operations and simple labour-intensive processes were quite typical in the early years. Many investors began by testing the waters and introduced more advanced technologies at later stages. Some of the latecomers decided to begin with capital intensive, advanced technologies. As a result, the manufacturing activities of both existing and new automotive producers are increasingly capital intensive. Still, automation levels definitely remain lower than in Western Europe.

New modes of production organisation such as lean manufacturing, an alternative to the highly integrated Fordist model, and modularisation have been widely applied. In fact, their implementation may be more comprehensive in Poland than in the traditional manufacturing core of Europe. There are two possible explanations. Firstly, it has been easier to introduce new models of organisation at a time of overall radical change in Central and Eastern Europe than in stable Western societies. Secondly, implementation of such new solutions is always easier in greenfield locations than in existing plant. For example, the General Motors factory in Gliwice is regarded as the pinnacle of achievement in lean manufacturing within the corporation based on the experience of Japanese plants, the Californian NUMMI joint venture with Toyota and Eisenach. The Polish factory produces three completely different cars (Zafira, Astra and Agila) on a single assembly line and is a destination of study visits by GM staff from other countries.

Labour

The fundamental feature attributed to post-communist economies is the low cost of labour. This, however, cannot fully account for the good economic performance of foreign subsidiaries and domestic producers. Skills and motivation are vital labour qualities, which underlie the enhanced production competencies of Polish plants. The process of continuous upgrading of skills is under way.

The gap in productivity levels between Poland and Western Europe has been closing. The managers of major OEMs and component manufacturers maintain that their productivity is already higher. The advantage of automotive production in Poland and other Central European countries in terms of simple wage adjusted labour productivity is demonstrated in Eurostat data.

The good reputation of automotive employers allows them to select the best employees on the local labour market: mainly young, highly motivated people, who can be easily trained, adjust well to new production methods and lack 'bad' work habits. This lies at the root of good product quality and productivity.

Job security and working conditions in the Polish automotive industry generally compare favourably with other sectors, for example electronics and clothing. The automotive sector offers higher than average manufacturing wages and foreign corporations tend to offer significantly better rates than their local counterparts, although wages are obviously lower than in Western Europe. Moreover, contrary to common belief, employment standards do not normally diverge from those found in West European countries, and working conditions in new factories may be superior. The Polish labour code is less conducive to arrangements for flexible working hours than, for example, the new German contracts, but employees are generally eager to work overtime due to the high wage rates on offer.

A broad labour market, where a workforce with various skills is available, constitutes an important advantage of the core countries. Difficulties are faced in Poland in recruitment to certain specialist positions.

Market

The small size and volatility of demand may limit the scale of operations and undermine the profitability of production in less-developed countries. There is no doubt that market size remains one of the elementary advantages of the core countries in comparison to Central and East Europe, even though the latter has better growth prospects. In addition, fluctuations in new car sales in Poland are considerable. Nevertheless, this has a limited impact on the current development of automotive manufacturing in the country since 85 per cent of output is exported.

A vast range of components and vehicles is exported to various West European countries. Germany (28 per cent) and Italy (15 per cent) are the top export destinations. Thus, there is limited dependence on a single market. In contrast to the Czech Republic and Slovakia, these exports are also far more diverse in terms of the number of companies involved.

Intercity bus operators and public transport companies in many Polish towns are busy modernising their fleets while at the same time tourist operators are expanding. This creates a growing market for bus and coach manufacturers.

The rapid overall growth of production of final assemblers and first-tier component manufacturers together with their search for local suppliers provides an expanding market for second- and third-tier producers. Moreover, there is increasing demand for spare parts as a result of a rising number of newly registered vehicles in the country (from 12.5 million in 2000 to 16.8 million in 2005).

As most sales take place through transnational corporations, they are usually based on long-term contracts and conventional short-term subcontracting plays a rather limited role.

Suppliers

Initially, foreign car assemblers and component manufacturers relied on imported inputs to a large extent. Nowadays the local content in most automotive products manufactured in Poland is relatively high. The prevalent strategy of foreign corporations investing in Poland has been to increase local sourcing in order to lower their costs. This has been increasingly possible due to the rapid expansion of the entire automotive sector creating a larger and diversified pool of local suppliers. Three processes combine to produce the effect: modernisation and expansion of former state-owned factories acquired by foreign investors, mushrooming greenfield plants, and the development of new indigenous producers.

The increase in local purchases is underpinned by the reliability and quality of local suppliers. The confidence in Polish suppliers has grown gradually as foreign companies learnt that their requirements could be met in parallel with the fast learning of local firms. This is illustrated by the opinion of the American manager: 'At the beginning it was a disaster; the suppliers had no idea about QS, there were just a few with ISO certificates. There was a queer attitude to quality and productivity. Nowadays they are totally different companies. The Polish suppliers have made enormous progress; they provide good quality, flexibility and fast decisions'.

In the perspective of the global commodity chains approach, we can point out that a substantial part of the Polish automotive industry is linked with the 'lead firms', such as General Motors, Volkswagen, Fiat, Toyota, Volvo Bus. It is also linked to the major 'system integrators' – the companies that are capable of designing, manufacturing, and delivering complete modules, e.g. Delphi, Valeo, TRW. This provides access to leading product design, new technologies and markets.

The weakness of the automotive sector in Poland, heavily dominated by foreign manufacturers, is the limited role of indigenous producers. On the one hand, this reflects a history of isolation and the technological backwardness of the industry under communism resulting in the acquisition of major firms by foreign investors. On the other hand, it reflects global trends towards a reduction in the numbers of suppliers and the shift to the supply of complete systems or modules. The highly capital intensive character of production and the lack of R&D capabilities are formidable barriers to Polish-owned direct suppliers of OEMs. More commonly they figure among the manufacturers of pressed metal parts and plastic and rubber products, and there is an extensive network of second-tier and third-tier Polish suppliers. The positive trend is an increasing number of indigenous firms allied with transnational corporations which are beginning to supply specialised products for foreign markets. New markets for local businesses consisting of the delivery of tools and specialist machinery and its maintenance have been created by large automotive companies. In addition, there are numerous suppliers to the aftermarket.

Non-production Competence Areas

Strategic decision-making, marketing, purchasing and R&D form some of the key competence areas of core industrial countries. Their presence in the periphery tends to be rare. The majority of foreign-owned affiliates in Poland lack such competencies.

Purchasing has recently appeared as a new function of large subsidiaries. Some can choose their own suppliers within the country, others have become responsible for finding and auditing suppliers located in Central and Eastern Europe for the entire corporation. Fiat Auto Poland is an example of the latter. General Motors has gone even further and has globalised its direct materials purchasing. Managers responsible for purchasing particular commodity groups for the corporation are situated in various countries. In this intra-corporation division of labour, high-level decision-making competencies in purchasing certain commodities have been assigned to Gliwice in Poland.

In research and development, on the one hand, there is inertia of large established centres in Western Europe with strong competencies among the personnel and expensive laboratories. On the other hand, the increasing manufacturing activity in Poland, where the main plants for the production of numerous products are now located, may be an argument for integrating production and design in one place. Some of the major manufacturers (system integrators) have already established their R&D units in Poland. Valeo has created its R&D department at an engine cooling systems plant, TRW has opened its engineering centre for safety systems, Faurecia and Tenneco are following a similar path. Delphi has built a large technical centre specialised in the design of suspension systems (its staff complement reached 850 engineers in 2007). Thus, there are subsidiaries that are moving towards a 'product specialist' role within the corporation, combining the design and production of certain components.

Other Issues

What the survey material particularly emphasised is the greater flexibility of factories in Poland in response to customer demand in comparison to West European plants. The problem-solving focus among local managers and engineers is cited as a primary reason for this advantage. Thus, suppliers from Poland can provide not only high quality and low cost, but also fast response, which makes them attractive business partners.

Most large corporations appear to be following long-term strategies in Poland. They develop local networks of suppliers and create demand for various services, generating direct and indirect multiplier effects. The extensive, stable relationships entail trust in employees, suppliers and other partners, which constitute vital elements embedding foreign investors in the economy.

Still, the overall dependence of the Polish economy on the automotive sector is smaller than in some other European countries. The sector accounts for 19 per cent of Polish exports. It is heavily dependent on transnational corporations, but not on one or two companies. Outward foreign investment is insignificant, although some successful Polish producers have established subsidiaries in Ukraine.

Conclusion: From Periphery to Where?

The analysis shows that the automotive industry in Poland increasingly departs from several of its peripheral attributes. It does not fit the classic periphery model, where companies and subsidiaries build their strategies on low labour costs as the primary component of their profitability and function as Taylorist global outposts employing unskilled workers.

The enhanced role of Poland's automotive manufacturing in the European division of labour is most evident in its upgrading through the value chain into higher value-added products and first-rate product quality. Poland is now a place of production of a broad range of products: from sophisticated to simple, from specialised to standardised. There is the integrated production of vehicles and components with a high local content, as well as simple assembly. A somewhat larger gap, although closing, is found in production processes. Models of production organisation may already be more advanced than in Western Europe. An increasing importance is being placed on non-routine, skill-based operations. The adaptability and flexibility of producers located in Poland, including their willingness and ability to meet the specific requirements of customers and their fast response to customer demands are considered vital advantages. One has to bear in mind that the assessment is based on the opinions of managers of companies operating in Poland. However, in many cases those opinions are supported by quantitative indicators, e.g. in terms of quality and productivity. A less diversified supplier base and smaller domestic markets are characteristics of the periphery. There is some progress in R&D and some in purchasing functions, but on the whole Poland lags significantly behind the core countries in terms of non-production competencies, which is related to the overall dependence on foreign TNCs.

All in all, the good economic performance of Polish producers rests on high quality, reliability, adaptability, and fast response, as well as on low cost. It leads to success in intra-corporate competition over successive rounds of investment. Polish subsidiaries and companies win new projects (becoming suppliers to new car models) and are assigned to undertake new activities and new products. This results in an increase of production capacity and the widening of the range of manufactured products in existing and new factories.

Evidence of a trend towards complex upgrading processes in automotive subsidiaries in Poland is supported by other empirical studies (Fuchs 2005). Collective learning is similarly evident in developing countries of Asia and Latin America (Lara and Carrillo 2003; Ivarsson and Alvstam 2004). Lung (2004) claims that the learning processes are operating much more rapidly now in Central European automobile production than they did in Spain and Portugal in the 1980s.

The question is how can these processes be explained? According to Layan and Lung (2004, 69) manufacturing of innovative products in the periphery may result from: 1) a chronological convergence of various projects: greenfield plants host new products, and 2) the managing of risk on a territorial basis by relocating new products with unknown demand away from the central regions.

A broader explanation can be found in two other concepts: complementary specialisation and localised capabilities. The development of automotive production in Poland largely corresponds to the strategy of complementary specialisation identified by Kurz and Wittke (1998). Wage-cost advantage and market access do not sufficiently explain the involvement of TNCs in post-communist Europe. What we are seeing is a fundamental reorganisation of all tasks and functions within the value chain, which provides a basis for improved efficiency. The industrial capacities of Central Europe increasingly include capital-intensive production and complex, high-skill processes, which is most evident in the automobile industry. 'In this approach, not only would manufacturing processes be assigned to Central Europe, but so would functions of engineering, marketing, and innovation' (Kurz and Wittke 1998, 88).

This is consistent with the view that the European integration trajectory is a process of continued reconstruction of localisation advantages and specialisations within Europe and challenges 'a static vision which draws premature conclusions as to an alleged .... spatio-functional division that is featured in a stable centre-periphery model' (Layan and Lung 2004, 73-74). This supports a 'Europeanisation', rather than a 'hollowing-out' thesis.

The features which constitute the advantages conferred on subsidiaries and firms in the automotive industry in Poland can be interpreted as the creation of localised capabilities, which result from the interaction of corporations and the local environment. Investors have gradually learnt how to combine local assets, especially labour-force attributes, into desired localised competencies such as product quality and economic efficiency (Domański 2005). Thus, they both modify their earlier strategies and significantly change the economic environment of the country in which they are investing, rather than simply conforming to the conditions they met when they arrived. There has been a fast learning process among local staff and establishments. This is underlain by good general education, technical skills and industrial traditions, the motivation of employees and managers, cultural proximity to West European and American investors, political stability and EU membership. Moreover, geographical proximity to Germany is another advantage in comparison to Spain and even more to Asian countries, particularly in the case of just-in-time deliveries.

We must not forget that Poland is part of an emerging Central European agglomeration of the automotive industry comprising the Czech Republic, western Slovakia, Hungary and south-west Poland. Future prospects will significantly depend on the synergies and interconnections in this region (Pavlinek et al. 2009).

All things considered, the former post-communist periphery has gone a long way from its initial situation in the early 1990s, especially in terms of manufacturing capabilities and competencies. On the other hand, the technological, financial and strategic decision-making dependence on the companies from the core demonstrates the semi-peripheral position of the Polish automotive sector in the international division of labour and creates barriers and uncertainties about its future. Can the local capabilities created provide the basis for further growth and enhanced competencies of automotive firms in Poland in the years to come? Can Poland follow the path of the automobile industry in Spain, which has developed a relatively strong sector of indigenous component manufacturers and bus makers? Can growing economies of scale in manufacturing, and pressure from the local Polish managers who run foreign-owned subsidiaries, encourage further development of R&D and other services?

References

Bordenave, G. and Lung, Y. (1996), 'New Spatial Configurations in the European Automobile Industry', *European Urban and Regional Studies* 3, 305-321.

Domański, B. (2001), 'Poland: Labour and the Relocation of Manufacturing from the EU', in Gradev, G. (ed), *CEE countries in EU companies' strategies of industrial restructuring and relocation* (Brussels: European Trade Union Institute), 21-49.

Domański, B. (2005), 'Transnational Corporations and the Post-socialist Economy: Learning the Ropes and Forging New Relationships in Contemporary Poland', in Alvstam, C. and Schamp, E. (eds) *Linking Industries across the World: Processes of Global Networking* (Aldershot: Ashgate), 147-172.

Domanski, B. and Gwosdz. K. (2009), 'Toward a More Embedded Production System? Automotive Supply Networks and Localized Capabilities in Poland', *Growth and Change* 40 (forthcoming).

Dunning, J.H. (2000), 'The Eclectic Paradigm as an Envelope for Economic and Business Theories of MNE Activity', *International Business Review* 9, 163-190.

Freyssenet, M., Shimizu, K. and Volpato, G. (eds) (2003), *Globalisation or Regionalisation of the European Car Industry?* (Basingstoke: Palgrave Macmillan).

Fuchs, M. (2005), 'Borders to the Internationalisation of Knowledge? Engineering in Automobile Components Supply Companies in Poland', in Wever, E. and van Vilsteren, G. (eds), *Changing Economic Behaviour in a Unifying Europe* (Assen: Van Gorcum), 43-61.

Gereffi, G. (1999), *A Commodity Chains Framework for Analysing Global Industries*, <http://www.yale.edu/ccr/gereffi.doc>, accessed 20 May 2006.

Hudson, R. and Schamp, E. (eds) (1995), *Towards a New Map of Automobile Manufacturing in Europe? New Production Concepts and Spatial Restructuring* (Berlin: Springer).

Humphrey, J., Lecler, Y. and Salerno, M.S. (2000), 'Introduction', in Humphrey, J., Lecler, Y. and Salerno, M.S. (eds), *Global Strategies and Local Realities; the Auto Industry in Emerging Markets* (Basingstoke: Palgrave Macmillan), 1-15.

Humphrey, J. and Memedovic, O. (2003), 'The Global Automotive Industry Value Chain: What Prospects for Upgrading by Developing Countries', *UNIDO Sectoral Studies Series*, Vienna.

Hymer, S. (1976), *The International Operations of National Firms: a Study of Direct Foreign Investment*. Cambridge: MIT Press.

Ivarsson, I. and Alvstam, C.G. (2004), 'International Technology Transfer to Local Suppliers by Volvo Trucks in India', *Tijdschrift voor Economische en Sociale Geografie* 95, 27-43.

Lara, R. and Carrillo, J. (2003), 'Technological Globalisation and Intracompany Coordination in the Automotive Sector: The Case of Delphi-Mexico', *Journal of Automotive Technology and Management* 3, 101-121.

Layan, J. and Lung, Y. (2004), 'The Dynamics of Regional Integration in the European Car Industry', in Carrillo, J., Lung, Y. and van Tulder, R. (eds), *Cars, carriers of regionalism?* (Basingstoke: Palgrave Macmillan), 57-74.

Lung, Y. (2000), 'Is the Rise of Emerging Countries as Automobile Producers an Irreversible Phenomenon?', in Humphrey, J., Lecler, Y. and Salerno, M.S. (eds), *Global Strategies and Local Realities; The Auto Industry in Emerging Markets* (Basingstoke: Palgrave Macmillan), 16-41.

Lung, Y. (2004), 'The Changing Geography of the European Automobile System', *International Journal of Automotive Technology and Management*, 4, 137-64.

Maskell, P. (2001), 'The Firm in Economic Geography', *Economic Geography* 77, 329-344.

Pavlinek, P. (2002), 'Restructuring the Central and Eastern European Automobile Industry: Legacies, Trends, and Effects of Foreign Direct Investment', *Post-Soviet Geography and Economics* 43, 41-77.

Pavlinek, P., Domański, B. and Guzik, R. (2009), 'Industrial Upgrading through Foreign Direct Investment in Central European Automotive Manufacturing', *European Urban and Regional Studies* (forthcoming).

Porter, M.E. (2000), 'Location, Competition and Economic Development: Local Clusters in a Global Economy', *Economic Development Quarterly* 14, 15-34.

Richet, X. and Bourassa, F. (2000), 'The Re-emergence of the Automotive Industry in Eastern Europe', in von Hirschhausen, C. and Bitzer, J. (eds), *The Globalisation of Industry and Innovation in Eastern Europe* (Cheltenham: Edward Elgar), 59-94.

Sadler, D. (1997), 'The Role of Supply Chain Management Strategies in the 'Europeanization' of the Automobile Production System', in Lee, R. and Wills, J. (eds), *Geographies of Economies* (London: Edward Arnold), 311-320.

Sadler, D. (1999), 'The Internationalisation and Specialisation in the European Automotive Components Sector: Implications for the Hollowing-out Thesis', *Regional Studies* 33, 109-119.

Tulder van, R. (2004), 'Peripheral Regionalism: The Consequences of Integrating Central and Eastern Europe in the European Automobile Space', in Carrillo, J., Lung, Y. and van Tulder, R. (eds), *Cars, Carriers of Regionalism?* (Basingstoke: Palgrave Macmillan), 75-90.

Vernon, R. (1996), 'International Investment and International Trade in the Product Cycle', *Quarterly Journal of Economics* May, 190-207.

Worrall, D., Donnelly, T. and Morris, D. (2003), *Industrial Restructuring: The Role of FDI, Joint Ventures, Acquisitions and Technology Transfer in Central Europe's Automotive Industry,* Paper presented at Regional Studies Association Conference Reinventing Regions in a Global Economy, <http://www.regional-studies-assoc.ac.uk/events/pisa03/worrall.pdf>, accessed 20 May 2006.

Chapter 9

Toyotaism Travels to Poland:
A Case Study of a Hybrid Factory

Tomasz Majek and Roger Hayter

Introduction

During the 1970s, the explosion of Japanese exports of automobiles (and other products) led American and European governments in particular to introduce various forms of protection for their domestic industries. In response to trade barriers, including the 'voluntary export quota' system in the US, Japanese auto manufacturers began to invest in branch plants in the US, Canada, the UK and, later, in continental Europe. Among western countries, there was a widespread view – one that helped rationalise protectionist measures – that Japanese export success was based on low labour costs, government support and domestic market protection in Japan. Increasingly, however, the organisational efficiencies of Japanese production systems became recognised, heralded by Womack et al.'s (1990) landmark study that summarised the competitive advantages of the Japanese model in terms of 'lean production'. In the US, Toyota's first manufacturing investment in California in 1984, a joint venture with GM to resurrect a failed factory, was encouraged to help reveal the basis of lean production (to American manufacturers).

According to Womack et al. lean production is a global 'best practice' that could and should be copied and transferred around the world. Yet, numerous studies in geography have shown that Japanese foreign direct investment (FDI) in the auto and other industries has created branch plants that, to various degrees, modify Japanese practice in accordance with 'local cultures of production' (Edgington 1990; Kenney and Florida 1993; Mair 1992, 1994; Gatrell and Reid 2001). Within the business and related literatures these adaptations are often expressed as processes of hybridisation and the hybrid factory (Abo 1994; Itagaki 1997). This chapter contributes to geography's understanding of the hybridisation (and adaptation) of lean production by a case study of Toyota's new factory at Walbrzych (TMMP) in Poland. Within geography, previous studies along these lines have emphasised North American and West European contexts and have focused on the external relations of branch plants, especially with respect to firm-supplier relations. In contrast, this study focuses on the internal operations of Toyota's factory in Poland, a transitional economy within Eastern Europe.

The specific objective of the chapter is to assess the extent to which TMMP duplicates or differs from company operating norms in Japan. While this assessment is fundamentally qualitative, it is organised by a six-fold classification of factory organisation, developed in the business literature, that emphasise: work organisation, production control, procurement, group consciousness, labour relations and parent-subsidiary relations (Kumon 2004). Each attribute is qualitatively assessed in terms of departure from Toyota norms, based on information obtained via a visit to the factory, an in-depth interview with a senior representative of the company, and through a review of information provided by the company and government sources. The discussion is briefly framed by reference to the problems of assessing hybridisation.

Japanese FDI: Assessing Branch Plant Hybridisation

In general, the assessment of branch plant hybridity is not straightforward. So-called 'national' models of production in donor countries invariably exhibit internal differences, and changes over time. Corporate and factory cultures are highly nuanced and not readily measured by a common metric. Similarly, host economies vary in terms of their traditional practices by industry, stage of development and locality. Moreover, the nature of hybridisation is affected by the method of entry of FDI, for example, as green field investments facilitate adoption of parent company practices compared with entry achieved via acquisition of existing local operations.

In the specific context of Japanese FDI, there is both acknowledgement of a nationally distinctive 'Japanese model' of the organisation of production systems and of (dynamic) variations around this model among firms, even in the same industry (Liker et al. 1999). Moreover, the Japanese model is itself a hybrid. In an account of exceptional clarity, Odaka (2001) outlines the origins of Japanese manufacturing methods in a manner that implicitly privileges the concept of the hybrid factory both with respect to the evolution of lean production within Japan and its transfer beyond. For Odaka, the combination of manufacturing practices that emerged in Japan in the latter half of the twentieth century has resulted from the attempt to introduce American engineering and production methods into the unique social and economic environment of the Japanese economy. He illustrates the role of American FDI in the transfer of technology and knowledge and highlights the role of specific individuals as well as corporate benefactors in the development of Japanese methods of production. Moreover, Japanese FDI is a source of learning for parent companies as Japanese best practices are shaped by foreign experiences to some extent. From this perspective, learning processes are recursive, not simply from multinational corporations (MNCs) to the local economy.

Yet the dynamic, interactive nature of hybridisation and the problems of measurement are not meant to undermine its importance either to corporate performance or local development. Indeed, hybridisation is a vital issue from the perspective of local development, especially apparent in transitional economies such as Poland. Thus, the Polish government welcomes FDI, including from Japan, *precisely* to encourage transformation of traditional practices to global benchmarks, in order to become internationally competitive. In practice, this encouragement is dependent on the behaviour of individual firms as well as their ability and willingness to introduce parent company practices into Poland. Local groups, such as established unions, may resist change and so parent company practices may not be appropriate for all local situations. In such circumstances, adaptations may be required, and the efficiency of branch plants is dependent on the success of the modifications made to this now localised production system. Success on the 'shop-floor' is clearly significant for the host country; it is the objective of importing best practice in the first place.

If hybridisation processes are difficult to assess and locally contingent, Kumon's classification of six key attributes of factory organisation, already applied in studies of Japanese factories in America, East Asia and Western Europe (Abo 1994; Itagaki 1997), permits a quasi-systematic comparison of hybrid factories with parent company norms. In the following case study, the localised practices of Toyota, a particularly well-known (albeit evolving) 'template' model of the Japanese production system are assessed in the context of the company's first factory in Poland.

Toyota's Walbrzych Factory in Poland

Since the 1960s, but principally from the 1980s, Toyota has expanded its operations in Europe significantly and now operates facilities in the UK, Belgium, France, Portugal, Turkey as well as Poland (Kamiyama 2004). In 2001, TMMP was established in Walbrzych, a former mining town in a traditional industrial area of the country, to manufacture gearbox assemblies. This factory employs over 2,000 workers, represents an investment of over 400 million euros, and is a lynch-pin of Eastern European investments by Toyota which, like other Japanese companies, is a relative latecomer to the region. Following the establishment of TMMP, a second factory was opened to produce diesel engines further north in Jelcz-Laskowice. Generally considered a success, Toyota's factory in Walbrzych is assessed here according to the similarities and differences with parent company norms in the context of the following six attributes of factory organisation.

Work Organisation

Work organisation, intimately tied to local cultures, is arguably the most difficult element of a production system to implement in a foreign setting (Abo 1994). However, the Walbrzych factory closely approximates Toyota's work organisation principles, as defined by the wage system, job rotation, as well as training.

With respect to job classification, there are ten team members for each team leader, and three to four teams per production group at the facility. Each group has a group leader, who in effect fulfils supervisory functions and the supervisor-to-worker ratio is typical for Toyota (but relatively high in comparison to the auto factories of other Japanese manufacturers in Poland). For job classification and the range of tasks performed by workers on the shop floor, the sharpest distinction at Walbrzych is between production and maintenance workers, with the former rotated on the assembly line at two-hour intervals. This practice is similar to Toyota norms, although the rationale for job rotation in terms of worker training and reduction of monotony varied at Walbrzych. Thus, given the simple nature of existing operations at the facility in Walbrzych, the main objective of job rotation is to reduce monotony, rather than the training of workers in a diverse set of tasks. Nonetheless, as the company introduces more complex workflows at the factory in Walbrzych, job rotation will provide workers with skills in multiple areas of production.

Most workers training at Walbrzych is 'on-the-job', although advanced training for selected workers is undertaken in facilities in Japan. Internal training begins with a week-long initial period, required by Polish law, to familiarise new workers with health and safety regulations, basic operating procedures, house rules, and the Toyota production system philosophy. Subsequently, on-the-job training is designed to meet license-related obligations, to develop the competencies of team leaders and group leaders, and to explore the potential of individual workers. In this regard, the workers sent to Japan for training play key roles. At Walbrzych, as with Toyota as a whole, on-the-job training is an ongoing, continuous process that does not terminate after any pre-determined length of time. In contrast, our research shows that other Japanese auto manufacturers in Poland are less committed to ongoing training.

Typically, length of service as well as performance evaluations are used in order to determine wage levels in Japan (Abo 1994; Liker et al. 1999). At Walbrzych, although in operation for a relatively short time, length of service is important, as the first intake of workers are now earning more than newer workers from subsequent intakes. Over time, following parent company practices, it is the intention of Walbrzych's management to put more weight on performance evaluation in setting wages. Job category is also an important determinant of wage levels and maintenance workers consistently earn higher starting wages than production line workers, since young workers suited to maintenance tasks are in greater demand in the region as a whole.

The training and wage systems tie into the broader system of promotion used at Toyota's factory in Walbrzych. Promotion is largely internal, occurs across job categories and up the hierarchy from production worker to team leader and group leader. While management states that the company prefers to promote internally, sometimes experienced workers are recruited from outside the factory, especially in the case of new production projects. Nonetheless, in line with practice in Japan, the criteria used for promotion include a combination of length of service and performance evaluations, in addition to the particular requirements of experience in a specific production process.

Production Control

Toyota in Walbrzych acquired the vast majority of its factory equipment and machinery directly from Japan. Sources of equipment and machinery used in production are important for several reasons. First, the use of Japanese equipment means that maintenance activities become tied to specialists from Japan and local involvement is therefore limited. Second, the level of automation is in large part determined by the choice of equipment and machinery, and these must match the skill sets of local workers and the chosen work organisation practices at the facility. At TMMP, the recognition of these implications has led the company to decrease the level of automation in the facility, compared to what is in place in its factories in Japan. Since a high level of automation requires considerable maintenance expertise not available locally in Poland, the reduction in automation is aimed at reducing the facility's dependence on maintenance experts from the company's Japanese facilities.

Assurance of product quality is a complex, difficult operation and is key to Toyota's operation and overall image. The stated objective of management is to realise 100 per cent quality control within the production process and to completely eliminate quality control checks at the end of the line. This type of 'quality-built-in' is in line with practice in Toyota facilities in Japan, but is not yet possible in Poland because of limited experience among production workers. Even so, to instil the philosophy of quality-built-in at Walbrzych, Toyota employs quality control circles and other such tools to empower production workers with respect to the quality control process. Evidently, these tools are working, as data presented by the company indicated that defect rates at the facility were on par with similar facilities in Japan as of the spring of 2004. In contrast, other Japanese auto manufacturers interviewed in Poland have not made a similar commitment.

Maintenance practices relate to the level of automation, hiring and training philosophy and systems of quality control. TMMP hired a distinct (inexperienced) pool of maintenance workers based on their specific technical knowledge and skills. To complement this pool, engineers from Toyota Japan were used on a temporary basis. Indeed, there is a strong emphasis on developing in-house competencies in maintenance through a skills transfer and training programme with Japan. Some of Toyota's maintenance experts have more than 40 years experience in Japan and the plan is to transfer their knowledge to the new workers in Walbrzych. Nonetheless, at present, the facility lacks the depth of skill and experience required for all maintenance operations. Thus Toyota has modified the automation levels in the facility rather than adapt maintenance practices by out-sourcing servicing of (some) machinery, for example. The relatively low wage levels in Poland have meant that the impact of reduced automation and the corresponding increase in the number of production workers has not significantly affected TMMP's profitability.

With respect to the production line setup itself, as well as arrangements to cope with line failures or defects, Toyota brought in manuals from its UK operations when setting up the factory in Walbrzych in order to ease the implementation of Japanese practices into a European setting. Although the current product mix within the facility is low (the plant produces one type of transmission in several models), the facility employs typical elements in a Toyota production facility including fail-safe devices and production signal boards. Additionally, die change times and other measures of production efficiency were according to management, meeting expectations and comparable to facilities in Japan.

Procurement

Reliance on suppliers and the just-in-time philosophy are widely recognised aspects of Toyota's production system. Because Toyota's supply chain in Eastern Europe is relatively underdeveloped, however, the Walbrzych factory manufactures most of its components in-house, with little external sourcing and none from local suppliers. The analysis of relationships between the branch plant in Poland and Toyota's supplier network is complicated by the fact that Poland is becoming integrated within a broader European economic system. Over the years, Toyota has expanded its supplier base in Europe and increasingly draws on European components in the assembly of its European automobile models.

With respect to the procurement method by which the subsidiary relates to its suppliers, an interesting situation arises at Toyota in Walbrzych. While the company is philosophically dedicated to Toyota production system, including just-in-time, the operation is practically constrained by European logistics. In addition, the facility produces only one type of product in six models at present and production volume is limited, so that TMMP's thinking about logistics is not yet fully realised. Management intends to develop just-in-time and related practices as production volumes and product mix increase. Presently, since there are no local suppliers for the Walbrzych factory to use and logistics are uncomplicated, less than 30 per cent of supplies are sourced outside the plant and these are supplied from other Toyota factories. Here too, procurement logistics are constrained more by the limited maritime transport connections between Japan and Poland than by the unwillingness of the subsidiary to implement just-in-time practices.

Group Consciousness

The widespread practice among Japanese manufacturing companies to foster a shared worker consciousness and a sense of unity within their facilities is difficult to implement in foreign environments (Alston 1986). The goal of Japanese information-sharing practices among management, team leaders and workers, as well as between production, maintenance, and engineers is the company-wide dissemination of information regarding the current status of production, defect rates, and so on, as well as long term company strategy. The rationale for such practices is to foster shared commitments to problem solving and loyalty to the company.

Toyota has attempted to introduce Japanese-style small group activities, information sharing, and other practices to foster worker unity at Walbrzych. However, these initiatives have been limited to date and involve only about half of the shop-floor members. According to TMMP management, the short time that this facility has been operating is more of a factor than is Polish worker culture as such. Management argues that communication is only possible through 'informal channels' at Walbrzych because of the relatively small size of the workforce and the nature of production. The company believes that once the workforce grows beyond 1,000 a more formal information sharing system will need to be enacted within the facility.

The presence of small group activities, information-sharing and sense of unity within the workforce is indicated by the presence of uniforms, corporate logos, shared parking and cafeteria facilities and social events as well as in more formal elements of the production system. At TMMP, workers wear Toyota-badged uniforms[1] and cafeteria facilities are common and open to anyone, although there are no morning songs or exercises, nor common parking for management and employees. Interestingly, Toyota tries to involve the community in events and hosts a summer festival linked to the opening of the factory in Walbrzych, where interested workers and their families tour the plant and enjoy food and festivities.

Labour Relations

In Poland, unions have played a powerful role in industry and the political scene as a whole. Indeed, a clash between Polish trade unions and Toyota's preference for corporate unionism might be expected, for example, with respect to hiring policy, job security and grievance procedures.

In contrast to most other Japanese firms, who typically locate branch plants in 'new economic spaces', TMMP is located in a traditional industrial area, albeit within a newly designated special economic zone. Yet, despite a locally available large, unemployed, industrially experienced workforce, TMMP specifically seeks out young workers who are looking for their first job. In addition, beyond basic educational requirements (in most cases secondary school graduation), the selection process is geared to identify individuals who are seen as flexible, accepting of change and open to cultural exchange. According to management, the most important characteristic of new employees is their willingness to develop themselves eagerly, through continuous personal improvement. For TMMP the successful hiring of a young generation of flexible, keen, Polish workers has also meant the avoidance of the ingrained attitudes of older, formerly unionised workers from the country's industrial past.

---

1 Polish health and safety regulations require uniforms, but these do not have to carry the corporate logo of the employer (i.e. workers cannot wear jeans and a t-shirt).

With respect to grievance procedures, TMMP emphasises informal, open-style conflict resolution, with little formalised routine and external arbitration. In addition to specific resolution procedures outlined in the human resources management system documentation, workers are encouraged to speak directly with any level of management whenever problems arise. Management at TMMP highlights that this type of grievance resolution is 'a typical Toyota way', and so far, there have been no drives to unionise.

In the context of long-term job security, unlike at Toyota's facilities in Japan, there is no explicit 'no layoff' policy at TMMP in Walbrzych. Management stated that such a policy is currently not required, since the company was hiring more than 50 new people per month in Walbrzych alone in 2004. Nonetheless, the company recognises an unwritten commitment to long-term employment, along with the on-the-job training and continuous improvement programmes within TMMP.

Parent-Subsidiary Relations

Important indicators of Polish subsidiaries' relationships with the parent company include; 1) the ratio of Japanese expatriates, 2) the delegation of decision-making authority, and 3) the appointment of locals to managerial positions. In this regard, Toyota's facility in Walbrzych demonstrates a low percentage of Japanese expatriates, with only the most senior positions being held by personnel from Japan. Furthermore, management is committed to enabling Polish managers to run the facility without any assistance from Japanese expatriates, preferably within three to four years, and there is the intent to see a Polish president in the future. This commitment is reflected in company efforts to train two general managers, one in human resources and administration and the other in production.

Regardless of the nationality of management, the decision-making autonomy of the branch plant is an important issue with respect to foreign direct investment. At present, TMMP controls decisions at the factory and especially at the shop-floor level on work organisation, production control, labour and community relations, and to some extent procurement. However, decisions related to product mix, choice of suppliers and production volume are made at Toyota headquarters in the UK and in Japan. These strategic decisions cannot be made locally at present as TMMP has limited resources, knowledge and experience to manage and consolidate this type of operation beyond the shop-floor level.

Summary Scores of Hybridisation

To develop summary measures of branch plant hybridisation, Abo (1994) and others (Itagaki 1997, Boyer 1998, Liker et al. 1999) have assigned scores to attributes of factory organisation to reflect the closeness of branch plant practices to those of the parent company. Usually, five point nominal scales have been used, with a score of five indicating virtually identical practice in both the parent company and the branch plant, and a score of one indicates little or no similarity between branch plant and parent company practices. While the assignment of scores is judgemental, and there is no question of statistical significance, such hybridisation scores are a useful summary of real differences in practices.

Using this approach, hybridisation scores have been developed for each of six attributes of factory organisation at Walbrzych (Figure 9.1, Majek 2005). As Figure 9.1 shows, all six elements of Toyota's production system are implemented to some extent at Walbrzych, but only production control and work organisation elements are near equivalents of practice in Japan. Other elements showed considerable adaptation to the local circumstances. Just-in-time procurement and contracting practices are clearly constrained by the lack of local suppliers and limited transport connections with Japan. However, the reasons for the adaptation of labour relations, group consciousness and parent-subsidiary relations reflect managerial discretion rather than the imposition of hard constraints. These findings resonate with recent empirical studies of hybrid factories in Europe (Abo 2004), and demonstrate that Toyota has adapted operations to suit local conditions and maintained the overall integrity of its operations. Other hybrid branch plants in the US and East Asia have been less successful (Abo 1994, Itagaki 1997). In this regard, TMMP's relative success is at least consistent with management's claim to be committed to 'learning the patterns' of productive organisation that work locally (Interview, May 2004). In this sense, Toyota's Walbrzych plant highlights the importance of what Gatrell and Reid (2001) identify as 'cultures of production', wherein the competencies of the firm are leveraged using the strengths of the local community.

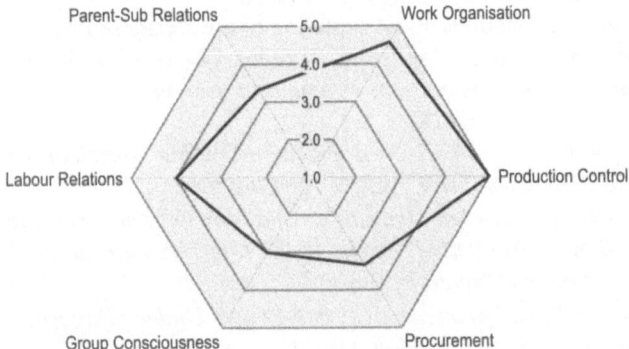

Figure 9.1   Hybridisation of the production system at Toyota's factory in Walbrzych

Conclusions

The implications of this study are twofold. First, contrary to what has been determined in earlier studies (most significantly, Abo 1994), the hybridisation of lean production does not necessarily result in performance penalties. Even with highly modified practices spanning procurement, group consciousness, parent subsidiary relations and labour relations, Toyota's Walbrzych factory matches the company's Japanese facilities in many important measures of performance, including production defect rates (Interview May 2004). In this regard, the hybridisation of factory organisation is a proactive measure taken to safeguard operational efficiency. Thus, while average wages in the region are lower than in Japan, the company has implemented a lower level of automation in Poland not to capitalise on this difference but to avoid maintenance problems in a context where experienced technicians are not available.

Second, and perhaps more importantly, evidence gathered in this study supports the idea of a global *Toyotaism* fuelling localised development as proposed by Fujita and Hill (1995). A substantive case can be made that hybrid practices are the result of a learning processes to find locally functional equivalents of practices that work in Japan. To this extent, our findings resonate with calls for the recognition of multiple co-existent models of productive organisation, although perhaps in a more fluid framework than that suggested by Boyer et al. (1998). An interesting question is whether or not it is possible to identify hybrids that offer the most potential for local development in the long run.

References

Abo, T. (1994), *Hybrid Factory: The Japanese Production System in the United States* (New York: Oxford University Press).

Abo, T. (2004), 'Applications and Adaptation of the Japanese Production Systems in Europe', in Kumon, H. and Abo, T. (eds), *The Hybrid Factory in Europe: The Japanese Management and Production System Transferred* (New York: Palgrave Macmillan), 52-74.

Alston, J.P. (1986), *The American Samurai: Blending American and Japanese Managerial Practices* (New York: W. de Gruyter).

Boyer, R. (1998), *Between Imitation and Innovation: The Transfer and Hybridisation of Productive Models in the International Automobile Industry* (New York: Oxford University Press).

Edgington, D. (1990), *Japanese Business Down Under: Patterns of Japanese Investment in Australia* (London, New York: Routledge).

Fujita, K. and Hill, R. (1995), 'Global Toyotaism and Local Development', *International Journal of Urban and Regional Research* 19:1, 7-22.

Gatrell, J. and Reid, N. (2001), 'The Cultural Politics of Local Economic Development: The Case of Toledo Jeep', *Tijdschrift voor Economische en Sociale Geografie* 93:4, 397-411.

Itagaki, H. (1997), *The Japanese Production System: Hybrid Factories in East Asia* (London: Macmillan Business).

Kamiyama, K. (2004), 'Toyota', in: Kumon, H. and Abo, T. (eds), *The Hybrid Factory in Europe: The Japanese Management and Production System Transferred* (New York: Palgrave Macmillan), 231-239.

Kenney, M. and Florida, R.L. (1993), *Beyond Mass Production: The Japanese System and its Transfer to the U.S.* (New York: Oxford University Press).

Kumon, H. (2004), 'Analytical Perspectives on Japanese Factories in Europe', in: Kumon, H. and Abo, T. (eds), *The Hybrid Factory in Europe: The Japanese Management and Production System Transferred* (New York: Palgrave Macmillan), 1-31.

Liker, J.K., Fruin, W.M. and Adler, P.S. (1999), *Remade in America: Transplanting and Transforming Japanese Management Systems* (Oxford: Oxford University Press).

Mair, A. (1992), 'Just-in-time Manufacturing and the Spatial Structure of the Automobile Industry: Lessons from Japan', *Tijdschrift voor Economische en Sociale Geografie* 83:2, 82-92.

Mair, A. (1994), *Honda's Global Local Corporation* (London: Macmillan).

Majek, T. (2005), *The Hybridisation of Lean Production: The Case of Japanese Subsidiaries in the Polish Auto Manufacturing Industry.* Unpublished MA thesis (Burnaby, Canada: Simon Fraser University).

Odaka, K. (2001), 'American Factory - Japanese Factory', *Social Science Japan Journal* 4:1, 59-75.

Womack, J.P. (1990), *The Machine that Changed the World: Based on the Massachusetts Institute of Technology 5-million Dollar 5-year Study on the Future of the Automobile* (New York: Rawson Associates).

Chapter 10

Spatial Division of Competencies and Local Upgrading in the Automotive Industry: Conceptual Considerations and Empirical Findings from Poland

Johannes Winter

Introduction

The automotive industry is Europe's leading industrial sector and transmitter of innovations and cyclical trends. Due to the sector's downturn in the 1990s, car manufacturers have begun to focus on their core competencies, such as product development, process engineering and co-ordination of the value chain. Initially, unprofitable functions such as manufacturing were outsourced to subsidiaries and suppliers. Today internationalisation also affects knowledge-intensive capacities. As a result, new patterns emerge in the global economy. A shift can be observed away from the traditional division of competencies, where complex and knowledge-intensive capacities are based in the Triad and labour-intensive ranges in developing countries (Dunford 2003, 842). Instead, greater variety emerges in the international division of competencies, which opens up upgrading opportunities for emerging markets.

Industrial upgrading enhances competitiveness of locations and gives impetus to regional development. But what do the upgrading opportunities of locations depend on? There is broad understanding of the impact of trade on regional development, but there is far less understanding of the impact of how that trade is organised (Schmitz 2004). This chapter suggests that the organisational structure of industries is instrumental in understanding the upgrading opportunities of locations and their regional effects. It is argued that upgrading depends on the way value chains are co-ordinated. While the resources for this process mostly emerge locally (Humphrey and Schmitz 2004, 95), decisions on upgrading are mostly taken centrally within value chains. Nevertheless, there is scope for some locally triggered development. The aim of this chapter is to highlight divisions of competencies and to provide two cases for local upgrading in an emerging market using the automotive industry in Poland as an example.

Methodology and Conceptual Framework

This contribution is empirically based, relying on primary data collected through 55 qualitative guided interviews with executives of the automotive industry, branch representatives and stakeholders in Poland and Germany. The interviews were conducted in 2005 and 2006. Subsequently, they were transcribed and analysed using 'Qualitative Content Analysis' (Glaeser and Laudel 1999). The study forms part of an ongoing research project at the University of Cologne/Germany, funded by the German Research Foundation.

The two cases selected for analysis are Polish subsidiaries of the German original equipment manufacturer (OEM) Volkswagen Commercial Vehicles and the American supplier Delphi Automotive Systems. Both exemplify local upgrading at different levels. Volkswagen Poznan was initially one of many 'extended work benches', but has now been transformed into a competitive company with extended technical and organisational competencies (Winter 2006, 46-52). In contrast, Delphi Cracow not only built up labour-intensive, but also knowledge-intensive capacities in Poland. Admittedly, both corporations are unique cases. Volkswagen has its own internationalisation strategy, where export not only refers to technology but also to organisational principles such as strong union representation on the supervisory board. Delphi is unique because of its critical financial situation which has required radical cost reduction and has led to reallocation of capital-intensive competencies such as R&D. But, what are the differences between these two specific cases and the Polish subsidiaries of companies such as the regionally well-connected Fiat Corporation, which has produced licensed products in Poland since the 1970s, or Bosch, with its special ownership structure in which 90 per cent of the shares are held by a charitable foundation? This chapter argues that corporate singularity should indeed be considered in any analysis of the division of competencies. However, the individual circumstances exemplified in the two case studies also mirror certain general trends that can be observed in the division of competencies and in local upgrading in the automotive industry.

Conceptual Framework

Competencies are defined in two different ways. Firstly, the term refers to any individual and organisational skills, capabilities or resources that are bound to staff. Employees acquire knowledge of intra-corporate rules of conduct as well as the capability to implement them within the specific context (Kolk and Van Tulder 2005, 3). Secondly, competencies refer to all forms of technical and organisational responsibilities that are necessary for meeting intra-corporate requirements and the needs of the market (Dosi and Teece 1998, 284). Such competencies are assigned to the company as a whole by corporate management. To specify the interlocational division of competencies which may result in local upgrading – or local downgrading if competencies are forfeited – a conceptual framework is proposed here suggesting four levels of corporate competencies (Figure 10.1):

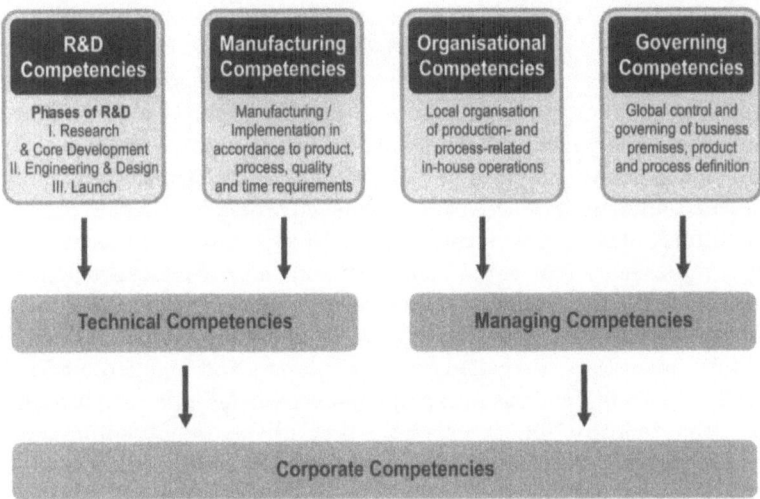

Figure 10.1   Levels of corporate competencies

R&D and manufacturing competencies can be summarised as technical competencies, whereas organisational and governing competencies are defined as managing competencies. In economic geography, the recent debate on the internationalisation and division of competencies has mostly focused on the allocation of technical and organisational competencies rather than how governing competencies are divided. This is due to the fact that regions and their residing companies have always been a central focus. Many companies are keen to maintain their highly developed competencies in core economic regions. A key factor is to what extent knowledge is tied to these specific regions by untraded interdependencies (Storper 1999) or to what degree they may undergo ubiquitification (Malmberg and Maskell 1999). Emerging markets are also confronted with the question of how to extend and build up competencies (Humphrey et al. 2000). These are markets which have already passed through import-substitution industrialisation and now need globally marketable knowledge in order to obtain access to core markets.

Spatial Division of Competencies and Local Upgrading

The main arguments for centralising competencies are the retaining of strategic, firm-specific knowledge at company headquarters and proximity to customers and core markets. Driving forces for decentralisation are cost advantages, increases in labour flexibility, access to skilled workforce, market exploitation and the need to meet local content requirements. Thus, corporate management is faced with a variety of reasons both for and against centralisation. Decisions on the spatial division of competencies are accompanied by continuous evaluation processes, clarifying whether selected paths of internationalisation should be abandoned or continued (Fuchs 2005).

In order to understand the division of competencies and their impact on upgrading opportunities, the governance mechanisms in the value chain need to be considered. Chain governance refers to the exercise of control along the chain, including the inter-firm relationships and institutional mechanisms through which non-market co-ordination of activities is achieved (Humphrey and Schmitz 2004, 98). Two features of chain governance emerge: first, co-ordination across company boundaries represents a relevant influence; and second, processes are increasingly dominated by key drivers of the global economy (Gereffi et al. 2005, 5). Gereffi (1999) has differentiated between buyer-driven and producer-driven chains. Buyer-driven chains are dominated by large retailers and branded marketers. In producer-driven chains, such as in capital- and knowledge-intensive branches like the automotive industry, OEMs operate as key drivers which co-ordinate their intra- and inter-corporate production networks (Dicken et al. 1995, 10). Such governance mechanisms heavily influence the trajectories open to locations.

Acquisition of competencies can take place through top-down processes, where decisions to hand down capabilities and responsibilities are taken by central management, or bottom-up-processes, that result from local management initiating co-operations with research institutes. Once acquired, competencies are relevant in different fields. The subsidiary may introduce an innovative product, implement a superior technology or gain new divisions for his firm. Upgrading enables emerging markets to build up and acquire additional competencies, to link up with global production networks and to achieve a better position in the chain. Humphrey and Schmitz (2004, 95) differentiate between three categories of upgrading and these are outlined in Figure 10.2.

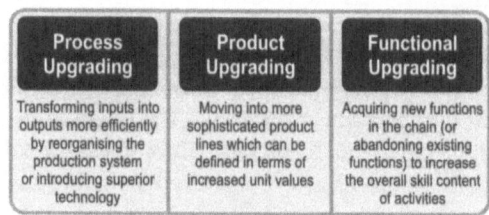

Figure 10.2   Categories of upgrading

Upgrading in value chains is subject to well-targeted influences by corporate and regional actors. These interactions include OEMs, suppliers, R&D centres and various stakeholders. Schmitz (2004, 7) identifies four types of relationship in value chains: a) market-based, in which enterprises deal with each other through arms-length transactions; b) balanced networks, in which enterprises co-operate and have complementary competencies but no control over each other; c) quasi-hierarchies, in which a lead firm sets the parameters within which others in the chain operate; and d) hierarchies, where enterprises are vertically integrated and the parent company controls its subsidiaries. The present chapter focuses on upgrading in hierarchically organised chains (type d) which represent the dominant mode of operation in the automotive industry.

The Automotive Industry in Central Eastern Europe

Central Eastern Europe (CEE) is one of the most important growth regions for FDI worldwide. According to UNCTAD (2005, 84) approximately US$20 billion FDI were allotted to locations in CEE. Of these, Poland, the Czech Republic, Slovakia and Hungary benefit most. Although the focus of FDI is likely to shift to the external borders of the EU, numerous OEMs and mega-suppliers are still located in CEE. The attractiveness of CEE is based on low costs, competitive productivity and qualified labour. Labour costs in the Polish manufacturing industry amount to approximately five euros per hour, while in Western Europe the figures are six times higher. This wage differential was economically justified in the past, when many affiliates were still in their start-up phase and a long way from economic standards such as 'zero-error-production'. Today however, some CEE locations approximate EU-standards in terms of profitability and productivity. Official statistics document a strong rise of labour productivity – defined as gross output per employee – and product quality in CEE. Yet, there are marked regional disparities: whilst front runner Slovenia already exceeds the average EU-15 productivity level, Polish and Czech workers are currently only half as productive as their Western European colleagues. Thus, due to high productivity and low costs, the automotive industry in CEE is highly cost competitive, even leaving some scope for future wage increases (Urban 2005, 18). As far as the spatial division of FDI in the automotive industry is concerned, the majority of capital appropriation flows into the automotive belt, which extends from south western Poland, the Czech Republic and Slovakia to north western Hungary. Currently, approximately seven per cent of global car production takes place in CEE, with further growth expected.

The Automotive Industry in Poland

Poland has four large car manufacturers: Fiat, GM, Volkswagen and the tarnished national manufacturer, FSO. In total, approximately 580,000 cars were produced in 2007 (CSM 2007). In addition, there are producers of engines and gear boxes as well as more than 150 component suppliers. The Polish automotive industry, which represents an important national industry, has approximately 90,000 employees. About 25 per cent of all FDI in manufacturing are allocated to this branch. Beyond low costs, Poland's spatial proximity to Western Europe is a facilitating factor in terms of logistics, marketability and socio-cultural closeness. Poland's accession to the EU in 2004 has stabilised political and economic conditions. Another important benefit is a ready pool of highly skilled people with experience in car production: from the 1970s until the end of the cold war, the state-owned companies FSO and FSM produced cars in Poland under a Fiat licence.

Upgrading in the Polish Automotive Industry

Since the 1990s, Poland has attracted mainly the labour-intensive capacities of the global automotive value chain. OEMs built up assembly plants, whilst numerous suppliers strengthened their manufacturing presence in close proximity to their customers. Recently, R&D competencies also began to move eastwards. Whilst most OEMs keep their R&D departments organisationally central, mega-suppliers such as Delphi, Remy and TRW have decentralised engineering competencies to Poland. The following sections give two examples of local upgrading in Polish subsidiaries.

Upgrading at Volkswagen Poznan

Volkswagen Commercial Vehicles is the market leader in Western Europe within the segment of light commercial vehicles. About 22,000 employees work at the three production sites in Germany (Hanover), Poland (Poznan) and Brazil (Resende), turning out 440,000 vehicles per year. Because of increasing global competition and cost pressures, Volkswagen began to reorganise its production process by implementing a new technology called virtual concept development. The cost-intensive and complex process of producing vehicle prototypes, which is part of application engineering, was substituted by virtual applications such as 'Computer Aided Design' of parts and components. Time saved amounts to 30 per cent; cost saved up to 75 per cent. The prototype is only built when the conceptual design phase has been completed. This example illustrates process upgrading at the corporation's central location which is responsible for product and process development. At non-central locations, the initiation of upgrading processes depends on the interplay between corporate decentralisation strategies and local efforts, as the following example illustrates.

The Volkswagen plant in Poznan was established in 1993 as a limited company. This was made possible by a joint venture between Volkswagen and the state-owned producer Tarpan. Since 1996, all shares have been owned by Volkswagen. In its early stages, the Polish subsidiary employed around 500 people, assembling 5,000 vans annually in CKD-modus ('completely-knocked-down'). Working through transnational production networks, cars built in a low-cost-country are imported from core regions as CKD-kits (modules) and assembled locally. The objective of this form of organisation is to avoid high import taxes levied on complete cars and to receive tax allowances for providing local employment (Winter 2006, 46). This production system corresponds to the extended workbench: labour-intensive and low-skilled operations are cut out of the existing production processes at central locations and moved across the border to subsidiaries.

The upgrading process at Volkswagen Poznan began in 2002 and included centrally instructed investments amounting to 600 million euros. The transition process can be characterised as organisational development from a highly dependent car assembler to a fully integrated manufacturer operating as a profit centre. Changes affected both product and process. Today, 6,700 employees turn out 167,000 vehicles per year and supply some 4.5 million components, such as pipe modules and cylinder heads to sister plants. About 95 per cent of vehicles produced in Poznan are exported, as are almost 75 per cent of components and modules.

To attain competitiveness and to reduce costs and quality problems in the medium-term, the Polish subsidiary implemented a single new assembly line instead of separated assembly areas. It also used 'just-in-time' logistics, created a local supplier park and introduced new car models. These measures are based on top-down decisions. Central management hands out the corporate strategy and defines the scope for action and decision-making for its affiliates. But, the hands of local management are not completely tied. Volkswagen Poznan won a high-profile global tender which included the design and manufacturing of steering gear housings, now produced at a corporate foundry in Poznan. Another bottom-up process was initiated by Polish management by promoting local investment strategies to the headquarters investment committee. This successful assertion enabled a profound transformation of the plant including the construction of new plant departments, such as a car body and painting shop. In addition, local management built up a public-private vocational school in co-operation with the municipal government to improve the professional education of job applicants. As previously stated, local upgrading is strongly influenced by the structure of the value chain and its key drivers. However, this case study shows that central locations not only set out corporate standards and govern peripheral trajectories, they also give room to local initiative and efforts.

Upgrading at Delphi TCK Cracow

Delphi Automotive Systems is one of the largest automotive suppliers worldwide and produces of electronics, transportation components and systems technology. The American company has 172,000 employees in about 40 countries. Since its 1999 spin-off from General Motors, Delphi has struggled financially, requiring it to look for cost-effective and more efficient solutions for product design and manufacturing. Cutting several hundred jobs in Western Europe, Delphi concurrently attempted to consolidate operations in CEE, where many of its customers are located. Presently, Delphi is embarking on an aggressive expansion strategy which, for example, includes the establishment of R&D competencies beyond the Triad. Several technical centres have been opened in emerging markets, and the Technical Centre Kraków (TCK) was the most recent of 31 centres spread across 11 countries.

In Poland, Delphi started its business in 1994, building up six production sites for shock absorbers, exhaust systems and electrical systems. Its regional headquarters, located at Cracow, not only manages the country's organisation but also CEE-wide operations in terms of logistics, purchasing, tax and legal operations, and human resources. The location in Cracow has passed through different types of upgrading. In 2000, Delphi was the first automotive corporation to develop an engineering centre in Poland. The TCK was established at the regional headquarters. Today, about 450 engineers and technicians support the development and design of components and modules for half a dozen OEMs. Eighty per cent of the staff are graduates of regional universities, such as the Cracow University of Technology. Initially, TCK started out by providing engineering support for the production of shock absorbers. As a result of restructuring at Delphi's Western European plants, the Polish location extended its range of activities. New R&D competencies were acquired to provide engineering support for exhaust systems, electrical systems and software development for electronic control units. The design department and the laboratory carry out simulation, modelling and testing for component development. Additionally, a workshop for prototyping was created and a local supplier network established to reduce delivery times and costs.

Evidently, the TCK already possessed a high level of competency, acquiring R&D competencies without recourse to manufacturing competencies. Nevertheless, all peripheral activities represent support functions for triadic R&D centres. The TCK is involved in application engineering and design but not in core development. The strategic and most of the operational decisions on development, planning and control of production and processes are taken centrally. The Polish location can thus be characterised as an extended workbench at a knowledge-intensive level. In contrast to the Volkswagen case, the Delphi parent company decided to establish not only manufacturing and organisational, but also R&D competencies locally. Through its stronger presence in CEE, Delphi attempts to combine high technology with lower engineering and manufacturing costs. There were plans for upgrading the TCK to a global centre of expertise for shock absorbers. Yet, corporate restructuring as well as local growth in terms of engineering experience led to extended local upgrading. These extensions of technical and managing competencies were initiated top-down. Local influence on the strategic development of the TCK is much lower than in the Volkswagen case. This can be explained by the corporate structure of Delphi, which is characterised by strong hierarchies and centralisation of governing competencies. However, since business expectations are being met and the region continues to be attractive for FDI, highly dependent affiliates can recommend themselves for upgrading (Fuchs and Winter 2008).

Comparison of Upgrading Processes in Poland

The scope for upgrading in Polish subsidiaries is closely related to the organisational patterns of the global automotive value chain. There is, however, still some scope for locally initiated processes. These local processes depend on the commitment of local management and regional stakeholders. In the case of Volkswagen, the Polish subsidiary acquired additional manufacturing and organisational competencies in several fields during the transition process from a CKD-assembly plant to a competitive firm with extensive vertical integration. In contrast, Delphi TCK acquired competencies to support engineering for Western European plants. It has maintained advanced competencies due to successful business development and increasing experience (see Figure 10.3). This leads to the conclusion that upgrading in hierarchical production networks is mainly initiated by central locations, but can be amplified as a result of local efforts and regional incentives. High quality engineering resources linked to low labour costs and a well-prepared work force were crucial factors for Delphi to further invest in TCK.

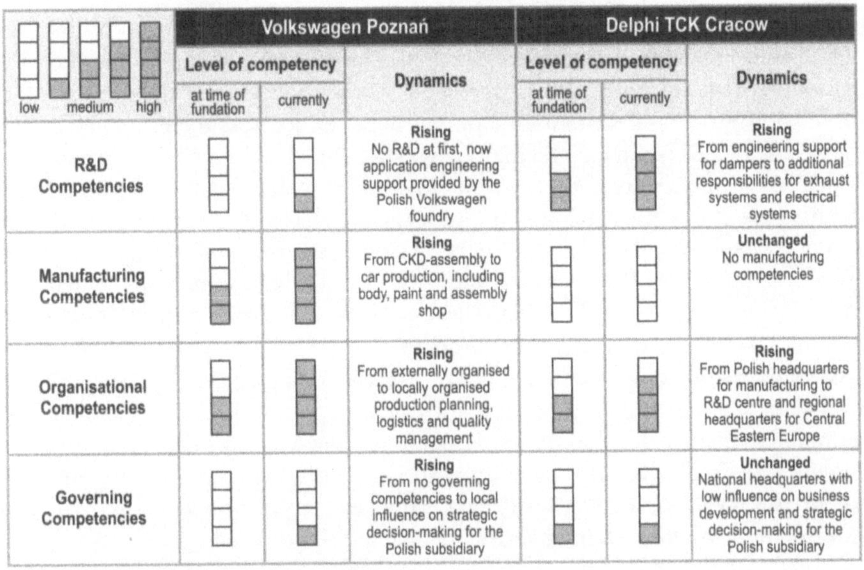

	Volkswagen Poznań			Delphi TCK Cracow		
low medium high	Level of competency		Dynamics	Level of competency		Dynamics
	at time of fundation	currently		at time of fundation	currently	
R&D Competencies			**Rising** No R&D at first, now application engineering support provided by the Polish Volkswagen foundry			**Rising** From engineering support for dampers to additional responsibilities for exhaust systems and electrical systems
Manufacturing Competencies			**Rising** From CKD-assembly to car production, including body, paint and assembly shop			**Unchanged** No manufacturing competencies
Organisational Competencies			**Rising** From externally organised to locally organised production planning, logistics and quality management			**Rising** From Polish headquarters for manufacturing to R&D centre and regional headquarters for Central Eastern Europe
Governing Competencies			**Rising** From no governing competencies to local influence on strategic decision-making for the Polish subsidiary			**Unchanged** National headquarters with low influence on business development and strategic decision-making for the Polish subsidiary

Figure 10.3   Dynamics of upgrading in the selected Polish subsidiaries

Acquisition of competencies and local upgrading not only affect plant level, but also the regional level activity. Upgrading stimulates regional development in terms of not only purchasing power, employment and tax revenue, but also through the improvement of educational systems and regional skills in manufacturing and engineering. Stimulating factors also include the initiation of applied research in R&D centres and technical universities, and the development of organisational competencies through changes in the supply chain. Whilst, in the case of Volkswagen, the Polish plant initially had to import most assembly components from abroad, today more than one-third are delivered by local firms. Companies are searching for new suppliers locally, which should be upgraded to competent partners. The main objective in integrating local suppliers is to save costs, to ensure 'just-in-time' production and to reduce the impact of mega-suppliers on the value chain. The impacts generated by transnational corporations therefore enable local firms and the surrounding regions to acquire additional competencies and make better use of their upgrading opportunities.

Conclusions

Based on two empirical case studies from Poland, this chapter has shown that upgrading in the automotive industry may extend the operational scope of integrated locations whilst also allowing them to map out more independent local business development strategies. The chapter points to the critical roles played by organisational structures and the way value chains are co-ordinated in maximising the upgrading opportunities of places and locations. In the producer-driven automotive chain, OEMs operate as key governors, managing their global production networks even though suppliers become increasingly influential through the acquisition of technical and managing competencies. In both OEMs and mega-suppliers, headquarters set the parameters under which subsidiaries operate: by defining the corporate standards for products, processes and administration as well as the interlocational division of capabilities and responsibilities. Nevertheless, some niches exist for locally triggered development processes when local management and regional stakeholders provide positive business development and successfully acquire additional competencies. Figure 10.4 is a diagram of current upgrading opportunities for emerging markets based on empirical findings. From this diagram it is clear that the greatest opportunities for upgrading in the automotive value chain exist in production-related areas such as engineering, logistics and manufacturing. Although these require technical and organisational know-how, tasks are standardised and barriers to entry are much lower than in core development which is defined as a core competency of triadic producers.

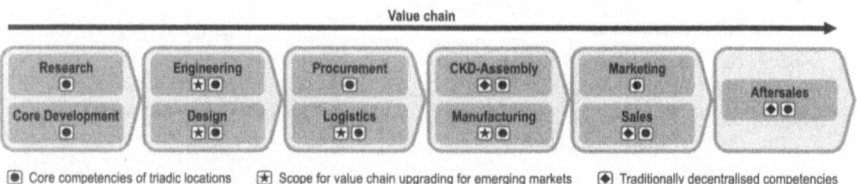

Figure 10.4 Division of competencies and upgrading opportunities in the automotive value chain

Thus, emerging markets such as Poland benefit from the reshaping of the spatial division of competencies by attracting FDI and additional capabilities and responsibilities. These regions are, however, heavily dependent upon only a very small number of foreign corporations and their predominantly centrally-positioned decision-makers. Thus, it is essential for emerging markets not only to acquire competencies, but also to anchor them locally and complement them by self-generated (in-)tangible assets. This could be achieved for instance by building up local firms, enabling them to become trading partners of established companies, by creating human capital, generating highly skilled employees in vocational schools and technical universities in the region, or by improving the infrastructure, connecting local firms to their customers and markets. Without an appropriate focus on independent local development and upgrading strategies, there is a risk of these places becoming locked into a race to the bottom, competing globally by providing lower costs and regulations, while being increasingly threatened by still cheaper and more deregulated competitors.

Acknowledgements

I am grateful for the constructive feedback on earlier drafts of the article by Martina Fuchs, Michael Plattner, Heiko Faust, André Scharmanski and two anonymous reviewers.

References

CSM Worldwide (2007), *CSM Global Automotive Production Survey* (Detroit: CSM).

Dicken, P., Hudson, R. and Schamp, E. (1995), 'New Challenges to the Automobile Production Systems in Europe', in Hudson, R. and Schamp, E. (eds), *Towards a New Map of Automobile Manufacturing in Europe? New Production Concepts and Spatial Restructuring* (Berlin: Springer), 1-20.

Dosi, G. and Teece, D. (1998), 'Organisational Competencies and the Boundaries of the Firm', in Arena, R. and Longhi, C. (eds), *Markets and Organisation* (New York: Springer), 281-301.

Dunford, M. (2003), 'Theorising Regional Economic Performance and the Changing Territorial Division of Labour', *Regional Studies* 37, 839-854.

Fuchs, M. (2005), 'Internal Networking in the Globalising Firm: The Case of R&D Allocation in German Automobile Component Supply Companies', in Alvstam, C. and Schamp, E. (eds), *Linking Industries across the World. Processes of Global Networking* (Aldershot: Ashgate), 127-146.

Fuchs, M. and Winter, J. (2008), 'Competencies in Subsidiaries of Multinational Companies: The Case of the Automotive Supply Industry in Poland', *The German Journal of Economic Geography* 4, forthcoming.

Gereffi, G. (1999), 'International Trade and Industrial Upgrading in the Apparel Commodity Chain', *Journal of International Economics* 48, 37-70.

Gereffi, G., Humphrey, J. and Sturgeon, T. (2005), 'The Governance of Global Value Chains', *Review of International Political Economy* 12, 78-104.

Glaeser, J. and Laudel, G. (1999), *Theory-guided Analysis of Texts? The Potential of a Variable-oriented Qualitative Content Analysis* (Berlin: VS Verlag).

Humphrey, J. and Schmitz, H. (2004), 'Governance in Global Value Chains', in Schmitz, H. (ed.), *Local Enterprises in the Global Economy: Issues of Governance and Upgrading* (Cheltenham: Edward Elgar), 95-109.

Humphrey, J., Lecler, Y. and Salerno, M. (eds) (2000), *Global Strategies and Local Realities: The Auto Industry in Emerging Markets* (New York: St. Martin's Press).

Kolk, A. and van Tulder, R. (2005), 'Setting New Global Rules? TNCs and Codes of Conduct', *Transnational Corporations* 14, 1-28.

Malmberg, A. and Maskell, P. (1999), 'The Competitiveness of Firms and Regions. Ubiquitification and the Importance of Localised Learning', *European Urban and Regional Studies* 6, 9-25.

Schmitz, H. (2004), 'Globalised Localities: Introduction', in Schmitz, H. (ed.), *Local Enterprises in the Global Economy: Issues of Governance and Upgrading* (Cheltenham: Edward Elgar), 1-19.

Storper, M. (1999), 'The Resurgence of Regional Economics. Ten Years Later', in Barnes, T. and Gertler, M. (eds), *The New Industrial Geography. Regions, Regulations and Institutions* (London: Routledge), 23-53.

UNCTAD (2005), *World Investment Report 2005: Transnational Corporations and the Internationalisation of R&D* (Geneva: UNCTAD).

Urban, W. (2005), *The Vehicle Industry in the New Member States* (Vienna: WIIW).

Winter, J. (2006) ‚Kompetenzerwerb in der Automobilindustrie. Das Beispiel Volkswagen Poznan' [‚Acquisition of Competencies in the Automotive Industry: The Case of Volkswagen Poznan], *Geographische Rundschau* 58, 46-52.

Chapter 11

Analysis of the Automobile Cluster in a Multi-dimensional Spatial Framework: The Case of Ulsan

Jeong Hyop Lee

Introduction

Ulsan, an automobile agglomeration in Korea, can be described as a creation of the developmental state which is built on strong ties between large Korean conglomerates (chaebols) and government. The formation and evolution of the Ulsan local production system strongly reflects this national regime. The nationally embedded local system has now extended into global production. This spatial extension of trading structures of the Ulsan complex is beyond the reach of traditional cluster concepts, which focus on geographically confined innovation activities. At the same time, the concept of *global production networks* (GPNs) seeks to explain theoretically the relationship between global dimension and local dimension of economic activity. However, neither the cluster concept nor the global production networks concept is an adequate theoretical platform to explain the current functioning of the Ulsan automobile cluster. What is needed is the mediation of local, national and global dimensions of the economic relationships that support, maintain and enhance this cluster.

This chapter examines the Ulsan automobile cluster using a multi-dimensional spatial framework. This framework is discussed against the background of a critical review of both cluster and GPN-related theoretical debates. The historical background and challenges of the Ulsan automobile cluster will be explored, and the cluster will be analysed using the proposed multi-dimensional spatial framework. The characteristics of each spatial dimension of the framework will be described and explained in qualitative terms and measured through a quantitative analysis of its spatial networks. Finally, the possible path of the cluster's development and its policy implications will be discussed in terms of the case study of Sejong, one of the large first tier vendors of the Hyundai Motor Company in Ulsan.

A Multi-dimensional Spatial Framework

The system and mechanism of a regional cluster can be best understood from the wider national and global perspective of the related value chain of industrial activities of which it is part, including the national and global division of labour. This is summarised succinctly in the definition of *Regional Innovation Systems* as "interacting knowledge generation and exploitation sub-systems linked to global, national and other regional systems for commercialising new knowledge" (Cooke 2004, 3).

Asheim and Coenen (2005) also recognise that the regional level is not sufficient for firms to sustain their innovativeness and competitiveness and that their learning processes are increasingly related to the multiple spatial levels of innovation systems. However, they also emphasise the importance of the regional level in the innovation process as has been confirmed repeatedly in comparative analyses of clusters. The multiplicity of spatial scales in an innovation system is now recognised as an important issue in the configuring of a regional cluster.

The extensive body of research on *global production networks (GPNs)* has expanded our understanding of the spatial dynamics of industries and enterprises, especially in the current era of globalisation. It focuses on the integration of the ideas of global value chain approach and those of the new regionalism (Coe et al. 2004). It throws light on how leading companies and regions organise their production networks, highlighting the region. Those regions that have been successfully adopted by global companies or have entered the networks of global companies are able to sustain their competitiveness in the competitive global economy.

However, the dynamic integration of regional assets into global production networks misses a specific and important spatial dimension; the national character of industrial activity. As it currently stands, the GPN approach focuses only on the strategic coupling of global production networks and regional assets. It does not provide for and incorporate these national level issues and differences into the ties that engender value creation, enhancement, and capture. Bathelt (2003) argues that the level of the nation state is important in determining the configuration of a regional production system while that region might host a large part of a global economic value chain.

Borrus et al. (2000) also introduced the concept of *cross-border production networks* in which the characteristics of the lead firm and the lead firm's home country are imprinted in the structure and operation of production networks. Comparative studies have shown that the American production systems contrast strongly with Asian production systems, under the control of Japanese, Taiwanese, Korean and other Asian firms (Borrus et al. 2000).

In the same way, the characteristics of a home country might influence the way production networks are organised and operated. Henderson et al. (2002) argue that national institutions, such as national ministry of finance, a national environmental group, and labour unions, for example, impact on the nature and operation of production networks. However, national influences on production networks are believed to go beyond the impact of national institutions. Historical, cultural and economic legacies embedded in the organisation and transactions of business activities in a country also mediate the "glocal" integration of industrial enterprises.

From this discussion it can be argued that regional clusters are spatially multifaceted. The spatial multiplicity of regional clusters, however, is not a static or uniform phenomenon. It is dynamic in its formulation. It can be thought of as having been configured in a certain spatial dimension but that this configuration evolves to incorporate resources in other spatial dimensions. This re-configuration process is also quite distinctive according to a region's spatial legacies. The Third Italy was initiated locally within Italy, while the formation of Hsinchu Science Park in Taiwan was triggered by global forces and the major influx of businesses from Silicon Valley.

The Korean economy has been widely identified as a successful example of a developmental state. The legacies of the developmental state have a strong influence on the way large companies organise their production networks locally, nationally, and globally. The characteristics of the national level have dominated the local and global dimension of most industrial agglomerations in Korea. In particular, the relationship between large companies and their subcontractors in the developmental state impacts strongly on the formation of production networks and their spatial configurations. The relationships generated at this scale have a distinctive impact on the other spatial dimensions of global production networks. As time passes, the regionalised, national innovation system (Asheim and Coenen 2005) of Korea will incorporate other resources, globally and locally. This process will result in a unique combination of innovation and economic activity in Korean clusters embedded within global production networks.

Formation and Challenges of the Ulsan Automobile Cluster

Ulsan is located at the southeast corner of the Korean peninsula and is well-known for being the hometown of Hyundai Motor Company (Figure 11.1). It was a rural town before the government built the heavy and chemical industrial complex there during the 1970s. Currently, the major industries in Ulsan are automobiles, petrochemicals, and shipbuilding. In 2002, the automobile industry employed 28.2 per cent of the workforce and produced 27.4 per cent of the cluster's total production.

The Ulsan automobile industry began in 1968 when Hyundai built its production plant there. The first cars Hyundai produced were the *Cortina* and the *D-750 truck* by way of CKD (completely knocked down). At that time, most of the components and parts were imported from Ford. Hyundai also had 40–50 domestic suppliers, most of whom were located in the Capital region, and these domestic components and parts were delivered to Ulsan by railway.

Since 1975, when Hyundai established a production line of 80,000 cars per year and changed its production from CKD to the original car model, *Pony*, the production base for automobile components and parts began to develop in Ulsan. As production capacity increased, Hyundai needed suppliers located close by. By the mid 1970s, several automobile components and parts companies were already established in Ulsan, for example Hanil e-hwa, Duckyang and Sejong, amongst others. Most were started by Hyundai-related personnel. A few were established by former Hyundai employees after they retired.

In the 1980s, Hyundai suffered from price increases for components and parts from Japan. As it increased its production capacity to 300,000 per year, it became necessary for Hyundai to build close co-operative networks with suppliers in

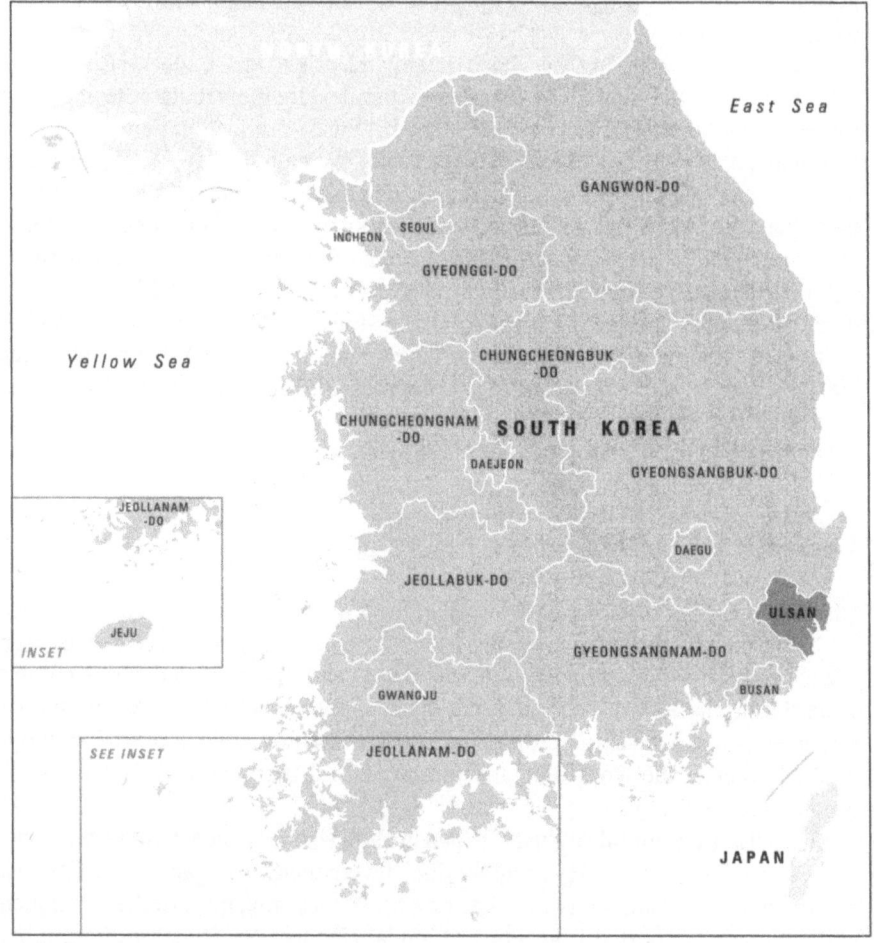

Figure 11.1   The location of Ulsan

Ulsan. As a consequence, around 100 companies in the Capital region have moved to Ulsan since 1986.

At the moment, Hyundai produces 1.53 million passenger cars per year in Ulsan, which is the largest production facility in the world. Automobile parts and components suppliers that have various types of business linkage with Hyundai's automobile production constitute the Ulsan automobile cluster. There were around 167 parts and components suppliers in Ulsan in 2002, and the geographical boundary of the automobile cluster may expand beyond the Ulsan administration area. There are also many other suppliers in places adjacent to Ulsan, including Gyeongju, Yangsan.

The Ulsan automobile cluster faces inter-regional and international competition to host the production and R&D facilities of the Hyundai Motor Co. In Korea, the middle region around Namyang and Asan is emerging as an R&D centre and an alternative production site for Hyundai automobiles. Hyundai is also continuously expanding its overseas investment, especially in China, India and the USA. This expanding multi-locational production by Hyundai may undermine and erode the production foundation of Ulsan to which it is currently strongly connected.

Multi-dimensional Analysis of the Automobile Industry

Ulsan and the National Production Network of the Korean Automobile Industry

The characteristics of the innovation activities of the Ulsan automobile cluster have been dominated by Korea's national regimes and institutions. Korea's economic growth has been built on a well-known catch-up model based on reverse-engineering, conditioned by "developmental state" planning. These characteristics, which were formulated during the period of rapid economic growth since the 1960s, also shaped the automobile production system in Korea. The Korean automobile industry has not developed its own original technologies but has borrowed them from advanced countries. The Korean automobile industry has been based on a very hierarchical structure of business linking car makers and suppliers, and has suffered from militant labour union demands since the mid 1980s. In this environment, only "value cars" could be produced, but not "premium cars".

Hyundai automobiles' power brand ranked third in the USA market in 2006. The ranking had been 34th in 2000. The question for Hyundai is how to meet customers' requirements in the USA market. The Korean system was not nurtured by the global production networks of major car makers.[1] It does not have the

---

1 According to many GPN researchers, the superior systems of advanced countries have been transferred to less developed countries and dominate the GPNs, as multinational companies have organised the diverse regional assets. The cluster models of advanced countries have been exported to industrial clusters in developing countries such as China and South America (Depner and Bathelt 2005; Ivarsson and Alvstam 2005).

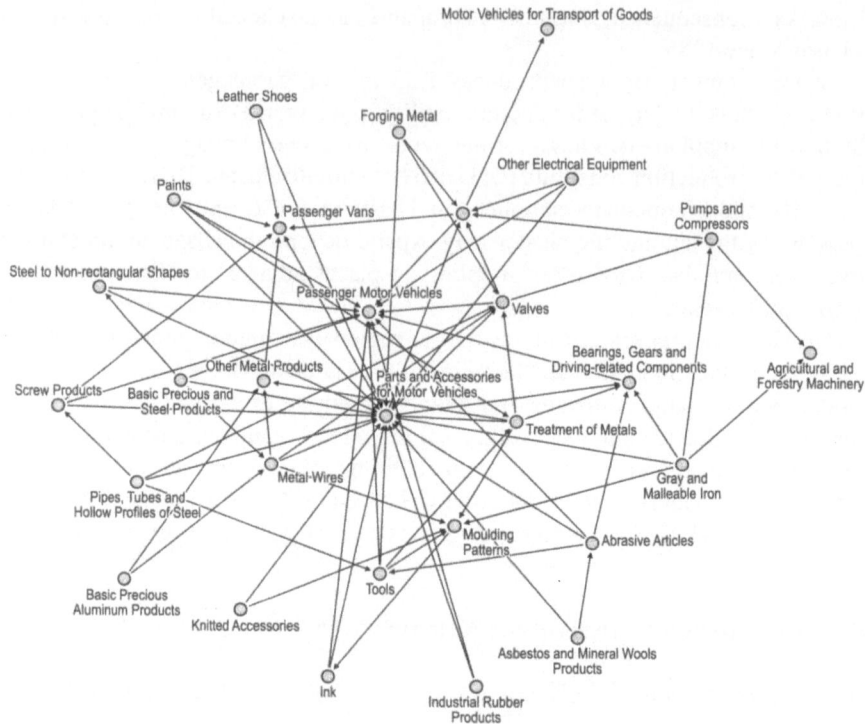

Figure 11.2   The national input-output structure of the Korean automobile industry

Source: Kim et al. (2005)

western type of innovative milieu in its automobile clusters, which is characterised by continuous innovation based on interactive learning between automobile companies, suppliers, and others. The issue is where does the Korean system's competitiveness come from?

As the Korean automobile industry has evolved, it has created its own, distinctive, domestic production network. Figure 12.2 illustrates the input-output structure of the Korean automobile industry.[2] Parts and accessories for motor vehicles, passenger motor vehicles and passenger vans occupy the central positions in the network. The balanced development of forward and backward linkages around these positions has provided the necessary conditions for the Korean automobile industry to build its price and quality competitiveness.

---

2 This network was drawn by social network analysis using the industrial group data drawn out from Korean input-output table 2000. The industry groups were extracted by the principal components and factor analysis. The automobile group was the second important group in the Korean industry (Kim et al. 2005).

Reverse engineering and hierarchical structure of the system also helped the Korean automobile industry to equip itself with the resources to innovate in the industry. In particular, engineer networks of engineers between Hyundai and other suppliers enabled Hyundai to sustain and increase its competitiveness in the global market. These networks have emerged as engineers moved out to suppliers from the Hyundai Motor Company. When Hyundai promoted quality management over competing schemes among suppliers, these engineer networks proved to be very effective. These characteristics of production networks formulated at the national level have constrained local and global level activities.

In the analysis of regional spatial change in the automobile-related employment of the production chain[3] between 1995 and 2004, Ulsan has the largest portion of employment in final automobile productions, although the total numbers employed fell from 38,175 to 27,298 during the period. This fall also occurred in Gyeonggi in the Capital region, the second-largest agglomeration of car makers. These falls reflect the impact on employment of the expansion of the modularised production system based on out-sourcing.

Ulsan does not dominate the automobile-related materials industries. Instead, these industries are concentrated along the western and southern coasts of the Korean peninsula. At the same time, the manufacture of parts and components for the automobile industry is particularly strongly concentrated in the Capital region. The industries are also localised in the South East region, Daegu, Ulsan and Southern parts of Gyeongnam.

The analysis of spatial change in Korea's automobile industry between 1995 and 2004, shows the Ulsan automobile cluster as having expanded beyond the administrative region, where it first began, and to be functionally integrated with the Capital region. It is now a part of the whole production network of the Korean automobile industry. Drawing on Bathelt (2003), the Ulsan automobile cluster is part of a national and global automobile production chain, performing assembly line and parts and components-making functions.

The New Path of Korean Automobile Globalisation

Very recently, Korean automobile companies have begun to develop their own GPNs by investing in advanced countries and other developing countries. The way they organise their own GPNs are quite different to previous mechanisms and this may be a new path to develop their unique GPNs targeting global markets. Companies from advanced countries are built on their soft competencies, and are mostly

---

3 The Korean automobile input-output network was reorganised into three important parts of the automobile production chain: materials, parts and components, and final automobile products. Materials are industrial plastic products, steel, metal wires, etc. Parts and components include valves, pumps and compressors, parts and accessories for motor vehicles, etc. And, final products cover passenger motor vehicles, passenger vans, motorcycles, etc. (Lee et al. 2005).

interested in acquiring diverse regional assets. Consequently, they organise their GPNs using local sourcing (Borrus et al. 2000). In contrast, engineering networks provide a core competence around which Korean companies can organise their GPNs, giving them a very distinctive potential. Because their core competencies are in their domestic engineering networks, Korean companies making overseas investments need to duplicate these relationships with the consequence that they are often accompanied by their major domestic subcontractors in their overseas investment ventures.

Hyundai started global production in the 1980s, later than its foreign competitors, by establishing large manufacturing plants and expanding production volumes in Canada, India, Turkey, China, and USA. Its global strategy was to mass-produce a small number of products, targeting new markets. According to an interview with one of Hyundai's board members, Hyundai plans to build ten large production facilities (production capacities of about 300,000 vehicles) across the world by 2010.

Hyundai has developed its competitive advantages in mass production and it exports low-to-mid price cars based on technology adoption and reverse engineering. As mentioned previously, the close relationship the company has with its domestic suppliers has been the major source of Hyundai's competitiveness. This domestic production system is required to be transferred in the new foreign production sites of Hyundai. So, Hyundai is usually accompanied by major domestic suppliers when it establishes production facilities in foreign countries. At the same time Hyundai allows its plants and R&D centres less autonomy in foreign countries than do its USA and European competitors.

This globalisation of Hyundai Motor Co. threatens the Ulsan automobile cluster through the possible de-territorialisation of the local production system. The local knowledge creation and dissemination system of the Ulsan cluster has not yet matured. Continuous globalisation may lead to the relocation of the Hyundai assembly line. This would seriously challenge the Ulsan cluster because it would also require the relocation of other suppliers.

The Spatial Evolution of Sejong in the Multi-dimensional Spatial Framework

In 1976, Sejong was established by seven people who retired from the Hyundai Motor Company. Sejong was guaranteed the contract to supply mufflers because of the contribution made by the Sejong founders to Hyundai. At that time, Sejong was simply a muffler maker based on the design provided by Hyundai. As Hyundai increased its production, Sejong also grew to have second tier suppliers in Ulsan.

In 1993, the founders of Sejong established their own R&D centre and, after an approximately five year trial, the company acquired the ability to produce its own muffler designs. The designs drawn by Sejong were put out to second and third tier local manufacturers helping to increase their production capabilities. In the mid

1990s, Hyundai established another assembly line at Asan in the Central region of Korea. Sejong was also asked to set up a new plant in Asan to provide mufflers to the plant. The Asan Sejong plant was provided with designs by the Ulsan R&D centre and the parts and components that were required were also provided by Ulsan suppliers.

At the end of 1997, Korea was hit by a severe financial crisis. Since 1998, four engineers have moved from Hyundai to Sejong. One became a director, another became general manager, and the others became design team leaders in the R&D centre. They had gained experience in Hyundai in muffler design, engine design, car architecture design, test and examination. These complementary skills and experience significantly enhanced the muffler design capability of the R&D centre. The design capacity of Sejong was also nurtured by co-operation with foreign companies, such as Japanese and USA muffler producers. Local universities, such as Ulsan University, also provided ideas on muffler design and analysis.

Since the financial crisis, Hyundai has adopted a "competition strategy" to promote the quality of parts and components it receives from domestic suppliers. Sejong is one of the domestic suppliers that has successfully met Hyundai's requirements. Sejong is now also competing with other domestic muffler suppliers, but it still fulfils around 80 per cent of Hyundai's orders, including muffler orders in Hyundai assembly lines in China and the USA. It also has around 40 second tier suppliers in Ulsan and adjacent areas.

Sejong was invited to invest in China, India and the USA as Hyundai established their production lines in those countries. Sejong established production plants in China and the USA, and is considering investing in India. After failure in Canada, Hyundai believed that it was essential for its domestic Korean partners to accompany them as they set up new production lines overseas. The domestic production system is copied in the new overseas production systems, with suppliers provided with the designs of parts and components, and machines from the Ulsan base.

Sejong is now competitive in muffler design and analysis. But, to lead the whole Ulsan production system is a complex task. As a result, the Ulsan metropolitan government and the central government initiated a special project "Auto Valley" to upgrade the capabilities of the local suppliers of parts and components. The project consists of the establishment and operation of the Automotive Parts Innovation Centre, the Industrial Complex for Module Producers, the Industrial Complex for Component and Parts, and Graduate School of Automobile and Shipbuilding.

A second goal of the "Auto Valley" project is to promote Ulsan as a global sourcing base for automobile parts and components. To achieve this goal, first tier suppliers, such as Sejong, are required to supply their products to other car makers, such as GM and Ford. As global car makers expand their production lines in China, the opportunities for Ulsan automobile parts makers to supply their products are increased. This might help second and third tier suppliers in Ulsan to generate profits to invest in their production technologies. And, if Hyundai moves from the production of "value cars" to the production of "premium cars", it will also help suppliers enhance their production capabilities through the profitable business opportunities it will create.

The Ulsan automobile cluster is quite weak in terms of local knowledge-creation and dissemination. It is not an innovative milieu because it has been formulated as a local element of a much bigger national automobile production network. Policy interventions that promote the competitiveness and innovativeness of local suppliers can lead to the Ulsan automobile becoming a global automobile parts sourcing base in a changing global production network of the automobile industry. Sejong, one of the few companies that have sustained their R&D facilities in Ulsan, even after the relocation of Hyundai's R&D facilities from Ulsan to Namyang,[4] can play an important role in changing the developmental path of Ulsan.

Conclusion

In this chapter, the Ulsan automobile cluster has been analysed within a multi-dimensional spatial framework. It has evolved since the 1960s as the home town of the Hyundai Motor Company. The evolutionary path of the Ulsan automobile cluster shows the strong influence of a national spatial dimension. The whole production network of the Korean automobile industry dominates the characteristics of the cluster because it has strong functional integration with other regions in the country. This has also created a distinctive path for the development of global production networks by the Korean automobile industry, with vehicle producers always being accompanied by the domestic subcontract system. The legacies of national production systems, especially the engineer networks, also condition the possibilities for changing the developmental path of the cluster. As shown through the case study of the first tier supplier, Sejong, successful transformations can significantly influence the success of other second and third tier suppliers at the regional scale in Korea. Although Sejong's production system is strongly dependent on national and global production and R&D networks that are not embedded in Ulsan, the regional networks of related cluster actors may be developed as a result of policy interventions by central and local governments.

References

Asheim B.T. and Coenen, L. (2005), 'Knowledge Bases and Regional Innovation Systems: Comparing Nordic Clusters', *Research Policy* 34:8, 1173-1190.
Bathelt, H. (2003), 'Geographies of Production: Growth Regimes in Spatial Perspectives 1 – Innovation, Institutions and Social Systems', *Progress in Human Geography* 27:6, 789-804.

---

4 This relocation was in the process since the mid 1990's, and completed very recently.

Borrus, M., Ernst, D. and Haggard, S. (2000), *International Production Networks in Asia: Rivalry or Riches?* (London: Routledge).

Coe, N.M., Hess, M. Yeung, H.W., Dicken, P. and Henderson, J. (2004), 'Globalising' Regional Development: A Global Production Networks Perspective, *Transactions of the Institute of British Geographers* 29:4, 468-484.

Cooke, P. (2004), 'Regional Innovation Systems: An Evolutionary Approach', in Cooke, P., Heidenreich, M. and Braczyk, H. (eds), *Regional Innovation Systems: The Role of Governance in a Globalised World*, 2nd Edition (London: Routledge), 1-18.

Depner, H. and Bathelt, H. (2005), 'Exporting the German Model: The Establishment of a New Automobile Industry Cluster in Shanghai', *Economic Geography* 81:1, 53-81.

Henderson, J., Dicken, P., Hess, M., Coe, N. and Yeung, H. (2002), 'Global Production Networks and the Analysis of Economic Development', *Review of International Political Economy* 9:3, 436-464.

Ivarsson, I. and Alvstam, C.G. (2005), 'The Effect of Spatial Proximity on Technology Transfer from TNCs to Local Suppliers in Developing Countries: The Case of AB Volvo in Asia and Latin America', *Economic Geography* 81:1, 83-111.

Kim, S.B., Jeong, J.H. and Lee, J.M. (2005), *Linkage Structures and Spatial Patters of Industrial Clusters in Korea* (Seoul: Korea Institute for Industrial Economics & Trade).

Lee, J.H., Kim, H.J. and Sohn, D.W. (2005), *Korean Models and Strategies of Regional Innovation Systems 1: Multi-layered Spatial Frameworks of Regional Innovations* (Seoul: Science & Technology Policy Institutes).

Chapter 12

The Identification of Potential Cluster Areas Using Local Indexes of Spatial Autocorrelation

Michael C. Carroll, Bruce W. Smith and Joseph P. Frizado

Introduction

In recent years, there has been a growing interest in cluster-based economic development (CBED) as an alternative economic development strategy. CBED is an attempt to take advantage of new global economic configurations. This concept has been popularised by the work of academics, such as Porter (1998). Also, it has gained acceptance among practitioners. For example, in his 2003 survey of states in the US, Akundi (2003, 2) reported that "40 states had boarded the cluster bandwagon". Despite the increasing popularity of the idea, there is a good deal of confusion and misinterpretation about cluster definitions, appropriate cluster identification methodologies, and the like. For example, Akundi (2003) reviewed several definitions adopted by practitioners. Owing to the vagaries surrounding CBED, it has been characterised by Martin and Sunley (2003) as being popular, but problematic. They (2003, 7) suggest: "... the rush to employ 'cluster ideas' has run ahead of many fundamental conceptual, theoretical and empirical questions".

The definition currently in vogue was offered by Michael Porter in 1998 and states that: "Clusters are geographic concentrations of interconnected companies, suppliers, service providers, firms in related industries, and associated institutions (for example universities, standards agencies, and trade associations) in particular fields that compete but also co-operate" (Porter 1998, 197). Others have attempted to refine the definition but most are essentially a restatement of Porter. Unfortunately, in practice and in the literature, the concept of a cluster has expanded and taken on multiple meanings. The term is no longer limited to external economies (Bathelt and Taylor 2002). It now is used to describe a wide variety of firm behavioural characteristics and policy prescriptions. Martin and Sunley (2003) offer a list of ten definitions that have appeared in the economics, policy, and geography literatures. The term cluster has morphed with the presupposition that clusters create innovations; they represent "learning regions" (Glaeser 1999); and increase the transference of knowledge simply because the firms are co-located. Furthermore, Gordon and McCann (2000, 515) observed: "... there has been a tendency to use terms such as "agglomeration", "clusters", "new industrial areas", "embeddedness", "milieu" and "complex" more or less interchangeably, with little concern for questions of operationalisation, which are actually far from straightforward, and should be different for each".

In empirical work, researchers have used a variety of methodologies to identify clusters. For example, some have used location quotients to identify spatial concentrations of industry and high location quotients are interpreted as an indicator of a cluster (Miller *et al.* 2001). Other approaches have relied, at least in part, on expert opinions to guide the process (Roberts and Stimson 1998). Statistical analyses of input-output tables to gauge the interdependences in the regional economy as the basis for empirically deriving clusters have been another strategy (Hill and Brennan 2000.) For example, Feser and Bergman (2000) used principal components analysis on the 1987 US input-output accounts to derive 23 industrial clusters, which they believe can be used as templates in subsequent regional analyses to develop a strategic view of a regional manufacturing economy. A critical review of the methodologies used in the identification of clusters has been published by Martin and Sunley (2003).

In this chapter, the focus is only on identifying the "spatial footprint" of potential cluster areas (PCAs) in a three state region. However, it is important to understand that we believe CBED is more than the co-location of firms. Spatial concentration is a necessary, but not sufficient, condition of a CBED policy. We believe CBED is a network-driven strategy stressing two-way communication between firms in the core industry, the core's local supply-chain, local government and support institutions such as universities, think-tanks and development agencies (Reid and Carroll 2006). In essence, a PCA has the potential to be a cluster due to the co-location criterion. Whether the cluster will be viable or not is a function of the other aforementioned elements, such as communication, local firms in the supply-chain, support institutions, etc. In this context, examination of industry location patterns to delineate concentrations is the first phase of a cluster-based development policy. Elimination of areas without sufficient concentrations of firms will reduce the likelihood of failed cluster projects due to the lack of critical mass.

The purpose of this chapter is to illustrate the use of Moran's I in the identification of PCAs. The benefits and limitations of using Local Indicators of Spatial Association (LISA) to identify clusters will be discussed. Floriculture was selected because of an ongoing cluster-based project, known as "Maumee Valley Growers". The project has been developing since 2003 to assist greenhouse growers in northwestern Ohio cope with increasing Canadian imports and other competitive challenges. The original cluster was composed of growers in northwestern Ohio. However, there has been recent discussion about expanding the cluster to southwestern Michigan, and perhaps, in the longer term, to more of Lower Michigan and northeastern Indiana. Thus, the identification of greenhouse PCAs is of more than academic interest.

Data and Methodology

The data for the greenhouse industry was obtained from the US *Census of Agriculture 2002* (US Department of Agriculture 2002). More specifically, information on floriculture was collected, with the floriculture industry including bedding/garden plants, cut flowers and cut florist greens, foliage plants, and potted flowering plants. Floriculture, instead of the more generic greenhouse industry, was used because it included those crops most commonly produced in the study area. County data were employed because counties are the smallest areal units for which data are reported.

Utilisation of *Census of Agriculture* data creates some problems. One problem is that the census does not publish data when those statistics could reveal sensitive information on individual farm operations. This non-disclosure rule results in data on the number of operations being reported more frequently than information on the nature of production. For example, it would be desirable to measure the amount of production occurring under glass versus open fields. However, such data for over 40 per cent of the counties are not reported. Another issue is that no sales data were reported in the 2002 census even though such data were published in previous censuses. Due to these data constraints, the measure of production in each county is the number of floriculture operations. Also, changes in the 2002 census preclude direct comparisons with earlier years.

The three states in our study area (Indiana, Michigan, and Ohio) are major floricultural producers in the United States. Between 1992 and 2002, the value of sales of floricultural products in the region increased slightly from 10.7 per cent of the national total to 11.4 per cent of the nation's output (USDA 2005). In 2002, Michigan and Ohio ranked fourth and fifth in the nation in production, following California, Florida, and Texas. Moreover, Michigan contains 10 of the top 5 per cent of the leading floricultural counties in the US, and Ohio contributes another 8 of the top 5 per cent. Viewed from another perspective, only California has more counties in the top 5 per cent of sales than do either Michigan or Ohio.

One of the advantages of the study region is that the counties are more homogenous in areal extent than counties at the national level. The mean size of counties in the study area is 479 square miles, ranging from a maximum of 957 square miles to a minimum of 81 square miles. In comparison, at the national level, the mean size is 966 square miles, which varies from 20,174 square miles to approximately 2 square miles. This wide variability in county size has important implications for the construction of spatial weights matrices, as will be discussed later.

In this project, one of the Local Indicators of Spatial Association (LISA), Moran's I, is used to identify clusters of floricultural operations. Feser and various colleagues have applied different spatial statistics to identify clusters, including Moran's I, Getis and Ord's Gi*, as well as the local G statistic. All of these measures can be used to identify "hot spots" or "cold spots" in spatial distributions (Mitchell 2005). "Hot spots" are pockets of concentration of phenomena where areal units and their neighbours have similar values, which are indicative of a cluster. Feser et al. (2001) used Gi* in conjunction with county data and national inter-industry data in a 2001 study. In 2002, Feser and Sweeney, used the local G statistic to identify clusters in manufacturing value chains among 14 US metropolitan areas. Again Feser et al. (2005) applied Gi* to identify discrete industrial complexes in the US in 1989 and 1997 using county level data (Feser et al. 2005). In an unpublished study, Feser and Isserman (2005) used Moran's I to identify industrial clusters in rural counties in the US.

Moran's I was selected because it possesses some desirable characteristics not inherent in the other measures. For example, Wong and Lee (2005, 389) have argued that, "... the distribution properties of the local Geary statistic are not as desirable as those of the local Moran statistic". One disadvantage of the local G is that it is not sensitive to low value spatial clustering (Lin 2004). That problem is important because there are many counties in the study area with few or no floriculture operations. Gi* does not possess the weaknesses of Geary's local C and the local G statistic, but Moran's I has some characteristics which make it more desirable for this project than Gi*. Moran's I measures elements of spatial association not described by Gi*; i.e. the case of a county with low values being surrounded by high neighbouring values and a county with high values being surrounded by neighbours with low values (Anselin 1995).

The local version of Moran's I is an adaptation of the global Moran's I. The global statistic calculates a single measure across the study area, whereas the local statistic calculates a statistic for each areal unit based on its similarity to its neighbours (Mitchell 2005). In order to determine the level of spatial autocorrelation locally, a spatial autocorrelation value is computed for each areal unit, which enables the user to determine how similar or dissimilar each county is to its surrounding neighbours.

One of the decisions involved in using LISA measures is to define the local neighbourhood or the spatial extent of a county's neighbours to be included in the calculations. The concept of the neighbourhood around each county is implemented by the spatial weights matrix. Alternative specifications of that matrix will lead to differing neighbourhoods and, therefore, perhaps the value obtained. Methods of neighbourhood definition that can be utilised include rook or queen's measures of adjacency, distance between county centroids, inverse distance function, inverse distance squared, stochastic weights, etc. (Getis and Aldstadt 2004; Wong and Lee 2005; Mitchell 2005).

Selection of a particular spatial weights matrix ideally should be based on theoretical considerations of the nature of spatial interactions between counties. In the case of cluster development, there is no such theoretical rationale. Moreover, there is little consensus in the cluster literature as to the appropriate spatial extent of a cluster. According to Porter (2000, 16), "[t]he geographic scope of clusters ranges from a region, a state, or even a single city to span nearby or neighbouring countries". In contrast, May et al. (2001) suggest that a cluster is characterised by firms agglomerating in a region up to 50 miles in radius. Martin and Sunley (2003,

12) strongly criticised this lack of geographic specificity: "[t]he key weakness is that there is nothing inherent in the concept itself to indicate its spatial range or limits, or whether and in what ways different clustering processes operate at different geographical scales". In the only cluster study using Moran's I, Feser and Isserman (2005, 26) used a "first-order, rook spatial weights matrix", i.e. they looked at only adjacent counties.

Identifying Potential Floriculture Clusters

The measure of floriculture production is the number of farms, and Figure 12.1 shows the distribution of floricultural operations in the study area. There are three main belts of floricultural operations. One is in southwestern Michigan consisting of at least six counties containing 443 farms. Another belt containing 395 floricultural operations comprises approximately nine counties in northwestern Ohio and southeastern Michigan. The final belt is roughly a 12-county area in northeastern Ohio containing 395 floricultural operations. In addition, there are lesser concentrations in the Columbus and Cincinnati, and Indianapolis regions.

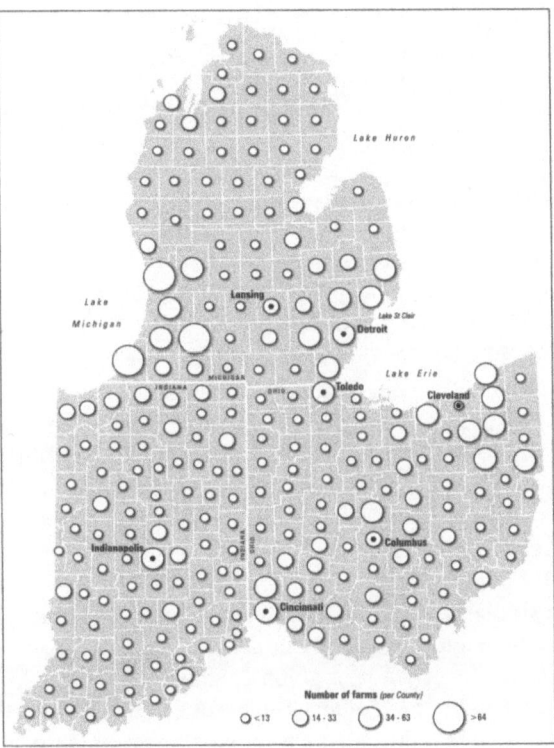

Figure 12.1   Floricultural operations 2002

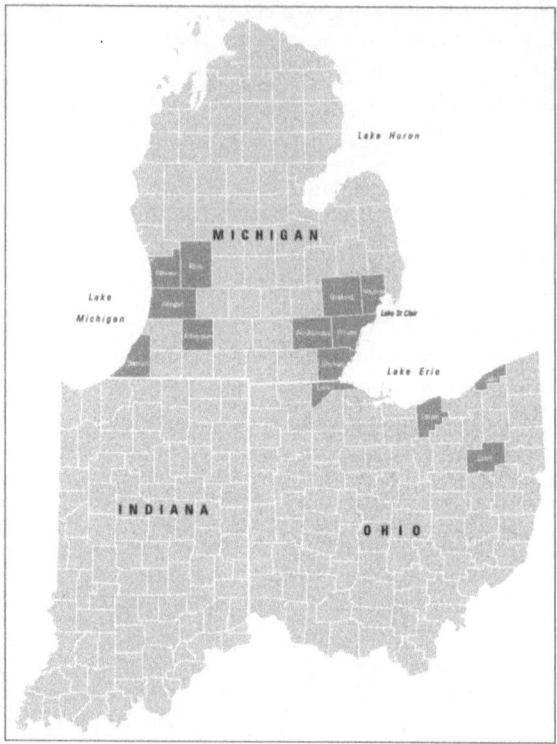

Figure 12.2   Potential cluster areas

Identification of possible clusters around these areas is problematic because it is not possible to be certain as to how many counties should be included. The purpose of using Moran's I is to make that decision more objectively.

In this study, an inverse distance spatial weights matrix of Euclidean distances has been used. The Cluster and Outlier Analysis tool in the ArcGIS 9.1 Spatial Statistics Tools was used to compute the value of the Moran I. The Cluster and Outlier Analysis Tool required a threshold distance, as a cut-off, beyond which counties were ignored. A distance of 150 miles was selected for the present study, this being large enough to include all possible counties in a potential cluster, given the geographic extent of the study area.

A large positive value of Moran's I indicates that a county is surrounded by neighbours with similar values, but those values could be high or low. A negative value for Moran's I indicates that a county is surrounded by neighbours with dissimilar values. To determine if a given I value, either positive or negative, is indicative of a clustering of high or low values, one must examine the spatial lag of each county. To accomplish this, all counties having z scores for Moran's I and the number of farms that were significant at the 0.10 level of confidence were mapped

(Mitchell 2005). It is these "high-high" counties that constitute a cluster (Figure 12.2, Table 12.1).

Three potential cluster areas (PCAs) can be identified within the study region. The most evident one is the southeastern Michigan and northwestern Ohio PCA, consisting of six counties. This cluster runs approximately 100 miles north to south and about 65 miles east to west at it widest point. Within this cluster are 304 floricultural operations, or 9.1 per cent of the study area's total. A second PCA is located in southwestern Michigan, comprising four contiguous counties and one non-adjacent county. This is the largest centre of production within the study area, containing 407 floricultural operations, or 12.1 per cent of the total. The third potential PCA is located in northeastern Ohio near Cleveland. It is composed of three non-contiguous counties which are situated within 65 miles of each other. These three counties contain 146 floricultural operations, or 4.3 per cent of the region's total.

Table 12.1   County z scores and probabilities

County	Z Scores	
	Moran's I	Farms
Allegan, MI	13.606***	2.435**
Berrien, MI	8.664***	4.224***
Kalamazoo, MI	10.909***	3.916***
Kent, MI	11.183***	2.743***
Monroe, MI	7.978***	2.188**
Macomb, MI	12.815***	3.052***
Oakland, MI	9.926***	2.188**
Ottawa, MI	12.815***	3.052***
Washtenaw, MI	6.954***	1.812*
Wayne, MI	8.873***	1.694*
Lake, OH	10.400***	2.620**
Lorain, OH	8.688***	2.558**
Lucas, OH	3.491***	2.805***
Stark, OH	5.795***	1.965**

*$p < 0.10$, ** $p < 0.05$, *** $p < 0.01$

In general, Moran's I does identify clusters of floricultural operations based on patterns of co-location. Moreover, the geographic extent of those clusters seems reasonable given the characteristics of the regional industry. Most floricultural operators are small and the owners most likely would not be inclined, or have reasons, to interact with other growers over a large geographic area.

Discussion of Moran's I

Moran's I is useful for identifying clusters, particularly since it examines patterns of co-location. This measure distinguishes concentrations of phenomena, or clusters, across county boundaries within a specified neighbourhood. In contrast, other measures, such as location quotients, examine only the value for a single county without reference to spatial lag.

The output of Moran's I provides a quick method of searching large numbers of areal units for PCAs. In addition, it does not rely on a fixed, predetermined spatial area. It is adaptable to varying sizes of areal units, be they counties or metropolitan areas. At the same time, however, this methodology does not indicate the processes behind the spatial concentration of industry. Nonetheless, the "hot spots" identified by the spatial statistics provide the bases for further analyses.

A concern in using Moran's I or other spatial statistics is the selection of the spatial weights matrix. Since the spatial weights matrix defines the spatial extent of a given areal unit's neighbourhood, it has an obvious impact on the degree of similarity or dissimilarity in a given neighbourhood. If cluster-based economic development had a theoretical base to guide the selection of a spatial weights matrix, that would be helpful. Therefore, the only possible approach is an empirical "trial and error" strategy. As ESRI's White Paper (2005) suggested, various distances can be tried using alternative distance functions and then selecting the combination that demonstrates the greatest amount of clustering. This strategy may work in empirical studies with a limited purpose, such as this study, but it does not make any substantive theoretical contribution to the cluster literature.

Directly related to the distance issue is the problem of great variations in the sizes of counties in the US. The largest county in the continental US is San Bernardino County in California, covering over 20,000 square miles. In contrast, there are 58 counties, primarily in the eastern US, which cover less than 100 square miles. If one is using some type of distance function, these differential distances are likely to distort the results. If one uses an adjacency definition of neighbours, then one must be aware of the implied differences in distance. This is an issue that the authors are continuing to study.

Conclusion

Identifying regions that have potential for industrial clustering is often a laborious sort through county data tables. This chapter shows how a Local Index of Spatial Autocorrelation (LISA) can be used to identify potential cluster areas (PCAs). This method is much faster than the traditional county data ranking or using some form of factor analysis. Further, it spatially identifies the unique footprint of the cluster. As can be seen from Figure 12.2, clusters rarely display any set geometric pattern. Clusters are a function of market areas and proximity of the industry's supply chain.

The LISA techniques described in this chapter are the first step in any cluster project. As described in the introduction of this chapter, co-location is a necessary but not sufficient condition for clustering. There are other necessary conditions such as joint action leading to collective efficiency which must occur before true cluster activity can take place. However, without the necessary industrial concentrations the process cannot start. The methodology described in this chapter can help policymakers develop a cluster strategy for existing industries. Embryonic or emerging clusters would obviously not show in the existing data. Also LISA could be used to locate primary suppliers. Economic development practitioners often attempt to form development strategies around potential or even aspirational clusters. LISA quickly shows if the necessary conditions exist, therefore, helping to shape policy in its early stages.

Future research will investigate the use of alternative measures of spatial autocorrelation, such as Getis-Ord's G and Geary's C. In addition, the utility of a non-symmetric spatial weights matrix incorporating measures of spatial interaction will be investigated. Finally the feasibility of using Geographically Weighted Regression in this application will be investigated.

References

Akundi, K. (2003), *Cluster-Based Economic Development, Part 1: A Survey of State Initiatives*, Texas Economic Development, Business and Industry Data Centre <http://www.bidc.state.tx.us/Cluster%20Based%20EconDev%20PARТ1.pdf>, accessed 28 January 2007.

Anselin, L. (1995), 'Local Indicators of Spatial Association-LISA', *Geographical Analysis* 27:2, 93-115.

Bathlelt, H. and Taylor, M. (2002), 'Clusters, Power, and Place: Inequality and Local Growth in Time-Space', *Geografiska Annaler B* 84:2, 93-109.

ESRI (2005), *Spatial Statistics for Commercial Applications, An ESRI White Paper* <http://www.esri.com/library/whitepapers/pdfs/spatial-stats-comm-apps.pdf>, accessed 28 January 2007.

Feser, E. and Bergman, E. (2000), 'National Industry Templates: A Framework for Applied Regional Cluster Analysis', *Regional Studies* 34:1, 1-19.

Feser, E. and Isserman, A. (2005), *Clusters and Rural Economies*. <http://www.urban.uiuc.edu/faculty/feser/Pubs/Clusters_and_rural_economies.pdf>, accessed 28 January 2007.

Feser, E., and Sweeney, S. (2002), 'Theory, Methods and a Cross-Metropolitan Comparison of Business Clusters', in McCann, P. (ed.), *Industrial Location Economics* (Cheltenham, UK: Edward Elgar), 222-257.

Feser, E., Koo, K., Renski, H., and Sweeney, S. (2001), *Incorporating Spatial Analysis in Applied Industry Cluster Studies* <http://www.urban.uiuc.edu/faculty/feser/PUBS/ EDQ%20Revised.pdf>, accessed 28 January 2007.

Feser, E., Sweeney, S. and Renski, H. (2005), 'A Descriptive Analysis of Discrete US Industrial Complexes', *Journal of Regional Science* 45:2, 395-419.

Getis, A. and Aldstadt, J. (2004), 'Constructing the Spatial Weights Matrix: Using a Local Statistic', *Geographical Analysis* 36:2, 90-104.

Glaeser, E. (1999), 'Learning in Cities', *Journal of Urban Economics* 46: 2, 254-277.

Gordon, I. and McCann, P. (2000), 'Industrial Clusters: Complexes, Agglomeration and/or Social Networks', *Urban Studies* 37:3, 513-532.

Hill, E. and Brennan, J. (2000), 'A Methodology for Identifying the Drivers of Industrial Clusters: The Foundation of Regional Competitive Advantage', *Economic Development Quartery* 14:1, 65-96.

Lin, G. (2004), 'Comparing Spatial Clustering Tests Based on Rare to Common Spatial Events,' *Computers, Environment and Urban Systems* 28: 6, 691-699.

Martin, R. and Sunley, P. (2003), 'Deconstructing Clusters: Chaotic Concept or Policy Panacea?', *Journal of Economic Geography* 3:1, 5-35.

May, W., Mason, C., and Pinch, S. (2001), 'Explaining Industrial Agglomeration: The Case of the British High-Fidelity Industry', *Geoforum* 32: 3, 363-376.

Miller, P., Botham, R., Martin, R., Moore, B. (2001), *Business Clusters in the UK: A First Assessment* (London: Department of Trade and Industry).

Mitchell, A. (2005), *The ESRI Guide to GIS Analysis, Volume 2: Spatial Measurements and Statistics* (Redlands, CA: ESRI Press).

Porter, M. (1998), *On Competition* (Boston: Harvard Business School Press).

Porter, M. (2000), 'Location, Competition, and Economic Development: Local Clusters in a Global Economy', *Economic Development Quarterly* 14:1, 15-34.

Reid, N. and Carroll, M. (2006), 'Cluster Development: The Case of the Northwest Ohio Greenhouse Industries from Concept to Implementation', in Reid, N. and Gatrell, J. (eds), *Enterprising Worlds: A Geographic Perspective on Economics, Environment & Ethics* (Dordrecht, The Netherlands: Springer), 42-56.

Roberts, B., and Stimson, R. (1998), 'Multi-Sectoral Qualitative Analysis: A Tool for Assessing the Competitiveness of Regions and Formulating Strategies for Economic Development', *Annals of Regional Science* 32:4, 469-494.

US Department of Agriculture, Economic Research Service (2005), *Floriculture and Nursery Crops Yearbook, 2005* <http://jan.mannlib.cornell.edu/data-sets/specialty/FLO/>, accessed 28 January 2007.

US Department of Agriculture, National Agricultural Statistics Service (2002), *Census of Agriculture, 2002* <http://www.nass.usda.gov/Census_of_Agriculture/index.asp>, accessed 28 January 2007.

Wong, D. and J. Lee. (2005), *Statistical Analysis of Geographic Information with ArcView GIS and ArcGIS* (Hoboken, NJ: John Wiley & Sons, Inc.).

Chapter 13

The Role of an Antecedent Cluster, Academic R&D and Entrepreneurship in the Development of Toledo's Solar Energy Cluster

Frank J. Calzonetti

Introduction

Altering a 'development trajectory' is a major challenge when promoting innovation-based development in declining manufacturing regions, as few regions are successful in creating *new* clusters in emerging technological areas. Toledo, Ohio, not unlike other industrial cities, is looking for new technology industries to offer opportunities for economic diversification and high paying jobs through the support and 'creation' of technology clusters in selected areas that offer future promise. In 2003, Toledo's economic development organisations sponsored a study that resulted in the identification of technology clusters. Alternative energy was one of these clusters. The alternative energy cluster includes a range of technologies with solar energy as the premier area of focus. The solar energy cluster faces serious challenges but has important strengths. The first challenge is that the cluster is developed around emerging technologies that have serious technical issues to overcome through research and development (R&D) before they are competitive in the marketplace. Second, the technology under development may be undercut if market conditions change in favour of traditional energy sources or if national policy supports another alternative (e.g. nuclear). Third, entrepreneurship is considered by local leaders as a local weakness. Finally, the region lacks venture capital and groups to support technology investment.

A strength of the cluster is that it builds upon an antecedent cluster in the glass industry. Toledo has deep experience in research, development and production of glass that is a component of solar cells. Second, the thin-film programme at The University of Toledo (UT) provides research support, a technical workforce, and access to federal funding to support industrial initiatives in solar energy. Third, despite the dearth of entrepreneurial talent in the region, the region has produced some exceptional entrepreneurs that provided leadership in creating a local solar energy industry. Fourth, the region has the strong advocacy of the local congresswoman who has been successful in steering congressionally-directed funds to support the cluster.

Thus, while many community leaders view the solar energy as a new 'emerging' cluster, looking deeper shows that this new technology-based industry grew from technology developed in an earlier era when Toledo reigned as the 'glass city' of North America. Thus, if Toledo is successful in 'creating' a solar energy cluster, it will have been built upon a strong foundation from an earlier era. Through most of the last century Toledo had a number of firms in the glass industry with major local R&D facilities that employed scientists and engineers who produced many innovations. This industry was largely hollowed out late in the twentieth century as glass R&D centres were dispersed or closed, business units were sold, and bankruptcies occurred. With recent local efforts to advance a cluster in solar energy and with UT supporting programmes to serve as a source and supporter of innovation, the community is in a position to alter its trajectory from a declining manufacturing city to one competing in a rapidly growing technological field.

This chapter shows that Toledo's new solar energy cluster owes much to its antecedent glass cluster and that the new cluster has taken many years to emerge. Although the cluster has been supported in many ways, it grew with little public policy attention or nurturing. The emerging cluster has many of the attributes associated with technology clusters, including sources of innovation, entrepreneurship, ability of the region to absorb innovations, university support, and sharing of information and people. Following a brief review of the relevant literature, the chapter provides information on the growth of the solar energy industry, describes the growth of the cluster in Toledo, and provides information on prospects of the cluster's ability to sustain its growth.

Innovation-based Cluster Development

It is not necessary to review the vast literature on technology clusters here (see Cooke 2002; Taylor 2005; Martin and Sunley 2003). Case studies indicate that some clusters develop without direct public policy intervention or direction and in other cases, state or regional leaders work to create a cluster in a region that does not have much activity related to the planned cluster. 'Created' clusters are often associated with rapidly expanding knowledge-based sectors that are heavily dependent upon universities, entrepreneurship and skilled labour. A common strategy is to create clusters around core activities such as local industry, a major federal centre, or university research centres. A rare path is the development in an emerging technological area that needs to overcome both technical and competitive market conditions before profitable returns are realised (e.g. Lester 2005). Cortright and Mayer (2001, 8) conclude that 'economic development efforts should be tailored to build on or extend existing strengths or emerging local competence; trying to create a totally new high tech centre where none currently exists is likely to be a lengthy and probably fruitless endeavor'.

A top-down approach has a questionable record of success when the task is to transform an 'old economy' (industrial plants, hierarchical, skill-based, etc.) into a 'new economy' (lifelong learning, risk-taking, teaming, etc.) (Cooke 2002, 131-156). Many regions try to create clusters in areas lacking a tradition of entrepreneurship. The approach requires that leaders redirect thinking away from the view that competing on the basis of low-cost manufacturing or attracting new manufacturing operations will revive the economy (Boschma and Lambooy 2000; Asheim and Isaken 2002). Even if a new development model is endorsed, the challenge is great if the area does not have local sources of innovation creation, entrepreneurs with technology experience, and investment funds or venture companies who have experience dealing with the uncertainty of technological risk in addition to the expected risk for all start-up businesses. 'Technological entrepreneurs' can develop and introduce new technical ideas into the marketplace (Feldman and Francis 2004, 131). For instance, a cluster can provide learning experiences that can assist new entrepreneurs overcome barriers to entry. For the most part, entrepreneurship is viewed in the context of new firm formation and not other variants of entrepreneurship such as that within firms, universities, or research institutes (Fromhold-Eisebith 2006).

The expectation is for newly-created knowledge to spill over into the local region if that region can 'absorb' the knowledge created (Breschi and Malerba 2001; Feldman and Massard 2002). This means that barriers to the exchange of information are reduced (restrictions on sharing information or post employment covenants) and that the region has individuals and organisations with the capacity to capitalise on locally available knowledge. A good example is provided by Saxenian (1994) who showed how the movement of engineers and technicians between jobs among firms in Silicon Valley helped to spread technical knowledge within that region. The growth potential and 'youth' of the industry has also been shown to be important to both clustering and technology spillovers (Glaeser et al. 1992).

If innovation is central to successful long-lasting clusters, a local source of continuous innovation is imperative. The availability of specialised knowledge and the creation of new knowledge may attract firms from elsewhere into the region, may attract entrepreneurs into the region, and may create new firms through spin-off activities (Maskell 2001). The most significant source of new knowledge creation is from research and development (R&D) activities with universities as major R&D producers (Audretsch 1998, 20). University R&D expenditures have been shown to be significantly related to new firm formation and a contributor to economic growth (Kirchhoff et al. 2002). Other important sources of innovation are industry R&D laboratories, federal R&D laboratories, or even 'decentralised industrial creativity' within the collective capacity of small firms within a region.

Growth of the Photovoltaic Industry

Photovoltaic (PV) cells or solar cells are devices that can generate electrical energy under exposure to light. The technology can be traced to 1839 when the photoelectric effect was identified. Some forty years later, the first experimental solar cell was constructed. It was not until the mid-twentieth century that the first solar cell was patented based upon discoveries at Bell Laboratories. Soon after, commercial activity began and solar cells found use in space applications. As photovoltaics became a critical source of power for satellites and other spacecraft, the first oil shock of the early 1970s spurred interest in the development of terrestrial applications of PV technology.

Solar energy is one of the fastest growing technology industries in the world with growth expected at an annual rate of 35 per cent through 2020. In 2000, worldwide solar cell production stood at 288 MW. In 2005, production worldwide soared to 1,759 MW (Maycock 2006). The three major producing areas are Japan (47.5 per cent), Europe (26.7 per cent), and the US (8.7 per cent).

There are different solar cells technologies in use and under development, and different companies and different research groups have focused their investment and interest to advance certain technologies. The major types are silicon crystal and thin-film solar cells. Major issues driving market potential include the efficiency of the cell (how much solar energy is transformed into electricity), production cost of the cells; aesthetics, life cycle environmental costs, and balance of costs to install the entire system. Silicon crystal solar cells (often seen on highway signs) accounts for over 90 per cent of the total solar cell market (SolarBuzz 2006). The high cost of silicon wafers, which accounts for up to 50 per cent of the total cost of the cell, is leading manufacturers to seek less expensive alternatives. Silicon crystal cells account for almost the entire output of Sharp, Q-Cells, and Kyocera, the world's top three solar cell producers (Maycock 2006). Thin-film technologies have taken many years to develop. Thin-film technologies are promising because the material costs are driven down since the films are exceedingly thin. Systems that provide for large area deposition leading to high volume production further reduce the cost of the cells. Toledo's PV technology is focused almost entirely on thin-film technology. First Solar, Unisolar (nearby in Michigan), start-up companies Solar Fields and Xunlight, as well as University of Toledo's research focus are in thin-film PV.

Foundations of the Toledo Solar Energy Cluster

Before technology clusters were a topic of discussion, Toledo had a vibrant glass industry cluster that included some of the leading glass technology companies: Owens-Illinois, Owens-Corning, Libbey-Owens-Ford, and many others. Toledo's advantages included its natural gas reserves, labour, and access to silica beds. Toledo was a source of major glass industry innovations with the source of innovation being entrepreneurs and local industry. In 1903, Michael Owens, of Owens Bottling Machine Company, invented the first automatic bottle-making machine, recognised as one of the most significant developments in glassmaking since the invention of the blowpipe some 2,000 years before. With this invention, a plant in 1905 could produce over 17,000 bottles a day compared to 2,880 bottles a day in a plant using hand-blown labour (Brickey 1990b). By 1912, the plant with the Owens machine could produce over 50 thousand bottles a day. The inventions continued. Libbey and Owens assisted Irving W. Colburn, who was working on a continuous-drawing flat sheet glass machine to form a new company in 1916, the Libbey-Owens Sheet Glass Company, whose main product was window glass (Ward M. Canaday Centre). In 1930, Libbey-Owens-Ford Glass Company was formed through the merger of Libbey-Owens Glass Company and the Edward Ford Plate Glass Company. Glass from this company was used in the construction of the Empire State Building. In 1929, Owens Bottle Company merged with Illinois Glass Company to form Owens-Illinois Glass Company. In 1935, Corning Glass and Owens-Illinois started working together in the development of glass fibre. By 1938, Corning and Owens-Illinois created a separate company, Owens-Corning Fiberglass from this joint venture, changing its name to Owens Corning.

Although UT was primarily a teaching institution in the 1950s, the University did support the industry, but not as a source of innovations. In 1952, the University formed The Institute for Silicate Research for a 'purely scientific purpose' that included a number of industry sponsors.

Thus, the 'glass city,' held a leadership position in glass technology and manufacturing. Over the decades, local industrial R&D declined as companies shed their research operations and moved to administrative and management centres or as production units. The interest in silicate materials waned at UT. Some of the industrial laboratories spun out of local companies and moved into the community as university centres or new companies, but much of this talent was lost forever. For instance, in the mid-1980s Owens-Illinois, under increasing pressure from the plastics and aluminium industry, scaled down its research efforts in materials in favour of work on processing and manufacturing (to drive down manufacturing costs). In 1987, Owens-Illinois, University of Toledo, and the State of Ohio agreed to place the Owens Illinois R&D laboratory into the University. Likewise, Libbey-Owens-Ford had a major R&D laboratory in nearby Perrysburg that produced leading technology in plasma display systems. As late as 1990, Libbey-Owens-Ford had about 275 people in its Toledo Technology Center, and its R&D centre was viewed as a 'profit centre' (Brickey 1990a). This activity was dispersed throughout other units and much of the expertise lost. By the mid-1990s, most of the local industrial and university R&D relating to glass was a chapter in Toledo's history.

Although a lack of entrepreneurship is frequently mentioned as a Toledo problem, the origins of the cluster are based solely on the initiatives of Toledo entrepreneurs. Few would dispute that Harold A. McMaster is the father of Toledo's solar energy cluster. McMaster worked as a physicist for automotive glass maker Libby Owens

Ford on wrap-around auto glass bending and tempering. Frustrated that his novel ideas were not receiving company endorsement, McMaster left Libby-Owens-Ford in 1948 and started experimenting with glass in his basement, paying particular attention to ways to bend glass to meet specific needs (e.g. glass for dime store racks). As he continued to produce new inventions, he formed Permaglass in 1948, which merged into Guardian Industries in 1969, then, with his partner Norman Nitschke (another prolific inventor) formed Glasstech Inc. in Toledo's neighbour city Perrysburg. They quickly constructed a new machine for 'tempering' glass that increased its strength, and on breaking caused it to crumble instead of breaking into shards. Glasstech grew into a leader in the manufacture of furnaces for tempered glass. An estimated 80 per cent of the world's automotive glass and 50 per cent of its architectural glass is manufactured using machines developed by the work at Glasstech (*The Toledo Blade* 2003). In 1984, with the support of 57 investors, McMaster formed Glasstech Solar Inc. to produce efficient solar cells by coating glass with thin layers of chemicals. Glasstech initially worked on thin-film silicon technology at its Wheatridge Colorado location. Continued research led to thin films based on cadmium telluride. This 'thin-film technology' would reduce the amount of material required by standard polycrystalline photocells and reduce manufacturing time. Three years later, Glasstech announced that it would be constructing a new US$15 million plant to produce 100,000 21-square-foot solar panels annually. This plant under the new name Solar Cells was located in the former Owens-Illinois Technical Center that was now on the UT campus. Thus, Toledo's first solar plant was located in a laboratory that supported R&D for the glass industry of a previous generation.

With Solar Cells locating on the University of Toledo campus, UT worked to support solar energy technology development. As McMaster and his colleagues at Glasstech advanced their technology, they sought out assistance from the University. In 1987, the University hired Dr Alvin Compaan and, with Compaan's support, two State of Ohio Edison awards were won to address processing issues in thin-film solar cell development. These awards brought sophisticated thin-film deposition systems to the region. From this base of collaboration, and with the new instrumentation, Compaan and Solar Cells Inc. won a competitive award in 1989 from the US Department of Energy's Solar Energy Research Institute, the first federal research award to Toledo in the area of solar energy. Since then, UT has received continuous funding from the US Department of Energy in support of its solar cell research.

In 2001, the University selected advanced thin-film materials as a premier area of research within the University. With this standing, the University dedicated an endowed chair position to this area, provided favourable cost-sharing arrangements, and approached Congress for support. With the active support of Representative Marcy Kaptur, UT received US$6 million from the Department of Defence and Department of Energy for thin-film PV research over the 2002-06 period. University solar energy spin-off companies are often subcontractors on these awards. In addition to these funds, UT continued to win competitive awards from the federal government and state with external funding in support of PV research now at more than US\$4 million per year. Providing additional support to the cluster, the University created an alternative energy incubator in 2005 that houses small companies and university research centres.

Formalisation and Local Recognition of the Cluster

Recognition of the cluster came with the winning of an NSF award, public promotion by the local congressional representative, the designation as a local technology cluster, the creation of a solar energy business council, and the winning of Ohio technology development grants. In 2002, UT won a National Science Foundation award to create a Partnership on Alternative Energy Systems. This award brought commitments from local government, local political leaders, and local industry. Significant awards from the State of Ohio provided boosts to the cluster. The UT photovoltaics group won a US\$2 million Ohio Third Frontier Award in 2004 and in December 2006 won an US\$18.6 million Third Frontier Award for its Center for Photovoltaics Innovation and Commercialisation. Evidence of the recognition of the cluster in Toledo was provided when Governor Taft recognised alternative energy research in Toledo in his 2005 State of the State address.

McMaster's vision that Toledo would emerge as a leader in solar cell production is ringing true. Bringing an infusion of cash and management expertise and with its new Perrysburg location, True North Partners, LLC of Scottsdale Arizona (including Wal-Mart heir John Walton) took control of McMaster's company in 1999 (Pakulski 2000) renaming the company First Solar LLC. After a dispute with the new investors, McMaster's ownership stake in the company ended. The McMaster family continued work in solar energy independent of First Solar LLC, and in 2001 at the age of 84 Harold McMaster created McMaster Energy Enterprises. First Solar produced about 100,000 4'x 2'panels in 2004 (compared to 30,000 in 2003) and produced over 300,000 panels in 2005. The company benefits greatly from incentives in Germany that allow property owners to sell power generated from solar back to the utilities for 49 cents a kilowatt-hour.

Until recently, public promotion of the cluster was almost entirely based upon university support. Recently, however, local companies, under the leadership of the CEO of McMaster Energy Enterprises have formed the 'Northwest Ohio Alternative Energy Business Council'. Business Council leaders have mounted an aggressive campaign in the state's capital to push state government to invest in solar energy and related technologies.

Table 13.1   Major firms in the Toledo solar energy cluster

Company	Contribution
First Solar LLC	CdTe thin-film photovoltaics
UniSolar	a-Si thin-film photovoltaics
McMaster Energy Enterprises	Thin-film photovoltaics (atmospheric deposition system)
Xunlight	Thin-film photovoltaics and hydrogen production from solar cells
Innovative Thin Films	Glass Superstrates and substrates
Advanced Distributed Generation	Design and installation of solar energy arrays
Pilkington	Tech Glass for solar energy industry
Owens Corning	Building materials and building integrated PV

Table 13.1 lists the major players in the cluster. This includes UniSolar of Auburn Hills, Michigan, who collaborates with UT on projects. The cluster is vertically integrated and includes major corporations, such as Owens-Corning and Pilkington with entrepreneurial managers who have recognised the opportunities of the industry. The cluster includes entrepreneurs who are working to move their business from the incubation stage into the acceleration stage. The University, with its specialised equipment, expert faculty, skilled students and Clean and Alternative Energy Incubator is a central focus of the cluster. In addition to serving as a source of innovations, the University is able to increase networking and communication through regular workshops, seminars, visitors and conferences on alternative energy offering the community valuable interpretative benefits.

Challenges and Next Steps

Despite a bright outlook, there are major challenges to face including overcoming technical questions; expansion of markets; retaining talent and retaining firms; attracting talent and attracting investment; and competition from other regions. A major research goal is to increase the efficiency of solar cells. To drive prices down, it is necessary to increase the efficiency of solar cells and to reduce the cost of the entire system (invertors and installation). Commercial solar cells now have efficiencies of about 15 per cent. Improving the efficiency of solar cells will reduce the size of cell needed to produce electricity, thus lowering costs for the entire system. Overcoming efficiency limits is important to an expanded commercial and residential market for PV.

Although PV systems are cost competitive in remote locations, such systems are not competitive with grid-connected power from existing conventional generation sources without federal or state incentives. The US commitment to solar energy is uneven, and although some states have taken steps to stimulate development; the nation is far behind other counties in taking a national approach to stimulate markets for new technology. Germany's 2000 'feed-in law' required utilities to pay 48.1 cents per kilowatt-hour for solar energy. Not surprisingly, solar energy installations increased dramatically and now stand at over 400 megawatts. In Japan, incentives were introduced to help stimulate production and installation of systems to drive down prices. As more experience was gained, the price of solar systems declined as did the incentive. Without the international market, or the promotion in certain US states, the outlook for local companies would be quite uncertain.

Conclusions

Toledo appears to be approaching the completion of a full circle in its position as a technology leader. Once a global leader in glass, with major industry R&D laboratories supported by a well-focused university centre, the region went into a period when it lost its competitive position. Now, the emerging alternative energy cluster that had its origins with people who contributed to the earlier era is emerging as a national or even international leader on a very specific solar cell technology: thin-film photovoltaics. Compared to regions that are building clusters around more technically mature and commercially successful activities, there still exists much technical and market uncertainly for the core technologies involved in solar energy. The gamble is that Toledo may be able to focus attention and develop momentum in an emerging area of technology that could position the region for greater returns. The downside is that market conditions may change, federal priorities may change, competing PV technologies will prevail or technical problems continue to thwart commercialisation of the technology.

Even though the solar energy cluster is now just emerging and gaining recognition, the foundations for the cluster can be traced to the mid-1980s as Harold McMaster moved into the solar cell arena. As his company continued to make steady but slow progress toward commercial sales, UT was building research capacity that was gaining momentum. With continued development of UT's research in PV systems, the recent expansion of research into other alternative energy areas throughout campus, and the growing interest in alternative energy by local industry, regional branding is beginning to occur. The cluster now is recognised in the region and increasingly in Ohio, as indicated by recognition in newspaper stories and other state economic reports.

The case study provides further indications that clusters may help develop and encourage entrepreneurship. The existence of the cluster and the focused university support has provided opportunities for entrepreneurs to partner with university faculty in attracting grants for technology development. These include State of Ohio grants, federal grants from the National Renewable Energy Laboratory, Small Business Innovation Grants, and congressionally-directed projects. In addition, the existence of the incubation facility assists small companies. The cluster also helps to draw national and international attention to the cluster, such as articles in national magazines and newspapers that highlight the cluster.

The technology cluster literature brings forward very few examples of lagging or declining areas that have 'created' technology clusters in new areas (Cortright and Mayer 2001). The solar energy cluster may appear to some to be a new 'created' cluster, but it has a foundation in an antecedent industry. Perhaps other newly-emerging technology clusters elsewhere have roots in an unidentified antecedent cluster. Lester (2005, 18) goes so far as to say that '... the emergence of an industry that is entirely without antecedent in the region is actually a very rare event'. An interesting point is the long gestation period for the emergence of this new industry as difficult technology questions are addressed, investors come forward, and market conditions change to open the door for the new cluster. Of course it is too early to say if the solar energy cluster will take hold in Toledo and be an example of a city moving from a position of national technological leadership in one technology (glass) that is able to take technological leadership in a new technology (solar energy). The 1987 *The Toledo Blade* quote predicting future fortunes with Toledo as a leader in new energy technology may be correct: 'Glasstech's plant could help to give Toledo an international reputation as the global centre in solar-energy technology. That easily could serve as a catalyst for attracting other solar energy-related industry, as well as research and development money and facilities to area universities'.

Acknowledgements

This chapter has been supported by NSF Grant #0227899.

References

Anon, (2003), 'Inventor became Philanthropist', *The Toledo Blade.* (26 August).

Asheim, B.T. and Isaksen, A. (2002), 'Regional Innovation Systems: The Integration of Local 'Sticky' and Global 'Ubiquitous' Knowledge', *Journal of Technology Transfer* 27, 77-86.

Audretsch, D.B. (1998), 'Agglomeration and the Location of Innovative Activity,' *Oxford Review of Economic Policy* 14, 18-29.

Boschma, R. and Lambooy, J. (2000), 'The Prospects of an Adjustment Policy Based on Collective Learning in Old Industrial Regions', *Geojournal* 49, 391-399.

Breschi, S. and Malerba, F. (2001), 'The Geography of Innovation and Economic Clustering: Some Introductory Comments', *Industrial and Corporate Change* 10, 817-833.

Brickey, H. (1990a), 'O-I, LOF Close, But Worlds Apart', *The Toledo Blade* (14 August).

Brickey, H. (1990b), 'His glass machine broke a bottleneck.' *The Toledo Blade* (14 August).

Cooke, P. (2002), *Knowledge Economies: Clusters, Learning and Cooperative Advantage* (London: Routledge).

Cortright, J. and Mayer, H. (2001), *High Tech Specialisation: A Comparison of High Technology Clusters* (Washington, D.C.: The Brookings Institution Centre' on Urban and Metropolitan Policy).

Feldman, M.P. and Francis, J.L. (2004), 'Homegrown Solutions: Fostering Cluster Formation', *Economic Development Quarterly* 18, 127-137.

Feldman, M.P. and Massard, N. (2002), 'Location, Location, Location', in Feldman, M.P. and Massard, N. (eds), *Institutions and Systems in the Geography of Innovation* (Boston/Dordrecht/London: Kluwer Academic Publishers), 1-20.

Fromhold-Eisebith, M. (2006), 'Which Mode of (CLUSTER) Promotion for which aspect of Entrepreneurship: A differentiating view on institutional support of automotive clusters', in Gatrell, J. and Reid, N. (eds), *Enterprising Worlds* (Dordrecht: Springer Science), 13-28.

Glaeser, E.L., Kallal, H.D. Scheinkman, J.A. and Shleifer, A. (1992), 'Growth in Cities', *Journal of Political Economy* 100, 197-207.

Kirchhoff, B.A., Amington, C., Hasan, I. and Newbert, S. (2002), *The Influence of R&D Expenditures on New Firm Formation and Economic Growth* (Maplewood, N.J.: BJK Associates).

Lester, R.K. (2005), *Universities, Innovation, and the Competitiveness of Local Economies: A Summary Report from the Local Innovation Systems Project—Phase I*, Working Paper 05-010. (Cambridge, MA: MIT Industrial Performance Centre').

Martin, R. and Sunley, P. (2003), 'Deconstructing Clusters: Chaotic Concept or Policy Panacea?', *Journal of Economic Geography* 3, 5-35.

Maskell, P. (2001), 'Toward a Knowledge-based Theory of the Geographical Cluster', *Industrial and Corporate Change*, 10, 921-941.

Maycock (2006), 2005 World PV Market Data.

Pakulski, G.T. (2000), 'Solar Cells Readies Production of More Economical Power Source', *The Toledo Blade* (1 February).

Saxenian, A. (1994), *Regional Advantage: Culture and Competition in Silicon Valley and Route 128* (Cambridge, MA: Harvard University Press).

Solar Buzz (2006), <http://www.solarbuzz.com/Technologies.htm>.

Taylor, M. (2005), *Clusters: The Mesmerising Mantra*, Paper Presented at the Meeting of the IGU Commission on the Dynamics of Economic Spaces. Toledo, OH. 2-5 August.

Ward M. Canaday Center, *Libbey-Owens Ford Glass Company Records, 1851-1991* (Toledo, OH: The University of Toledo).

Chapter 14

The Restructuring of Metal Manufacturing in the West Midlands Region of the UK: An Emerging New Geography

Michael Taylor and John Bryson

Introduction

The West Midlands region, centred on the major conurbation of Birmingham and the Black Country (including Coventry), is the UK's industrial heartland, where manufacturing is more important within the economic structure than in any other region in the country. It has long been recognised as having a distinctive structure of production based on the local linkage of essentially small firms – a 'locational integration' founded on the functional inter-linkage of small metal manufacturing businesses (Florence 1948, 1961). This distinctive structure developed from the late eighteenth century and was founded upon the activities of thousands of small workshops rather than large factories. Within Birmingham, strongly clustered manufacturers constantly innovated as they identified new markets and applied established techniques and technologies to new products. The system relied on the activities of factors who travelled around the country collecting orders which they executed by subcontracting production to the region's independent manufacturers. The factors carried samples and pattern books or catalogues. The importance of pattern books in selling products to consumers led to the development of a new trade in the city associated with the commercialisation of engraving and copper-plate printing. In Birmingham, 'the execution of the copper-plates, the impressions from which form the pattern books of the various trades, has extended the occupation of the engraver and copper-plate printed to a degree elsewhere unknown out of the metropolis' (Hawkes Smith 1836, 21). The development of a 'thousand trades' also led to the growth of a whole series of other occupations, for example, manufacturers of cases and boxes for the display of jewellery and other ornamental ware.

The West Midlands' distinctive economic structure resonates with Porter's (1998, 2000) ideas on clusters and clustering processes that are argued to involve processes of information mobilisation and transfer, learning and innovation, with the potential to generate internationally-significant competitive advantage. In recent years, however, the role of manufacturing, regionally in the West Midlands as nationally in the UK, has been in decline in terms of both employment and gross value-added. As companies have made productivity improvements, production has moved to low-wage countries and UK companies have moved off-shore. It is only too clear that globalisation has taken a significant toll on this region's competitive advantage. Associated with this decline of manufacturing in the West Midlands has been a reconfiguring of the geography of metal manufacturing production within the region – a reconfiguration that parallels changes in other elements of the region's economy, especially in business and professional services (Bryson 2007; Bryson and Daniels 2007). This chapter explores the spatial consequences of manufacturing change in the West Midlands that has resulted from globalisation. The focus of the chapter is on the processes of polycentric development that are reshaping the region's metal manufacturing activities.

The West Midlands Region and the Decline of Manufacturing

The West Midlands region, once 'the workshop of the world', is focused on two conurbations: the Birmingham and Black Country conurbation (including Coventry), and the much smaller North Staffordshire Conurbation ('The Potteries') centred on Stoke-on-Trent. It has a population of 5.3 million, of which 3.2 million are in the Birmingham and Black Country conurbation. This major industrial engine exists in a broader regional setting within which substantial areas are devoted to agriculture, horticulture and orchards. The region's workforce of almost 2.6 million is employed across a very wide range of activities. Traditionally, the region's economy is associated with metal-based industries as well as high-profile food-manufacturing. The economy now includes traditional manufacturing, hi-tech industries and increasingly a diverse range of business and professional services. Currently, the main sectors are automotive; plastics and rubber; software; food and drink; electronics and telecommunications; and business services.

The West Midlands is the region of the UK in which the implications of the end of old economic certainties are at their most pronounced. It was once the UK's most prosperous region and an economic powerhouse based on manufacturing. In 1950, Walker and Glaisyer were optimistic in their examination of the future of the Birmingham economy when they suggested that 'the industries of the district rest secure ... upon that facility for work in metal which the local folk have inherited from their forefathers ... [and] which the rest of the world demands' (1950, 259). But, the region's economy suffered very badly from the late 1960s as economies began to internationalise and, latterly, in the face of neo-conservative, Thatcherite deregulation of the UK economy, the intensifying pressures of economic globalisation, corporate restructuring and the growing impact of the processes and pressures of the knowledge economy. Available statistics paint a picture of a regional economy still dominated by manufacturing, but with high unemployment, low skill levels, and an enterprise deficit. Incongruously, though, it is still innovative at a regional level.

The pace of change in the region's manufacturing base has also accelerated in recent years. While the West Midlands regional economy as a whole has continued to grow year on year in terms of gross value-added, growth in the region's total employment continued only until 2003, after which time it began to decline, picking up again in 2005. In manufacturing in the region, the picture is very much worse. Gross value-added began to fall consistently after 1998, while the accompanying fall in manufacturing employment has accelerated markedly, bringing in its wake serious enterprise and skills deficits. It is worth noting that it is important not equate employment shifts with alterations in regional industrial structure. Productivity improvements in manufacturing can enhance profits and competitiveness, but at the same time lead to a decline in employment. It would appear that manufacturing in the West Midlands was surviving through productivity improvement until 1998, but experienced some considerable difficulties thereafter.

The West Midlands Production System: Local linkage, External Economies and Behaviouralism

The current pattern of manufacturing decline in the West Midlands stands in stark contrast to the model and mechanisms of growth that had long been recognised in the region, especially in the works of Allen (1929), The West Midlands Group (1948), and Florence (1948, 1961). This body of literature raised the West Midlands to the status of a type-example of a linked enterprise structure. Within this structure, the output of firms in metals production (from foundries to the manufacture of cars and jewellery) were seen as feeding into firms engaged in later-stage, specialised production involving metal forming, the production of components and sub-assemblies and, finally, finished and consumer products (especially vehicles, and other regionally distinctive products) (also see Berkeley et al. 2005). The close linkages between firms within this system were argued to generate local *external economies of scale*, making the region's industry more *efficient and competitive* than that in other regions. The whole system was seen as a 'locational integration' in which, according to Florence (1948), separate small firms were seen as operating much as the separate departments of larger enterprises, but with the powerful profit motive brought into play at every stage in the production sequence.

Such an interpretation sits easily with current ideas on clusters. It meant that manufacturing was tied to the region, and especially to the Birmingham and Black Country conurbation. The hypothesised need for proximity to fuel locational integration underpinned approaches to economic planning in the region that attempted to limit relocation outside the conurbation while at the same time, at least until the 1980s, encouraging some relocation of growth to other regions within the UK that were experiencing economic and particularly employment problems.

However, other research on the region's economy in the 1970s and 1980s was unable to identify the cost efficiencies of locational integration in the West Midlands metal manufacturing system and argued, instead, that the region's locally linked enterprise structure owed more to behavioural processes that made business in the West Midlands 'easier' rather than more 'cost effective' and 'cheaper' (Taylor 1973, 1975, 1978; Taylor and Thrift 1982a, 1982b; Taylor and Wood 1973; OECD 2004). This latter interpretation fits well with the region's economic history. Historically, the region's 'metal clusters' were founded on socially constructed relationships between firms that were formally and informally organised as part of a 'group contracting' system, and within which the region's principal 'market-makers' were factors and agents (see Taylor and Bryson 2006). However, although the linked enterprise system of the West Midlands might show signs of continuation to the present, the system has also undergone radical restructuring at various times, and especially since the 1940s. The region's industrial dynamism was constructed upon thousands of small innovative firms that were able to respond relatively rapidly to changing market conditions. As one set of products became unfashionable or less profitable these firms shifted to exploit new market niches. This dynamic economic structure was partially undermined by the development of large firms and the introduction of supply chains that were controlled by market makers located outside the region. It is to these changes that we now turn our attention.

Change and the Pressures of Restructuring

From the 1920s, Fordism and mass production brought major changes to the West Midlands regional economy: the limited liability company replaced the family firm and the entrepreneur; mergers brought a loss of local control and corporate concentration; the scale of production increased as did the extent of vertical integration. Fordist mass production created boom conditions in the West Midlands through the 1950s and 1960s, but left the region's economy with significant vulnerabilities. The number of 'market makers' within the region, those brokering its products into the markets of the world, was radically reduced, as indicated by the fact that 48 per cent of West Midlands manufacturing employment was in just 25 companies in 1977 (Spencer et al. 1986). That concentration, especially in the car industry, had significant implications. Many small, independent engineering firms were transformed into subcontractors, and the large employers became both powerful and vulnerable – powerful enough to elicit government assistance, while simultaneously being vulnerable to union pressure. The outcome was that the region's metal and engineering industries lost their resilience (Bryson et al. 1996). In 1976, Wood estimated that the proportion of West Midlands' output and employment that depended on the car industry was approaching 30 per cent. The increasing dependence of suppliers on the performance of a limited number of large companies was a structural weakness in the West Midlands economy. Increasing competition from overseas producers led to the collapse of many of these inter-firm linkages as major firms were forced to renegotiate conditions of trade and product quality with local suppliers. Competition, combined with the increasing external control of West Midlands companies, led to the replacement of local subcontractors by non-local and even overseas suppliers.

After 1970, in an emerging era of mass customisation rather than mass production, the need for economic restructuring left the West Midlands region striving to cope with change. The region's economy was a weak player in the service economy (Bryson et al. 1996). It was labelled as having non-innovative small firms, especially in the 1990s (OECD 2004). And, it was and is considered to have serious and problematic skills gaps (Bryson and Daniels 2008).

What has been distinctive about the West Midlands post-Fordist restructuring has been the growth of firm registrations in finance and business services: especially in service industries; also in wholesaling, retail and repair; hotels and restaurants; and construction. However, the 'new firm formation rate in the manufacturing sector was a little higher than in the UK but much lower than the employment the sector accounted for in the region' (OECD 2004, 39). The suggestion is, therefore, that the West Midlands regional economy is attempting to restructure by responding to the processes and pressures of change, but not responding rapidly enough.

Unfortunately, the ethnic communities of the West Midlands, which are the largest outside London and the Southeast region, do not appear to have contributed proportionally to the new firm formation rate in manufacturing. The data on this issue provide only a very partial picture, but it would appear that the ethnic communities of the West Midlands have created new enterprises not in metal manufacturing and engineering, but predominantly in health care activities and retailing (OECD 2004), and also in clothing manufacture (although this sector has declined sharply in the early years of the present century (Berkeley et al. 2005).

Although the most recent restructuring of production in the region would appear to indicate the existence of an 'enterprise deficit' in the West Midlands, there would appear to be no parallel innovation deficit. The 2004 OECD report considered, on the information that was assembled, that in terms of innovation performance, the region had 'not fared badly'. More recent time-series data on regional variations in the granting of patents, registered trademarks and registered designs suggest that, relative to gross value-added, *the West Midlands is one of the more innovative regions in the country* behind the East of England (Cambridge's 'Silicon Fen'), the South East (the M4 corridor), and the South West (Bryson and Taylor 2006). This is a point that is too easily overlooked. It is a potential driver of regional economic growth, with the capacity to stimulate enterprise, entrepreneurship and new firm formation.

Recent interview evidence from firms in niche metal manufacturing (forging, pressing, rolling and shaping) and lock making in the West Midlands suggests very specific processes of change are occurring in what remains of the region's metal manufacturing system (Bryson, Taylor and Cooper 2008). Large companies are disengaging from the local production system to shift high volume, low value production overseas. At the same time, they are maintaining skills-based specialist production in the region, and also research and design. SMEs in niche metal manufacturing are focusing on quality and good customer service, using CAD CAM to produce small batches with a rapid turnaround (see Bryson, Taylor and Cooper 2008). They also play to wider markets to reduce their commercial vulnerability. Many pride themselves in having the most advanced production facilities in Europe. Great care needs to be taken in assuming a uniformly-declining metals manufacturing sector in the West Midlands. High-end, niche metal manufacturers appear to have the potential to be the 'seedbed' from which may grow a very different metal manufacturing sector in the region.

Localisation and the Shifting Spatial Pattern of Production

The metals and metal manufacturing sectors that have been the traditional heart of West Midlands manufacturing are strongly localised in the districts of Birmingham and the Black Country, and significant nodes of these activities have developed elsewhere in the region, creating a new and emergent polycentric system extending beyond the bounds of the principal metal manufacturing and engineering system in the Birmingham and Black Country conurbation. This pattern of change is outlined schematically in Figure 14.1.

In Figure 14.1, the 'locational integration' of metal manufacturing and engineering that has for so long typified the West Midlands is portrayed as four stages in the sequence of production starting with raw materials and moving through to finished products. The foundation of the sequence is the production of basic metals. Firms in metal manufacturing work on the output of that sector and, in turn, supply products to equipment manufacturing in firms making machine tools, for example. Finally, the firms of the region's metal manufacturing and engineering complex make finished products, most notably motor vehicles.

Traditionally, proximity has been seen as vital to the efficiency of this system of production with firms necessarily locating with the region's conurbations, and especially within the Birmingham and Black Country conurbation. What Figure 14.1 demonstrates is that the localisation of firms engaged in metal manufacturing and engineering have shifted to towns and districts well beyond the confines of the conurbation and are now forming a belt around it. Based on an analysis of location quotients (see Isard 1960) there are now notably strong concentrations of basic metals production to the east of the conurbation, in the contiguous districts of Nuneaton and Bedworth, Rugby and Warwick. A somewhat stronger localisation is evident in the contiguous districts of Worcester, Wychavon, Redditch and Hereford. Also notable in this context are the localisation of basic metals production in Bridgnorth.

Metal manufacturing, interpreted in the past as the sectors working on the output of the basic metals production sector, has also spread beyond the conurbation (Figure 14.1). As illustrated by location quotients for the forgings and pressing sector, the core of metal manufacturing has now extended beyond the Birmingham and Black Country conurbation to the south (to Kidderminster, Bromsgrove, Redditch and Stratford-upon-Avon) to the northern periphery of the conurbation (to Cannock Chase, Lichfield for example), and to the emergent node of Telford.

Moving upwards again in the metals and engineering sequence, location quotients Machine tools manufacturing (Figure 14.1) shows significant polycentric development in the region. Machine tools manufacture has extended from the Birmingham and Black Country core, to Telford and Cannock in the north and west, to Tamworth, Nuneaton, and Rugby in the east, and to Bromsgrove, Redditch and Worcester to the south.

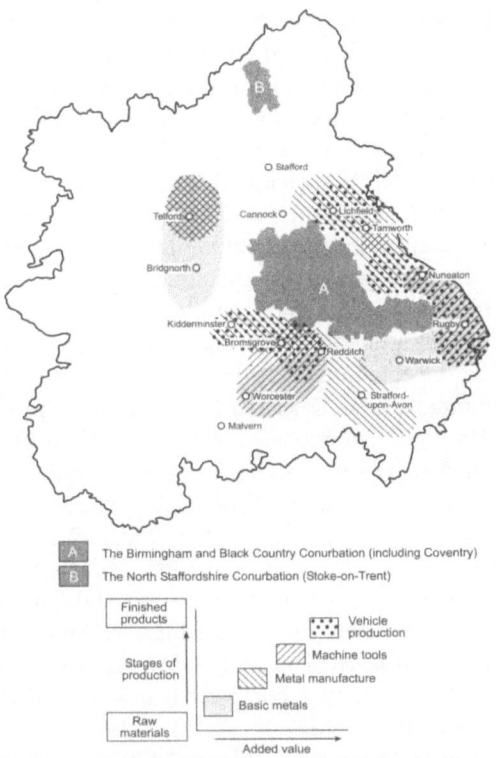

Figure 14.1   Geographical   extension   of   the   West   Midland   metal manufacturing complex

The peak of the West Midlands production system is dominated by vehicle production, and Figure 14.1 also shows that the localisation of this aspect of production has extended principally to the east of the conurbation, to North Warwickshire, Nuneaton and Rugby. Lesser concentrations of vehicle production are also found in Redditch, Kidderminster and Lichfield. However, notably strong concentrations have now emerged in Telford and Cannock Chase. This pattern suggests that the peak of the West Midlands production system has spread from its origins in the West Midlands conurbation and Coventry to embrace the whole of the eastern part of the region.

Extending this issue of the spatiality to embrace innovation, a strong regional pattern also emerges from a location quotient analysis of the locations of production that can be considered likely to underpin and guide the future of West Midlands metal manufacturing and engineering. In this context, two sectors of current production can be used as indicators. The first is R&D activity and the second is 'miscellaneous manufacturing n.e.c'. The 'not elsewhere classified' elements of a sectoral economic classification were considered by Florence (1961) in the context of the West Midlands to contain those elements of a production system that are emergent and do not fit readily with static classificatory systems. It is this interpretation that is applied here. What is clear from Figure 14.2 is that these two sectors have distinctive intra-regional distributions within the West Midlands, as measured by location quotients. R&D activity (Figure 14.2) is very strongly localised in Solihull, Warwick and Malvern, is strongly localised in the adjacent Stratford-upon-Avon district, and has an above average representation in Lichfield. These are all affluent, desirable places to live that are strategically close to the main centres of production and equally strategically placed on main motorway connections affording ready national access. Malvern is especially interesting as the concentration of R&D firms in the locality was a direct result of a wartime strategy to relocate radar research to the Midlands (Bud and Gummett 1999). The Telecommunications Research Establishment (TRE) was established near Swanage in May 1940 and relocated to Malvern College, Worcestershire in August 1942. TRE had some very specific locational requirements that involved access to an aerodrome and a site on a hill. TRE was combined with the Army Radar Establishment in 1953 to become the Radar Research Establishment. It became the Royal Signals and Radar Establishment when the Army Signals Research and Development Establishment moved to Malvern in the 1980s and in 2001 the British Government partly privatised this R&D and renamed it QinetiQ. In February 2006 QinetiQ was listed on the London Stock Exchange with a market capitalisation of £1.3 billion and the company is now one of the largest defence research organisations in the world.

Miscellaneous manufacturing n.e.c. (Figure 14.2) shows a different pattern of localisation, more strongly linked to parts of the Black Country but extending beyond. Essentially, it is the districts of the south west of the West Midlands region that specialise in these activities, and outside Dudley in the conurbation, there are notably strong localisations in Bridgnorth, Bromsgrove and Malvern Hills.

This shift in manufacturing to exploit new materials, processes and technologies has been an important feature of the region's industrial economy. Manufacturing companies that have survived the various gales of creative destruction have responded by proactively exploiting new markets. For example, Walter Somers Ltd, a metal forging company established in 1866 and still operating, constantly shifted markets, moving from the production of rickshaw axles and railway buffers in the nineteenth century to the production of difficult, high integrity forgings in

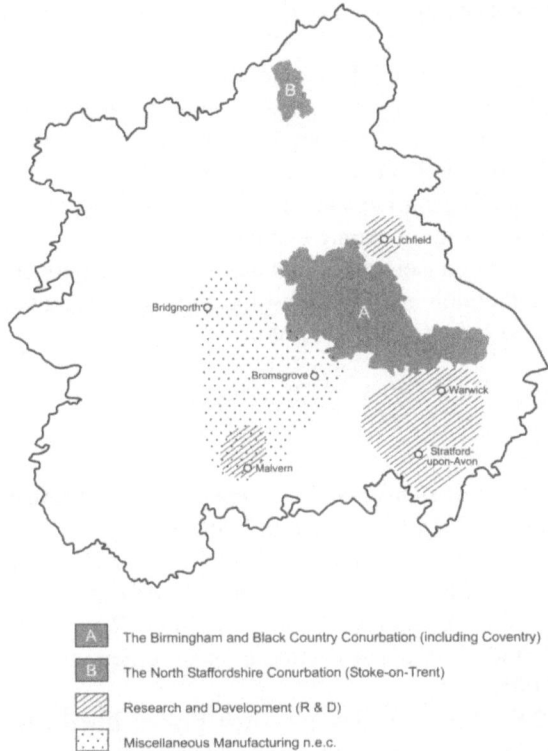

A. The Birmingham and Black Country Conurbation (including Coventry)

B. The North Staffordshire Conurbation (Stoke-on-Trent)

Research and Development (R & D)

Miscellaneous Manufacturing n.e.c.

Figure 14.2   Localisations of R&D and miscellaneous manufacturing

all types of metal in the late twentieth century (Gale 1987). New manufacturing companies continue to be established.

Conclusion: Polycentricity and the Emergence of the E³I Belt

From this analysis it is readily apparent that the spatial patterning of manufacturing activity within the West Midlands region is breaking away from Birmingham and the Black Country. The innovative and dynamic elements of manufacturing are shifting to a belt that encircles the conurbation; a belt that lies between around 20km to 40km from the conurbation which includes Stratford, Warwick, Rugby, Lichfield, Cannock, Bridgnorth, Worcester, Redditch and Malvern. Within the belt there is an important differentiation of production activities from centre to centre, bringing a new polycentric structure to the West Midlands region. Other research suggests that this newly emerging polycentric structure is being reinforced and intensified by the concentration of highly skilled people in some parts of the belt, together with high rates of new firm formation, localisations of producer service firms, and major concentrations of logistics activities in other parts (Bryson and Taylor 2006).

The emergence of the belt reflects a combination of factors including lifestyle, accessibility, and quality of environment. It combines '*economic, entrepreneurial, environmental and innovation*' factors and is more conveniently labelled the E³I belt. The newly emerging E³I belt is perhaps partially driven by the Birmingham city-region, but equally its drivers are elsewhere in the regional economy. It is to this belt, within which the dynamic elements and drivers of the West Midlands regional economy are concentrated, that academics, planners and policy makers need to turn their attention in order to fashion and facilitate an economic recovery that is congruent with the region's strengths.

References

Allen, G.C. (1929), *The Industrial Development of Birmingham and the Black Country: 1860-1927* (London: George Allen and Unwin).

Berkeley, N., Jordane, C. and Jarvis, D. (2005), *The Role of Mature Sectors in Promoting Regional Economic Development: The Case of the High Value Added Consumer Products Cluster in the West Midlands Region*, Paper Presented at the Regional Studies Association International Conference, Aalborg, Denmark, 28-31 May 2005.

Bryson, J.R. (2007), 'Lone Eagles and High Flyers: Rural-based Business and Professional Service Firms and Information Communication Technology', in G. Rusten and S. Skerratt (eds), *Being Rural in a Digital Age* (London: Routledge), 36-60.

Bryson, J.R. and Daniels, P.W. (2007), 'A Segmentation Approach to Understanding Business and Professional Services in City-regions: Shifting the Horizon beyond Global Cities', in Rubalcaba L., Kok H. and Baker P. (eds), *Business Services in European Economic Gro*wth (London: Palgrave Macmillan), 251-262.

Bryson, J.R. and Daniels, P.W. (2008), 'Skills, Expertise and Innovation in the Developing Knowledge Economy: Business and Professional Service Firms in the West Midlands, UK', *International Journal of Services Technology and Management*, 9(3/4), 249-267.

Bryson, J.R., Daniels, P. and Henry, N. (1996), 'From Widgets to Where? A Region in Economic Transition', in Gerrard, A.J. and Slater, T.R. (eds), *Managing a Conurbation: Birmingham and Its Region* (Birmingham: Brewin Books), 156-168.

Bryson, J.R. and Taylor, M. (2006), *The Functioning Economic Geography of the West Midlands Region, Advantage West Midlands* (Birmingham: West Midlands Regional Observatory).

Bryson, J.R., Taylor, M. and Cooper, R. (2008), 'Competing by Design, Specialisation and Customisation: Manufacturing Locks in the West Midlands (UK)', *Geografiska Annaler* 90(2), 173-186.

Bud, R. and Gummett P. (ed.) (1999), *Cold War Hot Science: Applied Research in Britain's Defence Laboratories: 1945-1990* (London: Science Museum).

Florence, P. (1948), *Investment, Location and Size of Plant* (Cambridge: CUP).

Florence, P. (1961), *The Logic of British and American Industry* (London: Routledge and Kegan Paul).

Gale, W.K.V. (1987), *Walter Somers Ltd – A History: 1866-1986* (Halesowen: Walter Somers Ltd.).

Hawkes Smith, W. (1836), *Birmingham and its Vicinity as a Manufacturing and Commercial District* (London: Charles Tilt).

Isard, W. (1960), *Methods of Regional Analysis* (Cambridge MA: MIT Press).

OECD (2004), *OECD Local Entrepreneurship Reviews: West Midlands, United Kingdom,* Report undertaken by Advantage West Midlands as part of the OECD Leed programme, < http://www.advantagewm.co.uk/downloads/oecd-final-full-report-aug-2004.pdf>

Porter, M. (1998), *On Competition* (Boston MA: Harvard Business School Press).

Porter, M. (2000), 'Locations, Clusters, and Company Strategy', in G. Clark, M. Feldman and M. Gertler (eds), *The Oxford Handbook of Economic Geography* (Oxford: Oxford University Press), 253-274.

Spencer, K., Taylor, A., Smith, B., Mawson, J., Flynn, N. and Batley, R. (1986), *Crisis in The Industrial Heartlands: A Study of the West Midlands* (Oxford: Clarendon Press).

Taylor, M. (1973), 'Local Linkage External Economies and the Ironfoundry Industry of the West Midlands and East Lancashire Conurbations', *Regional Studies* 7, 387-400.

Taylor, M. (1975), 'Organisation Growth, Spatial Interaction and Location Decision-making', *Regional Studies* 9, 313-23.

Taylor, M. (1978), 'Linkage Change and Organisational Growth: The Case of the West Midlands Ironfoundry Industry', *Economic Geography* 54: 4, 314-336.

Taylor, M. and Bryson, J. (2006), 'Guns, Firms and Contracts: The Evolution of Gun-making in Birmingham', in M. Taylor and Oinas, P. (eds), *Understanding the Firm: Spatial and Organisational Dimensions* (Oxford: Oxford University Press), 61-84.

Taylor, M. and Thrift, N. (1982a), 'Industrial Linkage and the Segmented Economy: 1. Some Theoretical Proposals', *Environment and Planning A* 14, 1601-1613.

Taylor, M. and Thrift, N. (1982b), 'Industrial Linkage and the Segmented Economy: 2 An Empirical Reinterpretation', *Environment and Planning A* 14, 1615-1632.

Taylor, M.J. and Wood, P. (1973), 'Industrial Linkage and Local Agglomeration in the West Midlands', *Transaction of the Institute of British Geographers* 59, 127-154.

Walker G. and Glaisyer E. (1950), 'A Survey of the Industrial Population of Birmingham and the Black Country', in Kinvig, R., Smith, J. and Wise, M. (eds), *Birmingham and its Regional Setting* (British Association), 249-260.

West Midland Group (1948), *Conurbation: A Planning Survey of Birmingham and the Black Country* (London: Architectural Press).

Wood P. (1976), *The West Midland* (London: David and Charles).

Chapter 15

Labour Market Intermediation

James W. Harrington and Nicholas Velluzzi

Introduction

The normative model of atomised labour markets is powerful in its simplicity: so powerful that its use in positive analysis of labour markets persists despite its obvious inconsistencies with the way people acquire occupations, employment, self-employment, and wage/salary/benefit levels. These inconsistencies have many sources and manifestations, including: asymmetries in the information and legal power of employees and employers, labour market segmentation, and the geographic specificity of labour markets. The resultant discrepancy between theoretical expectations and labour market outcomes has led to a rich study of social institutions that govern the operation of labour markets (Solow 1990), and a recent view that employers exert wage discretion against labour supplies that are less than perfectly elastic (Burdett and Mortensen 1998; Manning 2003). This chapter focuses on another complication of labour market operation: the use of companies, agencies, and information networks to mediate the employer-employee relationship.

The chapter has two purposes. The first is to improve the conceptualisation of labour market intermediation (LMI) by applying theories of intermediation and commodity chains. Specifically, intermediaries exist to reduce and to shift uncertainty and risk: the reductions increase overall productivity, while the shifting increases the returns for those actors who control scarce and valuable assets. The second is to outline and exemplify the sources of local/national specificity in the operation and effects of LMI.

Intermediation

Commercial intermediation is the process of connecting suppliers and buyers via a third party, the intermediary. Wu (2004, 67) defined intermediaries as 'economic agents who co-ordinate and arbitrate transactions in between a group of suppliers and customers'.

We may gain insight into the broad range of intermediaries and intermediation by recognising their varied roles in managing uncertainty (ignorance of future states) and risk (potential for gain or loss resulting from making decisions in the face of uncertainty). 'Market making' intermediaries (such as retailers and most wholesalers) select suppliers, buy, sell, and hold inventory (Spulber 1996;

Wu 2004). They thereby accept the risk of misreading or changes in the market. Banks, as financial intermediaries, accept and manage credit risk, interest risk, and temporal risk (of short-term debts and long-term assets). Their scale, scope, and expertise reduce the overall risk in the system, even as they accept some risk. Brokers are intermediaries who 'provide co-ordination services without buying and selling goods' (Spulber 1996, 145). Thus, they do not take on the liquidity, inventory, and other risks of retailers, wholesalers, and the like. Their value-added is in the centralisation and consolidation of buyers' and sellers' searches, creating scale economies and reducing uncertainty by:

- determining a price that allows suppliers and buyers to interact on either side of the intermediary;
- identifying a wide variety of sellers and buyers for heterogeneous goods or services;
- guaranteeing the quality of suppliers, based on experience with given suppliers; and
- monitoring suppliers, especially suppliers of services – more cheaply than individual buyers because of scale.

Thus, they attempt to reduce uncertainty for sellers and buyers, for which buyers or sellers are willing to pay.

Labour Market Intermediation

Definitions

Most simply, labour market intermediaries (LMIs) are 'mechanisms or institutions that intercede between job seekers and employers' (Autor 2004, 1). These can range from the passive information channels of a newspaper or online listing to the intensive activity of a state-sponsored 'one-stop shop' that provides screening, training, and job placement for a targeted population.

The phrase LMI invites comparison to financial intermediation. 'Financial intermediaries lend to large numbers of consumers and firms using debt contracts and they borrow from large numbers of agents using debt contracts as well' (Gorton and Winton 2003, 2). LMIs attempt to fill employment needs by large numbers of clients through their information about large numbers of employment seekers (or to fill the needs of seekers through their information about vacancies). However 'the obligations of firms (to a financial intermediary) and the claims ultimately owned by investors are not the same securities; (financial) intermediaries transform claims' (Gorton and Winton 2003, 5). Thus, contracted production rather than LMI may be the labour market analogue to financial intermediation, because this is when the worker is paid for labour while the intermediary is paid for product. Rather, the major reasons for LMIs' existence are to ease the flow of information between/

about job vacancies and job seekers and to reduce producers' long-term liabilities to workers.[1]

Functioning

In the search for workers, employers can trade off: search costs (length of search, intensity of search, geographic scope of search, or use of intermediaries), wage rates (which would elicit more and better qualified applicants), training costs (which would allow hiring less immediately qualified applicants) (Mills 1978), and labour productivity (higher with better skills, lower with higher turnover based on poor matches). In opposite fashion, job seekers make tradeoffs among search costs, wage rates, and self-financed training.

Producers or job-seekers will find it worthwhile to engage LMIs to the extent that these brokers can reduce search costs, lower producers' wage rates, increase job seekers' wage rates, or vet applicants' skills. 'Intermediaries have traditionally addressed these objectives with four basic types of assistance:

- improved access to labour supply and demand;
- improved quality and access to labour market information;
- special assistance to help people prepare for and adapt to job opportunities and labour market transitions; and
- interpretation of and adaptation to imbalances in the interaction of labour supply, demand, information, and price' (Cassell and Rodgers 1978, 118).

This list emphasises the uncertainty reduction provided by LMIs. Each side, then, would be more likely to use intermediaries when greater wages are at stake, and when intermediation is less expensive.[2] A rationalist view would assume that workers who expect (based on experience, networks, or socialisation) a high return to a longer search, and whose living costs are covered during the search, will make use of longer or more expensive searches.

Types and Roles

Table 15.1 provides a typology of LMIs, using Cassell and Rogers's categories of uncertainty reduction as the row headings, and the aforementioned variations in

---

1 We use the word 'producer' rather than 'employer' to designate a public or private organisation that produces a good or service for a market or client base, and 'worker' rather than 'employee,' because one important role of some LMIs is to become the employer of record, so that producers can reduce their official 'employment.'

2 Autor (2001) observed that the nearly trivial costs of posting résumés and job announcements on on-line job boards (a form of intermediary) has created a market for further intermediaries (some automated, some with personal interaction) that will sort through the millions of seemingly relevant resumes and announcements.

risk management as the column headings. In general terms, uncertainty reduction is a form of or means toward risk reduction: more uncertainty implies more risk of an unforeseen outcome. However we can maintain orthogonality between the two concepts by recognising that 'risk' is borne by the party that gains or loses if the outcome is different from the most expected outcome.

In labour markets, uncertainty surrounds: the worker's decision to seek certain types of training; the employer's decision to recruit a worker at a given wage/salary; the cost an employer incurs when a new worker must be subsidised while being trained on the job. How do employers and workers shift or reduce their risk? Which of those methods entail an intermediary, either to bear or shift risk, or to reduce information so that whoever bears the risk is less likely to have an unforeseen upside or downside? Note that this typology implies some possible intermediary functions that are not commonly assumed by LMIs – especially those functions that assume risk. We have included internal labour markets as forms of intermediation – institutional practices that allow producers to assess and train employees over time, so that some employees are deployable in new positions within the organisation (Doeringer and Piore 1971).

Table 15.1   Using uncertainty and risk to develop a typology of LMIs

Reduce uncertainty by:	Transfer risk		Accept risk	No transfer of risk
	From worker to producer	**From producer to worker**		
Increasing mutual access		TS agencies		Executive search firms Alternative TS
Increasing info to employers	Internal labour markets with long-term contracts	Short-term internships or trial periods		Online résumé posting services
Increasing info to workers				Want ads Job matching services
Improving LF transitions	Internal labour markets with employer-provided training			Alternative TS
Reducing supply/ demand imbalances			Contracted production; government supported training & placement programmes	Internal labour markets in which employees are promoted or moved as needed Private/ public workforce development boards

Temporary and Project-based Employment

The US National Research Council (2001) defined a broad category of 'project-based employment' as: (1) regular employees whose continued employment is contingent on the organisation matching them with subsequent projects; (2) self-employed contractors; (3) third-party contractors, employed by (a) temporary services firms that do not oversee the actual work done or (b) 'business services firms' hired by clients on a project basis.

Temporary employment, whether direct or through LMIs, assists producers to realise labour costs as truly variable (Ciscel and Smith 2005). Doeringer (1994) recognised the growth in what he calls 'the employment-at-will sector', which indeed has become the dominant form of employment growth for highly skilled technical workers, many trades people, as well as low-skilled clerical, service, and assembly workers. These entail the most rapidly-growing occupations (with the exception of clerical occupations) in some of the most rapidly growing sectors within the US. These 'firms want to hire workers who are job-ready, and they deliberately place the burden of human capital formation on workers and schools' (Doeringer 1994, 102) – essentially shifting the risks of selecting and paying for training to the individual worker (or the state). Producers can externalise risk by maintaining a flexible, contingent workforce and gaining the ability to rapidly respond to changes in labour demand, screen workers, decrease search costs, and externalise employee training. By shifting risk, client firms reduce their level of uncertainty, knowing that their labour demand requirements will be met at a specific price and lower cost. Ciscel and Smith (2005) noted the use of LMIs to move the culpability for work conditions from the retailer or brand name, and even from the contract manufacturer, to a third-party labour supplier (or to individual contract workers).

Temporary Services Agencies

Temporary services (TS) agencies can provide producers with quantitative flexibility (especially important in clerical and support employment), economies of scope (for specialised workers, not needed by individual producers at all times, especially important in technical and professional workers), and/or regulatory slipperiness (especially important in blue-collar employment) (Theodore and Peck 2002). For 'basic' labour, TS agencies provide economies of scale in search, from the employer's and employee's perspective. For highly specialised labour, TS agencies provide the benefits of established networks of employers and employees.

Gannon (1978) cited early 1970s figures that the US temporary services industry had about 65–70 per cent of its employment in clerical workers, 25–30 per cent in industrial occupations, and 2–5 per cent in professional/technical. By 2004, the TS industry had seen rapid growth in the US. Kilcoyne (2005) noted annual TS growth rates eight times greater than overall employment in the US, representing 13 per cent of net new jobs in 2003. Clerical occupations (clerks, secretaries, data entry, and word processing) comprised only 17 per cent of those employed through TS agencies, while industrial occupations (labourers, packers, assemblers) comprised 42 per cent. Professional and technical occupations still comprised only 6 per cent of these workers, largely registered and licensed practical nurses, nursing aides, computer support specialists, and computer programmers.

Intermediation in Commodity Chains

The configuration of a commodity chain reflects deliberate attempts by each constituent producer, provider, and intermediary to regulate the entire chain (not just the relationship with its immediate clients and suppliers) to its advantage. Gereffi (1999) emphasised that market power is conferred by the ability to enforce a near monopoly or monopsony on the supply of or demand for a product or service. In any commodity chain, the greatest market power is by the component of the chain with the most sustainable monopoly and monopsony. This component is no longer manufacturing except in the most specialised contexts. This component is seldom labour except in its most specialised forms.

'The equilibrium bid-ask spread, which separates buyer willingness to pay and supplier costs, is a consequence of transaction costs (searching and matching), asymmetric information (monitoring quality) and the returns to intermediation activities (including the value of time and effort that would go into direct or non-mediated procurement or sale)' (Spulber 1996). Generally, TS firms have maintained a 30–40 per cent spread between wages paid and revenues/worker. However, this margin has tightened as competition has increased and as TS firms began using less- and less-skilled labour for the low-end industrial positions – where producers have begun to see TS firms as a source of low-wage labour, as opposed to primarily as a source of temporal and regulatory flexibility (Peck and Theodore 2002).

Forde and Slater (2005, 259) illustrated that temp agencies in the UK have been somewhat successful in upgrading their services to the provision of managers and 'welfare associate professionals (nurses, therapists, welfare workers)', but that knowledge-intensive and IT workers are hired through other, non-mediated temporary arrangements. This suggests that higher skilled and highly specialised workers are more able to rely on their specific resumes and networks, and their employers recognise the need to search directly for a particular set of qualities and capabilities. Meanwhile, less-skilled clerical workers and operatives, and more certificated health-care workers (and some IT workers) are able to be 'bundled' and hired through intermediaries.[3] Producers would rather negotiate temporary or

---

3 'The TSI has restructured [by increasing its] occupational, industrial, and geographic 'reach,' while the basic business model of the temp agency – the wholesaling of industrial

project contracts directly with specialised labour, rather than going through a TS firm. The search for possible candidates may be helped by a search firm, especially one that focuses on such specialised labour.

Using 2004 Bureau of Labour Statistics data, Kilcoyne (2005) showed that the average hourly wage paid to TS employees was 25 per cent (or five dollars) below overall average wages. However, there was a striking variation across occupations. All industrial and clerical occupations had lower average hourly wages in TS firms than in the economy overall. Of the 48 occupational titles studied, the average hourly wage of TS employees was higher than the overall wage in only five titles: computer programmers (US$7.85 higher), registered nurses (US$4.93 higher), licensed practical and vocational nurses (US$3.88 higher), nursing aides, orderlies, and attendants (US$1.25 higher), and home health aides (US$0.60 higher). These occupations share two key characteristics: skills that are transferable across producers and sectors, and generally tight labour markets. In addition, programmers, registered nurses, and licensed nurses require years of specialised training and experience to claim those occupations. These three characteristics make producers willing to pay a premium for LMI, and provide intermediaries and employees with limited monopsony in labour supply. However, these five occupations represented only 132,000 TS employees, less than six per cent of all TS employees.

We conclude that TS firms, and LMIs in general, seldom play dominant or market-shaping roles in the supply of labour. Their presence increases the ability of workers and producers to connect across sectors, extends the information available to workers about positions and to producers about workers, and increases the employment flexibility of producers and the job mobility of workers.

Labour Market Intermediation across Economic Spaces

LMIs increase information flow and worker mobility across regions, as well. Different forms of labour market intermediation play different roles in these processes. Search firms, and online services that categorise and publicise job vacancies and seekers, increase information flow across regions. For occupations that face inelastic supply because of long and specialised training, interregional flows of information may serve to reduce supply or increase wages in low-wage regions. Another form of LMI, internal labour market arrangements within large employers, increases information flow and worker mobility across regions. Most TS agencies focus on individual local labour markets. Truly national or international TS agencies could aid in interregional information flow and worker mobility.

Local and national labour-market institutions determine the types of LMI that are feasible, and the roles available to intermediaries. Strong national regulation

---

and clerical labour under tight profit margins and relentless competitive pressures – has remained stubbornly impervious to fundamental change' (Theodore and Peck, 2002: 484).

of the direct employment relationship encourages the development of indirect employment arrangements, *if* these are allowed by labour regulation (e.g. union agreements). Peck and Theodore (2002, 144) noted that temporary employment arrangements account for a much smaller proportion of total employment in the US than in other OECD countries: 'The potential demand for temps may be dampened in the US because the "mainstream" employment relation is already relatively "deregulated" and because the labour market as a whole is already substantially "flexible"'. Relevant regulatory dimensions include requirements of justification or notice for layoffs, and non-wage benefits.

National labour regulation and local or national rates of unionisation affect the ability of LMIs to shift risk from producer to worker or to establish a lower tier of wage rates by occupation. Peck and Theodore (2002) reported that the proportion of temporary employment handled by separate LMIs (specifically, temporary services firms) is much greater in the US than in many other OECD countries (where producers are more likely to hire temporary workers directly). Within the US, the proportion of employment handled by temporary services firms is much greater in non-unionised, relatively deregulated regions (Peck and Theodore 2002). Their observation that the actual use of separate LMIs reflects their regulatory capability to shift risk and lower wages, supports risk shifting and wage differentials as substantial motivations for the use of these types of LMI.

Labour Market Intermediation and Workforce Development

Workforce intermediaries (WIs) illustrate another type of LMI to emerge in response to the overarching institutional changes in the labour market. Many types of organisations qualify as WIs; in the United States these include Workforce Investment Boards (WIBs), community colleges, One Stop Career Centers, vocational and adult schools, literacy providers, welfare-to-work programmes, community-based organisations (CBOs), employer associations, and the Employment Service. WIs typically operate on a regional scale defined by commuting sheds. The vast majority of WIs are non-profit organisations. Although WIs are present throughout most of the US, their greatest concentrations are in the West (56 WIs), Midwest (52 WIs), and Northeast (43 WIs) (Marano and Tarr 2004). Unlike TS firms, WIs have a dual customer approach. They view both job seekers and producers as clients and seek to provide services to both sides of the labour market. Many workforce intermediaries attempt to improve labour market opportunities for disadvantaged workers. In these cases, WIs will combine basic education, job training, and soft skills training (workplace etiquette, resume preparation, interviewing skills, communication skills, etc.) with the purpose of building career ladders that can assist disadvantaged individuals advance from their current positions into higher paying ones. In addition, WIs may also co-ordinate the provision of social service supports, such as child care, transportation, and counselling, to reduce exogenous barriers that prevent disadvantaged individuals from participating in the labour market. In this regard, workforce intermediaries attempt to bridge structural gaps in the labour market.

The Walla Walla Community College (WWCC) in Walla Walla, Washington, illustrates how an institution can effectively respond to a changing regional economy and address shortages in skilled labour supply through the provision of education and sector-specific training. In the mid- to late-1990s, the wine industry in the Walla Walla Valley began to expand at a rapid rate. Although the population of wineries and vineyards was relatively small, it more than doubled from 10 wineries in 1995 to 29 wineries in 2000.[4] More wineries were expected to start up in the upcoming years. As existing wineries grew and the population of wineries increased, winery owners were experiencing a shortage in local labour supply, especially during the fall harvest and crush. The community college and the wine industry formed a partnership from which they created a vocational certificate and degree programme in enology and viticulture. The programme, founded in 2001, was designed to provide individuals with a classroom education and experiential learning opportunities in commercial vineyards and wineries. In the initial stages of the programme, a majority of the coursework took place in the wineries and vineyards. Students were required to complete a number of practicum hours each quarter as well as an internship. The close ties between the WWCC and the local wine industry also facilitated job matching. Many of the programme graduates secured entry level employment in local wineries, as assistant winemakers or working as cellar assistant. The college also serves as a supplier of contingent labour during the fall harvest and crush. Depending on the winery, crush can last anywhere from one to three months. Crush is a critical learning experience for students because this is the moment when you assist in the process of making wine. It is also an opportunity for producers to meet their short-term labour requirements and screen potential permanent workers.

The partnership between the WWCC and the wine industry is the cornerstone of this programme. Close communication and interaction minimises risk and uncertainty, facilitates workforce training, and matching programme graduates in jobs. The WWCC is a partner on the local WIB and training provider for all workforce development needs in the region. The programme was initially funded through workforce development funds. As a result, a small number of enology and viticulture students applied and secured childcare support and tuition funding through the Washington State Employment Security Department, which administrates the Workforce Investment Act.

---

4 As of this writing, the expansion has continued to over ninety commercial wineries in the Walla Walla region.

Trade Unions as LMIs

Trade unions provide yet another variant of labour market intermediation. In contrast to the institutional environment of individually based employment relations in which TS firms and WIs are embedded, unions operate within an institutional framework of workforce or job class-based labour relations. This framework permits workers of specified job classes and employment status to collectivise their power and voice into a bargaining unit and negotiate over the terms and conditions of employment with the employer. Unions have substantial experience building career ladders for workers. Union-sponsored apprenticeship programmes in the building trades[5] have created access to training and high paying jobs for many years. More recently, unions have negotiated contract provisions that establish career ladders in such industries such as health care, child care, and hospitality (Fitzgerald 2006).

In a similar example, the Wisconsin Regional Training Partnership (WRTP) is cited as an example of how partnerships between organised labour, higher education, and industry can effectively address industrial upgrading, competitiveness, and workforce competences (Fitzgerald 2006; Garmise 2006; Leete et al. 2004). The WRTP was established in 1992 to address the regional contraction in manufacturing employment. The WRTP would offer a suite of sequential services: work with local employers to identify and implement technology upgrading strategies, retrain incumbent workers, and recruit and train new employees. The training curriculum was attached to newly created job classes that would form a career ladder for employees while attempting to increase the competitive position of the firm (Garmise 2006).

These examples illustrate how institutional arrangements are shaped by the environment in which they are embedded. In many respects, these modes of labour market intermediation reflect the institutional and geographical contexts in which they operate. In Walla Walla, the lens of workforce intermediary broadens our understanding of the forms LMI can take in specific sectoral and geographical contexts. As a community college, the mission of WWCC includes workforce and economic development. Institutional flexibility and responsiveness to local needs, in combination with industry involvement formed the building blocks of an effective sectoral initiative. As a regional actor in workforce development, WWCC accepts the risk for reducing workforce imbalances. However, as labour supply increases and the labour market become tighter, producer risk is reduced through increased information flow and the ability to screen workers. Consequently, worker uncertainty increases as the labour market tightens. In Milwaukee, where organised labour has had a significant presence in industrial relations, unions had a key role in the formation of the WRTP. The WRTP illustrates a mix of organisational creativity and leadership to address problems of industrial decline

---

5 Historically, union apprentice programmes in the building trades actively excluded women and people of colour, primarily African Americans.

and job loss. In a conventional unionised setting, risk is transferred to employers as the contract sharply reduces worker uncertainty. However, the WRTP illustrates a fairly even distribution of risk and uncertainty though the division of labour among the participating partners. Although the WRTP is a private sector entity, it accepts risk in similar ways of public entities by addressing issues of labour supply and demand, and attempting to increase industrial competitiveness.

Conclusions and Implications

LMIs take many forms. Most forms entail the reduction of producer and worker uncertainty through the LMI's ability to gain information about large numbers of producers and workers, despite the short-term experience of individual producers and workers with each other. Some forms encourage the transfer of risk from producers to employees. However, the control that LMIs can hold over workers is generally insufficient to allow LMIs to play dominant or market-making roles in the supply of labour.

The wide variety of LMIs leads us to develop an institutional approach to studying their operation. As such, the development and operation of LMIs must fit into the mix of economic, demographic, and political characteristics of each specific location and industry. The examples above suggest that local context plays an active and dynamic role in shaping institutions, including LMIs. In addition, public policy can have a large effect on the nature and operation of LMIs available to producers and workers in a locality.

References

Autor, D. (2001), 'Why Do Temporary Help Firms Provide Free General Skills Training?', *Quarterly Journal of Economics* 116:4, 1409-1448.

Autor, D. (2004), 'Labour Market Intermediation: What it is, Why it is Growing, and Where it is Going', *NBER Reporter (Fall)* <http:www.nber.org/reporter/fall04/autor.html>, 24 May 2006.

Burdett, K. and Mortensen, D.T. (1998), 'Wage Differentials, Employer Size, and Unemployment', *International Economic Review* 39:2, 257-273.

Cassell, F.H. and Rodgers, R.C. (1978), 'The Public Employment Service as a Labour Market Intermediary', in National Commission for Manpower Policy (ed.), *Labour Market Intermediaries*, Special Report 22 (Washington, DC: National Commission for Manpower Policy), 105-154.

Ciscel, D.H. and Smith, B.E. (2005), 'The Impact of Supply Chain Management on Labour Standards: The Transition to Incessant Work', *Journal of Economic Issues* 39: 2, 429-437.

Doeringer, P.B. (1994), 'Can the US System of Workplace Training Survive International Competition?', in Asefa, S. and Huang, W.C. (eds). *Human*

Capital and Economic Development (Kalamzoo, Michigan: Upjohn Institute), 91-107.

Doeringer, P.B. and Piore, M.J. (1971*), Internal Labour Markets and Manpower Analysis* (Lexington, Mass.: DC Heath).

Fitzgerald, J. (2006), *Moving Up in the New Economy: Career Ladder for US Workers* (Ithaca: Cornell University Press).

Forde, C. and Slater, G. (2005), 'Agency Working in Britain, Character, Consequences and Regulation', *British Journal of Industrial Relations* 43:2, 249-271.

Gannon, M.J. (1978), 'An Analysis of the Temporary Help Industry', in National Commission for Manpower Policy (ed.), *Labour Market Intermediaries*, Special Report 22 (Washington, DC: National Commission for Manpower Policy), 195-226.

Garmise, S. (2006), *People and the Competitive Advantage of Place: Building a Workforce for the 21st Century* (Armonk: M.E. Sharpe).

Gereffi, G. (1999), 'International Trade and Industrial Upgrading in the Apparel Commodity Chain', *Journal of International Economics* 48, 37-70.

Gorton, G. and Winton, A. (2003), *Financial intermediation*. Wharton Financial Institutions Center working paper 02-28. Reprinted in Constantinides, G., Harris, M. and Stulz, R. (eds), *The Handbook of the Economics of Finance* (Amsterdam: North Holland), 431-522.

Kilcoyne, P. (2005), 'Occupations in the Temporary Help Services Industry', *Occupational Employment and Wages*, May 2004, 6-9.

Leete, L., Benner, C., Pastor, M. and Zimmerman, S. (2004), 'Labour Market Intermediaries in the Old and New Economies: A Survey of Worker Experiences in Milwaukee and Silicon Valley', in Giloth, R.P. (ed.), *Workforce Intermediaries for the Twenty-first Century* (Philadelphia: Temple University Press), 263-290.

Manning, A. (2003), *Monopsony in Motion: Imperfect Competition in Labour Markets* (Princeton, NJ: Princeton University Press).

Marano, C. and Tarr, K. (2004), 'The Workforce Intermediary: Profiling the Field of Practice and its Challenges', in Giloth, R.P. (ed.), *Workforce Intermediaries for the Twenty-first Century* (Philadelphia: Temple University Press), 931-123.

Mills, D.Q. (1978), 'Labour Market Intermediaries: An Overview', in National Commission for Manpower Policy (ed.), *Labour Market Intermediaries*, Special Report 22 (Washington, DC: National Commission for Manpower Policy), 13-54.

Peck, J.A. and Theodore, N. (2002), 'Temped Out? Industry Rhetoric, Labour Regulation and Economic Restructuring in the Temporary Staffing Business', *Economic and Industrial Democracy* 23:2, 143-175.

Solow, R.M. (1990), *The Labour Market as a Social Institution* (Cambridge, MA: Basil Blackwell).

Spulber, D.F. (1996), 'Market Microstructure and Intermediation', *Journal of Economic Perspectives* 10:3, 135-152.

Theodore, N. and Peck, J. (2002), 'The Temporary Staffing Industry: Growth Imperatives and Limits to Contingency', *Economic Geography* 78, 463-493.

US National Research Council, Committee on Workforce Needs in Information Technology (2001), *Building a Workforce for the Information Economy* (Washington DC: National Academy Press).

Wu, S.D. (2004), 'Supply Chain Intermediation: A Bargaining Theoretic Framework', in Simchi-Levi, D., Wu, S.D. and Shen, Z.M. (eds), *Handbook of Quantitative Supply Chain Analysis: Modeling in the E-Business Era* (Boston: Kluwer Academic Publishers), 67-116.

Chapter 16

Globalisation, Skilled Migration and the Mobility of Knowledge

Christine Tamásy

Introduction

In the New Zealand policy arena, as in many other OECD countries, the development of a knowledge-based economy is seen as essential for international competitiveness and economic growth. Attracting and retaining skilled labour are central elements of knowledge-based strategies (OECD 2004) as the number of skilled migrants has risen dramatically in recent decades. Docquier and Rapoport (2004, 3) state that the number of foreign-born individuals (expatriates) residing in OECD countries increased by 51 per cent between 1990 to 2000, "a figure that jumps to 70 per cent for high-skilled migrants against only about 28 per cent for low-skilled migrants". The answer to the question, "can one really talk of the globalisation of migration flows?" (Tapinos and Delaunay 2000, 35) is emphatically "yes", due to increased people-flows and structural changes of the international migration system that have occurred over the last decades (IOM 2005; OECD 2005). Beaverstock and Boardwell (2000, 279) point out, however, that migration has to be conceptualised as, "an integral globalisation process, which has produced new geographies of migration with respect to both process and pattern".

In New Zealand, a country built on migration, the percentage of foreign-born citizens is one of the highest in the world – close to 20 per cent. At the same time, the analysis of Dumont and Lemaître (2005) shows that, expressed as a percentage of the total population, almost 16 per cent of people born in New Zealand are currently living as expatriates in OECD countries, and many of them are high-skilled, with tertiary education. In most OECD countries, the percentage of the skilled who are expatriates is below 10 per cent, but in New Zealand it is 24 per cent. This phenomenon is the result of pull factors in receiving countries, but also of push factors in the New Zealand economy. A survey of New Zealand expatriates found that salaries and career opportunities were the main reasons that keep them abroad (Inkson et al. 2004). Shortages of skilled workers are one of the most challenging obstacles faced by New Zealand's businesses at the beginning of the twenty first century (DoL, 2006). Thus, "Migrants quitting NZ in bigger numbers" (23 June 2006), "Skill shortage forces closure" (13 May 2006), "IT skills shortage still causing pressure" (22 February 2006), "Skills shortage reaches the acute stage" (15 December 2005), "Labour shortage problem for horticulturists" (9

December 2005), "Telecom looks overseas as skill shortage bites" (17 November 2005), and "Expats called home to 'fizzing and buzzing' NZ" (2 November 2005) are examples of headlines that can be found in the *New Zealand Herald*, the largest national newspaper.

Increased people-flows have led New Zealand, like several European countries, to find improved ways to regulate the cross-border movements of skilled immigrants and, more recently, to facilitate the international recruitment of foreign professional workers and expatriates.[1] However, scientific knowledge about skilled migration to guide political practices is much less developed than for migration in general. Freeman (2006) is critical of the peripheral status accorded to people-flows in debates on globalisation, although the interplay between migration, capital and trade is essential to understanding globalisation processes. The aim of the chapter is, therefore, to look deeper into the black box of 'skilled migration'. It cautions against simplistic interpretations based on fuzzy concepts or wobbly data and reconceptualises skilled migration as one element of the international mobility of knowledge. The chapter argues that the mobility of skilled workers enables the spread of tacit knowledge in intra- and inter-firm (globalising) networks, because organisational and relational proximity is integral to the transmission of knowledge that is difficult to communicate.

Fuzzy Concepts and Wobbly Data

The concept of "skilled migration" is fuzzy because the phenomenon possesses different meanings (Markusen 1999). The theoretical and empirical literatures provide no coherent framework for defining who is a "skilled" migrant and who is not.[2] The level of human capital is often central and skilled workers are usually defined as having completed tertiary education (OECD 2005). However, formal qualifications may provide signals about individuals' skills, but they do not measure practical knowledge gained through work experience and on-the-job training. A meaningful comparison of different national qualifications is itself a difficult task. In New Zealand, for example, the New Zealand Qualifications Authority (NZQA) evaluates the overseas secondary and tertiary qualifications of applicants for immigration purposes. However, such a focus entirely ignores aspects of human capital *formation* as an important dimension of international skilled migration. According to Skeldon (2005, 14), "education is at the heart of human capital formation". Studying abroad for a limited period of time promotes

---

1 New Zealand's immigration policy is based on the implicit assumption that new, skilled migrants support national economic growth. Moody (2006) discusses the ways through which immigration can influence GDP per capita growth via higher labour productivity and utilisation.

2 Migration can be broadly defined as the process of moving, either across or within an international border (IOM 2005).

long-term migration, as students often move to their host country with the intention of staying after graduation (IOM 2005).

The International Organisation for Migration (IOM 2005, 461) follows a more contextual approach and defines a skilled migrant as "a worker who, because of his/her skills, is usually granted preferential treatment regarding admission to a host country". While this definition indicates that institutional settings are important factors in understanding international people-flows, it excludes skilled migrants who might migrate to a country as refugee, partner or spouse. Surveys often take a pragmatic approach to define skilled workers by focusing on one or a few selected professional groups (e.g. teachers in higher education, scientists, engineers; see Ackers 2005; Findlay et al. 1996). In the context of skilled migration, therefore, "how do I know it when I see it?" (Markusen 1999, 870). As yet we have no answer.

If the concept of skilled migration is fuzzy, the available secondary data is even more problematic. New Zealand, for example, keeps record of international people-flows based on the intentions expressed by travellers on their arrival or departure cards. These data, however, are not available by level of education or work experience. Flow data are collected on immigrants by age, qualification and work experience, but it is only a partial picture because comparable data on emigrants and Australian immigrants is missing or incomplete.[3] Although there are significant developments taking place in New Zealand linking the arrival/departure data with information on immigrant approvals, the overall migration picture still remains fragmented and incomplete. Census data (e.g. for 1996, 2001) provide information on place of birth, length of stay in New Zealand, school and other qualifications, which can be used to calculate stocks of immigrants for particular points in time. However, only census data compiled from other countries allow stock estimates for expatriates to be made to produce a more complete picture of international migration from a New Zealand perspective.[4]

Census birthplace information is also the basis of the most detailed data sets on migration by educational attainment to date, created and analysed by the OECD (2005) and the World Bank (Docquier and Marfouk 2005). But, largely unknown is whether an expatriate obtained a degree in higher education in the country of birth or in the host country (Skeldon 2005). Another methodological problem results from changing national boundaries, e.g. the break-up of the Soviet Union or the unification of Germany. Former intra-national people-flows count now as international migration and vice versa. Furthermore, in the New Zealand 2001 census, for example, 4 per cent of the population did not specify a place of birth (OECD 2005). Although empirical information is most helpful to understand the

---

3 Since 1973, the Trans-Tasman Travel Arrangement (TTTA) between Australia and New Zealand has allowed free movements of citizens (no application for permits required).

4 However, not all countries worldwide have reliable census information and/or identify systematically countries of birth.

phenomenon of skilled migration, wobbly data can only produce shaky evidence that is neither helpful for advancing migration theories nor for developing recommendations to guide policy and practice.

Towards a Reconceptualisation

The "brain drain" debate dominated early discussions on skilled migration in the international arena, focused mainly on the loss of skilled workers and potential negative economic consequences. The story is complex (Iredale 1999) and has reached a stage where the mobility of skilled individuals is seen as often temporary and even sometimes beneficial for origin countries due to, for example, increased remittances and relieved labour market pressures. Thus, the international mobility of skilled workers has been discussed more recently in terms of a "brain exchange" and "brain gain" rather than "brain drain" (Bedford et al. 2002; Findlay 2002; World Bank 2006). Hugo's (1999) analysis of cross-border migration patterns to and from Australia illustrates, for example, an increasing importance of non-permanent settlement movements. Empirical studies of international migration also show a strong presence of return migration which may have been planned as part of a life cycle in a residential location sequence (Borjas, Bratsberg 1996). Thus, it can be expected that relocation activities are often *not* a single event in the biography of some individuals ("multiple migrants"). This highlights the importance of analysing international migration from a process perspective. The emphasis on interrelated phases (and types) of mobility implies that a set of research questions concerns what individuals and their families *do* over time when they initiate, refine and realise ideas to settle in another country, region or locality. The migration decision and its realisation do not occur as sudden events. The migration process includes at least three separate steps that can be called "idea generation", "opportunity identification", and "opportunity realisation".[5] The whole process, however, is not a linear sequence of events but a dynamic phenomenon with break-ups, iterations and feedback loops.

The international mobility of highly qualified individuals is primarily one from and to large urban areas (Skeldon 2005),[6] whereas in particular the interactions between the mobility of skilled workers and the emergence of "global cities" have received attention (Koser and Salt 1997; Beaverstock and Boardwell 2000). Cities with "global control capability" (Sassen 2001, 6) are attractive locations for transnational corporations, which move skilled workers on an international

---

5 The author's interest in entrepreneurship research may have influenced the use of terminologies to describe the migration phenomenon. Opportunities are defined in a constructivist way as perceived possibilities to settle in another country, region or locality.

6 In Auckland, New Zealand's primate city, 31 per cent of the resident population in 2001 were foreign born.

scale to balance the competitive needs of the company at different locations.[7] Berset and Crevoisier (2006) argue that the movements of highly skilled labour play a determining role in the competitiveness of any regional production system competing on a global scale.

A way forward to better understand the phenomenon from a globalising (firm) network perspective is, hence the main argument of this chapter, to *reconceptualise skilled migration as one element of the international mobility of knowledge*. Leamer and Storper (2001) argue that the international flow of people as a cause and consequence of globalisation processes is becoming increasingly important. While goods may be transported cheaper than ever before over long distances, humans are most important "containers" for transferring non-codified, tacit knowledge. Explicit or codified knowledge refers to knowledge that is transmittable and documented in publications, blueprints and databases. On the other hand, tacit knowledge is complex, context-dependent and difficult to communicate.[8] As a consequence, the internet age is more likely to affect the transmission of codified than of tacit knowledge (*ibid.*). Skilled workers embody explicit knowledge as well as tacit knowledge, acting as agents for the production and diffusion of know-what, know-why, know-how, and know-who (Lundvall and Johnson 1994). The empirical findings by Beaverstock and Boardwell (2000, 297), for example, clearly illustrate how transnational banks use professional and managerial migrants to transfer knowledge over time and space.

The more easily explicit knowledge can be accessed, the more crucial tacit knowledge becomes for firms to succeed in a globalising world economy (Maskell and Malmberg 1999). As tacit knowledge is embodied in people and its transfer depends on shared understandings (Gertler 2003), its transmission over distance requires organisational and relational proximity. While organisational proximity refers to "thick" intra-firm structures, relational proximity describes a wider framework consisting of intra- *and* inter-firm networks. What matters is the existence of socio-economic relationships in which individuals develop so-called "learning communities" (Amin and Cohendet 2000). Thus, the mobility of workers enables tacit knowledge to transcend boundaries within network structures of shared understandings: "Tacit knowledge is non-transferable without the exchange of key personnel and all the systems that support them" (Nonaka et al. 2000, 5). In this view, the transfer of tacit knowledge within spatially

---

7 International recruitment agencies are another important "channel" for international labour migration, which handles large numbers of self-generated people-flows to satisfy demands in areas of skill shortages (Iredale 1999, 2001).

8 Explicit knowledge and tacit knowledge should be seen as two ends of a continuum, rather than discrete categories. Explicit knowledge is transmittable in formal, systematic language, while tacit knowledge has a personal quality, which makes it hard to communicate (Nonaka 1994). Gertler (2003), however, criticises the sloppiness in the economic geography literature to conflate "uncodified" with "uncodifiable" tacit knowledge (see also Cowan et al. 2000).

dispersed globalising firm networks is possible. These may include a company's relationships with "external" customers, suppliers and research institutes (e.g. universities, technical colleges). Neither the production of tacit knowledge nor its transmission are necessarily localised. This is because shared beliefs, values and norms may span political and organisational boundaries. They enable a process of individual and collective learning behaviour that allows the production and flow of tacit knowledge over distance.

This line of argument, however, does not negate the transmission of tacit knowledge between economic actors located in geographical proximity to each other. Moreover, it does not deny arguments of Porter's (1998) industrial organisation model. Firms located within industrial clusters have clear advantages that are not available for firms located outside a shared knowledge basis. However, in recent years it has become increasingly recognised that most firms build "global pipelines" (Bathelt et al. 2004; Maskell et al. 2006) as channels of international knowledge flows. According to Faulconbridge (2006, 537) firms create knowledge processes "that stretch beyond scale-defined boundaries" into global spaces of learning. These literatures are most helpful for a reconceptualisation of skilled migration processes, although the discussions often (explicitly or implicitly) ignore that individuals or groups of individuals engage in learning processes within and across firms. The creation and transmission of tacit knowledge depend on a successful transfer of professionals (and their embodied expertise) as it is best illustrated by the mobility of workers who make knowledge transfers and learning processes work (Bastian 2006). However, the availability of tacit knowledge is not sufficient for effective learning processes in globalising firm networks. Firms and skilled workers have also to be prepared (and willing) to engage in learning by doing and interacting. The so called "absorptive capacity" (Cohen and Levinthal 1990) is a prerequisite for the production and diffusion of knowledge in intra- and inter-firm relationships.

Expatriation is one important firm strategy that refers to inter-firm skilled migration on a temporary or permanent basis through international assignments. In his book *Globalisation: the people dimension – human resource strategies for global expansion* Perkins (1999, 50) defines the "international cadre" as executives, "capable of transferring the enterprise's commercial and operational philosophies and systems into every location". Based on their career paths, three types of mobile professionals can be identified (*ibid.*). First: domestic staff hired locally who are unlikely to work outside the domestic market other than for infrequent events (e.g. training, one-off projects). Second: parent company staff who may work at some point in their career overseas for skills transfer reasons, but will subsequently return to the domestic base. Third: international staff who begin their careers with multinational companies from anywhere in the world, and have skills that can be used in a variety of different markets. Expatriates are executed by transnational corporations for specific staffing needs (e.g. to fill vacancies in foreign units), for the development of an internationally experienced management team, and for organisational development purposes such as effective knowledge transfers (Riusala and Suutari 2004). Findings by Beaverstock (2004) clearly illustrate that international assignments are a major globalisation strategy of London-based professional legal service firms, executed to develop, manage and diffuse idiosyncratic knowledge in globalising intra-firm networks in order to increase profitability and market share.

From a New Zealand perspective, the analysis by Dumont and Lemaître (2005) yielded an estimated total of 410,663 expatriates worldwide (in 2001). Kiwi Expat Association Inc. (Kea) has recently set up a website (www.keanewzealand.com), which offers a number of channels to connect the global "Kiwi" talent community and to facilitate the sharing of knowledge of New Zealanders overseas. From 1 April 2006, the New Zealand government has introduced a temporary tax exemption on foreign income for returning expatriates and new migrants. The tax exemption is for four calendar years and can only be granted once in a life-time. Apart from these fiscal incentives to re-attract expatriates and attract new, skilled migrants, permanent movements to New Zealand are subject to a points-based system with an increasing emphasis on the migrant's skills (formal qualification, work experience, skilled employment in areas of absolute skills shortages). 48,815 people were approved for residence in 2004/05, of which 61 per cent entered New Zealand through the Skilled/Business Stream, 29 per cent through the Family Sponsored Stream and 10 per cent through the International/Humanitarian Stream (DoL 2005). Migration processes are, therefore, constrained by regulatory forces at the national scale that influence cross-border mobility practices of firms and individuals.[9]

Conclusions

In this chapter, aspects of skilled migration in a globalising world economy have been discussed. In recent years, there has been a growing recognition of the role of skilled migration as a component and a "driver" of economic globalisation processes. Since the beginning of the 1990s, policy instruments have been established in most OECD countries for regulating the cross-border movements of skilled migrants and for facilitating – more recently – the international recruitment of foreign professional workers and expatriates. However, scientific knowledge about skilled migration is much less developed than for migration in general. The chapter cautions against simplistic interpretations based on fuzzy concepts or wobbly data and reconceptualises skilled migration as one element of the international mobility of knowledge. I would like to make two main concluding comments.

---

9 A comprehensive overview on migration literatures in the New Zealand context can be found in Trlin, Spoonley and Watts (2005). New Zealand's skilled migration policy is discussed in Bedford (2006).

Firstly, the chapter has shown that in order to analyse global skilled migration processes, researchers cannot solely rely on official migration data, if at all available, for intra- and inter-firm global mobility patterns. Most helpful are qualitative case study approaches that are concerned with how the world is viewed, experienced and constructed by socio-economic actors who practise and/ or manage skilled migration processes in globalising firm networks. Investigations by Beaverstock (2004), and Beaverstock and Boardwell (2000), for example, show the particular value of using in-depth case study evidence for an analysis of intra-firm mobility at a global scale. Mixed-methods triangulation in particular can provide a basis upon which different insights upon the migration phenomenon can be sensibly combined through a mixture of quantitative *and* qualitative methods. Downward and Mearman (2006) explore from the perspective of economics how mixed-methods approaches can even help to break down disciplinary boundaries in order to establish interdisciplinary social sciences, of which human geography (as economics) is a part.

Secondly, and more importantly, the chapter argues that the mobility of skilled workers enables the spread of tacit knowledge in globalising firm networks, because organisational and relational proximity is integral to the transmission of knowledge that is difficult to communicate. This line of argument breaks out of the simple "tacit knowledge is local" and "explicit knowledge is global" dichotomy which can often be found in the literature. Both tacit knowledge and explicit knowledge can be exchanged globally and locally. Large, multinational firms which operate on a global scale may have competitive advantages in being able to combine knowledge from different locations. However, Beaverstock (2004, 157) criticises that despite intensive research examining transnational corporations, "geographers have been slow to address how such firms manage the complexity of knowledge management". There is clearly considerably more work to be done before we can claim to have a theory of globalising networks and understand skilled mobility patterns and processes at a global scale.

Acknowledgments

This research is supported by a Heisenberg Fellowship of the German Research Foundation (Grant No Ta 277/2-1).

References

Ackers, L. (2005), 'Promoting Scientific Mobility and Balance Growth in the European Research Area', *Innovation* 18, 301-317.
Amin, A. and Cohendet, P. (2000), 'Organisational Learning and Governance through Embedded Practices', *Journal of Management and Governance* 4, 93-116.

Bastian, D. (2006), 'Modes of Knowledge Migration: Regional Assimilation of Knowledge and the Politics of Bringing Knowledge into the Region', *European Planning Studies* 14, 601-619.

Bathelt, H., Malmberg, A. and Maskell, P. (2004), 'Clusters and Knowledge: Local Buzz, Global Pipelines and the Process of Knowledge Creation', *Progress in Human Geography* 28, 31-56.

Beaverstock, J.V. (2004) '"Managing across Borders": Knowledge Management and Expatriation in Professional Service Legal Firms', *Journal of Economic Geography* 4, 157-179.

Beaverstock, J.V. and Boardwell, J.T. (2000), 'Negotiating Globalisation, Transnational Corporations and Global City Financial Centres in Transient Migration Studies', *Applied Geography* 20, 277-304.

Bedford, R. (2006), 'Skilled Migration in and out of New Zealand: Immigrants, Workers, Students and Emigrants', in Birrell, B.; Hawthorne, L. and Richardson, Evaluation of the General Skilled Migration Categories <http://www.immi. gov.au/media/ publications/research/gsm-report/index.htm>, 224-251.

Bedford, R.; Bedford, C.; Ho, E. and Lidgard, J. (2002), 'The Globalisation of International Migration in New Zealand: Contribution to a Debate', *New Zealand Population Review* 28, 69-97.

Berset, A. and Crevoisier, O. (2006), 'Circulation of Competencies and Dynamics of Regional Production Systems', *IJMS: International Journal on Multicultural Societies* 8, 61-83.

Borjas, G.J. and Bratsberg, B. (1996) 'Who Leaves? The Outmigration of the Foreign-born', *The Review of Economics and Statistics* 78, 165-176.

Cohen, W.M. and Leyinthal, D.A. (1990), 'Absorptive Capacity: A New Perspective on Learning and Innovation', *Administrative Science Quarterly* 35, 128-152.

Cowan, R., David, P.A. and Foray, D. (2000), 'The Explicit Economics of Knowledge Codification and Tacitness', *Industrial and Corporate Change* 9, 211-253.

Department of Labour (DoL) (2005), *Migration Trends 2004/2005* (Wellington: DoL).

Department of Labour (DoL) (2006), *Skilled Migrants in New Zealand: Employer's Perspectives* (Wellington: DoL).

Docquier, F.; and Marfouk, A. (2005), *Measuring the International Mobility of Skilled Workers (1990–2000): Release 1.0*. Research Working Paper, 3381 (Washington: The World Bank).

Docquier, F.; and Rapoport, H. (2004), *Skilled Migration: The Perspective of Developing Countries*. Policy Research Working Paper 3382 (Washington: The World Bank).

Downward, P. and Mearman, A. (2006), 'Retroduction as Mixed-methods Triangulation in Economic Research: Reorienting Economics into Social Science', *Cambridge Journal of Economics* 13 April 2006.

Dumont, J.-C.; and Lemaître, G. (2005), *Counting Immigrants and Expatriates in OECD Countries: A New Perspective.* Social Employment and Migration Working Papers, 25 (Paris: OECD).

Faulconbridge, J.R. (2006), 'Stretching Tacit Knowledge beyond a Local Fix? Global Spaces of Learning in Advertising Professional Service Firms', *Journal of Economic Geography* 6, 517-540.

Findlay A. (2002), *From Brain Exchange to Brain Gain: Policy Implications for the UK of Recent Trends in Skilled Migration from Developing Countries.* International Migration Papers, 43 (Geneva: ILO).

Findlay, A.M.; Li, F.L.N.; Jowett, A.J. and Skeldon, R. (1996), 'Skilled International Migration and the Global City: A Study of Expatriates in Hong Kong', *Transactions of the Institute of British Geographers* 21, 49-61.

Freeman, R.B. (2006), *People Flows in Globalisation.* NBER Working Paper, 12315 (Cambridge: National Bureau of Economic Research).

Gertler, M.S. (2003), 'Tacit Knowledge and the Economic Geography of Context, or the Undefinable Tacitness of Being (There)', *Journal of Economic Geography* 3, 75-99.

Hugo, G.J. (1999), 'A New Paradigm of International Migration in Australia', *New Zealand Population Review* 25, 1-2, 1-39.

Inkson, K.; Carr, S.; Edwards, M.; Hooks, J.; Jackson, D.; Thorn, K. and Allfree, N. (2004), 'From Brain Drain to Talent Flow: Views of Kiwi Expatriates', *University of Auckland Business Review* 8, 29-39.

International Organisation for Migration (IOM) (2005), *World Migration 2005. Costs and Benefits of International Migration* (Geneva: IOM).

Iredale, R. (1999), 'The Need to Import Skilled Personnel: Factors Favouring and Hindering its International Mobility', *International Migration* 37, 89-123.

Iredale, R. (2001), 'The Migration of Professionals: Theories and Typologies', *International Migration* 39, 7-26.

Koser, K. and Salt, J. (1997), 'The Geography of Highly Skilled Migration', *International Journal of Population Geography* 3, 285-303.

Leamer, E.E.; and Storper, M. (2001), *The Economic Geography of the Internet Age.* NBER Working Paper, 8450 (Cambridge: National Bureau of Economic Research).

Lundvall, B.A.; and Johnson, B. (1994), 'The Learning Economy', *Journal of Industry Studies* 1, 23-42.

Markusen, A. (1999), 'Fuzzy Concepts, Scanty Evidence, Policy Distance: The Case for Rigour and Policy Relevance in Critical Regional Studies', *Regional Studies* 33, 869-884.

Maskell, P.; and Malmberg, A. (1999), 'Localised Learning and Industrial Competitiveness', *Cambridge Journal for Economics* 23, 167-185.

Maskell, P.; Bathelt, H. and Malmberg, A. (2006), 'Building Global Knowledge Pipelines: The Role of Temporary Clusters', *European Planning Studies* 14, 997-1013.

Moody, C. (2006), *Migration and Economic Growth: A 21st Century Perspective.* Working Paper, 06/02 (Wellington: New Zealand Treasury).

Nonaka, I. (1994), 'A Dynamic Theory of Organisational Knowledge Creation', *Organisation Science* 5, 14-37.

Nonaka, I.; Toyama, R.; and Nagata, A. (2000), 'A Firm as a Knowledge-creating Entity: A New Perspective on the Theory of the Firm', *Industrial and Corporate Change* 9, 1-20.

OECD (Organisation for Economic Co-operation and Development) (2004), *Global Knowledge Flows and Economic Development* (Paris: OECD).

OECD (Organisation for Economic Co-operation and Development) (2005), *Trends in International Migration 2004* (Paris: OECD).

Perkins, S.J. (1999), *Globalisation: The People Dimension – Human Resource Strategies for Global Expansion* (London: Kogan Page).

Porter, M.E. (1998), 'The Adam Smith Address: Location, Clusters, and the 'New' Microeconomics of Competition', *Business Economics* 33, 7-13.

Riusala, K. and Suutari, V. (2004), 'International Knowledge Transfers through Expatriates', *Thunderbird International Business Review* 46, 743-770.

Sassen, S. (2001), *The Global City: New York, London, Tokyo*, 2nd Edition. (Princeton [et al.]: Princeton University Press).

Skeldon, R. (2005), *Globalisation, Skilled Migration and Poverty Alleviation; Brain Drains in Context.* Working Paper, T15 (Sussex: Development Research Centre on Migration, Globalisation and Poverty).

Tapinos, G. and Delaunay, D. (2000), 'Can One Really Talk of the Globalisation of Migration Flows?', in OECD (ed.), *Globalisation, Migration and Development*, (Paris: OECD), 35-58.

Trlin, A.D.; Spoonley, P. and Watts, N. (eds) (2005), *New Zealand and International Migration. A Digest and Bibliography, No. 4* (Palmerston North: Massey University).

World Bank (2006), *Prospects for the Global Economy: Economic Implications of Remittances and Migration* (Washington: The World Bank).

Chapter 17

Rural Development and Social Embeddedness: Banks and Businesses in Thuringia, Germany

Sabine Panzer

Introduction

Business finance plays a key role in processes of industrial production, and access to capital and financial services is central to most business operations. Therefore, the flow across space of financial capital from suppliers of money to buyers of money is an important issue in the discussion of the geographies of commodity chains. Since the majority of finance for small and medium-sized enterprises (SMEs) is still provided by the banking sector (Binks and Ennew 1998; Pieper 2005), the structure and geographical reach of banks, as well as the relationships between businesses and their banks, have to be considered an essential aspect in the shaping of the economic landscape. Although research on the geography of finance has grown in recent years (Martin 1999; Clark, Thrift and Tickell 2004; Leyshon and Pollard 2000; Leyshon and Thrift 1997; Dow and Rodríguez-Fuentes 1997; Grote 2004), Pollard (2003) points out that business finance has not been considered in any depth in economic geography – in general, it has been regarded as a 'black box' and a 'largely taken-for-granted aspect of production' (Pollard 2003, 430).

This chapter seeks to redress this shortcoming by analysing an empirical study on business finance in rural areas of Germany. Against the background of the transformation of the banking industry and a steady withdrawal of bank infrastructure from rural areas, the chapter explores whether trust-based bank-business-relationships can be created and developed over greater geographical distances and help SMEs to access external finance more easily. On the basis of the model of relationship banking (see, for example, Binks and Ennew 1998), the chapter assesses how German businesses in rural areas cope with the spatial concentration of financial institutions and whether the development of close working relationships founded on trust is affected.

The remainder of the chapter is divided into three main sections. Section 2 outlines the changes the German banking sector has undergone during the last three to four decades and the role trust plays in the relationships between banks and businesses. Section 3 focuses on the nature of bank-business-relationships in two rural areas in the federal state of Thuringia both from the viewpoint of local businesses and from the perspective of banks. Section 4 concludes the chapter.

Changes in the German Banking Sector and the Role of Trust in the Relationships between Banks and Businesses

Structural Changes and Spatial Effects in the German Banking Sector

Since the 1970s the financial systems in most western countries have undergone a fundamental process of restructuring that has led to radical organisational changes in the banking industry worldwide (see, for example, Leyshon and Thrift 1999). This transformation has been driven by three main factors. First, the progress in modern technology and telecommunication permitted the automation of financial services and induced a constant cost reduction of banking processes through standardisation. Second, market liberalisation and globalisation have led to a greater level of competition between financial service firms as foreign banks, as well as non- and near-banks, have pushed into national markets and the traditional scope of duties of the banking sector. Third, bank customers, especially business clients, have become much more confident and demanding in dealing with financial institutions. They tend to have multiple bank connections – a behaviour often referred to as 'cherry picking'. Thus, the financial market transformed from a seller's to a buyer's market (French and Leyshon 2004; Pieper 2005).

At the beginning of the 1990s, O'Brien (1992) argued that the technological revolution would lead to an economic dispersion and decentralisation in the financial industry and thus to the end of geography; a state where geographical distances do not matter anymore. Cairncross (1997) similarly predicted the 'death of distance'. Interestingly however, a centralisation of banking operations has taken place and has been analysed in a number of research studies (see, for example, Sassen 1991; Grote 2003; Beck, Demirgüç-Kunt and Levine 2006). The withdrawal of bank infrastructure from rural areas and a spatial concentration of bank operations can be observed in all western countries. Table 17.1 illustrates this change through an international comparison of branch office densities for the years 2003 and 2004. The Netherlands has the lowest branch office density (3,952 inhabitants per branch office in 2004), followed by Sweden and the US. Belgium and Italy have the highest branch office densities, while Germany, with 1,731 inhabitants per branch office in 2004, is one of the middle-ranking countries. However, there has been a marked reduction of branch offices in Germany (-4.1 per cent), making it an interesting study from which to assess the effects of bank infrastructure withdrawal on business finance in rural areas.

Table 17.1 **International comparison of branch office density (inhabitants per branch office)**

Country	2003	2004	Change (%)
Netherlands	3,582	3,952	- 10.3
Sweden	3,130	3,442	- 10.0
United States	2,349	2,357	- 0.3
Canada	2,207	2,251	- 2.0
Japan	2,138	2,178	- 1.9
Great Britain	1,889	1,996	- 5.7
Germany	1,663	1,731	- 4.1
Belgium	1,587	1,691	- 6.6
Austria	1,551	1,570	- 1.2
France	1,562	1,554	+ 0.5
Switzerland	1,370	1,427	- 4.2
Italy	1,298	1,298	0.0

Source: Deutsche Bundesbank (2006)

The German banking system, as a traditionally decentralised system, has seen a decline of bank branches from 49,186 in 1993 to 35,041 in 2004 (-28.8 per cent). While the greatest numbers of branch closures were registered in the sectors of savings banks and credit unions – mainly driven by mergers and acquisitions – these two groups still have the highest branch office densities (14,530 and 12,978 branches respectively in 2005). In the sector of private banks, there has also been an impressive consolidation process, where branch office networks were reduced by 49 per cent between 1998 and 2003, leading to much thinner networks of branch offices (Deutsche Bundesbank 2006).

Figure 17.1 illustrates how the four major banks (Deutsche Bank, Dresdner Bank, Commerzbank, HypoVereinsbank) have reorganised their operations over space by showing branch closures and openings between 1996 and 2003. While branch closures were predominantly realised in major urban areas (Berlin, Ruhr, Frankfurt etc.), many rural areas have been affected as well. Taking into consideration that bank infrastructure had been scarce in these areas before, closures of bank branches often meant a total withdrawal from these regions. There were also a small number of branch openings; however, they did not compensate for branch closures.

The development of the banking industry in Germany shows, similarly to other western countries, a moving away from the concept of branch offices to a multi-channel system, with branches being only one option amongst others such as tele-banking and direct banking that enable banks to deliver financial services more cheaply to customers. Against the background of the New Basel Capital Accord (Basel II), computer-based credit-rating systems have been developed during the last decade that allow banks to analyse their customers' credit-worthiness on the

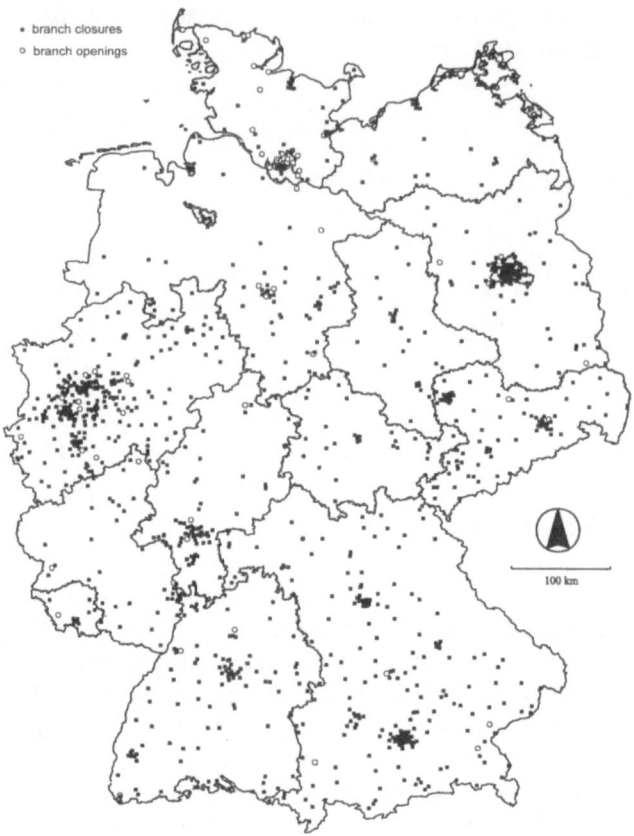

Figure 17.1   Reconfiguration of German major banks: branch closures/ openings

Source: Pieper (2005, 107)

basis of a number of assessment criteria in centralised back-office locations, so that customer contact can be reduced to a minimum (Leyshon and Pollard 1999).

The Role of Trust in Bank-Business-Relationships

Yet, the financing of SMEs, particularly in rural areas is relatively unattractive for financial institutions as they experience considerable transaction costs in gathering the information necessary to estimate the risk that the potential borrower will not repay (Pollard 2003; Leyshon and Thrift 1999). Additionally, SMEs' financial needs are not restricted to standardised retail products. Rather, they are characterised by individualised corporate banking services and money advice – a field of banking that is associated with the exchange of tacit knowledge that cannot easily be transferred across distances. These services are traditionally founded on close working relationships between banks and their customers which are based on face-to-face communication and trust. Against this background, a continuous withdrawal of bank infrastructure and centralisation processes of banks may lead to financial exclusion (Leyshon and Thrift 1999) in rural areas when SMEs face increasing problems in gaining access to the financial services that require trust.

For these businesses, the role of the local bank manager becomes important. It is the SME's bank manager who takes on the position of an intermediary between the business and the bank's back-office decision-makers. Thus, although the role of the individual bank manager is limited concerning the approval of borrowing requests, a close relationship between the bank manager and the SME can improve information flow significantly and, therefore, yield benefits for both parties. On the one hand, transaction costs for banks decline; on the other hand, SMEs located in rural areas might be able to access the financial system more easily.

This interpretation of bank-business relationships requires that SMEs and their bank managers are embedded in networks of reciprocity, interdependence and mutual trust at a personal level. The concept of embeddedness as suggested by Granovetter (1985) has been broadly studied in economic geography in a local context presuming that spatial proximity promotes the development of social networks (see, for example, Porter 1998; Storper 1997; Agnes 2002). This chapter, however, goes beyond this perspective and analyses how the increasing distance between banks and their business clients might affect close working relationships and thus the geographies of capital allocation.

Personal trust is seen as the foundation of successful and sustainable relationships between banks and SMEs as it can help to reduce uncertainty in the field of corporate banking. It is characterised by the willingness of the two parties to increase their vulnerability to economic action that is outside their sphere of influence (Hosmer 1995). Although numerous theoretical approaches to the term 'trust' exist (see, for example, Luhmann 1989; Coleman 1991; Giddens 1995) this chapter focuses on Glückler's (2004) concept which differentiates personal trust between expertise trust (referring to the trust-giver's anticipation in the trust-taker's qualities, expertise and techniques) and good-will-trust (meaning the anticipation that the partner of co-operation does not renege on that trust).

In an empirical study, business finance and the relationships between bank managers and their business clients in the federal state of Thuringia – which with 144 inhabitants/sq. km (Germany: 231 inhabitants/sq. km) is one of the least densely populated German federal states – were analysed to assess the development of trust in the light of banks retreating from rural areas. In the following sections, the results of this empirical study are presented.

The Relationships between Banks and Businesses, Illustrated by Examples in Thuringia

Design and Methodology of the Empirical Analysis

Two different administrative districts in Thuringia were selected to investigate rural bank service provision – one which is characterised by a low density of bank branch offices ('Kyffhäuserkreis': 2.5 branch offices/100 sq. km) and one characterised by a relatively high density of bank branches ('Kreis Sonneberg': 9.7 branch offices/100 sq. km).

To investigate trust as a mechanism of social embeddedness in the rural banking context, a survey of enterprises was carried out in both study districts via an internet platform as well as interviews with a number of businesses and banks.

Of 292 SMEs contacted in the two selected districts, 83 participated in the online survey – 31 from 'Kyffhäuserkreis' and 52 from 'Kreis Sonneberg' – a 28 per cent response rate. In addition, staff at two SMEs from 'Kyffhäuserkreis' and three SMEs from 'Kreis Sonneberg' were interviewed. Finally, interviews were undertaken with seven banks – four of which do not operate branch offices in the investigation areas.

Trust, Spatial Proximity and Local Businesses

The survey data show very clearly that SMEs prefer to be as close as possible to the banks with which they have commercial business. For most businesses (83 per cent in 'Kyffhäuserkreis' and 71 per cent in 'Kreis Sonneberg') this distance is no more than 5 to 10 km, which means that they keep their major banking connections with financial institutions which still have relatively well-structured branch office networks. These are mostly savings banks or credit unions, and they benefit from the radical withdrawal of private banks from rural areas as a number of SMEs have changed their bank connection for proximity reasons.

About two thirds of businesses in both study areas have, in addition to their major bank connection, one or even more extra bank accounts. Although these accounts mainly function to provide retail services that are characterised by low specificity and high market transparency, and for this reason are generally not sensitive to distance, they are nonetheless often kept at local banks. However, nationwide-operating banks with a thin network of branch offices and direct banks without any branches also play an important role in SMEs' access to retail services. Altogether, it can be said that in both study areas spatial proximity to a bank is a crucial criterion for bank selection by local businesses.

To assess whether there is any link between the distances separating SMEs and their major banks and the quality of their working relationships, the trust businesses have in their banks was measured and correlated with the level of geographic separation between them. Drawing on Glückler's (2004) concept, businesses were asked to estimate the extent to which they trust their bank manager's expertise and to what extent they believe that their bank manager will give them the most honest money advice, by rating them on a six-point scale (1 = very good trust; 6 = no trust). Thus, the levels of subjective 'expertise trust' and 'good-will trust' were evaluated.

The results presented in Table 17.2 show clearly that most businesses have great confidence both in their bank's expertise and in its good-will. However, at the same time it can be observed that the trust in the bank's expertise is usually higher than the confidence in its goodwill. All in all, 32.3 per cent of the businesses surveyed in 'Kyffhäuserkreis' and 21.2 per cent of the enterprises surveyed in 'Kreis Sonneberg' ranked their trust in their banks' good-will at 5 or 6.

Table 17.2   Expertise trust and good-will trust

Rank 1 = very good trust to 6 = no trust	Expertise trust – proportion of respondents in %		Good-will trust – proportion of respondents in %	
	KYF	SON	KYF	SON
1	16.1	15.4	16.1	7.7
2	38.7	48.1	29.0	36.5
3	25.8	17.3	12.9	26.9
4	16.1	7.7	6.5	5.8
5	3.2	7.7	22.6	13.5
6	0.0	1.9	9.7	7.7

KYF = Kyffhäuserkreis; SON = Kreis Sonneberg

To explore the relationship between 'trust' and proximity, a general index of trust has been constructed from the survey data, combining the indices of 'expertise trust' and 'good-will trust'. This index related to proximity (measured as average spatial distance in each category of 'trust') is shown graphically for the two study areas in Figure 17.2. Interestingly, this visual presentation shows that there is no positive correlation between spatial proximity and trust in the German case study regions. Levels of trust are unrelated to distance, and high levels of trust can be built at distance and in the absence of proximity. It would appear, therefore, that the branch office density of a region does not play a vital role in the development of trust-based bank-business relationships.

However, the relatively low levels of trust by businesses that are only a short distance from their major bank can be explained by the fact that many SMEs do not recognise bank-customer relationships as mutual and reciprocal. Most of the surveyed businesses (64.5 per cent and 46.2 per cent) do not have any social contact with their bank manager that goes beyond the usual business contact. Only 12.9 per cent of the enterprises surveyed in 'Kyffhäuserkreis', and 17.3 per cent of those in 'Kreis Sonneberg', reported that they have such contact often or regularly. These results show clearly that the possibility of personal contact is seldom associated with measures of trust building. Instead, many businesses describe

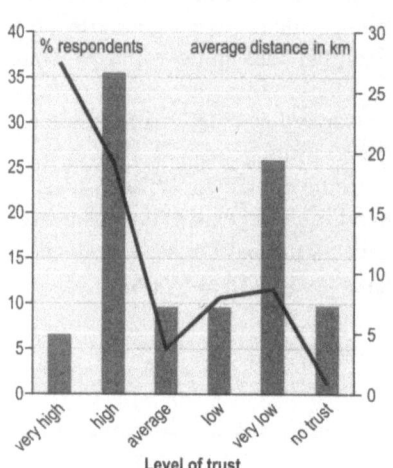

Figure 17.2a
Trust and spatial proximity
between banks and businesses
a) Kyffhäuserkreis

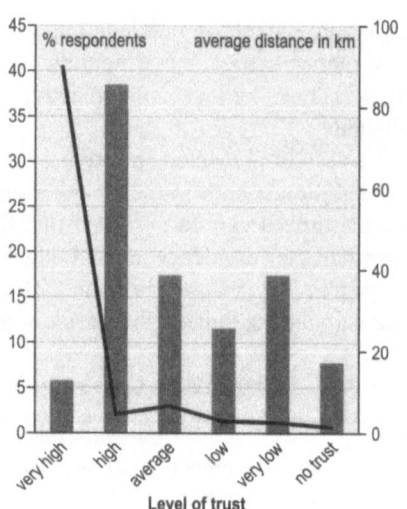

Figure 17.2b
Trust and spatial proximity
between banks and businesses
b) Kreis Sonneberg

their relationships with their bank as instrumental. It is, therefore, unsurprising that 54.8 per cent of the businesses in 'Kyffhäuserkreis' and 32.7 per cent of the businesses in 'Kreis Sonneberg' say that they experience severe problems in accessing external finance.

Nevertheless, from the detailed interviews with local enterprises, it was clear that, despite the low interest in personal contact with the bank manager, social nearness is considered an essential factor for good bank-customer relationships. However, social nearness is in most cases equated with spatial proximity. In contrast, banks with a centralised network of branch offices are associated with anonymity.

Banks' Strategies to Build Trust

The previous section explored the relationship between trust, proximity and social nearness from the perspective of banks' SME business clients in rural areas. The focus of this section turns to the banks themselves to explore banks' strategies to maintain social nearness to their rural business clients despite the continuous closure of bank branches and thus greater spatial distance that this imposes on bank-client relationships. The analysis is built on detailed interviews with banks. The analysis clarifies the fact that both banks with a well-structured network of branch offices and nationwide-operating banks with a thin network of branch offices actively seek to establish reciprocity, interdependence and trust in bank-business relationships. It emphasises the importance of continuously maintaining customer contacts and developing those relationships over time. However, the strategies banks use to achieve this aim differ considerably.

From the interviews it is clear that banks with decentralised branch office networks still consider spatial proximity a fundamental element in the development of trust-based bank-business relationships. They use their relatively well-structured networks in the regions in which they operate to create social nearness to their customers. With the strategy of 'one face to the customer', they ensure that each business client is assigned to only one bank manager. However, turnover of bank personnel complicates this aim. The branch office is still the central location for meetings and discussions. Thus, decentralised banks pursue, in many respects, the traditional way of banking, ignoring their own structural changes.

In contrast, banks with thinner networks of branch offices normally do not consider spatial proximity as essential for building and maintaining trust-based bank-business relationships. Nevertheless, the maintenance of client contacts is central to their work, too, and they achieve this goal primarily through mobile sales by approaching businesses in their vicinity. This enables both parties to transfer tacit knowledge even more easily than in branch offices as the functioning of the business can be demonstrated and assessed more easily. While face-to-face contacts help to establish trusting relationships, they become less important as the relationships evolve. Then, even non-codified information can be communicated via modern technologies such as telephone or email. Therefore, from the perspective of financial institutions with a thin network of branch offices, knowledge-intensive corporate banking services are not at all sensitive to spatial distance.

The idea of one-stop finance in order to improve customer loyalty can be observed at both banks with well-structured networks of branch offices and banks operating nationwide. The strategy aims to offer the whole range of financial services to customers by selling packages including a number of service goods. In this manner, banks try to establish interdependence between the bank and business clients to prevent the customers from drawing on competing banks' offers. However, all the banks interviewed seemed to struggle with the problem of 'cherry picking'.

The analysis also shows that many banks organise customer events to generate trust in bank-business relationships and to build up a positive reputation to attract new customers. Customer events can include business dinners, rounds of golf or workshops in which ongoing financial topics are discussed. Although trust cannot be built up strategically at such events, they can help banks to create a good environment in which it can easily be established. As soon as trust in bank-business relationships exists, it is possible to bridge spatial distances, so that well-structured networks of branch offices are not integral to the process. Therefore, SMEs in rural areas do not necessarily suffer a shortage of financial services from corporate banks when those banks retreat from these areas.

However, the interviews show that not all the banks arrange such customer events, and neither are all kinds of businesses offered these activities. On the one hand, banks with decentralised branch office structures do not normally organise events for their business clients. On the other hand, only 'profitable' customers are included; 'profitability' being measured not only as sales volumes but also by the quality of the relationship. Therefore, 'cherry pickers' are usually not on the banks' invitation lists.

As the analysis demonstrates, social nearness and thus good working relationships can be, but need not be, realised by spatial proximity. In fact, spatial proximity can be substituted by social nearness.

Conclusion

This chapter has explored the role of proximity and social nearness in the creation and maintenance of trust in the relationships between banks and their SME clients in rural Germany as banks have closed branches and reduced their networks. Based on the concept of relationship banking (Binks and Ennew 1998) and Glückler's (2004) model of trust, the analysis of empirical data on trust in the relationships between banks and businesses from two rural districts in the federal state of Thuringia ('Kyffhäuserkreis' and 'Kreis Sonneberg') shows three key results. First, from the perspective of local businesses, spatial proximity to a major bank seems to be the basis for building trusting relationships. For them, spatial proximity means social nearness, and banks with a thin network of branch offices are associated with anonymity. Second, however, many of these businesses do not have a high level of trust in their bank. At the same time they do not see any need to actively maintain relationships. Accordingly, a high percentage of SMEs experience severe problems in accessing external finance. Third, the data show that the building and developing of close working relationships based on trust does not depend on spatial proximity – in fact, spatial proximity can be substituted by social nearness. Therefore, the ability to create trust in bank-business relationships has a significant impact on the shaping of the economic landscape. It can be suggested that if SMEs in rural areas do not committedly co-operate with their banks they will experience more problems concerning adverse prices and conditions for financial services and even financial exclusion in the future. Against the background of many SMEs' reluctance to do so, it is assumed that the banks' retreat from rural areas will further deepen uneven development.

Acknowledgements

I would like to thank three anonymous referees for their valuable comments and suggestions on earlier drafts of this chapter.

References

Agnes, P. (2002), 'Local Embeddedness and Global Financial Services: Australian Evidence on "The End of Geography"', in Taylor, M. and Leonard, S. (eds), *Embedded Enterprise and Social Capital. International Perspectives* (Aldershot: Ashgate), 127-150.

Beck, T., Demirgüç-Kunt, A and Levine, R. (2006), 'Bank Concentration, Competition, and Crises: First Results', *Journal of Banking and Finance* 30, 1581-1603.

Binks, M.R. and Ennew, C.T. (1998), 'Smaller Businesses and Relationship Banking. The Impact of Participative Behaviour', *Entrepreneurship – Theory and Practice* 21, 83-92.

Cairncross, F. (1997), *The Death of Distance: How the Communications Revolution will Change our Lives* (Boston: Harvard Business School Press).

Clark, G.L., Thrift N. and Tickell, A. (2004), 'Performing Finance: The Industry, the Media and its Image', *Review of International Political Economy* 11: 2, 289-310.

Coleman, J.S. (1990), *Foundations of Social Theory* (Cambridge, MA: Harvard University Press).

Deutsche Bundesbank (ed.) (2006), *Entwicklung des Bankstellennetzes im Jahr 2005* <http//:www.bundesbank.de/download.bankenaufsicht/pdf/bankstellenbericht06.pdf> accessed 27.03.2007.

Dow, S. and Rodríguez-Fuentes, C.J. (1997), 'Regional Finance: A Survey', *Regional Studies* 31: 9, 903-920.

French, S. and Leyshon, A. (2004), 'The New, New Financial System? Towards a Conceptualisation of Financial Reintermediation', *Review of International Political Economy* 11: 2, 263-288.

Giddens, A. (1990), *Consequences of Modernity* (Cambridge: Polity Press).

Glückler, J. (2004), *Reputationsnetzwerke. Zur Internationalisierung von Unternehmensberatern. Eine relationale Theorie* (Bielefeld: transcript).

Granovetter, M. (1985), 'Economic Action and Social Structure: The Problem of Embeddedness', *American Journal of Sociology* 91: 3, 481-510.

Grote, M. (2004), *Die Entwicklung des Finanzplatzes Frankfurt. Eine evolutionsökonomische Untersuchung* (Berlin: Duncker & Humblot).

Hosmer, L.T. (1995), 'Trust: The Connecting Link between Organisational Theory and Philosophical Ethics', *Academy of Management Review* 20, 379-403.

Leyshon, A. and Pollard, J. (2000), 'Geographies of Industrial Convergence: The Case of Retail Banking', *Transactions of the Institute of British Geographers* 25, 203-220.

Leyshon, A. and Thrift, N. (1997), *Money/Space: Geographies of Monetary Transformation* (London, New York: Routledge).

Luhmann, N. (1989), *Vertrauen: Ein Mechanismus zur Reduktion sozialer Komplexität* (Stuttgart: Enke).

Martin, R. (1999), 'The New Economic Geography of Money', in Martin, R. (ed.), *Money and the Space Economy* (Chichester *et al.* Wiley), 3-27.

O'Brien, R. (1992), *Global Financial Integration: The End of Geography* (London: Council on Foreign Relations Press).

Pieper, C. (2005), *Banken im Umbruch. Strukturwandel im deutschen Bankensektor und regionalwirtschaftliche Implikationen* (Münster: Lit).

Pollard, J. (2003), 'Small Firm Finance and Economic Geography', *Journal of Economic Geography* 3, 429-452.

Porter, M.E. (1998), *On Competition* (Boston: Harvard Business School Press).

Sassen, S. (1991), *The Global City. New York, London, Tokyo* (Princeton: Princeton University Press).

Storper, M. (1997), *The Regional World: Territorial Development in a Global Economy* (New York: Guilfort Press).

Chapter 18

A Spatial Investigation of the Governor's Opportunity Fund in Virginia

John R. Lombard

Introduction

In the ever increasing competitive environment for attracting footloose capital and new jobs, political jurisdictions at all spatial scales have been driven to offer economic development incentives. Indeed, governments from advanced, emerging and lagging economies have long battled over inward investment dollars. Furthermore, in the European and US contexts, sub-national quasi-government institutions are taking a larger and more direct role in facilitating investment through the use of incentives (Phelps and Wood 2006). Even localities are pursuing investment by offering incentives on a global scale as they institute mechanisms to cope with increasingly competitive pressures from jurisdictions throughout the world. To a large extent, however, localities' abilities to provide incentives and establish linkages to the global economy are mediated by state and national institutions.

For example, in the US, all states offer some form of economic development incentive programme. The bulk of incentives offered by US states are non-discretionary tax-based and are awarded to all firms that meet statutory eligibility or performance requirements (Poole et al. 1999) These incentives are known as 'as of right' and are provided to businesses meeting the qualifying criteria for the incentive. Many US states have expanded the traditional tool box of 'as of right' economic development incentives to include discretionary economic development incentives. Discretionary incentives can be distinguished from 'as of right' incentives in that the former may be awarded at the discretion of the administrating office. Typically, these programmes are offered on a first come, first served basis owing to annual caps on the amount of total taxes foregone or by the level of programme funding by the state. Unlike place-based incentives such as state enterprise or federal empowerment zones that provide benefits based on locating within the zone, discretionary incentives are not bound by geography. As an example, many states offer infrastructure improvement grants that are associated with specific project transportation requirements even if the project is not located in a designated incentive zone.

While providing incentives in exchange for capital investment has been institutionalised throughout the world, an interesting trend appears to be emerging in the US. More recently, some US states have expanded the realm of discretionary incentive programmes to what can be considered as 'deal closing' funds (Moret 2002; Ellison 2004). These funds are typically cash-based, in most instances require a competitive bidding situation, and usually necessitate rapid approval (days or weeks rather than months or legislative sessions). They are administered by the state's executive office or designee and are awarded on top of any 'as of right' incentive. These incentives can be awarded directly to a business, locality, or economic development related institution. The express purpose of awarding the incentive is to insure that the economic development project will locate within state boundaries when the project is in a competitive situation with other states.

One such discretionary 'deal closing' incentive fund is the Commonwealth of Virginia Governor's Development Opportunity Fund (GOF). The GOF is used to attract or retain businesses that create jobs or provide new investment. A main characteristic of the GOF is that it is a cash-based discretionary fund administered by the office of the Governor and allocated to a locality or institution to help secure an investment, particularly for Virginia communities suffering economic and fiscal stress. Its purpose is to provide the proverbial 'icing on the cake' or the final incentive that solidifies a competitive economic development project for Virginia.

This chapter is an exploratory investigation of the spatial distribution of GOF awards across Virginia at the sub-county or city level. The research question is simple: are these awards going to those locations most in need of economic development? It is based on the assumption that economic development incentives should be targeted to distressed communities (for supporting arguments see Bartik 2005; Persky, Felsenstein and Carlson 2004; Nizalov and Loveridge 2005). Unfortunately, research in Virginia examines economic development incentives from a broad state-wide perspective employing fiscal impact and return on investment analyses. However, as Dewar (1998) argues, the public story as presented is incomplete. Explaining the success or failure of economic development needs to include economic, political and, as is argued here, spatial considerations.

Economic Development Incentives: A Critique

There are several reasons why an examination of the distribution of regional economic development incentives is an important research question. Firstly, this study embraces the notion that human and physical capital is unevenly distributed across space and, consequently, any evaluation of economic development incentives needs to examine the characteristics of the places and people where these awards are allocated (Greenbaum and Bondonio 2003). Also, the literature on the targeting of incentives suggests that poorer places with higher unemployment can benefit more from incentives than places that are relatively better off (Bartik 2005; Persky et al. 2005). Furthermore, research on spatial mismatch suggests that neighbourhoods are an important determinant of employment (Houston 2005) and that physical job proximity is a major determinant of local employment (Immergluck 1998, 185). Recent research, that overcomes the endogenous neighbourhood choice bias, indicates a positive association between job access and employment outcomes (Aslund, Osth, and Zenou 2006). For the purpose of this investigation, it is argued that the distribution of economic development incentives should be investigated at the local or neighbourhood scale, and that an important first step in presenting a more complete public story of economic development programmes is to understand the spatial distribution of the awards at a local or neighbourhood scale.

Secondly, evaluating economic development incentives has been criticised for both the techniques and methods that have been employed, as well as for the scope and breadth of the questions that have been asked (Persky et al. 2004). Studies tend to focus on aggregate employment growth at the regional, metropolitan or industry level, and evaluate programme success in terms of simple ROI calculations and cost-benefit analyses. This traditional tallying of jobs and investment dollars fails to answer important questions on what kinds of jobs are created and for whom these jobs are intended. In a summary of existing literature examining the research on incentive effectiveness, Fisher and Peters (2004, 35) concluded that after years of policy experimentation and hundreds of scholarly studies 'the standard justifications given for incentive policy by state and local officials, politicians, and many academics are, at best, weakly supported'. This study adds to the body of literature calling for critical perspectives on economic development and associated policy programmes.

Thirdly, given that existing literature on the effectiveness and efficiency of economic development incentives in general is mixed and does not prioritise one specific economic development policy initiative over another, it is important that policy research in this area continues. In addition, while the literature on tax-based economic development incentives is quite extensive, the literature examining fiscal incentives is limited (Norris and Higgins 2003). Moreover, 'deal closing'-based discretionary incentives are relatively new phenomena and are under-researched. Virginia's GOF provides a valuable case study of a discretionary incentive.

Fourthly, this chapter contributes to the general call within geography for more applied geographical research in the policy area (e.g. Pacione 1999; Martin 2001). Descriptive analysis using GIS can contribute to a better understanding of the spatial distribution of GOF awards and better inform policy decision-makers as to the likely effectiveness of their business incentive programmes.

State 'Deal Closing' Incentives

It is no easy task to compile information on incentives across states. Prior cataloguing attempts have been notorious for their lack of comprehensiveness. More recently, the Council for Community and Economic Research has compiled an on-line listing of state incentives that is updated on a regular basis which offers a typology of incentives. However, there is no information on 'deal closing' programmes. Poole et al. (2004) surveyed state offerings and found that varying definitions and methodological differences have contributed to the uncertainty surrounding the information on economic development incentives. Some states compile inventories of all programmes that provide business assistance and label these as economic development incentives while other states only count those programmes that provide direct tax or fiscal assistance as economic development incentives.

According to Ellison (2004), economic development incentive grant programmes exist in 25 states. Some of these programmes can be classified as mainly deal closing funds housed within traditional economic development agencies while others are hidden in other administrative agencies such as transportation departments. Ellison's research uncovered eight states that have instituted deal closing funds financed by general tax revenues: Florida, Georgia, Louisiana, Maryland, North Carolina, Pennsylvania, Texas, and Virginia. Each state funds its programme in various amounts. Aside from Texas, which had the largest funding amount (biennial) in 2005 at US$295 million, Virginia's funding level at US$10 million is more indicative of the relative small scale of these 'deal closing' incentive programmes. Both North Carolina and Virginia require local participation through matching funds and other states require application through local government. While all states have some form of codified investment or job growth numbers as a precursor to an award, most states provide exceptions to their stated requirements.

Moret (2002) criticises deal closing funds on several fronts: most funds are not designed to assist with major projects; they are subject to political and private interest and manipulation; the state is picking industrial winners; and these incentives only help at the margins. For example, a North Carolina Department of Commerce-funded assessment of its incentive programme found that most awards were provided to businesses in the state's wealthiest counties (Fain 2003).

Previous Research on Virginia's Governor's Opportunities Fund

In a review of the literature and legislative reports, only three studies examining the GOF were found. The first study was conducted by the Virginia Joint Legislative Audit and Review Committee (JLARC) in 2002. This study evaluated the awarding of incentives in Virginia to determine their cost effectiveness. The JLARC study selected two years of incentive awards that included activity and awards for workforce services, another large grant programme offered by the Department of Business Assistance Workforce Services Division. JLARC compared announced job and investment amounts to actual job and investment amounts two and three years out from the award year of the GOF. Overall, JLARC found that more jobs were created than were announced. However, the study revealed that several projects vastly exceeded announced job creation numbers which more than made up for projects that underperformed their stated job creation numbers. Furthermore, JLARC constructed several conservative scenarios to estimate the likely personal income tax benefit to the state associated with actual new jobs. Their estimates suggested an average payback on GOF and Workforce Services awards of three to five years. One conclusion of this report is that if the GOF and Workforce service awards programme were eliminated then the State would suffer longer term consequences of fewer new jobs and reduced tax revenues. A fundamental assumption the JLARC study was that the awarding of the GOF was essential to the location of a particular project in Virginia. However questionable this assumption may be in light of counterfactual evidence, this study did provide confirmation of the importance of incentives to elected officials. While the JLARC study can be criticised in several respects, one principal area of concern is the failure to ask basic evaluative questions such as where were the jobs created and to whom were they most likely go.

Another study assessed the impact of incentives to attract new business to Virginia (Alwang, Peterson and Mills 2001). As part of the general investigation, the research examined the overall effectiveness of GOF awards at the planning district level for Virginia from 1998 to 2000. It was concluded from this research that GOF programme funds appear to be randomly scattered across Virginia and show no relationship to the goals of the programme as presented in the enabling legislation. The authors suggested that the effectiveness of the GOF could be improved by stating clearly the objective of the fund. They criticised the return on investment (ROI) criterion used by VEDP to assess effectiveness because the calculation did not take into account the costs of services provided to the firm or the negative impact of the GOF on existing firms.

A third study examined Virginia's incentive programmes and their impact on land use and sustainable growth (Environmental Law Institute 2001). GOF activity from 1998 to 2000 was one of the economic development programmes examined. Their study criticised existing state evaluations on the use of incentive programmes because they did not take into account their impact on patterns of growth. They called for more stringent data reporting requirements to allow a broader assessment of incentive impact.

While existing research on the GOF is informative, it only examines the impact and distribution of incentives at an aggregate level. This accounting mindset is prevalent across the economic development hierarchy in Virginia. Regional marketing organisations, as well as local city and county economic developers, are guilty of the same short-sightedness in evaluating their programmes. Certainly the data are available to allow more detailed analysis than economic impact and the technology (GIS) to facilitate spatial analysis is ever-present in government. Therefore, the research reported here begins to fill an important gap in evaluating the impact of the GOF by examining the spatial distribution of awards at the local and neighbourhood level.

A Re-analysis of the Governor's Opportunity Fund (GOF): Data and Procedures

For this study, details on GOF awards were collected through press release announcements tallied by the Division of Legislative Services of the Virginia General Assembly. This report contained for each recipient, the company name, locality, amount of GOF dollars approved, investment dollars, jobs created and their average wage, and locally matched dollar funding. The first step in the analysis was to examine the distribution of awards at the city/county level. Each city/county was rank ordered and classified into deciles using the 2000 fiscal stress index computed by the Department of Housing and Community Development Commission on Local Government. Data on GOF awards were aggregated by city/county deciles to provide a snapshot of GOF awards.

To determine the spatial distribution of GOF awards at the neighbourhood level, a convenient sample of all GOF projects awarded between 1996 and 2004 was selected. This time period was chosen because it coincides with the 1996 legislation as well as the first year of capturing detailed data. Street addresses of each GOF award were obtained in one of three ways; (1) searching the web, (2) printed telephone directories, or (3) contacting local county or city economic development representatives. Each address was verified by telephoning the company directly to ensure that the operation was still in business and to obtain contact information for future research. Of the 189 GOF awards announced during this period, 16 were no longer in business or we were unable to verify address information.

Demographic data was obtained for all Virginia census tracts (1530) using the Neighbourhood Change Database (NCDB) for the 2000 census. The 2000 census was chosen because it provides the richest data at the tract level. Each project was geo-coded to its corresponding street address and census tract for analysis purposes. Selected socioeconomic data were analysed using simple difference of means tests to assess differences in census tracts receiving or not receiving GOF awards. The difference of means test was conducted again using a spatially smoothed average value taking into account every tract that bordered a GOF award tract. For analysis purposes, several census tracts received multiple GOF awards during this period and therefore the sample size for socioeconomic comparisons was reduced to 124 census tracts.

To determine census tracts most in need of economic development assistance, three proxy measures were used: (1) the level of unemployment, (2) the proportion of families below the federal poverty level, and (3) median household income. Each census tract was rank ordered on each dimension of need and those tracts in the bottom five per cent (n=76), and ten per cent (n=153) were identified. Using GIS, distances were calculated from the location of the geo-coded street address for each GOF award to the centroid of the nearest high need census tract. This serves as a proxy (albeit weak) for job accessibility associated with the GOF projects.

Results and Discussion

This chapter has attempted to explore the spatial distribution of Governor's Opportunity Fund awards in Virginia to assess whether or not the awards are located near neighbourhoods suffering from economic or fiscal stress. Table 18.1 presents a summary of allocation of GOF awards by number, jobs created, and investment dollars at the city/county level by fiscal stress deciles classification. It is interesting to note that the relatively well-off counties in the top deciles or 10 per cent (the least fiscally stressed) accounted for 15.2 per cent of GOF funds, 28.4 per cent of new jobs, and 34.7 per cent of total investment which contrasts with the bottom 10 per cent (the most fiscally distressed) which received approximately the same total GOF dollar award (15 per cent) over this time frame but these awards leveraged much less in terms of jobs or investment, accounting for only 15 per cent of jobs and only 12.2 per cent of investment. The top and bottom 20 per cent of fiscal stress counties and cities again received approximately the same per cent of GOF dollars, but once again, the least stressed localities received 38.1 per cent of new jobs and 51.5 per cent of total investments while the most stressed localities received only 20 per cent of new jobs and only 15.2 per cent of investment.

Table 18.1   GOF awards by county/city fiscal stress (ranked by deciles)

Deciles	GOF count	%	GOF funds	%	New jobs	%	Retained jobs	%	Investment	%
1	26	14.3	11150	15.0	9735	16.3	210	5.0	1149	12.2
2	36	19.8	15705	21.1	11915	20.0	210	5.0	1435	15.2
3	50	27.5	19150	25.8	13939	23.4	290	6.9	1535	16.3
4	89	48.9	31662	42.6	21076	35.4	1960	46.7	2209	23.5
5	106	58.2	37912	51.0	25623	43.0	2596	61.9	2536	26.9
6	117	64.3	43947	59.2	28863	48.4	2669	63.6	3134	33.3
7	138	75.8	49607	66.8	34200	57.4	3599	85.8	3564	37.9
8	153	84.1	53822	72.5	36890	61.9	3934	93.8	4565	48.5
9	169	92.9	62979	84.8	42685	71.6	4149	98.9	6149	65.3
10	182	100.0	74279	100.0	59605	100.0	4194	100.0	9411	100.0

Note: GOF funds and investment are in $ thousands.

Characteristics of GOF award census tracts in Virginia were compared to non-GOF census tracts across select socioeconomic measures. Table 18.2 presents the results of a difference of means test. On average, GOF award census tracts had higher population levels, lower mean travel time to work, and lower median household income. Surprisingly, the GOF award tracts exhibited no significant difference in the proportion of families below poverty or the unemployment rate.

To obtain a sense of the proximity of GOF awards to stressed census tracts, three indicators of perceived need were examined; (1) median household income, (2) percentage of families in poverty, and (3) the employment to population ratio. The first indicator of proximity was the share of GOF awards located in one of the census tracts ranked in the lowest 5 per cent, 10 per cent, and 20 per cent of all census tracts ranked across the three indicators of stress. As Table 18.3 shows, stressed census tracts across the three indicators of need are not receiving a higher share of GOF awards. In all cases, the share quotient is less than one.

Table 18.2   Select socioeconomic characteristics for non-GOF and GOF census tracts

	Non-GOF census tract (N=1406)	GOF census tract (N=124)	Mean difference	Non-GOF v GOF t-test significance	Confidence interval
Mean population (16 years and older)	3550	4339	-789	p < .01	-1122 to -454
Mean travel time to work (minutes)	27.3	24.8	2.5	p < .01	1.5 to 3.4
Mean % below poverty level	10.91%	11.42%	-0.51	NS	---
Mean % Unemployed	2.90%	2.85%	0.05%	NS	---
Median household income	$49,768	$41,103	$8,662	p < .01	$5,347 to $11,984

Table 18.3   GOF share of stressed census tracts by need indicator

	Census tract need indicator		
Census tract	High unemployment number (per cent)	Median household income number (per cent)	Per cent of households below poverty level number (per cent)
Top five per cent (n=77)	3 (2.4)	5 (4.0)	4 (3.2)
Top ten per cent (n=153)	8 (6.5)	10 (8.1)	8 (6.5)
Top twenty per cent (n=306)	24 (19.4)	21 (16.9)	23 (18.5)

For the second indicator of proximity, the distance from each GOF award location to the nearest census tract centroid, the average distances were calculated from award locations to the nearest census tract ranked in the bottom (for household income) or top (for the proportion below poverty and unemployment rate) 5 per cent, 10 per cent, and 20 per cent of all Virginia census tracts. As Table 18.4 shows, the distance to the nearest census tract ranked in the lowest 5 per cent ranged from 19.4 miles for median family income to 28.9 miles for the poverty indicator. The median distance for the same group of census tracts ranged from 10 to 15 miles. The average distances for those census tracts ranked in the bottom 10 per cent and 20 per cent are reduced substantially from the distance in the bottom 5 per cent.

Table 18.4   Average distance to nearest census tract by need indicator

	Census tract need indicator								
Distance to nearest Census tract	High unemployment number (per cent)			Median household income number (per cent)			Per cent of households below poverty level number (per cent)		
	Mean	Median	IQR	Mean	Median	IQR	Mean	Median	IQR
Top five per cent	19.9	10.5	31.6	19.4	14.8	29.3	28.9	15.3	42.4
Top ten per cent	12.1	6.9	14.6	13.6	7.5	17.5	20.3	9.0	26.2
Top twenty per cent	7.1	3.9	7.2	6.4	4.2	6.9	6.3	4.0	6.4

Conclusions

Returning to the original question posed in this analysis: are Governor's Opportunity Fund (GOF) awards in Virginia going to those locations most in need of economic development? The preliminary investigation on the distribution of GOF awards reported here suggests that most awards are not allocated to communities or neighbourhoods deemed to be most in need. Simple statistics indicate that for high unemployment and high poverty census tracts, the null hypothesis, that there is no difference between census tracts in their receipt or non-receipt of GOF awards, has to be rejected. Certainly the distribution of existing industry in Virginia influences the location of subsequent investment. Clearly, GOF projects are located in those census tracts with lower median household incomes. However, if the GOF programme allocates incentives to take into consideration those areas most in need then we would expect to find a relationship. Of course, this suggests that all communities have the same ability to bid for incentives. Unfortunately, not all locations can muster resources to meet the fund-matching requirements of the GOF scheme. Indeed, the equity issue looms large and there is a need for further inquiry into local government's ability to meet the local participation requirements of the Governor's Opportunity Fund. Unfortunately, this question is not answered in our preliminary investigation.

Underpinning the research reported here is the normative assumption that some neighbourhoods are more in need of economic development than others. Knowing that a particular business in a region, or even city or county, has benefited from an economic development incentive award is not enough. There is a need also to know which neighbourhoods benefit and, more importantly, which individuals are benefiting from the creation of new jobs. While this research does not answer the 'who' question, fuller evaluation must consider not only the site and situation of investment but, importantly, the type of jobs being created and their accessibility.

In this investigation census tracts have been used as a proxy for neighbourhoods and have shown that, on average, the nearest high distressed neighbourhood was more than 19 miles away. While this distance may seem trivial for those who have cars and routinely endure commutes in excess of 20 miles, for some distressed neighbourhood residents without access to private or public transportation, this distance may seem insurmountable. From a state-wide policy perspective, there is a need for future investigations of incentive impact to go beyond simple fiscal impact. Spatial investigations, along with economic impact analyses, must be considered in the context of local neighbourhoods.

Moreover, this chapter presents only a spatial perspective. There is a need to understand more fully who is getting the jobs that are created and linking this evidence to a job-chain framework as suggested by Persky et al. (2004). If economic development incentives are provided to enhance the welfare of all Virginia residents, then taking into consideration the specific location of activities that benefit business should provide a better sense of accomplishment and impact. Amendments to the GOF act put forward in the 2006 legislative session have tightened eligibility requirements and even instituted a prevailing wage requirement. However, a main evaluative guide is that the GOF should be geared toward communities with unemployment rates above the state average. Until state policy makers acknowledge the tremendous disparity between neighbourhoods, even within counties and cities that are relatively well off, incentive awards will continue to hit the target but miss the bull's eye.

References

Alwang, J., Peterson, E., Mills, B. (2001), *Assessing the Impacts of Incentives to Attract New Business: A Case Study of the Scrap Recycling Industry*, A report to the Virginia Legislative Subcommittee on Industrial Incentives.

Aslund, O., Osth, J., and Y. Zenou (2006), *How Important Is Access To Jobs? Old Question-Improved Answer*, Institute for Labour Market Policy Evaluation.

Bartik, T. (2005), 'Solving the Problems of Economic Development Incentives', *Growth and Change* 36:2, 139-166.

Dewar, M. (1998), 'Why State and Local Economic Development Programs Cause So Little Economic Development', *Economic Development Quarterly* 12:1, 68-87.

Ellison, C. (2004), *Survey of State 'Deal-Closing' Funds and Other Incentive Grant Programs for Job Creation*, Texas Legislative Council.

Fain, J. (2003), *William S. Lee Act 2003: Assessment of Results*, State of North Carolina Department of Commerce; Division of Policy, & Strategic Planning.

Fisher, P., and Peters, A. (2004), 'The Failures of Economic Development Incentives', Journal *of the American Planning Association* 70:1, 27-37.

Greenbaum, R., Bondonio, D. (2003), 'Losing Focus: A Comparative Evaluation of Spatially Targeted Economic Revitalisation Programmes in the US and the EU', *Regional Studies* 38:3, 319-334.

Houston, D. (2005), 'Methods to Test the Spatial Mismatch Hypothesis', *Economic Geography* 81:4, 407-434.

Immergluck, D. (1998), 'Neighborhood Economic Developments and Local Working: The Effect of Nearby Jobs on Where Residents Work', *Economic Geography* 74:2, 170-187.

Martin, R. (2001), 'Geography and Public Policy: The Case of the Missing Agenda', *Progress in Human Geography* 25:2, 189-210.

Moret, S. (2002), *Considering a 'Deal Closing Fund' for Louisiana*, Public Affairs Research Council.

Nizalov, D. and Loveridge, S. (2005), 'Regional Policies and Economic Growth: One Size Does Not Fit All', *The Review of Regional Studies* 35:3, 266-290.

Norris, D.N., and Higgins, E. (2003), 'The Impact of Economic Development Incentives and Programs: A Review of the Literature', A Component of the *Biennial Unified Economic Development Budget Report* Provided to the Louisiana Department of Economic Development. Louisiana Tech University.

Pacione, M. (1999), 'Applied Geography: In Pursuit of Useful Knowledge', *Applied Geography* 19, 1-12.

Phelps, N. and Wood, A. (2006), '"Lost in Translation?" Local Interests, Global Actors and the Multi-scalar Dynamics of Inward Investment', *Journal of Economic Geography* 6:4, 493-515.

Poole, K., Erikcek, G.A., Iannone, D., McCrea, N., and Salem, P. (1999), *Evaluating Business Development Incentives*. Report prepared for US Department of Commerce Economic Development Administration by the National Association of State Development Agencies (Cleveland, Ohio: W.E. Upjohn Institute for Employment Research, and The Urban Center, Cleveland State University).

Persky, J., Felsenstein, D., and Carlson, V. (2004), *Does 'Trickle Down' Work? Economic Development Strategies and Job Chains in Local Labor Markets* (Kalamazoo, MI: W. E. Upjohn Institute for Employment Research).

Virginia's Economic Incentives: Missed Opportunities for Sustainable Growth, (2001), Environmental Law Institute, Washington, D. C.

Chapter 19

A New Alternative to Air Travel in Malaysia: Low-cost Carriers and Their Impacts on Mobility and Full-Service Airline

Abd Rahim Md Nor and Nor Ghani Md Nor

Introduction

In the aviation industry, the term 'low-cost carrier' refers to airlines which offer alternative air passenger transport to that offered by conventional airlines at significantly lower fares but without the luxury of business and first class services and in-flight complementary extras commonly found in a full-service airline (Doganis 2001; Francis et al. 2006; Lawton 2002). Globally, low-fare carriers have reshaped the competitive environment within liberalised markets, and have made significant impacts in the world's domestic passenger markets which had previously been largely controlled by full-service network airlines. However, low-cost air travel is a relatively new phenomenon in Southeast Asia, introduced in the region less than five years ago. Malaysia joined the bandwagon when its first low-cost airline, AirAsia, made its maiden commercial flights to a few domestic destinations in late 2001. This chapter examines the relevance and importance of the low-cost airlines in an increasingly globalised Malaysian economy, traces its origin and expansion, looks into its impacts on mobility for domestic and international passengers, and explains how it has affected the incumbent national flag carrier.

Achieving Lower Cost: Economies and Business Models

Globalisation has transformed the world economy into what might reasonably be called a new geo-economy. Although globalisation is a set of tendencies and there is neither a single predetermined trajectory nor a fixed endpoint (Dicken 2003), the hyperglobalists argue that we now live in a borderless world in which the state, the border and the national are no longer relevant (Ohmae 1995). In this era of increased globalisation, the economic system has become more integrated or globalised in many aspects, and the world is tied together into a single globalised marketplace in which the majority of the people and countries are able to partake in today's globalised economy and information networks, and to be affected by them (Friedman 1999).

The impacts of globalisation on the air passenger industry can be seen from at least two perspectives. First, it has far-reaching impacts on the role of the government in the industry in which the balance shifted decisively towards greater deregulation of the nationally-based systems and the privatisation of the state-owned companies. In Malaysia, the opening up of the commercial aviation industry began in the mid-1990s when the national airline, MAS, was privatised. In late 2001, for the first time, the government opened the air passenger market in a big way when the first low-cost airline in the country was allowed to operate by the private sector. It was quickly followed by the award of another two privately-operated low-cost carrier licences to two companies to operate no-frills airlines called Fly Asian Xpress (FAX) in August 2006, and AirAsia X and FireFly both in early 2007 (Jayaseelan 2007; Gunasegaran 2007). In Southeast Asia, there is significant pressure to deregulate the commercial aviation industry in view of the decision made by member countries of the association of South East Asian nations (ASEAN) to open up the international air passenger markets in the region by year 2008 under an open-sky policy (*New Straits Times* 2 August 2006, 7). Also, two innovations that have been especially significant in the development of the global economy, namely satellite communications and optical fibre technologies (Henderson and Castells 1987), are the prime movers in facilitating the business of low-cost passenger airlines, in the form of e-ticketing and internet booking, two prime factors influencing the success of the industry (Doganis 2001). In Malaysia, its first no-frills airline, AirAsia, jumped onto the bandwagon in April 2002, less than six month after starting operations, by introducing internet booking for its tickets. By late 2005, approximately 65 per cent of its flight ticket reservations were made through the internet (*New Straits Times* 7 December 2005, 9).

One of the issues discussed in this chapter is the way in which the first low-cost airline in Malaysia achieved cost advantages greater than the incumbent full-service airlines. In the commercial aviation industry, a number of studies have examined and discussed the factors that contribute to or impede the success of the low-cost carriers. Porter (1985) identified three generic strategies adopted by low-cost airlines for them to compete on price and achieve cost advantages relative to their competitors, namely leadership in cost reduction, product differentiation, and market segmentation. He also showed that low-cost airlines have been predominantly short-haul operations, partly because some aspects of low-cost operations have been seen as less compatible with long-haul operations such as the need for food, seat pitch and in-flight entertainment. Williams et al. (2003), suggest that shorter routes offer the greatest potential to achieve cost competitiveness over the full-service carriers. Mason et al. (2000), Doganis (2001) and Williams (2001), among others, have shown that the strategies adopted by the majority of low-cost airlines to lower costs and achieve success are; minimum cabin crew, one class of seating allowing more seats per aircraft, no seat allocation, passengers having to pay for food and drink, internet booking, a simple fare structure and pricing strategy, e-ticketing, high aircraft utilisation, short 'on the ground' turnaround times, no cargo carried to slow down turnaround times, use of secondary airports, lower wage scales, lower rates of unionisation among employees, flexible working terms and conditions for employees, point-to-point services, and no offer of connections. Combined, these factors have enabled airlines to reduce their costs and lower fares.

Gillen and Lall (2004) suggest that choice of business model for point-to-point services is the source of competitive advantage. The example of Southwest Airlines shows that organising its crewing 'team' organisation and simplicity of information flows result in greater relational co-ordination. In contrast, Ryanair sought lower costs through lower fares. They argued that the Southwest Airlines model is not generic and is difficult to duplicate, whereas the Ryanair model can be more easily duplicated. In a recent study on the relationship between ownership and corporate governance, Carney and Dostaler (2006) found that low-cost carriers seem to best fit the pattern of entrepreneurial governance; a pattern characterised by a more direct control of management decisions which enables quicker and more risky decision-making. Alves and Barbot's (2007) study of different business and management models adopted by airlines found that the low-cost carriers organise their boards differently from full-service carriers to achieve the lower costs and faster decision-making required by their business model. It was also revealed that these carriers solve their potential agency cost problems differently to full-service carriers who have more board monitoring committees and a closer interest between shareholders and executive directors.

The Origin of Low-cost Air Transport in Malaysia

When commercial aviation started in Malaysia[1] in 1947, the industry was dominated by a single, state-owned airline offering full-service flights to the public. The low-cost concept began in December 2001, when Tune Air officially acquired 99.25 per cent of the equity of AirAsia[2] from DRB-Hicom, one of Malaysia's giant conglomerates. Tune Air became the holding company, while AirAsia remained as the operating company. With this acquisition, AirAsia became Malaysia's first low fare, ticketless airline. From its humble beginning with just two medium-sized airplanes plying domestic routes, the low fare airline has expanded at an unprecedented pace: by mid-2005 it was already using 21 aircraft, each with a seating capacity for 148 passengers. After remarkable success in the capital, Kuala Lumpur, AirAsia opened a second hub in 2003 at Senai International Airport in Johor Bahru near Singapore, and launched direct flights to Sabah and Sarawak on Borneo Island.

---

1 Known as Malaya before independence from Britain in 1957.

2 At that time AirAsia was an ailing Malaysian full-service airline using small aircraft.

In January 2004, AirAsia set a milestone in Asian aviation history when the low fare airline formed a partnership with an aviation company in Thailand to develop a low fare carrier named Thai AirAsia in that country with a 49 per cent stake. In November 2004, a company that was 99.8 per cent owned by AirAsia successfully concluded a sale and purchase agreement with the Indonesian private airline PT AWAIR to acquire a 49.0 per cent stake in the company. Currently, together with its regional partners, Thai AirAsia from Thailand and AWAIR from Indonesia, Air Asia ferries passengers to many cities in the region, ranging from Bali in the south to Manila in the east, Chiang Mai in the northwest, and Xiamen in the north.

AirAsia and its partners in Thailand and Indonesia currently operate point-to-point domestic and international flights from its hubs at Kuala Lumpur in Malaysia, Bangkok in Thailand, and Jakarta in Indonesia. AirAsia offers a simple 'no frills' service at fares that are on average significantly lower than those offered by traditional full-service airlines. Modelled on successful low fare airlines such as the US-based Southwest Airlines and Dublin-based Ryanair (O'Connell and Williams 2005), AirAsia was established to create a new aviation product in Malaysia, one that would revolutionise air travel, and grow the local aviation market by providing low fares to enable more people to fly.

Table 19.1   Malaysia's LCC domestic and regional routes, 2006

Origin cities	Domestic destinations	International destinations
Kuala Lumpur International Airport, Malaysia	14	Thailand (3), Indonesia (8), Macau (1), Philippines (1), Cambodia (2), Brunei (1)
Johor Baharu, Malaysia	10	-
Kota Kinabalu, Malaysia	3	Indonesia (1), Philippines (1)
Penang, Malaysia	10	Indonesia (1), Thailand (1)
Kuching, Malaysia	5	-
Bangkok, Thailand	9	Singapore (1), Malaysia (3), Macau (1), China (1), Cambodia (1), Vietnam (1)
Jakarta, Indonesia	6	Malaysia (1)

Factors Contributory to Success

With the tagline 'now everyone can fly', AirAsia's philosophy of low fares is aimed to make flying affordable for everyone (Air Asia 2006). AirAsia also aims at making travel easy, convenient and fun for its passengers. AirAsia's operations are based on the following key strategies. The low fare, no frills policy means its fares are significantly lower than those of other operators. This service targets the passengers who will do without the frills of meals, frequent flyer miles or airport lounges in exchange for fares up to 80 per cent lower than those currently offered with equivalent convenience. No complimentary drinks or meals are offered on board. Instead, it offers a range of snacks and drinks available on board at very affordable prices and prepared exclusively for its passengers. Passengers have the choice of purchasing food and drinks on board. AirAsia's high frequency service ensures passenger convenience is met. The airline practises a quick turnaround of 25 minutes, which is the fastest in the region, resulting in high aircraft utilisation, lower costs and greater airline and staff productivity.

Passengers can make bookings through a wide range of facilities including a nationwide call centre providing convenient telephone booking services; ticketless service providing a low-cost alternative to issuing printed tickets; easy payment channels to enable their passengers to pay for their telephone bookings by credit card or by cash; internet booking; reservations and sales offices facilities available at airports and town centres for the convenience of walk-in customers; and authorised travel agents that make immediate payment via a virtual AirAsia credit card. It attracts potential passengers by constantly looking for ways to improve its services and increase savings for its passengers. AirAsia is the first airline in Asia to have a multi-lingual website, with seven languages. The airline also practises cost optimisation operations by striving to maximise profit and provide low fares with quality service. The airline optimised costs by operating a faster turnaround time, improving aircraft utilisation and crew efficiency, providing a 'no frills' service, using one type of aircraft to save training costs, all of which result in savings which are passed back to consumers in the form of low fares.

Impacts on Mobility

Air Asia began operation by introducing a fare policy attractive to passengers, just as major airlines worldwide were under pressure to cut costs in an industry with high fixed costs and low margins. The airline company brought down domestic air fares in Malaysia to a level previously considered impossible. Suddenly it became cheaper to fly from Penang from Kuala Lumpur than to drive by car on the North South Expressway. This attracted new classes of passenger, particularly those with low incomes, students, and tourists on budget trips. A year and some 1.1 million passengers after commencing operation, the airline was operating smoothly and was not affected by volatility in the industry, not even the worldwide slow down in air passenger traffic after the September 11 terrorist bombing in the United States. From a humble beginning with a few thousand monthly passengers, the airline and its partners in Thailand and Indonesia carried nearly 10 million passengers up to early 2006. Currently, the airline makes a total of 521 weekly flights and serves 31 cities both in Malaysia and in the neighbouring countries in South East Asia, thus boosting mobility of the population.

Apart from the convenience and simplicity of travel which attracts new passengers and regular travellers, fares are the most important determinant of air travel. Fares are structured in multiple tiers, reflecting demand and supply. On a typical journey, the cheapest fare is 80 per cent lower than the market fare, and the highest is 20 per cent lower. Depending on the time of booking, the customer will get the cheapest, compared to rival airlines. The earlier the customer books, the lower the fare.

The demand for air travel in the region is growing as a result of growing urban concentration, increased incomes, improving access to the internet and the opening of national economies. The level of the region's incomes and trading activities is ideal for starting single-service airlines offering flights of up to four hours' duration. The region also provides opportunities for low-cost airlines flying international leisure travellers.

In Malaysia, until recently, domestic flights were served only by the national flag carrier Malaysian Airline (MAS), which offers only full-service facilities, and AirAsia. Both contribute to the total number of domestic passengers carried in the country. When overall domestic and international passenger growth at all Malaysian airports showed a decline between 2002 to 2003, domestic passengers at Kuala Lumpur International Airport, the main hub both for MAS and AirAsia, showed a positive and relatively high growth of approximately 32 per cent, indicating the contribution of the low-cost airline to air travel at the country's premier airport.

A recent study (Abd Rahim Md Noor 2006) revealed that the low-cost airline was a popular choice for many passengers, especially those with low incomes, the young, and those who were travelling for leisure. The study also showed that gender, education and income contributed affected the frequency of use of the airline. Males were likely to make more trips than females. Education also had a positive effect, with the more highly educated making more trips than those with middle and lower levels of education. Personal income, too, was positively correlated with trip frequency, as was age. Many young people, under the age of 25, had travelled with the airline, and some 60 per cent had made either one or two trips with them. The study also found that a large majority of passengers used the airline for short distance trips to domestic destinations while a small percentage travelled to destinations in the neighbouring countries in South East Asia.

Impacts on the State-owned, Full-service Airline

Several studies outside South East Asia have found that the entry of a low-cost carrier impacts on routes it enters. Windle and Dresner (1995) found that entry of a low-cost airline onto a route reduces fares by 48 per cent and increases passenger numbers by as much as 200 per cent. In low-income countries, like those of South East Asia, with a relative lack of land transport infrastructure, Lawton and Solomko (2005) predicted that low-cost airlines are likely to achieve the greatest market stimulation.

In Malaysia, the low-cost carrier started operation when the state-owned full-service carrier MAS was facing losses, declining patronage, and a deteriorating image. The introduction of the low-cost airline by AirAsia further threatened the national airline by taking its passengers. From 2001, when AirAsia started operation, until 2004 when the low-cost airline was sharing domestic routes with the national flag carrier, passenger volumes for this sector in Malaysia showed an overall downward trend (Figure 19.1). Mismanagement problems in MAS culminating in its take-over by the government in 2001, and the spread in Asia of the viral respiratory illness SARS in 2003 and avian flu in 2004, contributed to the decline in passenger volume on the national airline. Despite the shrinking in the total domestic air passenger market from 19.3 million passengers to 15.7 million between 2001 and 2004, AirAsia managed to increase its total annual passengers from approximately half a million to 3 million (Figure 19.1), increasing its share of passengers from only 3.1 per cent to 19.1 per cent (Figure 19.2).

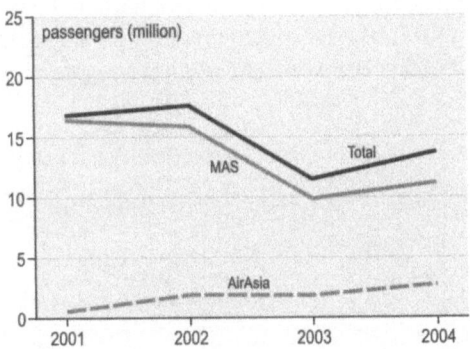

Figure 19.1 Air Asia and MAS's share of domestic air passengers (%), 2001–2004

Source: Abd Rahim Md Nor (2005)

In competing for passengers, both airlines cut prices. For example, MAS introduced a Super Saver fare scheme in which it was possible to buy a return flight ticket from Kuala Lumpur to Bangkok at a cheaper fare than AirAsia (Tan 2005). Another sign of threat to MAS was in early 2005 when Penerbangan Malaysia, a wholly owned company of Malaysia's Ministry of Finance Inc., which controls as much as 69 per cent of the national carrier, held discussions with Singapore Airline's low-cost carrier, Tiger Airways, on the possibility of setting up a joint-venture budget airline that would operate out of Malaysia from Subang Airport in Kuala Lumpur (Barrock 2005). The plan was for Penerbangan Malaysia to lease aircraft to the joint venture company in return for a stake in the company. The joint venture would enable the Malaysian company to address its losses from MAS's domestic operations, while Tiger Airways, which operated routes between Singapore and

Thailand, would have a foothold in the more lucrative Malaysia market. However, the plan never came to fruition.

At around the same time, MAS also launched a tripartite code-sharing agreement with Singapore Airline and its sister airline Silk Air, ending decades of rivalry between the two major national flag carriers. Under the pact, the foreign airlines would have code-share flights with MAS for the busy routes from Singapore to Malaysian cities such as Kuching, Kota Kinabalu and Penang. Since both airlines are government-owned, investors could see the pact as a strategy to block competition especially from a private entity like AirAsia. MAS claimed that the pact would allow all three carriers to further enhance their competitiveness and tap new opportunities flowing from the ASEAN Open Skies Policy to be implemented in 2008 which would allow national carriers of ASEAN countries to enjoy unrestricted intra-ASEAN access between the ASEAN capitals. It was to pave the way for the airlines to pool resources and to better position themselves in the rapidly changing landscape of aviation industry. AirAsia was baffled by the move, contending that the code-share pact contradicted the national aspiration of turning Kuala Lumpur into a regional aviation hub, as the code-shared flights between Singapore, Sabah, Sarawak and Penang would bypass the country's premier airport.

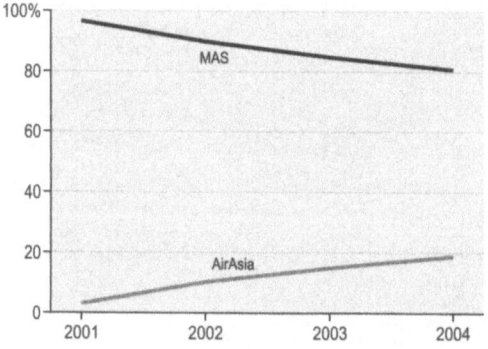

Figure 19.2   Shares of domestic air passengers (%), 2001–2004 in Malaysia
Source: Abd Rahim Md Nor (2005)

Despite the measures to enhance competitiveness against AirAsia, MAS continued to deteriorate financially. In late 2005, MAS (which had been losing money since 1997) announced a further loss of RM280.7 million for the first quarter of the year, and replaced its management team. In March 2006, after examining a joint report by both airlines, the government announced that, from August 2006, MAS would operate only 19 domestic 'trunk' routes (*New Straits Times* 25 May 2006, 36), and Air Asia would be given 96 routes, so-called 'non-trunk routes'. Both companies were given the right to decide on the frequency of flights, capacity distribution, type of aircraft used and airfares, under the supervision of Ministry of Transport. On the domestic routes, MAS would not be allowed to sell tickets at a discount and the prices would not be lower than the full economy fare. Air Asia was asked to build up capability to operate all domestic routes, including ensuring connectivity with places that MAS does not fly to. MAS would reduce aircraft numbers in its domestic sector and reduce its manpower from 23,000 to 16,500. AirAsia will expand operations at its hubs in Kuala Lumpur and Johor Baharu on the peninsular part of Malaysia, and open new hubs in Kota Kinabalu and Kuching in East Malaysia in the anticipation of increased frequencies of domestic flights in the future.

Concluding Remarks

Malaysia's first low-cost airline, AirAsia, was introduced in the country's economic landscape at a time when the country was undergoing rapid economic development and becoming integrated into a new geo-economy in an increasingly globalised world. Globalisation has affected the development of this type of air passenger service through the deregulation of the commercial aviation industry resulting in the opening up of the air to privately operated airline companies, particularly in the low-cost sector, and through the use of communication technologies in facilitating its business operations. The emergence of the low-cost airline and its partnership with operators in neighbouring countries have revolutionised the way people travel in the region, introducing a new concept of air travel to a wide range of people previously denied access owing to the high cost of air travel offered by full-service airlines. The impact of the low-cost airlines on the economy and business can be seen from the net income that the low-cost airline had accumulated, the employment that it has created, and the tax it has paid to the government. The expansion of operations to neighbouring cities in the region brings in international visitors and boosts the country's tourism industry. The different business model adopted by the low-cost airline enables it to increase efficiencies, offer much lower fares, and to attract more passengers, making it more profitable and attractive than the government-owned national airline, especially for low income, non-business travellers.

References

Abd Rahim Md Nor (2005), *Air Transport and Regional Cross-border Passengers: The Roles of Malaysia's No-frills Airlines in South East Asia.* Paper presented at 2nd International Malaysia-Thailand Conference on Southeast Asian Studies jointly organised by Universiti Kebangsaan Malaysia (National University of Malaysia) and Mahidol University International College, Thailand, 29 Nov –1 Dec.

Abd Rahim Md Nor (2006), *Low Cost Air Transport: Preliminary Study on its Impacts on Tourism in Malaysia.* Research Report No. SK-002-2005, Faculty of Social Science and Humanities, Universiti Kebangsaan Malaysia (National University of Malaysia), Bangi, Malaysia.

AirAsia Berhad (2006), *Annual Report 2005* (Kuala Lumpur: AirAsia Berhad).

Alves, C.F. and Barbot, C. (2007), 'Do Low Cost Carriers Have Different Corporate Governance Models?', *Journal of Air Transport Management* 13:2, 116-120.

Barrock, J. (2005), 'Competition Heightens for AirAsia', *The BizWeek*, 29 January: 3.

Carney, M. and Dostaler, I. (2006), 'Airline Ownership and Control: A Corporate Governance Perspective', *Journal of Air Transport Management* 12, 63-75.

Dicken, P. (2003), *Global shift: Reshaping the Global Economic Map in the 21st Century.* (London: Sage Publications).

Doganis, R. (2001), *The Airline Business in the 21st Century* (London: Routledge).

Francis, G., Humphrey, I., Ison, S. and Aicken, M. (2006), 'Where Next for Low Cost Airline? Spatial and Temporal Comparative Study', *Journal of Transport Geography* 14, 83-94.

Friedman, T. (1999), *The Lexus and the Olive Tree* (New York: Harper Collins).

Gillen, D. and Lall, A. (2004), 'Competitive Advantage of Low-cost Carriers: Some Implications for Airports', *Journal of Air Transport Management* 10:1, 41-50.

Gunasegaran, P. (2007), 'It's the Turnaround Time, Stupid', *The Edge*, 8 January, 14.

Henderson, J. and Castells, M. (eds) (1987), *Global Restructuring and Territorial Development* (London: Sage Publications).

Jayaseelan, R. (2007), 'Fly Asian Xpress's Game Plan', *The Edge*, 8 January, 1.

Lawton, T.C. and Solomko, S. (2005), 'When Being the Lowest Cost not Enough: Building a Successful Low-fare Airline Business Model in Asia', *Journal of Air Transport Management* 11, 355-362.

Lawton, T.C. (2002), *Cleared for Take Off: Structure and Strategy in Low Fare Airline Business* (Aldershot: Ashgate).

Mason, K., Whelan, C. and Williams, G. (2000), *Europe's Low Cost Airlines* (Cranfield University Air Transport Group, Cranfield).

O'Connell, J.F. and Williams, G. (2005), 'Passengers' Perceptions of Low Cost Airlines and Full Service Carriers: A Case Study Involving Ryanair, Aer Lingus, Air Asia and Malaysia Airlines', *Journal of Air Transport Management* 11, 259–272.

Ohmae, K. (ed.) (1995), *The Evolving Global Economy: Making Sense of the New World Order* (Boston: Harvard Business Review Press).

Porter, M. (1985), *Competitive advantage* (New York: Free Press).

Tan, M. (2005), 'Trial by Flight', *The Edge*, 14 February, 8.

Williams, G. (2001) 'Will Europe's Charter Carriers Be Replaced by "No Frills" Scheduled Airlines?', *Journal of Air Transport Management* 7:5, 277–286.

Williams, G., Mason, K., and Turner, S. (2003), *Market Analysis of Europe's Low-cost Airlines: An Examination of Trends in the Economics and Operating Characteristics of Europe's Charter and No-frills Scheduled Airlines*, Air Transport Group Research Report, Cranfield University.

Windle, R.J. and Dresner, M.E. (1995), 'The Short and Long Run Effects of Entry on US Domestic Air Routes', *Transportation Journal* 35, 14-25.

Chapter 20

Globalisation and Local Flavour in Business Organisations: The Case of Norwegian Elite Football Clubs

Stig-Erik Jakobsen, Arnt Fløysand and Hallgeir Gammelsæter

Introduction

Globalisation is influencing a growing number of sectors of social and economic life. Two aspects of this development are of specific interest in this chapter. National systems of organisation are increasingly exposed to global processes and influences, and norms and values of the market are invading sectors that traditionally have been based on non-profit values. One such sector is professional football, where the process of commercialisation has been accelerating since the beginning of the 1990s. This has raised the prominence of economic institutions, market transactions, and development in an institutionalised community of organisational actors that establish, maintain, and transform the rules of the football business across nations and political boundaries (Holt et al. 2005).

This chapter explores these changes through an analysis of Norwegian professional men's football. It begins with a discussion of approaches to understanding globalisation and local responses, and then moves to explore these responses to globalisation in the context of professional football clubs in Norway. The analysis starts by discussing how external factors, such as international player migration and international regulations, are affecting the organisation of clubs in Norway. It proceeds to discuss the specific way in which the corporate form (public limited companies) has been implemented within Norwegian elite football. This is followed by an examination of how business principles are influencing the operation of clubs, using the transformation of the governance structure of two selected clubs as examples. In our discussion, we examine how the processes of globalisation are met by responses at the national and local levels. Using this approach, we are able to discuss whether the clubs become internationalised with a purely business form of organisation or if they develop more hybrid forms. We argue that, although organisational practices of Norwegian professional football are developing towards homogenisation, the local flavour remains noticeably prominent.

Data for the analysis have been drawn from publicly available documents and literature, and interviews with representatives of the Norwegian Football

Association (NFA). We also conducted case studies of two elite clubs. Each case study is based on several interviews with people who are connected to, or have been part of, the club, in such capacities as managing directors, marketing consultants, accountants, board members, investors, sponsors, and coaches. The data were collected during 2005 and 2006.

Globalisation: Towards Homogeneous or Diversified Organisations?

Within the literature on globalisation and local response, there is a school of thought that argues that globalisation and commercialisation lead to a homogeneous form of organisation across territories (Ohmae 1990; Hall 1995). This approach argues that these two processes are interwoven, and globalisation accelerates the commercialisation process and the invasion of market principles into new sectors and geographical areas. Drawing on the knowledge of neo-institutional organisational theory, it can be argued that organisations in a field become more homogeneous in terms of formal structures, organisational culture, goals, or mission when the field is institutionally defined (DiMaggio and Powell 1983). Greater institutionalisation increases both the density of organisations and the similarity of their structures. It also increases the extent of interaction among organisations, the clarity of patterns of domination and co-operation, the information load with which they must contend, and the mutual awareness of being involved in the same type of activity. The main objective of this literature has been to explain the emergence of fairly standardised templates within organisational fields (DiMaggio and Powell 1983; Greenwood and Hinings 1996). Departing from this argument we expect to find growing similarities across and between nations in modes of organising professional football clubs in the global era.

Another school of thought argues that organisations are still diversifying despite globalisation and commercialisation (Maskell and Malmberg 1999; Cooke 2001). This position is rooted in the classic work of Polanyi (1944) and especially in the analysis of Granovetter (1985). Polanyi demonstrated how economy is enmeshed or embedded in non-economic institutions. The central insight of Polanyi's work is that the market is socially constructed and not a naturally occurring form. Thus, organisation in such a system will be shaped by its institutional environment. While Polanyi offered a macro-perspective, Granovetter (1985) managed to scale down the embeddedness concept towards an emphasis on the agency of individuals and organisations. For Granovetter, embeddedness refers to the ongoing contextualisation of economic activity in existing patterns of social structure and relations. In a more recent contribution, Dicken and Thrift (1992) stated that organisations are produced through a historical process that embeds practice in specific cognitive, cultural, social, political, and economic characteristics of the organisations' home territories. There has been a special emphasis on the national and local anchoring of organisation, for instance on how embeddedness within national systems and local networks influences organisational form (Maskell and Malmberg 1999). These lines of thought imply that the organisation of a football club will be strongly influenced by its national and even local context despite the processes of globalisation and increased commercialisation.

Development Within Norwegian Elite Football

The Process of Commercialisation

At the founding of the Fédération Internationale de Football Association (FIFA) in 1904 seven nations were represented. More than 100 years later (2006), its membership has grown by exactly 200 national associations. However, just as important as the specific football associations in constituting professional football in an increasingly dense and interconnected and globalised field, has been the entry of big sponsors and media companies. The interests of sponsors and media corporations have converged to the extent that the income from TV rights has become the most important revenue source for many professional soccer leagues in Europe. Whereas the gate receipts and commercial product sales in the English Premier League increased by an impressive 342 and 365 per cent, respectively, between 1992 and 2003, revenue from broadcasting increased by a stunning 3520 per cent (Michie and Oughton 2004). Pay TV broadcasters BSkyB and Setanta propose to pay a combined €2.5 billion for TV rights for the English premier league for the three-year period from 2007–08, with BSkyB paying €1.9 billion for its four packages and Setanta €0.6 billion for its two packages (www.sportbusiness. com). Modern communications have shrunk the sports world, and it is possible for people to view events all over the world and place bets on the outcomes. To deal with the demand placed on them by broadcasters, sponsors, advertisers, and spectators requiring entertainment, sports organisations have had to develop more business-oriented practices and new modes of organisation (Robinson 2003).

Commercialisation is also a distinctive feature of Norwegian elite football. The 1990s were the most successful period in the history of Norwegian football, with the national team qualifying for the World Cup in 1994 and 1998 and for the European Championship in 2000, and Rosenborg qualifying for the Champions' League every year since 1995, except for 2003 and 2006. This boosted interest in football in the popular media and among the public (Jakobsen et al. 2005). In 2005, the Norwegian commercial broadcaster, TV2, paid a total of €75 million (600 million NOK) for the TV rights for the Norwegian top league for the three-year period from 2006 to 2008 (www.vg.no). The development has triggered a growth in turnover among clubs, the emergence of larger and more complex organisations, and a more prominent position for economic institutions and market transactions. It is also more common to speak of clubs as 'corporations' competing in the 'entertainment industry'. In 2004, the total turnover of the 14 clubs in the Norwegian top league was approximately €70 million. The total number of employees in these clubs, including players, coaching staff, and administration, is about 500 (Jakobsen et al. 2005). In the next section we will look closer into this recent transformation at club level.

International Influences on Norwegian Football

A striking feature of modern football, underlining the increased international influence at club level, is the increased mobility of football players (Maguire 2004). The Bosman ruling, which caused the near elimination of restrictions on player transfers in Europe, and the increased wealth of football clubs, have fuelled the movement of players across national borders. The English premier league, for instance, is marked by a growing number of acquisitions of players from outside the UK, while the number of English-born players who progressed to the premier league decreased from 134 in 1998–99 to 63 in 2004–05 (Littlewood and Richardson 2006). A similar tendency can be observed in the Norwegian top league. In 1995, there were only 13 foreign players in the Norwegian top league, but during the second half of the 1990s Norwegian clubs started to import foreign players, mainly from other minor European leagues. This rise has continued though the 2000s. A degree of diversity appears at club level, but the trend of increased mobility among football players is clearly illustrated in Norwegian elite football. The number of foreign players in the Norwegian premier league increased from 55 in 2000 to 110 in 2005, which is close to one third of the total number of players (see Table 20.1). These players come from a diverse range of countries, demonstrating that football is becoming a world game. In 2000, there were only seven players from non-European countries; in 2005, there were 45.

Table 20.1   Numbers of foreign players in the Norwegian top league

	2000	2001	2002	2003	2004	2005
Total number in top league clubs	55	70	78	63	76	110
Percentage of total numbers	15%	19%	22%	18%	21%	29%
Average number in top league clubs	3.9	5.0	5.6	4.5	5.4	7.9

Sources: www.vg.no, www.aftenposten.no

Another dimension of the internationalisation process is that practices at club level have to conform to international regulations. Organisations such as FIFA, the regional confederations (such as UEFA), and the European Union have used their power to coerce clubs to comply with regulations such as stadium manuals and licence obligations. This can trigger a process of homogenisation in club organisations. In its club licensing system, UEFA has allowed national associations flexibility when developing their own licensing systems based on the UEFA master document (Morrow 2003). In Norway, the NFA has introduced a club licence to play at the elite level that is co-ordinated with the UEFA licence. A vital part of these provisions are the demands related to stadium facilities. All stadiums must be licensed as suitable for UEFA tournaments or at the elite level, and at the beginning of the century these facility requirements for a football stadium were strengthened. There have been differences in clubs' stadium projects, related to differences in the kind of constituents they have been able to mobilise to invest in a stadium. However, most clubs have recently put considerable resources into upgrading their stadium facilities. Of the 14 clubs that comprised the top division in 2005, nine had either finished building a new stadium during the 2000s, were in the process of constructing a new stadium, or had carried out considerable modernisation of an existing stadium. To pay their costs, many of these new stadiums have a multitude of facilities such as meeting/convention areas, health facilities, premises for business and trade, and even apartments for rent. Despite the potential for debt, most of the new and modernised stadiums in Norwegian football have generated increased gate revenue. New or modernised stadiums have also made the clubs more attractive to sponsors and business partners. The upgrading of Norwegian football stadiums was triggered by the NFA, which introduced a rather strict interpretation of the original UEFA licence. For the NFA, the modernisation of stadium facilities has been important in developing football as a 'product' and in strengthening its commercial foundation. Investors have been involved in some stadium projects by offering equity or even long-term loans to the property companies that own and operate the stadiums. Together with the general increase in turnover among clubs in Norwegian elite football during the past decade, the development of stadium facilities and the infrastructure implies new challenges in organising a professional football club. This will be discussed in the following sections.

The Dual Model

As already mentioned, the commercialisation of football has triggered a growth in turnover and the development of larger and more complex organisations: football clubs are frequently said to have become more 'businesslike' during this process. By examining two selected clubs (Lillestrøm and Brann), we will elaborate on these issues, emphasising the governance structure of these clubs. Before this discussion, we will present the dual model of Norwegian football, which is essential for organising at club level.

Over the past 20 years, following the accelerating commercialisation of European soccer, the corporate form (notably the public limited company [plc]) has gained ground. In the UK, football started to change in 1983 when Tottenham Hotspur listed publicly on the London Stock Exchange (LSE) with a holding company owning the club. Several more clubs were listed during the 1990s, and by October 2002 there were 20 UK clubs listed on the LSE (Morrow 2003). Danish Brøndby was in fact the second European club to be listed, in 1987, and during the 1990s Denmark became home to the largest concentration of listed clubs after the UK, when a total of six clubs were listed before the turn of the century (Morrow 2003). In Spain, most of the elite clubs were turned into joint stock companies by government intervention in 1992 in an effort to ameliorate their recurrent financial problems. In the Netherlands, Germany, and Sweden the national football associations have lifted their bans on the plc model as a valid way of organising professional clubs, although with the restriction in the two latter countries that the majority of the shares must be held by the association.

Unlike the elite clubs in the abovementioned countries, those in Norway must still be organised as voluntary sports clubs. Coaches and players must be members of association clubs that in turn must be members of the NFA to be licensed for the soccer league. Corporations cannot be members of the NFA. This does not mean that the commercialisation of Norwegian men's soccer has lagged behind that of its Scandinavian counterparts. On the contrary, the salaries of players have on average been higher in Norwegian top clubs than in Danish and Swedish clubs (Goksøyr and Olstad 2002), a trend that seems to be continuing. It is believed that this has been made possible, at least to some extent, by the setting up of limited companies that have established contractual relationships with their respective association clubs, thus forming a dual governance structure.

The first club to invite investors to form a limited company was Molde FK, which experienced great financial problems in 1991–92, notably because the club could not service its debt after having constructed a new stand at the stadium. In 1992, investors set up a limited company that established a contractual relationship with the association club. In return for the exclusive rights to utilise the club's name and the team in sponsorships, advertising, public relations, hospitality, and merchandise, the plc took over the administrative staff and the personnel costs, including the players' salaries, although the players and the sporting staff were still employed by the association. During the 1990s, several clubs at the top level adopted the model. By 2000, 12 out of 14 clubs had established a dual governance structure. In 1997, the NFA introduced a provision for 'commercial co-operation between a club and a plc', stating that any agreement between a club and a plc thereafter had to be approved by the NFA. A more comprehensive provision followed in 1999. The intention of the regulation was to restrict the influence of the plc and 'protect the interests and the autonomy of the association club', according to an NFA spokesman (in an interview). Investors were displeased with the restrictions. One investor stated that: '… the top flight clubs are large companies and they have to be operated by business principles. This will make it easier for those providing the capital' (*Aftenposten* 3 February 2000).

Transforming the Organisational Structure: The Cases of Lillestrøm and Brann

The dual model has institutionalised business principles in Norwegian elite football. Both of our two selected cases, the Lillestrøm and Brann football clubs, have a history of contractual co-operation with a plc. This implies the introduction of corporate for-profit values in organisations that traditionally have been based on voluntary democratic non-profit values (one member, one vote). The following discussion will illustrate how the introduction of such business principles has been instrumental in transforming the organisational structures of the clubs.

Lillestrøm has been the overall leading club in the greater capital region of Norway for the past 30 years. However, during the past couple of years, the club has been marked by stagnation. For three successive seasons, the club finished seventh in the top league (2002, 2003, and 2004). It also encountered huge economic difficulties: revenues did not keep up with expenditure, and its affiliated plc accumulated a significant debt. During the last few months of 2003, a local businessman took over the plc and became the sole proprietor. The takeover was seen as friendly, since the businessman was a former Lillestrøm player. However, the club was in desperate need of fresh capital to remain competitive with other top clubs.

The new owner believed that the organisation had for too long been living on the reputation of earlier successes. He wanted to restructure the club by establishing a more efficient and transparent mode of organisation. He immediately started to reorganise the club:

> When I started to look into the figures for the club there were a lot of errors in the budget, related to both expenditure and income. There was no system for cost control, and it was difficult to find out which people in the organisation were to be held responsible for attaining the budgeted figures.

Key issues in this reorganisation have been the appointment of a new sports director and a new chief financial officer. The former has a longstanding reputation as a player, both at Lillestrøm and abroad, and is a well-known figure in Norwegian football. The chief financial officer is a 'non-football' person, but has strong knowledge of managing and reorganising businesses. He is a business partner of the owner. Their mission was to 'leave no stone unturned' to find ways in which the club could improve. They started with the financial state of the club. A new accounting system was introduced, responsibility for income and expenditure in the budget was delegated to selected departments and individuals, and there was a reshuffling of the staff. Some staff members also left the organisation.

After establishing a new financial management structure at the club, the administrators proceeded to examine the general governance structure of the club. The management attempted to make the responsibilities and tasks of every department (for instance Finance, Marketing, Sport, Arrangement, Youth) more explicit, and selected roles were defined. They also tried to implement a sort of 'team organisation', with tight relationships between teams and the individuals involved. A club representative stated:

> We try to introduce a 'pressure to achieve' in every part of the organisation. It is not only our strikers that should be measured and evaluated every week. This also goes for the marketing consultant.

The new management wants to make the organisation more 'business minded.'

The Lillestrøm case illustrates that establishing a system for cost control is an increasingly important issue related to the governance of football clubs. The economic difficulties of SK Brann underline this. SK Brann is one of the biggest clubs in Norway, being the uncontested representative of Bergen, the second largest city in the country. However, the club also has a history of economic difficulties. Annual deficits during the period 1998–2002 resulted in an accumulated loss of NOK 80 million (€10 million). A key figure in the administration of Brann says the following:

> When I arrived at the club in 1999, there was no system for cost control. But the plc got hold of a controller, and he introduced new models for cost control and economic planning. Even if the period of the plc was characterised by huge economic difficulties, there was a time of learning for the organisation. Particularly during the last couple of years, they (the plc) put a strong emphasis on budgeting, solvency, and cost control.

Brann plc was established in 1997 with the target of attracting money to develop an elite team, but in the 2003 pre-season the club withdrew from the agreement with its indebted plc, which resulted in the latter's liquidation. The responsibility for running the club reverted to the club organisation. Thanks to the Norwegian model that prohibits the plc from being licensed to play in the league, Brann was not relegated as a result of the liquidation. Partly as a consequence of success on the field, generating a rise in income, a restructuring of debts, and new investors providing money for new players, the finances of Brann seem to be reasonably healthy at the time of writing. Despite not being an economic success, the plc years entailed that modern principles for running a business were introduced into the organisation. A representative for the club stated that it now has a well-adapted system for cost control:

> We have learned to be level-headed when budgeting, and some of our business partners have even criticised us for not being expansive.

Each department within the club has its own budget and a nominated person responsible for not exceeding that budget. Combined with systems for income control, it is now easier for the club to monitor its economic status and to predict the annual economic results. Thus, the financial room-for-manoeuvre can be estimated.

Concluding Discussion

This chapter began by recognising two distinctive processes related to globalisation; first, that national and local systems of organising are increasingly being exposed to global processes and influence, and second, that the norms and culture of the market are invading sectors that traditionally have been based on non-economic principles. These aspects of globalisation have been explored through a discussion of the organisation of professional football in Norway, a sector that is characterised both by the integration of a national system into a global system, and by the introduction of economic principles into a sector that traditionally has been distinguished by the characteristics of the civic or the voluntary sector. The question is then whether the outcome of this process is the development of professional football clubs as business organisations, or if organisational practice within football clubs still has idiosyncratic (sector-specific) characteristics that have a distinctive local flavour.

To elaborate on this it is necessary to look into the ways in which market principles are integrated into the football domain. This discussion has illustrated that there has been a growing market for the trading of football players in Norwegian elite football. In addition, the analysis has demonstrated that Norwegian elite football clubs have introduced practices that traditionally have been associated with business organisations. There is a clear trend towards developing formal principles and procedures for organisational practices. This is related to the development of governance structures and transparent systems of cost control. The employment of players, coaching staff, and administrators also implies that the relationship between the organisation and individuals has been monetarised and conducted under market conditions.

However, elements of the voluntary sector are still viable within Norwegian professional football. The specific rule of regulation within the NFA has prevented the development of a purely business-oriented organisational model, since only membership-based voluntary organisations can participate in the league. Despite the introduction of investors through a contractual relationship between plcs and the clubs, the activities of the clubs are not directed towards seeking rental income or a certain rate of return for their members or shareholders. Their main target is to develop football teams that are the pride of their organisations and their communities. Profit can of course be a motivation for the investors in the plcs, but few if any of the investors in Norwegian football have had any sustained positive returns on their investment in monetary terms. This can partly be explained by the fact that high expectations both within and around a club often result in high-risk strategies for obtaining success. There are several examples of clubs having invested beyond their capacity, and ultimately only a minority of participating clubs bring home trophies at the end of the season. Also, a large number of volunteers remain who contribute to the club. These volunteers are especially important on match days in their roles as guards, ticket sellers, and refreshment vendors. Nevertheless, the clubs feel the pressure from external stakeholders, such as investors, to portray themselves as businesslike organisations. To achieve external legitimacy, organisations must adapt and evolve in ways that convey appropriate messages to other actors (Scott 1995).

This means that recent developments within Norwegian football do not reflect the establishment of business organisations with consistent goals, targets, and performance measurements. Instead, they illustrate the emergence of a hybrid type of organisation that seeks to resolve various dilemmas and tensions. To achieve coherence the organisation must bridge different values within the organisation and reduce the potential tension. The dual model development within Norwegian professional football certainly involves what Granovetter (2005) describes as the interpenetration of economic and non-economic action. Norms that are developed within each sphere of social life are merged, creating a hybrid organisational form combining values from different spheres. Williams (2002) has stated that even in a global modern society there exist large economic spaces that combine various principles and have a certain resistance to pure market logic. The dual model, combining business principles and those of the voluntary sector, illustrates that despite increased external influence on club practice, an extension of their geographical scope, and the spreading of universal organisational principles, club organisations are still marked by a distinct local flavour.

References

Cooke, P. (2001), 'Regional Innovation Systems, Clusters and the Knowledge Economy', *Industrial and Corporate Change* 10, 945-974.

Dicken, P. and Thrift, N. (1992), 'The Organisation of Production and the Production of Organisation: Why Business Enterprises Matter in the Study of Geographical Industrialisation', *Transactions of the Institute of British Geographer* 17, 279-291.

DiMaggio, P.J. and Powell, W.W. (1983), 'The Iron Cage Revisited: Institutional Isomorphism and Collective Rationality in Organisational Fields', *American Sociological Review* 48, 147-160.

Goksøyr, M, and Olstad, F. (2002), *Fotball! Norges Fotballforbund 100 år* (Oslo: Norges fotballforbund).

Granovetter, M. (1985), 'Economic Action and Social Structure: The Problem of Embeddedness', *American Journal of Sociology* 91, 481-450.

Granovetter, M. (2005), 'The Impact of Social Structure on Economic Outcomes', *Journal of Economic Perspectives* 19, 33–50.

Greenwood, R. and Hinings, C.R. (1996), 'Understanding Radical Organisational Change: Bringing Together the Old and the New Institutionalism', *Academy of Management Review* 21, 1022-1054.

Hall, S. (1995), 'New Cultures for Old', in Massey, D. and Jess, P. (eds), *A Place in the World? Places, Cultures and Globalisation* (Oxford: Oxford University Press), 205-209.

Holt, M., Michie, J., Oughton, C., Tacon, R. and Walters, G. (2005), *The State of the Game: The Corporate Governance of Football Clubs 2005*. Research paper 2005, No. 3 (London: Birbeck Football Governance Research Centre, University of London).

Jakobsen, S-E., Gammelsæter, H., Fløysand, A., and Nese, G. (2005), *The Formalisation of Club Organisation in Norwegian Professional Football*. SNF-working Paper 83/05 (Bergen: Samfunns og næringslivsforskning).

Littlewood, M. and Richardson, D. (2006), *Football Labour Migration: Player Acquisition Trends in Elite level English ProfessionalFfootball 1990/91–2004/04*. Paper presented at the 14th EASM Congress (European Association for Sport Management), 6–9 September 2006, Nicosia, Cyprus.

Maguire, J. (2004), 'Sport Labour Migration Research Revisited' *Journal of Sport and Social Issues* 28, 477-482.

Maskell, P. and Malmberg, A. (1999), 'The Competitiveness of Firms and Regions. Ubiquitification and the Importance of Localised Learning', *European Urban and Regional Studies* 6, 9-25.

Michie, J. and Oughton, C. (2004), *Competitive Balance in Football: Trends and Effects*. Research Paper 2004, No. 2 (London: Birkbeck Football Governance Research Centre, University of London).

Morrow, S. (2003), *The People's Game? Football, Finance and Society* (New York: Palgrave Macmillan).

Ohmae, K. (1990), *The Borderless World* (London: Collins).

Polanyi, K. (1944), *The Great Transformation. The Political and Economic Origins of Our Time* (Boston: Beacon Press).

Robinson, L. (2003), 'The Business of Sport', in Houlihan, B. (ed.), *Sport & Society. A Student Introduction* (London: Sage Publications), 165-183.

Scott, W.R. (1995), *Institutions and Organisations* (Thousand Oaks: Sage Publications).

Williams, C.C. (2002), 'A Critical Evaluation of the Commodification Thesis', *The Sociological Review* 50, 525-542.

Chapter 21

Commodity Chains, Natural Disasters, and Transportation Infrastructure: From Kobe to Katrina

Julie Cidell

Introduction

Though largely unacknowledged in the literature, cheap, reliable transportation is a key component of global commodity chains. Global transportation facilities have become more and more concentrated in fewer and fewer locations, funnelling commodities through a shrinking number of terminals. The vulnerability of many of those places to natural disasters means that the commodity chains of which they are a part are vulnerable along their lengths, whether or not the actual sites of production are vulnerable themselves. This chapter explores the implications of localised natural disasters for the geographies of transportation and global commodity chains by examining two particular natural disasters, the Great Hanjin Earthquake of 1995 and Hurricane Katrina of 2005.

Commodity chains may be disrupted by natural disasters in two main ways. First, sites of activity along the chain may be damaged. Crops in the field may be wholly or partially destroyed; processing or manufacturing plants may be damaged; or markets can be severely depressed due to evacuation or economic losses suffered by households. In all of these cases, alternate locations must be found to either source raw materials, process them, or sell the final product. Many economic analyses have considered the regional costs of a natural disaster, taking into account these different sites of activity.

However, there is another type of disruption that has to do with the links, not the nodes, in the chain. Transportation facilities such as ports and roads are vulnerable to natural disasters, and if they are damaged, the disruptions can reverberate throughout the commodity chain to places far removed from the disaster itself. Additionally, there is the question of how commodity chains or networks adapt when their links are severed by a sudden event such as a natural disaster. Because these links rely on transportation, studying these disruptions in the commodity chain can highlight the importance of transportation within commodity chains. It also demonstrates the flexibility of these chains and how specific places can be inserted or removed from the chain.

The chapter consists of three main sections. First, the literature on commodity chains, networks, and circuits is briefly reviewed. Secondly, the effects of the Great Hanjin Earthquake on Kobe and some of the commodity chains of which that city is a part are described. Finally, the effects of Hurricane Katrina on New Orleans, the Gulf states of Louisiana and Mississippi, and other locations, metaphorically upstream and downstream, are similarly discussed. The latter two sections are based on a qualitative analysis of media coverage, research reports, and briefing papers on the two events.

Commodity Chains and Networks

There are generally considered to be four different approaches to the commodity *chain* concept. As many authors have explicated, one branch originated with the world systems theory of Wallerstein (Hopkins and Wallerstein 1994). The focus of commodity chains here has been on single commodities, linking together a series of places from raw materials to the point of final sale. The concept has expanded to include the processes of globalisation, being renamed global commodity chains (GCC). Another approach has been the French filière model, which derived from work on contract agriculture in both developed and developing countries (Raikes et al. 2000). Third, Porter's value chain concept (Porter 1985) has been included by some as an example of commodity chains within the business literature, although its almost exclusively local focus has led to criticism that it ignores larger structures and institutions. Finally, the new political economy of food and agriculture is considered within that field as a separate strand of commodity chain literature (Jackson 2006).

Criticisms of the linear nature of GCCs, as well as its focus on product rather than process, have led to a second main approach: commodity *networks* or *circuits*. Developed along with the cultural and network turns in economic geography, this approach considers connections between all elements of the system in multiple directions instead of flowing from producer to consumer (Henderson et al. 2002). This allows for a better analysis of flows between multiple scales as well as territories.

No matter whether commodity chains, networks, or circuits are the spatial model under discussion, the various nodes are still largely treated in isolation from one another, without consideration of the material connections among them. Smith et al. (2002) briefly discuss this as follows: 'Flows of value are our starting point, but underpinning these ... are relations of production (and we would add distribution and consumption)' (p. 55). While relations of consumption have been addressed by commodity networks to a considerable extent, relations of distribution continue to be neglected (Schwarz 2006). In particular, the transportation infrastructure and vehicles that move the goods and people that comprise commodity chains and networks are almost completely taken for granted in the literature, perhaps because of the growing though unfortunate distance between economic and transport geography (Rodrigue 2006).

One way to approach the question of the difference transportation makes to commodity chains is to look at what happens when it is disrupted. How do chains adapt to breaks between raw materials and processing, or between the factory and final market? Examining the consequences of natural disasters, where transportation links can be damaged or broken in a matter of hours or even minutes, can illuminate the importance of transportation in commodity chains and networks.

Kobe

On 17 January 1995, the Great Hanshin Earthquake shook Kobe, Japan, for nearly fifteen seconds at a magnitude of 6.9 (Tanaka 2000; USGS 2005). Approximately five thousand people were killed and over 200,000 buildings were destroyed (ibid.). The city also suffered damage to two of its major transportation facilities. First, a partial collapse of the Hanshin Expressway closed that major artery for well over a year (Ghasemi et al. 1996). More significant, however, was the damage to the Port of Kobe. The port was largely built on reclaimed land, which is prone to liquefaction under earthquake conditions (Tanaka 2000). Of the 187 berths at the port, 181 were left unusable, quay walls were displaced by an average of two to three meters, and nearly all of the port's cranes experienced some kind of vertical or horizontal displacement (ibid.). The port was closed in its entirety for two months, and it wasn't until two years later that all of the damage was repaired (Landers 2001).

Kobe's recovery *should* serve as an example of a city quickly and successfully responding to a major disaster. Within a year, exports were at 80 per cent of what they had been before the disaster, and within 18 months, manufacturing had almost completely recovered at 98 per cent (Postrel 2005); additionally, the number of shipping routes leaving the port was up to 80 per cent within just a year's time (*Straits Times* 1996). However, in the ten years after the quake, the port went from being the sixth busiest in the world to twenty-seventh (ibid; *Financial Times* 2004). In an era when the number of containers being handled at Asian ports continues to increase dramatically, Kobe suffered a sharp downturn. From 1994 to 2000, while Tokyo increased by 46 per cent and Pusan, South Korea, increased by nearly 100 per cent, Kobe declined by 22 per cent (Landers 2001).

Initially, local officials blamed the media for failing to publicise the port's quick recovery. For example, the manager of the city's Port Harbor Bureau argued, 'now that the port has largely gone back to normal, CNN is not interested' (*Straits Times* 1996). Then, high labour costs and excessive bureaucracy were blamed for the failure of traffic to return to the port (Becker 1997). In other words, pre-existing trends of neo-liberal competition (particularly decreasing the costs of labour and regulation) were given a boost by the temporary disruption to Kobe's infrastructure. Disasters tend to accelerate economic changes already underway, particularly for economic sectors in strong competition with other locations (Postrel 2005). One shipper noted that before the quake, the cost of using Kobe as a trans-shipment point was almost ten times that of Pusan, but the reliability of the port had kept it competitive. After the two months when the port was unusable, shippers found that Pusan was sufficiently reliable, and they never transferred their business back (Landers 2001).

Although local industries had their commodity chains disrupted by the earthquake, most disruption occurred in the area of trans-shipment. Here, the physical breaking of the links between nodes in commodity chains ranging from clothing to electronics resulted in a shift of economic activity, but not at the production or consumption nodes. Instead of using Kobe as a trans-shipment point, many firms turned to Pusan, Taiwan, Hong Kong, or Singapore. Kobe, therefore, serves as a lesson of why relying on trans-shipment can be dangerous for a region; any disruption of that transportation infrastructure can lead to a competitive loss that is difficult to recover from, as the city-as-link is replaced by another within the commodity chain.

Katrina

Hurricane Katrina was a different kind of natural disaster, and demonstrated in a different way how transportation matters in commodity chains. In particular, the disaster was spread over a much wider area, and it affected transportation infrastructure and commodity chain links in a variety of ways. The damage to the Port of New Orleans, as well as smaller facilities along the Gulf Coast, prevented raw materials from getting to processing plants, prevented processing and manufacturing from taking place, damaged warehouses storing goods at intermediate processing stages, and delayed or prevented manufactured goods from getting to market.

In 2003, the Port of New Orleans was ranked fifth in the country in terms of tons of cargo. However, in terms of value, it was only ranked twelfth, indicating the low-value nature of much of the cargo that passes through this port (AAPA 2006). Furthermore, the port is unusual in its emphasis on exports. Of the top twenty US ports, only New Orleans, the Port of South Louisiana (upstream from New Orleans), and Anchorage have a higher dollar value of exports than imports. This reflects Louisiana's location at the mouth of the Mississippi, as well as the chemical industry located just upstream from New Orleans. It also reflects the importance of bulk goods such as grain whose point of origin is difficult to trace and so is assigned to the final point of export (Russell 2006).

On 29 August 2005, Hurricane Katrina made landfall as a Category Three storm. Having been a much stronger Category Five over the Gulf of Mexico, there was still a considerable storm surge in addition to torrential rainfall and heavy winds. As has become well known in media accounts, the actual hurricane was not the most damaging part of this particular natural disaster: the overtopping and breaking of the levees holding back Lake Ponchartrain is what devastated New Orleans. Nevertheless, the Gulf Coast states of Mississippi and Louisiana were also damaged severely from the storm itself.

As with Kobe, the transportation facilities that were the most disrupted were maritime facilities, specifically the Port of New Orleans. The Mississippi River was closed to all traffic for five days, and the Port of New Orleans was closed for nearly two weeks (Russell 2006). However, as of 15 February 2006, the number of ships docking at the Port of New Orleans was back up to pre-Katrina levels (AP 2006a). For the port itself, the value of exports fell 3.5 per cent from 2004 to 2005 (Russell 2006), with a 23 per cent decline in agricultural products partially offset by the rising prices of steel imports (White 2006).

Transportation vehicles are as much a part of the network as is fixed infrastructure. For example, 60 per cent of the grain exported from the US leaves through New Orleans. In this case, it was not the damaged port facilities that were a problem, but the hundreds of barges that were contaminated with floodwaters and had to be cleaned of spoiled grain before they could be reused (White 2005a). Because barges are one-third the cost of truck transport and one-ninth the cost of rail, shifting to a different mode meant higher production costs. The US Department of Agriculture spent nearly US$2 million to help offset the costs of changing modes, although only a small percentage of the total flow of grain exports was redirected to other ports (ibid.).

One industry which is heavily concentrated in Louisiana along the Lower Mississippi is the chemical industry. While the factories were far enough upstream that they did not suffer severe damage, they were strongly affected by the damage to transportation infrastructure. The chemicals produced in these plants generally have one of two destinations: another plant in the immediate area for further processing, or export to another part of the country or globe (Boone 2005). In the first case, damaged rail lines and interstate highways limited local movements, while international exports were limited by damage to the Port of New Orleans. Similarly, the damage to the Port of Gulfport in Mississippi left DuPont Chemical unable to bring in the titanium ore it processes into titanium dioxide at its Pass Christian, MS, plant (AP 2006b).

As an interview with the executive director of the Greater Baton Rouge Industry Alliance noted, for firms in the region 'the biggest issues have been logistical: ensuring a supply of raw materials and the ability to ship products to consumers' (Boone 2005). Companies with international markets had to find other export locations, and as of March, many of those had not returned (Russell 2006). At the same time, as the *Times-Picayune* stated, 'International exports actually helped many companies survive following the hurricanes. As local markets dried up and traditional delivery methods became nonexistent, companies found alternative ways to get their products to their international customers' (Russell 2006, 1). Houston and Baton Rouge, for example, served as temporary export facilities for companies until New Orleans was up and running again. For these firms, where logistics was the only problem and not actual damage to production facilities, the international nature of their market was a benefit in that demand was still there.

Another affected area was the warehousing of bulk goods. The shift to just-in-time manufacturing has not occurred in all sectors; many commodities remain as break-bulk cargo. At the Port of New Orleans in particular, about 50–60 per cent of the volume of cargo is break-bulk, not containers (White 2005b). For example, New Orleans is the second largest coffee importer in the US, in large part because of its proximity to the Latin American coffee-producing regions. For break-bulk commodities, including coffee and steel, warehousing is still an important part of the commodity chain (*Times-Picayune* 2005), and so flooding damage to warehouses played a major role in the viability of the firms who used them.

Despite all of these impacts, the most dramatic influence on transportation and thus commodity chains as a result of Hurricane Katrina is not likely to be the Port of New Orleans itself, but the Mississippi River-Gulf Outlet, known as MR-GO. 'Mr. GO', as it is colloquially known, was established in 1965 as a shortcut for ships to avoid the frequently-shifting channel of the Mississippi River delta. The Gulf Outlet is a 76-mile channel from the Port of New Orleans to the Gulf of New Mexico (Sayre 2005). Unfortunately, the straight line of the canal acted as a funnel for the storm surge, increasing its intensity and leading to the breaching of levees in St. Bernard Parish and in New Orleans along the Industrial Canal (Brown 2005). Businesses located along the Industrial Canal were some of the most damaged by floodwaters (White 2006).

The canal itself acquired enough sediment during the storm that it is currently unusable for deep-sea ships. Because of the role the canal played in worsening the effects of the storm, the US Army Corps of Engineers has put an eighteen-month moratorium on dredging MR-GO while they study the costs and benefits of keeping it in place. In the meantime, while the Port of New Orleans suffered approximately US$80 million in damages from the storm itself, the estimated cost to relocate the businesses that are dependent on MR-GO is estimated to cost another US$385 million (Stinson 2006).

At least temporarily, many different kinds of activities that rely on MR-GO were moved out of Louisiana and Mississippi to facilities untouched by the storm. For example, Maersk Sealand shifted its operations to Houston, TX, and Port Everglades, FL. While the former relocation is considered to be temporary, the shift to Florida is permanent (White 2006). As with Kobe, this natural disaster may have exacerbated existing competitive trends for container shipping. Prior to the storm, Zim Integrated Shipping Services had already pulled out of New Orleans in favour of Houston and Mobile, AL, in large part because of the automobile manufacturing at the former and larger distribution facilities at the latter. The locational advantage that the Mississippi River confers on New Orleans of cheap shipping throughout the Midwest is not such an advantage for containers (White 2005b). New Orleans Cold Storage (NOCS) is one business whose ships are too wide to fit up the Mississippi, and port authorities are worried they and other companies that relied on MR-GO will leave for Houston (White 2006). NOCS is one of the largest poultry exporters in the US, and its ships are currently unable to reach its cold storage facilities near the central city. Trucking the poultry to a point on the river that the large ships can reach increases the cost to the point that the company is considering a move to a different city or state (Santora 2006). The company may able to keep its commodity chain intact, but only at the cost of moving its site of production.

Conclusion

While recovery from a natural disaster might appear to be quick, some disruptions to economic activity may, in fact, be permanent. The contrasting experiences of Kobe, Japan, and New Orleans, LA, show some of the different ways in which natural disasters affect commodity chains by disrupting the transportation infrastructure on which flows of commodities and people depend.

In the case of Kobe, the main impact was the year-long disruption of the port. While it was the largest export facility in Japan, most of that activity was trans-shipment rather than goods manufactured within the country. For the places connected to Kobe on either side of the chain, the disruption simply meant a shift to another port such as Pusan or Taiwan. In many cases, this shift may have actually resulted in a cost savings for the firms involved. The negative impacts were, therefore, largely felt within the disaster zone itself, as Kobe was removed as a link in many different commodity chains.

Hurricane Katrina, however, impacted a much larger area: not just in terms of the physical storm itself, but because of the many different roles that transportation infrastructure plays, from transporting intermediate chemical products from one factory to another, to getting agricultural or manufactured products from local and regional hinterlands to foreign markets. As a link, the Port of New Orleans is less likely to suffer long-term economic decline because its absolute location at the mouth of the Mississippi River has not changed, whereas Kobe's relative location shifted dramatically as compared to other ports in East Asia. However, the damage to MR-GO has the potential for longer-term impacts on the city of New Orleans, which may lose its place as a node in several commodity chains.

The spatial implications for commodity chains of this preliminary analysis are threefold. First, the concentration of traffic at fewer and fewer ports means that they are more vulnerable to disaster; some redundancy may therefore be desirable in shipping traffic (against current consolidation trends). Secondly, places that are vulnerable to natural disasters should be aware of their location within particular commodity chains, as well as the vulnerability of specific transportation links. Finally, producers and manufacturers should also be aware of the vulnerability of the links and nodes throughout the chains of which they are a part, including their potential alternative suppliers and/or markets.

The increasing interconnectedness of economic activity means that places at great distances from each other are more and more likely to be affected by seemingly local events. Looking at how the disruption of transportation links affects commodity chains is one way of broadening this type of analysis to reflect that commodity chains (and networks and circuits) are composed of links as well as nodes. Future work on GCCs should broaden the analysis to include the places that commodities pass through on their way from producer to consumer as part of the distribution network. As these two examples show, commodity chains can be shaped by events along their length as well as at their ends, something which needs to be considered in how we conceptualise chains, networks, and circuits.

References

AAPA (American Association of Port Authorities) (2006), *Port Industry Statistics* <http://www.aapa-ports.org/industryinfo/statistics.htm>, accessed 25 May.

AP (Associated Press) (2006a), 'Activity Back to Normal Levels at Port of New Orleans', 15 February.

AP (Associated Press) (2006b), 'DuPont Looking at Other Facilities to Handle Cargo', 5 January.

Becker, D. (1997), 'Failure to Change Would Give Added Impetus to the Tendency of Japanese Ports Moving to the Status of Feeder Ports', *Journal of Commerce*, 30 September, 12A.

Boone, T. (2005), 'Industrial Recovery Faces Headwind', *The Advocate* (Baton Rouge, LA), 6 September, 3-C.

Brown, M. (2005), 'MR-GO'ing, Going, Gone?' *Times-Picayune*, 27 November, 1.

Financial Times (London). (2004), 'Port City Seeks to Nurture New Enterprises', 27 April, 3.

Ghasemi, H., Otsuka, H., Cooper, J., and Nakajima, H. (1996), 'Aftermath of the Kobe Earthquake', *Public Roads* 60:2, 17-22.

Henderson, J., Dicken, P., Hess, M., Coe, N., and Yeung, H. (2002), ' Global Production Networks and the Analysis of Economic Development', *Review of International Political Economy* 9:3, 436-464.

Hopkins, T. and Wallerstein, I. (1994), 'Commodity Chains: Construct and Research', in Gereffi, G. and Korzeniewicz, M. (eds), *Commodity Chains and Global Capitalism* (New York: Praeger), 17-20.

Jackson, P. (2006), 'Commercial Cultures: Transcending the Cultural and the Economic', *Progress in Human Geography* 26:1, 3-18.

Landers, P. (2001), 'Kobe Disaster Offers Clues on Rebuilding', *Wall Street Journal*, 19 October, A19.

Porter, M. (1985), *Competitive Advantage: Creating and Sustaining Superior Performance* (New York: Free Press).

Postrel, V. (2005), 'When Disasters Act as Accelerators of Change', *New York Times*, 6 October, C2.

Raikes, P., Friis Jensen, M. and Ponte, S. (2000), 'Global Commodity Chain Analysis and the French *Filière* Approach: Comparison and Critique', *Economy and Society* 29:3, 390-417.

Rodrigue, J-P. (2006), 'Challenging the Derived Transport Demand Thesis: Geographical Issues in Freight Distribution', *Environment and Planning A* 38, 1449-62.

Russell, P. (2006), 'State's Export Business Weathers Hurricanes', *Times-Picayune*, 4 March, 1.

Santora, T. (2006), 'N.O. Cold Storage Frosted by Move to Clean up MS River Gulf Outlet', *New Orleans CityBusiness*, 17 April.

Sayre, A. (2005), 'Hurricane-hit Port Recovers Half of Business', *Associated Press State & Local Wire*, 27 December.

Schwarz, G. (2006), 'Enabling Global Trade above the Clouds: Restructuring Processes and Information Technology in the Transatlantic Air-cargo Industry', *Environment and Planning A* 38, 1463-85.

Smith, A.M., Rainnie, A., Dunford, M., Hardy, J., Hudson, R. and Sadler, D. (2002), 'Networks of Value, Commodities and Regions: Reworking Divisions of Labour in Macro-regional Economies', *Progress in Human Geography* 26:1, 41-63.

Stinson, S. (2006), 'Port of New Orleans Expects Huge Expense Stemming from Closure of Mississippi River Gulf Outlet', *New Orleans CityBusiness*, 27 February.

Straits Times (Singapore) (1996), 'Port Almost Back into Shape, but not as Competitive as Asian Rivals', 14 April, 4.

Tanaka, Y. (2000), 'The 1995 Great Hanshin Earthquake and Liquefaction Damages at Reclaimed Lands in Kobe Port', *International Journal of Offshore and Polar Engineering* 10:1.

Times-Picayune (2005), 'Picking up Speed', 10 December, 1.

USGS (United States Geological Survey). (2005), *Earthquake Hazards Program* <http://earthquake.usgs.gov/regional/world/events/1995_01_16.php>, accessed 12 December.

White, J. (2005a), 'Katrina Cuts La. Grain Exports', *Times-Picayune*, 28 October, 1.

White, J. (2005b), 'Port Struggling to Regain Business after Katrina', *Times-Picayune*, 11 November, 1.

White, J. (2006), 'Port Tries to Keep Key Firms Afloat', *Times-Picayune*, 30 April, 1.

Chapter 22

Reconsidering Commodities in International Trade and Economic Growth: The Case of New Zealand

David Hayward

Introduction

New Zealand's merchandise trade remains heavily dependent on exports of land-based production. Prevailing beliefs in public policy discourses regard this as undesirable and hold it accountable for the country's poor comparative trade performance. This chapter re-examines the evidence and offers an alternative interpretation of the potential for economic growth based upon primary sector exports.

A particular stimulus for this inquiry lies in a policy document that criticised New Zealand's recent trade performance and restated the common wisdom that land-based industries offer limited potential for sustained growth (Skilling and Bowen 2005). Although intended for local policymakers the document echoed recent OECD analysis and appeared to be informed as much by ideology as by reason and empirics. This chapter is a response and rebuttal to that report but it also challenges the binary notion of simple and complex commodities, in which the former are deemed to contain little added value.

Context and Critique

There are at least three components to the context of this inquiry. The first is the comparative performance of national economies in respect to their trade and national income. Typically, these are represented by tabular rankings and time series analysis of national economies according to derived statistics – the values of which are deemed to measure economic performance. The second is the conventional wisdom on the long term decline in international prices for simple commodities in comparison to complex manufactured goods. The third is the fashionable attention to technology and innovation in the production of traded commodities, and their positive association with economic growth. In each case sound, logical arguments are invoked but the critique developed here will expose the weaknesses in aggregate data and arbitrary industry and commodity classification schemes.

In December 2005, the New Zealand Institute, an independent policy analysis organisation, released a report that sharply criticised New Zealand's export performance over thirty years (see Skilling and Bowen 2005). The publication was widely discussed in New Zealand, and cited in public policy pronouncements. Its analysis revealed New Zealand's poor export growth performance in comparison to other OECD countries. However, alert readers will be intrigued by the observation that, 'Only Switzerland generated slower export growth than New Zealand over this period' (Skilling and Bowen 2005, 7). For a small, peripheral economy such as New Zealand, whose perspective on the global economy is normally filtered through a post-colonial cringe, being compared to a high income European country with a historically open economy would normally be considered an achievement. Indeed, the paradox exposes the weakness of time-series growth measures, wherein the growth rates tell only part of the story and the initial starting positions are overlooked. Thus, while New Zealand and Switzerland may have 'lagged' in terms of export growth rates they were of course starting from far more advanced positions than their peers, although this is overlooked in the report.

Cross-sectional analysis too suggests that New Zealand's export performance has lagged. Exports are equivalent in value to 29 per cent of gross domestic product (GDP), which places the economy in the lower rankings of OECD members (Skilling and Bowen 2005). However, it remains notably higher than the OECD average of 20 per cent and so New Zealand is doing 'poorly' in comparison only to its smaller OECD peers. As with the time-series record, these measures require careful interpretation, and in this case at least two caveats apply. The first is that exports include a substantial 'import component'. That is to say that if exports are to be used as a measure of economic performance then we must take into consideration the value of the imports that went into their making. Unlike most of the small OECD economies with which it is routinely compared, the import content of New Zealand's exports is comparatively low – estimated to be half that of the smaller European countries, for instance (Black et al. 2003). The second caveat is the effect of geographical and political-economic context. In almost all cases, New Zealand's smaller OECD peers are either European Union members or geographically contiguous with it. In contrast, New Zealand is geographically and politically distant from its trading partners; the most similar 'peer' may be found in Australia, which has a much lower export 'performance' statistic. Nonetheless, the Institute concluded that New Zealand, 'has not kept pace with … other developed countries in terms of expanding their export activity' (Skilling and Bowen 2005, 9) – a conclusion that was widely echoed in national media.

The New Zealand Institute concludes that, 'The perception of New Zealand as a small trading nation is unfortunately well out of date' (Skilling and Bowen 2005, 9) with the primary reason being its dependence on a small number of land-based industries. The dominance of agricultural exports is well known but interpreting this as a 'problem' opens up a conceptual and politically-charged debate. Two factors are cited in particular; the long-term price decline for land-based commodities, and the low levels of technology development and adoption within land-based industries. The first derives from the Prebisch-Singer thesis, which describes a long-term decline in the terms of trade for agricultural products in favour of manufactured products (Prebisch 1950; Singer 1950). Since the utility value of simple commodities changes little over time it is argued that their value will decrease in comparison to complex manufactured goods and services. For this reason, land-based industries are conventionally regarded as a poor basis for long-term, export-based economic growth. However, it is erroneous to presume that all land-based products conform to the model of a 'simple' commodity, and moreover that they are unable to incorporate intellectual property. The integrity of the Prebisch-Singer thesis is not at question, but rather its application and the analysis of comparative statistics that allow little scope for differentiation among outwardly similar commodities and industries.

A further example of the inherent weakness in utilising standard commodity and industry classification schemes is the OECD's recent analysis of the export performance of industries in relation to their levels of technology development and innovation (OECD 2005). This document figures prominently in the Institute's report, and reveals that New Zealand's manufacturing exports rank second to last in their technological composition. However, caveats apply that are pertinent to the study of land-based exports. First, the OECD's analysis relates only to manufacturing industries; and second, the information is derived from only selected, and generally large OECD member states. Furthermore, the measurement of technology is derived from formal expenditures on research and development (Hatzichronoglou 1997) and so overlooks other kinds of innovation, such as quality assurance, logistics and marketing. The OECD report is a valuable contribution to our understanding of the importance of technology in manufacturing industries, but its scope does not encompass land-based production and so overlooks its innovative, value-adding potential. Like the Institute's report, the OECD report was greeted with enthusiasm and little criticism because it appears conceptually rigorous and empirically sound. The following analysis offers an alternative interpretation.

Commodity Chains and Commodity Value

The notion of an industry as a sequence of value-adding activities – the 'commodity chain' – presents a valuable analytical framework for examining industrial dynamics, national and regional economies and the context of international trade. The concept may be found in different incarnations, such as: global commodity chains (Gereffi and Korzeniewicz 1994; Gibbon 2001), global commodity systems (Friedland 2002), commodity chains analysis (Hughes and Reimer 2004; Leslie and Reimer 1999), filières (Lewis et al. 2002; Raikes et al. 2000), industrial production systems (Dicken 2003; Walker 1988), supply chains (Doel 1999; Le

Heron et al. 2001), systems of provision (Watts et al. 2005), value chains (Porter 1985; Sturgeon 2000; Wood 2001) and global value chains (Gereffi et al. 2005). The great advantage of commodity chains analysis ('CCA' – in its many guises) is the reduction of specific economic features to abstract elements. Thus, both the simplest and the most complex industries may be described equally in terms of commodities, their transformation, discrete technologies, flows of information and flows of value. Commodities are necessarily transferable and substitutable, as are labour, capital, land and technology. In adopting commodity chains analysis, this chapter places primary attention on the notions of commodity and value, rather than on the peculiar geographies of nodes and flows. In CCA terms, international trade is almost irrelevant. It occurs at arbitrary points in the chain and in itself says little about the overall structure and performance of an industry. For countries, however, trade is an important measure of value transfer, and differentiated commodity values are a better indication of trade performance than gross measures.

The source of commodity value is an especially interesting and relevant topic. The value chains version of CCA puts great emphasis upon the processes of adding value to commodities and the capture of value at nodes within the chain (Wood 2001). The value-adding sequence describes the supply side of commodity value while the demand side includes consumers' preferences and the utility of the commodity. From this perspective, commodity value may be usefully deconstructed into types: a first order, utility value; and a second order, symbolic value. This approach was developed in social semiotics (Gottdiener 1995) and applied to a commodity chain analysis by Hartwick (1998). The utility component of a commodity is an objective attribute; the second order, aesthetic or symbolic value, however, is subjective and can be determined only by the consumer. The distinction between these two enables us to explore differential prices for commodities that may in outward appearance seem to be good substitutes. In both Hartwick's analysis of the gold commodity chain and Le Heron and Hayward's (2002) analysis of the breakfast cereal industry, this concept was employed to examine the industries' efforts to nurture and manipulate the second order value. This perspective extends the conventional notion of value chains and allows for the possibility that even unprocessed commodities may incorporate added value; where that value is second order, symbolic value. This may be evident as product differentiation is revealed in differential prices; and the productive elements to create this may include brand marketing, quality assurance, the attachment of intellectual property, and the matching of the product to specific consumer requirements. Indeed, anything that appeals to the irrational human consumer and leads to differentially higher market prices may be interpreted as realised added commodity value. In the following analysis, evidence of added value will be interpreted as an alternative measure of trade performance.

The Structure of New Zealand's Exports

New Zealand is a comparatively small economy with a small number of export-oriented industries, a modest range of domestically-oriented industries, and a dwindling number of import-substitution industries. The general pattern is well known although obscured by the conventional categories of primary, secondary and tertiary sectors. Commodity chain analysis transcends this typology through tracing the sequence of commodity production through various transformation nodes. From this perspective both raw agricultural products as well as the manufactured products derived from them would be considered to be outputs of the same industry; for instance, raw milk and pharmaceutical grade lactose would both be considered products of the same industry/commodity chain.

The data summarised in Table 22.1 are derived from four digit Harmonised System data for New Zealand's commodity exports and excludes 're-exports' that often contaminate international trade databases. The two-tiered industry groupings reported here were derived from 1,145 HS categories and are organised according to commodity chains. At the first tier, a distinction is made between animal-based and plant-based industries, in recognition of the commodity and technology linkages between them. For instance, the leather, live animal and carpet (which in New Zealand are wool floor-coverings) industries are functionally linked to the meat, dairy and wool industries. In this way, these 'animal-based industries' are grouped together to form a second tier of analysis. Individually, the second tier industries each combine several HS commodity categories. As with any classification scheme, this is not without weaknesses, but the intent here is to present a commodity chain informed account of New Zealand's merchandise trade. This presentation further accentuates the well-known dependence on Animal- and Plant-based industries which, along with Fisheries, Wood, Metal and Mineral industries, account for three quarters of merchandise exports.

Table 22.1   New Zealand merchandise export industries, 2004

Industry	Exports (NZ$ millions)	%	Main Products
Animal-based industries			
Dairy	6,764.5	23.0	Milk powder, butter, cheese, casein, whey
Meat	5,208.6	17.7	Mostly beef and lamb
Wool	799.4	2.7	Raw wool fibre
Leather	489.0	1.7	Raw and tanned hides
Animals	266.1	0.9	Horses and cattle
Carpets	113.2	0.4	Wool carpets
Other animal products	11.0	0.0	
	13,651.8	46.%	

Industry	Exports (NZ$ millions)	%	Main Products
Plant-based industries			
Fruit	1,165.8	4.0	Kiwifruit and apples
Wine	469.3	1.6	Table wine
Vegetables	357.6	1.2	Squash, onions, peas, sweet corn (incl. frozen)
Prepared foods	277.5	0.9	Mostly sauces and preserved vegetables
Other plant-based products	278.6	0.9	
	2,548.8	8.6	
Wood products	2,877.5	9.8	Logs, timber, particle and fibre boards, plywood, pulp, papers and furniture
Fisheries	1,232.9	4.2	Shellfish and fish (fresh and frozen)
Other food products	512.4	1.7	Includes confectionery, baked goods, juices and preserves
Metal products	2,173.8	7.4	Aluminium, gold and steel
Mineral products	963.0	3.3	Petroleum, chemicals, ceramics and glass
Basic & Elaborate Manufacturing			
Machinery	1,076.8	3.7	Industrial and agricultural machinery
Electrical	886.6	3.0	Boards, transformers and other devices
Plastic and rubber	523.4	1.8	Packaging, articles, tyres and gumboots
Whiteware	357.5	1.2	Refrigeration, dishwashing and laundry
Boatbuilding	207.2	0.7	Recreational and sports yachts
Motor parts	178.4	0.6	Mostly alloy wheels
Pharmaceuticals	190.1	0.6	Medicaments from animal sources, mostly

Industry	Exports (NZ$ millions)	%	Main Products
Misc manufacturing	1,071.2	3.6	Including medical and navigational instruments
	4,491.2	15.2	
Other – not elsewhere specified	1,020.4	3.5	
Total Merchandise Trade	29,471.7	100.0	

Source: INFOS database, Statistics New Zealand, (cited 8 June 2006).

The Performance of New Zealand's Exports

The New Zealand Institute and OECD reports consider New Zealand's trade performance in aggregate terms. Similar data may be used to explore comparative trade performance in value-added terms. Specifically, the use of ratio statistics comparing the value of exported products in comparison to their volume may be interpreted as measures of revealed added value. These measures are most appropriate for comparatively undifferentiated commodities. For instance, New Zealand beef may be presumed to be equal in utility value to Argentine beef. Thus, any difference in per unit value may be interpreted as revealed added value obtained through brand identity, quality assurance, packaging, or some other consumer-recognised attributes. Furthermore, these value-to-volume ratio statistics may be compared to the world averages in order to derive standardised statistics of the form recognised by geographers as location quotients, and by some economists as export efficiencies.

Table 22.2 includes comparative value: volume measures for eight of New Zealand's leading export commodities from its Animal- and Plant-based industries. The standardised value: volume ratios presume an underlying world market price for each commodity and percentage statistics are used to express each country's deviation from this. Note that intra-European Union trade is excluded, since this is presumed to be distorted by Common Agricultural Policy and is in any case less inter-national trade than extended inter-regional trade.

Table 22.2   Export value-to-volume ratios for selected commodities, 2004

Commodity	Leading exporters	Value: Volume ratio, expressed as a percentage of Rest of the World Total (excluding Intra-EU trade)	% of World Total (excluding Intra-EU trade)
Butter	European Union	115	38
	Canada	111	2
	Australia	98	8
	New Zealand	96	36
	Belarus	85	5
	Ukraine	74	5
Cheese	Switzerland	224	4
	European Union	149	35
	United States of America	95	4
	Australia	76	17
	New Zealand	69	18
	Ukraine	68	6
	Belarus	62	4
Milk Powder	European Union	118	26
	Australia	105	11
	Argentina	96	6
	New Zealand	89	30
	United States of America	88	9
	Belarus	85	2
	Ukraine	77	3
Beef (boneless)	Australia	137	25
	United States of America	130	4
	Canada	120	11
	New Zealand	110	10
	Nicaragua	100	1
	Uruguay	90	6
	Argentina	89	8
	Brazil	69	25
	Paraguay	68	2
	European Union	60	5

Commodity	Leading exporters	Value: Volume ratio, expressed as a percentage of Rest of the World Total (excluding Intra-EU trade)	% of World Total (excluding Intra-EU trade)
Sheep meat	New Zealand	141	51
	Bulgaria	135	1
	Uruguay	105	1
	Chile	102	1
	Argentina	78	1
	Australia	75	38
	India	53	1
	China	47	3
	United States of America	45	1
Apples	New Zealand	183	8
	United States of America	164	11
	South Africa	117	7
	Brazil	92	4
	European Union	89	14
	Chile	88	17
	Argentina	85	5
	China	65	18
Kiwifruit	New Zealand	227	53
	United States of America	108	2
	European Union	76	14
	Chile	46	24
	China	37	1
	Iran	25	5
Wine	New Zealand	217	1
	European Union	201	34
	Australia	112	17
	South Africa	71	7
	United States of America	66	10
	Chile	60	12
	Argentina	48	4
	Moldova	32	6
	Bulgaria	30	2

Source: FAOSTAT database, Food and Agriculture Organisation of the United Nations. Available at: http://faostat.fao.org/ (cited 12 June 2006).

For each commodity the comparative value: volume measures are reported for New Zealand and the other leading exporters. The patterns are revealing and warrant some interpretation in respect to the characteristics of the associated industries. In the case of butter, cheese and milk powder, New Zealand's value: volume ratios are below the world average – markedly so, in the case of cheese. Of the three, cheese has the widest range of value: volume ratios, which suggest that it is more qualitatively differentiated and that Switzerland and the European Union are achieving realised added value. For these dairy commodities it appears that New Zealand trades largely on price competitiveness founded upon low-cost production.

The New Zealand meat industry, however, appears to be achieving added value – as indicated by the above average value: volume ratios for beef and sheep meat. For the latter it is notable that New Zealand earns 41 per cent more per unit than its competitors, even while accounting for over half the total world trade by volume. The value: volume ratio for beef is lower but still noticeably higher than South American competitors. One can infer from these statistics that the New Zealand meat industry appears to be successfully achieving some added commodity value.

The evidence of realised added value is more convincing for the three principal exports of the 'plant-based industries': apples, kiwifruit and wine. In each case, New Zealand's exports earn revenues markedly above the rest of the world averages. It is well known that New Zealand wine trades at relatively high price points internationally and the industry is oriented towards the production of wine imbued with second order, symbolic value. This is expressed materially in a value: volume ratio twice that of the rest of the world average, comparable to the European Union and markedly ahead of other so-called 'New World' wine producers. New Zealand appears to be performing well with the less differentiated kiwifruit and apple commodities, too. The global kiwifruit market is dominated by a very small number of national producers, and New Zealand has succeeded in maintaining a dominant position in extra-EU trade as well as a significant price premium. Consistent and assertive brand marketing by the monopoly exporter, efficient logistics and effective quality control has translated into added value for New Zealand producers. The utility value of kiwifruit is broadly the same for fruit from any source and so this price premium may be interpreted as second order, symbolic value. Even the humble and ubiquitous apple reveals marked differences between global producers. New Zealand is a comparatively small producer and one of many international traders. Nonetheless, it has achieved a value: volume ratio almost twice that of its main competitors – the other southern hemisphere producers – owing to seasonality. It appears that for even this relatively undifferentiated commodity there is scope for exacting second order commodity value. And, here again, this has been realised through innovations that relate to product form and aesthetics, plus branding, logistics and quality assurance.

A longitudinal analysis affords an examination of the durability of these differences. Figure 22.1 illustrates New Zealand's comparative value: volume ratios for five of the selected commodities – dispensing with the dairy industry products that for New Zealand appear to be undifferentiated commodities. For beef, sheep meat and apples the record extends back to 1961, but for kiwifruit and wine the record is shorter owing to their more recent emergence as export

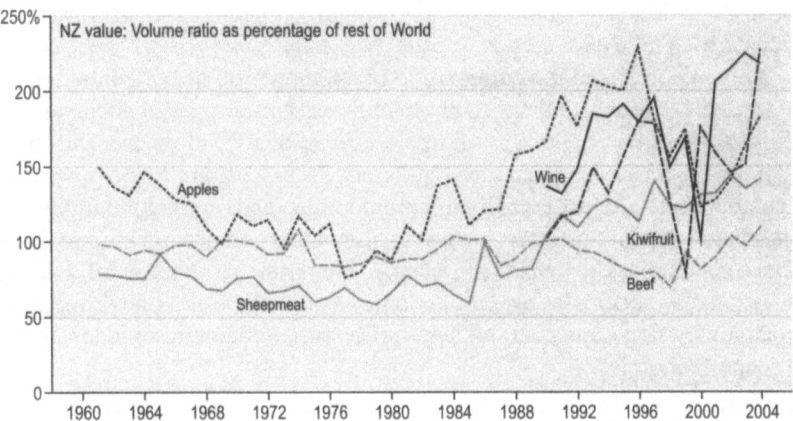

Figure 22.1 Value-volume ratios for selected New Zealand exports, 1961–2004

Source: FAOSTAT database, Food and Agriculture Organisation of the United Nations. Available at: http://faostat.fao.org/ (cited 12 June 2006)

commodities. The time-series record allows some refinement to the interpretations made on the cross-sectional data. For instance, the long-term record for beef shows minor deviations around the world average and so the present, comparatively high values cannot be viewed as anything other than a normal fluctuation. However, the record for sheep meat indicates two distinct periods: the first from 1961 through to 1990 during which below-average commodity values were obtained; and the recent period of increasing above-average earnings. This may be taken as evidence of the sheep meat industry successfully achieving sustained, realised added value. A similar pattern is evident for apples although less stable: from the early 1980s through 1996 apples earned realised added value. This was followed by a severe decline and then a recent recovery. The shorter longitudinal records for kiwifruit and wine are similar and the underlying trend for both has been for increasing average value: volume ratios, apart from an interruption around the millennium.

Interpreting Commodity Values and Trade

The analysis has considered eight agricultural commodities, each of which is a significant export commodity for New Zealand. The analysis of value: volume ratios reveals distinct differences in the comparative performance of these as traded commodities. The three dairy products achieve below average prices; the two meat products indicate some realised added value; and the fruit products all achieve substantial realised added value. To interpret these findings we should consider them in respect to the industries' characteristics – summary details of which are included in Table 22.3.

The New Zealand dairy industry is highly integrated both vertically and horizontally. The 2001 merger of the two largest processors and the former monopoly exporter was the culmination of a long period of horizontal integration and created a national industry able to fully exploit scale and scope economies (Willis 2004). The 14,000 dairy farms are functionally integrated with the three remaining co-operative processor-marketers. New Zealand dairy products trade as comparatively undifferentiated commodities and compete largely on the basis of low production costs owing to the natural environment, and scale and scope economies. The largest processor, Fonterra, has horizontally integrated internationally, accumulating production assets in Australia and Chile, and is a comparatively large transnational enterprise with a competitive focus upon scale and scope economies.

Table 22.3   New Zealand animal- and plant-based export commodities and industry characteristics

Industry	Main Export Commodities	Industry Characteristics	Commodity Characteristics
Dairy	Butter Cheese Milk powder	14,000[a] dairy cattle farms *Near* monopoly processor and exporter (Fonterra) Vertical integration well developed for scale and scope economies Horizontal integration nationally and overseas	Qualitatively undifferentiated Price sensitive Incipient brand-marketing
Meat	Beef Sheepmeat	28,000[a] sheep and beef cattle farms 150 processor/ exporters Vertically and horizontally disintegrated	Qualitatively undifferentiated Price sensitive Some brand-marketing
Fruit	Kiwifruit	2,300[a] orchards in regional concentrations Monopoly exporter Horizontal integration with overseas producers	Qualitatively differentiated – strong brand marketing Price sensitive but earning premium New product innovations

Industry	Main Export Commodities	Industry Characteristics	Commodity Characteristics
	Apples	1,400[a] orchards in regional concentrations Horizontal and vertical disintegration nationally Limited horizontal integration with overseas producers	Qualitatively differentiated – weak brand marketing Price sensitive New product innovations
Wine	Table wine	1,200[a] vineyards in regional concentrations; approximately 44% [b] grape volume from vertically integrated wineries. 530[c] wineries: Larger enterprises, horizontally-integrated nationally and internationally Smaller enterprises, regionally focused	Qualitatively differentiated – strong region and brand marketing Comparatively price insensitive New product innovations

Sources: Various industry and statistical sources, including:

[a] Figures for 2002 from Ministry of Agriculture and Fisheries (http://www.maf.govt.nz)

[b] Figures for 2002 from Grape and Wine Industry Statistical Annual 2003, New Zealand Winegrowers, Auckland.

[c] Figures for 2006 from New Zealand Winegrowers Annual Report 2006 (http://www.nzwine.com)

The New Zealand meat industry has around 150 processing plants, operated by a small group of multi-plant enterprises and a larger group of independents. In comparison to the dairy industry, the meat industry is markedly disintegrated, and in recent years, plant sizes have decreased in size – in many respects conforming to a post-Fordist transition. There has been a concomitant shift in products from moderately processed, undifferentiated commodities to more highly processed forms targeted to specific markets. The value: volume ratios for sheep meat provide evidence for realised added value based upon second order, symbolic value.

The three plant-based industries are markedly different in organisation but in each case production and marketing strategies have been employed effectively to raise the second order commodity values. The apple industry's plummeting value: volume ratios in the late 1990s may be explained by the coincidence of depressed world market prices and horizontal and vertical disintegration of the domestic industry. Having recovered, the New Zealand apple is once again a product differentiated by quality assurance and brand identity, in which the premium prices earned represent realised added value. New Zealand kiwifruit has been even more successful in achieving added value through product differentiation, and remaining integrated with a single exporter. The global market was pioneered by New Zealand growers in the 1970s and for some time scarcity ensured premium prices. Subsequently, with the arrival of competitors the commodity has been marketed as a qualitatively differentiated product that appears to be achieving second order commodity value. The wine commodity is evidently a highly differentiated product and one in which strategies to add second order, symbolic value are well developed. The New Zealand industry has reoriented itself towards premium product and has recently been extremely successful in achieving realised added value even in the face of the present global over-supply and depressed world market prices. New Zealand wine features as a premium product in the global portfolio of transnational distributors and comparative scarcity has underpinned the realised added value of its exports.

Summary

This analysis explores the realised added value that may be achieved in the products of land-based industries. This is interpreted to be second order, symbolic value and thus additional to the first order utility value to which these would be limited by conventional expectations over the trade performance of agricultural commodities. It is argued that premium, above average prices for some agricultural products may be interpreted to be evidence of symbolic commodity value. As realised added value, this symbolic value is comparable to that achieved by complex, technology-driven exports; in this case the innovations lie in quality assurance, logistics and brand marketing. This is not the case for all of the examples studied and it is likely that the specific economic history of New Zealand and its contemporary economic-political context would be relevant factors, too. Nonetheless, the longitudinal record of these industries in New Zealand suggests that realised added value can be sustained for agricultural exports.

The analysis presented here challenges the common wisdom on the growth potential for land-based commodity exports. In particular, the crude application of these ideas may overlook the significant innovation going on already and the potential for further value-adding in the land-based industries. The perspective taken by New Zealand Institute in its report is aligned to notions promulgated by the OECD, which are derived from observations of technology development and export performance in the manufacturing industries in global core economies. New Zealand's land-based industries are experienced in trading in global markets, and some industries are especially successful in extracting premiums above the simple, utility value of their commodities. However, the success of these industries tends to be overlooked through a conventional approach that arbitrarily separates the economy into primary and secondary sectors, and then applies notions such as the Prebish-Singer thesis too simply. Commodity chain analysis dismisses the arbitrary primary-secondary (and tertiary) distinction and focuses on the flow of commodity value. From its perspective the specific forms of product are irrelevant and one avoids the fetish of production innovations and novel technologies. Innovation leading to realised added value is equally important, and thus even 'simple' commodities offer the potential to sustained economic growth.

References

Black, M., Guy, M. and McLellan, N. (2003), 'Productivity in New Zealand 1988 to 2002', *New Zealand Economic Papers* 37:1, 119-50.

Dicken, P. (2003), *Global Shift: Reshaping the Global Economic Map in the 21st Century* (London: Sage).

Doel, C. (1999), 'Towards a Supply-Chain Community? Insights Form Governance Processes in the Food Industry', *Environment and Planning* A 31, 69-85.

Friedland, W. (2002), 'Reprise on Commodity Systems Methodology', *International Journal of Sociology of Agriculture and Food* 9:1, 82-103.

Gereffi, G. and Korzeniewicz, M. (eds) (1994), *Commodity Chains and Global Capitalism* (Westport: Praegar).

Gereffi, G., Humphrey, J. and Sturgeon, T. (2005), 'The Governance of Global Value Chains', *Review of International Political Economy* 12:1, 78-104.

Gibbon, P. (2001), 'Agro-Commodity Chains: An Introduction', *IDS Bulletin* 32: 3.

Gottdiener, M. (1995), *Postmodern Semiotics: Material Culture and the Forms of Postmodern Life* (Oxford: Blackwell).

Hartwick, E. (1998), 'Geographies of Consumption: A Commodity-chain Approach', *Environment and Planning D: Society and Space* 16, 423-37.

Hatzichronoglou, T. (1997), *Revision of the High-Technology Sector and Product Classification*, OECD Science, Technology and Industry Working Papers, 1997/2 (Paris: OECD Publishing).

Hughes, A. and Reimer, S. (eds) (2004), *Geographies of Commodity Chains* (New York: Routledge).

Le Heron, K. and Hayward, D. (2002), 'The Moral Commodity: Production, Consumption, and Governance in the Australasian Breakfast Cereal Industry', *Environment and Planning A* 34:12, 2231-51.

Le Heron, R., Penny, G., Paine, M., Sheath, G., Pedersen, J. and Botha, N. (2001), 'Global Supply Chains and Networking: A Critical Perspective on Learning Challenges in the New Zealand Dairy and Sheepmeat Commodity Chains', *Journal of Economic Geography* 4:1, 439-56.

Leslie, D. and Reimer, S. (1999), 'Spatialising Commodity Chains', *Progress in Human Geography* 23:3, 401-20.

Lewis, N., Moran, W., Perrier-Cornet, P. and Barker, J. (2002), 'Territoriality, Enterprise and Réglementation in Industry Governance', *Progress in Human Geography* 26:4, 433-62.

OECD (2005), *OECD Science, Technology and Industry Scoreboard* (Paris: Organisation for Economic Co-operation and Development).

Porter, M. (1985), *Competitive Advantage: Creating and Sustaining Superior Performance* (New York: Collier Macmillan).

Prebisch, R. (1950), *The Economic Development of Latin America and its Principal Problems* (New York: United Nations).

Raikes, P., Jensen, M. and Ponte, S. (2000), 'Global Commodity Chain Analysis and the French Filière Approach: Comparison and Critique', *Economy and Society* 29:3, 390-417.

Singer, H. (1950), 'U.S. Foreign Investment in Underdeveloped Areas: The Distribution of Gains between Investing and Borrowing Countries', *American Economic Review* 40:2, 473-85.

Skilling, D. and Bowen, D. (2005), *Dancing with the Stars?: The International Performance of the New Zealand Economy* (Auckland: New Zealand Institute).

Sturgeon, T. (2000), 'How Do We Define Value Chains and Production Networks?', *IDS Bulletin* 32:3.

Walker, R. (1988), 'The Geographical Organisation of Production Systems', *Environment and Planning D: Society and Space* 6, 377-408.

Watts, D., Ilbery, B. and Maye, D. (2005), 'Making Reconnections in Agro-Food Geography: Alternative Systems of Food Provision', *Progress in Human Geography* 29:1, 22-40.

Willis, R. (2004), 'Enlargement, Concentration and Centralisation in the New Zealand Dairy Industry', *Geography* 89:1, 83-88.

Wood, A. (2001), 'Value Chains: An Economist's Perspective', *IDS Bulletin* 32:3.

Chapter 23

Influencing Global Economic Participation? 'After-neoliberal' Policy Practices and Discursive Alignments in Auckland's Regional Governance

Steffen Wetzstein

Introduction

New Zealand has become [in]famous for its rapid and extensive neoliberal political reforms of the 1980s and 1990s, which resulted in intensive restructuring of the national economy (Le Heron and Pawson 1996). A new political economic governance framework was put in place that focused on the creation of competitive markets as the key strategy for achieving allocative efficiency (Deeks and Enderwick 1994). In recent times, sources of economic value have increasingly been sought in the areas of knowledge-rich activity and the deeper integration of local actors and activities in the globalising economy. In the context of New Zealand's peripheral, small, and largely resource-based economy, policy makers today face the key challenge of influencing local-global connections in investment processes in value-adding activities in ways that make private sector decision-making reconcilable with wider objectives for New Zealand's sustainable development. Auckland, as New Zealand's largest regional economy and traditional gateway to the world, plays a central role in these policy considerations. It is, therefore, a relevant site for analysing the ways current policy practices are able to influence the global economic participation of local economic actors.

It is recognised in the literature that increasing global economic integration, with the increasingly rapid circulation of capital and information flows, poses particular challenges for state and territorial intervention. The 'global' constitutes an emerging governance domain with particular difficulties for actors to guide capital accumulation (Jessop 1998). In this context, urban and regional spaces of governance take on particularly important roles as centres of co-ordination of global economic flows under contemporary conditions of 'globalising capitalism' (Florida 1995). The challenge to exert regulatory control at the urban scale has been made visible in the well-documented emergence of speculative modes of governance under 'urban entrepreneurialism' (Harvey 1989). Less well known, however, is how expansionary economic processes and globalising economic relations are thought to be facilitated from and through city-region political spaces under 'advanced' neoliberal political-economic conditions.

This chapter seeks to interpret the work of political and policy interventions in Auckland's regional economy in the post-restructuring period between the mid-1990s and mid-2000s. By focusing on the work of particular regional actors, such as the Auckland Regional Growth Forum (RGF), the Auckland Regional Economic Development Strategy (AREDS) and the 'Competitive Auckland' business initiative, answers are sought to two questions:

- How, and through what practices, is governing performed in this political-economic moment, and
- can an expansionary effect on economic processes be discerned?

This type of analysis calls for a multi-method investigative framework incorporating interviews with key actors (such a local politicians, government policy analysts and business people), re-interpreting policy texts (such as strategy documents and minutes of officer meetings), and the critical observations and experiences of the author who himself has worked in Auckland's local government. The central claim is that in an 'after-neoliberal' moment, emerging governing practices are of a largely discursive nature with, to date, unknown material effects. The remainder of this chapter first utilises a range of literatures to interpret recent changes in Auckland's economic governance before outlining in more detail key governing practices that aim to change the behaviour of political and economic actors in Auckland. After exploring and discussing some of the observable effects of current governing approaches, the concluding discussion answers the question raised in the chapter title.

After Neoliberal Restructuring: the Re-worked State, 'at a Distance' Governing and Public Policy Complexity

As a consequence of neoliberal reform, the New Zealand state was organised according to private sector principle. Its policy functions were redesigned on the basis of an output-focused policy framework that separated development from delivery structures. The central state removed itself from direct economic management in the regions, while local state interests responded by shifting to a more facilitating mode of engagement with private investors. In Auckland, a fragmented and competitive local state had been further consolidated under new local government legislation. Today, Auckland's 1.4 million people are governed by seven local territorial councils and an overlaying regional council (see Figure 23.1). Increasingly, the regulatory local state apparatus has included many quasi-autonomous actors that, in their entirety, constitute a complex institutional system of local governance. Auckland's businesses and labour responded to the new conditions by competitive upgrading and increased flexibility. Overall, regional

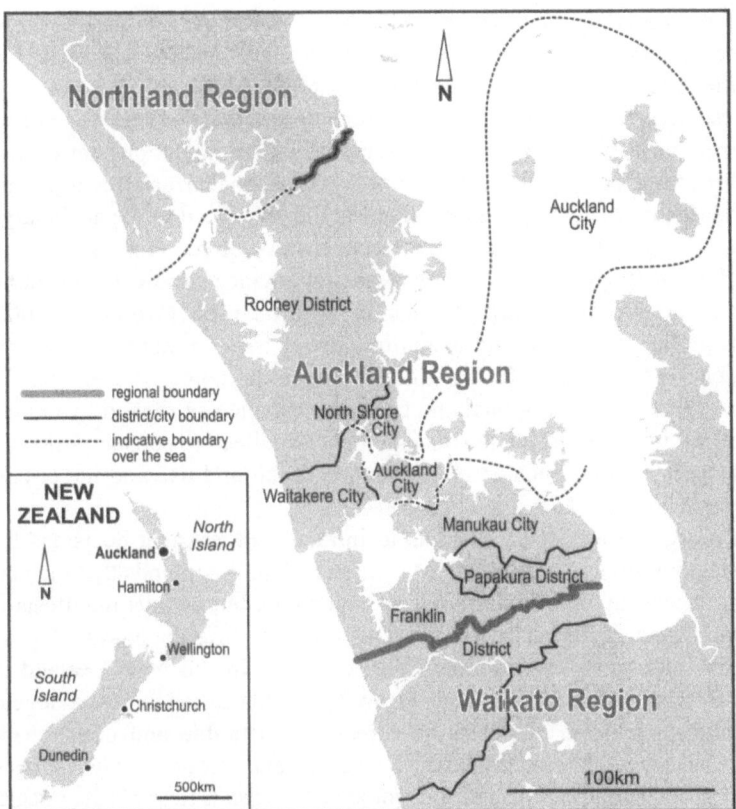

Figure 23.1   Local councils in the Auckland region

development has been driven by consumption-led and land-based investment strategies. While international migration and rising imports of goods have become an expression of intensifying local-global links, recent analyses have highlighted the increasing economic marginalisation of Auckland in terms of exporting activity, its role as a physical gateway between New Zealand and the world, and its position in servicing the country as a business hub (Le Heron and McDermott 2001). Other policy challenges include pressures on land and infrastructure, social polarisation, rapid cultural diversification, environmental degradation and heritage loss.

A series of public and public-private initiatives has emerged in reply to the new problems in Auckland's development. A first region-wide governance arrangement was the RGF, a co-operative partnership between Auckland's councils to examine the options and alternatives for future population growth, and to manage its effects on the environment, infrastructure and local communities. The activation of local business interests in the 'Competitive Auckland' initiative aimed at both raising awareness of Auckland's globally referenced economic underperformance and influencing state institutions to promote Auckland globally. This project became part of the founding platform for AREDS, a public-private policy project that produced a twenty-year strategy for facilitating Auckland's growth based on intensified global economic connectivity (AREDS 2002).

These diverse initiatives on the state-economy interface can be understood as an expression of the widespread institutional experimentation that characterises this political-economic moment. Ever new institutional arrangements are invented in search for more closely aligned interests that may lead to expansionary effects on economic processes. Auckland's recent history of economic governance also highlights the importance of networked governance arrangements rather than purely hierarchical or market forms of co-ordination (Wetzstein 2006). Yet, interest group politics around a multiplicity of issues remain a core ingredient of regional governance. This highlights the importance of conceptualisations that view processes of regional and urban governance as shaped by conflicts and political struggles regarding the management of collective consumption and social reproduction in areas such as infrastructure, local land-use and the environment (Ward and Jonas 2004; While et al. 2004).

An intriguing aspect of attempts to influence investment decisions has been the ways governing is performed. MacKinnon (2000) contends that the neoliberal state is not absent from governance arrangements but always manifests itself in different forms, different representations, and in different ways of governing. Thus, the state acts 'at a distance' (Rose 1999). In the New Zealand context, policy aspirations are now re-oriented from outputs to outcomes, a development that contrasts with an emphasis on directly measurable and often quantifiable outputs that prevailed over the last two decades, and a concern with inputs in prior decades. Policy complexity is further enhanced by trends towards the integration of outcomes in current discourses. Terms such as 'sustainable development', 'quality of life' and 'economic wellbeing' link together separate policy objectives and create phrases that produce new meanings (Dalziel and Saunders 2004). The shift towards outcome-focused policy frameworks aiming to align policy and resource allocation goals of a multitude of actors has produced both more complex economic governance and a more complicated public policy world over recent years. Larner and Le Heron (2002) contend that emerging practices of governing 'from afar' increasingly entail incorporations of spaces of calculation and measurement such as indicators and benchmarking. They use the phrase 'imaginaries' to describe how discourses and practices are constitutive of new globalising spaces and subjects. They propose a shift in globalisation accounts from those premised on flows to the 'imaginaries' that constitute the global as made up of flows. The latter can be usefully understood as actor networks, which include both discursive and technical dimensions.

Recently, Larner and Craig (2005) refer to changes in New Zealand's political economy as an 'after-neoliberal' political project. This political moment and strategy is linked to New Zealand's fifth Labour government which, elected in 1999, has sought to rebuild economic and social institutions. Larner et al. (2007) argue that 'after-neoliberalism' is made up of projects such as globalisation, knowledge society, creative industries and social development which are interlinked and co-constitutive in nature. This political programme can be viewed as an attempt to align a series of political projects that are in different ways designed to re-invigorate economic and social participation in the context of a globalising and knowledge-driven economy. Part of the 'after-neoliberal' regulatory moment seems to be the emergence of discursive practices of governance that have received increasing attention recently in research on Sydney's global city discourse and its metropolitan planning processes (McGuirk 2004).

Attempts to Re-orient Actors' Goal Settings: 'Story-telling' and Calculative Practices

Narrating Auckland's Performance

A key objective of current policy initiatives is the mobilisation of actors and resources by focusing on actors' goal-settings in relation to governing and investment. 'Story-telling' has become a central practice in this context. The stories construct Auckland's performance in economic, social, cultural and environmental arenas. They are now central ingredients of political strategies to intervene in the regional economy. Developing narratives to inspire and mobilise other actors has become a key governing technique to achieve desired governance effects in complex and interdependent policy arenas. Important tools in telling stories that others can buy into are an appeal to visions and broader aspirations, the use of persuasive and emotive language, and the mobilisation of imagination and new imaginaries. The differing narratives on the performance of New Zealand's largest city-region are embedded in distinct political discourses. These can be understood as both the expressions and the media of evolving and co-constituted political projects. Auckland's policy world is a particular governance arena where these wider political projects are played out and given content.

The narration of a coherent and appealing story is perceived as effective in changing actors' behaviours. As a 'Competitive Auckland' member points out:

> ... there are two change agency models: the integrated story and a set of recommendations. The later will be shortened, let's say four out of twelve recommendations are implemented. The former makes people sign up, makes them compromise with their own objectives. (Private Sector Manager, 2004)

In Auckland's economic intervention space, multiple stories are activated and circulating at any time. One discourse constructs Auckland as a coherent city rather than a fragmented region. In this context, a diverse coalition of interests including Auckland City Council politicians and local business leaders have placed an emphasis on the Central Business District, heritage questions, global economic comparisons and ceremonial moments in public life that together represent Auckland as a unified, urban, sophisticated and global city-region. In contrast, Auckland is continually being depicted as a non-coherent, divided and competing place of many sub-regions, by local growth coalitions consisting of local councils and developers. Another policy discourse constructs Auckland as a centre of the knowledge economy and creativity, in which innovative industry sectors and knowledge workers abound. However, there are also absent discourses and discursive silences such as the fates of the so-called 'sunset industries' and small to medium-sized businesses.

A central part of telling stories in governance arenas is the enrolment of knowledges that provide content to, and authorise, these narratives. In particular, globally sourced knowledges played an important role in recent initiatives. Global consultancy knowledge created for 'Competitive Auckland' shaped thinking in the AREDS initiative. Embodied knowledges such as those of overseas 'experts' – mostly from the USA – influenced conceptualisations on growth issues in a range of local sites. These findings not only confirm that New Zealand's public policy history has been strongly influenced by ideas from other parts of the world (Craig and Porter 2005), but that enrolling knowledges for specific governing projects, and commissioning studies on socio-spatial topics, state policy-makers, politicians and other local interests effectively talk a city's identity into existence (McNeill et al. 2005).

Creating Imaginaries through Benchmarking and Indicators

Emerging practices of economic governing are directed towards the strategic self-management of actors. A key governmental technique in this regard is the constitution of particular governance domains through the use of calculative practices aimed at creating desired imaginaries among actors. Two broad orientations can be discerned: the imagining of globalising links, flows and relationships through global benchmarking, and an increasing trend towards stimulating thinking about what sustainability may mean in practice through the widespread use of indicators. Both practices, besides audit and contractualism, are part of the calculative practices that are 'emerging as a generic term for an entire family of conceptually related comparative techniques' (Larner and Le Heron 2002, 762).

According to Larner and Le Heron (2004) it is the potential to assemble and translate measurements relating to the performance of other actors in near and far places that makes benchmarking a globalising practice. It has been vital in the constitution of governing thought in Auckland's recent policy initiatives. For example, the rationale for 'Competitive Auckland' to start its campaign was the realisation among its members that Auckland's economic performance over the past decade had been poor relative to that of benchmark cities. Their research on problems and potential solutions to Auckland's perceived institutional deficits had been based on international benchmarking derived from case studies conducted in the United Kingdom, Canada, USA and Australia. AREDS has used benchmarking as a means to construct Auckland as part of the Asia-Pacific region (NZIER 2002). Finally, policy work in the Auckland Regional Council has been increasingly guided by benchmarking exercises and global case studies to inform local decision-making processes.

Benchmarking in a global comparative sense includes two dimensions. As a policymaker explains: 'we benchmark ourselves against cities we admire but we won't catch-up to, and against cities that we are competing against for investments looking for a new home' (local government manager 2004). The management literature on benchmarking confirms that this notion is a mixed metaphor, emphasising both collaboration between benchmarking partners as well as principles and language that convey notions of competition (Cox et al. 1997). Benchmarking is now transforming into a governmental strategy. Larner and Le Heron (2004) link this tendency with the increasing application of this practice to the public sector. They point out that benchmarking has become a 'buzzword' for industry, government and individuals who aspire to be 'world-class'. However, it can also fuel and give content to inter city competition for investment and people (Wetzstein and Le Heron 2003). By naturalising an outward-focused mentality, benchmarking contributes to a new globalising governmental rationality that aids to align diverse actors such as individuals, organisations, industry, cities and regions (Larner and Le Heron 2004).

A related trend has been the increasing use of indicators in governing practice. They have become widely accepted as tools to measure strategic action in New Zealand's and Auckland's policy environment. As Table 23.1 shows, they emerge in various governing sites such as central government entities, organisations of Auckland's local state and private consultancies. The objects that are measured through indicators are observable dimensions of economic, social, environmental, cultural and institutional processes. New indicator developments increasingly aid in making associations between these different processes. As indicator-based practices are constitutive of new governance domains, they help to bring into being new spaces of calculation that make sustainable economic activity thinkable. Indicators representing economic processes and economic relations are now constructed as part of an integrated framework of development that has in its centre the objective of sustainability. At the heart of this notion, overlapping economic, social, cultural and environmental processes are thought to be measurable in their interrelatedness and integration; and therefore imaginable in this way by actors.

Table 23.1   Benchmarking and indicator reports in New Zealand's and Auckland's policy world

No.	Policy Report	Author	Date	Pages	Report Content Overview
1	Quality of life in New Zealand's six largest cities (Auckland City, Manukau City, Christchurch City, North Shore City, Waitakere City, Wellington City)	150+ Forum, CEO's project group	2001	119	55 indicators in 9 indicator areas of demographics, housing, health, education, employment and economy; findings inform participating councils' strategic and annual planning and will be used for advocacy purposes
2	Auckland regional economic development strategy (AREDS): Measures and benchmarking	New Zealand Institute of Economic Research, NZIER (2002)	July 2002	45	7 economic, 5 social and 1 environmental measure for 11 Pacific Rim cities: Sydney, Brisbane, Melbourne, Adelaide, Hobart, San Francisco, San Diego, Seattle, Portland, Vancouver, Singapore
3	Our Changing Environment	Auckland City Council	2002	166	Environmental monitoring in 12 indicator areas of air quality, water quality, hazardous substances, solid waste, natural hazards, contaminated land, noise, heritage, climate, transportation, energy, growth and amenity

No.	Policy Report	Author	Date	Pages	Report Content Overview
4	Performance Measures: Auckland Regional Council, Annual Report	Auckland Regional Council	2003	32	Strategic and key performance indicators in the areas of regional direction, environmental quality, transport, parks and recreation, heritage, business units, Ericsson stadium, Rideline (Auckland transport information system), residual functions, support functions
5	Benchmark Indicators Report 2003 - Growth and Innovation Framework	Ministry of Economic Development	Aug 2003	72	Indicators on material standard of living, productivity, supply of skills and talent, changes in investment, innovation, entrepreneurship and technological change, global connectedness and others
6	Quality of life in New Zealand's eight largest cities (Dunedin City, Hamilton City and Local Government New Zealand (LGNZ) in addition to cities in Policy Report 1)	Local Government New Zealand Metro Sector Forum, CEO's project group	2003	172	56 indicators in 11 indicator areas of people, knowledge and skill, economic standard of living, economic development, housing, health, natural and built environment, safety, social connectedness, civic and political rights; findings inform participating councils' strategic and annual planning and will be used for advocacy purposes

No.	Policy Report	Author	Date	Pages	Report Content Overview
7	Export Development and Promotion - Lessons From Four Benchmark Countries	Boston Consulting Group	May 2004	231	Benchmarking case studies on export promotion and development activities in Denmark, Malaysia, Chile and the United Kingdom
8	Economic Development Indicators 2005	Ministry of Economic Development and Treasury	2005	104	Indicators on material standards of living, labour utilisation, productivity, investment, savings and financial development, innovation and enterprise, international connections, skills and talent, economic foundations
9	The social report: indicators of social wellbeing in New Zealand	Ministry of Social Development	2005	180	42 indicators in 10 indicator areas of health, knowledge and skills, paid work, economic standard of living, civil and political rights, cultural identity, leisure and recreation, physical environment, safety, social connectedness

Discursive Alignments of Interests in 'After-neoliberal' Governance Moment

Current governing practices of 'story-telling', benchmarking and indicatorisation aim to re-direct actors' imagination and aspiration to particular aspects of economic activity. These practices are constitutive of new spaces of governance in which value creation is linked with global networks and comparisons as well as a high content of knowledge and creativity. Economic activity is now constructed in ways that makes the reconciliation with wider societal objectives in social, environmental and cultural realms imaginable. But what have emerging discursive forms of governing actually been able to achieve in regard to addressing Auckland's development challenges? The investigation of recent policy and political projects reveals that the understandings of actors about the economy and interventions are diverse and multiply constructed. However, there are increasing discursive alignments of interests across institutional and geographical spaces around particular narratives. The latter include stories of Auckland's globally referenced economic development, the importance of urban-sustainable initiatives and heritage considerations for its future growth as well as the assumption that economic activity must be creative to create value. But alignment has limits. While some actors have changed their assumptions, governing intentions and expectations as a result of recent engagement processes, others have been unaffected.

What then is the regulatory nature of this 'after-neoliberal' moment in Auckland's and New Zealand's political-economic history? It can be claimed that New Zealand's biggest regional economy is largely discursively regulated. The widespread use of discursive governing practices such as the use of stories and calculative practices influence actors' goal-settings and aspirations with regard to governing and investment decisions. They contribute to increasing (but by no means fully) alignments of interests around particular political projects and thus mobilise actors and the resources they posses. As further research indicates, the most immediate effects of these emerging practices are changed processes, institutions and policies of the state. This, for example, involves changed assumptions about the nature of interventions under globalising conditions among some central and local state actors. But so far, neither large private sector organisations such as Auckland's international airport and its seaport nor bundles of small investors were mobilised in any significant way through new governing techniques (Wetzstein 2006). 'After-neoliberal' governance seems to largely stop short of changing private investment behaviours. Differences in the way regulators and private actors construct their worlds – their differing languages, assumptions and work modes – have not been challenged in any significant way.

Concluding Reflections: Globalising Orientation of Auckland's Policy World and Unknown Regulatory Effects

This chapter argues that in New Zealand's neoliberalising context and under globalising conditions, Auckland's regional economic development policy practices are about influencing other actors' perceptions and assumptions in regards to investment objectives, and not about direct economic management. Key emerging discursive practices of governance encompass a combination of 'story telling' to inspire, motivate and mobilise other actors, the use of benchmarking to create globalising imaginaries for local actors, and the proliferation of indicators to constitute 'self-reflexive' actors that pursue sustainability through the balancing of investment goals. These governmental techniques used in recent political interventions and policy initiatives facilitated discursive alignments of interests leading to transformative effects on the state-regulatory apparatus. Altered private investment behaviours, however, have not been detected so far. They are unknown. Thus, economic processes involving actors and activities in Auckland are largely discursively mediated by political and policy efforts.

To what degree then can Auckland's global economic participation be influenced politically under current conditions? At best, proliferating discursive practices of governance provide a very modest hope for achieving this ambitious regulatory goal. Given the challenges posed by the global integration of the world economy for the future of a small and geographically isolated country such as New Zealand, the extent and the nature of current interventions may be not sufficient to influence its sustainable economic transformation. A more supportive, targeted and co-ordinated economic development approach might be a better way of addressing Auckland's and New Zealand's economic challenges. A stronger-resourced and increasingly locally present central state may be a key actor in this process.

Future research projects could assist the development of better suited policy. They must look more critically at the ways investment conditions are actually reshaped by political and policy interventions by tracing regulatory effects in multi-method case studies. Importantly, there needs to be a feeding-back of the results into policy learning processes. More investigations are needed in contemporary political economy and policy-directed research that put the individual actors – firms, organisations and people – in the centre of their analyses. Finally, new research could ask more specifically whether current changes in Auckland's economic governance allow for a better reconciliation between economic and social-environmental objectives as stressed by current policy discourses. In order to understand how to assist the re-positioning of people and firms in a globalising world, knowledges about their multiple relationships and interactions in the emerging global economic configurations and regulatory spaces are paramount.

In conclusion, the discursive nature of public policy and economic governance changes means that current intervention patterns may not affect private investment decisions and outcomes, and thus make little impression on the materiality of globalising economic processes. A case can be made, however, that through narrating stories about a global and competitive Auckland and by constructing global imaginaries using numbers and benchmarks, Auckland's public policy world becomes more globally oriented. But despite this, it is very obvious that influencing Auckland's and New Zealand's global economic participation under current political-economic conditions remains a formidable political and policy task.

Acknowledgements

The author is very grateful to the contributions of Richard Le Heron and two anonymous referees in shaping this chapter. I also gratefully acknowledge Igor Drecki for his cartographic work.

References

AREDS (2002), *Growing Auckland: Auckland Regional Economic Development Strategy 2002-2022* (Auckland: Auckland Regional Economic Development Strategy <www.areds.co.nz/>.

Cox, J.R.W., Mann, L. and Samson, D. (1997), 'Benchmarking as a Mixed Metaphor: Disentangling Assumptions of Competition and Collaboration', *Journal of Management Studies* 34:2, 285-314.

Craig, D. and Porter, D. (2005), 'The Third Way and the Third World: Poverty Reduction and Social Inclusion Strategies in the Rise of 'Inclusive' Liberalism', *Review of International Political Economy* 12:2, 226-63.

Dalziel, P. and Saunders, C. (2004), 'Regional Economic Development Planning in New Zealand: Who Owns it?', *Sustaining Regions* 4:1, 16-29.

Deeks, J. and Enderwick, P. (1994), *Business and New Zealand Society* (Auckland: Longman Paul).

Florida, R. (1995), 'Toward the Learning Region', *Futures* 5:27, 527-36.

Harvey, D. (1989), 'From Managerialism to Entrepreneurialism: The Transformation in Urban Governance in Late Capitalism', *Geografiska Annaler* 71B:1, 3-17.

Jessop, B. (1998), 'The Risk of Governance and the Risks of Failure: The Case of Economic Development', *International Social Science Journal* 155, 29-45.

Larner, W. and Craig, D. (2005), 'After Neoliberalism? Community Activism and Local Partnerships in Aotearoa New Zealand', *Antipode* 37:3, 402-24.

Larner, W. and Le Heron, R. (2002), 'The Spaces and Subjects of a Globalising Economy: A Situated Exploration of Method', *Environment and Planning D* 20, 753-74.

Larner, W. and Le Heron, R. (2004), 'Governmentality, Geography and Globalising Processes: The Role of Calculative Practices', in W. Larner and Walters, W. (eds), *Global Governmentality* (London: Routledge), 212-232.

Larner, W., Le Heron, R. and Lewis, N. (2007), 'Co-constituting 'After Neoliberalism': Globalising Governmentalities and Political Projects in Aotearoa/ New Zealand', in England, K. and Ward, K. (eds), *Neoliberalisation: States, Networks, People* (London: Blackwell), forthcoming.

Le Heron, R. and McDermott, P. (2001), 'Rethinking Auckland: Local Response to Global Challenges', in Felsenstein, D. and Taylor, M. (eds), *Promoting Local Growth* (Aldershot: Ashgate), 365-386.

Le Heron, R. and Pawson, E. (1996), *Changing Places. New Zealand in the Nineties* (Auckland: Longman Paul).

Local Government Manager (2004), Interview with Author on 26 April 2004 .

MacKinnon, D. (2000), 'Managerialism, Governmentality and the State: A Neo-Foucauldian Approach to Local Economic Governance', *Political Geography* 19, 293-314.

McGuirk, P.M. (2004), 'State, Strategy, and Scale in the Competitive City: A Neo-Gramscian Analysis of the Governance of "Global Sydney"', *Environment and Planning A* 36, 1019-43.

McNeill, D., Dowling, R. and Fagan, B. (2005), 'Sydney/Global/City: An Exploration', *International Journal of Urban and Regional Research* 29:4, 935-44.

NZIER (2002), *Auckland Regional Economic Development Strategy: Measures and Benchmarking/ Report for AREDS* (Auckland: New Zealand Institute of Economic Research.

Private sector manager (2004), Interview with author on 18 November 2004.

Rose, N. (1999), *Powers of Freedom: Reframing Political Thought* (Cambridge: Cambridge University Press).

Ward, K. and Jonas, A.E.G. (2004), 'Competitive City-regionalism as a Politics of Space: A Critical Reinterpretation of the New Regionalism', *Environment and Planning A* 36, 2119-39.

Wetzstein, S. and Le Heron, R. (2003), 'A Research Project as Narrative: The Making of a 'Creative Auckland' Story', *Proceedings from New Zealand Geographical Society 22nd conference 'Windows to a Changing World'* (Auckland: New Zealand Geographical Society), 160-163.

Wetzstein, S. (2006), 'Networked Governance for Global Participation: The Case of New Zealand's Largest City', in Daniels, P.W. and Harrington, J.W. (eds), *Perspectives on Services and Development in the Asia-Pacific* (Aldershot: Ashgate), forthcoming.

While, A., Jonas, A.E.G. and Gibbs, D.C. (2004), 'Unblocking the City? Growth Pressures, Collective Provision, and the Search for New Spaces of Governance in Greater Cambridge, England', *Environment and Planning A* 36, 279-304.

Chapter 24

Auckland's Metro Project:
A Metropolitan Governance Strategy for
Regional Economic Development?

Richard Le Heron and Philip McDermott

Introduction

New Zealand joined the international trend for major cities and regions to focus on strategies to improve international economic competitiveness during the 1990s. This chapter provides a critical discussion of the most recent phase in the development of Auckland's Regional Economic Development Strategy (AREDS), the Metro Auckland Project (MAP). In doing so, it argues that assessment of a civic development project, such as the MAP, must explicitly recognise the "before", "during" and "after" politics of policy formation.

When New Zealand Prime Minister Helen Clark launched the Metro Project Action Plan in October 2006 (Auckland Regional Council 2006), which signalled the culmination of the MAP, she stated that "we have the *development* of a *platform* for the *delivery* of a *single action plan*" (emphasis added). The Prime Minister's participation in the launch was itself notable. Her speech affirmed three issues from the point of view of the Labour government, that: Auckland's economic performance is fundamental to New Zealand's economic transformation, Auckland politicians and policymakers have bridged a long-standing Auckland-Wellington (capital city) divide; and apparent support by Auckland's local governments for a single regional economic governance framework meant Auckland-Wellington conversations about resources for Auckland were likely to be productive.

Yet, the Prime Minister's statement hints at the politics and contingencies of *actually* performing regional economic governance. This chapter takes up this theme, asking three questions: "What political imaginaries or energising rationales went into constituting the MAP? What political imaginaries collided during the MAP? And, what political work became possible as the MAP proceeded?" The chapter focuses on the politicised nature of the MAP policy.

This emphasis on politics and political imaginaries is a departure from much regional economic governance research by economic geographers in several respects. First, few studies reflect on actual involvement *in* specific policy processes, informed by recent geographic writings *on* politics.[1] If, as J-K Gibson-Graham (2006, xxxiii) contends, "politics is a process of transformation instituted by taking decisions on an ultimately undecidable terrain", then, following Amin and Thrift (2005, 236), we should attempt to learn from "a politics of working through inevitably difficult coalitions".

Second, the chapter explicitly situates the MAP in terms of the New Zealand and Auckland contexts (Auckland Regional Economic Development Strategy 2002, 2002-2022) and positions the authors as members of the MAP policy development team. In particular, discrete policy processes construct and maintain intensity in interactions that have "important political resonances which we need to examine both theoretically and empirically [because] very small spaces and times ... can have very large consequences" (Thrift 2004, 730).

Third, the attention given to the political projects as imaginaries involving the pursuit of resources draws directly from theoretical work in New Zealand about neo-liberal governance. This work has identified a distinctive moment around joined-up government and partnerships between the private and public sector, referred to as *after-neo-liberalism* (Larner and Le Heron 2004; Larner et al. 2007; Wetzstein 2007).

The chapter is organised to disclose dimensions of the political associated with the MAP. It opens with a contextual account of regional economic policy processes in Auckland in relation to national policy frameworks since the early 1990s. It argues that prior alignments amongst key regional and local actors in Auckland and with Wellington-based central government were important pre-conditions for the MAP.

The MAP is then outlined in conventional policy terms. Illustrative examples of events during the policy process show how the MAP was fashioned through the politics of contingent situations. Discussion then shifts to the broader political effects of the MAP. The concern is to see how the MAP was harnessed by particular political projects. The final section re-situates policy projects in the New Zealand and Auckland regional governance context.

A few acronyms...

ARC – Auckland Regional Council
AREDF - Auckland Regional Economic Development Forum
AREDS – Auckland Regional Economic Development Strategy
ARGF – Auckland Regional Growth Forum
ARGS – Auckland Regional Growth Strategy
AUT – Auckland University of Technology

---

1 For exceptions, at least in terms of recognising the politics of urban and regional policy processes under neo-liberalism see Jessop (2002), Jonas and Pincetl (2006), Le Heron and McDermott (2001), McGuirk (2000, 2007), Neill et al. (2005), Painter (2005) and While et al. (2004).

CA – Competitive Auckland
CA2 – Committee for Auckland
GIAB – Growth and Innovation Advisory Board
GIF – Growth and Innovation framework
GUEDO – Government and Urban Economic Development Unit
MAP – Metropolitan Auckland Project
MPAP – Metro Project Action Plan

Contextual Trajectories and Contingencies

The genesis of the MAP lies in developments from two decades of New Zealand's neo-liberalising reforms (Britton et al. 1992; Le Heron and Pawson 1996). The reforms were intended to stimulate economic restructuring by business and government. They established new ground rules for organisational actors in the political landscape. Where these actors have taken regional economic governance in Auckland is our starting point.

A round of reform-driven local government amalgamations in 1989 created four cities and three predominantly rural districts in Auckland, from 29 territorial local authorities. The regional planning framework of the Auckland Regional Authority was dismantled and replaced by the Auckland Regional Council (ARC). The new cities and districts gained local economic development functions, instituted through Local Authority Trading Enterprises, which had responsibility primarily for delivery of infrastructure services and, subsequently, by establishing Economic Development Agencies.[2]

The new ARC had its roles limited to environmental planning, regional parks, transport planning, and managing public transport. Public transport was, however, to be delivered via contract through the private sector. The Resource Management Act 1991 restricted *regional* governance in environmental matters to the quality of water, soil and air, and pest control, leaving land use planning firmly in the domain of *local* city and district councils.

In this landscape, those with a potential interest in Auckland's economic development became: local government (because they were mandated), the ARC (because questions within its purview always seemed to end up involving economic development issues), the business community (often concerned about regulations and compliance costs) and central government (which saw Auckland as a problem and Auckland actors as responsible for the problem). Over the next 15 years this mix of actors struggled to create a regional economic governance

---

2 One result has been the proliferation of economic development offices in Auckland. Today, there are five enterprise or development agencies supported in whole or in part by local councils, two economic development offices within local councils, and a regional promotions agency in the form of a business unit of the Regional Council.

framework. Their individual and collective efforts were energised in part by their relative autonomy and in part through lines of competition and collaboration.

The first development of major significance was a common commitment to a regional growth forum in the mid 1990s. This resulted from a stand-off between the local and district councils on the one hand, and the ARC on the other, over the limits of regional intervention in land use. The Draft Regional Policy Statement (1994), through which the ARC was to fulfil its environmental planning commitment under the Resource Management Act 1991, continued to reflect the wider scope of the defunct Auckland Regional Planning Scheme (1988), promulgated under an earlier planning regime (the Town and Country Planning Act). The Draft Regional Policy Statement placed strict expansion limits on councils through Metropolitan Urban Limits, and was vigorously contested by them. The Ministry for the Environment sympathised with their contention that the Policy Statement exceeded the regional mandate.

Consequently, to avoid having the Environment Court resolve the differences and recognising the interdependence of the environmental outcomes across the statutory divisions in regional and local governance, the ARC promoted the establishment of the Auckland Regional Growth Forum (ARGF) to develop an Auckland-wide consensus on growth management.

The ARGF acted as a vehicle for discussion by mayors and officers about regional land use and environmental matters. Its purpose became progressively broader with social and economic objectives introduced to complement the core focus on land use and transport. The Auckland Regional Growth Strategy (ARGS) was published by the ARC on behalf of the Forum in 1999 as "a visioning and conceptual process". As a vehicle for collaboration among local government agencies, the Forum placed emphasis on the process rather than on the strategy itself, which was characterised by a high degree of generality and a commitment to flexibility (Warren, Chairman, ARGF 1999).

During this period, individual councils continued to promote local economic development, establishing or restructuring their respective development agencies. In 2000, a group of business interests established a charitable trust, Competitive Auckland (CA), to explore and respond to their perception that the region had performed poorly in economic terms and was losing business and talent. CA's objective was to "deliver a well-articulated competitive strategy to enhance Auckland as an internationally-competitive location to undertake business", a project in which it progressively engaged a range of interests, including the region's councils. Among other things, CA highlighted the need for a "world-class" city in which the business and educational sectors were fully engaged, and for cost-effective and efficient infrastructure, especially transport.

The Committee for Auckland (CA2) was formed in early 2003, from largely the same interests that made up CA, with an emphasis on promotion and leadership.

Partly in response to this private sector initiative, and particularly to its initially critical approach of the regulatory and infrastructure performance of local government and to its apparent lack of equity considerations, the Auckland

Regional Economic Development Strategy Strategic Leadership Group (AREDS) was established in late 2001. This comprised a coalition of CA, district, city and regional councils, Maori and other stakeholder representatives, and the education sector. With funding by the ARC, the local councils, and Industry New Zealand, it aimed at a model for regional collaboration following the precedent of the ARGF.

The AREDS Founding Document acknowledged that Competitive Auckland had developed "a business-growth strategy for Auckland" and that its "strategy work will form an important part of AREDS" (November 2001). However, AREDS also acknowledged, "the increasing ... importance of city\regions in national economic development ... the need for central government intervention to improve the structure of the economy ... and acceptance by local government that there was a regional gap in economic development activity" (Anon 2004).

The AREDS Strategy was published in October 2002. It was welcomed by the Minister for Economic, Industry and Regional Development as "a partnership that can bring about real growth in the Auckland region". He commented that it was remarkable that developing the strategy led to business, local and central government, Maori, Pacific Peoples and the communities of Auckland coming together in a partnership, and highlighted the importance of Auckland speaking with one voice, as a single city. The Minister also committed to working with Auckland to reduce barriers to development (Anderton, Beehive Press Release, 16 October 2002).

At the same time, the Minister for Auckland (a position established by the Labour government, recognising the importance of the region to New Zealand's economic progress and political governance) acknowledged that the Strategy contained a message for central government departments: "that a whole-of-government approach is essential if Auckland is to achieve its economic potential" (Tizard, Beehive Press Release, 16 October 2002).

The strategy was directed at a vision for Auckland as "an internationally competitive, inclusive, and dynamic economy", calling for "an outward focus", building on a platform of "exceptional people, cultures, environment and infrastructure". AREDS produced a number of reports (Rowe 2004; Wetzstein 2006) that sought to clarify Auckland's economic role and its economic challenges (Fairgray 2006). This was within a setting, though, in which the legitimacy of its own governance continued to be challenged and evolve.

The AREDS strategy can be interpreted as aligning regional with central government thinking under the Labour government's two mega developmental visions; growth and innovation framework (GIF) supported by a Growth and Innovation Advisory Board (GIAB) and sustainability, although this was not the intention at the time.

AREDS lost momentum after development of the strategy, as it struggled with governance and management issues. The Establishment Group canvassed options among stakeholder organisations to resolve the conflict between accountability and participation in 2003 and 2004.

As a result of this exercise (Johnston and Norgrove 2004), the ARC adopted a leadership role by establishing the Auckland Regional Economic Development Forum (AREDF). This was constituted as a committee of the ARC comprising elected and invited members (including representatives of local councils). Its role was to oversee a new economic development agency, Auckland Plus, situated within the ARC. This centres regional *economic development* responsibility on the ARC.

It is against this background of economic governance, transforming from a mosaic of industry and local council initiatives to a regionally-focused local government initiative, that the MAP can be located.

The MAP initiative arose initially as a collaboration between the Auckland University of Technology (and particularly the Head of the School of Public Policy, a member of the AREDF) and the ARC. The aim was to, "bring together local, national and international expertise through an OECD review which will assess the region's performance in terms of productivity ... It is intended that an action plan will be developed ... which will drive the work programme for the recently-established Auckland Economic Development Forum" (Regional Partnership Programme Funding Application, ARC, November 2005).

Some 40 per cent of project funding was sought from government. A condition of government support was that the Ministry of Economic Development and the Department of Trade and Enterprise became active partners in the project, together with the Department of Labour and Ministry for the Environment, all working out of the recently established Government Urban and Economic Development Office (GUEDO). Raising Auckland's productivity was seen as a key to New Zealand regaining a more acceptable ranking in the OECD country economic performance stakes. Viewed in this manner, the MAP was a revitalisation of AREDS, but understood and imagined differently by Auckland and Wellington actors.

Making a Policy Project through Political Imaginaries[3]

At the early meetings[4] the broad architecture of the MAP policy project[5] was outlined, along with discussion about the MAP's formative politics. The project

---

3 The content and partiality of this section reflects the particular engagements of the authors at various stages in the MAP process and is thus only a vehicle to explore the play of different political imaginaries. The description of events and processes is, in places, greatly abbreviated.

4 We were among the authors of the MAP proposal and then part of the initial MAP development team that first met in November 2005.

5 The background team was drawn together on the basis of "expertise", not "representation". The membership included three from regional government, three academics, five from business and consultancy, one from local government and three from

had three targets; regional capability (focusing on the quality and supply of infrastructure), business capability, and people and skills development.

The original suggestion that the MAP encompass an official OECD review foundered when AUT and Trade and Enterprise differed over which section of the OECD should guide the MAP, revealing different personal experiences and preferences.

Instead, the MAP policy process was shaped to include an international panel that would offer suggestions and reflections on a re-activated AREDS-informed project. Thus, from conception, the form that the MAP was eventually to take was both unclear and indeterminate. Indeed, the only obvious logic to the MAP team at the outset was the internal logic of the work streams and timetable, which were to culminate in an action plan which would enable implementation of the AREDS.

The work programme was geared to the preparation of background papers to brief the visiting panel, the organisation of the panel's weeklong visit, the panel's subsequent report, the development of an action plan (informed by the background papers and the panel's report) and, finally, the launch of the action plan (signalling the conclusion of the MAP policy process). While these policy steps seemed at the time unproblematic, the politics encountered by the team in the further development of the regional economic governance framework proved to be complicated and challenging.

Bearing in mind the wider political environment in New Zealand, the omission of both governance and sustainability from the agenda of the Metro team was somewhat surprising.[6] This closed off several lines of investigation. Subsequent discussion identified the multiple understandings of Auckland by those involved.

The MAP team found two aspects especially hard to work with. First, while the idea of productivity was repeatedly stressed as the MAP's modus operandi, the idea of productivity itself proved to be very illusive, as did thinking about Auckland's economic performance from perspectives that embrace productivity. This stemmed partly from the decision to use existing AREDS and other reports, many of which drew on old if not dated information and long-established themes that really required interrogation and perhaps reinterpretation. These were supplemented with interviews in the business sector which by their nature were

---

central government. The membership rationale of expertise broke down several months into MAP.

6 Governance issues relating to the region were considered by the ARC project manager to have been settled as MAP was being overseen by the AREDF, a standing committee of the ARC. The MAP, with its principal outcome of an action plan, was merely another step in the implementation of AREDS. This can not be said for sustainability, which was expressly excluded from consideration. The MAP was thus firmly located within the GIF framework, despite the reasonably holistic goals of AREDS. It is significant however that the ARC initiated a parallel collaborative process, Sustaining the Auckland Region Together (START) that was dedicated to sustainability.

bound to be cautious if not reactionary with respect to the role of local government in economic development.

These sources had minimal information relevant to examining productivity. They set bounds on discussion, making it hard, for example, to consider whether metropolitan form or an enlarged Auckland region might stimulate productivity gains. Notions such as the economic benefits of agglomeration economies were taken as given, despite significant structural changes which might favour decentralisation and in contradiction of the belief held by many participants in the MAP and its predecessors that high infrastructure costs and especially traffic congestion were penalising Auckland business.

Early scepticism that the policy process might produce a plan with little connection to productivity grew as the panel's visit approached. We presented a position paper dealing with regional dimensions of productivity on the day the membership of the MAP team was extended to include every territorial local authority.[7] Almost the first question asked of us by the new arrivals was "Where is sustainability in your document? Everything we do has to be sustainable under the 2002 Local Government Act!"

Second, the objective of producing an action plan posed special challenges, in that it was the end point of professionals engaging in policy making, yet the possible content of the action plan always seemed to be remote from investment decisions. While the work stream format guided the arrangements of the panel's visit, the focus of overall effort began to shift from identifying technical issues about productivity that might be held out as questions to the panel, to consideration of how to enrol as many strategic stakeholders as possible to support the MAP's agenda.

At this point, productivity as a rationale began to disappear. This shift meant the panel's visit, in its arrangements and conduct, began to take on instead the twin functions of attempting to demonstrate the unity of Auckland and encouraging champions of such a conception (especially from business) to step forward. The government's "single voice - single city" response to AREDS rose above the issues. The pursuit of unity or at least consensus which had marked the development of the Auckland Regional Growth Strategy once more became the focus, with contentious issues consequently downplayed in the work streams or quarantined within the panel's deliberations and conclusions.

The MAP international review team programme included separate Pacific Island peoples and Maori Economic Development Forum meetings to examine key issues, concerns, priorities and opportunities; presentations by the Southern, Central and North-West sectors of the Auckland region; a public forum "Making Cities Work"; three work stream workshops around Regional Capacity and Civic Infrastructure, Skills and Innovation; meetings with the central government's Growth and Innovation Advisory Board, key central government CEOs and

---

7 Auckland City via one team member had been the sole local government contributor in the first months of MAP.

GUEDO; a day long "Symposium for Auckland"; and de-briefings to the central government (GIAB and GUEDO) and regional government entities, as well as the MAP project team.

At the de-briefing of the visiting panel to the MAP project team in May 2006, some sense of the bigger picture politics of MAP surfaced. The panel indicated, for example, that the CEOs of the local authorities told them that they were committed to making a change if the initiative was about leadership rather than governance, changes in the latter being seen as a path to a threatening unitary type governance structure.[8]

Much discussion around the edges of the week-long Metro panel programme was reported to have dealt with governance organisation – to oversee the AREDS process, versus the championing of causes relating to regional economic development, versus the contribution of political capital from local authorities to endorse the AREDS process. The panel observed that while there was great energy at the symposium, the event was not representative of the future stakeholders of the region, saying that this concern must be a permanent agenda item.

The most telling remarks were made by the chair, reflecting on his dinner conversation with the Prime Minister during the week, who intimated that Auckland was being thought about in terms of what it could do for the nation, what it could do right and what help was needed to achieve its goals. This was translated into the view that MAP needed to offer a clear path and articulate its contribution to national economic development and that these expectations meant Auckland could argue in effect for freedom and flexibility. Put another way, if MAP outlined a clear, unitary vision, then government was ready to respond.

Leveraging Political Imaginaries from a Policy Project

The published Metro Project Action Plan (MPAP, ARC 2006) contains 5 objectives, 17 strategies and 31 actions. The abbreviated details summarise the developments in the regional economic governance framework (number of actions relating to each objective are shown in brackets):

- take effective and efficient action to transform Auckland's economy to become world-class (3),
- infrastructure and world-class urban centres (5),

---

8 Later in the year regional government was to surface as an issue at least twice – first, as a proposal by four mayors to form a unitary governmental body to replace the ARC (which flopped after a short time), and second, as a proposal with wider political support to create a regional council from seven cities (which continues to simmer). The former initiative reflects a continuing dissatisfaction with the role of the ARC, despite the latter's best efforts to bring the councils into the fold. The latter reflects a technocratic view that efficiencies will be achieved by consolidation.

- transform Auckland into a world-class destination (4),
- develop a skilled and responsive labour force (8), and
- increase Auckland's business innovation and export strength (11).

Thus, at the launch the MPAP stood as the successful conclusion of a highly visible policy intervention.

Two comments are nevertheless pertinent. Productivity is mentioned in only one strategy and one action point. Elements of strategy that directly address economic transformation (interpreted as structural changes in the profile of activities, occupations, movement up value chains, extracting more value from off shore networks) took the form of:

- facilitative actions centred on governance arrangements rather than outcomes (e.g. deliver a single plan for the Auckland city-region, plan all infrastructure within the wider context of a single vision for the Auckland city-region),
- supporting strong centres (regarded as a key source of productivity gain and incidentally aligning with a node-based view of urban consolidation that underlies the ARGS) rather than any illustration or analysis of potential output or productivity benefits),
- developing a distinctly Auckland regional brand,
- improving co-ordination between innovation programmes and agencies,
- strengthening university/CRI/business collaboration,
- building an information base,
- completing CBD and waterfront development in Auckland City.

These all relate in one way or another to pre-existing strategies or the agenda of different participants.

Hearing the Prime Minister state that the 20-page brochure of the MPAP was "a deceptively simple document, a lot lies behind it", was reassuring in its (perhaps unintended) accuracy. At least two of the fifth Labour government's political projects – getting Auckland local authorities to talk to each other and talking in a single voice – can be said to have been formative in MAP outcomes, if not constitutive in its genesis. In contending that "the challenge for New Zealand is the development of a significant city… only one metropolis of international scale…major international and business hub", the Prime Minister reminded her audience of the realities of the government's view on inter-city competition, "there are 400 cities in the world larger than Auckland; five are in Australia.[9]

To suggest, however, as Michael Barnett (Chair of AREDF) said in his opening speech that the "thinking to fuel the journey has been done" is to ignore the politics running through the MAP. Noting that leadership was "not originally on

---

9 In fact, only four metropolitan areas in Australia are larger than Auckland – Sydney, Melbourne, Brisbane and Perth.

the agenda", he staked his claim by calling for a different brand, publicly rejecting the City of Sails, Auckland's present moniker. Within a week his effort to rally support for a new Auckland brand had failed, however. A *New Zealand Herald* (2006) editorial captured the tensions: "If a change is to be made ... let's hope that it will enhance Auckland's image to the world, and not be some mishmash of political correctness in an attempt to appease every interest group in the region".

The launching of the project results was itself instructive. Despite the recommendations pertaining to one plan for a single region, the prevailing division was sustained when the Chairman of the AREDF convened two workshops to launch the action plan in October 2006, separating officials, politicians and advisors from business interests.

Only a few weeks after the launch of the MPAP, the GUEDO in Auckland hosted a meeting aimed at facilitating the development of a five-year research plan centred on Auckland's economic transformation. The circulated email of invitation stated that "Local and central government initiatives need a strong evidence base to underpin effective development paths" (Marra 2006). The Chair remarked that while Auckland had shifted to being on the government agenda, the government actually had little to look at, a telling reflection on the content of the MAP.

The MAP was described as "not having a strong research focus" (consistent with our earlier comments and frustrations) and was "standing on thin ice when making recommendations". Discussion started by identifying the desirability of building on the MAP and START. The depreciation of both the MAP and START was salutary. In the case of the MAP it was asked, but "Will it catalyse courageous decisions?" The contribution of the START process was reduced to the claim that because sustainable cities are the rich cities (unverified), Auckland must attract highly-skilled/high-income individuals and attract firms that employ them.

Conclusions

The question posed in the chapter's title raises issues about the contribution of the MAP to Auckland's regional economic governance. The chapter has briefly outlined the paths and politicised contingencies of economic governance in Auckland in relation to a single, potentially unifying project. In the event, the MAP was a meeting ground for a range of political projects but it is not clear whether it was a distinctive framework for integrating them so much as appropriating them into the work of other political projects.

The contest between the public and private sector within the region to make economic policy was submerged in the initial steps, but the division nevertheless persisted and was manifest at the end of the MAP process. The contest between local and regional councils over matters affecting the economic well-being of regional communities continues, although the MAP clearly highlighted the central role assumed by the ARC and, within the process, subordinated the role of local councils. The division between central and local government has been narrowed, superficially at least as, through the MAP, the ARC has succeeded in melding a single document on which to claim a single economic voice for Auckland. The ARC is clearly positioning itself as the central partner in the government's often diffuse and fraught relationship with its Auckland constituencies.

The ARC's voice has been strengthened subsequently by the ill-advised and failed attempt to promote an alternative regional governance structure by the four cities, an attempt that nevertheless confirmed a lingering suspicion over the ARC's claims to powers in economic and land use areas that local councils still see as theirs. This action suggests that the governance relativities defined by the MAP process are not accepted unequivocally by local politicians.

The fragility of the apparent regional unity and convergence between central and local agenda in the MAP were also revealed when in October and November 2007 the Prime Minister and the Minister for the Rugby World Cup promoted the development of an iconic waterfront stadium as part of a drive to promote Auckland as a world-class city. In the event, the Auckland City Council endorsed the proposal but it was rejected by the Auckland Regional Council, and collapsed accordingly.

While the merits of the waterfront stadium proposal were never proven, and were probably far outweighed by its costs, the lack of direction available from the MAP and the absence of collaboration among the MAP partners prior to its promotion raise doubts over the likely efficacy of the MPAP.

References

Amin, A. and Thrift, N. (2005), 'What's Left? Just the Future', *Antipode* 37, 220-238

Anon (2004), *Overview of AREDS Working Group Progress*. Report to the Auckland Regional Economic Development Strategy Establishment Group. Auckland.

Auckland Regional Council (2006), *Metro Project Action Plan. Implementing the Auckland Regional Economic Development Strategy*. Auckland.

Auckland Regional Economic Development Strategy 2002. Growing Auckland. Auckland.

Auckland Regional Economic Development Strategy 2002-2022. Policy Report. Auckland.

Britton, S., Le Heron, R. and Pawson, E. (eds) (1992), *Changing Places in New Zealand: Geographies of Restructuring* (Christchurch: New Zealand Geographical Society).

Fairgray, S. (2006), *Auckland's Role in the Australasian Economy: Implications for AREDS*. Unpublished MSc thesis, School of Geography and Environmental Science, University of Auckland.

Gibson-Graham, J.-K. (2006), *A Postcolonial Politics* (Minneapolis St Paul: University of Minnesota Press).

Jessop, B. (2002), 'Liberalism, Neoliberalism and Urban Governance: A State-theoretical Perspective', *Antipode* 34:3, 452-472.

Johnston, A. and Norgrove, K. (2003), *Proposal for the Governance of the Auckland Regional Economic Development Strategy.* Policy Report. Auckland.

Jonas, A. and Pincetl, S. (2006), 'Rescaling Regions in the State: The New Regionalism in California, *Political Geography* 25:5, 482-505.

Larner, W. and Le Heron, R. (2004), 'Global Benchmarking: Participating "at a Distance" in the Globalising Economy', in Larner, W. and Walters, W. (eds), *Global Governmentality. Governing International Spaces* (London: Routledge), 212-232.

Larner, W. Le Heron, R. and Lewis, N. (2007), 'Co-constituting "After Neoliberalism": Globalising Governmentalities and Political Projects in Aotearoa/New Zealand', in England, K. and Ward, K. (eds), *Neoliberalisation: States, Networks, People* (London: Blackwell), 233-247.

Le Heron R. and Pawson E. (eds) (1996), *Changing Places. New Zealand in the Nineties* (Auckland: Longman Paul).

Le Heron, R. (2006), 'Towards Governing Spaces Sustainably – Reflections in the Context of Auckland, New Zealand', *Geoforum* 37:1, 441-446.

Le Heron, R. and McDermott, P. (2001), Rethinking Auckland: Risks and Issues for Industry Policy and Governance', in Felsenstein, D. and Taylor, M. (eds), *Promoting Local Growth: Process, Policy and Practice* (Aldershot: Ashgate), 365-386.

Le Heron, R. and McDermott, P. (2006), *Productivity Issues in the Regional Economy: A Framework for Auckland's Economic Development.* Prepared for the Metro Auckland Project Team. Auckland.

Marra, L. (2006), 'GUEDO Workshop', October 16.

McGuirk, P. (2000), 'Power and Policy Networks in Urban Governance: Local Government and Property-led Regeneration in Dublin', *Urban Studies* 37:4, 651-672.

McGuirk, P. (2007), 'Political Construction of the City-region: Notes from Sydney, *International Journal of Urban and Regional Research* 31, in press.

Neill, D., Dowling, R. and Fagan, B. (2005), 'Sydney/global/city: An Exploration', *International Journal of Urban and Regional Research* 29:4, 935-944.

Painter, J. (2005), 'Governmentality and Regional Economic Strategies', in Hillier, J. and Rooksby, E. (eds), *Habitus: A Sense of Place* (Aldershot: Ashgate), 131-157.

Rowe, J. (2004), 'A Case Study in Metropolitan Regional Economic Development', *Sustaining Regions* 4:1, 30-40

Thrift, N. (2004), 'Transurbanism', *Urban Geography* 25:8, 724-734.

Wetzstein, S. (2006), *Economic Governance for a Globalising Auckland? Political Projects, Institutions and Policy.* Unpublished Ph.D. thesis, School of Geography and Environmental Science, University of Auckland.

Wetzstein, S. (2007), 'Networked Governance for Global Participation: The Case of New Zealand's Largest City', in Daniels, P. and Harrington, J.W. (eds),

Perspectives on Services and Development in the Asia Pacific (Aldershot: Ashgate), forthcoming.

While, A Jonas, A and Gibbs, D (2004),'Unblocking the City? Growth Pressures, Collective Provision , and the Search for New Spaces of Governance in Greater Cambridge, England', *Environment and Planning A* 36, 279-304.

Chapter 25

Christmas in a Box:
Unravelling the Christmas Catalogue
Commodity Chain

Juliana Mansvelt and Caroline Miller

Introduction

Despite attempts to acknowledge the significance of consumers and consuming practices in commodity networks, there is still some way to go in unpacking the 'black box' of consumption (Lockie and Kitto 2000). We argue that catalogue shopping – a long established retailing mode which has survived spatial and social change – provides a useful context within which to explore, in a preliminary manner, how production and consumption relationships are shaped through a single association. A textual analysis of catalogues and advertisements reveals how meanings surrounding 'a New Zealand Christmas' are transformed through commodity relations, influencing the practice and performance of retailer and consumer subjectivities and the construction of value.

This chapter examines what catalogue shopping can reveal about the assembly of commodity networks, examines the emergence of the Christmas hamper catalogue in New Zealand, and concludes with a discussion of how catalogue shopping practice shapes, and is shaped by, wider socialities and spatialities.

Making the Link: Producer, Consumers and Catalogues

Literature on global commodity chains has emphasised political economy of linkages between consumers and producers by highlighting global change (Raikes et al. 2000). Gereffi (2001) proposes that since WWII there has been a general shift from manufacturer-led chains, to chains governed by retailers and brand name marketers, reflecting 'consumer pull'. While this research has revealed consumers' social and spatial connections to production, encouraging political mobilisation of consumers (Slocum 2004), there is an enduring challenge with regard to how best to recognise all aspects of the integration of production and consumption without reifying the divide, (Hughes and Reimer 2004).

This chapter focuses on the consumption end of the chain, understanding it as a network of social and material relations, in order to examine the representations made available to consumers. We reflect on how the discourses of 'family' and of 'saving' are used not only to brand and fetishise relatively mundane products, but how this normative valuation of the symbolic works as part of the democratisation of desire (Zukin 1998) promising inclusion, freedom of choice and prescribing appropriate and moral forms of consumer citizenship (Jayne 2006). Rather than seeking to unveil the commodity fetish to critique the representations imbued in commodities, the chapter seeks to understand how such representations may shape notions of value in the catalogue commodity network.

Our 'unravelling' of the Christmas catalogue chain is informed by actor-network theory (ANT) approaches which enable commodity linkages to be realised as 'webs of interdependence' (Hughes 2000, 188). By focusing on connections, that is 'how things are 'stitched together' across divisions and distinctions' (Murdoch 1997a, 322), ANT approaches provide a framework for exploring how commodities, people, places, material and non material relationships are assembled, governed and manifest across space. It also allows us to begin to reveal how discourses may operate as modes of ordering, helping stabilise network relations.

Focusing on one association within an actor network can help understanding of how power oscillates through networks (Hitchings 2003). Catalogue shopping provides an ideal medium through which to examine these connections, given the relatively enduring relationship between retailers and consumers. The relationship between consumers and their purchases is made explicit in the case of catalogue shopping, with consumers' purchases, physical addresses, and other personal details being specifically linked to retailers' databases, providing information which may be used for marketing and adjusting inventory and stocking levels. The retailer-consumer relationship may also be more easily sustained than for many other forms of purchase, with returns policies, ongoing mail-outs, emails, special offers and loyalty schemes building this connection.

The catalogue is also a tangible link between material production systems and the selection and use of commodities as material culture, a repository of social and spatial imaginaries surrounding both production and consumption of commodities. Retailers which use catalogues capitalise on 'vicarious exploration' (Stell and Paden 1999, 332), the ability to examine and compare a range of new or unfamiliar commodities in a manner not open to other forms of retailing. The catalogue may be viewed in spaces and times chosen by the recipient and can remain in use as a site of exploration and imagining.

Catalogue retailers may consequently operate as centres of calculation (Latour 1990), inscribing and circulating particular meanings in the consumption of commodities, with the catalogues forming 'immutable mobiles' moving between retailer and consumer with their material form creating stability within networks. Although catalogues are material objects, the network associations are complex. The possibility of retailer and consumer enrolment via purchase, exists alongside material and symbolic arrangements which may work to refuse the capacity to assemble such an association (Callon and Law 2005). This is marked with the Christmas catalogues, where the commodities are explicitly presented as part of 'giving' a magical Christmas, a construction which embodies a tension between values inscribed in 'the gift' and the economic relation. This tension may result in a failure to establish a material association between seller and consumer, or a deformation, and subversion of the normative construction of the value relation. Thus, consumption must be understood as being created through a diverse range of structures, practices and commodity imaginings between companies and consumers in specific time and place contexts (Barnett et al. 2005), with these heterogeneous elements together co-constructing the association between retailers and consumers.

The catalogue's material form may give stability to the network association, with its textual and symbolic representation adding 'value'. Existing research on direct mail catalogues tends to be orientated towards the actantality of the retailer, emphasising marketing, logistics and supplier practices (Earl 1998). Literature focused on the consumer has tended to examine motives for purchase that is; convenience, greater merchandise selection, unique products and low prices, while noting the increasing recreational and experiential aspect of mail order shopping (Eastlick and Feinberg 1999). It is the symbolism of commodities presented in Christmas catalogues which forms a critical part of the performance and practice of the retailer-consumer association shaping moral economies of purchase.

Catalogue Shopping: Putting NZ in Context

The tyranny of distance and isolation has meant catalogue shopping has always been popular in 'settler societies of the New World', emerging in the United States with Montgomery Ward in 1872. Others followed and while the big book format has survived, they have increasingly shifted to more specialised markets offering commodities at better quality and higher pricing than mainstream retailers. More than 80 per cent of the US mail order business comprised of smaller companies offering niche products (Morganosky and Fernie 1999, 275-277). In Britain, the development of catalogue selling was adapted to existing cultural arrangements of savings clubs, which saw catalogues mailed to agents rather than consumers. Agents were responsible for selling the products and for collecting orders and payments on commission. A number of these catalogues offered credit, and subsequently catalogue shopping came to be seen as a working class phenomenon. The most prominent British mail order catalogue company, Littlewoods, commenced in 1932. New players are now challenging the 'down-market' imagery of catalogue shopping by targeting the income-rich and time-poor, a trend already identified in the US.

In New Zealand, catalogue shopping emerged through a different set of political, economic and social relations. Initially they were a means of extending markets spatially, allowing distant customers access to a range of commodities. While New Zealand's populations were spatially diverse the market for goods was always small and all the catalogues attempted to serve both rural and urban dwellers' needs. Catalogues which originated from large department stores have disappeared, replaced by a range of more specialised catalogues. It is perhaps surprising that catalogues have survived amongst the range of other shopping spaces from malls, the internet, to retailer clubs.

In New Zealand, catalogue shopping has adapted and grown often as part of multi-channel retailing strategies to build brand awareness, increase store traffic, and create intimacy with customers, providing consumers with 'lower levels of subjective uncertainty when facing competing offers' and increasing consumer satisfaction (Liebermann 1999, 291). Mail order retailers such as Ezibuy and Pumpkin Patch, all offer commodities via retail outlets, online, and in postal formats. In New Zealand 34 per cent of consumers prefer addressed mail for receiving sales and advertising promotion, with 64 per cent believing 'advertisers who use this medium care more about reaching me personally' (NZ Post 2005). An estimated 86 per cent of New Zealanders start their shopping experience with some form of mail catalogue or flyer (Mediacom Marketing Digest 2004). Mail order companies have continued to consolidate their market, with the major retailers now complemented by a range of specialist catalogue firms such as Gourmet Meats and Maruia Nature Eco-store.

Christmas has always been a site for mail order sellers to exploit the 'vicarious exploration' aspect of catalogues, with special Christmas catalogues quickly becoming the source of consumers' 'wish lists'. In 1995 a new form of Christmas catalogue emerged in New Zealand – from Chrisco,a Christmas hamper company which offers credit-based purchasing. In 2004, it was joined by the more 'up market' Mrs Christmas. These catalogues represent an excellent example of the ways in which the material form of the catalogue can influence the shaping of the retailer and consumer association, the practices and performances of this relation, and the spaces in which they are shaped.

Christmas: The Ultimate Retail Event

Christmas is a key moment in the gift economy, one in which purchase of the commodity must be transformed into the social context of gift exchange (Carrier 1993). It involves elements of risk – confronting anxieties about one's own aesthetics and taste, and the uncertainty of making purchases for significant others (Gregson and Crewe 2003, 181). Negotiating these uncertainties is more difficult if purchases are second hand, so buying new can become a matter of bargain purchasing, investing in saving schemes, or paying in instalments. Christmas is also a season which induces significant social, spatial and cultural expectations about the production of 'the Christmas event'. In New Zealand, these expectations revolve around the production of food 'at home' and of social obligation which may include extended family, friends and the wider community. The advent of specialist Christmas food catalogues demonstrates how retailers' shaping of cultural and economic imperatives surrounding Christmas has allowed them to circumvent the established link between wholesaler and in-store retailer, creating new networks and more direct associations with consumers.

The Commodity Chain Players

Christmas food hampers arrived in New Zealand in 1995 with Chrisco, a mail order catalogue still predominantly owned by Richard Bradley. Bradley came to New Zealand in 1978, with his business centrally directed through a private company, Hats Holdings Ltd. It has expanded to include a finance company (Hopscotch Money Ltd), an assurance company (Hopscotch Assurance Ltd) and the Australian (established 1997) and Canadian (in 2003) Chrisco mail order catalogue companies. These have all been developed as private companies with a small group of associates, primarily Simon West and family members. This private company structure limits access to information on Bradley or his companies that even the business press has been unable to breach.

Chrisco: 'Helping Everyone Save for a Magical Christmas'

Chrisco's catalogue, which has now morphed into Chrisco Christmas Hamper Club Catalogue, is a combination of direct marketing to consumers and elements of the British agents' approach. Payments are made on a weekly basis, by direct credit, allowing the company to stress 'small weekly payments' and first call on purchaser's income. Consumers can place their orders directly with the company or through an agent, with customers being encouraged to become agents. Agents work hard for rewards, only receiving a 'Distributor Discount' after they have sold 'at least NZ$750 worth of hampers to a minimum of three customers' (Chrisco Make Money Catalogue) with the reward being a discount on the distributor's own Chrisco purchases. This integration strengthens network stability and is a key to the Chrisco performance of 'family owned company' working for families. Consumers are encouraged to view themselves as 'members' of the Chrisco 'family' which assists them in creating a 'magical Christmas' for their families. Television advertising has consistently featured a pleasant 'motherly' Mrs Christmas, with Christmas fulfilled and made magical by the arrival of a Chrisco hamper which 'you and your family will have fun going through'. Buyer's testimonials are a feature of both print and television advertising, highlighting the importance of meeting family expectations and the stress-free and value aspects of Chrisco hampers. Family provision and obligation is strongly gendered with the use of feminine images to portray an ethic of care.

Since 2005, Chrisco has offered 'the ability to save for the 'big ticket' items you've been dreaming of', through the Chrisco Saving Club which extends purchases to a range of goods from handycams to children's toys. Should debt be a problem Hopscotch Savings and Loan, which in 2005 utilised Chrisco's mailing list, can 'combine your current debts into one loan' with both 'family' firms, caring for members – 'the Chrisco team like Hopscotch's commitment to helping families'. The shaping of commodity value in catalogues is linked to budgeting, saving and debt reduction rather than price, with consumer debt being constructed as part of 'other forms' of Christmas shopping. This branding continues despite evidence that the hampers are more costly than buying products 'off-the-shelf'. Perhaps mindful of the negative publicity surrounding the economic value of Hamper commodities, Chrisco commissioned its own survey into Christmas shopping, noting impulse spending in supermarkets was a significant cause of 'blowing the budget' and recommending 'our hamper and savings club' to save in advance for Christmas and other special times of the year (Dol 2007, 38).

Mrs Christmas: 'New Zealand's Best-value Hamper Range'

Mrs Christmas Ltd began selling in 2004, fronted by Annabelle White, a celebrity cook whose motto, 'the Cuddly Cook', is emblazoned on the apron she wears in all advertisements. Annabelle White, as the maternal, generous and familial face of Mrs Christmas is also a minority shareholder in another private company, Mrs Christmas Ltd, which is owned by a group of four investors. The Mrs Christmas hampers are directed at slightly higher socio-economic groups with some hampers including balsamic vinegar and Belgian chocolates. In these catalogues, generosity and excess is given acceptability with encouragement to join the 'huge success' of The Jackpot Hamper which is the 'long-awaited (from a company which is only two years old) solution to every family's Christmas needs', at a mere NZ$15.78 a week or NZ$789 in total (Mrs Christmas 2006 Catalogue). Here value is related to time, with the time-poor consumer offered the opportunity to 'trust' Annabelle with selecting those important Christmas goods, hence the motto 'Live Well, Pay Less' and the assurance given that the commodities shown 'are a selection of some of my must-have favourites for Christmas' (Mrs Christmas 2006 Catalogue). The 'select' nature of the goods is reinforced by Annabelle reminding you that she tests products for a Sunday paper and that these tests guide her selection of goods.

Like its competitors, the company offers vouchers including ones for Ezibuy, and Progressive Enterprises which controls a substantial part of the supermarket sector. With this catalogue there is less emphasis on payment, although it uses the same direct debit system that Chrisco employs. Again, the family ethic of care features strongly in the representations of the company: 'you are our customers, and we proudly accept the responsibility of looking after you, time after time' (Mrs Christmas 2005, 2006 Catalogues). Aspirations that can only be made real by a catalogue purchase are also apparent: 'make your wildest dreams come true with Mrs Christmas' (Mrs Christmas 2006 Catalogue). Like Chrisco, Mrs Christmas has also moved into offering other goods including top end home theatre systems.

Shaping Commodity Meanings for Consumers

The social and cultural importance of Christmas makes it a unique selling event, linking the individual to the intangible aspects of production associated with producing value via branding and marketing, something increasingly important for the profitability of lead firms in commodity chains (Bair 2005). Miller (1993) believes that Christmas is an example of syncretism, a dynamic form shaped globally, but constructed and expressed differently in place creating tension between local (the gift) and global (the commodity). Yet while this tension exists, the association between retailer and consumer is more nuanced than this. Our examination of New Zealand Christmas catalogues, for example, differs from Clarke's (1997) analysis of the British catalogue bulk-buy retailer Argos. Although Chrisco and Mrs Christmas, like Argos, are designed to appeal to groups precluded from more expensive forms of shopping, these commodities are not 'cut price' relative to other forms of provisioning, such as shopping in-store, nor does their arrangement in hampers provide 'maximum choice' (Clarke 1997, 74). It is the association with normative social and cultural obligations of the Christmas event, and the gendered, raced and classed subjectification of consumers that frames the potential association and the creation of value. The consumer, once enrolled as purchaser, becomes entangled in the heterogeneous commodity network of the seller, and becomes a largely passive recipient of the seller's hamper packages (although a modicum of choice has gradually crept into catalogues where the purchaser is allowed to add or substitute a small range of goods). However, while discourses of 'family' and 'saving' form an organising mode (Law 1994) in the Christmas commodity chain, influencing the performance of catalogues companies and the representations of commodities in catalogues, notions of value may be shaped and performed quite differently by the retailers and purchasers.

Offering small weekly payments enrols some consumers in financial and commodity networks which they might not normally be able to enter. All sellers stressed the value and economy of their approach, despite an analysis in 2005 showing that a Chrisco Ham Supreme Hamper sold at NZ$372 could be bought for NZ$265 (Stock 2005, D7). The direct link between consumer and seller allows the latter to create an enduring and complex relationship. As mentioned previously, consumers are also invited to identify with the producer as part of the wider familial network. The retailers stabilise the commodity network by repeated mailings, prize draws and purchaser points schemes. Sales agreements stitch together the retailer-consumer association as the penalties for withdrawal are quite severe. Chrisco, for example, offer no refunds and will 'reduce your order to the amount you have already paid'.

Christmas catalogue retailers attempt to enhance the symbolic value of their commodities by mobilising place-based commodity biographies and narratives (see Lyons 2005). The companies make a clear effort to promote themselves as 'local' companies with 'New Zealand' brands which the customer can know and trust. They also manipulate the naming, content and size of hampers to create 'culturally' appealing New Zealand hampers. For Pacific Islanders and Maori, providing food is at the heart of celebrations that are family-focused and Chrisco in 2007 has upgraded its existing Pacific Hamper to the 'Pacific Mega Combo' (at a cost of NZ$1637.50), presumably to cater for the traditionally larger Pacific families.

Consumers may be sentient beings but, ironically, they may be captivated by actorial roles offered by catalogue companies and what they sell. These companies reproduce particular social and cultural meanings of Christmas commodity acquisition in which the moral economy of Christmas is constituted through purchase, shaped through discourses of family and familial obligation, saving and thrift. This represents a 'spatialisation of virtue' (Voyce 2006, 2811), located ostensibly in and around the home, premised on ideologies of heterosexual 'family', a gendered division of labour founded upon notions of consumer citizenship located in the economic and cultural constructions of obligation, gift giving and caring at Christmas.

Conclusions

Despite the development of new forms of retailing, direct mail catalogues have survived and evolved. Our case study of Christmas catalogues, shaped through retailer-commodity association reveals the way in which cultural and economic socialities, spatialities and subjectivities collide. Chrisco and Mrs Christmas both draw strongly on discourses of Christmas as material bounty, filial love through purchase, and familial responsibility. Family operates as a mode of ordering for the network, particularly Chrisco, with the altruistic retailer helping the consumer through a stressful time and simultaneously enrolling new 'family members' through its weekly payment system. The consumers, as part of the 'catalogue family', are constituted as 'empowered'. They are portrayed as clever shoppers, avoiding stress and anxiety by saving for a hassle- and debt-free 'magical' Christmas.

The chapter has highlighted some of the ways in which caring subjectivities are framed and performed in the retailer consumer association. The emphasis in this chapter has been on the production of consumption by catalogue retailers, and how the representations made available to consumers are shaped by, and continue to shape, the catalogue commodity chain. Although the catalogues connect retailers to particular and distanced commodity production networks through the combinations and organisation of commodity packages, these spatialities are obscured by the rhetoric of the 'local' family company offering trusted brands. Consequently, the retailer:consumer relationship discussed here must be understood as a single association which is a part of much wider commodity networks and shopping practices. It is vital to learn more about the chains that give rise to particular combinations of commodities in catalogues, how the commodities themselves shape the network (e.g. food versus homewares, toys and the like), and how the individuals themselves choose, purchase and consume items from these catalogues, in the social, discursive and material contexts which enable them to do so. Understanding how the normative shaping of moral dispositions, representations and encounters with commodities is experienced by consumers is also critical if understanding of catalogues is to move beyond their role as communicative texts (Selby 2004). The significance of the 'Christmas event' in the assembling of this network is also important, and there appears to be considerable tension in a relation in which dreaming, desire and the magical must be balanced with economic imperatives to purchase gifts and provide social occasions for others, giving rise to questions of whether such 'savings schemes' may actually contribute to household indebtedness.

This preliminary survey of two Christmas catalogue companies has highlighted the potential of this form of retailing for revealing how meaning, material and imagined geographies are shaped up, transformed and governed at a distance via a specific commodity chain. Although this is only a partial unravelling, it is nevertheless a telling one. The arrival of Christmas in a box, both materially and metaphorically brings more than the turkey home!

References

Bair, J. (2005), 'Global Capitalism and Commodity Chains: Looking Back, Going Forward', *Competition and Change* 9:2, 153-180.

Barnett, C., Cloke, P., Clarke, N, and Malpass, A. (2005), 'Consuming Ethics: Articulating the Subjects and Spaces of Ethical Consumption', *Antipode* 37:1, 1-23.

Carrier, J.G. (1993), 'The Rituals of Christmas Giving', in Miller, D. (ed.), *Unwrapping Christmas* (Oxford: Clarendon), 55-74.

Clarke, A.J. (1997), 'Window Shopping at Home: Classifieds, Catalogues and New Consumer Skills', in Miller, D (ed.), *Unwrapping Christmas* (Oxford: Clarendon), 73-99.

Dol, J. (2007), 'Planning Ahead for Xmas', *Herald on Sunday*, 7 January, 38.

Earl, V. (1998), 'Catalogue Marketing – The Attraction of Catalogues', *Marketing Magazine* July: 45.

Eastlick, M.A. and. Feinberg, R.A. (1999), 'Shopping Motives for Mail Catalogue Shopping', *Journal of Business Research*, 45 (3), 281-290.

Gereffi, G. (2001), 'Beyond the Producer-driven/Buyer-driven Dichotomy', *IDS Bulletin* 32:3, 30-40.

Gregson, N. and Crewe, L. (2003), *Second-Hand Cultures* (Oxford: Berg).

Hitchings, R. (2003), 'People, Plants and Performance: On Actor Network Theory and the Material Pleasures of the Private Garden', *Social & Cultural Geography* 4:1, 99-114.

Hughes, A. (2000), 'Retailers, Knowledges and Changing Commodity Networks: The Case of the Cut Flower Trade', *Geoforum* 31, 175-90.

Hughes, A., and Reimer, S. (eds) (2004), *Geographies of Commodity Chains* (London and New York: Routledge).

Jayne, M. (2006), *Cities and Consumption* (Abingdon, Oxon and New York: Routledge).

Latour, B. (1990), 'Drawing Things together', in Lynch, M and Woolgar, S. (eds), *Representation in Scientific Practice* (Cambridge, Mass, MIT Press), 19-68.

Law, J. (1994), *Organizing Modernity* (Oxford: Blackwell).

Liebermann, Y. (1999), 'Membership Clubs as a Tool for Enhancing Buyers' Patronage', *Journal of Business Research* 45, 291-97.

Lockie, S., and Kitto, S. (2000), 'Beyond the Farm Gate: Production-Consumption Networks and Agri-food Research', *Sociologia Ruralis* 40:1: 3-19.

Lyons, J. (2005), 'Think Seattle, Act Globally' Speciality Coffee, Commodity Biographies and the Promotion of Place, *Cultural Studies* 19:1, 14-34.

MEDIACOM, (2004), 'Press Release, Scoop Independent News', *Marketing Digest 04 February 2004* <http;//io.knowledge-basket.co.nz/magz/cma/cma.pl>, accessed 31 October 2006.

Miller, D. (1993), 'A Theory of Christmas', in Miller, D. (ed.), *Unwrapping Christmas* (Oxford: Clarendon), 3-37.

Morganosky, M.A. and. Fernie, J. (1999), 'Mail Order Direct Marketing in the United States and the United Kingdom: Responses to Changing Market Conditions', *Journal of Business Research* 45, 275-79.

Murdoch, J. (1997), 'Towards a Geography of Heterogeneous Associations', *Progress in Human Geography* 21:3, 321-37.

New Zealand Post (2006), *Consumer Media Preference 2005* <www.nzpost.co.nz/cultures/en-NZ/business/DirectMarketing/Consumermedia>, accessed 31 October 2006.

Raikes, P., Jensen, M.F., and Ponte, S. (2000), 'Global Commodity Chain Analysis and the French *Filiere* Approach: Comparison and Critique', *Economy and Society* 29:3, 390-417.

Selby, M. (2004), 'Consuming the City: Conceptualising and Researching Urban Tourist Knowledge', *Tourism Geographies* 6:2, 186-207.

Slocum, R. (2004), 'Consumer Citizens and the Cities for Climate Protection Campaign', *Environment and Planning A* 36, 763-782.

Stell, R. and Paden., N. (1999), 'Vicarious Exploration and Catalog Shopping: A Preliminary Investigation'. *Journal of Consumer Marketing* 16:4, 332-46.

Stock, R. (2005), 'It's Hardly the Spirit of Christmas' *Sunday Star Times*, 30 January 2005, D7.

Voyce, M. (2006), 'Shopping Malls in Australia. The End of Public Space and the Rise of 'Consumerist Citizenship'?, *Journal of Sociology* 42:3, 269-286.

Zukin, S. (2005), 'Urban Lifestyles: Diversity and Standardisation in Spaces of Consumption', *Urban Studies* 35, 825-839.

Index